www.brookscole.com

www.brookscole.com is the World Wide Web site for Brooks/Cole and is your direct source to dozens of online resources.

At *www.brookscole.com* you can find out about supplements, demonstration software, and student resources. You can also send e-mail to many of our authors and preview new publications and exciting new technologies.

www.brookscole.com
Changing the way the world learns®

MOONS & PLANETS

Apollo 12 landing module orbiting over the crater Herschel, prior to its lunar touchdown in 1969. (NASA)

Wynken, Blynken, and Nod one night,
Sailed off in a wooden shoe—
Sailed on a river of crystal light
Into a sea of dew.
"Where are you going and what do you wish?"
The old moon asked the three.
"We have come to fish for the herring fish
That live in this beautiful sea.
Nets of silver and gold have we,"
Said Wynken, Blynken, and Nod.
                    —*Eugene Field*

MOONS & PLANETS

FIFTH EDITION

WILLIAM K. HARTMANN

THOMSON

BROOKS/COLE

Australia • Canada • Mexico • Singapore • Spain
United Kingdom • United States

THOMSON
BROOKS/COLE

Executive Editor: *David Harris*
Acquisitions Editor: *Keith Dodson*
Development Editor: *Marie Carigma-Sambilay*
Assistant Editor: *Carol Ann Benedict*
Editorial Assistant: *Melissa Newt*
Technology Project Manager: *Samuel Subity*
Marketing Manager: *Erik Evans*
Marketing Assistant: *Leyla Jowza*
Advertising Project Manager: *Kelley McAllister*
Project Manager, Editorial Production: *Cheryll Linthicum*
Art Director: *Vernon Boes*
Print/Media Buyer: *Karen Hunt*
Permissions Editor: *Stephanie Lee*
Production Service: *Tom Novack/Upon A Mouse Publishing Service*

Text Designer: *Detta Penna*
Copy Editor: *Tom Novack, Patterson Lamb*
Illustrators: *Thompson Type, Catherine Brandel, Victor Royer, Carole Lawson, Pat Rogondino, Jill Turner*
Cover Designer: *Roy R. Neuhaus*
Front Cover Image: *Painting by William K. Hartmann*
Back Cover Images: *NASA/JPL/SSI. Saturn image at: http://ciclops.lpl.arizona.edu. Mars image at: http://marsrovers.jpl.nasa.gov/newsroom/pressreleases/20040311a.html*
Cover Printer: *Phoenix Color Corp—BTP*
Compositor: *Thompson Type*
Printer: *Phoenix Color Corp—BTP*

For more information about our products, contact us at:
Thomson Learning Academic Resource Center
1-800-423-0563
For permission to use material from this text or product, submit a request online at **http://www.thomsonrights.com**. Any additional questions about permissions can be submitted by e-mail to **thomsonrights@thomson.com**.

Library of Congress Control Number: 2004091448

ISBN: 0-534-49393-9

Brooks/Cole–Thomson Learning
10 Davis Drive
Belmont, CA 94002
USA

Asia
Thomson Learning
5 Shenton Way #01-01
UIC Building
Singapore 068808

Australia/New Zealand
Thomson Learning
102 Dodds Street
Southbank, Victoria 3006
Australia

Canada
Nelson
1120 Birchmount Road
Toronto, Ontario M1K 5G4
Canada

Europe/Middle East/Africa
Thomson Learning
High Holborn House
50/51 Bedford Row
London WC1R 4LR
United Kingdom

Latin America
Thomson Learning
Seneca, 53
Colonia Polanco
11560 Mexico D.F.
Mexico

Spain/Portugal
Paraninfo
Calle/Magallanes, 25
28015 Madrid, Spain

For Gayle and Amy

Preface

The preface to the 1972 first edition and the 1983 second edition of this book stated that "the last decade of space exploration has revealed a new solar system. This book is for any reader—scientist, amateur astronomer, or student—who wants a contemporary view of that new solar system." Clearly the solar system has expanded even more dramatically in the years since the early editions, in terms of both scientific detail and sheer natural wonder. Now we know the soils and rocks of Mars, the lavas of Venus, the active volcanoes on Io, the ice crust and probably-buried ocean of Europa, the smoggy atmosphere of Titan, the geysers of Triton, the faces of several asteroids, jetting vents of Halley's comet, and the coma of Chiron.

The first edition preface also noted that ". . . probably no one has a wide enough background to do full justice to such a broad subject." Today that is even more true. Perhaps it is crazy to try. Yet the need exists. Since planetary science is an interdisciplinary field, it may be first encountered by students at any level between freshman and first-year graduate student. This encounter may occur in an astronomy department, a geology department, or a newly created planetary science department. The teacher of such a course is likely to have his or her own specialty and emphasis. While our educational system easily produces specialists, it creates obstacles for individual students and readers who want a broad overview of the discoveries and terminology of our explorations into space. In order for all of us to share the excitement of these explorations, or to build a personal philosophy about our human role in the universe, we need to share a *first-order understanding* of the basic scientific facts.

I have been reminded of this while participating in international conferences where meteorologists, astronomers, geologists, spectroscopists, chemists, and physicists struggled with each others' terminology and world views while trying to share hypotheses and conclusions about the geochemistry, landscapes, internal evolution, volatiles, and magnetic fields of strange worlds. I was reminded of it again while listening to astronomers argue with paleontologists about the radical consequences of the asteroid impact 65 My ago, which apparently wiped out three-fourths of then-existing species on Earth.

For such reasons, my publishers and I have retained one of the best-liked features of the earlier editions, the basic framework for presenting planetary science. This book does not work its way through the solar system, planet by planet. A number of other books do that. One good example, which would make excellent supplemental reading for this course is *The New Solar System* (ed. Beatty and Chaikin, Cambridge, Mass., Sky Publishing Corp.), which has been reprinted in several updated editions. However, the basic principles of planetary science do not work planet by planet. They are universal. The equation for escape velocity, the principles of mantle and atmosphere convection, or the identification of basic mineral types are not unique to Mars, Jupiter, or Enceladus. When you master these basic ideas, you can apply them across the solar system, and understand phenomena on many worlds. We arranged this book according to some basic topics: origins, small bodies, planetary interiors, planetary surfaces, planetary atmospheres, etc. The latter chapters end with wrap-up sections that describe applications of the general principles to specific worlds.

In addition to this, several users have suggested a chapter that would apply the general principles of the book to a specific world. I have adopted this suggestion in this edition, and because of the striking number of planned Martian missions over the coming years, I end this edition with a chapter that applies the principles and nomenclature from the rest of the book to the case of Mars. This permits a sophisticated discussion of the recent Martian discoveries and the goals of Martian exploration.

The updating of this edition includes some exciting advances, such as the apparent discovery of ice at the lunar pole, the Pathfinder landing on Mars and the surveys from the Mars Global Surveyor and Mars Odyssey orbiters, Galileo photography and measurements of Jupiter satellites, the NEAR spacecraft landing on asteroid Eros, recent excitement about meteorites from Mars, and the discovery of a number of planet-sized bodies around other stars.

I have also abandoned the traditional, Victorian-era split between asteroids and comets. Numerous examples are now known of bodies cataloged as asteroids turning into comets, and vice versa. A single chapter emphasizes the continuum of various types among the small bodies.

The basic text material is a virtually nonmathematical presentation, starting with background on basic facts and planetary motions, and moving into the topical areas. Numerous equations are developed in isolated mathematical notes. Using the book without the math notes should give a good descriptive over view. Using the book with the math notes should allow a moderate level of physical sophistication.

Study aids at the ends of chapters include "problems" (descriptive questions) and "advanced problems" (more mathematical problems, usually based on the boxed mathematical notes in the preceding chapter or chapters). A hurdle in entering any new field is learning the relevant terminology, contributed in this case from astronomical, geological, meteorological, and physical traditions. We approached this problem by presenting important terms in **boldface** with adjacent definitions, and by repeating these terms in a concept list at the end of each chapter. A good studying technique for each chapter is to *review these concepts,* and *define or use each one in a sentence.* Most of these concepts are also listed in the index, which thus serves as a glossary to guide the reader to each term's definition.

Acknowledgments

I extend thanks to the staff of Wadsworth Publishing Company for cooperation, hospitality, and friendship (even during tight schedules and manuscript cutting), with special thanks to editor Keith Dodson and associate editor Carol Benedict, project manager Cheryll Linthicum, and production editor and copyeditor Tom Novack. I also thank many scientific colleagues in research and teaching who offered suggestions, criticisms, help with illustrations, and/or encouragements since the first edition, including Fran Baganal, Sasha Basilevsky, Jeff Bell, Mike Belton, Richard Binzel, Dominique Bockelée-Morvan, Leigh Broadhurst, both Robert Browns, Joe Burns, Dale Cruikshank, Don Davis, Mike Drake, Jim Elliot, Steffi Engel, Paolo Farinella, Ron Greeley, Richard Greenberg, Dave Grinspoon, Alan Harris, Jim Head, Floyd Herbert, Alan Hildebrand, Don Hunten, Bill Hubbard, John Kerridge, Baerbel Luchitta, Mike Malin, Al McEwen, Lisa McFarlane, Bill McKinnon, Carlé Pieters, Nicole Rappaport, Carl Sagan, Bob Singer, Peter Smith, Alan Stern, Dave Tholen, Joe Ververka, Stu Weidenschilling, George Wetherill, Charles Wood, and many other colleagues for their helpful conversations. Thanks also to my students, especially Michelle Gates, who helped locate typos and errors in earlier editions. Thanks to Elaine Owens of the Planetary Science Institute, a division of the San Juan Institute, for assisting me. Thanks to Kathleen Komarek, Gil Esquerdo, and Dan Berman of the staff of the Planetary Science Institute for help with image processing and transferral. I am grateful, too, to the Institute for giving me a flexible part-time status that allowed me to work on this project in the midst of other demanding projects.

Special thanks to Elizabeth Alvarez and Sara Smith for assistance in providing images from the Galileo and Mars Pathfinder missions, respectively. And to Jurrie van der Woude (Jet Propulsion Lab), and the Jet Propulsion Lab and NASA PIO staff for supplying up-to-date images from various spacecraft.

I also appreciate the knowledgeable craftsmanship and interest in space exploration of artists such as Michael Carrol, Don Davis, Jim Hervat, Pamela Lee, Ron Miller, Hiroki Morinoue, James Nichols, Andrei Sokolov, and other members of the International Association for the Astronomical Arts, for helpful conversations and for allowing me to represent their work in my Wadsworth textbooks.

Finally, special thanks to Gayle and Amy Hartmann, and Kelly Rehm, for putting up cheerfully (more or less) with the cosmic misfortune of having a scientist and writer around the house.

William K. Hartmann

Contents

Chapter One

Introduction: Planetary Science and the Cosmic Perspective 1

Why Study Planets? 1

A Cosmic Perspective 2

Socio-Environmental Issues on the Space Frontier 3

A View of Science 4

Planetary Science 5

Implementing Planetary Science 5

Evolutionary Viewpoint of this Book 6

Planetary Evolution: A Modern Conundrum 6

Summary 7

Concepts 7

Problems 7

Projects 7

Chapter Two

The Solar System: An Overview 9

What Is a Planet? 9

Bode's Rule 12

A Survey of the Planets 12

Mercury 12

Venus 13

Earth 15

Moon 17

Mars 18

Jupiter 23

Small Worlds of the Outer Solar System: A Simplified Overview 27

Callisto 28

Ganymede 28

Europa 28

Io 28

Saturn 30

Saturn's Satellites 32

Titan 33

Uranus 34

Neptune 35

Pluto 37

"Planet X?" 37

Telescopic Appearance of the Planets 39

Miscellaneous Basic Data and Terminology 39

Summary 43

Concepts 44

Problems 44

Projects 44

Chapter Three

Celestial Mechanics   47

**Historical Development Through
the Renaissance 47**

Kepler's Laws 48

Newton's Laws 49

Orbits 49

Circular Velocity 49

• **Mathematical Notes on Kepler's Laws 50**

Escape Velocity 50

**Astrometry, Orbit Determination, and
the Doppler Shift 51**

The Three Body Problem 52

Perturbations and Resonances 53

• **Mathematical Notes on the Law of Gravitation 54**

Lagrangian Points 54

• **Mathematical Notes on Circular Velocity 55**

• **Mathematical Notes on Escape Velocity 55**

• **Mathematical Notes on the Velocity Equation 56**

• **Mathematical Notes on the Doppler Effect 56**

Horseshoe Orbits 57

Tidal Effects 57

• **Mathematical Notes on Tidal Forces 58**

Tidal Evolution of Orbits and
Rotation Rates 58

Roche's Limit 59

Tidal Heating 59

• **Mathematical Notes on Roche's Limit 60**

Rings 60

**Dynamical Effects of Solar Radiation and
Solar Wind 65**

Radiation Pressure 65

• **Mathematical Notes on Radiation Pressure 66**

Solar Wind and Interplanetary Gas Motions 67

Poynting-Robertson Effect 67

• **Mathematical Notes on the Poynting-Robertson
Effect 68**

Yarkovsky Effect 68

Turbulence 69

• **Mathematical Notes on Turbulence 70**

Summary 70

Concepts 71

Problems 71

Advanced Problems 72

Chapter Four

The Formation of Stars and Planetary Material   75

Evidence That Stars Are Forming Today 75

The Interstellar Material 75

Molecules, Molecular Clouds, and Dust 77

**Gravitational Collapse: A Theory of Star
Formation 78**

The Virtual Theorem 79

Comment on the Formation of Just About
Everything 79

**Newly Formed Stars: Theory and Direct
Observations 80**

Bok Globules 80

• **Mathematical Notes on the Virial Theorem 81**

Cocoon Nebulae 81

Bipolar Mass Ejection and Herbig-Haro Objects 84

**The H-R Diagram: A Tool for Discussing Protostar
Evolution 84**

• **Mathematical Notes on Wien's Law
and Temperature Measurement 85**

Protostars and the H-R Diagram 86

Evolutionary Track of a Protostar as Observed
from Outside Cocoon Nebula 87

T Tauri Stars: A Later Stage in Star Formation 87

FU Orionis Stars: Infall of the Nebula 88

Beta Pictoris Systems: Residual Dust Disks 88

Binary and Multiple Stars 90

Rarity of Single Stars 90

Planetary Systems of Other Stars 91

Orbits as a Test of the Origin of Companions 93

Earth-sized Alien Planets? 93

Origins of Binary, Multiple, and Planetary
Systems 93

Summary 94

Concepts 96

Problems 96

Advanced Problems 97

Chapter Five

The Formation of Planets and Satellites   99

Date and Duration of Solar System Formation 99

Formation Age of the Solar System 99

Did a Nearby Supernova Explode Just Before the
Solar System Formed? 100

Formation Interval and Duration of the Formative
Process 101

**Solar System Characteristics To Be Explained
by a Successful Theory 101**

**The Solar Nebula: The Nebular Hypothesis
Confirmed 102**

Origin and Mass of the Solar Nebula 102

Composition and Shape of the Solar Nebula 103

Density and Pressure in the Solar Nebula 104

• **Mathematic Notes on the Shape of the Solar
Nebula 104**

Cooling of the Solar Nebula 105

Magnetic Effects 105

Was the Solar Nebula Highly Ionized? 106

Magnetic Braking of the Sun's Rotation 106

**First Planetary Material: Evolution of Dust
in the Solar Nebula 106**

The Condensation Process 106

Condensation of the High-Temperature
Refractories 108

Major Condensates: Nickel-Iron and Silicates 108

The Carbonaceous Condensates 110

More Major Condensates: The Ices 110

Gross Compositions of the Planets 111

Chemical Complexities 111

Further Evolution of the Solar Nebula Dust 111

Collisional Accretion of Subkilometer
Planetesimals? 112

Gravitational Collapse of Kilometer-Scale
Planetesimals? 113

• **Mathematical Notes on Collision Velocities 114**

**Collision Accretion of Full-Scale Terrestrial
Planets 114**

Forming the Giant Planets 116

Sweep-Up of the Last Planetesimals 117

Ring Systems as Clues to Planetary Origin 117

Origin of Satellites 118

Major Prograde Satellites: Miniature Solar Systems 118

Captured Satellites 119

Origin of Earth's Moon 121

Other Satellites of Catastrophic Origin: Charon, and Selected Asteroids? 122

Summary 124

Concepts 124

Problems 124

Advanced Problems 125

Chapter Six

Meteorites and Meteoritics 127

Fate of Planetesimals 127

Ejection from the Solar System 127

Collision with Planets 127

Capture into Satellite Orbits or Resonant Orbits 127

Fragmentation 127

Preservation until Today 128

Meteorites as "Free Samples" of Planetesimals 128

Meteoritics and Some Associated Definitions 130

History of Meteorite Studies 130

Phenomena of Meteorite Falls 132

Sound 132

Brightness 132

Train 132

Temperature 132

Velocity 132

Impact Rates: Meteoroid Flux 132

Present-Day Impact Rates 133

Primeval Impact Rates—The Early Intense Bombardment 133

Impact Rates on Other Planets 132

A Meteorite Classification System 134

Chondrules 136

Carbonaceous Chondrites 138

Chondrites 140

Gas-Rich Chondrites 140

Achondrites 141

Stony-Irons 142

Irons 142

Widmanstätten Pattern: Size of Parent Bodies 142

Neumann Bands: Evidence for Collisions 143

Brecciated Meteorites: Proof of Collisional Mixing 143

Meteorite Ages 143

• **Mathematical Notes on the Rubidium-Strontium System of Rock Age Measurement 144**

Cosmic Ray Exposure Ages and the Yarkovsky Effect 145

Clustering of Ages: Evidence for Specific Collisions 145

Date of Fall 146

Tektites 146

Where Do Meteorites Come From?—Orbits and Origins 146

Meteorites From the Moon and Mars 148

Siberia 1908: A Grand Meteoritic Explosion 149

Summary 152

Concepts 152

Problems 153

Advanced Problems 153

Chapter Seven

Interplanetary Worldlets: Asteroids and Comets 155

Comets: From Omens to Interplanetary Bodies 155

The Comet Discovery Process 156

Asteroids: Discovering a New Class of Solar System Bodies 156

Small Bodies: Confusion Among Types 157

Can Some Asteroids Be Comets and Vice Versa? 160

Expanding the Inventory: Centaurs and Kuiper Belt Objects 161

Compositional Trends: The Taxonomic Classes 162

Spectroscopic Properties and the Taxonomic Classes 162

Do the S-class Asteroids Equal Chondritic Meteorites? 163

The Small-Body Zones of the Solar System 163

How Asteroids and Comets Reached Their Present Locations 166

Trojan Asteroid Histories 167

Relating Small Moons to Asteroids and Comets 167

Evidence of Collisional Fragmentation: Hirayama Families 168

Physical Nature of Asteroids and Comets 169

Asteroids' Moons and Compound Shapes 169

Absolute Magnitudes, Sizes, Albedos, and Masses of Interplanetary Bodies 172

Densities of Interplanetary Bodies 173

When the Ice Sublimes: Phenomena of Active Comets 173

Tails and Comas: Appearance and Composition 174

Composition of Comet Solid Materials 174

Rotation and Shapes of Interplanetary Bodies 176

Feasibility of Irregular Shape as a Function of Size 176

Centrifugal Force Versus Gravity 176

Asteroid Rotations as a Clue to Planet Evolution 176

Number and Size Distributions of Asteroids and Comets 177

• Mathematical Notes on Maximum Rotation Rates 178

• Mathematical Notes on Maximum Sizes of Irregular Bodies 178

Mass Distribution as a Clue to Collisional History 178

Numbers and Size Distribution of Comets 179

Fragmentation and Outbursts of Comets 179

Meteors and Cometary Meteor Showers 180

Velocities and Orbits 180

Meteor Showers and Comets 181

Physical Nature of Cometary Meteoroids 182

Meteoroid Ejection Velocities 183

Zodiacal Light 183

Pristine Bodies 185

Exploring Asteroids 186

Visiting Asteroids 186

The Asteroid Threat Versus the Asteroid Opportunity 186

Summary 188

Concepts 188

Problems 188

Advanced Problems 189

Chapter Eight

Planetary Interiors 191

Four Basic Observations and a Basic Concept 191

Theoretical Techniques for Calculating Interior Conditions 191

The Equation of State 191

Computer Models of Planets 192

Plastic Flow Inside Planets 192

Changes of State 193

Minerals, Rocks, and Ices 193

Melting 193

• Mathematical Notes on Pressures Inside Planets 194

Solid State Phase Changes 194

Laboratory Experiments Versus Theory of State Changes 195

Pressure Ionization and Metallic States in Giant Planets 195

A Density-Mass Diagram for Planets 195

Differentiation 196

Additional Observational Checks on Planetary Models 197

Moment of Inertia 198

Geometric Oblateness 198

Gravitational Field and Dynamical Ellipticity 198

Rotation Rate 199

Surface Heat Flow and Temperature Gradient 199

Composition of Neighboring Planets and Meteorites 200

Magnetic Field 200

Deep Drilling and Direct Sampling 200

Seismic Properties 200

Seismology and Earthquakes 200

The Seismometer 200

Wave Types and Early Seismic Observations 201

Origin of Earthquakes 202

Earthquake Distribution Plates, Asthenosphere, and Lithosphere 203

"Moonquakes" and "Marsquakes" 204

Thermal Histories of Planets 205

Definition of the Problem 205

Theory of Heat Transport 206

Convection 206

Initial Thermal State of Planets 206

Principles of Crust and Lithosphere Formation 208

Heat Balance of Planets 208

Magnetism of Planets 209

Magnetic Interactions with the Solar Wind 209

Changes in a Planet's Magnetic Field 209

Paleomagnetism 210

Origin of Planetary Magnetic Fields 210

Strength of Planetary and Solar Magnetic Fields 211

Synthesis: Interiors of Specific Worlds 211

Earth: Core-Mantle-Crust Evolution 211

Earth: Unraveling the Dynamics of the Interior 212

Venus: Earth's Enigmatic Sister 217

The Moon: A Low-Density World That Cooled Rapidly 219

Mercury: Moonlike, but not Entirely Moonlike 220

Mars: the Intermediate Case 221

Icy Satellites of Outer Planets: Overview 221

Callisto 222

Ganymede 223

Europa 224

Io 227

Satellites of Saturn 227

Miranda 228

The Giant Planets 230

Summary 232

Concepts 232

Problems 234

Advanced Problems 234

Chapter Nine

Planetary Surfaces I: Petrology, Primitive Surfaces, and Cratering 237

Petrology 237

Minerals 237

Rocks and Rock Types 239

A Survey of Planetary Rocks 243

Early Evolution of Lithosphere Materials 244

Regolith and Megaregolith 246

Primeval Lithospheric Evolution: Rock Formation Versus Rock Destruction 248

Discovery of Meteorite Impact Craters 249

Mechanics of Impact Crater Formation 249

Features of Impact Craters 251

Simple Craters, Complex Craters, and Multiring Basins 255

Utilizing Impact Craters to Learn About Planets 256

Craters as Tools for Dating Surfaces 258

Stratigraphic Studies of the Earth-Moon System 258

Crater Saturation Equilibrium 260

Crater Counts and Isochrons 261

Crater Retention Ages 261

The D_L Method of Dating 261

Microeffects on Airless Surfaces 264

Micrometeorite Effects 264

Glasses and Iron Coatings 264

Albedo Effects and Sputtering Caused
by Irradiation 264

Compositional Effects 265

Production of Carbon and Colored Organics 266

The Dark Side of Iapetus: A Case Study 266

Space Weathering 266

Summary 266

Concepts 267

Problems 268

Advanced Problems 269

Chapter Ten

Planetary Surfaces 2: Volcanism and Endogenic Processes of Surface Evolution 271

Volcanism and Tectonics 271

Rocks Produced by Volcanism 271

Structures Produced By Volcanism 272

• **Mathematical Notes on Volcanic Eruptions 275**

What Causes Volcanism 278

Structures Produced By Tectonic Activity 279

**Volcanic and Tectonic Landforms
of the Moon and Planets 279**

Large-Scale and Global Lineament Systems 284

Tectonic Patterns Associated with Impact Basins 284

Mass Movements 285

Atmospheric Effects on Planetary Surfaces 287

Windblown Deposits on Mars 287

River Channels on Mars 291

Geochemical Cycles 291

Carbon Dioxide, Carbonates, and the Urey
Reaction 291

The Chemistry of the Venusian Surface 292

Chemistry and Red Color on the Surface
of Mars 292

Evaporite Minerals on Mars 293

Temperatures of Planetary Surfaces 293

**Summary: Synthesizing Planetary Surface
Processes 294**

Phobos, Deimos, Amalthea, and Other Small
Bodies 294

• **Mathematical Notes on Planetary Surface
Temperatures 297**

The Intermediate-Size Satellites of Saturn 298

Triton and Pluto 298

Io 299

Europa, Ganymede, and Callisto 301

Titan 305

The Moon 305

Mercury 308

Mars 309

Venus 311

Earth 313

Concepts 314

Problems 314

Advanced Problems 315

Chapter Eleven

Planetary Atmospheres 317

Origin of Planetary Atmospheres 317

Primitive Atmospheres 317

Inert Gases as Tracers of Early Conditions 317

Secondary Atmospheres of Earth, Mars, and Venus 318

Comparative Planetology of the Atmosphere of Venus, Earth, and Mars 318

Structure and Condensates of Planetary Atmospheres 321

Temperature Structure 321

• **Mathematical Notes on Atmospheric Structure 322**

Rayleigh Scattering, Blue Skies, and Pink Skies 322

Radiative Transfer of Heat Energy 323

The Greenhouse Effect 324

Condensable Substances—Moist Atmospheres 326

The Sulfuric Acid Condensate Clouds of Venus 327

Martian Clouds 328

Dynamics of Planetary Atmospheres 328

Vertical Mixing by Convection 328

Martian Dust Storms 328

Global Circulation of the Planetary Atmosphere 329

Atmospheric Levels and Upper Atmospheres 331

Troposphere, Tropopause, and Stratosphere 331

Temperature Irregularities and the Mesosphere 331

Thermospheres, Ionospheres, and Exospheres 332

Escape of Planetary Atmospheres 333

• **Mathematical Notes on Atmospheric Escape 334**

Summary: Atmospheres of Selected Worlds 335

Venus 335

Earth 337

Mars 338

Jupiter and the Giant Planets 338

Titan 342

Small Worlds and Thin Atmospheres 343

Concepts 347

Problems 347

Advanced Problems 347

Chapter Twelve

Life: Its History and Occurrence 349

The Nature of Life 349

The Origin of Life on Earth 350

From Organic Molecules to Living Cells 351

What Mars Tells Us 354

The Viking Landers' Experiments 354

Fossil Microbes in Martian Rocks? 355

Planets Outside the Solar System 356

Has Life Evolved Elsewhere? 356

Effects of Planetary and Astronomical Processes on Biological Evolution 358

Impacts as a Driver of Biological Evolution on Earth 358

Evolutionary Processes on Other Worlds? 359

Adaptability and Diversity of Life 359

Appearance of Alien Life 360

Effects of Technological Evolution on Biological Evolution 360

What Ends Civilization? 361

Alien Life in the Solar System? 362

Alien Life among the Stars? 363

Where Are They? 363

Evolutionary Clocks and the Explorative Interval 364

Radio Communication 365

Summary 366

Concepts 368

Problems 368

Advanced Problems 369

Chapter Thirteen

Martian Epilogue: Applying Planetary Science on a New Frontier 371

Brief Review of Present Martian Features 371
The Martian Geological Eras 372
Origin and Compositional Questions 372
Interior Structure and Related Considerations 373
Dust Storms and Dust Mantling 374
Water on Present-Day Mars 376
Condensation of Water and Carbon Dioxide on Mars 376
Processes at the Martian Polar Caps 378
Martian Mystery No. 1: The Channels, The History of Water, and Climate Change 379
Ancient Martian Seas and Lakes? 381
Long-Term Climate Change and the History of Water 382

Hillside Gullies, Glaciers, and Recent Climate Changes 383
Obliquity Cycles and Recent Martian Climate Variations 384
Obliquity Cycles and Pre-Tharsis Mars 385
Searching for Ancient Water and Ice—The Secret of Mars 385
Martian Mystery No. 2: Did Life Ever Form on Mars? 387
Summary 389
Concepts 389

Appendix

Planetary Data Table 391

References 396

Index 419

MOONS & PLANETS

Departure of Apollo 16 for the moon in 1972 symbolizes the direct exploration of the solar system by space vehicles, activity that has revolutionized our understanding of planets in the last two decades. (NASA)

CHAPTER ONE

Introduction: Planetary Science and the Cosmic Perspective

Why Study the Planets?

Fittingly, the first edition of this book was started on July 16, 1969. That morning in Florida, Armstrong, Aldrin, and Collins sat in their command module atop the 98-meter Saturn V. Lift-off came with a burst of smoke and brilliant flame. The first flight to the moon had begun. At that moment, we didn't know whether they would make it.

Apollo 11 achieved lunar orbit on July 20, 1969. The ship came around the limb of the moon and reestablished contact with Houston. The astronauts were "go" for the landing. Armstrong and Aldrin left Collins for their long voyage down. Newscasters who had been keeping up a commentary for hours, announced that they simply had nothing more to say during these historic minutes, so they let us listen to the dialogue between Houston and the lunar module: "Picking up some dust." Contact. "Tranquility Base here. The Eagle has landed." 1:18 P.M. M.S.T.

Six hours later, at 7:20 P.M. M.S.T., Armstrong and Aldrin opened the hatch. Armstrong came down the ladder, moving faster and surer than we had expected. He stepped down onto the landing pad and as quickly off, with his left foot, onto the moon, as if there was nothing to it. A wonderful moment of human fallibility ensued. Everyone had been wondering what the first spoken words would be as humans stepped onto the moon; Armstrong had kept mum about what he would say. What he planned to say, and thought he *did* say, is "That's one small step for a man; one giant leap for mankind." Unfortunately, the "a" was either left out or didn't transmit. Commentators celebrated the wonderful sentiment "That's one small step for man; one giant leap for mankind," and only later began to notice that it didn't really make much sense in that form; it is generally quoted in the form Armstrong intended. Anyway, the first human steps had been taken on the moon. 7:45 P.M. M.S.T., July 20, 1969.

That week won't be forgotten. The president called it the greatest event since Creation. We witnessed an evolutionary quantum jump: humankind's first step across the void from the world where we first appeared to another world.

Perhaps the only equivalent moment in earlier times is a date—fairly well known but universally ignored—probably in the summer of A.D. 1003 (Boorstein, 1983). Records of Viking explorers and settlers in Newfoundland reveal that they first made contact with Eskimo natives at about that time. Why is this day important in our planetary history? Humanity originated in Africa (or Asia?) and spread outward around the world. This spreading required hundreds of thousands of years. The fateful day around A.D. 1003 marks the culmination of the spread, the first moment when the two branches—west-migrating and east-migrating—first circled the entire globe and met each other on the other side. Strange that we don't mark it as an important event. The first step across the gulf to another world represents an equivalent defining moment for humanity. Yet it, too, is marked with less fanfare than we give Halloween, when imaginary spirits are supposed to come to life.

What can we hope to find on other worlds? We hope to find knowledge of how planets evolve, knowledge about what governs their crustal structure and their climates, knowledge that can help us live more successfully in our own environment. We expect to find abundant solar energy sources and new supplies of raw material such as nickel-iron. We may even find new worlds to inhabit or other life forms with whom we will have to co-exist. We hope to leave for our children better answers to the questions that people have asked for 10,000 years; questions about where we and our world came from—questions dealt with some 3,000 years ago in a book called Genesis, which three men read to us as they made the first human flight around the moon on Christmas Eve, 1968.

Our generation faces the problem of whether to turn inward on itself in response to what a congressman, as early as the year of the first moon landing, described as "a real questioning about whether life is really going to be better." Our generation also has begun—for the first time in history—operations at a scale that can affect the whole planetary environment. Consequences could affect all of us, whether they result from industrial activity (carbon dioxide emissions that accelerate climate change, radioactive wastes), accidents (oil spills or nuclear mishaps like Chernobyl), unforeseen by-products of consumer goods (fluorocarbon "spray-can" emissions that damage the ozone layer), warfare (nuclear fallout and effects of biological weapons), or purposeful environmental changes (projects to alter the climate or retard desertification).

The more we limit humanity and our attentions to the finite globe of Earth, the bleaker the future seems, as we lose our sense of frontier. We decrease our chances to learn how worlds work and our chances to escape our own mistakes. If we study Earth and other planets together, in both their present states and their (remarkably different) earlier states, we gain a better idea of the future possibilities for human life, on Earth and elsewhere. The activities of humanity off Earth replace the threat of an earthbound civilization with the promise and excitement of a civilization in which outward exploration and growth can resume productively.

The exploration of space has already given us a new perspective, as shown in two passages quoted by Oran Nicks (1970), associate administrator of the National Aeronautics and Space Administration (NASA). The first passage was written in 1948 by British astrophysicist Fred Hoyle:

> Once a photograph of the earth, taken from the outside, is available—once the sheer isolation of the earth becomes plain—a new idea as powerful as any in history will be let loose.

The second passage, written after the exploration of space had become reality, is quoted from a letter by John Caffrey:

> I date my own reawakening of interest in man's environment to the Apollo 8 mission and to the first clear photographs of the Earth from that mission. My theory is that the views of the Earth from that expedition and from the subsequent Apollo flights have made many of us see the Earth as a whole, in a curious way—as a single environment in which hundreds of millions of human beings have a stake. I suspect that the greatest lasting benefit of the Apollo missions may be, if my hunch is correct, this sudden rush of inspiration to try to save this fragile environment . . . if we still can.

The exploration of space may affect humanity's outlook in another way. The historian, Catherine Drinker Bowen (1963), wrote of the sense of optimism engendered in Europe by the Copernican revolution (which showed that Earth was not a stationary center of the universe but that the planets moved around the sun) and the exploration of the New World in the late 1500s:

> Were God's heavens then no longer immutable? And not only heaven but earth was shifting its geography. Francis Drake sailed home, having seen the limitless Pacific. . . . People spoke of America as today we speak of the moon, yet far more fruitfully. [Francis Bacon spoke of] "that great wind blowing from the west . . . the breath of life which blows on us from that New Continent." Columbus, he said, had made hope reasonable.

Surprisingly, scientists began to echo earlier science fiction writers in reflecting on the lunar landings. Physicist Freeman Dyson (1969), of Princeton's Institute for Advanced Study, wrote:

> We live on a shrinking and vulnerable planet which our lack of foresight is rapidly turning into a slum. Never again on this planet will there be unoccupied land, cultural isolation, freedom from bureaucracy, freedom for people to get lost and be on their own. Never again on this planet. But how about somewhere else?
>
> . . . Many of you may consider it ridiculous to think of space as a way out of our difficulties, when the existing space program . . . is being rapidly cut down, precisely because it appears to have nothing to offer to the solution of social problems. [But] if one believes in space as a major factor to human affairs, one must take a very long view.

Our **goals in exploring space** include not just a catalog of random new facts and third-order theories about the universe, but a basic first-order understanding of how our own world works, how other worlds work, and whether life exists elsewhere in the universe. These goals are symbolized in Figures 1-1 and 1-2. Perhaps just as important to us is an awareness that there really is a frontier in space—scientifically, psychologically, and physically.

A Cosmic Perspective

A long view, which I like to call the "cosmic perspective," goes a step beyond the environmental perspective that most of humanity began to take seriously (by foresight or necessity) in the 1970s. Partly, it grew out of space exploration and was fostered by scientific spokespersons, such as the late planetary scientist, Carl Sagan. The cosmic perspective recognizes that although we have evolved in the finite ecosystem of Earth, we really live in a much larger environment that we have just begun to probe. As our old geographical frontiers swept around the globe and closed in on themselves, in the millennium of A.D. 1000–2000, we finally realized (after the Copernican revolution of the 1500s–1600s) that we have been living all the time on the edge of a new frontier: Earth's surface, with its very thin atmosphere, is the edge of interplanetary space. The sky is no longer the limit for human operations.

Realizing that we live on a spherical frontier has forced us to face new issues of how to interact with our cosmic environment as well as the environment of the biosphere around us. An example comes from the recent realization of the last two decades that we live in a cosmic shooting gallery; scientists now generally agree that given the observed number of interplanetary bodies, impacts large enough to end civilization as we know it—by disrupting the fragile climatic/agricultural/economic balance—happen on a timescale of roughly every 10^4 years. In other words, a global technical civilization in a planetary system like ours is unlikely to last more than 10,000 years or so unless it learns to alter the orbits of asteroidal debris. To treat the same problem on a smaller scale, I was once asked to participate in a study of the stability of buried nuclear wastes. The engineering question: How do you create a chamber that will not be breached for a million years? (This number was chosen arbitrarily and is longer than the decay time of many nuclear waste products.) It turned out that one threat, which had not been evaluated, was the possibility of a modest-sized meteorite impact on the facility. It was a good example of a practical need for what had once been "purely academic" astronomical data. We really are beginning to live on a planetary scale that requires a cosmic perspective.

The natural tendency for a successful species to expand its numbers will intensify planetary-scale problems if we remain confined to Earth. In 1950 Earth's total population was about 2.5 billion people, in 2000 it was 6.1 billion, and by 2050 it is projected to be 9.4 billion (according to U.S. Census Bureau projections). The projection includes a declining world birth rate—the growth rate in the 1950s was about 20%/decade, in 2000 about 13%/decade, and in 2040 is expected to be about 6%/decade. The 2000 mean population density for Earth's land area is about 40 persons/km², including uninhabited regions. This compares to figures for Japan (335 persons/km²), India (317), United Kingdom (244), China (133), Mexico (52), the United States (29), Russia (8), and Canada (3). Since the days of economist Thomas Malthus (1766–1834), we have been warned that such population numbers can't keep increasing in a finite world without outstripping the raw materials needed for their support. In spite of continued mineral exploration, we know that the total terrestrial fossil fuel and ore supply is obviously finite.

In summary, we can confidently predict that over the next few generations, society will change dramatically, either from negative effects of our impact on the finite Earth or from our learning how to reduce that impact. The latter may involve reducing the birth rate, but may also—over 100 or 200 years—involve utilizing the material and energy resources of space. Carrying human operations into the interplanetary environment is not just an exciting scientific enterprise; it may ultimately buffer us from the dangers of a finite homeland with a closed frontier. A legitimate question is whether we can reach that kind of sustainable future before a resource-depletion-induced economic collapse.

Socioenvironmental Issues on the Space Frontier

Old science fiction and new engineering studies suggest the opportunity of utilizing interplanetary materials to further space exploration, establish space colonies, alleviate raw material and energy shortages on Earth, and protect our species against catastrophic collapse of civilization on the home planet. Through the Russian and American space programs, our generation has taken the first steps toward developing a human capability to operate in space. In 1981, the Soviet Union demonstrated the first modular space station by docking the 15-ton Cosmos satellite with the 21-ton

Figure 1-1. Two astronauts on the moon during the last lunar voyage in 1972, which was televised live on Earth, symbolize possible planetary explorations in the future. Lunar astronauts began the transfer of traditional geological field techniques to the surfaces of other planets. (NASA)

Salyut station. By 1988, individual cosmonauts had lived in the much larger *Mir* space station for as long as a year—time enough to fly to Mars or Venus. By 2000, full-fledged cooperation was underway and international teams inhabited the new International Space Station. The future of the international station remains somewhat uncertain.

What of the longer-term future interaction with our cosmic environment? In principle, we could continue lunar and Martian exploration, explore asteroids as a source of metals and other raw materials, and utilize solar energy resources in space to process asteroid materials. Learning how to use resources in space might reduce the pressures caused by ever-expanding processing of lower-grade ores within Earth's biosphere. As later chapters show, asteroids contain metals and other resources. Studies suggest that acquiring such asteroid resources could become economically viable in the future, as costs of processing terrestrial resources increase (Gaffey and McCord, 1977; Hartmann and others, 1984; Lewis and Lewis, 1987; Lewis, Matthews, and Guerrieri, 1993). Instead of digging ever deeper into Earth for lower-grade ores and dumping the waste products into our ecosphere, a future society based on interplanetary operations could allow Earth to begin to relax back toward its more natural state.

A dreamy idea? Perhaps, but who in 1776 would have predicted railroads surging across the American West within a century, or jets spanning intercontinental skies a century after that? As if to emphasize the seriousness of the idea, the wealthy entrepreneur, James Benson, backed by a number of asteroid experts, held a press conference in September 1997 to announce a serious attempt to fly a mission, funded by private venture capital, to a nearby asteroid, prospect for mineral value, and sell scientific data to NASA for less than NASA's cost to fly the mission.

Figure 1-2. Great public excitement was generated in 1997 during the successful lander of Mars Pathfinder on Mars, and operation of the rover, Sojourner, which left the lander and measured chemistry of nearby rocks. The NASA Web Site for this mission, which brought daily news and views from Mars into homes and schools, had the largest number of web "hits" in history. This view, from the wide-angle camera on Sojourner, shows the lander sitting on the deflated balloons that cushioned its landing. Martian rocks and soils dominate the foreground. (NASA)

On the other side of the issue, asteroid scientist George Wetherill (1979) proposed setting aside asteroids as a scientific preserve instead of exploiting their resources. In the big picture, progress toward human capability in space increases our chances of surviving natural, accidental, or purposeful destruction of Earth's environment, but it also underscores the age-old question of how we should best handle our natural resources. The era of public-science motivation for solar system exploration may be culminating, and the era of private-profit motivation, with its mixture of dynamism and excess, may be beginning.

In any case, the current growth rates in population and resource consumption cannot be sustained over the next few generations on the finite Earth; significant social change of one sort or another will come in the 2100s. This book is not the place to imagine future history, but I do invite you to consider, as you read on, the challenging opportunities and issues that are emerging as we take our first steps into our cosmic environment.

A View of Science

Many people erroneously envision science as a tree branching outward; in time the divisions of science branch finer and finer so that science becomes more and more complex, more and more specialized. Science is like this only in the sense that data proliferate.

Science should really be viewed in the opposite way, as a process that synthesizes many observations into theories that enable a person to understand a wide range of natural phenomena. By "understanding" I mean predict the right answers to problems. Take mechanics, for example. During the Renaissance, scientists made endless observations of levers, falling objects, inclined planes, rotation, orbital motion, and so on. Yet even the most learned people were not able to predict very many experimental results until Newton and his successors saw the relationships among the observations. Today, a college sophomore studying physics can understand (that is, quantitatively predict) most of the phenomena that Newton studied, and a senior's understanding surpasses that of Newton.

Thus, science is like working *down* a tree instead of up. First, we can understand only individual events—the twigs. Then we see how they all join together, and we can grasp a branch of knowledge. Eventually, we can understand whole portions of the tree. (Nobody has yet reached the trunk!)

In this book, I hope to emphasize that by mastering a few basic, unifying concepts from several different fields, the reader can understand a remarkable variety of processes that affect Earth and other planets.

Planetary Science

Planetary science is the study of the origin, evolution, and current environments of planetary bodies, whether they orbit around the sun, as in our solar system, or around other stars. **Stars** can be defined as masses of cosmic gas larger than about 75–85 Jupiter masses (designated $M_{Jupiter}$), so large that the temperature and pressure in their centers are great enough to force the nuclei of atoms together, thus causing sustained nuclear fusion reactions that consume hydrogen atoms. These reactions release the energy that the stars radiate. **Planetary bodies,** on the other hand, can be defined as bodies generally composed of silicates, ices, and high-pressure forms of hydrogen-rich material that orbit the sun or other stars and are not large enough to generate nuclear fusion reactions in their interiors; thus, they do not give off visible radiation or very large amounts of infrared radiation of their own. Based on criteria involving internal structure and forms of high pressure hydrogen, the upper limit of planet size is placed by some astronomers at about 2 to 13 Jupiter masses. Thus, there is a class of objects between planets and stars, called **brown dwarfs.** Brown dwarfs are too small to sustain H fusion reactions, but are big enough to create substantial heat by their gravitational compression, and thus to give off substantial radiation. One definition of brown dwarfs includes objects between 13 $M_{Jupiter}$ and 75 $M_{Jupiter}$ (Kulkami, 1997). In this range, the objects may have short-lived nuclear reactions that consume lithium and deuterium (^{2}H, the heavy isotope of hydrogen), but not H itself. Such objects might resemble Jupiter, or dull-glowing red-hot versions of Jupiter.

Planetary science, sometimes called planetology, can be defined as the physics of planetary bodies, using "physics" in a broad sense. Originally, **physics** was defined as "the branch of knowledge treating the material world and its phenomena." In that sense, all science is physics. The discipline now called physics concerns the behavior of matter and energy, usually in idealized circumstances. Physicists may study atoms, or friction, or electromagnetic fields, but they usually try to isolate a specific phenomenon in order to determine the basic laws involved.

Other sciences deal with the physics of more complicated systems, in which many phenomena may operate at once and specific subjects can't be isolated. Chemistry, for example, is the physics of matter interacting at the electron level (that is, without atomic nuclei interacting). Astronomy is the physics of cosmic matter, mostly in the form of stars and systems of stars. Meteorology is the physics of the atmosphere; geophysics, the physics of Earth as a body; geology, that of Earth's (and nowadays all planets') outer layers. Biology is the complicated physics of living material.

Planetary science is more integrative than geology, meteorology, and biology, for it combines them all. An analyst trying to understand the peculiar riverbed-like channels on Mars may deal with geology (could water or some other fluid have eroded these channels?), geochemistry (could water have been produced on Mars in the past?), ordinary chemistry (under what conditions would the water freeze instead of flow?), or meteorology (what was the Martian air pressure in the past as gases erupted from volcanoes and dissipated?).

Similarly, a planetary scientist may have to deal with a wide variety of data sources. An analyst trying to understand the satellites of Jupiter will deal with theoretical data (the dynamics of satellite orbits), conventional astronomical data (spectrophotometric data that indicate composition), and data sent back from space vehicles (digitized and computer-enhanced photographs and magnetic-field measurements).

Implementing Planetary Science

Planetary research in America is supported primarily by grants from the National Aeronautics and Space Administration (NASA) and, to a lesser extent, the National Science Foundation (NSF). NASA supports research that fulfills its mission of exploring the planets for the benefit of humanity. This mission was set by Congress and is funded according to the level of public interest as perceived by Congress. In the post-Apollo 1970s and 1980s, the funding dropped somewhat. In Europe, the European Space Agency (ESA) provides a similar function and serves as a role model for technical cooperation between different countries; about 1980, European planetary science exploded into a major influence in the field.

To apply for research funding, researchers write proposals (descriptions of planned studies). The proposals are sent to funding agencies, such as NASA, and are ranked for quality by panels of scientists appointed by the agency. (Membership of the review panels comes from the scientific community and rotates year by year to ensure fairness.) Available research money is apportioned among the proposals ranked best.

Planetary scientists must often make a creative compromise between being a specialist and being a generalist. Too general an education can leave one too far from the cutting edge to make a contribution. On the other hand, too specialized a background can convert one into a narrow scientist, the type who never sees the "big picture" that comes from data in different fields, but instead argues only over details. Such a scientist may become too much the narrow type of expert described by a variant of Newton's third law: "For every Ph.D. there is an equal and opposite Ph.D."

Thomas Kuhn (1962) pointed out in a famous essay on scientific revolutions that a new field of scientific inquiry becomes mature when the participants agree on a **paradigm,** a body of basic facts, theories, and approaches to their work. Planetary science is a new field, and its agreed-on body of knowledge is only now emerging. Journals, such as *Icarus* and *Meteoritics and Planetary Science,* consolidate planetary studies; in the meantime, planetary science papers are still being published in such varied journals as *Science, Nature,* the *Journal of Geophysical Research,* and the *Astronomical Journal.* Certain areas of vigorous study have developed names of their own. **Space science** has been used to designate the study of the interplanetary medium, uppermost atmosphere, solar-planetary interactions, and cosmic rays; **astrogeology** (something of a misnomer) designates geological and stratigraphic studies of lunar and planetary surfaces and is a popular term within the U.S. Geological Survey. **Exobiology** refers to hypothetical life not native to Earth and the attempt to detect such life.

Evolutionary Viewpoint of This Book

Because planetary scientists must understand *processes* as these apply to a variety of worlds, they cannot merely study descriptions of the present-day planets one at a time. Therefore, we do not describe the planets one by one. Also, we stress broad planetary processes over methods of measurement. We describe processes, such as the origin of planets and the evolution of their surfaces, and the planets as four-dimensional, extending in three dimensions of space and a fourth dimension of time. This process-oriented approach also allows us to address basic principles of planetary evolution and then compare how various planets fit into larger patterns. This approach, comparing the planets to learn about general planetary processes, is called **comparative planetology** and has been a cornerstone of the NASA planetary exploration program.

An example of this approach comes from modern studies of Earth. Until recent decades, geological investigations were mostly limited to travel over two-dimensional landscapes, with results plotted on two-dimensional maps. A generation or two ago, few techniques were available to study deeper crustal structure or to measure ages of formations. Gradually, we have extended our geophysical techniques (seismic, drilling, etc.) to measure subsurface structure and our techniques to get radiometric ages of formations all over the Earth. An excellent example of the new approach is the contemporary geologic study of the volcano Kilauea in Hawaii. Researchers have fitted the whole mountain with seismographs, tiltmeters, and other geophysical instruments whose data output is transmitted to the Kilauea Volcano Observatory on the rim of the main crater. You can stand in one room and watch an array of graphs that are recording the workings of the volcano's interior as well as of its surface. Motions of molten lava, for example, can be traced from deep below the mountain's summit as the lava works its way toward the surface.

An interplanetary example comes from comparative studies of Venus and Earth, two planets of similar size. The study of the carbon dioxide (CO_2)-rich atmospheres of Venus calibrates our understanding of how the CO_2 greenhouse effect might work on Earth, and comparisons of the surfaces lead to analysis of the different paths of crustal development on the two planets. Venus and also Mars offer us nature's grand experiments in building Earth-like planets, but with different parameters that affected each planet's evolution. The chapters are arranged from a chronological or evolutionary viewpoint. Early chapters describe basic matters: an overview of planets, definitions, and matters of celestial mechanics. We then begin with a view of the origin of the solar system (Chapters 4 and 5). Small bodies of the solar system are discussed next, with a reminder that they provide evidence of the earliest state of the planets (Chapters 6 and 7). Planet interiors and surfaces are then discussed (Chapters 8 through 10), followed by their atmospheres (Chapter 11). We then consider the development of life (Chapter 12), whose flowering (modestly referring to us) was extremely recent. Finally, as a way of summing up, we utilize material from all chapters to discuss what may be the most provocative extraterrestrial planet, Mars (Chapter 13).

Planetary Evolution: A Modern Conundrum?

As a final introductory note, consider this sticky philosophical problem. Most people would not quarrel with the assertion that we live in an evolving planetary system. Yet this belief is not universal, as the British philosopher and cosmologist H. Dingle (1960) reminds us:

> Nearly 100 years ago Philip Grosse, in order to reconcile the facts of geology with the Hebrew scriptures, advanced the theory that . . . "there had been no gradual modification of the surface of the earth, or slow development of organic forms, but that when the catastrophic act of creation took place, the world presented, instantly, the structural appearance of a planet on which life has long existed." The beginning of the universe on this theory occurred some 6000 years ago. There is no question that the theory is free from self-contradiction and is consistent with all the facts of experience we have to explain; it certainly does not multiply hypotheses beyond necessity since it invokes only one; and it is evidently beyond refutation by future experience. If, then, we are to ask of our concepts nothing more than that they shall correlate our present experience economically, we must accept it in preference to any other. Nevertheless, it is doubtful if a single person does so.* It would be a good discipline for those who reject it to express clearly their reasons for such a judgment.

---

*Here the philosopher is optimistic. Although this theory is hardly popular in our society, American fundamentalists and their evangelical radio programs insist on precisely this view—that all the seemingly evolved strata, fossils, and radioisotopes were put into their complex pattern just to tempt scientists into error. By 1981, two states passed legislation to require that this "creationist" theory be presented alongside any evolutionary theory of biology, geology, or astronomy. These laws are being tested in court; in other states, creationists have brought still more lawsuits to require that their beliefs be taught in science courses. In my own state of Arizona, the State Board of Education in 1998 proposed to drop the term "evolution" from the teaching of biological and physical sciences! We thus begin our planetary studies with our feet planted in an interesting sociological soil.

Can we, as "enlightened" planetary scientists, express clearly our reasons for utilizing the evolution of the cosmos as a meaningful concept? In nature, we find landscapes showing ordered strata, where fossils appear in an orderly sequence terminating in the present, and where rocks contain patterns of radioactivity that correspond to the fossil sequence in age. Our physical theories allow us—with our computers—to construct models that begin with hypothesized ancient conditions and evolve into the present conditions. These models have successfully *predicted* effects in nature (such as the volcanoes on Io)—a claim no other system of human thought has been able to make.

The success of our evolutionary viewpoint and scientific method does not prove that it is the ultimate truth. However, it does assure us that our philosophy is pragmatic and that future discoveries will continue to refine present beliefs. The beauty of our science is that it puts us in a direct question-and-answer dialog with nature. No other system of knowledge does that. This is the strength of the scientific method: It continues to converge on a more and more detailed, consistent, satisfying, and practical view of the universe. With a profound interest and a rather religious sense of wonder, we pursue the evolutionary thread that leads backward to what seems—according to recent findings—to have been the creation of the universe and leads forward into our own future.

SUMMARY

Planetary science is an integrative field that combines knowledge from other fields of science. Planetary scientists try to understand the origin and evolution of planets and their surface environments. Planetary scientists are, to borrow Barbara Novak's (1980) description of naturalist painters on the Western frontier in the last century, "archaeologists of the Creation." At the same time, planetary scientists try to apply their results to a better understanding of humanity's present and future environment.

In the next decade or so, more milestones of planetary science are expected through international space probes. Many robotic missions to Mars are planned, including landers that will probe below the surface. The 2004 Cassini probe is expected to parachute instruments through the clouds of Saturn's largest moon, Titan, to see what lies on the hidden surface. A Pluto probe is being discussed (Stern, 2002). Robotic lunar exploration will continue, and flights of astronauts in near-Earth space are planned to continue expanding human capabilities to operate in our cosmic environment.

CONCEPTS

goals in exploring space	physics
science	paradigm
planetary science	space science
stars	astrogeology
planetary bodies	exobiology
brown dwarfs	comparative planetology

PROBLEMS

1. How can data obtained on other planets help illuminate environmental processes and evolutionary processes on Earth?

2. Discuss how data on the history and conditions of our own planetary system can (or cannot) shed light on what types of planet systems exist around other stars.

3. Meteorites of pure nickel iron (which can probably be seen in a museum near you) testify that some asteroidal fragments are pure metal. Suppose the economic value of iron is of the order of ten cents per kilogram. Calculate the economic value of a 1-km-diameter asteroid of iron.

PROJECTS

1. See whether your school library receives any journals devoted to planetary science, such as *Icarus* or *Meteoritics* and *Planetary Science,* or journals in which planetary science results are published, such as the *Journal of Geophysical Research*. If so, report on some areas of current planetary science research.

2. In recent issues of general scientific journals, such as *Science, Nature,* and *Scientific American,* determine the percentage of articles related to planetary science. Does planetary science make a contribution to the modern scientific outlook?

A view of Jupiter's satellite Io from Voyager 1 spacecraft in 1979 provides an example of the variety of planetary bodies. Mottled patches on Io result from active volcanism that constantly modifies the sulfur-rich surface. Doughnut-shaped ring at right center is a cloud of debris being erupted from an active volcanic caldera (black area inside the doughnut). (NASA)

CHAPTER TWO

The Solar System: An Overview

The solar system is commonly said to have nine **planets,** shown in Figure 2-1. In order from the sun, they are Mercury, Venus, Earth, Mars, Jupiter, Saturn, Uranus, Neptune, and Pluto. A traditional mnemonic device for remembering this sequence is "*Men Very Early Made Jars Stand Upright Nicely (Period).*" (Surely today's students can do better than this!) Because confusion often sets in at *Saturn, Uranus,* and *Neptune,* remember that the *SUN* is in the system too.

The first four planets are sometimes called the **terrestrial planets** because of their nearness to Earth and the similarity of their rocky, metallic compositions. The four planets from Jupiter through Neptune are sometimes called the **giant,** or *Jovian,* **planets** because of their size and similarity to Jupiter.

What Is a Planet?

You might think that the term *planet* is so clearly defined that any body can be labeled either a planet or a nonplanet. We grew up being told there were nine planets moving around the sun, and nine bodies do seem uniquely distinguished by size, position, orbit, and so on. However, the situation is not really so clear. The word "planet" comes from the Greek term for "wanderer," referring to the fact that planets drift in the night sky among the stars, from night to night. Among the ancients, a "planet" was any of seven bodies that changed position from day to day with respect to the stars. These seven included the sun and the moon, as well as the first five planets (excluding Earth) out to Saturn. The rest had not yet been discovered.

A contemporary definition is that a planet is any body except a **comet, asteroid,** or **meteoroid,** orbiting the sun, but this definition is ambiguous unless comets, asteroids, and meteoroids are defined. We beg the question by calling those objects "interplanetary" bodies. And scientifically, how do we justify excluding the asteroid **Ceres,** which orbits between Mars and Jupiter and has the sizable diameter 1000 km, nearly twice as big as the other asteroids in that region?

If we try to define planets by the regular spacing of their nearly circular orbits, we face the fact that Pluto— which is generally pictured as the most remote planet— has a markedly noncircular orbit that crosses inside Neptune's. In fact, Pluto spent much of the 1990s inside Neptune's orbit! This trait, of crossing planet orbits, is generally associated with asteroids and comets, not planets themselves.

If we try to define planets by size alone, we are confounded by the fact that some satellites are larger than the planets Mercury and Pluto, as shown in Figure 2-2. Pluto is only about half the size of our own moon.

Many astronomers are moving toward the view that it makes more semantic sense to say that our own solar system has eight planets, not nine! When Pluto was discovered in 1930, it was hailed as a newly discovered planet. However, today's evidence is that it should be demoted to an asteroid. The arguments include the following:

- Since the 1990s, many bodies up to half the size of Pluto have been discovered in the same region. Called the "Kuiper belt" objects, they are cataloged as asteroids or comet nuclei.
- Pluto crosses inside Neptune's orbit, like an asteroid or comet.
- Pluto is now known to be only about half the size of our moon, and a bit more than twice the size of the next-largest known asteroid, Ceres. Seven satellites are larger than Pluto.
- Pluto's orbit is considerably more inclined to the plane of the solar system than the orbit of any other planet, but well within the range for asteroids and comets.

Generally, the whole distinction between planets and "asteroids and comets" is vague, because planets formed by the aggregation of smaller bodies; the asteroids and comets are these smaller bodies or their broken debris, so it is hard to say what is the smallest planet and what is the largest asteroid or comet.

To confuse matters more, a number of bodies, mostly around the size of Jupiter, were detected during the 1990s, orbiting around other stars. They are too small to be stars themselves—that is, to have nuclear reactions inside. Are they planets, in the same sense as Jupiter or Earth? Most astronomers initially said yes, but we might ask if they

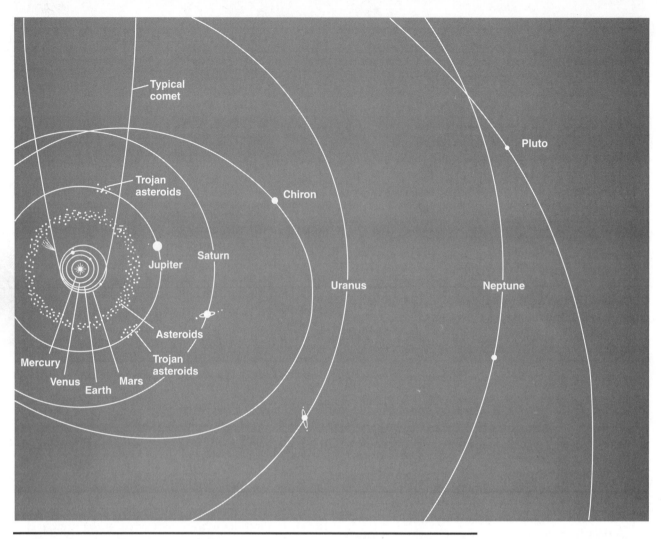

Figure 2-1. The arrangement of the nine principal planets and other lesser bodies of the solar system with orbits drawn to scale. This figure illustrates the relative closeness of the four terrestrial planets, the relatively vast distances in the outer solar system, the noncircular orbits of the small bodies Pluto and Charon, and the presence of comets and different classes of asteroids.

formed in the same way as the planets in our system. This is more than a semantic question because planet-sized bodies might form in different ways, and some of the extrasolar systems may be different from ours. For example, some roughly Earth-sized "planets" have been found around neutron stars—remnants of stars that exploded as supernovae. These might not be planets like ours but may have something to do with the debris from the explosion.

Operationally, we now define a **planet** as a nonstellar body larger than a certain size (1,000–3,000 km?), orbiting around a star. If we knew more about the orbits and other characteristics of recently discovered extrasolar planets, we might have a better answer to the question of whether they formed in the same way our system did.

Some additional terms can give more flexibility. **Planetary body** or **planetary material:** Includes planets and smaller objects, as used in Chapter 1. Planetary bodies are usually solid (with some liquid or gas components possible) and mostly composed of silicates or ices. Planetary bodies include the dust grains observed around some stars, asteroids, comets, and the planets orbiting other stars.

Planetesimal: One of the relatively small planetary bodies. Usually this word refers to one of the small planetary bodies in the *early solar system,* which aggregated into planets. The term usually implies a size between that of a microscopic dust grain (less than 1 mm in diameter) and a large asteroid (about 1,000 km in diameter).

Satellite: Any planetary body in orbit around a larger planetary body. At least 63 are known.

Asteroid: A small planetary body composed mostly of rock or metals; most orbit the sun between Mars and Jupiter. All known asteroids have diameters of less than 1,000 km. Usually the term is used only for bodies down to about 10 m in diameter, but there is no firm cutoff.

Comet: An ice-rich planetesimal that can emit an observable gaseous halo when its ice is warmed by the sun. Most comets spend most of their time in the outer solar system far beyond the asteroid belt. The term *comet* is

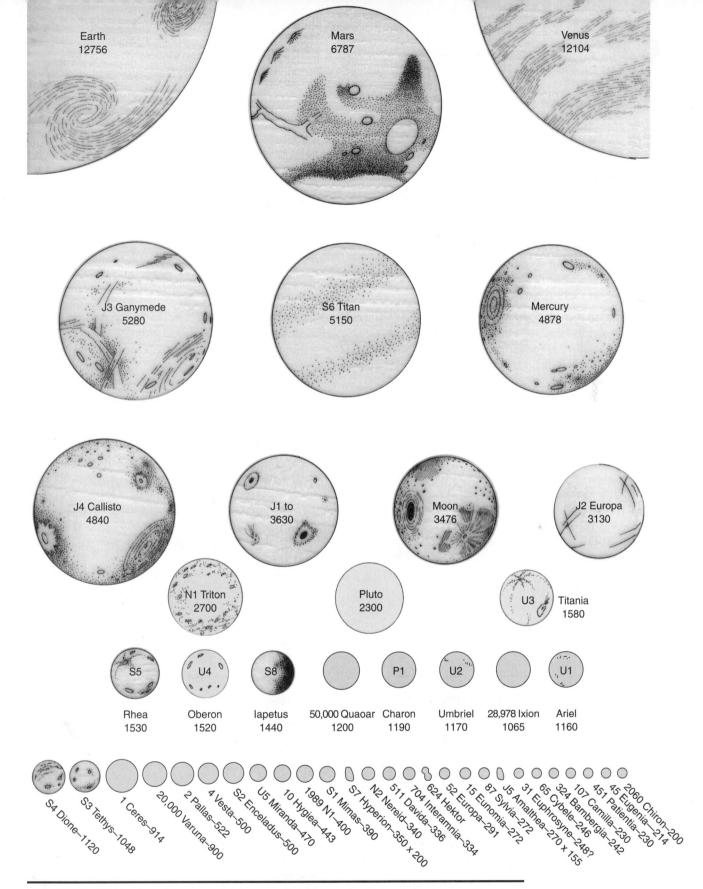

Figure 2-2. Relative sizes of the terrestrial planets and selected smaller bodies, drawn to scale. Diameters are given in km. Giant planets' satellites are indicated by letters J, S, U, and N and a number code, as well as names. (The number code is a traditional usage, indicating either the order outward from the planet, usually for larger numbers, or the order of discovery.) Asteroids have a number and a name. The three objects with numbers ≥ 20,000 are the largest known Kuiper Belt "asteroids" in the outermost solar system.

TABLE 2-1 Bode's Rule and Planetary Symbols

	Mercury	Venus	Earth	Mars	Asteroids	Jupiter	Saturn	Uranus	Neptune	Pluto
	4	4	4	4	4	4	4	4		4
	0	3	6	12	24	48	96	192		384
Predicted distance (AU)	0.4	0.7	1.0	1.6	2.8	5.2	10.0	19.6		38.8
Actual distance (AU)	0.4	0.7	1.0	1.5	2.8	.52	9.5	19.2	30.0	39.4
Symbol	☿	♀	⊕	♂	ⓝ[a]	♃	♄	⛢	♆	♇
No. known satellites	0	0	1	2	?	16[b]	18[b]	17[b]	8[b]	1

[a] The well-observed asteroids (roughly 2000 known) are numbered; the symbol is the encircled number.

[b] Voyagers 1 and 2 discovered several small satellites on the outskirts of the ring systems of Jupiter, Saturn, Uranus, and Neptune. The smallest may grade into the largest ring particles. At least four more probable moons were seen by Voyagers near Saturn.

ambiguous in that region because those comets are too far from the sun to be heated enough to make the ice turn into gas; thus they are **inactive** or **dormant comets** and are hard to distinguish observationally from asteroids. Many of the inactive comets have been cataloged as asteroids. In a sense, the "comet-asteroid" language breaks down beyond Saturn, where the bodies are surely icy but rarely give off gas. *Planetesimals* might be a better general term in that region.

World: A term useful in referring to any planetary body—planet or satellite—that is larger than 1,000 km in diameter. Bodies larger than that typically have enough internal energy to have individual geological "personalities." This term gained increased use in 1979 when the Voyager probes revealed that the four large satellites of Jupiter, originally assumed to be dead iceballs, instead had four distinct geologic characters, ranging from craters to smooth ice to erupting volcanoes.

Bode's Rule

In 1772, the German astronomer Johann Bode* popularized a simple empirical rule, **Bode's rule,** which is a helpful tool for remembering the distances of the planets from the sun. This rule should be memorized. Write down a sequence of 4s, one for each planet except Neptune and one for the asteroids, and add the sequence 0, 3, 6, 12, 24, . . . (see Table 2-1). Dividing by 10 gives the relative distances of the planets from the sun in terms of the mean distance from Earth to the sun. The latter distance is called the **astronomical unit** (AU). Thus, Jupiter is 5.2 AU from the sun.

Bode's rule lacked any theoretical justification, but it passed its first test in 1781 when the English astronomer William Herschel discovered Uranus at 19.2 AU. Afterward, a search was made for the "missing planet," which was supposed to lie between Mars and Jupiter. This led to the discovery of the first asteroid, Ceres, on the first night of the new century, January 1, 1801, by the Italian astronomer Giuseppe Piazzi, exactly at the predicted solar distance of 2.8 AU. Discoveries of more asteroids followed in subsequent years, most having solar distances around 2.8 AU.

Notice that from Bode's rule, planets are not evenly spaced, but each one tends to be nearly twice as far from the sun as the previous one. Modern work suggests that this spacing has to do with the way the aggregating planetesimals partitioned the solar system into zones, with each zone dominated gravitationally by one large body that grew by sweeping up the rest of the bodies.

A set of symbols for designating the planets is the one useful contribution of astrology. The symbols, listed in Table 2-1, form convenient subscripts and datum-point symbols in theoretical discussions and graphs.

*The rule was apparently first discovered by another German astronomer, Johann Titius; thus, it is often called the Titius-Bode rule.

A Survey of the Planets

Before proceeding to the chapters describing the dynamics, origin, and nature of the planets, we give a thumbnail sketch of each planet in order from the sun. These sketches emphasize a basic observational description of each planet; the interpretations appear in later chapters.

Mercury

Mercury is the second smallest planet and the closest to the sun. It has essentially no detectable atmosphere. Its surface has a slightly pinkish cast; Earth-based telescopes reveal only faint mottlings reminiscent of the appearance of the moon to the naked eye. Most of these dark regions are believed to be lava-covered areas similar to the dark lava plains of the moon. Like the dark regions on the moon and Mars, these dark regions are called *maria* (singular

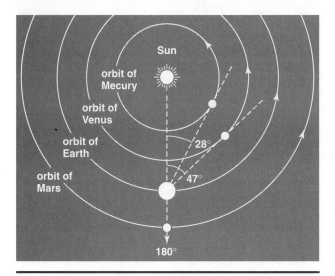

Figure 2-3. Because Mercury and Venus orbit between Earth and the sun, their maximum angular separations from the sun are 28° and 47°, respectively. Because the 28° angle is so small, solar glare makes Mercury difficult to see. Exterior planets, however, can appear at opposition—that is, at 180° from the sun.

mare) from the Latin word for sea, because early astronomers erroneously thought they were oceans.

Because Mercury's orbit is so close to the sun, Mercury is usually hidden by the sun's glare, as seen from Earth, as shown in Figure 2-3. It can be seen only at twilight or in the daytime sky, and even with a telescope it is hard to observe.

On three occasions in 1974, the American spacecraft *Mariner 10* sailed within 6,000 km of Mercury and made the first close-up photos of the planet, showing roughly half the planet in detail. The pictures, such as Figure 2-4, show the planet to be moonlike and crowded with **impact craters,** which are circular depressions caused by impacts of interplanetary debris. They are virtually identical with the impact craters on the moon. The lava plains are not as clearly defined as are those on the moon. Rugged cratered areas on planets like Mercury are called **uplands.** As seen in Figure 2-5, they dominate the planet. They also contain many areas of crater-damaged, dusky plains that may be very old lava flows.

Before the 1960s, astronomers tried to map the dusky markings to clock the rotation of Mercury. Most of them concluded that Mercury kept one side toward the sun during its 88-d (day) orbital period, just as the moon keeps one face toward Earth. To their embarrassment, radar work in the 1960s showed that Mercury rotates not in 88 d but in 59 d, and it does not keep one side toward the sun (Pettengill and Dyce, 1965)! The 59-d rotation period is exactly two-thirds of the 88-d revolution period; this relationship of two rotational or orbital periods by a simple fractional ratio, such as 2:3, 1:2, or 1:3, is called a **resonance** and is maintained by certain dynamical forces (see Chapter 3).

Venus

Venus is sometimes called Earth's sister planet. It most nearly matches Earth in size, and its orbit is closest to Earth's orbit. Its very dense atmosphere is composed mostly of carbon dioxide (CO_2). Opaque white or yellowish clouds hide its surface.

Because of the highly reflecting clouds and its close approaches to Earth, Venus is very bright when it appears in our sky. Like Mercury, it moves only a small angular distance from the sun as seen from Earth (see Figure 2-3). When it moves into a position to be our **evening star** or **morning star** (setting or rising a few hours after or before the sun), it is brighter than any other planet and is some 15 times brighter than the brightest star (Sirius); it can even cast shadows.

Through a telescope, Venus appears as a nearly blank disk, occasionally mottled by barely visible dusky bands or patches. These cloud patterns are much more prominent in ultraviolet light because some of the cloud or atmospheric constituents absorb the ultraviolet (Figure 2-6). Spacecraft photographs taken with ultraviolet filters clearly reveal the roughly banded cloud patterns, which lack the great cyclonic whorls characteristic of Earth's low-level clouds.

Because of Venus's cloud cover, we cannot watch the rotation of the solid surface of the planet, and until the 1960s, the rotation period was a mystery. Repeated attempts

Figure 2-4. A mosaic of Mariner 10 photos of Mercury shows a heavily cratered planet that resembles the moon. (NASA)

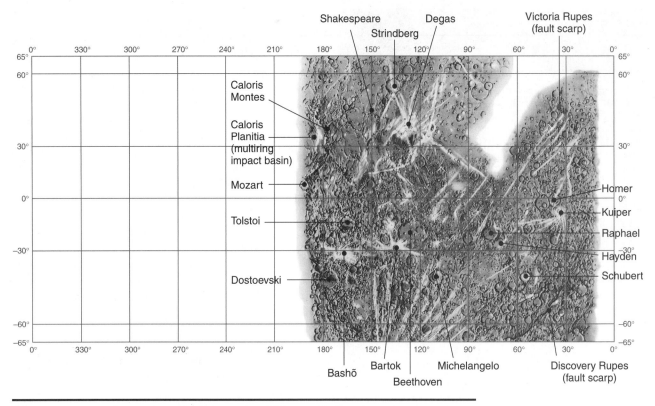

Figure 2-5. Map of the portion of Mercury mapped by Mariner 10, the only portion presently known in detail. Most features are named after historical figures in the arts. (Base map courtesy R. M. Batson, U.S. Geological Survey)

to determine the rotation by watching motions of the cloud markings on Earth-based ultraviolet photographs were frustrated by rapid changes in the clouds and their ill-defined patterns.

In the early 1960s, radar techniques yielded the totally unexpected finding that Venus rotates in 243.1 d, not in the same direction as Earth and most other bodies rotate, but backward, from east to west (Dyce, Pettengill, and Shapiro, 1967)! This peculiar situation may be abetted by dynamical resonance between Earth and Venus (Goldreich and Peale, 1967), although the resonance is apparently not exact. This east-to-west motion (clockwise as seen from the north side of the planet), whether in a planetary body's orbit or spin, is called **retrograde motion,** as contrasted to the usual west-to-east or **prograde motion** (counterclockwise as seen from the north).

Ultraviolet photos show that the high-atmosphere clouds circulate east to west in roughly 4 d (Smith, 1967), a result confirmed by spectroscopic studies from Earth (Betz and others, 1977) and by spacecraft photography (Murray and others, 1974). This movement corresponds to a circulating wind near the cloud tops (40 to 60 km above the surface) of about 100 m/s (meters/second). Other winds have been measured by spacecraft and spectroscopy, ranging from very gentle surface breezes to a 125 m/s wind from daytime to nighttime hemispheres at an altitude of around 115 km (Betz and others, 1977).

What lies hidden beneath the clouds of Venus? Decades of speculation have produced many theories. Many early 20th century scientists thought the clouds were composed of water droplets and that the surface must be a steamy jungle of vegetation swept by torrential rains. Others suggested a windy, dry desert or an ocean of oily liquid. Decades of speculation were ended by spacecraft encounters with the planet and new techniques of intensive Earth-based observation in the 1960s. Venus was first observed at close range on December 14, 1962, by the U.S. spacecraft Mariner 2, which passed 38,854 km (21,645 mi) from the surface of the planet. The first contact with the planet was achieved October 18, 1967, when the Soviet spacecraft Venera 4 parachuted into the atmosphere and radioed back data. On December 15, 1970, **Venera 7** made the first successful unmanned landing on Venus, sending back more atmospheric data. Finally, on October 22 and October 25, 1975, Veneras 9 and 10 sent back the first photos of the surface (Chapter 10), which reveal a desolate, dry, rock-strewn landscape (Florensky, Basilevsky, and Pronin, 1977; Keldysh, 1977). The atmosphere at ground level is surprisingly clear. Gravel and rounded bedrock outcrops give evidence of erosive and weathering processes.

Earth-based instruments and the Soviet landers showed that the surface temperature and pressure are extremely high, about 750 K (890° F) and 90 atmospheres (90 times the sea level air pressure on Earth), respectively.

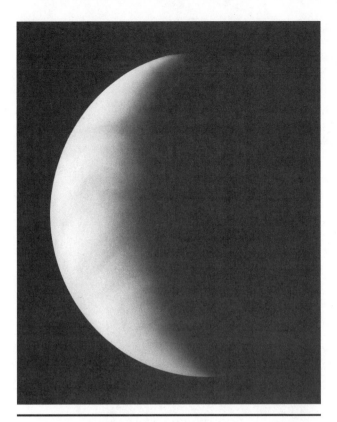

Figure 2-6. Ultraviolet image of Venus from Pioneer Orbiter in 1978 shows faint patterns of cloud bands in the opaque atmosphere. (NASA)

The mystery of the cloud composition was solved when Sill (1973) and Young (1973) independently found that the cloud properties could be explained by droplets of concentrated sulfuric acid (H_2SO_4 dissolved in water). The droplets are typically 2 μm across, with an acid concentration of 78% to 90% (Pollack and others, 1978). Venus is thus a fearsome place—hot and dry with a dense CO_2 atmosphere and sulfuric acid clouds.

Space probes orbiting around Venus have bounced radar waves through the clouds and off the surface, yielding detailed images and topographic maps of surface features (Figure 2-7). Global mapping was first done by the Soviet Venera 15 and 16 probes starting in 1983, and with more detail by the American Magellan probe in 1990–1992. The results reveal a volcanically active sister to planet Earth. Venus's volcanic plains are broken by a few Australia-size uplands 2 km to 5 km high, with a few volcanic peaks as high as 11 km above the plains, exceeding Mt. Everest's 8 km above sea level (Figure 2-8). Meteorite impact craters, sparsely scattered on the plains, suggest a surface nearly as young as Earth's, with a fairly uniform age of about 500 My to 800 My. With some surprise, geologists have interpreted this as suggesting a global resurfacing event at that time, perhaps due to major turnover in the planet's mantle. Fractures testify to seismic unrest, and some of the volcanoes may be active today. Circular features called coronas, a few hundred km across, may be caused by local upwelling currents in Venus's mantle (Phillips, Grimm, and Malin, 1991; Saunders and Pettengill, 1991).

Earth

The next planet from the sun is characterized by shifting white clouds and a bluish color resulting from sunlight scattered by molecules in its atmosphere. Different from the CO_2-dominated atmospheres of its neighbor planets, Venus and Mars, its atmosphere is mostly molecular nitrogen (N_2) and molecular oxygen (O_2), with variable traces of water vapor (H_2O). Its surface is very unusual, being temperate and 71% covered with liquid water.

The most important and unique characteristic of this planet—**Earth** (Figure 2-9)—is its widespread life, both on land and in the oceans. Protecting this life appears to be a major challenge to Earth's current civilization.

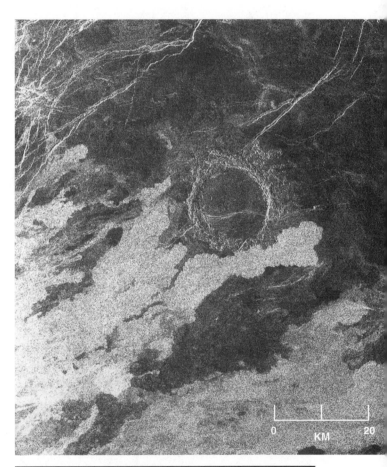

Figure 2-7. Radar image of impact crater on volcanic terrain of Venus. Dark background material is an old lava plain, cut by tectonic fractures (bright lines), typical on Venus. The impact crater (center circle) was formed on this surface, but was partly filled in by lavas. Tongues of new lava flows extend toward the crater from the southwest; they are bright because rough, young lavas reflect more radar than old, eroded lavas. (NASA Magellan mission image)

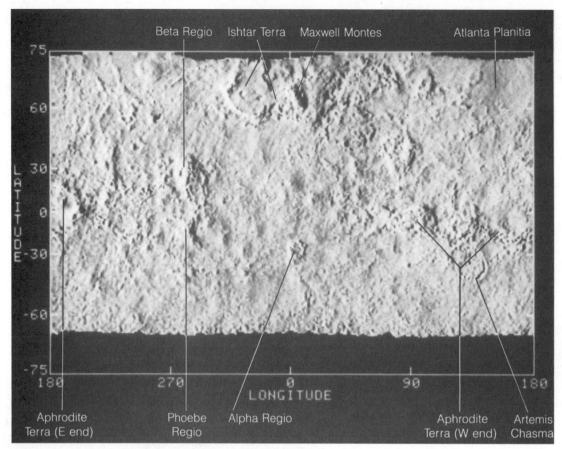

a

Beta Regio   Ishtar Terra   Maxwell Montes   Atlanta Planitia

Aphrodite
Terra (E end)   Phoebe
Regio   Alpha Regio   Aphrodite
Terra (W end)   Artemis
Chasma

LATITUDE

LONGITUDE

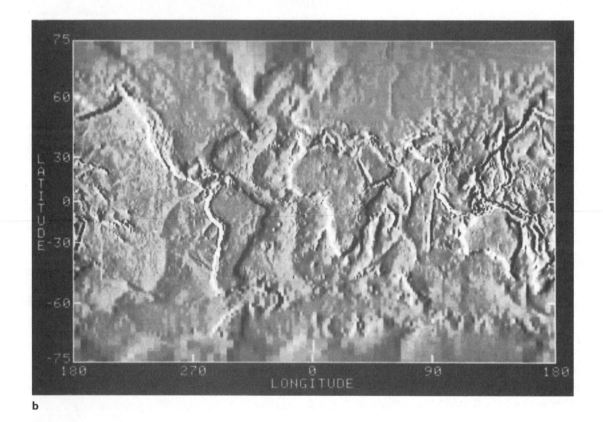

b

LATITUDE

LONGITUDE

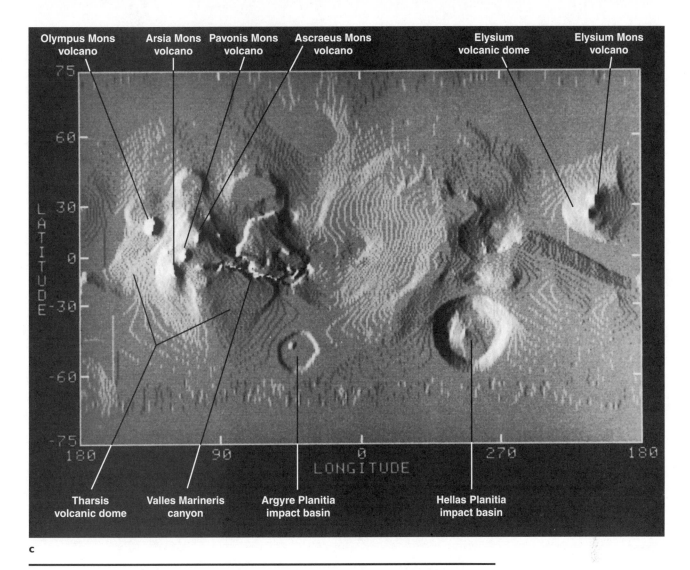

Olympus Mons volcano · Arsia Mons volcano · Pavonis Mons volcano · Ascraeus Mons volcano · Elysium volcanic dome · Elysium Mons volcano

Tharsis volcanic dome · Valles Marineris canyon · Argyre Planitia impact basin · Hellas Planitia impact basin

c

Figure 2-8. Comparative relief maps of (**a**) Venus, (**b**) Earth, and (**c**) Mars. Altitude data have been digitized and presented as shaded relief with light from the left. Venus has broader expanses of rolling plains than Earth but also has some elevated, continentlike masses (such as Aphrodite Terra) and features resembling sea-floor trenches (such as Artemis Chasa). Mars retains a more primitive topography with traces of old impact basins (such as Hellas). Martian relief is also dominated by large domes of volcanic lavas surmounted by high volcanic peaks (such as Olympus Mons atop the Tharsis dome). (Images courtesy Michael Kobrick, Jet Propulsion Laboratory)

Moon

Earth has one natural satellite, the **moon,** whose character remained rather mysterious during centuries of telescopic study. The broad dark patches composing the "man in the moon" were first thought to be seas and were given the Latin name *maria* (MAH-ree-a; singular *mare,* pronounced MAH-ray); these dark patches were finally found to be plains formed by lava flows about 3,500 My ago. Early observers gave these plains Latin names such as Mare Tranquillitatis (Sea of Tranquility) and Mare Imbrium (Sea of Rains). As on Mercury, the cratered regions of lighter color are called uplands.

After a century of debate, researchers concluded that most of the craters are caused not by volcanism, but by meteorite impact. As shown in Figure 2-10, craters and lava plains have been mapped in detail on both sides of the moon.

Unmanned landers in the 1960s first revealed that the lunar surface is covered by a dusty layer called a **regolith**—a layer of powdery soil and scattered rock fragments created by eons of bombardment by meteorites.

The moon's history remained enigmatic until 1969, when astronauts explored the ghostly lunar landscape and brought back the first rock and soil samples. In 1969–1972, lunar samples were returned from six Apollo

Figure 2-9. Photo of the Earth and the moon, taken from the Galileo space probe as it left the Earth-moon system on its way to Jupiter. (NASA)

landing expeditions and three sites visited by Soviet unmanned vehicles. These samples indicate that the moon has a bulk chemical composition roughly similar to that of Earth's outer mantle layers, with many rock types known to terrestrial geologists. However, the moon lacks water and other volatile compounds. These chemical differences are clues to the moon's origin and history and are the subject of much vigorous research. The rocks show that the moon, like Earth, formed about 4500 My (million years), or 4.5 Gy* ago. The mode of formation is uncertain. The early moon underwent intense bombardment by meteorites, forming most of the many craters shown in Figure 2-11 by 4.0 Gy ago. The early moon was volcanically active, and about 3.5 Gy ago, massive lava flows created the dark plains or maria. Most lunar volcanism ceased about 2 Gy ago. Most of the Apollo astronaut landings occurred on the maria, whose dusty, rock-strewn landscapes became familiar when the lunar exploration was broadcast live from the moon's surface into our living rooms.

Mars

If Venus is Earth's sister planet, **Mars** is a smaller brother, half the size of Earth. It has many Earthlike features, including an atmosphere with clouds and ice deposits at its poles, called **polar caps.** It rotates in 24½ h (hours), about the length of Earth's day. Unfortunately for the proposed family resemblance as well as for future manned exploration, the atmosphere is very thin, cold, and dry. Compared to Earth's sea level surface pressure of 1000 mbar (millibars), Mars's surface pressure in most regions is about 5 to 8 mbar and would require visitors to wear a spacesuit. Martian ground temperature exceeds freezing only on the warmest summer days (though the air temperature usually stays below freezing), and the air has only minuscule traces of water vapor. There is no liquid water; all the water of the planet is in the form of ice and water molecules trapped in the minerals. The famous red **surface coloration** is caused by oxidized iron minerals, or rustlike

---

*In the SI (International System) system of metric units, the prefix "giga," abbreviated G, stands for 10^9. The figure 4.5 Gy is also commonly referred to as 4.5 billion years.

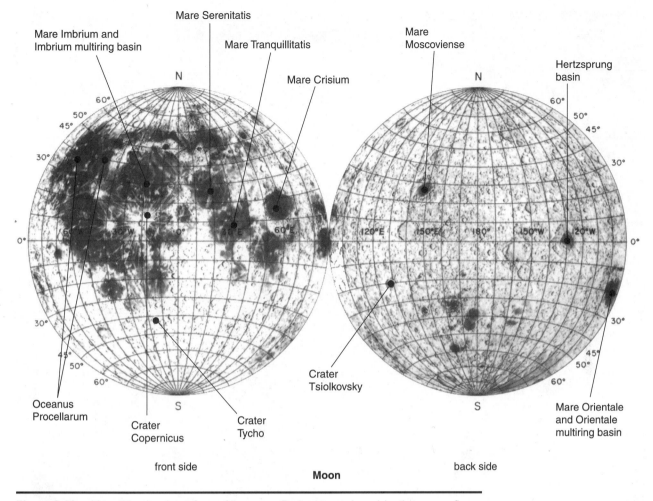

Mare Imbrium and
Imbrium multiring basin

Mare Serenitatis

Mare Tranquillitatis

Mare Crisium

Mare
Moscoviense

Hertzsprung
basin

Oceanus
Procellarum

Crater
Copernicus

Crater
Tycho

Crater
Tsiolkovsky

Mare Orientale
and Orientale
multiring basin

front side

Moon

back side

Figure 2-10. Map of front and rear faces of the moon. Dark areas are lava plains that occupy floors of large impact basins. (Base map courtesy R. M. Batson, U.S. Geological Survey)

material. This color is prominent even when Mars is viewed with the naked eye from Earth.

As shown in Figure 2-12, dusky markings have been mapped on Mars since 1659. They intrigued early astronomers because they change shape and intensity from season to season and year to year. Observers in the 1800s interpreted the seasonal changes as strong evidence that the dark markings were Martian vegetation.

When observers such as William Herschel discovered clouds and white polar ice caps around 1800, they thought these features bolstered the philosophical idea called the **plurality of worlds**—the idea that other planets were like Earth, and probably inhabited with other of God's creatures. By the late 1800s, astronomers began to realize that the Martian climate was inhospitable, but Darwin's 1859 theory of evolution and adaptation made it reasonable to suppose that Martian creatures might have evolved and adapted to their planet's conditions.

During the 1877 approach of Mars, the Italian observer Giovanni Schiaparelli called attention to the streaky

Martian markings, which he called by the Italian term, *canali*. Contrary to common belief, Schiaparelli did not discover these streaks; they had been drawn earlier, as shown in Figure 2-12b. However, Schiaparelli drew them differently, as straight or curved narrow lines, as can be seen in Figure 2-12c.

In the 1890s, Percival Lowell, a flamboyant American astronomer, pulled these discoveries into a radically new conception of Mars that influenced many 20th-century ideas. Lowell saw the lines, which he called **canals**, as very sharply defined, like steel engraving lines. He said they really were canals, built by a civilization of intelligent Martians. The Martians, said Lowell, were having trouble surviving because the weak gravity of their planet was allowing its water to leak off slowly into space.* Therefore, the Martians had constructed a vast network of canals to

---

*This idea of gases escaping into space more easily on weak-gravity planets is scientifically correct, as we see in Chapter 12.

Figure 2-11. The moon is dominated by rugged regions covered by craters mostly formed during intense meteorite bombardment about 4.5 to 3.5 Gy ago. In the upper left are dark, smooth lava plains called *maria*, dating back to about 3.5 Gy. (NASA, Apollo 16)

carry the spring runoff from the melting polar ice down to the warmer equatorial regions, where they could cultivate the vegetation.

Lowell's theory was wonderful, but it was all wrong. Its underpinnings, the canals, essentially do not exist as he drew them; they are only wispy streaks and splotches, as shown in Figures 2-13 and 2-14. The Martian canal affair occurred because the human eye tends to interpret rough alignments as linear streaks. Lowell was especially inclined toward this and even saw lines on Venus.

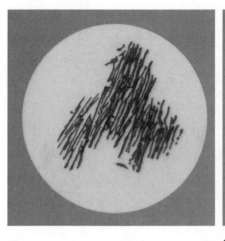  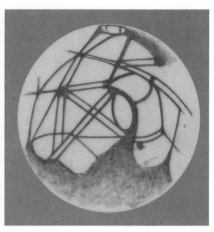

a   b   c

Figure 2-12. Drawings of Mars through telescopes. (**a**) One of the earliest known sketches of Mars was drawn by Christian Huygens on November 28, 1659. The northward-extending triangle is probably a dark region known as Syrtis Major. (After Huygens) (**b**) In a drawing by English observer W. R. Dawes during the 1864–1865 opposition, Syrtis Major extends northward but with a streaky extension of a type later called a canal. Clouds covering the north polar ice cap are outlined by a dotted line at the top. (After Dawes) (**c**) Italian observer Giovanni Schiaparelli drew Mars in 1888 with the dark triangle of Syrtis Major again visible with a streaky, dark tail. Schiaparelli first popularized the canals, shown by him here as nearly straight lines and line pairs covering Mars. However, spacecraft show that the canals do not exist in this form. Because we see Mars from different angles in different years, Syrtis Major may appear above or below the disk center. (After Schiaparelli)

Figure 2-13. Mars viewed from the Mars Global Surveyor spacecraft in 1997 shortly before it went into orbit. This image, sharper than any Earth-based pictures, shows bright clouds (right) and the types of dark streaky markings (lower third) that were mistaken for linear "canals" by many early observers. (NASA and Malin Space Science Systems)

Lowell's ideas, though disproven, forced scientists and intellectuals to recognize that theories of planetary and biological evolution raise the possibility of life on other worlds if the conditions are suitable. This idea shook intellectuals and the public at large and remains exciting today.*

On July 15, 1965, the first close-up spacecraft photos of Mars were taken by the American spacecraft Mariner 4. There were no signs of dying Martian cities or canals—only a moonlike cratered surface. In 1969, Mariners 6 and 7 flew by, showing more craters and jumbled, collapsed depressions that geologists called **chaotic terrain.** Although the nature of the changing dark markings was still unknown, many astronomers now concluded that Mars was dead and moonlike—similar to the moon but with a little wind to blow the dust around.

On November 14, 1971, Mariner 9 became the first spacecraft to go into orbit around Mars. This mission mapped the whole planet in detail for the first time, as shown in Figure 2-14, and proved that the vision of a moonlike Mars was too pessimistic. To everyone's surprise, Mariner 9 photographed young volcanoes, sand dunes, and many examples of dry riverbeds! Examples of the

---

*A more detailed, nontechnical history of the changing theories is given by Hartmann (2003). An excellent biography of the colorful Percival Lowell is given by Hoyt (1976).

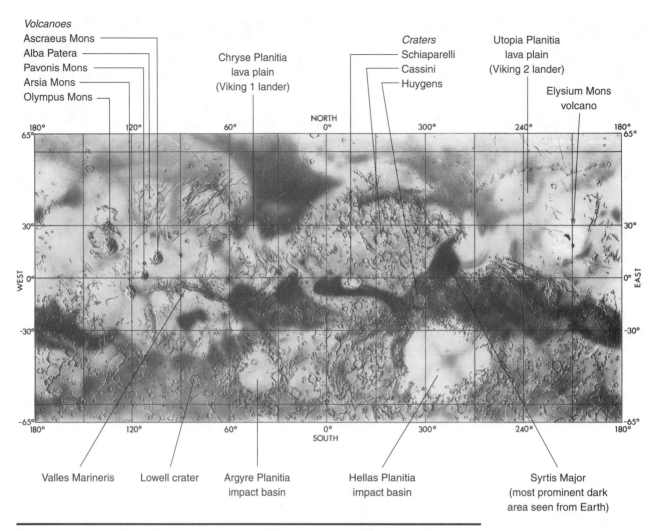

Volcanoes
Ascraeus Mons
Alba Patera
Pavonis Mons
Arsia Mons
Olympus Mons

Chryse Planitia
lava plain
(Viking 1 lander)

Craters
Schiaparelli
Cassini
Huygens

Utopia Planitia
lava plain
(Viking 2 lander)

Elysium Mons
volcano

Valles Marineris

Lowell crater

Argyre Planitia
impact basin

Hellas Planitia
impact basin

Syrtis Major
(most prominent dark
area seen from Earth)

Figure 2-14. Map of Mars. The prominent dark and light markings are semipermanent features seen from Earth, created mostly by dust deposits. The underlying topographic features, such as craters and mountains, were first mapped in the 1970s by spacecraft. (Courtesy R. M. Batson, U.S. Geological Survey)

riverbeds appear in Figure 2-15. Some of them emanate from the previously discovered chaotic terrain. After much controversy, most planetary geologists concluded that these features, called **channels** (not to be confused with the semi-illusory canals!), really were carved by running water. As discussed in more detail in Chapters 11 and 12, this finding suggests that the climate of Mars may have been more clement in the ancient past.

Mariner 9 also indicated that substantial amounts of H_2O are frozen at the poles under a transient winter cap of CO_2 snow and accompanied by stratified polar sedimentary deposits. The evidence for abundant ice and past liquid water raised profound questions. If liquid water existed on Mars in the past, did life evolve at that time? If so, what happened to that life? If not, why not?

Because no one had yet seen Martian details smaller than a couple hundred meters, astronomers still wondered

if the surface supported plants or even animals! The next step was to land cameras on the surface to see what was there. The first human-made devices on Mars were Soviet probes that failed before or after touchdown. Mars 2 crashed in November 1971; Mars 3 landed but failed 20 s (seconds) after touchdown in December 1971; Mars 6 failed moments before touchdown in February 1974.

July 20, 1976, the seventh anniversary of the first lunar landing, was a better day. The Viking 1 lander made the first successful touchdown on the planet (Figure 2-16). In the following months, Viking 1 and its sister ship, Viking 2 (which landed September 3, 1976) found a striking desert landscape devoid of any obvious life. There was not even any organic material in the Martian soil, measured to a precision of a few parts per billion. Although three experiments designed to seek life found peculiar reactions in the soil at both landing sites, most

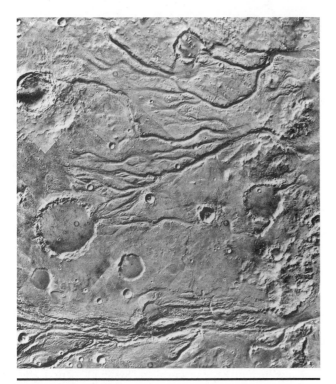

Figure 2-15. Dry riverbeds on Mars. These features, named channels (not canals), were apparently cut by running water. Like arroyos in Earth's terrestrial regions, they have tributary systems and get wider and deeper in the downhill direction. (NASA, Viking orbiter)

scientists relate those reactions to unusual soil chemistry. Mars Pathfinder landed at a third site on July 4, 1997, and photographed a similar terrain strewn with lavalike rocks. In the 1980s–90s, scientists identified meteoritic rocks on Earth that were blown off Mars by asteroid im-

pact. These priceless rocks confirm excitingly young volcanic and magmatic activity on Mars—as recently as 0.17 Gy ago.

Mars has two small satellites, Phobos and Deimos, discovered in 1877 by the American astronomer Asaph Hall. Curiously, literary works such as Swift's *Gulliver's Travels* (1720) and Voltaire's *Micromegas* (1750) referred to two moons of Mars years earlier. This led some pseudo-science writers to suggest occult knowledge of the two moons centuries ago, but the explanation is simpler. Johann Kepler, who discovered the laws of planetary motion around 1610, believed in numerology and suggested that if Earth had one moon and Jupiter four (the four discovered by Galileo at that time), then Mars should have two to maintain the progression. Later spacecraft have searched for smaller satellites and found none down to diameters of 1.6 km (Pollack, 1973).

Close-up photos of Phobos and Deimos show irregularly shaped satellites about 20×28 km and 10×16 km, respectively, with heavily cratered surfaces and brownishblack surface material, as shown in Figure 2-17 on page 24. In color and appearance, these two satellites resemble certain dark asteroids composed of carbon-rich minerals.

Jupiter

Jupiter is by far the largest planet, having more mass than all the other planets put together. It has more than three times the mass of the next largest planet, Saturn, and about a 20% larger diameter. Dynamically the solar system can be thought of as composed of two main bodies, the sun and Jupiter; Jupiter has 0.001 the mass of the sun, and the other bodies are negligible.

Jupiter has a dense atmosphere of molecular hydrogen (H_2, 79% by mass), helium (19%), and trace amounts of

Figure 2-16. The first photograph transmitted from the surface of Mars showed soil near the footpad of the Viking 1 lander. During the touchdown, some soil was blown into the footpad (lower right). The camera scanned vertically, beginning at the left and proceeding to the right, within minutes after touchdown. Light and dark streaks at left, therefore, may be light haze and shadows caused by dust clouds kicked up during the landing. The largest rock (center) is about 10 cm (4 in.) across. (NASA)

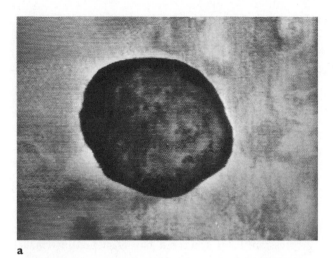

Figure 2-17. Views of Mars's satellite, Phobos. (**a**) Phobos hanging in front of craters and dusky markings of Mars, as seen from the Russian PHOBOS-2 probe in 1989. Phobos and Mars are equally illuminated by the sun; the view emphasizes how black Phobos's surface soil is, compared to Mars. (Courtesy B. Zhukov, IKI [Institute for Cosmic Investigations], Moscow) (**b**) Crescent view with the right side faintly illuminated by Mars. The sun is out of the picture to the left, and Mars is out of the picture to the right. The irregular shape of Phobos can be traced around the entire outline. (NASA, Viking Project; courtesy T.C. Duxbury)

water vapor (H_2O), methane (CH_4), and ammonia (NH_4). The planet is covered by clouds, arranged in dark **belts** and bright **zones,** parallel to the equator (Figure 2-18). The belts and zones are semipermanent and have been named, as shown in Figure 2-19. The clouds show a variety of colors—browns, tans, yellows, and reds. Large irregularities,

such as dark, oval-shaped clouds that may last for years, are called **disturbances.**

Another long-lived feature is the **Great Red Spot,** prominent in Figures 2-18 and 2-19. Probably first seen by Robert Hooke in 1664 or Giovanni D. Cassini in 1665, it was named in 1878 when it became very prominent and was rediscovered (Peek, 1958; Chapman, 1968). Time-lapse movies by Voyagers 1 and 2 in 1979 showed that the Great Red Spot is characterized by circulating currents—small clouds caught in the Red Spot will spiral around counterclockwise like a leaf caught in a whirlpool. It is a giant whirlpool indeed; somewhat variable in size, it can reach 4 Earth diameters!

Telescopic observers in the last century estimated the rotation rate of the planet by watching the cloud features as the planet turned, but they found that different cloud features circulate at different rates, due to zonal winds. The basic rotation rate of the underlying planet is probably about 9h (hours) 55m (minutes) 30s (seconds), a value associated with radio pulses associated with the planet's magnetic field. The equatorial clouds move faster than this because of jet streamlike winds, and give a rotation period of only 9h 50m 30s, the so-called "System I" period. High latitude clouds also show a "System II" rotation, or circulation period of 9h 55m 41s, closer to the planetary value. The Great Red Spot drifts along with a variable period, drifting sometimes ahead of and sometimes behind other features in System II; Peek (1958) gives values $9^h 55^m 31^s$ to $9^h 55^m 44^s$ over a 76 y (year) period.

The first spacecraft to Jupiter was the American probe Pioneer 10, which made some crude images and measurements during a fast flyby on December 4, 1973. It was followed by a similar probe, Pioneer 11, in December, 1974. The first detailed images of Jupiter and its four large moons were made by the U.S. probes Voyager 1 (March 1979) and Voyager 2 (July 1979) as they flew through the Jupiter system on their way to the outer planets (Figure 2-20). These were followed by the U.S. Galileo probe that went into orbit around Jupiter in 1995 and spent several years cruising from one moon to another, returning a wealth of data about the system. These vehicles took superb photos of Jupiter and the moons and gathered abundant additional data. Voyager 1 also discovered a ring around Jupiter, fainter and narrower than the famous rings of Saturn and probably composed of fine, rocky particles.

Jupiter has a complex system of at least 28 satellites outside the rings, somewhat resembling a small solar system. The four large moons are called the **Galilean satellites** because they were discovered by Galileo. The discovery date was January 7, 1610; on the next night they were independently discovered by the German astronomer Marius, who named them *Io, Europa, Ganymede,* and *Callisto* (in order outward from Jupiter) after associates and paramours of Zeus, the Greek version of the Roman god Jupiter. Jupiter's moons are also labeled with numerals; the Galilean satellites are J1, J2, J3, and J4 in the order given above. Subsequent numbers are assigned in order of discovery.

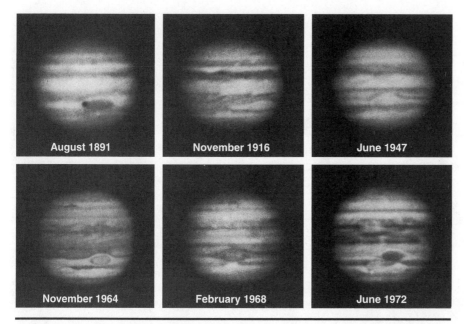

Figure 2-18. Photographs of Jupiter spanning 81 years, showing the changing array of Jupiter's belts and zones. The dark north temperate belt is relatively permanent, but the equatorial zone changes from bright (1891) to dark (1972). Many of these images show the Great Red Spot. (Lowell Observatory)

Within the orbits of the Galilean moons is the largest remaining moon, Amalthea (numbered J5), about 155 × 270 km in size, dark reddish-gray in color, and heavily cratered. The Voyagers discovered three other small moons in this region. Two are about 30 to 40 km across, very dark in color, and located between the rings and Amalthea

(Jewitt, 1979). The other is about 75 km across and is located between Amalthea and Io.

Beyond the Galilean moons are two peculiar groups of small moons. The inner group includes J13, J6, J10, and J7, in orbits from about 11,000 km to 12,000 km from Jupiter, and all inclined between 24° and 29° to the

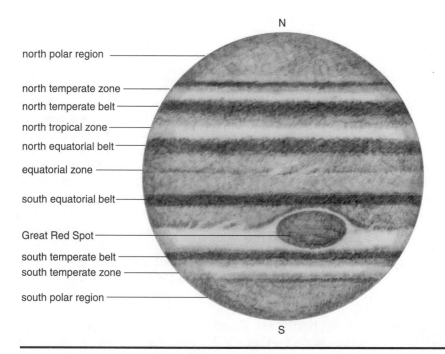

Figure 2-19. Semipermanent cloud formations of Jupiter, visible from Earth through small telescopes.

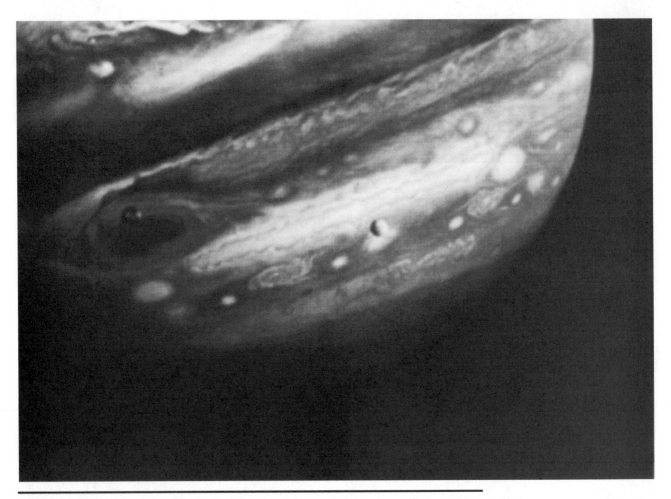

Figure 2-20. Jupiter looms beyond two of its satellites. Europa is the lighter moon (right), and Io the darker moon silhouetted against the Great Red Spot. Numerous other cloud formations on Jupiter are prominent in this Voyager 1 view from a distance of 20 million kilometers. Europa and Io are about the size of our moon. (NASA)

planet's equator. The second group, including J12, J11, J8, and J9, all lie at distances of about 21,000 km to 24,000 km with inclinations of 147° to 164°. (The use of an inclination figure greater than 90° is a convention for indicating a retrograde direction of revolution. Thus, an inclination of 147° is equivalent to an inclination of 180° − 147° = 33° but with retrograde orbital motion.) Why the outer satellites of Jupiter should be arranged in two such tidy groups rather than randomly scattered is a mystery to which we return in Chapter 5. The surfaces of at least some of these moons are black, owing to a soil rich in carbonaceous minerals like those in certain dark asteroids and meteorites. In addition to the eight moons mentioned above, many other captured moons have been found since 2000.

The "total number of moons" of a giant planet has become a semantic question. Ultra-sensitive telescopes routinely turn up new kilometer-scale moonlets, and 100-meter-scale objects may exist by the hundreds. Many are not original, but captured. This book concentrates on "world-class" examples.

Orbital calculations (for example, Carusi, Valecchi, and Kresak, 1981) suggested that cometary or asteroidal bodies are sometimes captured by Jupiter into temporary satellite orbits lasting months or years. A moonlet photographed several times in 1975 by Charles Kowal may have been one of these. The concept of temporary captures by Jupiter was confirmed in spectacular fashion in 1993 when astronomers Carolyn and Gene Shoemaker and David Levy discovered a cometary body caught in a loose, temporary orbit around Jupiter. Because of dynamical forces during the close pass by Jupiter, when it had been captured into orbit, the weak cometary body had been broken into a string of fragments that were destined to make a wide orbit around Jupiter and then crash into the planet. The spectacle climaxed in 1994 when the string of fragments struck Jupiter, making explosions that were visible from Earth, as seen in Figure 2-21 (Beatty and Goldman, 1994). This episode reminds us that throughout geologic time, the planets have been interacting with the smaller interplanetary bodies, resulting in explosions, impact craters, and occasional climatic disturbances.

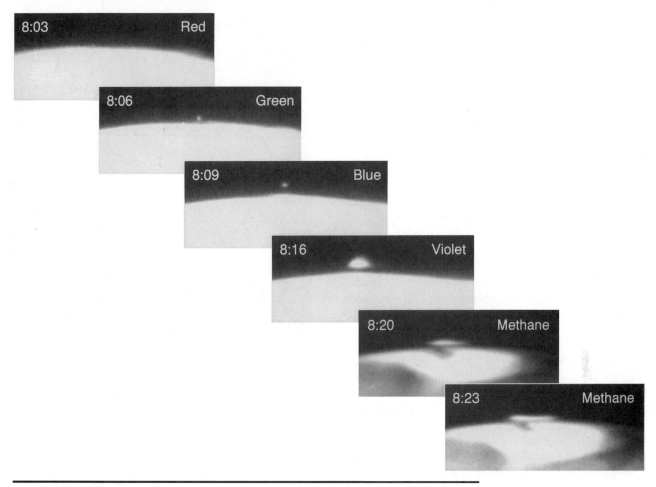

Figure 2-21. Sequence of photos showing the fireball rising and spreading over the edge of Jupiter as a fragment of comet Shoemaker-Levy 9 collided with the planet and exploded above Jupiter's clouds. Sequence covers a 20-minute period on July 22, 1994. (Hubble Space Telescope images)

Small Worlds of the Outer Solar System: A Simplified Overview

Telescopic views of even the largest satellites of the outer solar system revealed only pinhead-sized disks (less than 2 arc seconds across) on which visual observers in the mid-20th century reported dusky shadings. Spectroscopists, by the 1970s, reported H_2O ice with admixtures of soil. Astronomers tended to assume that these worlds would be rather alike—cratered, dead iceballs resembling icy versions of our own moon. Close-up photos of Jupiter's and Saturn's moons by Voyagers 1 and 2 revealed the opposite—an astounding variety from one world to the next, as if each were attempting to assert its own uniqueness!

The variety might make theorists throw up their hands and conclude that each world is governed by laws unto itself. But we can make some sense of the pattern by keeping in mind the following simple model. Theories of mineral condensation in the primordial solar system (which we discuss in more detail in Chapter 5) suggest that the main materials formed in the cold, outer parts of the system would be frozen water and a type of sooty-black dust rich in carbon minerals. Spectrophotometric observations seem to confirm that either ice or dark soil, or a mixture of the two, are the main constituents visible on the surfaces of most moons and all interplanetary bodies from the outer asteroid belt outward. Note that a chunk of frozen water is relatively stable against sublimation by solar heating only beyond about 4 AU (Lebofsky, 1975). If mixtures of ice and black soil are finely powdered and mixed in a regolith soil layer, the resulting material will look dark if the soil content is more than a few percent, because the opaque black minerals efficiently absorb the light. Thus, a primitive surface of, say, 50% ice and 50% carbonaceous soil would look dark. Similarly, any surface in which the carbonaceous component had been concentrated would look like dark soil (even though it might have some ice content). On the other hand, if heating occurred, the denser soil would sink and watery "lava" could coat the surface with regions of relatively pure bright ice or ice/soil layers (Hartmann, 1980). If enough heating occurred to evaporate most of the water, components such as sulfur compounds could be left on the surface (Fanale, Johnson, and Matson, 1974).

As we now see, this "salt-and-pepper" model ("salt"—bright ices; "pepper"—black soils) gives a first-order explanation of the range of surfaces that have evolved in the outer solar system. It is easiest to apply this idea if we describe the four Galilean moons, starting with the outermost and working our way inward.

Callisto

Of the four large Galilean moons of Jupiter, the outermost one, **Callisto,** most closely resembles a terrestrial planet.* At 5,000 km across, it is about 2% larger than Mercury. The surface is moderately dark and composed of silicatelike soil mixed with 30% to 90% H_2O ice by mass (Clark, 1980). The soil component is probably black carbonaceous material that may contain further water chemically bound in its minerals. The Voyager flybys in 1979 revealed a heavily cratered surface resembling the uplands of the moon or Mercury, as is seen in later chapters. The low mean density, 1790 kg/m^3 (compared with 1,000 for water and ice, and about 3,000 for rock) is the lowest of the Galilean moons, indicating that the interior consists of about one-fourth ice by mass. The lack of volcanic/tectonic features suggests little internal heating. Similarly, magnetic and dynamic measurements of the Galileo orbiter as it passed Callisto in 1996–1997 indicate no magnetic field and a homogeneous interior that never melted or formed a dense core. Bright rims and rays of some craters suggest that impacts have blown away a dark soil surface and exposed brighter icy material underneath. The largest feature is an enormous multiring bull's-eye, with outer rings spanning 2,400 km. This and other smaller, similar features are probably caused by impacts on the icy crust, later modified by isostatic leveling involving a watery substrate (McKinnon and Melosh, 1980).

Ganymede

Ganymede, the largest and next moon inward from Callisto, has different properties. With a 5,270-km diameter, it is 8% larger than Mercury and 75% the size of Mars. It has a moderately bright surface containing an average of 90% frozen water or frost by mass (Clark, 1980). Voyager closeups reveal some provinces of old, cratered, dark, terrain like Callisto's but these are broken by broad, bright fracture zones that appear to have fresher ice. Offsets have occurred on some fractures, but overlying undeformed craters indi-

Figure 2-22. Jupiter's icy "billiard ball" moon, Europa. This world is nearly the size of the moon but covered with relatively featureless, fractured ice fields. The ice is believed to float on a global ocean of water. (NASA Voyager photo)

cate that the fracture zones are moderately old and inactive. The bulk density is 1,930 kg/m^3, a little greater than Callisto's. The Galileo orbiter discovered that Ganymede has a weak magnetic field and evidence of a denser core. These observations (plus theoretical work—see Chapter 10) suggest that Ganymede melted or partially melted in the past, allowing denser matter to sink to the center to form a dense core, and forming a thin icy crust floating on a watery interior. The watery mantle soon solidified, but early motions cracked the ice allowing eruption of fresher water that froze into the bright, fractured ice swaths.

Europa

Europa, the smallest Galilean moon, is different from the others. Twelve percent smaller than our moon, it has a bright, featureless surface except for a system of shallow, dusky grooves. It looks like an icy billiard ball with smudges and scratches (Figure 2-22). Paradoxically, the very blandness is cause for great excitement among scientists. The near absence of impact craters shows that the surface is relatively young and that resurfacing processes have operated in recent times. The bright surface is more than 90% ice by mass, but the bulk density of 3,030 kg/m^3 implies that the interior is composed mostly of rocky material. The Galileo orbiter found evidence of a denser core and a possible weak magnetic field. Photos from the probe showed clearly that the ice, like pack ice in Earth's arctic, is a thin layer (hundreds of meters thick?) floating on an

---

*Discussions of the various Galilean worlds include Voyager 1 and 2 results described by Smith and Voyager Imaging Team (1979), along with other papers in *Science, 204,* No. 4396 (June 1, 1979) and *Science, 212,* No. 4491 (April 10, 1979), and Galileo probe results in *Science, 274,* No. 5286 (October 18, 1996).

underlying ocean, and that it fractures and breaks into ice rafts, with new ice filling in the gaps. All the evidence, taken together, indicates continued heating by an uncertain mechanism; some heat source is keeping the underlying water melted. Two mysteries are (1) the source of the heat and (2) the nature of the underlying ocean. As Arthur C. Clarke, author of *2001,* suggested after Voyager, a water ocean on Europa could conceivably harbor life.

Io

Io, the innermost Galilean moon, is one of the strangest worlds in the solar system. At 2,640 km across, it is only 2% larger than our moon. Its density is 3,530 kg/m³, highest of the Galilean worlds, implying an interior of rocky material with very little or no ice content.

A history of mystifying discoveries indicated Io's strange nature even before the Voyager photos revealed Io's unique properties. In 1964, observers reported Io to be anomalously bright for about 15 minutes after some, but not all, eclipses (Binder and Cruikshank, 1964). This effect was seen occasionally in later years. Also in 1964, radio observers discovered that bursts of radio radiation from Jupiter were correlated with the position of Io in its orbit relative to Earth (Bigg, 1964). Spectral observers in the next few years found that unlike the other satellites, Io lacks water frost or ice on its surface. High-resolution telescopic photos taken in 1973 show reddish and white patches (Minton, 1973). Observers then discovered that Io is surrounded by a thin, yellow-glowing cloud of sodium atoms knocked off the surface (Brown, 1973; Trafton, Parkinson, and Macy, 1974). This cloud, many Io diameters across and extending along Io's orbit, is too faint to be seen through telescopes but is probably bright enough to be seen as an aurora by an observer on Io. Sulfur and oxygen atoms have subsequently been found spread along Io's orbit. Some months before Voyager 1 reached Jupiter and Io, telescopic observers noted mysterious flare-ups of infrared (5µM wavelength) emission, but the observers did not guess the real cause (Witteborn, Bregman, and Pollack, 1979).

Many of these findings have now been clarified. For example, Io orbits within Jupiter's intense magnetic field and is coupled to Jupiter by electric currents through this field. This connection explains its influence on the directions of radio emissions arising in Jupiter's magnetic field.

A notable triumph of the scientific method came in 1979 when California dynamicist Stanton Peale and colleagues calculated the heat produced inside Io by a certain dynamical effect called **tidal heating** (see also Chapter 3). They found that Io's interior must be hot and *predicted* that Io might have active volcanoes. The prediction was published a few days *before* Voyager 1 reached Jupiter, and a few days later, active volcanoes were discovered on Io by Linda Morabito, an engineer in the Voyager program (Figure 2-23). This was the first discovery of active volcanoes

Figure 2-23. Photo on which Io's active volcanoes were discovered. The bright hemisphere is lit by the sun, whereas the dark side is faintly lit by Jupiter. Bright clouds (left and right edges of the crescent) are plumes of debris shot as much as 260 km above Io's surface by active volcanic eruptions and falling back through a near vacuum onto Io's surface. (NASA)

outside Earth. The chapter opener photograph for this chapter also shows an active Io volcano from "above," with a black caldera at the center, surrounded by a doughnut-shaped bright ring of debris.

"Predicted" is the key word in the above discussion because it is the key strength of the scientific method. What other system of thought has this ability to predict phenomena around us?

The Voyager and Galileo close-up photos reveal an extraordinary world of mottled patches in colors of red, orange, yellow, white, and black. When the first photos came in, one Voyager investigator quipped that he didn't know what was wrong with Io, but it looked as if it might be cured by a shot of penicillin. Reporters compared Io to a pizza.

Temperature measurements confirm that the volcanoes are hot (500 K or 440° F) and their outbursts account for infrared flare-ups seen from Earth. The heating is severe enough that it has driven off water and other volatile compounds, leaving a rocky/sulfurous composition without the ice that characterizes other moons of the outer solar system. The Galileo orbiter discovered that Io has a large metallic core, interpreted to be a mix of iron and iron sulfide (Anderson, Sjogren, and Schubert, 1996). The surface is dominated by sulfurous lavas, explaining the high density and unusual colors. Ions striking the surface dislodge atoms of sulfur and sodium that are excited by sunlight, explaining the yellow auroral glow around Io. Condensations of compounds such as sulfur dioxide (SO_2) when

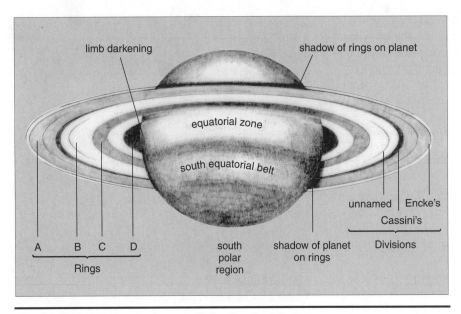

Figure 2-24. Features of Saturn. Telescopes with apertures as small as 5 cm (2 in.) will show the rings; telescopes larger than about 24 cm (10 in.) will sometimes show all these features.

cooling occurs during eclipses may explain some of Io's reported changes in brightness. In short, Io is an amazing world, with a colorful red, yellow-orange, and white surface, an intermittent yellow-aurora sky, and volcanoes ejecting vast fountains of gas and sulfur-ash debris.

Saturn

The globe of **Saturn** is rather like Jupiter but smaller, more flattened, with less-prominent cloud markings (Figure 2-24). Yellowish and tan cloud belts parallel the equator as seen in Figures 2-24 and 2-25. Occasional bright and dark markings disturb these belts, but a greater depth of overlying haze makes them less visible than on Jupiter. Spectroscopic studies from Earth and Voyagers 1 and 2 identified the main gases as molecular hydrogen (88.2% by mass) and helium (11%)—an atmosphere similar to that of Jupiter. Minor identified components include methane (CH_4), ammonia (NH_3), and ethane (C_2H_6).

Saturn has the lowest mean density of all the planets, 710 kg/m^3, indicating a very hydrogen-rich interior with few rocky materials. Given a sufficiently immense ocean, Saturn would float!

Saturn is best known for its **rings,** shown in Figure 2-25. When Galileo first turned his crude telescope on Saturn in 1610, he could not see the rings clearly. He saw only fuzzy appendages on each side of the globe, and drew Saturn as a triple planet. Saturn's true nature remained controversial until the 1660s, when the rings were clearly seen to encircle the planet (Alexander, 1962). In 1859, the Scottish physicist James Clerk Maxwell showed that they could not be a solid plate but must be made up of innumerable particles, each moving in an independent orbit around Saturn. The American astronomer James Keeler (1895) was the

first person to prove this when he used the Doppler shift to detect the varying orbital velocities of different parts of the rings; it was one of the first great successes of spectroscopic astronomy. Although the rings are 270,000 km (170,000 mi) in diameter, they are extremely thin, probably no more than a few hundred meters thick. Spectra show that the ring particles are composed of, or covered by, frozen water (Pilcher and others, 1970; Kuiper, Cruikshank, and Fink, 1970), possibly with traces of silicate or carbon minerals (Lebofsky, Johnson, and McCord, 1970; Pollack, 1975; Cuzzi and Pollack, 1978).* Various studies show that the common ring particles range from a few centimeters to a few decameters across; they are hailstones from the size of a Ping-Pong ball to the size of a house. A Voyager search for larger ring moonlets yielded negative results at sizes larger than a few kilometers.**

The rings are divided as sketched in Figure 2-24. The dusky outer ring, called ring A, is separated by a gap from the brighter ring B. The gap is called **Cassini's division.** A narrower division, called Encke's division, cuts ring A. On the inner side of ring B is a very faint, tenuous ring, C. The French observer Guerin (1970) announced a still fainter

---

*Measuring the composition of the rings illustrates the vicissitudes of planetary research. Kuiper, Cruikshank, and Fink (1970) obtained the first infrared spectrum of the rings, and although Kuiper as early as 1956 hypothesized that the rings are water ice, the spectrum matched lab samples of frozen ammonia at −20°C, and so they published the conclusion that the rings are made of ammonia ice. Immediately, Pilcher and others (1970) and Lebofsky and others (1970) pointed out that under the actual ring temperatures of −178°C, water ice spectra are altered, and *do* match the spectrum of Saturn's rings. Kuiper and his co-workers concurred at once that the spectra indicate water ice.

**Important Voyager results on Saturn, the rings, and the satellites have been published in *Science, 212,* no. 4491 (April 10, 1981) and *Science, 215,* no. 4532 (Jan 29, 1982). See also Beatty and others (1981).

Figure 2-25. Varying aspects of Saturn during its 29-y orbit around the sun, as photographed with Earth-based telescopes. Because the rings maintain a fixed tilt to the ecliptic, the Earth-based observer sometimes sees them from "above" and sometimes from "below" the ring plane. This series shows half of the 29-y cycle, including two views of the apparent disappearance of the rings as Earth passes through the ring plane. (Lowell Observatory)

Figure 2-26. Saturn's unusual satellite Iapetus has dark soil on its leading hemisphere (left) and bright icy material on its trailing hemisphere (right). A dark circular crater rim is seen on the right side. Cause of the strange soil/ice distribution is uncertain. (NASA, Voyager 1)

and more tenuous innermost ring, D, which extends from the planet toward ring C; the Voyagers confirmed a D ring here but not bright enough to have been seen from Earth (Smith, 1990). Rings A through D, together with the Cassini and Encke divisions, have been called the classical ring elements (Smith and others, 1981); they were charted from Earth. Pioneer 11, flying by in 1979, discovered a narrow ring just outside the A ring; it came to be called the F ring. Beyond it lies a thin G ring and a very diffuse E ring that extends beyond the orbit of Enceladus.

Saturn's Satellites

Thirty Saturnian moons were known by 2001. The system, like Jupiter's, includes a host of small moons near the rings, and a grouping of larger moons at intermediate distance, including the giant moon, Titan. As with Jupiter's Galilean system, we gain some insights by starting with the outermost moons and working inward.

The outermost sizeable moon is the small, dark satellite **Phoebe,** about 220 km across. Its orbit is retrograde and the most highly inclined of the satellite family. It is surely a captured interplanetary body rather than a native Saturn satellite. In 2001 another dozen, smaller, captured moons, both prograde and retrograde, were found (Gladman et al. 2001).

Next inward from Phoebe is the curious moon, **Iapetus.** It is a moderate-sized body, about 40% the size of our moon. Its diameter is 1,440 km. Its strangeness was recognized as early as 1671 when G. D. Cassini discovered that

he could see it easily when it was on the west side of the planet, but not when it was on the east side! This oddity was explained when large telescopes revealed that Iapetus keeps one face toward Saturn so that one hemisphere leads and one trails, and that the leading hemisphere is covered with very dark soil whereas the trailing hemisphere is covered with frost and nearly five times as bright. Voyager 2 photos in 1981 revealed that the boundary between the dark and light sides is irregular and sharp (Figure 2-26). One possible explanation is that dark dust knocked off Phoebe by meteorites spirals in toward Saturn and hits Iapetus's leading side, enriching that side's soil with dark material and altering the native material.

As if to reemphasize the rule that each world is unique, the next satellite is again different. **Hyperion** is a biscuit-shaped chunk about 350 km across and 200 km thick. Voyager 2 photos (Figure 2-27) showed its irregular, cratered shape. Its long axis does not remain lined up with Saturn, as would be expected if gravitational tidal forces

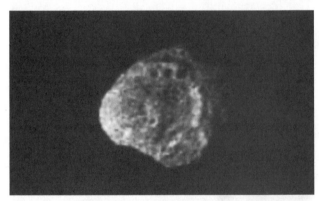

a

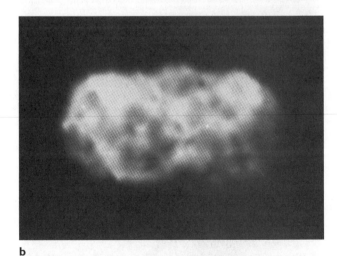

b

Figure 2-27. (a) Top view and side-on view of Saturn's biscuit-shaped satellite, Hyperion. The irregular shape may have occurred during impact fragmentation. Hyperion is about 360 km long. The side-on view (b) was made from a greater distance and is less sharp. (NASA Voyager 2 photo)

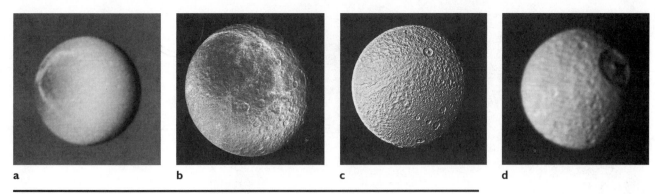

Figure 2-28. A rogue's gallery of Saturn's major icy moons. (**a**) An unusual view of Dione shows it against a background of Saturn's clouds; Dione, at 1,118 km diameter, has darker markings on its trailing hemisphere (left). (**b**) Rhea, at 1,530 km diameter, shows craters and dark patches. (**c**) Tethys, 1,048 km in diameter, shows a cratered surface and a linear canyonlike fracture. (**d**) Mimas, at 390 km, has an impact crater almost big enough to have shattered the satellite. (NASA Voyager photos)

had acted for a long time. Instead, Hyperion has an irregular and chaotic spin that is a result of gravitational forces and rotational dynamics acting in concert with Hyperion's irregular shape and eccentric orbit.

Because the next satellite inward, **Titan,** is unique and one of the largest moons in the solar system, we discuss it separately in a moment.

The next three moons, **Tethys, Dione,** and **Rhea,** are around 1,000 to 1,500 km across and have bright, icy surfaces, as seen in Figure 2-28a-c. Tethys, Dione and Rhea have densities of around 1,210 to 1,430 kg/m³, and probably have some rocky material mixed with their ice. These worlds are moderately to heavily cratered. Tethys's surface is broken by a huge canyon system and Dione's by swaths of lighter-toned material, reminiscent again of Ganymede. Cracking of the surfaces and resurfacing by water (quickly frozen to form bright ice plains and flows) may have occurred because of some source of internal heat, possibly tidal heating.

In 1980, Earth-based observers discovered interesting additional small moons perhaps 20 km to 50 km across in the orbits of Tethys and Dione. One orbits 60° behind Tethys, one is 60° ahead of Tethys, and the third is 60° ahead of Dione. At least two others, associated with Tethys and Dione, are suspected from Voyager photos. Gravitational forces make this 60° point—called a Lagrangian point—a stable location for small objects. Certain asteroids are known in similar 60° points ahead of and behind Jupiter. These discoveries expand the list of similarities between satellite families and the solar system as a whole—supporting the analogy of satellite families as miniature solar systems.

The next satellite, **Enceladus,** is especially fascinating as a "missing link" between Jupiter's moons Europa and Ganymede. Enceladus (Figure 2-29) has the linear grooves and bright, sparsely cratered, icy surface of Europa, but it also has some moderately cratered, fractured regions that resemble Ganymede. Voyager analysts suspect that Ence-

ladus has a fairly young surface created by eruptions of water and ice from an interior heated by tidal interactions with other Saturnian moons. This mechanism would resemble the heating that keeps Io's volcanoes erupting, but calculations have failed to show how the heating mechanism would be adequate. The bulk density of 1,200 kg/m³ indicates that Enceladus is mostly icy throughout, so that radioactivity from rock minerals (such as the process that heats Earth's interior) also appears minimal. A concentration of material in Saturn's E ring at the position of Enceladus's orbit also leads to suspicions of (volcanic or geyserlike?) eruptions of material off Enceladus. However, the mystery of Enceladus's heat sources and degree of geologic activity remains unsolved.

The next moon inward is **Mimas,** 390 km across, round, icy, and heavily cratered. Mimas's bright surface and low bulk density (1,190 kg/m³) indicate that it is composed mostly of ice. As seen in Figure 2-28d, Mimas has a crater so large that it was almost big enough to shatter the moon.

On the outskirts of the rings are at least five small moons with diameters of 30 to 330 km. Most were discovered by Voyager 1 and by intense Earth-based observations during a period of a few weeks when the rings were seen edge on and thus did not obscure the satellites by their glare. Voyager photos show that at least some of these moons are lumpy and cratered, but their composition is uncertain.

Titan

In the middle of the Saturn satellite system is the giant, **Titan,** with a diameter of 5,120 km. It is the solar system's second-largest moon, ranking just smaller than Ganymede. Titan rivals Io as the solar system's most bizarre world. It is the only moon with a thick atmosphere, discovered in 1944 when spectra of Titan revealed methane gas (CH_4). Observations in 1973 showed that Titan's sky is

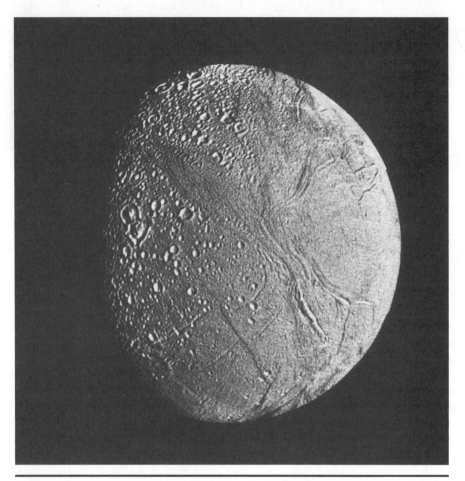

Figure 2-29. The surface of 500-km Enceladus is of special interest because, in the midst of heavily cratered areas, it shows smooth icy plains that indicate eruptions of water and resurfacing episodes. This may be a "missing link" that shows the conversion of a cratered world into an "icy billiard ball" like Jupiter's moon, Europa. (NASA, Voyager 2)

not clear,* but is filled with a reddish haze, creating a nearly featureless globe as seen from a distance. Later observations showed that this haze is photochemical smog produced by reactions of the methane and other compounds when they are exposed to sunlight—like the smog produced by the action of sunlight on hydrocarbons over Los Angeles. Titan is the smoggiest world in the solar system.

The Voyager probes showed that methane and smog are no more than 10% of Titan's atmosphere. The main constituent is nitrogen, as on Earth. The atmosphere is so dense that the surface air pressure is 1.6 times that on Earth! The main difference is that Titan's air is very cold, around 93 K (−292° F). Minor constituents include organic molecules such as ethane (C_2H_6), acetylene (C_2H_2), ethylene (C_2H_4), and hydrogen cyanide (HCN). Since Earth's air is also mostly nitrogen, Titan's atmosphere offers

fascinating comparisons; the abundance of organic molecules suggests that Titan offers a natural laboratory for research on the origins of life.

Under conditions on Titan's surface, pools of liquid methane and liquid nitrogen may exist. If the methane content exceeds some 8%, methane may also rain out of the clouds and exist as snow or ice, playing the same triple role of gas, liquid, and icy solid as water does on Earth. The smog may even produce gasolinelike compounds that rain out of the hazy clouds. The strange surface may be revealed by the Cassini/Huygens parachute probe in 2004.

Uranus

Now we come to the outer three planets, which were not known to the ancients. **Uranus** was discovered accidentally on March 13, 1781, by the English musician-turned-astronomer William Herschel. Herschel was observing star fields at the time with his telescope; later studies showed that other accidental observations of Uranus had been made earlier but the observers had mistakenly plotted the planet as a star.

---

*A clear sky had been fervently desired by a generation of astronomical artists, who love to depict Saturn hanging in a blue sky over the icy plains of Titan. At certain infrared wavelengths, telescopes can glimpse the surface.

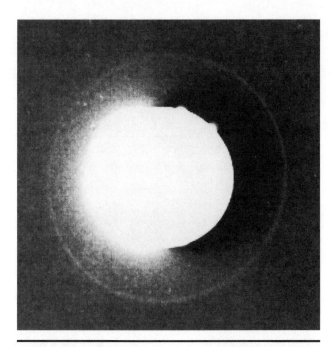

Figure 2-30. Overexposed photo of the globe of Uranus from approaching spacecraft Voyager 2 shows the brightest of the thin rings surrounding the planet. Because of Uranus's high axial tilt, Voyager approached from the poleward direction, giving a "face on" view of the rings. (NASA)

While Uranus is nearly four times the size of Earth, it is only 41% the size of Saturn. It is so remote that it presents only a tiny, somewhat bluish disk in large telescopes. Voyager 2 photos were the first high-resolution images, but revealed only a nearly featureless disk, with only faint traces of cloud bands and polar haze. Voyager 2 is the only spacecraft to have reached Uranus and Neptune; Voyager 1, after departing Saturn, flew on a different course out of the solar system. Spectroscopic studies indicate a dense atmosphere consisting of roughly the same mix of hydrogen to helium found on Jupiter and Saturn (Ingersoll, 1990). The planet rotates in 17.9 hours.

The bland appearance of Uranus (Figure 2-30) results from colder temperatures and lesser inner thermal activity, so that the atmosphere is less restless than those of Jupiter and Saturn. The color comes from Rayleigh scattering of blue light (see Chapter 11), as in our own atmosphere, plus a strong absorption of red light by methane gas (CH_4), which constitutes a few percent by mass. The net result is a bland, ethereal blue globe, with a slight greenish cast.

Uranus has a peculiar dynamical property. As shown in Figure 2-31 on page 36, its axis of rotation, instead of having only a slight tip to the plane of the solar system as is true of the other planets, lies mostly in the plane of the solar system. This means that sometimes the "north" pole of Uranus points toward the sun, whereas half a revolution later the "south" pole points toward the sun. The inclination of the axis to the orbit plane is 82° and the spin direction is retrograde; these two statements are usually com-

bined by saying that the inclination of the axis is 98°, or eight degrees *past* parallel. This is more than a curiosity; it is a property that must be explained by any theory of the evolution of planets.

The first edition of this book, in 1972, remarked that Uranus resembles a bluish Saturn without rings—a fine analogy until March 10, 1977, when many astronomers watched Uranus pass in front of a relatively bright star, and saw the star dim several times on both sides of Uranus's disk, indicating that it had been obscured by rings that go all the way around Uranus (Elliot, Dunham, and Mink, 1977, and many other papers published that year). When Voyager 2 arrived, it confirmed an elegant system of narrow, faint rings (Figure 2-30) totally unlike Jupiter's broad, tenuous ring, or Saturn's dramatic system of bright rings and narrow gaps. The faint rings are composed of black dust particles. Between some of the narrow rings are broader, still fainter rings. A flurry of subsequent studies indicated that the net gravitational actions of small satellites near the rings confine the ring particles into narrow, tightly defined rings.

Uranus has five substantial satellites discovered from Earth, and ten more small ones, close to the planet, discovered by Voyager 2 and two more small outer moons that may be captured. Voyager provided a surprise. As shown in Figure 2-32 on page 36, the satellites show amazing geologic features and individual personalities. For example, 1,170-km-diameter Umbriel is darker than the others, and 470-km Miranda has extraordinary swaths of grooves and faulted cliffs. The energy source for Miranda's ancient tectonic activity may be tidal heating. Other satellites, such as 1,160-km Ariel, also show tectonic signs of ancient activity (Figure 2-32b). The sixteenth and seventeenth moons were discovered in 1998 from photographs taken with the Palomar 200-inch telescope.

Neptune

After Uranus was discovered in 1781, dynamicists tried to determine its orbit. They found that Uranus's motions were being affected by the gravitational force of another planet beyond Uranus.

An interesting chapter in the history of science then ensued (Lyttleton, 1968). Unknown to each other, an English and a French astronomer set out in the 1840s to predict where the undiscovered planet must be. Based on Bode's rule, which seemed to be confirmed by Uranus, they assumed a solar distance of 38 AU—ironically, as **Neptune** is the one serious failure of Bode's rule. John Adams, just finishing his undergraduate work at Cambridge, had trouble getting his professors interested in the search. Urbain LeVerrier predicted a position in 1846, and the English astronomers began a desultory search, actually spotting it, but failing to recognize it. Meanwhile, LeVerrier got two young German astronomers interested, and they discovered it within half an hour of starting their search on September 23, 1846. Adams and LeVerrier are

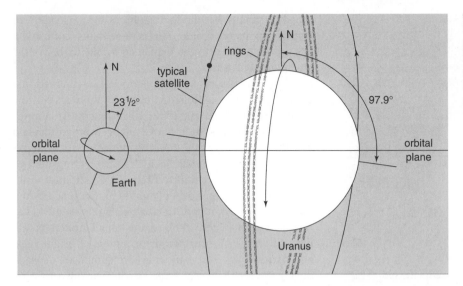

Figure 2-31. Comparison of the sizes and rotations of Earth and Uranus. Earth has an obliquity (or axial tilt) of 23 ½° and a prograde (west-to-east) rotation. Uranus has a much steeper obliquity and retrograde rotation.

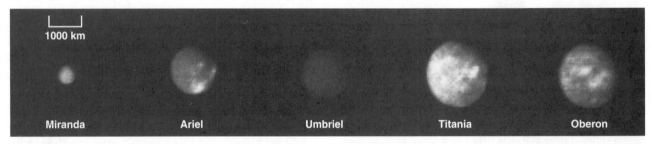

a

b   c

Figure 2-32. Satellites of Uranus (**a**) The five large moons in order outward from Uranus (left to right). They are reproduced at the same scale and with a uniform brightness scale. The unique, puzzling dark color of Umbriel is well seen, as is the small size of Miranda. (**b**) Close-up of part of Uranus's satellite, 1,160-km-diameter Ariel, shows old, heavily cratered surface, broken by numerous tectonic rifts. (**c**) Miranda, at 484 km diameter, surprised scientists by showing an intensely fractured and contorted surface. Origin of the sharply defined sectors of fractured terrain is unknown. (NASA Voyager 2 photos)

both credited with the discovery, but at the time it caused a scandal because of the failure of the British to grasp their opportunity.

Neptune is so far away that little was known of it before Voyager 2 climaxed its flight with a close flyby in 1989. Neptune is sightly smaller than Uranus, shares Uranus's blue color for the same reason, and rotates slightly more slowly, in 19.2 hours. Voyager 2 revealed a hazy disk (Figure 2-33 on page 38), but with a more active atmosphere than on Uranus. Neptune has faint blue cloud belts and a large, dark, oval storm system reminiscent of Jupiter's red spot. The atmosphere, as with the other giant planets, is roughly three-fourths hydrogen and one-fourth helium, with minor admixtures of methane (CH_4) and other gases.

Neptune, like the other three giant planets, has a ring system. The rings were first suspected during stellar occultation measurements in the 1980s (similar to the case of Uranus), but the measurements indicated a strange discontinuous structure to the rings. Voyager 2 confirmed narrow rings that aren't uniform but contain concentrations of particles called ring arcs. The ring particles are believed to be dark.

Once again, the satellites provided the big surprise. In addition to seven known small moons, about 50 to 400 km in diameter, Neptune has one big moon, **Triton** (Figure 2-34 on page 38). Its diameter is 2,700 km, about three-fourths the size of our moon. Triton's orbit is strange. It is retrograde and inclined, suggesting that Triton was captured from interplanetary space. During the 1980s, an atmosphere was suspected on Triton. Voyager 2 confirmed a very thin atmosphere of molecular nitrogen (N_2) and methane, with a high haze layer, and also a geologically young, uncratered surface of CH_4 and N_2 ices, along with a seasonal polar cap of transient frost (probably N_2). Most amazing of all were geyserlike plumes of darkish smoke, indicating active internal heating (see Chapter 10; also see Stone and Miner, 1989, and other papers in that special issue of *Science*).

The energy that drives the "geysers" is uncertain. Triton may have originated as a Pluto-like interplanetary body, been captured by Neptune, and heated by tidal forces (Goldreich and others, 1989; also, see next chapter).

Pluto

Irregularities in Neptune's motions suggested at least one more large planetary body beyond Neptune. Percival Lowell began a search for it in 1905. His assistant, Clyde Tombaugh, discovered **Pluto** in 1930 on Lowell Observatory photos. It was named for a god of the underworld, whose first two letters matched Lowell's initials. The opening section of the chapter gave some reasons to debate whether Pluto should really be called a planet or only the largest of a swarm of known interplanetary bodies in that region.

A satellite, named **Charon** (KEHR-on), was discovered in 1978 (Christy and Harrington, 1978). Pluto is the only "planet" not yet visited by a spacecraft, and we have no close-up photos. Pluto and Charon are so far away that even the Hubble Space Telescope shows only a fuzzy image of Pluto with the vaguest shadings. However, observations of spectra, star occultations, and a fortuitous set of eclipses (1985–1990) established many properties of Pluto and Charon. They keep the same face to each other and rotate every 6.39 days on an axis tipped 122° to the orbit plane. Pluto has an atmosphere thinner than that of Mars, probably composed of nitrogen and carbon dioxide (Stern, Weintraub, and Festou, 1993). Pluto's diameter is 2,302 ± 16 km, and Charon's is 1,186 ± 26 km. The daytime surface temperature is a frigid 36 ± 2 K (−396° F) (Stern and others, 1993). Pluto and Charon's mean densities are roughly 2,000 ± 200 kg/m^3, suggesting a composition of around 70% rock and 30% ice (Binzel, 1990; Stern, 1992).

An interesting observation is that Pluto is somewhat brighter and pinker (but not nearly as red as Mars) and has a surface rich in methane ice with blue geometric albedo (reflectivity) of 66%; Charon is more neutral gray and has a surface rich in water ice, with a blue geometric albedo of only 38%. Why would the two intimately linked worlds have different surface materials? Researchers think that the lighter molecules, such as methane, have escaped from Charon because of its lower gravity; once the molecules escaped into space within the Pluto/Charon system, some of them collided the surface of Pluto itself. Thus, over time, Charon has lost methane molecules and Pluto has gained some of them. This would explain the difference in surfaces.

Pluto and Charon are a frontier outpost of the solar system. From that distance, the sun would appear only as bright as a bright streetlight seen at night down the street.

"Planet X"?

Are there any other substantial planetary bodies beyond Pluto? Several astronomers have sought dynamical or photographic evidence of a planet there, sometimes called **"Planet X."** Clyde Tombaugh, the discoverer of Pluto, conducted a long search and ruled out any planet as large as Neptune near the plane of the solar system out to a distance of around 100 AU.

However, the mass of Pluto/Charon seems to be too small to account for all the irregularities in the motions of Neptune. This finding has prompted more searches; surveys in 1979, 1988, and following years found no more Pluto-sized objects, but they did turn up many smaller objects in Pluto's region and beyond, with

Figure 2-33. A portion of crescent-lit Neptune (bottom) sets off a view of its satellite, Triton. Crescent lighting and forward scattering of light through haze of Neptune makes its cloud features relatively invisible. (NASA Voyager 2 photo)

0   300 km

Figure 2-34. Two photo-mosaics of Neptune's sparsely cratered, largest moon, Triton. Due to the inclined orbit, lighting was mostly on the southern hemisphere, as shown by latitude/ longitude grid. Inset shows a view that includes more of the bright southern polar region. Dark streaky smudges (bottom) are associated with geyserlike vents. (NASA Voyager 2 mosaics)

diameters of a few hundred kilometers, approaching a fifth the size of Pluto. These discoveries convinced many astronomers that the outer fringe of the solar system is full of icy bodies, of which Pluto may be only the largest (or one of the largest). Probably no full-fledged Planet X, Earth-size or larger, will be discovered. However, additional Pluto-size bodies may lurk in the outermost solar system.

Telescopic Appearance of the Planets

Though much of the information we will discuss comes from spacecraft, it is still interesting to know what can be seen by examining the planets visually with an earth-based telescope. Except on rare nights, Earth's atmospheric turbulence blurs details with an angular size of less than 0.5 seconds of arc. Rare conditions with little turbulence are called good "**seeing.**" With even fair seeing, amateur astronomers' telescopes of 15 cm to 30 cm (6–12 in.) show most of the details shown in Figure 2-35. Experienced observers can always see much more detail than beginners.

The sizes of the sketches in Figure 2-35 indicate very roughly the apparent sizes of the planets seen from Earth. Note that the inner planets display the largest apparent size when they are in crescent phase between Earth and the sun. A good rule of thumb is that the east-west ("horizontal") thickness of Venus in most of its phases is roughly 10 (10 seconds of arc) across. Jupiter and the Saturn ring system are typically around 45 seconds of arc. When passing near Earth, Mars reaches as much as 25 seconds across. Jupiter's four Galilean satellites and some satellites of the giant planets are easily visible in small telescopes; but only the four large moons of Jupiter approach even one second in angular diameter, so that vague detail can be seen only with the largest telescopes under the finest conditions.

Miscellaneous Basic Data and Terminology

Readers intending to pursue scientific careers would be well advised to memorize a few numerical facts about the planets. In spite of popular arguments against "memorizing mere facts," a small investment in memorization can produce large savings in time spent looking up trivia and allow quick checks of ideas that might otherwise turn out to be blind alleys. Table 2-2 lists some data that are used in a large number of simple calculations. The **solar constant,** listed in Table 2-2, is defined as the rate at which energy is received from the sun by a 1-m^2 surface facing the sun at the mean distance of the earth's orbit (Joules per meter squared per second). In other words, it is the mean flux* of sunlight at the top of Earth's atmosphere. Helpful facts to remember are that the sun's mass is about 1,000 times that of Jupiter. Also, the sun's radius is about 10 times that of Jupiter, which is about 10 times that of Earth, which in turn is about 10 times that of the largest asteroid.

Figure 2-36 illustrates a number of terms common to planetary astronomy. Most of the terms in the diagram are self-explanatory. The term **apparition** refers to the period of several months when a planet is best visible—for example, an apparition of Mars (when it was on our side of the sun). **Elongations** (inner planets only) can be either eastern (appearing east of the sun in the evening sky) or western (west in the morning). A mnemonic device is the *e* in *east* and *evening*.

As shown in Figure 2-36, an eclipse occurs when one body enters the shadow of another (not necessarily when one appears to pass behind another, which is called an occultation). The shadow of any object (a planet or your hand) has two parts: the umbra, or dense inner region in which the light source is completely covered, and the penumbra, or lighter outer region in which the light source is only partly covered. An eclipse can be either umbral or penumbral, depending on which part of the shadow is occupied by the eclipsed body. Figure 2-37 shows a penumbral eclipse of the Viking 1 landing site on Mars by the Mars satellite Phobos (Phobos is not big enough to cover the sun totally, as seen from Mars).

Certain dynamical terms should be noted—for instance, the prefixes *peri-* (designating the point when a smaller body is closest to the primary body) and *ap-* or *apo-* (designating the farthest point). If one wants to speak in general terms, without specifying a particular primary body, one uses the suffix *-apsis;* thus, **periapse** and **apoapse** can be used to designate points in orbits around any planet or satellite. For referring to specific bodies, one uses specific suffixes (usually Greek), such as *-gee* for Earth (**perigee** and **apogee**), or *-helion* for the sun (**perihelion** and **aphelion**).

Certain other terms are of very general use. The **ecliptic** is the plane of Earth's orbit. **Inclination** is the angle between the plane of a planet's or satellite's orbit and the Earth's orbit. Earth's orbit, or the ecliptic plane, was chosen long ago as the standard of reference. Celestial mechanicians sometimes refer to the **invariable plane,** which is defined by the total angular momentum of the entire solar system. The distinction is small, because Jupiter's inclination is only 1.3°. The planets with the greatest inclinations are Pluto (17°) and Mercury (7°). Sometimes the "inclinations" given for satellite orbits are the angles between their orbit and the planet's equatorial plane, not the plane of the solar system. Thus, a listing under "inclinations" for Uranus's

---

*Flux is a general technical term for amount passing through unit area per second.

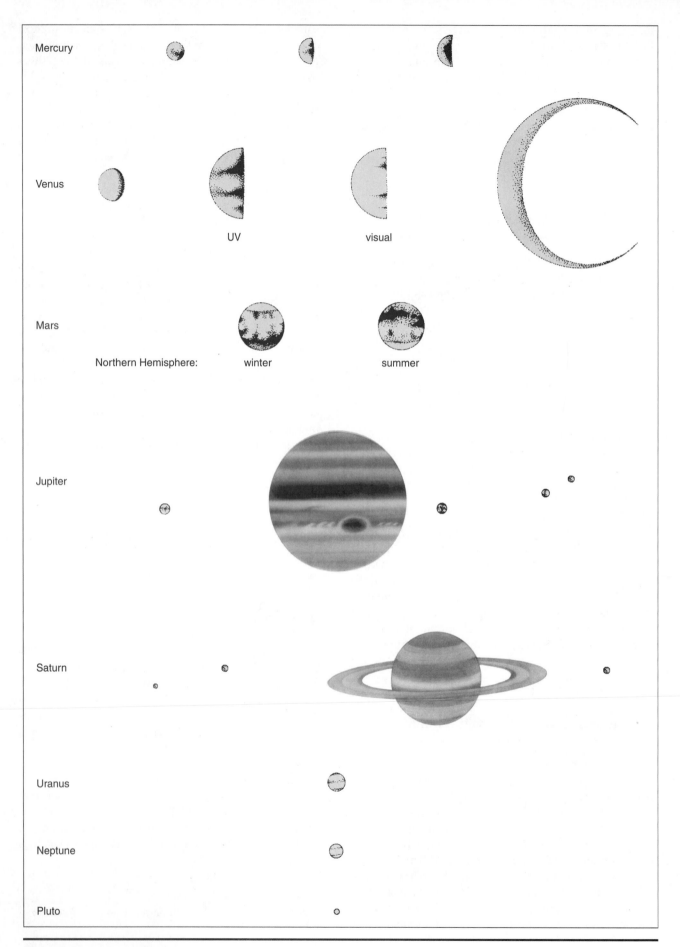

Figure 2-35. Appearance of the planets to a visually experienced observer using a moderate-sized telescope.

TABLE 2-2 Useful Facts to Memorize

Body	Important satellite	Mass (kg)	Radius (m)	No. known satellites
Sun	Jupiter	2×10^{30}	7×10^8	thousands
Mercury		—	—	0
Venus		—	—	0
Earth		6.0×10^{24}	6.4×10^6	1
	Moon	7.4×10^{22}	1.7×10^6	—
Mars		—	—	2
	Phobos	—	—	—
	Deimos	—	—	—
Asteroids		$\leq 10^{19}$	$\leq 5 \times 10^5$	?
Jupiter		2×10^{27}	7×10^7	≥ 28
	Io	—	—	—
	Europa	—	—	—
	Ganymede	—	—	—
	Callisto	—	—	—
Saturn		—	—	≥ 30
	Titan	—	—	—
Uranus		—	—	≥ 21
Neptune		—	—	≥ 8
Pluto		—	—	1
Comets		$10^{15}-10^{16}$	—	—

Gravitational constant	$G = 6.67 \times 10^{-11} \ N \cdot m^2/kg^2$
Astronomical unit	$1 \ AU = 1.49 \times 10^{11} \ m$ (~150 million km)
Boltzmann constant	$k = 1.38 \times 10^{-23} \ J/deg$
Stefan–Boltzmann constant[a]	$\sigma = 5.67 \times 10^{-8} \ J/m^2 \cdot deg^4 \cdot s$
Velocity of light	$c = 3.00 \times 10^8 \ m/s$
Mass of H atom	$M_H = 1.67 \times 10^{-27} \ kg$
Planck's constant	$h = 6.63 \times 10^{-34} \ J \cdot s$
Mass of sun	$M_\odot = 2.00 \times 10^{30} \ kg$
Luminosity of sun[a]	$L_\odot = 4 \times 10^{26} \ J/s$
Solar constant[a]	$F_\odot = 1.36 \times 10^3 \ J/m^2 \cdot s = 1.36 \ kw/m^2$

[a]Note that one Joule/second (J/s) is equal to one watt (W).

satellites might be given as about 1° or as 98°, and one must be careful to determine which definition is being used.

As shown in Figure 2-38, **revolution** is the motion of one body around a second body, whereas **rotation** is the spinning motion of a body around an axis within itself. Rotation and revolution are commonly confused, especially in everyday speech (to be entirely proper we should speak of rotating doors and rotators).

Obliquity is the tilt angle between a planet's axis of rotation and the pole of the orbit. For Earth, it has the familiar value 23½°. Most other planets have similar low values, except Venus (which has a value near 180°), Uranus (with

the unusual value, 98°), and Pluto (122°). The biggest planet, Jupiter, has the smallest obliquity, 3°, which may prove significant to theories of planetary origin. Obliquity is sometimes incorrectly labeled *inclination*. Obliquity is responsible for seasons on the planets, because it causes one hemisphere to be tipped toward the sun at a given point in a planet's orbit. As with inclinations, retrograde rotation is indicated by quoting an obliquity greater than 90°.

Albedo is a measure of the reflectivity of a surface—that is, the percentage of sunlight the surface reflects. An albedo can be calculated for each color, or an average albedo can be given for all the colors of sunlight (that is

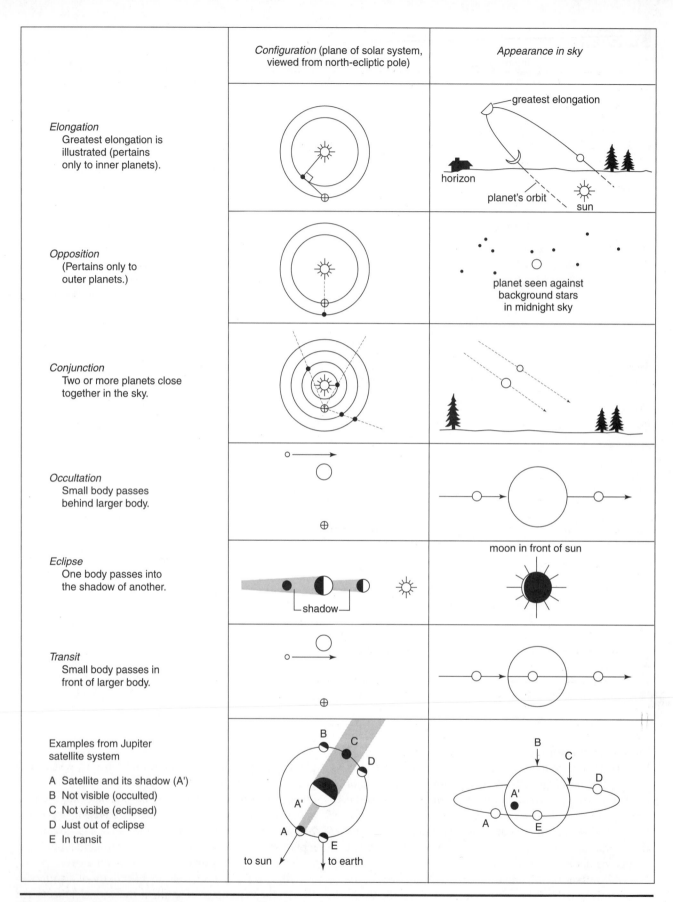

	Configuration (plane of solar system, viewed from north-ecliptic pole)	Appearance in sky
Elongation Greatest elongation is illustrated (pertains only to inner planets).		greatest elongation horizon planet's orbit — sun
Opposition (Pertains only to outer planets.)		planet seen against background stars in midnight sky
Conjunction Two or more planets close together in the sky.		
Occultation Small body passes behind larger body.		
Eclipse One body passes into the shadow of another.	shadow	moon in front of sun
Transit Small body passes in front of larger body.		
Examples from Jupiter satellite system A Satellite and its shadow (A') B Not visible (occulted) C Not visible (eclipsed) D Just out of eclipse E In transit	to sun to earth	

Figure 2-36.   Terms used in planetary astronomy. (See the text for discussion.)

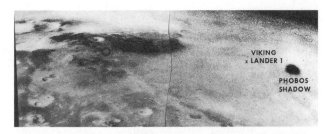

Figure 2-37. Three-minute sequence of pictures shows the penumbra shadow of Phobos passing across the Martian landscape, including the Viking 1 landing site. The shadow is about 90 km (56 mi) long. Timing of the event as observed here by the Viking orbiter and also on the ground by the Viking lander helped locate the lander. Contrast enhancement makes the shadow appear darker than it would otherwise. (NASA, Viking Project; courtesy T. C. Duxbury)

averaged over the wavelength range of sunlight), which is the more common practice.

A complication arises because of phase effects. **Phase** is defined as the angle between the sun and the observer as seen from the planet. The **Bond albedo,** the most commonly used definition of albedo, refers to the total percentage of sunlight reflected *in all directions.* To determine the Bond albedo, we have to observe a planet over a wide range of phases to determine how the amount of light reflected depends on direction, a dependence called the phase func-

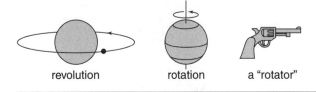

Figure 2-38. *Revolution* refers to the orbiting of one body around another. *Rotation* refers to the spinning of a body around its axis. A revolver, which has a rotating chamber, is misnamed.

tion. Different surface particles, such as fine dust, rocks, or cloud droplets, have different phase functions. For this reason the Bond albedo (A) is broken into two factors, p and q ($A = p$). The factor q is known as the phase integral and can be either observed or theoretically calculated. The factor p, known as the **geometric albedo,** gives the percentage of light reflected at zero phase angle.

In general, all these albedos are around 2% to 10% for dark rocks, 10% to 25% for lighter rocks, and about 40% to 70% for clouds and frosts. Albedos are thus indicators of surface and atmospheric properties. Venus, because of its nearly white clouds, has a Bond albedo of 76%. Earth has an intermediate value of about 36% (averaging over clouds, oceans, and land). The moon and Mercury have low values, in the range of 6% to 10%, depending on the region. Certain asteroids and most comets, whose surface soils are rich in carbon minerals, have albedos of only 2% to 5%.

The ages of the planets and older surface features are measured in units of 10^9 y, or billions of years (abbreviated "b.y."). To avoid problems of different abbreviations in different languages, the International System of Units (ISU or SI for *Systèm International*) has been adopted throughout science. The SI system employs the following prefixes to indicate various multiples of 10^3:

giga-means 10^9 (abbreviated G)
mega-means 10^6 (abbreviated M)
kilo-means 10^3 (abbreviated k)
milli-means 10^{-3} (abbreviated m)
micro-means 10^{-6} (abbreviated μ)

One may thus encounter 1 billion years = 1 b.y. = 1 Gy = 10^9 y. Similarly, 1 million years may be written 1 m.y. = 1 My = 10^6 y.

| SUMMARY |

Dynamical properties (such as orbit and size) and surface and atmospheric properties (such as composition and temperature) are now known to the first order for all principal planets and many satellites. The frontier of the solar system visited by spacecraft is now beyond Neptune. Earth has turned out to be unique, apparently being the only body in the solar system with substantial amounts of liquid surface water, a temperate climate, a breathable atmosphere, and probably life. The moon, Mercury, and most satellites present airless, or nearly airless, cratered landscapes with various degrees of tectonic or volcanic modification. Titan has a cold, dense atmosphere. Venus has a very hot surface with a high-pressure CO_2 atmosphere. Mars has the most earthlike surface, with a thin CO_2 atmosphere, blowing dust, temperatures occasionally above freezing, ice caps, seasonal change, and perhaps had some running water during ancient times. Io has active volcanoes. Europa is an icy billiard ball with a probable liquid water ocean beneath a thin ice crust. Ganymede (Jupiter), Enceladus (Saturn), and Miranda (Uranus) are notable for fractures and swaths of tectonically modified surface. Moons and planets have remarkably varied and eccentric "geologic personalities."

CONCEPTS

planet
terrestrial planet
giant planet
comet
asteroid
meteoroid
Ceres
planetary body
planetary material
planetesimal
satellite
inactive or dormant comets
world
Bode's rule
astronomical unit
Mercury
mare
impact craters
uplands
resonance
Venus
evening star
morning star
retrograde motion
prograde motion
Venera 7
Earth
moon
regolith
Mars
polar caps
surface coloration
plurality of worlds
canals (of Mars)
chaotic terrain
channels (of Mars)
Jupiter
belt
zone
disturbance
Great Red Spot

Galilean satellite
Callisto
Ganymede
Europa
Io
tidal heating
Saturn
Saturn's rings
Cassini's division
Phoebe
Iapetus
Hyperion
Titan
Tethys
Dione
Rhea
Enceladus
Mimas
Uranus
Neptune
Triton
Pluto
Charon
"Planet X"
seeing
solar constant
apparition
elongation
periapse, apoapse
perigee, apogee
(Martian) perihelion, aphelion
ecliptic
inclination
invariable plane
revolution
rotation
obliquity
albedo
phase
Bond albedo
geometric albedo

PROBLEMS

1. Why is the term *planet* difficult to define in a scientifically useful way?

2. How many "planetary bodies" are known in the universe if one defines them as (a) principal planets, (b) nonstellar bodies occupying nonoverlapping zones of the solar system (such as described by Bode's rule), (c) nonstellar bodies larger than 5,000-km diameter?

3. Why is Bode's rule not a "law of nature" with the same status as, for example, Newton's law of gravitation?

4. Why is Venus the planet most commonly associated with the phrase the "evening star"? Explain by referring to its distance, size, albedo, and angular distance from the sun as seen from earth. What planet would make the most prominent evening star as seen from Mars?

5. Explain why the "canals of Mars" were the most widely discussed Martian surface features at the turn of the century but are not discussed at all now.

6. What satellites have been photographed at a range close enough to reveal surface features, and how do these features compare with those of our moon?

7. Which planet, Earth or Pluto, is usually closer to Uranus?

8. Which planet, Earth or Saturn, comes closest to Jupiter?

9. Briefly describe the appearance or properties of each planet and of the satellite systems.

10. If Venus, Earth, Mars, and Jupiter are in a straight line on the same side of the sun, what phenomena does an observer on Earth see? What ones would an observer on Mars see?

PROJECTS

1. Observe as many planets as possible with the largest telescope available. Make sketches showing the appearance and any surface features seen (a disk 5 cm in diameter is often considered standard for sketching planetary images). Observe each planet for at least three nights if possible and observe changes from night to night. (Even experienced observers may need at least three nights before they can see substantial detail on the planetary disks, which usually seem disappointingly small to the inexperienced observer.)

2. By using an almanac, an astronomical guide such as the *American Ephemeris and Nautical Almanac,* or monthly listings in magazines such as *Sky and Telescope* and *Astronomy,* determine which planets will be reaching their best positions (oppositions and elongations) for observing in the next few months and plan observing programs similar to that in project 1 above for a week or two at those times.

3. If the Great Red Spot or other cloud details can be seen on Jupiter, monitor their position for an hour or so until you detect the planet's rotation. Monitor Jupiter's satellites for an hour or so until you detect their revolution.

The system of rings around Saturn, composed of independently orbiting particles, demonstrates the need to use celestial mechanics and dynamical theories of orbits in order to understand planetary phenomena. (NASA, Voyager)

CHAPTER THREE

Celestial Mechanics

Celestial mechanics is the science that attempts to describe and predict motions of objects—from planets and moons to asteroids trapped in resonances, broken satellites, twisted rings, and other curiosities in space. Archaeological and historical evidence shows that human societies long ago began to make observations of solar and planetary positions in order to calibrate calendars for agricultural and ceremonial purposes. This practice evolved into **astrology,** the superstition that planetary positions influence our lives on Earth. Astrologers from about 100 B.C. to A.D. 1500 created a demand for forecasts of planetary positions. Celestial mechanics as a science developed through efforts to improve humanity's ability to predict the positions of planets and to figure out the underlying principles of their motions as seen from Earth. This chapter gives a basic survey of many celestial mechanics phenomena—from basic orbital properties to peculiar effects that help explain the very existence of planetary systems.

Historical Development Through the Renaissance

As practical observations and astrology developed side by side, the Greeks discovered the **scientific method** of drawing conclusions about nature by making hypotheses on the basis of observations, and then testing the hypotheses by making more observations. In this way, some of the Greeks gained a surprisingly good conception of the solar system. Aristarchus of Samos (c. 270 B.C.) is said to have advocated the view that Earth rotated and revolved around the sun, and by using geometry he estimated the relative distances of the moon and the sun from Earth and the proportions of Earth and the moon. Eratosthenes (c. 200 B.C.), by a famous measurement of the elevation of the sun above the horizon at different latitudes, measured the spherical shape and size of Earth.

The most popular concept, however, was that of Ptolemy (c. A.D. 100), who viewed Earth as the stationary center of the planetary system, with the sun and other bodies revolving around it. This **Ptolemaic system** allowed approximate prediction of the positions of the planets in the sky, and it held sway for 15 centuries. Five names dominate the overthrow of the Ptolemaic system, a revolution that shook science, theology, philosophy, and perhaps our basic psychology. These people and their contributions to this revolution are as follows:

Nicolaus Copernicus	1473–1543. Theory of circular motion of Earth around the sun (1543)
Tycho Brahe	1546–1601. Observations of planet positions (c. 1600)
Johannes Kepler	1571–1630. Analysis, Kepler's laws (c. 1610)
Galileo Galilei	1564–1642. Telescopic observations supporting Kepler (c. 1610)
Isaac Newton	1642–1727. Laws of gravity (*Principia*, 1687)

The work of these five scientists produced a steady advance in our knowledge. During the rise of the spirit of free inquiry, the Polish astronomer Copernicus questioned the Ptolemaic assumption of Earth's central position. The motions of the planets across the sky could be explained much more simply, he found, by assuming that Earth and other planets revolved around the sun. Copernicus suspected that the paths of the planets were perfect circles with the sun at the center.

Inspired by the idea of making a new star catalog, measuring the planetary motions, and illuminating the controversy between Ptolemaic and the Copernican theories, the Danish astronomer Tycho Brahe built a private observatory and started recording planetary positions night by night. There were no telescopes in those days, but Tycho (as he is called) had elaborate instruments for naked-eye observations. (For historical reasons, Tycho and Galileo are generally called by their first names.)

Tycho hired a German assistant, Kepler, who inherited the stacks of observations. Kepler studied the data for years before discovering the empirical rules that describe how the planets move. He found that the planets move not quite in circles, but rather in ellipses. His three basic rules, known as Kepler's laws, are discussed in the next section.

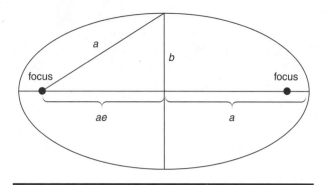

Figure 3-1. Geometric properties of an ellipse: (**a**) semimajor axis, (**b**) semiminor axis, and (**e**) eccentricity.

Meanwhile, the Italian observer, Galileo, became the first to use the telescope (probably invented a year or so earlier in Holland) for observing the planets. His discoveries include craters and mountains on the moon and spots on the sun, which overturned earlier, idealized conceptions of them as "perfect" spheres. More important, he found four large satellites orbiting around Jupiter, thus proving that not all bodies revolve around Earth. They are now called the Galilean satellites. Legend tells of some defenders of the Ptolemaic theory who refused to view this sight through Galileo's telescope. A black day in the history of formalized religion came when the Church forced Galileo to recant his work and sentenced him to house arrest for the rest of his life. Galileo remarked with just cause, "In questions of science, the authority of a thousand is not worth the humble reasoning of a single individual."

A second telescopic observation supporting the Copernican theory was Giovanni Cassini's 1666 description of a large spot on Jupiter (probably the Great Red Spot), which Cassini followed as Jupiter's rotation carried the spot across the disk. This discovery, showing that other planets rotate, helped convince skeptics that Earth also rotates.

In the year of Galileo's death, the English physicist Isaac Newton was born.* Newton, who said he made his discoveries "by always thinking about them," thought about the meaning of planetary motions. At this time, the motions of the planets were known empirically, but no one knew any theoretical reasons why planets should move in Kepler's ellipses instead of, say, circles or squares. Newton reasoned from his observations and experiments that all masses in the universe attract each other with a force proportional to the product of their masses but inversely proportional to the square of the distance between them. This discovery is known as **Newton's universal law of gravitation** and is discussed in his book *Principia*

(1687), probably the most important scientific book ever written. Through his law, Newton was able to derive all of Kepler's laws.

Newton's accomplishment exemplifies how science works. Science takes a lot of seemingly unrelated observations and combines them into theories that any person can use to predict correct answers to questions about nature. Of Newton, Alexander Pope wrote,

Nature and Nature's laws lay hid in night:
God said, "Let Newton be!" and all was light.

Kepler's Laws

Kepler's laws can be stated as follows:

1. Each planet moves in an ellipse with the sun at one focus.
2. The line between the sun and the planet sweeps out equal areas in equal amounts of time. (This is called the law of areas.)
3. The ratio of the cube of the semimajor axis to the square of the period is the same for each planet. (This is called the harmonic law.)

Newton's work showed that Kepler's laws apply to any situation in which a small body revolves around a much more massive body. The two pairs of bodies can be a planet and a satellite, the sun and a comet, or the moon and a spaceship, except that the ratio in the third law depends on the sum of their masses.

As shown in Figure 3-1, each ellipse has two **foci,*** or points of symmetry (one of which is occupied by the **primary body,** which is the more massive body). Ellipses have different shapes, the circle being a special case in which the two foci coincide in the center. The point closest to the primary is called the **periapse,** and the point farthest from it, the **apoapse.** The properties required to define the shape of a planet's orbit and its position in the orbit are called the **orbital elements.** Five elements define the size, shape, and orientation of the ellipse, and the sixth (the moment when the orbiting body passes through periapse) defines the orbiting body's position in the ellipse. The three most important elements, shown in Figures 3-1 and 3-2, are the following:

a (**semimajor axis**): greatest distance from the center of the ellipse to its periphery (for a circle, the radius)
e (**eccentricity**): a measure of the departure from circularity

---

*In a certain science fiction story, Galileo's spirit was transferred and reincarnated as Newton in 1642! This theory may be questionable, but it helps one to remember the historical framework of the scientific renaissance.

---

*The plural of focus is foci.

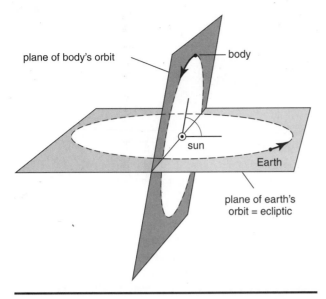

Figure 3-2. The inclination of a body's orbit is the angle between its orbital plane and the ecliptic plane.

i (**inclination**): the angle between the orbital plane and some reference plane, usually Earth's orbital plane (that is, the ecliptic)

Another useful property of an orbit is *P*, the **period,** or the time required for the body to complete one orbit.

Newton's Laws

As mentioned above, Kepler's laws describe *how* the planets move, but don't say *why.* Newton's laws gave a better framework for visualizing *why* the planets follow Kepler's laws. We should note, however, that science never ultimately answers "why" questions but merely moves us to more underlying relationships among phenomena, which had not earlier been seen to be related. Newton's idea of gravity, for example, revealed that all masses relate through their mutual attraction. As he realized, the motion of an apple falling from a tree is governed by the same "laws" that control the motion of the moon in orbit around the Earth. The falling apple is actually in orbit!

Newton clarified the meaning of the concepts of force, motion, energy, and so on. He realized that if any body is *not* moving in a straight line, some force must be acting on it to deflect it from straight-line motion. In the case of the planets, the principal force is **gravity,** the force by which any mass attracts any other mass. Newton determined that gravitational attraction between the sun and a planet must be proportional to the mass of the sun and to the mass of the planet, and he showed what mathematical form the force law has. This formulation is his **law of gravitation.**

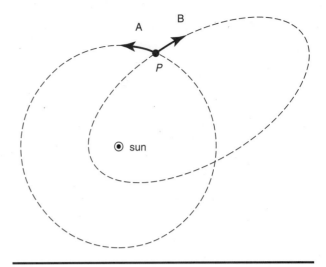

Figure 3-3. A body at point *P* can go into a variety of different orbits around the Sun, depending on its initial velocity. Two possible orbits are shown.

Orbits

A body anywhere in the solar system is under the gravitational influence of the sun. If the body is not too close to a planet, the sun's influence is dominant. If the body is given a velocity *v* in any direction, it will start to move in some orbit influenced by the sun. Body A in Figure 3-3 is pushed to the left and starts to curve downward toward the sun. Body B is pushed away from the sun but eventually turns back. Both bodies are in orbit around the sun. A falling apple or a football thrown by a quarterback is in orbit around Earth, but in this case the orbit intersects the surface of the ground.

Whether the object is in orbit around the sun, Earth, or some other body, the principle is the same. The primary body will be at one focus of the orbit. (Detailed analysis shows that the focus lies at the center of gravity of the two bodies, which is near but not exactly coincident with the larger body. The sun is so massive relative to the planets that it virtually coincides with the focus.)

Circular Velocity

Circular velocity is the speed of a satellite in a circular orbit about a primary body. The circular velocity decreases with increasing distance because the pull of gravity declines with distance.

It is instructive to imagine a body being inserted into orbit at various speeds. Suppose a spacecraft has reached a point *P* above Earth's atmosphere* (see Figure 3-4). Its

---

*If the spacecraft were in the atmosphere, atmospheric friction would produce a resistive force called *drag*, which would disturb the motions described here.

The laws can be simply stated mathematically:

1. $\quad r = \dfrac{a(1 - e^2)}{1 + e \cos \theta}$ (equation for an ellipse)

2. $\quad \dfrac{dA}{dt} = \text{constant}$

3. $\quad \dfrac{a^3}{p^2} = \text{constant}$

where r = distance from orbiting body to one focus

a = semimajor axis

e = eccentricity

θ = angular position in orbit measured from focus

A = area

P = period

In the most general case, with any small mass m that is not negligible and any large mass the third law takes the following form:

3. $\quad \dfrac{a^3}{p^2} = \dfrac{G}{4\pi^2} \cdot (M + m)$

Because every planetary mass is so much smaller than the solar mass (M here), application of the law to the system of sun and planets converts the right-hand side of the equation to approximately a single constant, $GM/4\pi^2$.

An important exercise in advanced courses is to derive Kepler's laws from Newtonian physical theory. This is very simple in the case of the second law, because the law of areas is a restatement of one of the basic conservation laws that are so important in physics: the law of conservation of angular momentum. Angular momentum is equal to the linear momentum times the radius, or mvr. Thus conservation of angular momentum says that

$$mvr = \text{constant}$$

or, substituting for v,

$$m\left(r\dfrac{d\theta}{dt}\right)r = \text{constant}$$

But from Figure 3-A we see that the area swept out in a time dt (being $\frac{1}{2}$ the base times the height of the triangle) is $dA = \frac{1}{2}r\, d\theta\; r$. Thus, substituting in the above equation,

$$m\, 2\, \dfrac{dA}{dt} = \text{constant}$$

Since the mass of the planet m and 2 are both constants, we now have the law of areas, dA/dt = constant. Another important exercise, beyond the scope of this book, is to derive Kepler's first and third laws from Newtonian physics.

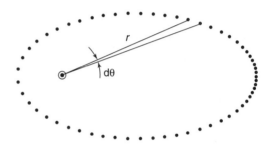

Figure 3-A. Orbital motion in an elliptical orbit. The dots are spaced to show motion during equal time intervals, indicating fastest motion at periapse and slowest motion at apoapse. Area swept out during the indicated interval is $\frac{1}{2}rd\theta r$.

---

attitude-control system has aligned it parallel with the ground, but its position is stationary with respect to the center of Earth. If a rocket motor is not fired, the spacecraft will fall directly to the ground on path A. If a short burst is fired, the spacecraft will fall into an elliptical orbit around the center of Earth, but the orbit, B, will intersect the ground like that of the football. With a higher speed the rocket could go nearly all the way around, striking the ground on the far side, on orbit C (similarly to an ICBM). A still higher speed will put the rocket into a complete orbit D, with the perigee on the far side. A still faster velocity is needed to put the rocket into a precisely circular orbit E. The description remains the same whatever the primary body and whatever the smaller body (Figure 3-5).

Nearly circular orbits were chosen for the first space flights because the rockets were barely powerful enough to climb to the top of the atmosphere, and the space capsules had to skim along at a constant height to avoid atmospheric drag.

Time-lapse movies from space probes dramatically record Keplerian orbital motion among satellites of planets. One of the most interesting cases exists within ring systems, such as the famous rings of Saturn. Each of the millions of small particles pursues its own nearly circular orbit around Saturn. As shown in Figure 3-6 on page 52, certain swarms of these particles can be tracked in their movement around Saturn on consecutive photos but the reason for the visibility of these swarms is unknown.

Escape Velocity

Escape velocity is the speed required to project a body completely free of a primary body. Like circular velocity, the escape velocity is less at larger distances from the primary body.

If our imaginary spacecraft in Figure 3-4 is accelerated to an even higher speed than that required for circular orbit E, it climbs into an elliptical orbit F, with the

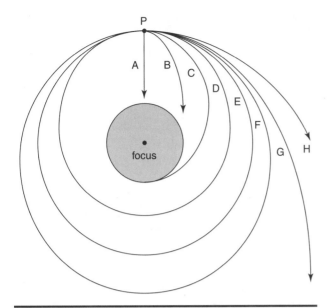

Figure 3-4. A rocket at point *P* above Earth can go into a variety of different orbits. Cases A through H correspond to firing the rocket parallel to the ground but with different velocities (see the text).

Figure 3-5. The condition called weightlessness in space is not due to any "lack of gravity" but merely to the fact that all objects in a spacecraft are co-orbiting together around some distant object (Earth, the moon, the sun, and so on) in nearly identical orbits. The condition is sometimes more aptly called "freefall" (if you jump off a high diving board and release a tennis ball during your fall toward Earth's center, the ball will appear to float weightlessly in front of you until you hit the water). Here, astronauts William Pogue and Gerald Carr clown to show the effects of weightlessness inside Skylab 4 during an 84-d flight in 1975. (NASA)

apogee on the far side of Earth. At the critical speed of escape velocity, the apogee is at infinite distance and the rocket never comes back. Instead, it keeps slowing down as it moves out on path G, the speed approaching but never quite reaching zero. Another way of saying the same thing is that the rocket with escape velocity has a kinetic energy just equal to Earth's gravitational potential energy. The orbit, G, has the shape of a parabola and is called a **parabolic orbit.** At each point on G, the speed will just equal escape velocity (sometimes called **parabolic velocity**) at that point. If the rocket is made to move faster than escape velocity, it goes into a different orbit H, with a slightly different shape from orbit G. Again the rocket never comes back, but in this case it has a kinetic energy greater than Earth's gravitational potential energy, and the rocket is always moving faster than the escape velocity at each point in its orbit. This path is called a **hyperbola** or a *hyperbolic orbit.* (Ellipses, circles, parabolas, and hyperbolas are a geometrically related family known as **conic sections,** as they can all be produced by taking cross sections of a cone at various angles.)

Another way of viewing escape velocity is to imagine a body dropped from rest at an infinite height. It falls toward the ground with increasing speed, and its speed at any point will be the escape velocity at that height.

Astrometry, Orbit Determination, and the Doppler Shift

If you could watch a spaceship returning from the moon night after night from your backyard, how could you determine what its orbit was in three-dimensional space?

Figure 3-6. Three photos of Saturn's rings, taken a half hour apart, reveal motion in circular Keplerian orbits. The rings are made of millions of separate particles (not visible here). Due to unknown causes, swarms of particles may become visible as spoke-like shadings, whose orbital motions can be tracked. In the top view, a double spoke approaches the tip of the rings. In the middle view, it rounds the tip, followed by another spoke. In the last view, this spoke has moved around the tip. As predicted by Kepler's laws, particles nearer the planet orbit faster, causing the spokes to shear from radial to diagonal patterns. (NASA, Voyager 1 sequence)

The task of measuring positions of objects in the sky is a branch of celestial mechanics called **astrometry.** In principle, only three observations of position, relative to background stars, are required to determine an orbit, but in practice many observations are used to determine the orbit as accurately as possible. In the 1800s, when many planets and asteroids were being discovered, the laborious computations needed to derive the orbit had to be done by hand and might require months; today they are done by computer in minutes.*

The science of **spectroscopy,** or analysis of the distribution of colors in light, is of extreme importance in studying all celestial bodies and can clarify motions as well as compositions. Each color corresponds to a certain wavelength. If a source, such as a distant satellite, is moving, then the wavelength of the light coming from it is shifted from that of the light received from a stationary source, and the amount of the shift is proportional to the velocity. This shift in wavelength is called the **Doppler shift,** after Austrian physicist Christian Doppler, who discovered it in 1842 while studying stars. The Doppler shift allows measurement of the velocity toward or away from the observer (but not the component of velocity perpendicular to the observer). Measurements of this **radial velocity,** as it is called, clarify the orbital motions of distant objects. If the object being tracked is a spacecraft emitting signals of precisely known frequency, the Doppler shift measures can be made with great precision and allow incredibly precise tracking. More recently, radar signals from Earth have been bounced off asteroids and other objects, allowing greatly improved orbit determination.

The Three-Body Problem

The earlier discussions exemplify the **two-body problem:** the description of motion in a system of two bodies with no other influences. Because the motion of a planet around the sun has only minor influences from other planets, two-body theory gives a good approximate description of planetary motion. This accounts for the success of Kepler's laws. The real universe, however, is not so simple; in many cases, we need to consider more than two bodies in a system.

The **three-body problem** is to describe the motions of three bodies when they are big enough to influence one another or at least when two of them are big enough to influence the third. An example would be a spacecraft moving in the Earth-moon system, where both Earth and the moon are big enough to influence the spacecraft's trajectory. There is no single, general analytical solution that

---

*However, the computer programmer may have to work for months to get the computer routine in working order, thereby illustrating the principle of Conservation of Human Labor, but that's a different story.

describes multibody motions, contrary to the known simple equations for two-body motions. Therefore, the only general way to predict motions in a three-body system is by **numerical integration,** the process of solving the problem in small steps. A computer could be set up to start with the spacecraft at position 1 and compute the forces on it from Earth and the moon, and then to let the spacecraft move a small distance under the influence of these forces to position 2, and so on, for hundreds or thousands of positions, depending on the accuracy needed.

The three-body problem applies not only in the Earth-moon-spaceship case but also in many other solar system cases, such as a sun-Jupiter-comet system. In addition, some special cases cause very interesting effects that help explain some puzzling properties of the solar system, as described in the following sections.

Perturbations and Resonances

Each planetary body in the solar system is affected not only by the sun but also by the other planets (such as massive Jupiter) and other smaller bodies. The sun dominates, but the interplanetary forces cause small effects, called **perturbations.** Because of these perturbations, the planetary and satellite orbits are not perfect ellipses, but ellipses with minor sinuosities. Uranus's motion is perturbed by Neptune, an asteroid's motion by Jupiter, and so on. A satellite of a planet may be perturbed by the sun and also by the mass contained in the equatorial bulge of the planet (which in turn is produced by centrifugal force due to the planet's rotation).

Effects of perturbations are interesting and varied. The elements of a perturbed orbit change with time. The changes may be **periodic** (varying smoothly between limits) or **secular** (tending to change in a certain direction). Perturbations of satellite orbits inclined to planets' equators typically cause the plane of the orbit and the line from periapsis to apoapsis (the line through the two foci) to swing around the planet on time scales much less than the solar system's age. In the lunar case, the line from perigee to apogee swings around Earth in only 9 y.

In many cases the perturbations cause only minor fluctuations in the orbital elements that look more or less random, but consider the case of one body whose period is $\frac{1}{2}$, $\frac{1}{3}$, $\frac{2}{3}$, or some other small-integer fraction of the period of a neighboring larger body, a condition in which the two bodies are said to be **commensurable.** A good example would be an asteroid whose period is $\frac{1}{2}$ of Jupiter's period, as shown in Figure 3-7. On every second trip of the asteroid around the sun, it will find itself again next to Jupiter (positions 1 and 9), and the disturbance by Jupiter will repeat.

In other words, commensurability causes a condition called **resonance,** the repetition of perturbations. If you push on a swing at random intervals, you may produce a small oscillatory motion, but if you push on the swing in a constantly repeated way that matches the swing's natural

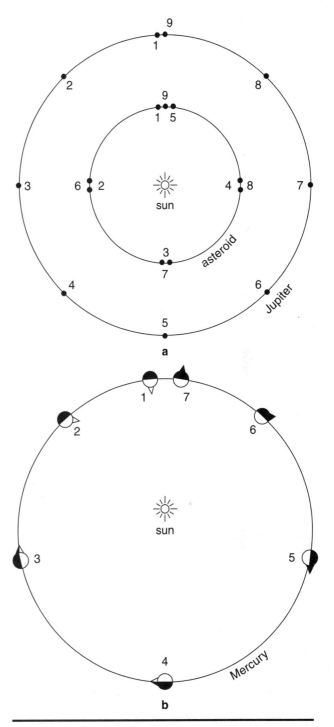

Figure 3-7. Examples of commensurabilities. Numbers give positions of bodies after equal intervals of time. (**a**) An asteroid in a 1:2 resonance with Jupiter completes two revolutions while Jupiter completes one. At position 9 they repeat configuration 1. Such a repeated perturbation has kicked asteroids out of their orbits, creating the Kirkwood gaps. (**b**) A 2:3 commensurability controlling the rotation of Mercury, illustrated with a mountain or tidal bulge on one side. Mercury, which completes 1 $\frac{1}{2}$ rotations in one trip around the sun (or 1 rotation in $\frac{2}{3}$ orbits), is locked into the commensurability by tidal forces strongest at perihelion (positions 1 and 7) in its elliptical orbit.

The law of gravitation states that

$$F = \frac{GMm}{r^2}$$

where F = force between two bodies
M = mass of larger body
m = mass of smaller body
r = distance between the two bodies
G = gravitational constant (see Table 2-2).

The equation gives the force exerted on a small body by a large body, or vice versa. Earth attracts familiar objects with a force that we call their weight. If you are given the value of the gravitational constant, the mass of Earth the radius of Earth, and the mass of a rocket, you could compute the rocket's weight, which equals the force necessary to lift it off the ground. This is why the force exerted by rocket engines is usually expressed by American rocket engineers as pounds of thrust.

Sample Problem

Prove that a person weighs only $\frac{1}{6}$ as much on the moon as on Earth, given that the moon's mass is $\frac{1}{81}$ of Earth's mass and the moon's radius is $\frac{1}{3.7}$ of Earth's.

Solution

The ratio of weight on the moon to the weight on Earth is

$$\frac{F_{\mathbb{C}}}{F_{\oplus}} = \frac{GM_{\mathbb{C}}m/r^2_{\mathbb{C}}}{GM_{\oplus}m/r^2_{\oplus}}$$

where M and r are the mass and radius of the planet involved, and m is the mass of the person. Canceling G's and m's, we have

$$\frac{F_{\mathbb{C}}}{F_{\oplus}} = \frac{M_{\mathbb{C}}r^2_{\oplus}}{M_{\oplus}r^2_{\mathbb{C}}} = \frac{1}{81}\frac{(3.7)^2}{1} \approx \frac{1}{6}$$

frequency, you can "pump up" the motion of the swing to dramatic, wide oscillations. Similarly, some resonances can have a dramatic pumping-up effect on a planetary orbit, causing a drastic change in the orbital elements, usually markedly increasing the eccentricity e.

Note that as stated in Kepler's third law, the period of the asteroid depends only on its semimajor axis and is not affected by eccentricity or inclination. Therefore, a resonance can happen not only to an asteroid in a circular orbit at a fixed distance from Jupiter (as in Figure 3-7), but also to any asteroid in an elongated orbit with that same period (and semimajor axis). Such a resonance is sometimes called a *mean motion resonance* because it depends on the period, or mean motion, of the body (see review by Froeschle and Greenberg, 1989).

In 1969, dynamicist James Williams discovered another type of asteroid-Jupiter resonance that involves more complex interactions with planetary motions; this type is sometimes called a *secular resonance*. Williams noted that at least one of these was efficient in throwing asteroids clear out of the belt into orbits that can approach Mars and subsequently Earth, helping to explain the supply of meteorites that fall on Earth (see Chapter 6). This resonance is designated the $\dot{v}_6$ resonance (read "nu dot six" resonance) after a term in the relevant dynamical equations. (See review by Scholl and others, 1989.)

Curiously, a few resonances, such as the 3:2 resonance with Jupiter, can actually have a stabilizing effect on orbits. Instead of destabilizing the objects and throwing objects out of that zone, they help hold objects in the zone.

The most dramatic manifestation of resonances is the structure of the asteroid belt. In 1866, Indiana astronomer Daniel Kirkwood discovered certain zones of semimajor axis within the asteroid belt that are not populated. He showed that they match up with Jupiter resonances (Cunningham, 1988). These zones are now called the **Kirkwood gaps.** Here are some of the important resonances and the positions in the belt (semimajor axis) where the effect is felt:

Resonance	Semimajor axis (A.U.)	Belt structure
v_6	~2.0	Kirkwood gap marking inner edge of asteroid belt; higher semimajor axis for higher inclinations.
3:1	2.50	Strong Kirkwood gap
5:2	2.82	Moderate Kirkwood gap
2:1	3.28	Strong Kirkwood gap; outer edge of asteroid belt (except for a few concentrations of additional objects)
3:2	3.97	Concentration of asteroids, known as the Hilda group, after largest member

Rotations of planetary bodies can also be affected by resonance when the effects involve tidal forces. An example is Mercury, whose rotation is sketched in Figure 3-7b.

Lagrangian Points

As mentioned earlier, predicting positions of bodies in three-body or n-body systems is very difficult. However, a specific analytic solution is known for a certain type of

It is easy to remember how to derive the circular velocity, because gravity must be just matched by centrifugal force:

$$\frac{GMm}{r^2} = \frac{mv_{circ}^2}{r}$$

Thus,

$$v_{circ} = \sqrt{\frac{GM}{r}}$$

Note that this gives v as a function of r and that as r increases, v decreases.

---

three-body system that corresponds to a condition of 1:1 resonance—namely, a system of two major co-orbiting bodies with nearly circular orbits (such as Earth and the moon), with a third small body nearby having the same revolution period P as the other two. In such systems, there are five points at which the gravitational forces of the two bodies plus the centrifugal force balance in such a way that a third body put in one of these points tends to remain in a fixed position with respect to the other two, or at least remains near that point.

The five positions are called **Lagrangian points** after the mathematician Lagrange (1736–1813), who first discovered their locations. The Lagrangian points are shown schematically in Figure 3-8. Of the five points, only two, L_4 and L_5, tend to be populated by stable groups of bodies.* The other three are only quasi-stable; if a small body were put at L_1, L_2, or L_3, it would stay only until an outside influence (such as a fourth body) moved it out of position, whereupon it would drift away. A body put at L_4 and L_5, if slightly disturbed, would oscillate around its original position.**

A key idea in considering Lagrangian points is that the whole system shown in Figure 3-8 rotates together, like horses on a merry-go-round, because the periods of all objects around the center of gravity are equal. Points L_4 and L_5 are 60° ahead of and behind the second-largest body, in its orbit. Objects "caught" near the L_4 and L_5 points would

---

*Greenberg and Davis (1978) review the Lagrangian points, and note that many authors have erroneously described the L_4 and L_5 points as stable potential minima, analogous to dimples in a plastic surface in which a steel ball could come to stable rest. They note that, in fact, the L_4 and L_5 points are potential maxima (analogous to bumps on the plastic surface), but that the Coriolis force is responsible for driving departing particles back toward these points. Greenberg (1978) emphasizes that in the presence of a resisting medium, such as a gaseous or dust nebula, energy loss could reduce the effectiveness of the Coriolis force and cause particles to drift away from the L_4 and L_5 points.

**This stability applies only if the mass ratio of the primaries M_1:M_2 is less than 0.0385.

It is easy to derive escape velocity, since the gravitational potential energy must just equal the kinetic energy:

$$\frac{GMm}{r} = \frac{1}{2}mv_{esc}^2$$

$$v_{esc} = \sqrt{\frac{2GM}{r}}$$

Note the important and useful fact that

$$v_{esc} = \sqrt{2} \times \text{circular velocity}$$

Sample Problem

Relative to the center of the sun, Earth's orbital speed is about 29.8 km/s. How fast would a rocket have to leave Earth's orbit in order to reach interstellar space?

Solution

This would be escape velocity. Using the $\sqrt{2}$ rule, we have

$$v_{esc} = \sqrt{2}\,(29.8) = 42 \text{ km/s}$$

---

seem to circulate aimlessly around them as seen by an observer rotating with the system (or riding on the merry-go-round). Of course, no part of the system is stationary with respect to the surrounding stars.

The Lagrangian points are practical realities. The sun and Jupiter form a system of two massive bodies, with the L_4 and L_5 points 60° ahead of and behind Jupiter in its orbit. Two groups of asteroids have been discovered occupying these positions. They are called the **Trojan asteroids** and are named after the Homeric heroes of the Trojan wars. Until 1906, when the first of the Trojans was discovered by the German astronomer Max Wolf, the Lagrangian

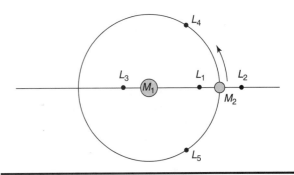

Figure 3-8. Location of the five Lagrangian points in an orbiting system.

Mathematical Notes on the Velocity Equation

Suppose that an orbit has been determined for a body and you want to know how fast the body is moving at a particular point relative to the primary body. A useful equation, sometimes called the *vis visa* or **velocity equation,** gives the velocity as a function of position (see Figure 3-B). With semimajor axis a and distance r to the object,

$$v^2 = GM\left(\frac{2}{r} - \frac{1}{a}\right)$$

This equation is a much more general form of our equations for circular and escape velocity, since it allows calculation of velocities required for any desired orbit.

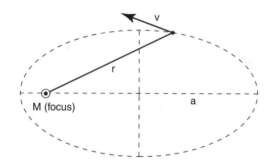

Figure 3-B. Orbital parameters used in the velocity equation.

Sample Problem

A comet in orbit around the sun is at aphelion 100 AU from the sun, and its perihelion is at Earth's orbit. Find its velocity at perihelion relative to the sun.

Solution

First, remember that all units must be in the same system, such as SI. A sketch of the orbit shows that the semimajor axis $a = 50.5$ AU $= 7.5 \times 10^{12}$ m. Substituting in the other quantities from Table 2-2 (page 41) in SI units, we have

$$v^2 = 6.67(10^{-11})\ 2.00(10^{30})$$

$$\left[\frac{2}{1.49(10^{11})} - \frac{1}{7.5 \times 10^{12}}\right]$$

$$\therefore v = 4.21(10^4)\ \text{m/s} = 42.1\ \text{km/s}$$

Note that this is essentially the escape velocity from Earth's orbit, where perihelion is located, as derived in the sample problem in the mathematical notes on escape velocity. In other words, 100 AU is so far out that the comet falls into the inner solar system essentially as if it had fallen in from infinity. Mathematically, we see that the second term in the brackets is almost negligible, reducing the equation to the equation for escape velocity.

Mathematical Notes on the Doppler Effect

The Doppler effect may be expressed as

$$\frac{\Delta\lambda}{\lambda} = \frac{v}{c}$$

where

$\Delta\lambda$ = change in wavelength
λ = normal wavelength
v = velocity along the line of sight
c = velocity of light = 3×10^8 m/s

For example, consider a spacecraft moving away from Earth at 3 km/s, which is 10^{-5} of the velocity of light. All its radio signals would be shifted in wavelength (or frequency) by 10^{-5} of their original amount. Monitoring the wavelength with great care thus gives information on the acceleration of the spacecraft. Such accelerations can be caused by local masses near which the spacecraft is moving. Using this fact, researchers have detected masses of small satellites, mass concentrations on the moon (Chapter 8), and effects of winds as they disturb the motions of spacecraft parachuting into the atmosphere of Venus and Jupiter (Chapter 11).

A body moving away from the observer has its light shifted toward longer wavelengths (the red end of the spectrum), an effect called a *red shift*. An approaching object has a decrease in the wavelengths of its light, called a *blue shift*.

points were mathematical abstractions, but now the two Trojan swarms are estimated to rival the main asteroid belt in numbers of objects. Trojans' orbits are seriously perturbed by other planets, and they may drift as far as 30° and more out of the L_4 and L_5 positions.

Another application of Lagrangian points came when Earth-based telescopic observations and Voyager photos in 1980–1981 revealed small moonlets near L_4 and L_5 Lagrangian points of Saturn's satellites Tethys and Dione. Mars also has at least one "Trojan" asteroid, 5261 Eureka. Searches for objects in the L_4 and L_5 points of Saturn and other planets have so far been fruitless (Gehrels, 1977).

The L_4 and L_5 points of our Earth-moon system gained attention as a result of proposals by Princeton physicist Gerard K. O'Neill (1974, 1975) and others that they are ideal locations for large, permanent space cities

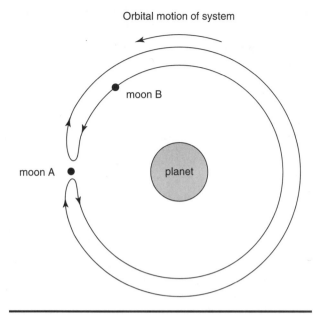

Figure 3-9. A horseshoe orbit: the apparent path of a small moon B as seen by an observer in a rotating coordinate system moving with moon A. (Both moons travel all the way around the planet, with respect to the stars. See text.)

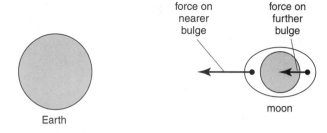

Figure 3-10. Body tides raised in the moon by the differential gravitational attraction of Earth.

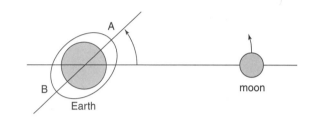

Figure 3-11. Tidal bulges (A and B) in Earth are dragged off the Earth-moon line by Earth's rotation. The effect of bulge A (which exceeds the effect of B) pulls the moon forward in its orbit, hence driving the orbital path slowly outward.

that could utilize at low cost construction materials and other resources from the moon. An advantage of these points is their fixed orientation relative to a lunar base, allowing easy ballistic launching of lunar material to them. These ideas are discussed in more technical detail in a NASA study (Johnson and Holbrow, 1977) and in a popular book (O'Neill, 1977).

Horseshoe Orbits

Another type of three-body orbital relationship, shown in Figure 3-9, has been found relevant to several solar system situations. Consider two small moons, A and B, almost in the same circular orbit. At the outset, B lags behind and is slightly closer to the planet. In keeping with Kepler's laws, B catches up with A. Figure 3-9 shows subsequent events as perceived by an observer riding on A. As B approaches on its "inside track," it is accelerated forward by A's gravity. This pulls it into a higher orbit, now farther from the planet than A. Following Kepler's laws, B now moves slower than A and begins to drop behind, in spite of A's gravitational attraction. B thus starts to recede from A on its new "outside track." In this sense, A and B have met, done a little dance, and exchanged orbits (since A is now closer to the planet). A and B continue to move apart until B drifts all the way around the planet (as seen by A), and a new encounter occurs in which B moves back to an orbit lower than A. The mean positions of the two orbits remain fixed over many encounters. As seen in Figure 3-9, the one moon (especially the smaller one) describes a horseshoe path relative to the other (Dermott and Murray, 1981).

This situation actually applies to two 200-km-scale inner moons of Saturn, 1980S1 and 1980S3, which have orbits only tens of kilometers apart. They don't collide, but approach to within about 16,000 km of each other, where one would loom about 1.5 times the apparent size of our moon in the sky of the other for a short while before shrinking and receding again into the distance (Harrington and Seidelmann, 1981). The encounters of these two moons occur roughly 4 y apart.

Tidal Effects

When two bodies orbit around each other, each exerts a force on the other, according to Newton's law of gravity. Consider Earth and the moon. The side of the moon facing Earth has a stronger force on it than does the far side, because the facing side is closer. This means that the body of the moon experiences a **net differential force** acting along the Earth-moon line at each moment. The moon stretches slightly along this line. The stretching forms approximately symmetric **tidal bulges** toward and away from Earth. These bulges, raised in the solid body of the moon, are called **body tides** (Figure 3-10).

The same situation arises in the case of Earth but with two differences. First, as shown in Figure 3-11, Earth has oceans that are free to flow and form **ocean tides**, bulges

Newton's law of gravity states that

(1) $F = \dfrac{GM_\mathbb{C}M_\oplus}{r^2}$

Defining masses as shown in Figure 3-C, we have

$M_\mathbb{C}$ = mass of moon
m = mass of tidal bulge
$M_\oplus$ = mass of Earth

Thus, the differential force on Earth caused by the moon is

$dF_\oplus \propto r^{-3}\, dr$

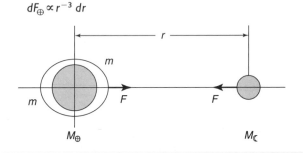

Figure 3-C. Properties related by Newton's law of gravitation in the case of tidal forces. $M_\oplus$ and $M_\mathbb{C}$ are the masses of Earth and the moon, which attract each other with force F; m marks the mass of each tidal bulge.

The mass m of the tidal bulge raised by this force must be proportional to the force, so

(2) $m \propto dF_\oplus \propto r^{-3}\, dr$

We want to know the net force on the moon caused by the two bulges. This is also a differential force, dF, proportional to the mass of the tidal bulge:

(3) $dF_\mathbb{C} \propto m r^{-3}\, dr$

obtained by differentiating (1). This becomes

(4) $dF \propto r^{-6}$

obtained by substituting (2) into (3). What we have shown is the important fact that *tide-raising forces are proportional to the inverse cube power of distance* and that the net effect of tidal forces depends on the inverse sixth power of distance. These are much stronger dependences on r than the inverse square law of gravity. Therefore, tidal effects can be very strong if the two bodies are close together but weak when the two bodies are somewhat farther apart. For this reason, tidal recession was very fast when the moon and Earth were close together, and is slow now.

A and B. Body tides form as well, but the ocean tides are much easier to observe, as beach people well know. Second, Earth rotates faster than the Earth-moon system revolves, so the side toward the moon is always changing. An observer at a given place on the turning Earth sees the ocean tide rise and then fall as the tidal bulge sweeps past him. In actual fact, high tide does not come when the moon is overhead because the shapes of shorelines and other complications modulate the tides.

Forces associated with the tidal bulges and acting between the bodies are called **tidal forces.** Tidal forces have come to be recognized as a fundamental phenomenon in planetary science and understanding some of their effects is important.

Tidal Evolution of Orbits and Rotation Rates

Because Earth is turning, the tidal bulges A and B, shown in Figure 3-11, are dragged off the Earth-moon line. This happens to both the body and ocean tides, although the amount differs because of the different amounts of friction involved. Bulge A is thus in front of the moon as it moves in its orbit, and bulge B is behind the moon. The bulges are, in fact, concentrations of mass, and so by Newton's law of gravity, they exert forces on the moon. Bulge A is closer to the moon and thus stronger; the net effect is to pull the moon ahead. It is as if the moon had a small rocket motor accelerating it steadily ahead in its orbit. As a result, the moon is slowly spiraling outward, away from Earth. This effect was actually detected late in the 19th century when observers noticed that the moon tended to depart from the position predicted by ordinary, nontidal, two-body Keplerian laws. Stated another way, tidal effects cause a transfer of angular momentum between two orbiting bodies.

In the same way, the moon has a net effect of pulling backward on bulge A, slowing Earth's rotation. Millions of years from now, the day will be noticeably longer than 24 h and the moon will be farther away. The angular momentum gained by the moon is given up by Earth, thus satisfying the law of conservation of angular momentum in the two-body system.

Similarly, Earth has tidally grabbed onto the longest axis of the moon and used it to slow the moon's rotation so that the moon now keeps its longest axis pointed along the Earth-moon line. This state is called **synchronous rotation** of a body; this is the state in which rotation and revolution rates are equal so that one side of a satellite always faces the companion body about which it orbits.

In the past, Earth turned faster, the moon was closer, and the month was shorter. Actual evidence of this comes from certain marine creatures whose shells have daily and monthly growth bands, allowing biologists to count the number of day bands in a monthly cycle in fossils of different ages. Whereas the present month of lunar phases (synodical month) is 29.5 d long, it was only about 29.1 d long 45 My ago (Kaula and Harris, 1975). Older fossils reveal monthly cycles as early as 2.8 By ago, when the month may have been as short as 17 d.

A linear extrapolation of the present recession would put the moon near Earth only 1 to 2 Gy ago, a result ruled out by the old fossils. In practice, it is not possible to compute from tidal theory the time when the moon was closest, because the extent and depth of oceans, and hence tidal effects, were different in the geologic past. Most researchers believe that the moon's closest approach was several billion years ago, probably during the planet-forming era. Mathematically, we can trace the motions backward and find that when the moon was closest, Earth's day (rotation period) was only 4 to 5 h long.

The same kind of tidal effects have acted on satellites throughout the solar system. Most satellites in the solar system, except for those farthest from planets, have had their long axes aligned toward the planet and are thus locked into synchronous rotation, keeping one side toward the planet.

The largest satellites are big enough to raise tidal bulges on the planets and thus are undergoing orbital evolution, usually spiraling outward like our own moon. In multisatellite systems, this has the interesting consequence that the largest satellites arrive in positions that put them in resonances with inner moons. A large satellite like Titan may have pulled smaller satellites with it as it moved outward. Systems of several major satellites, such as the Galilean moons and several pairs of Saturn's moons, may have arrived in their observed resonant relationships in this way (Weisel, 1981).

Tides raised on the planets by the sun may have slowed the rotations of Mercury until they were locked in resonant rotation states. Mercury is locked into a 2:3 resonant rotation with its orbital motion around the sun (Figure 3-7b). In the case of Venus, a solar tide was probably raised mainly in Venus's dense atmosphere, helping to slow the rotation (Lago and Cazanave, 1979). (An earlier theory involving resonant tidal forces raised on Venus by Earth has been disproven.)

Roche's Limit

In discussing the mathematical theory of tides, we showed that the force raising a tidal bulge obeys a $1/r^3$ law (whereas the net tidal interaction obeys a $1/r^6$ law). This means that as the distance r decreases, the stretching force gets very large. There is a critical distance between two bodies within which the tide-raising force on the smaller body is

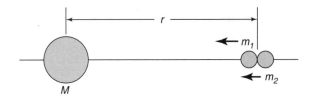

Figure 3-12. If two masses m_1 and m_2 orbit around a planet M, Roche's limit occurs where the excess attraction of m_1 toward the planet is equal to the gravitational attraction of m_1 and m_2 for each other.

strong enough to tear it apart. This critical distance is called **Roche's limit** after its discoverer Edouard Roche (1850), a French mathematician.

To understand Roche's limit more clearly, imagine the Earth-moon system. Suppose we represent the moon by two particles just touching each other (Figure 3-12). The only force holding the particles together is their own mutual gravity; the only force tending to separate them is the tidal force. If for any reason this moon approaches Earth, the tidal force will increase much faster than the gravitational force, and at some point it will exceed gravity, causing the two particles to drift apart.

In reality, of course, the moon is made of more than two particles. Its many particles are bonded together in the form of rock, with high **tensile strength** (strength against rupture by stretching), but the principle is the same. For mathematical convenience, Roche's limit is defined as the critical distance at which a body with no tensile strength would be torn apart by tidal forces. This is a good approximation for a large satellite, like the moon, where the internal pressure due to self-gravity exceeds the tensile strength of the rock. A kilometer-scale rocky satellite would have to pass well inside Roche's limit before it would actually fragment.

Tidal Heating

Probably the most profound effect of tidal forces in planetary science is the effect called tidal heating, which was studied by California physicist Stanton Peale and colleagues and led to their 1979 prediction of volcanoes on Jupiter's satellite, Io. Suppose Io (or any sizable satellite) had been formed close to Jupiter (or any large planet) in a circular orbit. A tidal bulge would be raised on Io, but if the orbit were circular, the distance between Io and Jupiter would be fixed and the size of the bulge would not change. Now suppose some other large satellites are added, nearby, as is really the case in the Jupiter system. They would perturb the orbit of Io, forcing it into a slightly variable, noncircular orbit. This forced eccentricity of the orbit would bring Io sometimes closer to Jupiter and sometimes farther away. In turn, the tidal bulge would rise and fall. Io would be *flexed,* in something like the way you might repeatedly flex a tennis ball in your hand. If you do this,

Mathematical Notes on Roche's Limit

The mutual gravitational attraction for two equal-sized touching particles is (see Figure 3-12)

$$(1) \quad F = \frac{Gmm}{(dr)^2}$$

The disruptive tidal force is the differential gravity force:

$$(2) \quad dF = \frac{GMm}{r^3} 2 \, dr$$

(see discussion of tides). At Roche's limit r_R, these two forces are equal. Equating (1) and (2) thus gives

$$\frac{Gm^2}{(dr)^2} = \frac{2GMm}{r_R^3} \, dr$$

Therefore,
Roche's limit (for two touching particles)

$$\equiv r_R = \left(\frac{2M}{m}\right)^{1/3} dr.$$

Since the mass of a body M is $4\pi R^3 \rho_M/3$ (where R is its radius and ρ_M its density) and since the radius of each body m is $dr/2$, this expression can be simplified to

$$r_R \simeq 2.5 \left(\frac{\rho_M}{\rho_m}\right)^{1/3} R$$

A somewhat more complex derivation gives the classical Roche expression for the breakup of a single zero-strength (or liquid) spherical body (instead of two touching spheres):

$$r_R = 2.44 \left(\frac{\rho_M}{\rho_m}\right)^{1/3} R$$

Aggarwal and Oberbeck (1974) have studied the case of the breakup of orbiting spheroidal bodies held together by the strength of their rocky or icy material. Generally, for bodies larger than about 40 km in diameter, orbiting icy or stony bodies of modest strength will break up at

$$r_{AO} = 1.38 \left(\frac{\rho_M}{\rho_m}\right)^{1/3} R$$

An incoming body about to impact a planet will get even closer to the planet before breaking up:

$$r_i = 1.19 \left(\frac{\rho_M}{\rho_m}\right)^{1/3} R$$

The actual limits depend on the size and strength of the bodies. Bodies smaller than about 30 to 60 km in diameter can penetrate much closer to a planet and will not break up at all if they are small and strong enough. Of course, aerodynamic stresses (treated by Melosh, 1981) may cause breakup of a different sort if the planet has a dense enough atmosphere.

you will notice that the tennis ball grows warm because of friction of the molecules inside. In the same way, the tidal flexing of the satellite creates internal heat.

In the case of Io, this tidal heating is so strong that it maintains molten conditions in parts of the interior and drives volcanoes. A provocative observation is that many other of the larger satellites close to giant planets, such as Europa, Enceladus, and Miranda, show highly fractured surfaces or signs of outright melting and resurfacing episodes—suggestive of tidal heating. Initial calculations seemed to indicate that tidal heating was not strong enough to do the job in these cases, but further studies have suggested past conditions in which perturbations by other satellites might have caused enough tidal heating to create melting, volcanism, or at least tectonic activity. The study of tidal heating, satellite resonances, and related dynamical effects is a fertile field.

Rings

Planetary **rings** are systems of small bodies moving in circular orbits in a planet's equatorial plane. As the French astronomer Andre Brahic commented, ring systems are the perfume of the solar system because they have negligible mass but enhance the character of the planets. The four known ring systems present fascinating variety, as shown in Figure 3-13. Figure 3-14 shows that Roche's limit and resonances are both important in shaping rings (Burns, 1990, provides more detail in a nontechnical review). The ring systems extend out to about Roche's limit, indicating that they are swarms of debris too close to a planet to coalesce.

As shown in close-up photos of rings and their shadows (chapter opening image, Figures 3-13 and 3-15), rings can have sharp or diffuse edges and well-defined gaps. These features are believed to be controlled primarily by resonant perturbation forces acting on the rings from the nearest satellites outside the rings, or from small moonlets inside the rings. For instance, as seen in Figure 3-14, the inner edge of Saturn's ring B, the inner edge of Cassini's division, and the outer edge of ring A fall very close to the 1:3, 1:2, and 2:3 resonances of the nearby satellite Mimas (Cuzzi, 1978). Particles lying at these resonances have their eccentricities pumped up until they collide with other particles outside the resonance, creating gaps and sharp edges.

In the Jupiter and Saturn systems, modest-sized moonlets are associated with the outer edges of the rings, near Roche's limit; some of the ring particles are probably being created by micrometeorite erosion of dust off these moonlets.

The dynamics of n-body motions among the ring particles produce fascinating situations. Consider the moonlet in Figure 3-16. As in any Keplerian motion, particles

Figure 3-13. Four different ring systems. (**a**) The ring of Jupiter. This ring is so faint it was seen only when backlighted by the sun, as in this view. (**b**) Saturn's ring system, the most extensive in the solar system. Note the visibility of the disk through parts of the ring system. (**c**) Composite view of Uranus's system of thin ringlets. This backlighted view shows the faint rings on either side of the over-exposed Uranus disk, which is blocked out in the center. (**d**) The ring system of Neptune, like that of Uranus, shows thin ringlets, but some of Neptune's rings show concentrations called ring arcs (bottom left). (NASA Voyager images)

on exterior orbits move slower, and particles on interior orbits move faster. Thus, small ring particles approach only from the outer leading quadrant (upper left in Figure 3-16) and the inner trailing (lower right) quadrant. But in the strange gravitational field inside Roche's limit, these particles are not pulled directly into the moonlet. As seen by an observer on the moonlet, approaching particles perform odd loops (related to horseshoe orbits) if the moonlet is large enough to have substantial gravity. Relative velocities among smaller interacting particles may be much slower.

Because rings are inside Roche's limit, you might suppose there is no chance for one of the incoming particles to come to rest on the moonlet's surface. But remember that the main Roche tidal disruptive force lies along the moonlet-Saturn line. Thus, at some point inside Roche's limit, a particle placed on the Saturn-facing side would

spontaneously levitate and drift away, but a particle that landed 90° from the sub-Saturn point could remain on the moonlet's surface if it did not rebound. Thus, the moonlet could accrete material in a band about 90° from the Saturn direction. It could grow, slowly changing shape and axis of rotation (Greenberg and others, 1977).

These results may explain the curious discovery that two opposing quadrants of Saturn's A ring (the quadrants preceding conjunction, as defined by the Saturn-Earth line of sight; that is, northwest and southeast of Saturn's disk) are a few percentage points brighter than the other two quadrants (Rietsema, Beebe, and Smith, 1976). This situation could arise if the larger ring particles are rotating synchronously and are struck by neighbors primarily only on the two quadrants, as shown in Figure 3-16, making the light-reflecting qualities of these sides different from those of the other two quadrants (Columbo, Goldreich,

CELESTIAL MECHANICS

61

Figure 3-14. Ring systems of giant planets. R = classical Roche limit below which zero-strength liquid sphere would fragment. S_E and S_U = Smoluchowski (1978) limit below which touching equal-size and unequal-size bodies, respectively, would fragment. AO = Aggarwal-Oberbeck (1974) limit below which large (D ≥ 40 km) solid spheres would fragment. Smaller solid bodies would have still closer AO limits. Rings tend to lie inside R; satellites, outside R.

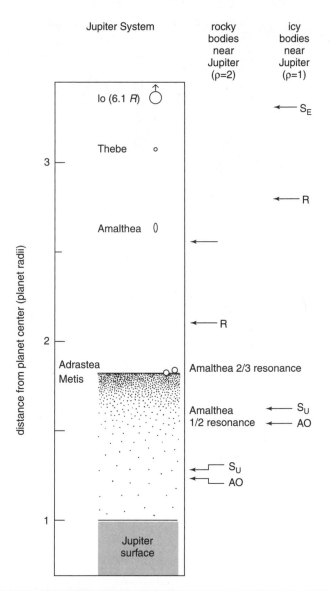

and Harris, 1976). Other explanations have also been proposed (Franklin and Columbo, 1978).

Another application of *n*-body ring dynamics is the attempt to understand unexpected details of ring structure observed by Voyagers 1 and 2 in the four-ring systems. For example, Goldreich and Tremaine (1979) and others developed theories in which narrow rings could be maintained by the gravitational action of "shepherding satellites"—small moons in or between rings, which force ring particles into narrow rings, like sheepdogs shepherding their flock. This theory seemed generally confirmed when Voyager 1 photographed two 100-km moons on either side of Saturn's narrow, outer F ring (Figure 3-17a), and later found two more apparent shepherds constraining the brightest, narrow ring of Uranus (Figure 3-17b).

Nonetheless, the detailed structures of rings remain puzzling. In addition to the question of how thousands of well-defined ringlets and gaps are created and/or maintained, scientists are pondering mysteries such as enhanced arcs within single rings, as found in Neptune's system (Figure 3-13d), and ringlets with kinks or braids (Figure 3-18). These curiosities probably involve disturbances of streams of ring particles as they pass by small moonlets, but the situation is unclear.

Figure 3-15. (**a**) The rings of Saturn, photographed at the intermediate range by Voyager 1, showed more complex structure than expected from Earth-based observations. Hundreds of fine divisions were found, in addition to the well-known Cassini division (main dark gap) and Encke division (thinner dark gap near outer edge). Box shows region of picture (**b**). (**b**) Detailed close-ups of certain ring regions showed more unexpected anomalies. This picture is a montage of two views of the leading and trailing sides. Comparison of thin bright ring inside dark gap at right shows that the thin ring is displaced from one view to the next. Dynamical forces causing such changes in fine structure are not well understood. (NASA, Voyager photos)

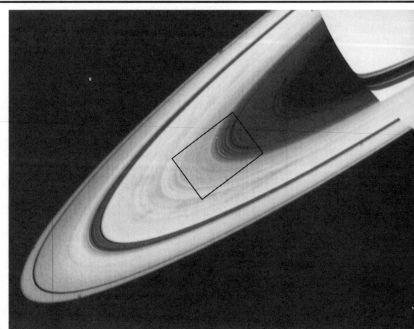

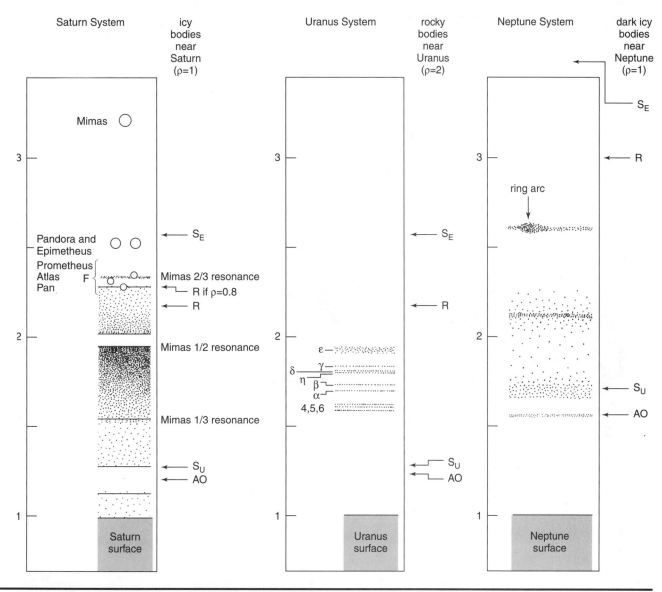

b

Figure 3-16. Motions of small, equally spaced ring particles passing an icy, 200-km diameter moonlet in Saturn's rings as seen by an observer moving with the moonlet and looking down on the ring plane. Particles approach from only 2 quadrants and collide or pass by with a velocity of about 60 m/s. They land only in a restricted band. Although this example exceeds the size of likely moonlets in Saturn's rings, it illustrates effects that would occur to a lesser extent among other major moonlets in ring systems. (Adapted from Greenberg and others, 1977; see text)

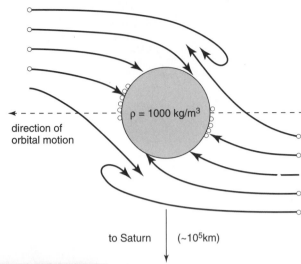

direction of orbital motion

$\rho = 1000$ kg/m³

to Saturn   (~10⁵km)

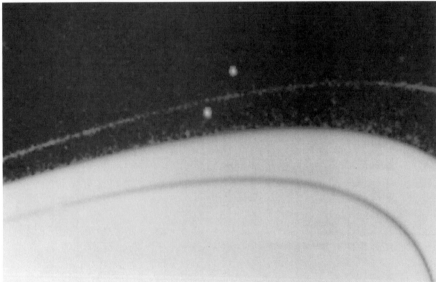

a

Figure 3-17. Shepherding satellites. (**a**) Two shepherding moons are seen on either side of Saturn's narrow F ring. The lower part of the picture is filled by the outer part of Ring A. In this view, the satellites are less than 1,800 km apart and will pass each other in about 2 hours. (**b**) Two more shepherding moons were found straddling the brightest ring in the Uranus system. Other thin rings may also be constrained by smaller, undiscovered moons. (NASA, Voyager 2 images)

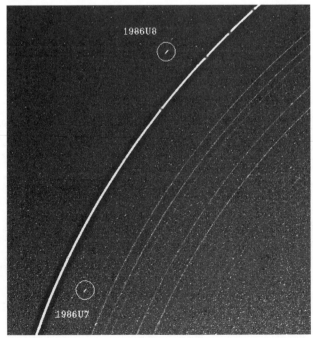

1986U8

1986U7

b

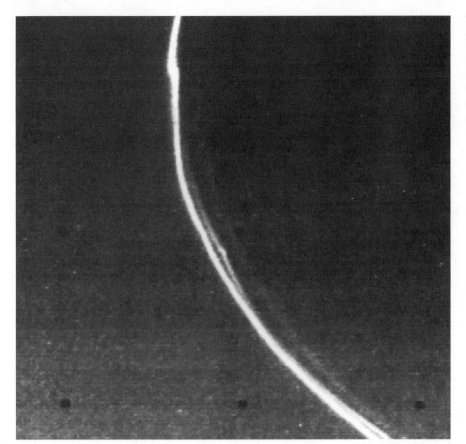

Figure 3-18. Unexpected twists or braids in Saturn's narrow F ring were discovered by Voyager 1. The diffuse ring segment to the right is about 35 km across. The knot near the center may be a local concentration of ring particles. (NASA, Voyager 1)

Whatever the solutions to ring dynamics, ring systems offer extraordinary environments. In Saturn's rings, the particles may be relatively close together, as the rings are no more than 100 m thick yet absorb considerable light. They probably move along together in virtually circular orbits; Goldreich and Tremaine (1978) suggested a random velocity dispersion of 0.1 cm/s or less! In the rings, we would be surrounded by a cloud of floating hailstones, perhaps occasionally jostling into one another, as imagined in Figure 3-19.

Dynamical Effects of Solar Radiation and Solar Wind

Radiation Pressure

Sunlight, like all light, is a form of electromagnetic radiation. Electromagnetic radiation consists of pulses transmitted through electric and magnetic fields; it has some of the properties of wave motion yet some of the properties of particles. We commonly speak of radiowaves or the wavelength of light, but the particlelike properties are not so familiar. One of these properties is that the pulses of light, called **photons,** carry momentum. Like BBs shot from an air rifle, they transmit this momentum to whatever they strike. When a photon strikes an object, an impulse is transmitted away from the light source.

Suppose a body is orbiting in space, exposed to the sunlight. The body is being struck by photons that transmit momentum (that is, exert pressure on the body). This phenomenon, called **radiation pressure,** has the effect of a force pushing the body away from the sun. If the cross-sectional area exposed to the sun is very large and the mass of the body is low, the radiation force can exceed the gravitational pull of the sun, causing the body to be "blown out" through the solar system.

One application of this has been the idea of a radiation-pressure-driven spacecraft, which would be propelled by an enormous thin sail. No such spacecraft has yet been built, although minor radiation pressure effects have been utilized in navigating existing space vehicles. True solar sailing would allow long solar system voyages with lower power and low fuel consumption.

Radiation pressure affects interplanetary dust particles in the solar system. Because the ratio of cross-sectional area to mass ($r^2 : r^3$) increases as the radius gets smaller, only small particles are strongly affected. (But if the radius gets much smaller than the wavelength of sunlight [5×10^{-7} m], the interaction with photons changes and radiation pressure again becomes negligible.)

Figure 3-19. An imaginary view within Saturn's ring system shows a sky full of floating hailstones. Saturn's cloudy surface lies below; the sun lies some degrees to the left of the ring plane. The densely populated portion of the rings may be as little as 100 m thick. (Painting by author)

Mathematical Notes on Radiation Pressure

Pressure can be expressed as the rate of transfer of momentum to a unit surface. The momentum carried by an individual photon is $h\nu/c$, where h is Planck's constant (see Table 2-2), ν is the frequency of the light, and c is the velocity of light. The total rate of momentum transfer is this momentum per photon times the number of photons per square meter per second for all frequencies:

$$P = \sum_{\nu} \frac{h\nu}{c} \left. \frac{dn}{dt} \right|_{\nu} = \frac{\mathscr{L}}{c}$$

where $\mathscr{L}$ is the total flux of radiation in Joules per square meter per second. The substitution of $\mathscr{L}$ can be made because $h\nu$ is the well-known energy per photon, so the total flux of energy is $h\nu(dn/dt)$ summed over all frequencies. If we want to know the force caused by radiation pressure on a small particle, we recall that pressure is force divided by area. Thus,

$$\text{radiation force on a spherical particle} = F = \frac{\mathscr{L}}{c}\pi a^2 Q$$

where a is the particle radius and Q is a correction factor on the cross section, since small particles may have an absorption cross section different from the geometric cross section:

$$Q = \text{correction factor}$$
$$= \frac{\text{effective absorption cross section}}{\text{geometric cross section}}$$

Q depends on particle size, particle composition, and wavelength of the light, but typically has values of 0.1 to 1.0.

Sample Problem

For particles of what size would the outward force of radiation from the sun be just balanced by the inward force of gravity? Assume that the particle is in Earth's orbit at solar distance r and that it is made of stone with density $\rho = 3 \times 10^3$ kg/m^3 and that $Q = \frac{1}{2}$.

Solution

Equating radiation and gravity force from solar mass we have

$$\frac{\mathscr{L}\pi a^2 Q}{c} = \frac{GMm}{r^2} = \frac{GM}{r^2}\frac{4\pi a^3 \rho}{3}.$$

Solving for a and substituting the values for the constants (from Table 2-2), we have

$$a \cong 1 \times 10^{-7} \text{ m}.$$

This is about 0.1 μm. Particles of this size and smaller would thus be blown out of the solar system when exposed to sunlight.

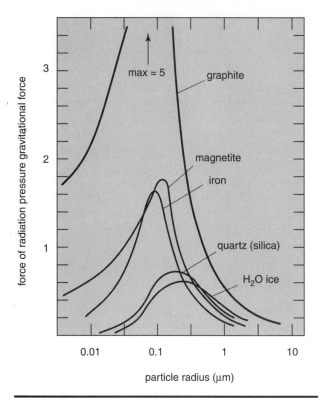

Figure 3-20. Ratio of outward radiation force to inward gravitational force for various-sized particles orbiting the sun. The force due to radiation pressure tends to maximize around sizes of 0.1 to 1 μm, but is greater for darker-colored particles. (Adapted from Burns, Lamy, and Soter, 1979)

Burns, Lamy, and Soter (1977) describe these effects and derive the results shown in Figure 3-20 for different kinds of particles. Radiation pressure acts strongly on micrometer- and submicrometer-sized particles (1 μm = 10^{-6} m). Because dark particles absorb the most light, the effect is strongest for them. Radiation pressure acts so strongly on submicrometer graphite and metal particles that they could be blown out of the solar system even if initially pushed toward the sun. For other submicrometer particles of stone or ice, radiation pressure merely decreases the net force toward the sun.

Solar Wind and Interplanetary Gas Motions

Analyses of comets, especially by Biermann (1951), showed that submicrometer material expelled from them is being accelerated away from the sun faster than can be accounted for by radiation pressure, suggesting that the interplanetary gas itself is moving away from the sun and carrying the cometary material with it. This idea led to the concept of the **solar wind,** an expanding, low-density **plasma** (ionized gas), emanating from the sun. Near Earth, this material is composed of a plasma with average electron densities around 2 electrons/cm³, temperatures around

200,000 K, and expansion velocities around 600 km/s, ranging occasionally up to 1,000 km/s. The solar wind can thus carry gas and fine dust from the inner to the outer solar system in only a month or so.

Theoretical models treat the solar wind in the first approximation as an expansion of the hot gases of the sun's outer atmosphere, or **solar corona** (temperature about 2 million K at three solar radii from the sun's center).

Spacecraft have returned good observations of the solar wind all the way out to the vicinity of Saturn. A current problem is the nature of the interface of the solar wind with the interstellar gas. One hypothesis places this beyond Pluto's orbit at around 50 AU from the sun. Space probes that leave the solar system, such as Pioneer 10 and Voyager 1, may clarify the location of this interface.

Poynting-Robertson Effect

The **Poynting-Robertson effect** is an interaction of light with (roughly) centimeter-scale particles. Unlike the solar wind and radiation pressure, this effect causes the particles to spiral *inward* toward the sun.

Consider a small particle moving in a circular orbit around the sun. Sunlight, which can be thought of as a steady stream of photons, flows outward from the sun while the particle moves at right angles to the photon stream. Thus, the photons strike the particle preferentially on its leading side, just as a car driven through a rainstorm is struck on its front side even though the rain may be falling vertically. This effect is shown in Figure 3-21. This makes a tiny change in the apparent direction from which the light is coming, an effect called the **aberration of light;** the apparent positions of stars seen from Earth are actually displaced up to 20.5 seconds of arc due to the aberration of light, as discovered by the English astronomer James Bradley in 1727.

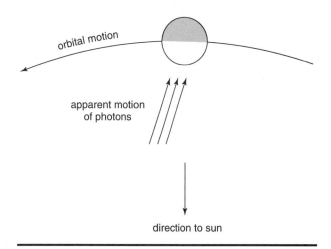

Figure 3-21. The Poynting-Robertson effect is caused by drag due to the apparent displacement of photons from the solar direction (see the text).

A concise discussion of the Poynting-Robertson effect (sometimes called the PR effect) is given by Burns, Lamy, and Soter (1979), who discuss several radiation effects. In general, if a particle starts on an eccentric orbit, the eccentricity will be reduced until the orbit is virtually circular. During the rest of its history, the particle will slowly spiral inward toward the sun (though any single trip around the sun will follow a nearly circular path). Most of the time may be spent in a fairly elliptical orbit. Once circularized, the orbit decays into the sun in a time given by

$$t_{PR} = 7.0\ (10^6)\ a\rho r^2/Q$$

where

$t_{\rho R}$ = decay time (y)
a = particle radius (m)
ρ = particle density (kg/m³)
r = orbit radius (AU)
Q = correction factor (typically 0.1 to 1.0)

The preferential absorption of photon momentum on the leading side is only one small part of a complex relativistic effect, whereby reradiation of absorbed photons causes a net loss of energy and the particle slowly settles into a smaller and smaller orbit, spiraling in toward the sun.

This effect was first predicted by the British physicist Poynting (1903) and later amended by the American physicist Robertson (1937) to take relativity into account. Wyatt and Whipple (1950) showed how this effect affects motions of interplanetary debris.

Mathematical theory enables us to calculate the history of a particle of any size, starting in any given orbit. Particles smaller than a few centimeters across will be swept from most orbits into the sun in a time less than the history of the solar system. Figure 3-22 summarizes the situation by plotting the time scale (solid lines) for particles with any initial eccentricity and perihelion distance to be swept into the sun. The dashed lines show the track followed by sample particles as they evolve across the diagram. For instance, a 1-cm particle starting at 1-AU perihelion with an eccentricity of about 0.7 will have a lifetime of only a few 10^7 y. Lifetime is proportional to diameter. In some size ranges, both Poynting-Robertson forces and radiation pressure combine to determine the particle's destiny.

Yarkovsky Effect

A curious effect with a curious history changes the orbits of larger bodies. The **Yarkovsky effect** is an orbital change on (roughly) meter-scale rotating particles, resulting from the radiation that they emit after they are warmed by the sun. If there were no rotation, the warmest spot would be the noon area, but rotation carries this spot around to the afternoon side. All bodies radiate photons in proportion to T^4 (T = surface temperature), so that the warmest area radiates the most photons. Thus, the most photon momentum is ejected on the afternoon side, and these photons act like a rocket exhaust. If the rotation and orbital revolution are both prograde, the "photon thrust" pushes the particle forward into an ever-expanding orbit. However, if either the rotation or the orbital motion is retrograde, the effect is reversed and the particle spirals inward. There is also a high-obliquity or "seasonal" version of the effect that arises from seasonal changes instead of diurnal rotation; it makes objects spiral inward, but rather slowly

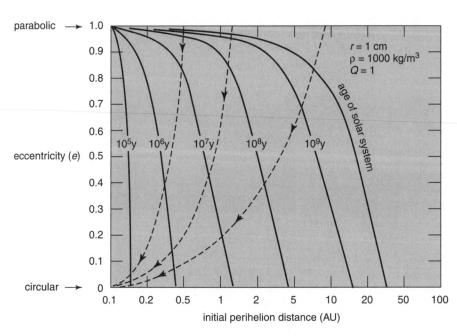

Figure 3-22. The time scales for a 1-cm-radius particle to spiral into the sun from starting orbits of different eccentricities and perihelion distances. Solid lines give the orbit-decay times; dashed lines give typical evolutionary tracks during decay. (Adapted from unpublished calculations by S. J. Weidenschilling)

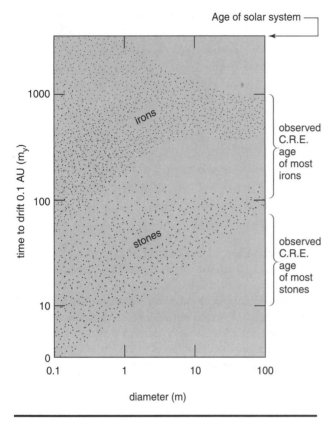

Figure 3-23. Drift of stone and iron fragments in the asteroid belt due to the Yarkovsky effect. This diagram shows the time for a fragment to drift 0.1 AU in semimajor axis, a characteristic distance from a random point in the belt to a resonance that might rapidly eject the fragment onto an orbit from which it could intercept Earth and become a meteorite. The observed duration of cosmic ray exposure (C.R.E.), which meteorites experienced in space as meter-scale fragments, is shown at the right. The diagram indicates that drift by the Yarkovsky effect may help explain the delivery of asteroid fragments to Earth as meteorites (see also Chapters 6 and 7). (After Farinella and others, 1998; Hartmann and others, 1998)

(Rubincam, 1995; Farinella and others, 1997). Thus, if the rotation direction of an asteroid or comet were occasionally changed by collisions, the planetesimal might wander about the solar system, sometimes moving outward and sometimes spiraling in toward the sun.

The Yarkovsky effect became widely known only in recent years after E. J. Öpik (1951) pointed out that it had been discovered by a Polish engineer, I. O. Yarkovsky, around 1900 and described in a paper that was lost. Öpik remarked that he remembered reading the lost paper in his youth around 1909; working from memory he re-derived the effect nearly half a century later! Peterson (1976) suggested that it helps deliver meteorites from the asteroid belt to Earth.

The effect is provocative because meteorites commonly come from meter-scale bodies, and if meter-scale bodies have their orbits moved around the solar system by the Yarkovsky effect, then the effect may play an important role in getting them out of the asteroid belt. Rubincam (1995) and Farinella and others (1997) pointed out that the Yarkovsky drift would not have to move an asteroid fragment from the belt all the way to Earth, but only from the belt to a nearby Kirkwood gap resonance, where it would be rapidly flung onto an orbit that approaches Earth. This is shown in Figure 3-23, which gives the time to drift 0.1 AU in the belt as a function of size. Asteroid fragments can be either stony or metallic, but the effect works better on stones, especially the black, carbonaceous chondrites. Figure 3-23 shows that the calculated drift times from random starting points in the asteroid belt (where an asteroid collision might produce a fragment) to resonances (where the fragment gets flung from the belt to Earth) actually match measured "cosmic ray exposure ages" of stone and iron meteorites—ages that measure their length of exposure in space (see also Chapters 6 and 7). Small stones would be cleared out of the asteroid belt much faster than small irons, and populations of 1 m to 100 m fragments in the belt, as well as properties of small meteorites hitting planets, may be strongly affected by these effects (Hartmann and others, 1999). Families of fragments from asteroid collisions would also be dispersed faster than otherwise expected (Bottke and others, 2001). Thus, the Yarkovsky effect, once a historical curiosity, may have an important role in the solar system.

Turbulence

Turbulence, the swirling motion of moving gas or liquid, is familiar to anyone who has watched the eddies in a flowing river, or an explosion, or the dust swirling behind a passing car. It is the opposite of streamline flow, or **laminar flow.** In planetary science, turbulence is important in a number of cases. The clouds of gas and dust from which stars form are turbulent; turbulence characterizes the motions of clouds in some parts of planetary atmospheres.

The **Reynolds number** is a useful concept in discussing turbulence; it tells whether turbulence will occur on a certain scale in a given gaseous medium. In the Reynolds number, which is a ratio, the numerator represents the forces that promote turbulence and the denominator represents viscous forces that tend to damp out turbulence. If the Reynolds number is greater than 1, turbulence will tend to increase in the medium; if it is less than 1, turbulence will decrease. Strongly turbulent media typically have Reynolds numbers above about 100.

In the mathematical theory of fluid mechanics there is a quantity known as **vorticity,** which is a measure of the amount of rotary circulation of the medium per unit area. The *vorticity equation* gives the rate at which vorticity increases or decreases. This equation has several positive terms that increase vorticity and a negative term that decreases it. The ratio of these terms gives the Reynolds number, which can be expressed as

$$R \equiv \frac{vL}{v}$$

where

$v = $ flow velocity in the medium

$L = $ dimension of possible turbulent eddies

$v = $ kinematic viscosity $= \dfrac{\text{absolute viscosity}}{\text{density}}$

In looking up viscosities, the reader should be aware of the difference between kinematic viscosity and absolute (sometimes called dynamic) viscosity, as given above. Some typical absolute viscosities are

$1.0 \times 10^{-3}\,Ns/m^2$ for water near room temperature
$1.8 \times 10^{-5}\,Ns/m^2$ for air near room temperature
$8.8 \times 10^{-6}\,Ns/m^2$ for H_2 gas near room temperature
$5.7 \times 10^{-6}\,Ns/m^2$ for H_2 gas at 160 K
$7.6 \times 10^{-6}\,Ns/m^2$ for CH_4 gas at 195 K

Sample Problem

With normal wind conditions, would you expect turbulence in Earth's atmosphere involving dimensions like those of a thunderstorm?

Solution

We must use SI units. Windy gusts of 36 km/h equal an air speed of 10 m/s for v. If we take 10 km for the storm dimension, $L = 10^4$ m. The kinematic viscosity will be the absolute viscosity of air (above) divided by air's density of about 1 kg/m³. Thus the Reynolds number comes out to be of the order 10^{10}, which is considerably greater than 1; and we correctly predict turbulence on this scale, as confirmed by satellite photos of storm clouds.

SUMMARY

The motions of the main bodies in the solar system, such as planets orbiting the sun or small satellites orbiting planets, are well described by Kepler's three laws. These in turn can be derived from Newton's more general laws of motion and law of gravitation. Motions of one small body around one large body (an example of the two-body problem) are easy to treat and well understood. The three-body or *n*-body problems are much harder and result in many interesting phenomena in the solar system, including small perturbations of orbits, resonance, and Lagrangian points. Many other effects cause major or minor alterations of Keplerian orbits.

Tidal effects are extremely important, playing many roles, from forcing moons to keep one face toward their planets, to causing volcanism on Io. Tidal forces cause the orbits of moons to evolve; they also create Roche's limit and help explain planetary rings.

Orbiting bodies interact with solar radiation and the solar wind in ways that tend to eliminate small particles from the solar system, either by blowing them out or by pulling them into the sun. The radiation-caused Yarkovsky effect probably mobilizes meter-scale bodies. As a function of size range, the effects are as shown in the diagram following.

Diameter (m) $10^{-9}\ 10^{-8}\ 10^{-7}\ 10^{-6}\ 10^{-5}\ 10^{-4}\ 10^{-3}\ 10^{-2}\ 10^{-1}\ 10^{0}\ 10^{1}\ 10^{2}\ 10^{3}$

Effect:	**Solar wind**	**radiation pressure**	**Poynting-Robertson effect**	**Yarkovsky effect**	**negligible force**
	blows out	blows out	spirals inward	drifts in or out	assumes Keplerian orbit

celestial mechanics	perturbations
astrology	periodic
scientific method	secular
Ptolemaic system	commensurable
Newton's universal law of gravitation	resonance
Kepler's laws	Kirkwood gap
foci	Langrangian points
primary body	L_4, L_5
periapse	Trojan asteroids
apoapse	velocity equation
orbital elements	net differential force
semimajor axis	tidal bulge
eccentricity	body tide
inclination	ocean tide
period	tidal forces
gravity	synchronous rotation
law of gravitation	Roche's limit
circular velocity	tensile strength
escape velocity	ring
parabolic orbit	photons
parabolic velocity	radiation pressure
hyperbola	solar wind
conic sections	plasma
astrometry	solar corona
spectroscopy	Poynting-Robertson effect
Doppler shift	aberration of light
radial velocity	Yarkovsky effect
two-body problem	turbulence
three-body problem	laminar flow
numerical integration	Reynolds number

PROBLEMS

1. Why did Kepler's, Galileo's, and Newton's celestial mechanics discoveries have such a profound philosophical impact on humanity?

2. Why are those discoveries called the Copernican revolution?

3. Consider three hypothetical solar systems:
   a. the sun with one planet of Earthlike size
   b. the sun with two planets, Jupiter and Earth
   c. the sun with two companions, a "planet" of 0.1 solar mass and Earth
In which system would the Earth-sized planet most nearly obey Kepler's original laws? Why?

4. In which of the above systems would the motion of the Earthlike planet be most disturbed and least well obey Kepler's laws? Why?

5. Compare the revolution periods of asteroids to the revolution period of Jupiter and explain why, in some cases, those orbital relations may result in asteroids being kicked out of the central asteroid belt into other parts of the solar system. What features of the asteroid belt's structure may arise from this process?

6. Suppose Jupiter had only one large satellite (as large as or larger than Ganymede). Where might you look in that system for evidence of possible smaller bodies or trapped meteoroidal debris? Why?

7. If Jupiter's semimajor axis is 5.2 and Jupiter has a period of 11.86 y, what is the typical semimajor axis and period for a Trojan asteroid?

8. Figure 3-11 shows how the tidal bulge raised on Earth by the moon accelerates the moon forward (like a small rocket pushing forward) and hence drives it slowly away from Earth. (a) Consider a system (such as Neptune's) in which the planet rotates prograde but the slowly orbiting large moon revolves retrograde. What is the result in terms of orbital evolution of the satellite? (b) What is the result if the planet rotates retrograde? (c) Explain recent studies suggesting that Neptune's satellite, Triton, will crash into Neptune some time in the future.

9. Why is tidal evolution ineffective if the satellite is much smaller than the planet, as in the case of Mars and Phobos?

10. Suppose a comet breaks apart while in an elliptical orbit, producing dark-colored silicate particles of various sizes traveling in the same orbit. The diameters are 0.01 μm, 1 μm, 1 mm, and 1 m. Predict what happens to each particle, and roughly estimate the timescale involved.

ADVANCED PROBLEMS

11. Show that an asteroid at $a = 3.28$ AU (just outside the orbit of the largest asteroid, Ceres) is in a 1:2 commensurability with Jupiter. Are many such asteroids known?

12. Show that the velocity of Earth in its orbit around the sun is about 30 km/s.

13. Show that the velocity equation reduces to the equation given for circular velocity (when $r = a$) and the equation given for escape velocity (under the appropriate condition in that case).

14. If a comet moving on a parabolic prograde orbit passes Earth, how fast would it move relative to Earth? What if it is moving on a retrograde parabolic orbit?

15. Escape velocity from a planet's surface is equivalent to the speed at which a falling object hits the ground if dropped with zero initial velocity from infinity (or very high altitude), neglecting atmospheric drag. What are the fastest and slowest velocities at which a large meteorite orbiting the sun could strike Earth's surface? Explain your reasoning.

16. (a) In the past, when the moon was 0.25 as far from Earth as it is now, estimate how much more massive Earth's average tidal bulge was than it is today. (b) How much stronger was the net accelerating effect of this bulge on the moon?

17. An early measurement of the outer radius of Uranus's rings was about 5.0×10^7 m from the center of the planet. Compare this to the Roche limit for icy bodies near Uranus, using the expression derived here for touching particles. What is the significance of this result? Would you expect Uranus's ring particles to aggregate into a satellite?

18. Show that the ratio between outward radiation force and inward gravity force is independent of the distance from the sun. If a bit of meteoritic dust, or a "solar sail" spacecraft, is moving away from the sun through the solar system by means of radiation pressure, would the acceleration diminish as it reached a large distance from the sun?

19. (a) Consider two particles in circular orbits in Saturn's rings 10^8 m from Saturn's center. One is located 1 m farther from Saturn than the other. By Kepler's laws, they have different periods and must occasionally pass each other. How fast do they pass by each other? (*Hint:* You could compute V_{circ} for each particle and subtract, but the difference in orbits is only one part in 10^8, so you would have to maintain accuracy to at least eight decimal places, which is unrealistic. Instead, use calculus and differentiate $V_{circ} = (GM)^{1/2} (r^{-1/2})$ with respect to getting

dV_{circ} = difference in velocity
$\quad = 1/2 \, (GM)^{1/2} r^{-3/2} dr$

where dr is the difference in distance.

(b) What if the particles are about 200 km apart, as in the example of Figure 3-17a? Compare your result with the 60 m/s figure given there (based on a more detailed calculation including gravitational attraction of the particles for each other).

(c) How fast would a shuttle in a 200-km-high circular orbit approach a space lab in a 201-km-high circular orbit? (Answer: about 0.6 m/s.)

20. During a Mars exploration expedition, a package is to be launched from Phobos to reach a party of astronauts on Deimos. Relative to Phobos, what launch velocity is needed for the minimum energy orbit, which has periapsis on Phobos and apoapsis on Deimos?

21. Two asteroids in the asteroid belt collide. They are on roughly circular orbits whose semi-major axes differ by 0.3 AU. At what speed do they collide?

22. A tidal stretching force of 1 lb is observed across a large space structure constructed in synchronous orbit around Earth. Later, it is decided to bring the structure down to a height of 200 km. What is the new stretching force?

23. During planet formation, a planet has acquired a satellite of rocky or icy material. The planet itself is still growing rapidly, due to accretion of meteoritic material. (a) How does the satellite's orbit change? (b) What will happen to the satellite as the satellite's orbit evolves? Specify distances for various events. (c) Describe the plausible final appearance of the system as compared to the appearance of, say, Saturn.

Central part of the Pleiades star cluster. This cluster illustrates that stars form in groupings. Surrounding the bright stars can be seen clouds of gaseous debris, about 20,000 AU across. The age of the Pleiades is estimated to be 80–150 My; the solar system would have been embedded in such a grouping during its first few hundred My. (Haute Provence Observatory)

CHAPTER FOUR

The Formation of Stars and Planetary Material

Having surveyed our planetary system and the forces that control the motions of particles in it (Chapters 2 and 3), we are ready to ask how this system formed. This question, as it turns out, boils down to how innumerable small particles around the early sun (the planetesimals) aggregated into a few big particles (the planets). To see this, we need to start with the general astrophysical question of how stars form. The interdisciplinary zone between astrophysics and planetary science is one of the most fertile areas for research on planetary origins (e.g., see conference volume edited by Black and Matthews, 1985).

Evidence That Stars Are Forming Today

Proofs that stars have continued to form throughout astronomical time come from both planetary and stellar research. The first involves the relative youthfulness of the solar system itself. The universe began in an outrushing of material, called the **big bang,** some 10 to 18 Gy ago. Galaxies or huge masses of typically 10^{11} stars aggregated from the outrushing material; our galaxy formed about 10 to 12 Gy ago. Yet from several kinds of dating methods based on the study of radioactive isotopes in lunar examples, terrestrial samples, and many different meteorites, we know that our solar system formed only 4.6 Gy ago. So our solar system has formed in only the last 38% of our galaxy's history; all stars did not form at the beginning.

The second proof of continuing star-forming activity comes from astrophysical analysis of stellar evolution. A star like the sun takes about 10 Gy to consume its supply of hydrogen fuel. But the more massive a star, the hotter it is and the faster it consumes its fuel. Consequently, a large star has a shorter lifetime as a normal, luminous star. A star of ten solar masses (a star of 10 M_{sun}), lasts only about 0.01 Gy. The most massive stars, of several tens of solar masses, may last only a few million years and then explode in dazzling **supernovas,** throwing much of their interior into neighboring interstellar space. Nonetheless, many of these massive stars exist, and we have historical records of supernovas. This evidence proves that stars have formed within the last few million years. Still more interesting is the fact that supernova explosions typically occur in **open star clusters**—groupings of several hundred stars that typically disband in a few hundred million years—and in **molecular clouds**—the densest, cool concentrations of interstellar dust and gas, proving that star formation tends to occur in these groupings. Open star clusters and molecular clouds are often associated with still larger diffuse clouds of interstellar dust and gas, called **nebulae,** such as shown in Figure 4-1.

A third line of evidence is that astronomers in the 1980s and 1990s have discovered numerous examples of stars surrounded by solar-system-sized clouds of dust and gas, exactly resembling the appearance deduced for our sun and planetary system as it was forming. In other words, we *see* other stars forming around us.

These findings have led to several conclusions about how stars form. First, they form in clusters rather than singly. Second, the clusters form from nebular concentrations of **interstellar material**—huge clouds of atoms, molecules, and dust grains. These findings suggest that interstellar material is the ultimate parent material of stars and planetary systems and that to understand star formation, we should look at the interstellar medium more closely.

The Interstellar Material

All evidence indicates that the overall composition of the interstellar material approximately matches that of the sun, supporting the idea that the sun formed from a concentration of such material. This composition is shown in Table 4-1. As shown in the table, spectroscopic studies show that some of the heavier elements, such as Ca, Al, Fe, Mg, and C are strongly depleted in the gas and concentrated in the dust. Provocatively, these are the

Figure 4-1. Dramatic nebular clouds of dust and gas lie in the star-forming region of the constellation Monoceros. The dark "cone" is a dust-grain and gas concentration estimated to be about 50,000 AU wide, silhouetted against more distant glowing clouds. This region lies in the outer part of the young star cluster NGC 2264, which is only a few million years old. Complex molecules such as formaldehyde (H_2CO) have been found in the region. (Lick Observatory)

TABLE 4-1 Composition of Interstellar Medium in Our Part of the Galaxy

Element	Atomic no.	Composition (% by mass)		Concentration factor (total ÷ gas)[c]
		Interstellar medium[a] (total)	Interstellar atomic gas[b]	
H	1	78.3	~78	1
He	2	19.8	~20	1
O	8	0.8	0.2	4
C	6	0.3	0.02	15
N	7	0.2	0.03	7
Ne	10	0.2	0.2	1
Ni	28	0.2	0.1?	2?
Si	14	0.06	0.03	2
S	16	0.04	0.03	1.3
Fe	26	0.04	0.001	40
Mg	12	0.015	0.0007	21
Ca	20	0.009	0.000002	4500
Al	13	0.006	0.00003	200
Ar	18	0.006	0.001	6
Na	11	0.003	0.0006	5
		99.98	~99.6	

[a] Includes atoms, molecules, and dust grains. Based on observed solar composition (which represents all interstellar material from which sun formed). Adapted from Gibson (1973).

[b] Does not include molecules and dust grains. Estimates based on abundances spectroscopically observed in interstellar gas between the solar system and the star ζ Ophiuchi.

[c] Probable indicator of concentrations in minerals of interstellar dust grains (quotient of preceding two columns).

planet-forming elements! These elements are especially prone to aggregate into dust grains and their absence in the gas probably means they are locked up in grains. The grains are the first steps on the road to rocky and icy planetary material.

Molecules, Molecular Clouds, and Dust

For many years, the formation of interstellar dust grains was a mystery. Early calculation showed that in most of interstellar space, atoms are too far apart to collide often enough to form molecules (Oort and van de Hulst, 1946). Where, then, could molecules form? Astronomical research in the 1970s showed that molecular clouds are the dense, cool environments where collisions occur frequently enough to allow many molecules and dust grains to aggregate. For example, when hydrogen atoms (H) collide, a certain fraction of them stick together and form molecular hydrogen, H_2.

Even larger particles form, as revealed by infrared spectroscopic observations (see Figure 4-2). Most molecules and mineral grains tend to absorb certain colors of light, not so much in the visible parts of the spectrum, but at **infrared wavelengths,** the portion of the spectrum with wavelengths too long for us to see. Detectors for measuring infrared light were developed about the time of World War II and were not widely applied in astronomy

until after 1960. An explosion in the discovery of interstellar molecules occurred in the late 1960s. One molecule (hydroxyl, OH) was discovered between 1963 and 1967; 24 molecules were discovered in the next 5 y. These include water (H_2O), ammonia (NH_3), formaldehyde (HCNO), and carbon monoxide (CO). A 1978 detection of methane (CH_4) in the Orion nebula was its first detection outside the solar system. Complex molecules such as methylamine (CH_3NH_2) and ethyl alcohol (C_2H_5OH) have also been found in interstellar clouds.

Note in Table 4-1 that the atoms involved in these molecules, such as C, O, and N, are among those most abundant in cosmic material, explaining why they collide often enough to make numerous molecules. Note also from the last column in the table that these atoms are among those depleted from the atomic gas, confirming that most of them have been tied up in molecules.

Heavier atoms have been even more depleted. They are tied up in the larger particles, the grains of **interstellar dust.** These are detected by the fact that the dust reddens the light of distant stars, much as dust in our own air reddens the light of the setting sun. The amount of reddening can be used to calculate the size of the interstellar dust grains, which turn out to be about 0.1 to 1 μm in diameter. Both spectroscopic and theoretical work indicates that these grains are mineral grains similar to ordinary rock-forming compounds. As shown in Figure 4-2, spectral **absorption bands near 10 and 20 μm,** found in starlight

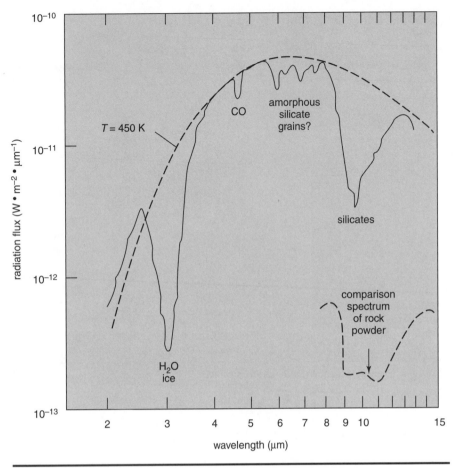

Figure 4-2. The infrared spectrum of NGC 7538, a compact infrared source associated with a large molecular cloud and nebula of ionized hydrogen. NGC 7538 is probably a star-forming region. The infrared radiation arises from dust whose temperature is about 450 K; the dotted line shows theoretical emission from a solid blackbody at this temperature. Absorption bands are labeled, including 3.1-μm absorption from microscopic silicate mineral grains. The dust may be heated by an embedded protostar. (Adapted from Willner, Puetner, and Russell, 1978)

transmitted through dust nebulae, match spectra of light transmitted through ordinary rock powders. In addition to silicate and carbon minerals, the dust probably also contains iron, a conclusion based on evidence that elongated grains, like little compass needles, are aligned with spin axes along the galactic magnetic field.

As we discuss in more detail in the next chapter, chemists have calculated the types of grains that ought to condense in a cooling cloud of interstellar gas. The process is like that by which snowflakes and raindrops condense in clouds in our own atmosphere as air cools. If a relatively dense nebula cools sufficiently, silicon, oxygen, magnesium, calcium, aluminum, iron, and nickel in this nebula form rocky and metallic mineral grains. This supports the interpretation of the spectra.

Besides spectroscopic and theoretical work, we have direct observations of samples! Cetain micron-scale grains found in meteorites and collected by high-flying aircraft have revealed oxygen-isotope ratios far different from solar system values, proving they come from other stellar environments. Those include mineral grains bound together by carbonaceous material. Measured compositions include minerals such as diamond (C), silicon carbide, graphite (another form of C), corundum (Al_2O_3), hibonite [(Ca, Fe)(Al, Ti, Mg)$_{12}$ O$_{19}$], spinel (MgA1$_2$O$_4$), and forsterite (Mg$_2$SiO$_4$) (Messenger and others, 2003).

Gravitational Collapse: A Theory of Star Formation

If a cool interstellar cloud is dense enough, the gravitational attraction of its particles for each other will be strong enough to make it start contracting or to keep it contracting if it gets an initial compression by external forces. The complete contraction process is called by astrophysicists **gravitational collapse** because once the process has begun, the cloud usually collapses to stellar sizes (less than a millionth of its initial size) in a relatively short time. During its collapse, but prior to onset of nuclear reactions, the object is called a **protostar.**

Because all particles have a gravitational attraction for other particles, why shouldn't all nebular clouds collapse and become protostars? The answer is that other forces compete with gravity by pushing outward on the cloud. Probably the most important as well as the most easily understood outward force is gas pressure caused by heat in the cloud. The hotter the cloud, the faster the molecules move, and the more the cloud tends to expand. In terms of atomic theory, each atom has a certain mean velocity that increases with the temperature of the cloud (remember that the temperature of a substance is defined as a measure of the kinetic energy of its atoms).

In a given cloud at a given temperature, the atoms are darting about, colliding with each other. This activity exerts an outward thermal pressure. If gravity is inadequate to hold the cloud together, the cloud rapidly expands away into space. The cloud can collapse only if the density of the gas becomes great enough that a typical atom's attraction to the cloud as a whole overcomes the outward pressure.

In addition to heat, other forms of energy in the cloud tend to affect contraction. Two of these are magnetic fields and turbulence. Detailed theories of star formation must consider those complications, which continue to attract current research. In this introductory discussion we consider only the dominant influences of gravity versus thermal pressure.

The Virial Theorem

The competition between gravity and thermal pressure is expressed by a crucial astrophysical theorem called the **virial theorem** or *Jeans theorem* (after one of its developers, English astrophysicist Sir James Jeans). This theorem, which can be derived by the methods of statistical mechanics, shows that gravitational collapse of a protostar will begin only if the total gravitational potential energy of the cloud exceeds twice the total thermal energy (if there are other energy sources, such as rotation, turbulence, or magnetic fields, these must be added to the thermal energy).

The fact that the entire interstellar medium is in motion—with turbulence, heated regions, and expanding supernova clouds adding to the chaos—helps assure that some portions of the interstellar gas will be compressed and become dense enough (that is, have enough mass and gravitational potential energy) for collapse to start.

The virial theorem also implies that as the collapse proceeds, energy is released in a specific way: About half the gravitational energy released by the infalling material heats the gas of the protostar, and the other half radiates away into space.

To see why gravitational energy gets converted into thermal energy that heats the cloud, try to imagine a collapsing cloud in which all the atoms are moving radially inward. In this case, there would be no collisions among them, and so no gravitational energy would be converted into heat. The gravitational energy released (that is, the change in gravitational potential energy) would be completely transformed into the kinetic energy of the atoms as they fall toward the center. In real life the atoms collide with each other, slowing the infall, but raising the temperature of the gas. The atomic collisions guarantee that some gravitational potential energy is converted into heat, and subsequent radiation by the gas ensures that some energy is radiated away into space.

The virial theorem has many important applications. It can be used to estimate the temperature and the amount of radiation coming from a collapsing protostar, and these estimates can be used to identify protostars in space and to calculate conditions during planet formation. We can also use the virial theorem to compute the size and density of interstellar clouds that could collapse to form a star such as the sun. Figure 4-3 shows an application of the virial theorem (in the absence of magnetic fields and turbulence) to predict the masses of clouds that would collapse under different temperature and density conditions. For example, the diagram shows that for cool ($T = 10$ K) interstellar clouds to collapse into objects ranging in mass from the largest to smallest stars (see right-hand scale), the initial cloud densities would have to be about 10^{-9} to 10^{-10} kg/m³.

Are nebulae of such mass and density common in interstellar space, so as to explain the continuing formation of stars? Paradoxically, the answer is no. Most nebulae are less dense.

How, then, do stars form? To resolve the paradox, remember that stars are observed to form in clusters. Looking along the bottom scale in Figure 4-3 to the densities typical of dense interstellar clouds (about 10^{-18} kg/m³), we find that such clouds, if reasonably cool, would contract into masses about the size of star clusters rather than individual stars. This explains the origin of the star clusters. A typical contracting cloud, as it collapses, subdivides to make several hundred or a thousand stars. Theoretical studies have revealed that during the collapse of these **protoclusters,** local condensations form within the cloud and begin to collapse as independent entities. These subcondensations then collapse to form stars, and the whole thing turns into an open star cluster similar to the Pleiades, the Hyades, or the region around the Orion nebula.

Comment on the Formation of Just About Everything

The theory of star formation is not restricted to masses the size of stars or clusters. That is why Figure 4-3 has been plotted to include such a large range of mass and density. It is provocative to consider the evolution of the whole universe from this point of view.

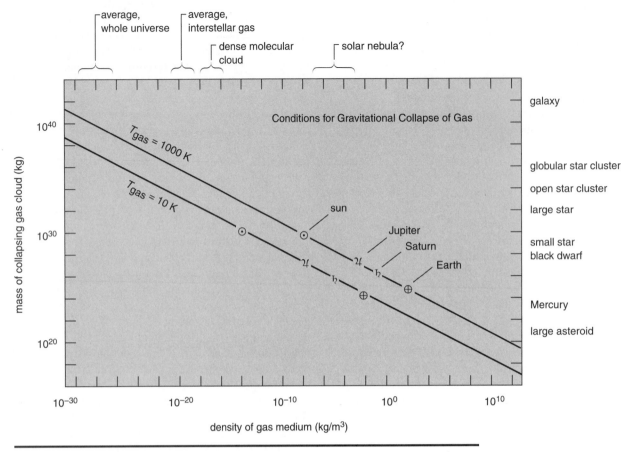

Figure 4-3. A guide to formation of astronomical objects by means of gravitational collapse. A diagonal line for each specified temperature shows the minimum density of gas (bottom scale) at that temperature required to initiate collapse in a gas cloud of specified mass (left scale). Clouds to the right of their appropriate temperature line would tend to collapse; clouds to the left would not.

For example, if all the material in the universe were spread out uniformly, a very low density of roughly 10^{-27} or 10^{-28} kg/m^3 would be obtained. Figure 4-3 shows that in a more or less uniform, relatively hot gas of this density, the condensations would have the mass of galaxies, about 10^{11} M$_{sun}$. In other words, even if no galaxies existed and only hot gas filled space, it is reasonable that galaxies should have formed. This deduction helps explain our earlier statement that galaxies formed out of the gas expanding from the big bang.

The same process applies to smaller, denser masses. Much evidence indicates that the newly formed sun, similar to other newly formed stars, was surrounded by a nebula with a density perhaps as high as 10^{-6} to 10^{-3} kg/m^3. Figure 4-3 shows that if this mixture of gas and dust got cool enough, masses the size of Jupiter or even smaller might have formed directly by gravitational collapse. One question, therefore, is whether Jupiter-scale giant planets and their satellite systems might have formed in this way, either in our solar system or in other planetary systems. Thus our star-forming theory has led us directly to a plausible planet-forming theory, to which we return in the next chapter.

Newly Formed Stars: Theory and Direct Observations

We have now seen evidence that (1) star systems are forming today, (2) solid "planetary" material can condense as dust grains in cool, dense environments, and (3) star formation is theoretically understood (approximately!) in terms of contracting dust and gas clouds. Consequently, we can now examine the actual properties of interstellar clouds and newly forming stars to understand the sequence of structures and processes that might lead to planet-forming activity beyond our own system.

Bok Globules

Dark, rounded clouds typically 1,000 to 100,000 AU across were first emphasized as possible collapsing stars by the Dutch American astronomer Bart Bok as far back as the 1950s. They are often seen silhouetted as dark spots in

A complete mathematical derivation of the virial theorem from first principles is beyond the scope of this text, but the basic relations can easily be set up from the verbal statement given earlier. According to that statement, a cloud will contract if the gravitational potential energy exceeds twice the thermal energy. The gravitational energy of a homogeneous sphere (which we can assume to represent the protostellar cloud) is

$$\text{potential energy} = \frac{3}{5}\frac{GM^2}{R}$$

where

G = gravitational constant
M = mass of cloud
R = radius of cloud

The thermal energy is the thermal kinetic energy of *one* particle, $\frac{3}{2}kT$, times the total number of particles, $M/\mu M_H$, where

k = Boltzmann constant
T = temperature of gas
μ = mean molecular weight $\simeq 1$
M_H = mass of hydrogen atom

The inequality stated by the virial theorem is thus

$$\frac{3}{5}\frac{GM^2}{R} > 2 \times \frac{3}{2}\frac{kTM}{\mu M_H} \quad \text{for collapse}$$

Canceling common factors and substituting for the mass of a sphere $M = \frac{4}{3}\pi R^3 \rho$, we have

$$\rho > \frac{15kT}{4\pi GR^2\mu M_H} \quad \text{for collapse to start}$$

where ρ = density of gas.

Once the collapse of an unstable object gets well under way (so that hydrostatic equilibrium is almost in effect), the inequalities expressed above become approximate equalities, so that

$$4\pi GR^2 \mu M_H \rho \approx 15kT$$
or
$$GM\mu M_H \approx 5kTR$$

This result allows approximate evaluation of the temperature inside a collapsing protostar or protoplanet of specified mass once it has reached a specified radius.

Sample Problem

Find the density required for collapse to begin in an interstellar cloud consisting mostly of atomic hydrogen at 50 K if the cloud has dimensions somewhat bigger than a typical open star cluster, say about 20 pc in diameter. What mass would be involved in this cloud?

Solution

We will substitute into the inequality for density ρ. Since the cloud is essentially hydrogen (like all interstellar clouds), the mean molecular weight $\mu = 1$. One pc is 3×10^{16} m. Other constants appear in Table 2-2. The density is found to be roughly 8×10^{-20} kg/m³, giving a total mass of roughly 5000 $M_\odot$, enough to make quite a respectable star cluster.

---

regions of bright nebulae. These **Bok globules,** as they are called, may contain several solar masses of dust and gas and are often millions of times denser than normal interstellar gas. Many may be objects beginning to collapse into cocoon nebulae, but recent work suggests that others are fragments of clouds blown apart by gas outrushing from star-forming regions. At the very least, they represent the kinds of discrete clouds that could collapse to form clusters of stars.

Cocoon Nebulae

Solar system studies, theoretical astrophysics, and direct observation have all indicated that a newly formed star is surrounded by a relatively dense cloud of gas and dust, some tens or hundreds of astronomical units across. When searches for newly formed stars were pursued in the 1960s, the Mexican astronomer A. Poveda (1965) correctly argued that new stars are likely to be obscured by this envelope of dust and gas. Davidson and Harwit

(1967) agreed and coined the term **cocoon nebula** for the dusty cloud predicted to surround newly born stars. The cocoon nebula was predicted to be shed later (probably by strong radiation pressure or solar wind) so that the star emerges into view like a butterfly emerging from a cocoon. We now review three lines of evidence for the existence of cocoon nebulae, the environments that may produce planets.

Solar System Evidence for Cocoon Nebulae: The Nebular Hypothesis. Even before stellar astronomers began to search for cocoon nebulae, planetary scientists realized that a cocoon nebula must have surrounded the early sun in order for the planets to have formed and to have obtained their present motions. After Kepler determined planetary motions around 1610, but before Newton published his analyses of the motions in 1687, the French philosopher René Descartes reasoned in 1644 that the initial gas in the universe broke up into rotating "vortices," much as the surface of a stream may develop rotating eddies. Each such rotating vortex produced a rotating star surrounded by gas, and Descartes thought that minor sub-vortices in the gas might have produced a family of planets near each star.

The German thinker Immanuel Kant correctly reasoned in 1755 (without much mathematical support) that gravity would make a circumsolar cloud contract and that rotation would flatten it. Thus the cloud would assume the general shape of a rotating disk, explaining the fact that the resultant planets form a disk-shaped system of bodies.

This view, that a disk-shaped nebula formed around the early sun and then spawned planets, came to be called the **nebular hypothesis.** It was further developed by the French mathematician Laplace, who proposed in 1796 that the rotating disk continued to cool and contract. Because of conservation of angular momentum, the sun would have spun faster as it contracted, just as a figure skater spins faster as she pulls in her arms. Laplace thought (incorrectly) that the outer rim of the spinning disk would shed concentric rings of material at various stages of the contraction, with each ring eventually condensing into a planet. In this way, Laplace explained the spacing of the planets in circular orbits.

These early ideas about the nebular hypothesis were important in two regards. First, they moved scientific thinking toward the modern view that a cocoon nebula developed as part of the normal star-forming process, but this movement was of limited value because it was not supported by either detailed calculations or direct observations. Laplace himself offered his discussion "with that diffidence that always ought to attach to whatever is not the result of observation or calculation."

Second, and more important, these early ideas helped clear away the tangle of confusion that connected Renaissance physical theories with theology. The work of Descartes, Newton, Kant, and others implied, contrary to prevailing thought, that even if God had created the original universe full of gas by some special, incomprehensible process, He seemed to be letting it evolve by cause-and-effect relationships. Thus, independent of theological speculations, physical laws derived from observations could be used to infer past events or to predict future ones.

The nebular hypothesis has been retained and expanded. Its modern version—that planets formed from dust embedded in a disk-shaped gaseous nebula—explains many properties of planets and meteorites as we will see in subsequent pages.

Theoretical Evidence for Cocoon Nebulae. A second line of evidence for cocoon nebulae is theoretical. Theorists follow the evolution of a contracting protostellar cloud in hopes of understanding not only the birth of the star but also the evolution of the cocoon itself—the placental medium in which planets grow. Essential to such investigations is correct estimation of various initial properties such as the total mass of the cloud, its temperature, its rotation rate, the degree of internal turbulence, and its magnetic fields.

The calculations made so far strongly indicate that a star forms as a central condensation in the disk-shaped collapsing cloud, as the inner part of the cloud collapses faster than the outer part. The outer part remains behind as the cocoon nebula. The same studies indicate that under many conditions of rotation, turbulence, and so forth, the inner star-forming nucleus may divide into two or three objects orbiting around each other, as shown in Figure 4-4 (left side). This phenomenon may explain why more than half of all star systems are binary (two co-orbiting stars) or multiple (three or more co-orbiting stars) rather than single stars like the sun. Slower rotation (or other conditions?) may produce single stars with planet systems (Figure 4-4, right side).

The process of fragmentation may also produce bodies too small to be true stars. The minimum mass of a star (as defined by the presence of nuclear reactions in its interior) is about 0.075-0.08 M_{sun}. Somewhat smaller objects could be large planets such as Jupiter (0.001 M_{sun}) or the larger planets that have been found in other systems. The whole process was reviewed by Cameron (1975); detailed calculations appeared in Bodenheimer and Black (1978), and Field (1978). The calculations indicate that the cocoon nebula would obscure the new star from outside view. Only the cocoon, a dusty, warm cloud emitting primarily infrared radiation, would be visible. Therefore, to detect these systems, infrared observations became the favored technique.

Telescopic Evidence for Cocoon Nebulae: From Infrared Stars to Proplyds. The third line of evidence for cocoon nebulae is direct observation. Prior to 1966, there was no firm observational proof that cocoon nebulae existed. However, the Mexican astronomer E. E. Mendoza (1966) observed an object cataloged as star R in the constellation Monoceros (called **R Mon** for short) and discovered that it was emitting large amounts of infrared radiation (peaking at wavelength around 2.4 μm). American astronomers Frank Low and Bruce Smith (1966) promptly interpreted R Mon as a 200-AU-diameter cocoon nebula surrounding newly formed stars. In this view, dust grains inside the nebula have been heated by the enclosed star to a temperature of 850 K, a temperature calculated from the spectrum of R Mon's infrared radiation. R Mon lies near the star-forming nebulosity shown in Figure 4-1.

Many similar systems have been discovered. The central star heats the dust that obscures it, and the outside observer often sees only the infrared thermal radiation from the warm dust, as shown in the sample spectrum in Figure 4-5. In some cases, the dust seems to be thinning and the visible light of the star peeks through, as seen in the figure. When the wavelength of the peak radiation is measured, the mean effective temperature of the dust can be calculated easily, by use of a physical relation known as Wien's law (see optional mathematical notes). By means

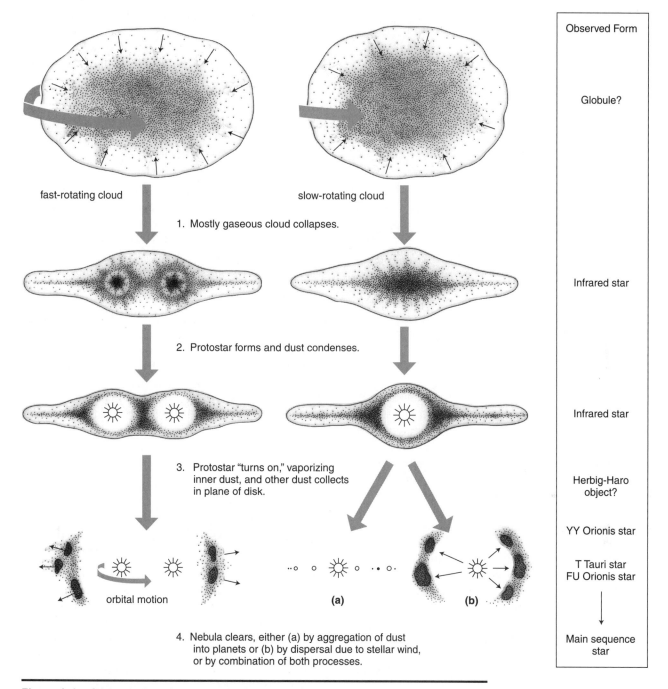

fast-rotating cloud

slow-rotating cloud

1. Mostly gaseous cloud collapses.

2. Protostar forms and dust condenses.

3. Protostar "turns on," vaporizing inner dust, and other dust collects in plane of disk.

orbital motion

(a)

(b)

4. Nebula clears, either (a) by aggregation of dust into planets or (b) by dispersal due to stellar wind, or by combination of both processes.

Observed Form

Globule?

Infrared star

Infrared star

Herbig-Haro object?

YY Orionis star

T Tauri star
FU Orionis star

Main sequence star

Figure 4-4. Schematic view of the possible sequence of evolutionary processes for a binary star (left) and a single (nonbinary) star (right). Depending on the processes that dominate stage 4, the potentially planet-spawning dust may aggregate into planets (case a) or be blown away (case b)

of detailed spectroscopic observations such as those shown in Figure 4-2, the composition of the dust has been identified as silicate and/or ice particles, thus confirming that cocoon nebulae actually do contain potential planet-forming material similar to that in our solar system.

The small infrared source, NGC 7538E, located in a probable star-forming cloud rich in large molecules and minerals, is a good example of an object in which potential planetary material has been identified (Willner, Puettner, and Russell, 1978). As shown in Figure 4-2, its infrared spectrum reveals the presence of ice crystals, carbon monoxide, and silicate mineral grains similar to rock-forming material. Such grains are microscopic, probably 0.1 to 1 μM across.

In a 1975 survey of 20 objects considered likely proto-stars or very young stars (including such objects as R Mon

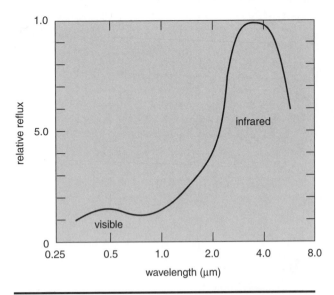

Figure 4-5. Spectrum of an infrared star. The dominant, infrared part of the spectrum is radiation emitted by dust particles of the cocoon nebula. In some cases, but not all, a little visible light from the central star leaks through the nebula.

and probable young stars T Tauri and FU Orionis), 19 had excess infrared radiation, presumably caused by circumstellar dust, and 9 actually had silicate-mineral features in their spectra. Such work was followed in 1983 by the launch of the Infrared Astronomical Satellite (IRAS), which surveyed the sky and discovered dozens of stars with excess infrared radiation indicating dust disks. Such stars are called **infrared stars.**

A dramatic advance in studying infrared stars and compact dusty nebulae came when the Hubble Space Telescope began taking extremely high-resolution images of star-forming regions, such as the Orion cloud (Greene, 2001). Here, the telescope revealed small, blob-shaped clots of gas that appear to be the protostellar disks or collapsing clouds—cocoon nebulae just forming. In 1994, these were given the name **proplyds,** standing for PROto-PLanetarY DiskS. In the Orion star-forming region, many of the proplyds have ionized surfaces because they are irradiated by one or more very hot young near-by stars, formed in a cluster at the heart of the Orion nebula (Figure 4-6). Typical diameters are several hundred AU, which again roughly matches the original size of our own system of planets, dust, and comets. In many of the proplyds, a central star can be glimpsed; one prominent example was interpreted from spectral data as having a mass of 0.2 M_{sun} and an age of 0.3 to 1 My. In one study region containing 110 stars deep in the Orion nebula, the Hubble Space Telescope netted 56 stars with proplyd structures. In other words, the Orion nebula displays many stars in their first million years or so; at least half of these display thick disks of dust in which planets might be forming.

Based on all these lines of evidence, we can construct a probable view of what a cocoon nebula looks like at close range (Figure 4-7 on page 86). The solar system itself must have looked this way in its first few million years.

Bipolar Mass Ejection and Herbig-Haro Objects

Various lines of evidence have shown that cocoon nebulae frequently are ejecting high-speed gas from the "top" and "bottom" of the central region—that is, along the polar axes of the central star. The mechanism is unknown but probably involves magnetic fields accelerating ionized material. The same thing has been observed in disks of active spiral galaxies, where the gas flow is often in tight beams. Figure 4-7 represents the gas flow in faint beams above and below the disk (Greene, 2001).

The high-speed ejected gas often collides with gas clouds in the vicinity of the newly forming star. Shock waves ionize some of the nebular gas, creating somewhat conical nebulae called Herbig-Haro objects, which were recognized in the 1960s and named after their discoverers. **Herbig-Haro objects** are often highly changeable on a timescale of a few years, testifying to the instability of the outflow process.

The H-R Diagram: A Tool for Discussing Protostar Evolution

Can we arrange the various types of infrared stars and dust-shrouded luminous stars in a sequence that represents stages in the formation of stars or planetary systems? More specifically, can we arrange them in order of age?

To do this, astrophysicists have long used a diagram called the **H-R diagram.** This diagram was first plotted around 1914 by Danish astronomer Ejnar Hertzsprung and American astronomer Henry N. Russell, after whom it is named.* The diagram plotted the luminosity of the star (total energy output expressed in Joules per second) against the temperature of the star. By historical tradition, luminosity scale increases upward, and temperature scale increases to the left.

The H-R diagram was a way to begin to make sense out of the observational data accumulating on stars early in this century. One of the first discoveries made with it is that most stars, including the sun, fall in a band called the

---

*Although some prefer the term *spectrum-luminosity diagram,* which describes the two observable quantities, H-R or Hertzsprung-Russell is the more common name. Originally, a parameter called "spectral class" was used instead of temperature, before the latter could be readily measured.

Wien's law tells which wavelength λ corresponds to the maximum amount of radiation at each temperature T. Using SI units, the wavelength is given in meters and the temperature in Kelvin. The law is

$$\lambda = \frac{0.00290}{T}$$

The number 0.00290 is a constant of proportionality and remains the same in all applications of the law. Thus, as T increases, λ decreases, giving shorter wavelengths and hence bluer light.

An example is instructive. The sun has a temperature of about 5700 K. Therefore, we can calculate the wavelength of the strongest solar radiation. Using SI units and powers of 10,

$$W = \frac{2.9 \times 10^{-3}}{5.7 \times 10^{3}} = 0.509 \times 10^{-6}$$
$$= 5.09 \times 10^{-7} \text{ m}$$

This, of course, is exactly in the middle of the range to which the eye is sensitive. (Otherwise we could not see sunlight!) Light of this wavelength corresponds to a yellow color.

Planetary matter, of course, is not so hot. The wavelength for maximum radiation from dust in a cool (500 K) nebula, for example, would be 5.8×10^{-6} m, or 5.8 μm. This figure is well into the infrared part of the spectrum. The infrared region of roughly 5 to 40 μm is often called the *thermal infrared,* as it is the region of emission caused by the body heat of nonluminous planetary sources.

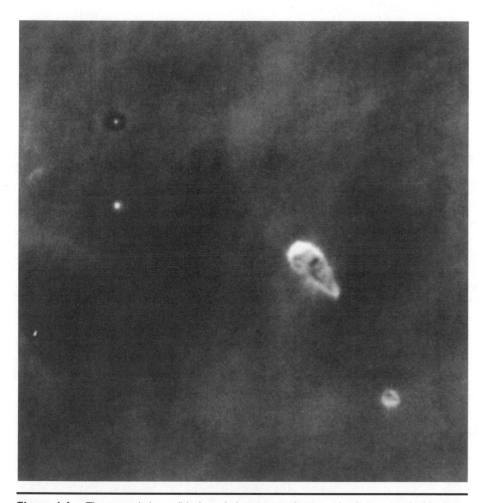

Figure 4-6. Three proplyds, or disk-shaped, dense protoplanetary clouds, are seen in this photo of the central region of the Orion nebula. The proplyd at the upper left is dark dust with a central star faintly visible. (Hubble Space Telescope photo; Space Telescope Institute)

Figure 4-7. Imaginary view of a late-stage planet-forming cocoon nebula at close range. The thick dust in the plane of the disk obscures the starlight, but the central star is faintly visible from this angle slightly above the disk. At an age of perhaps a few My, the system is still surrounded by star-forming nebulosity, visible in the background. (Painting by the author)

main sequence, as shown in Figure 4-8.* Main sequence are "normal," middle-aged stars, defined by the fact that their central nuclear reactions convert hydrogen to helium. As we say more informally, they "burn" hydrogen. (This common terminology is not strictly correct, because nuclear reactions are not the same as the chemical reaction of "burning.") Hot, luminous stars, in the upper left end of the main sequence, are the most massive; cool, faint stars, in the lower right, are least massive. The sun is in the middle of the main sequence.

Stars that do not generate energy by consuming hydrogen are not located on the main sequence; they appear in other parts of the diagram. These are of two types. Stars in the final stages of evolution run out of hydrogen and begin nuclear reactions that consume other elements, thus becoming **post-main-sequence stars.** They fall in various parts of the diagram and do not concern us here. The others are the **pre-main-sequence stars,** which are the topic of this chapter. These are protostars in which the normal H reactions have not quite started or are just beginning. Thus these stars are just settling into the main sequence. Some of them are undergoing brief series of nuclear reactions that consume other light elements, such as lithium.

---

*The conventional H-R diagram in most astronomy texts is just the upper left third of Figure 4-8, where most observable stars lie, but Figure 4-8 has been extended to include very cool stars, brown dwarfs, and planetary bodies.

Protostars and the H-R Diagram

To predict what a protostar looks like—how bright and hot it is—is to predict where it falls on the H-R diagram. A newly forming protostar begins life as a cool, dark cloud, and so it first appears in the lower right corner of the H-R diagram. Detailed astrophysical calculations by the 1970s followed the evolution of protostars across the H-R diagram (Larson, 1972, 1978; Bodenheimer, 1976; Bodenheimer and Black, 1978; Greene, 2001). In general, these results show that once the protostellar cloud becomes gravitationally unstable, it collapses very rapidly to a diameter of some tens of solar radii in only about a hundred years. The total luminosity of the central star goes up, bringing the protostar to a position above the main sequence. The protostar then contracts more slowly to produce a stellar core in the center of the cloud; this protostar is only a few solar radii across and forms within about a thousand years. On the H-R diagram, the star now moves downward toward the main sequence. At these rates of evolution, changes should be visible over the few decades that we have been observing suspected protostars; and indeed, such variations have been reported. This is the case with R Mon, though these changes may involve obscuration by clouds moving around the star rather than evolutionary changes in the star itself.

The stages followed by a star as it evolves define an **evolutionary track** across an H-R diagram. The early, high-luminosity part of the evolutionary track is called the **Hayashi track,** after Japanese astronomer Hayashi (1961),

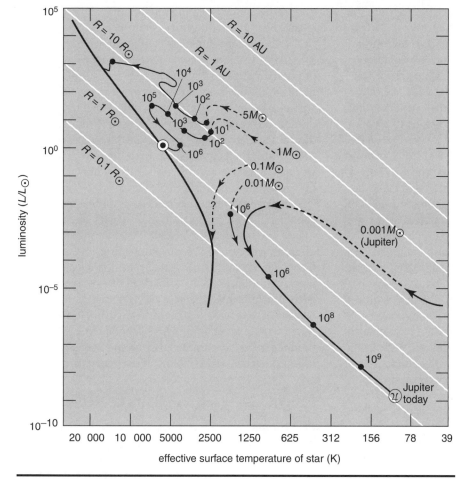

Figure 4-8. Calculated evolutionary tracks in the H-R diagram for stars of different masses. Note collapsing objects smaller than about 0.08 M$_{sun}$ do not initiate hydrogen-consuming nuclear reactions; hence, they do not reach the main sequence to become true stars. The planet Jupiter is such an object, yet it can be studied by the techniques of stellar astrophysics, as presented here. (Based on data from Larson, 1969, and Bodenheimer, 1976)

who recognized its properties. The duration and maximum luminosity of the Hayashi stage is important; it implies strong heating just before the planets form. Following the decline in luminosity along the Hayashi track, the evolutionary track tends to move erratically left toward the main sequence before dropping onto it. This erratic path is called the **Henyey track,** after Henyey, LeLevier, and Levee (1955), who first indicated its approximate position by means of simplified calculations.

Evolutionary Track of a Protostar as Observed from Outside the Cocoon Nebula

Most astrophysical calculations of protostar evolutionary tracks, as shown in Figure 4-8, focus on the history of the central protostar that forms at the core of the larger contracting cloud. However, although the protostar may radiate large amounts of visible light, this light does not get out of the surrounding cocoon cloud because it is absorbed by

nebular dust. It merely heats the nebula to an effective radiating temperature of some hundreds of degrees, so that the object as seen from the outside is an infrared source for 10^5 to 10^6 years (Larson, 1972). The *apparent* evolutionary track for a cocoon nebula thus lies well to the right of the main sequence.

T Tauri Stars: A Later Stage in Star Formation

Two processes in the nebula help to make the central star visible. First, the dust grains begin to aggregate; instead of countless tiny dust grains, we begin to have a smaller number of basketball-sized planetesimals, causing a clearing of the nebular fog. These planetesimals tend to settle into a thin layer along the plane of the nebular disk. Second, as the star heats up, it produces a "stellar wind" of gas blowing outward from the star, which helps disperse the rest of the gas and fine dust of the nebula.

The best-known stars representing this stage are the **T Tauri stars.** T Tauri stars, named after the star T in the constellation Taurus, were first recognized as an unusual group in 1943–1944 in work that went almost unnoticed because the author, K. Himpel, was publishing inside Nazi Germany. They were later emphasized by American astronomer A. Joy (1945). T Tauri stars represent the stage when the disk-shaped cocoon nebula begins to clear or blow away.

The star itself, at the center of the clearing nebula, draws closer to the main sequence during the Henyey track portion of its evolutionary path across the H-R diagram. Therefore, if we plot the H-R diagram of a cluster of newly forming stars, we find that most of the stars are T Tauri stars defining a band above and to the right of the main sequence. For example, the H-R diagram for stars in and near the Monoceros young clusters NGC 2264, shown in Figure 4-1, reveals a band of T Tauri stars whose position on the diagram indicates an average age of only a few million years.

T Tauri stars are highly and irregularly variable in brightness, accompanied by variable nebulae and strong magnetic fields and found in regions with abundant dust clouds and young, massive stars. They rotate fast (as predicted by Laplace's nebular hypothesis about contracting rotating stars) and are enveloped by remnants of cocoon nebulae with turbulent motions.

Many T Tauri stars have infrared excesses, as discovered by the Mexican astronomer E. E. Mendoza (1966), indicating that they probably have warm dust in their expanding nebulae. Many T Tauri stars have the telltale 10-μm and 20-μm silicate spectral features seen as emission bands, indicating silicate dust with temperatures on the order of 200 K. Doppler shifts in the spectra of most T Tauri stars often indicate expansion velocities of 70 km/s to 200 km/s for the surrounding gas, which may indicate the final outward dispersion of cocoon nebulae by strong stellar winds blowing out from the star. Cocoon nebulae up to a few tenths of a solar mass are apparently blown off before the star itself reaches the main sequence.

Not all T Tauri stars have a smooth outflow of gas. Redward displacements of absorption lines for many T Tauri stars indicate that perhaps a quarter or more of them have an infall of some matter, and some show both infall and outflow. Some that have been observed numerous times have shown an infall about 30% to 50% of the time (Kuhi, 1978). Probably they are disk-shaped rotating systems with outflow in some regions and infall in others. Views of the same T Tauri star at different times or from different angles can give different flow patterns, as turbulence, rotation, and evolution bring different gas-and-dust-cloud patterns into the line of sight.

The T Tauri stage presumably represents a stage near the end of any planet formation that may occur. The cocoon nebula is dispersing. The local conditions may have many ramifications for planets. Observations suggest strong magnetic fields and stellar flares that might send accelerated atoms crashing into the planetary material to create short-lived radioactive isotopes.

T Tauri stars dramatize the importance of astrophysical studies in gathering evidence on the formative stages of possible planetary systems elsewhere in the universe. Because T Tauri stars are common, they also show that common evolutionary processes, rather than rare catastrophes, cause the evolution from the state where planetary dust obscures the star in a cocoon nebula to a state where most of the planetary dust has become undetectable. The real questions are how often the dust is blown away, and how often it aggregates into planets.

In systems where planets form, the dispersal of cocoon nebulae is a crucial process. Planetary scientists speculate that the dispersal might produce all manner of interesting effects, such as stripping primitive atmospheres from planets; but there is little firm evidence on which to base such models. Ward (1981) pointed out that typical cocoon nebula models amount to 10 times the total planetary mass, so that the loss of the nebula may seriously alter the gravitational environment of the planet system, may produce resonances among planets, and may affect the orbital arrangement and stability of the planetary system.

FU Orionis Stars: Infall of the Nebula

FU Orionis stars are a subclass of T Tauri stars that exhibit extreme flare-ups. The prototype, FU Orionis, has been seen to brighten rapidly by a factor of about 250 in luminosity, remaining bright for years at a time. These stars have temporarily enhanced the rate of infall of material from the cocoon nebula into the star itself. Whereas T Tauri stars may typically absorb an average of 10^{-7} M_{sun}/year from infalling nebular material (i.e., a tenth of a solar mass in 1 My, which would more or less exhaust the nebula), FU Orionis stars have short-lived massive infalls as high as 10^{-4} M_{sun}/year. This causes a flare-up in brightness and a corresponding increase in material blowing out from the nebula, probably as bipolar outflow from the polar regions. The outflow rate can reach about 10% of the enhanced inflow rate. Eventually, the net flow in the disk becomes an outward stream, a stellar wind of thin gas, analogous to the solar wind rushing outward from the sun through the solar system.

Beta Pictoris Systems: Residual Dust Disks

Beta Pictoris systems are relatively normal stars around which thin disks of dust have been imaged. They emerged after the IRAS infrared satellite surprised observers in 1983 by revealing some solarlike stars with excess infrared radiation, suggesting surrounding dust. In 1984, Bradford Smith and Richard Terrile made ground-based telescopic images of some of these stars, to learn whether the disks

Figure 4-9. The Beta Pictoris edge-on dust system, about 50 light-years away. The black central disk is caused by a system used to block the glare of the star itself. If this were the solar system, Pluto's orbit would be near the edge of the black disk. (Photo by S. Larson and S. Tapia, 1987, at Cerro Tololo International Observatory, Chile)

could be seen. This project netted an image of the edge-on disk of thin dust around the star Beta Pictoris (Figure 4-9).

The disk reaches about 400 AU from the star—a radius intriguingly comparable to that of the disk-shaped "Kuiper belt" of cometary debris around the solar system, which appears to mark the size of the solar system's original cocoon nebula. In this context, note that the closest match to a broad spectral emission feature at 10 to 11 microns in the infrared spectrum of Beta Pictoris comes from silicate dust grains emitted by Halley's comet, as shown in Figure 4-10 (Artymowicz, 1997)! These grains are believed to be composed mainly of the minerals olivine and pyroxene, prime planet-forming minerals. In other words, there is direct evidence that the Beta Pictoris dust nebula is composed of the same type of dust that is spewing out of our own active comets.

The dusty disk around Beta Pictoris is not thick enough and does not have enough gas to be called a cocoon nebula, yet it probably has more total mass than our own interplanetary dust. These facts support the idea that it is a remnant of the cocoon nebula, perhaps partly accreted into unseen comet-sized bodies. A relatively dust-free zone appears to exist in the central region, and this may mark a zone where the dust aggregation has gone further, producing planetary bodies. Dust in the outer zone may be replenished by collisional erosion among planetesimals there. Analyses of the disk properties thus suggest that the whole system is less than 100 My old,

which fits with the idea that planet formation is still proceeding there.

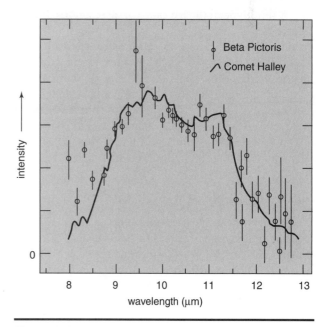

Figure 4-10. The infrared spectrum of the Beta Pictoris dust disk showing a broad emission caused by overlapping 9.6 and 11.2 micron bands of silicates, strikingly similar to that of the spectrum of silicate dust blown out of Halley's Comet. (Adapted from Artymowicz, 1997, based on data of R. Knacke and S. Fajardo-Acosta)

Binary and Multiple Stars

Our description of star formation has one major flaw: We have hardly considered the fact that at least half of all stars are binary or multiple. A **binary star*** is a pair of stars orbiting around a common center of gravity; a **multiple star** is a system of three or more co-orbiting stars. As binary and multiple star systems are so common, their formation must be a common process. Their properties, though poorly understood, must contain some messages about the formation of planetary systems, especially as some companions in such systems are nearly as small as planets.

One reason that binaries are poorly understood is our lack of good statistics. Selection effects create real difficulties. If we study only nearby stars, which we can see well, we have too small a number to have a good sample. If we study distant stars, the data are biased—for example, toward binaries with equal brightness because in very unequal pairs, the faint member is lost in the glare of the bright one. A **visual binary** is a binary with members far enough apart to be seen separately in a large telescope; their statistics are biased toward large separation distances. A **spectroscopic binary** has members so close together that their orbital velocities are very high (recall Kepler's third law), causing measurable, periodic Doppler shifts that identify the star as a binary. Statistics of spectroscopic binaries thus favor very close separations. Because of these selection effects in detection and cataloging, it is hard to guarantee statistical validity for lists of binary and multiple stars.

A third type of binary, called an **astrometric binary,** is detected by observations of stellar motions having incredible precision. These are called **astrometric observations.** If a star is accompanied by another star or planet, the two bodies orbit each other. Therefore, the star's motion through interstellar space will not be a straight line but will have a "squiggly" shape because it is always orbiting back and forth around its companion. The companion may be an unseen, tiny body; but if the star is close enough to the sun, the sinuosities will be detectable.

Rarity of Single Stars

These techniques have revealed that many stars have smaller companions, from low mass stars to objects roughly the size of Jupiter. A burning question in current research is how many of the small-mass companions are really planets and reflect planetary systems like our own solar system.

Some studies have suggested that if a star has a reasonably massive companion star, the companion may disrupt the motions of dust in the cocoon nebula in such a way as to inhibit aggregation of planets. Thus, the frequency of ordinary binary and multiple star systems is important if you want to talk about the fraction of stars that might have planets like ours.

Table 4-2 lists the nearest stars and shows the importance of binary and multiple systems. The largest member of each system is A, the second-largest, B, and the third-largest, C. Six of the 7 nearest stars and 16 of the 32 in the whole list are known or suspected binaries or multiples. Binary and multiple systems are hardly rare!

Several surveys have been made to determine the number of members in multistar systems. Table 4-3 lists some results. Once again, the selection effects create extreme difficulty because the farther away the star, the harder it is to detect small, faint companions by direct observation or astrometric means. Binary statistics suggest that about 50% to 80% of all star systems contain two or more companions (columns 2 through 4 in Table 4-3). Perhaps only a few percent to 50% of solar-type stars are really "single" stars that could have planets like ours.

Various observers, such as Abt (1978), have studied the mass distributions of binaries and multiples, leading to several interesting results. First, the mass distribution of systems with side separations (about 100 to 3,000 AU) is quite different from the mass distribution of systems with orbits about the size of the solar system. The wide systems seem to follow the size distribution found for ordinary field stars, suggesting that they may be field stars randomly paired in some way. The mass distribution of close binary companions suggests that they formed by a different process, probably a tendency for collapsing protostars to break up into two or more closely spaced, co-orbiting condensations that then became stars (Abt, 1978; Bodenheimer, Ruzmaikina, and Mathieu, 1993).

A second finding is that the oldest known stars—which happen to lack heavy elements—have the normal complement of widely spaced companions but do not have close companions! Here again, a separate origin for close companions is indicated, as is the possibility that these companions might be related to the dusty planetary material of cocoon nebulae that was absent during the formation of the most ancient stars (Abt, 1978). In addition, Fleck (1978) found a similarity of angular momentum between close binary systems and planetary systems, and suggests a similarity in origin.*

---

*The term **double star** applies to two stars at different distances that appear close together in the sky by fortuitous geometric alignment but have no gravitational connection.

---

*Note that the most ancient stars are deficient in heavy elements, which were mostly formed by nuclear reactions inside stars and then blown out into interstellar space by supernova explosions of large stars. Star-forming interstellar gas has thus grown richer and richer in heavy elements, which now amount to 2% to 3% of the total mass.

TABLE 4-2 Survey of 32 Nearest Star Systems out to 15 Light-years

Star or system	Distance (ly)	Mass of known or suspected components ($M_\odot$)			Approximate semimajor axis of orbit (AU)	
		A	B	C	AB	AC
Sun	0.0	1.0	0.001	0.0003	5.2	9.5
Alpha Centauri	4.3	1.1	0.9	0.1	24	10,400
Barnard's star	6.0	0.15	—	—	—	—
Wolf 359	7.5	0.1?	—	—	—	—
Lalande 21185 = BD + 36°2147	8.2	0.35	0.0015	0.001	10	2.5
Luyten 726-8 = UV Ceti	8.4	0.044	0.035	—	10.9	—
Sirius	8.6	2.31	0.98	—	20	—
Ross 154 = Gliese 729	9.4	0.2?	—	—	—	—
Ross 248	10.2	0.2?	—	—	—	—
Luyten 789-6	10.4	0.1?	—	—	—	—
Epsilon Eridani	10.7	0.7?	0.003??	—	—	—
Ross 128	10.8	0.2?	—	—	—	—
61 Cygni	10.8	0.6	0.6?	0.008?	85	?
Epsilon Indi	11.1	0.7	—	—	—	—
Procyon	11.4	1.8	0.6	—	16	—
Σ2398 = BD + 59°1915 = Gliese 725	11.5	0.4	0.4	—	60	—
Groombridge 34 = BD + 43°44 = Gliese 15	11.6	0.4?	0.2?	0.2?	156	?
Tau Ceti	11.7	1.0	—	—	—	—
Lacaille 9352 = CD − 36°15693	11.7	0.4?	—	—	—	—
G51-15	11.9	0.1?	—	—	—	—
BD + 5°1668 = Luyten's star	12.0	0.2?	0.1?	?	—	—
Luyten 725-32	12.3	0.1?	—	—	—	—
Lacaille 8760 = CD − 39°14192	12.5	0.5?	—	—	—	—
Kapteyn's Star	12.7	0.5?	—	—	—	—
Krüger 60 = DO Cep = Gliese 860	13.0	0.27	0.17	0.01	10	?
Ross 614	13.4	0.14	0.08	—	3.9	—
BD − 12°4523 = Gliese 628	13.7	0.2	0.1??	—	?	—
Wolf 424	13.9	0.2	0.1	—	3.0	—
van Maanen's Star = Wolf 28	14.0	0.3?	—	—	—	—
CD − 37°1549	14.5	0.2?	—	—	—	—
Luyten 1159-16	14.7	0.1?	—	—	—	—
Groombridge 1618 = BD + 50°1725	15.0	0.5?	—	—	—	—

Source: Data from Allen (1973; general data); Gatewood and Eichhorn (1973; Barnard's star); Harrington and Behall (1973; Luyten 726-8); Probst (1977; Ross 614); Campbell (1989, Epsilon Eridani and 61 Cygni).

In summary, the origin of some closely spaced binary and multiple systems may be related to the origin of planetary systems—planets being simply ultra-low-mass companions whose mass distribution is controlled by conditions such as angular momentum, turbulence, and mass distribution in the original collapsing protostellar cloud, as suggested as early as 1955 by G. P. Kuiper. However, questions remain: Do multistar systems grade smoothly into systems where the companions are planets? Can new, sensitive techniques actually detect alien planetary systems?

Planetary Systems of Other Stars

Astronomers have four techniques for searching for possible planetary companions of stars. The first is direct imaging, which is biased toward large, distant companions of nearby stars. The second is the spectroscopic technique that looks for slight changes in velocity, but this is biased toward large, close companions. The third is the astrometric technique, which is biased toward large companions of nearby stars. The fourth is the occultation technique—the

TABLE 4-3 Fraction of Systems Containing *n* Members

No. of members	19 systems within 12 ly	All systems		Solar-type main sequence stars[c]
		Average of 6 estimates[a]	Estimate favored by Batten (1973)[b]	
1	0.50	0.40	0.30	≤0.22
2	0.30	0.40	0.53	
3	0.15	0.15	0.13	
4	0.05[d]	0.036	0.03	≥0.78
5	—	0.01	0.008	
6	—	—	0.002	

[a] Estimates by various authors. Data from Batten (1973).

[b] Batten's estimates are averages over all stars on the basis of data from various sources.

[c] Data from Abt (1978). Abt's estimates are based on reviews of several sources, plus several comprehensive observational surveys of various types of stars by Abt and his coworkers.

[d] This figure represents one system, our solar system.

attempt to catch the slight dimming of a star if a planet passes in front of it; however, even Jupiter would block only 1% of the light of the sun, so the technique has to be very sensitive.

Until the 1980s, these techniques were too crude to detect objects as small as planets, but they did succeed in detecting some very small stars moving around other stars. Some examples include these:

- A faint star of about 100 $M_{Jupiter}$, moving around the very nearby, sunlike star Alpha Centauri at a distance about 10,000 AU from the star
- A faint star of about 170 $M_{Jupiter}$, moving around the cool, nearby star Kruger 60, at a distance around 10 AU from the star
- A faint star about 140 $M_{Jupiter}$, moving around the cool star BD +66°34, at a distance around 41 AU from the star

These detections barely approached the range of **brown dwarfs** (about 10 to 75–80 $M_{Jupiter}$), let alone Jupiter-sized objects. Indeed, some theories have suggested that bodies smaller than 10 Jupiter-masses may not be able to form, as the sun did, from a gravitationally shrinking cloud, so the statistics of bodies in that size range would be of great interest. Also, because hydrogen-rich giants larger than 2 $M_{Jupiter}$ have substantially different internal structure (in terms of hydrogen degeneracy and electron behavior), we might restrict the term planet to objects smaller than 2 $M_{Jupiter}$.

To claim that we have detected a planetlike object near a star, then, we need to be able to detect objects no bigger than 2 to 10 $M_{Jupiter}$. During the 1980s, a number of new instruments were built that could begin to detect objects in this range. First came some detections of possible brown dwarfs (Henry, 1996):

- A companion estimated at 4 to 80 $M_{Jupiter}$, having a temperature around 800 K, moving around the original T Tauri star, T Tauri itself, and detected by infrared observers (Hansen, Jones, and Lin, 1983)
- A brown dwarf of about 20 to 50 $M_{Jupiter}$, moving around the cool star Gliese 229, at a distance around 44 AU from the star (first imaged in 1994)
- A probable brown dwarf of at least 10 $M_{Jupiter}$, moving around the solar-type star HD 114762 B, at a distance of only 0.34 AU from the star, with a fairly high eccentricity of 0.33 (detected in 1988)
- An unseen companion of about 8 $M_{Jupiter}$, orbiting around the binary pair 61 Cygni, which is only 33 light years from the solar system

By the late 1980s, the race was on to continue using these techniques and be the first to discover objects actually in the planetary mass range, outside our solar system. Even before then, several claims of planet detections had been made, only to evaporate when other observers failed to confirm them. They were figments of noisy data. Instruments sensitive enough to probe into the planetary mass range—below a couple of $M_{Jupiter}$—were developed in the 1980s and began to come on-line, one after another. The payoff came in within a few years, with a variety of objects being detected at the rate of several per year.

- A possible object of 1 to 5 $M_{Jupiter}$, orbiting around the star Epsilon Eridani, a star only 10.7 light years from us (Campbell, 1989). Observations by 2003 give a semi-major axis $a = 3.3$ AU but a higher eccentricity than for solar system planets, $e = 0.61$, and also suggest a (dubious?) second planet as small as 0.1 $M_{Jupiter}$ at $a = 40$ AU and $e = 0.3$.
- A possible planet, with a mass as low as 1.6 $M_{Jupiter}$, orbiting about 2 AU from the star Gamma Cephei with $e = 0.2$ (Campbell, 1989)
- A possible planet, as small as 0.5 Jupiter masses, moving around the solar-type star 51 Pegasi, at a

distance of only 0.05 AU from the star (only a sixth of Mercury's distance from the sun!). It is more likely to have several Jupiter masses, depending on the factors involving the orbit geometry. It was discovered in 1995 and hailed as the first well-confirmed detection but subsequent studies have questioned its reality (Roth, 1997).

- A possible planet, as small as 6.6 Jupiter masses, moving around the solar-type star 70 Virginis at a distance of about 0.4 AU (about Mercury's distance from the sun), announced in 1996, with $e = 0.040$
- Two possible planets, as small as 2.5 and 0.8 $M_{Jupiter}$, moving in near circular orbits around the solar-type star 47 Ursae Majoris at a distance of about 2.1 and 3.7 AU, announced in 1996 to 2002
- Two planets, measured at 1.9 and 2.4 $M_{Jupiter}$, orbiting around the star Gliese 876 only 15 light years from Earth. The a values are 0.13 AU and 0.21 AU, respectively, and $e = 0.27$ and 0.12.
- A possible planet, about 1.6 Jupiter masses, moving around the solar-type star 16 Cygni B in an elliptical orbit from 0.6 to 2.8 AU, announced in 1996. The star is nearly identical to the sun, but has a companion star at roughly 700 AU.

Statistics in 2003 indicated a total of 94 known systems with planet-sized objects. These include 108 proposed planets, and twelve systems so far with more than one planet. We don't yet have a complete statistical census of the planetary citizens of our neighborhood, but it appears that of roughly 100 stars that have had at least preliminary study by techniques sensitive enough to detect Jupiter-scale objects, at least a few percent do have planets.

Orbits as a Test of the Origin of Companions

Are the "planets" that have been found precisely like the planets in our solar system? Did they form in the same way? Most of them have considerably more elliptical orbits than planets in our system, suggesting a different origin or history. Another puzzle is how giant planets, such as the half-Jupiter mass object orbiting 51 Pegasi B, ended up at distances from their star comparable to Mercury's distance from the sun. This again suggests some dynamical history different from that in our system.

The planets in our system have nearly circular orbits that lie nearly in a common plane, and this is almost certainly because they originated from finely dispersed material in a disk-shaped nebula around the early sun. Collisions among gas molecules, dust grains, and planetesimals would have damped out noncircular motions until everything traveled in parallel, circular, coplanar orbits. These features, then, may be diagnostic of this type of origin process.

Do binary and multiple star systems show these features? No. Binary orbits are frequently much more ellip-

tical than are the orbits in the solar system. As for relative inclinations, in the solar system, Pluto's orbit, the most inclined, lies at 17° to the system, and none of the rest is inclined more than 7°. However, in ten triple- and quadruple-star systems reviewed by Batten (1973), no orbit pair is reported at less than 19° inclination, and half are at more than 40°. One other system, BD + 66°34, may have three low-mass stars in coplanar orbits. Furthermore, most multiple systems are not smoothly and concentrically ordered as if by a Bode's rule relation. Instead, they usually have two closely spaced members and a very distant member, or two co-orbiting, closely spaced pairs.

As for Jupiter-scale planets very near stars, they create real problems for theorists because they are so unlike our system. Recent work (see Chapter 5) suggests multiple planetary bodies may have formed in each solar system zone before aggregating into single planets. Theorists Francesco Marzari and Stuart Weidenschilling (2002) proposed a "jumping Jupiters" model in which near-miss encounters among three such bodies could deflect one of them into an elliptical orbit very near the parent star—which would slowly circularize, as observed.

As more multiplanet systems are found, we will want to test the similarity of their origin to that of the solar system by determining whether the orbits of the planetary mass companions are circular, coplanar, and spaced in a way resembling Bode's rule.

Earth-sized Alien Planets?

Do "normal Earths" exist, which might have water and life-supporting conditions? As more sensitive instruments come on-line, the search will continue to tabulate the frequency of ordinary Earth-sized worlds that formed in the way our world formed.

It is too early to have good detections or statistics on Earth-sized objects for most candidate stars. However, since 1992, at least two Earth-scale objects have been found in orbit around an exploded star remnant—a pulsar—called PSR 1257+12. Most astronomers think the explosion should have altered or even destroyed any planetary system, and that the two Earth-sized objects must therefore have formed by a process different from the one that spawned our solar system. They may even have aggregated from debris left over after the explosion; in any case they do not seem very Earthlike in terms of history or environmental conditions.

Origins of Binary, Multiple, and Planetary Systems

As was suggested in Figure 4-4, a rotating protostar may fission into a binary pair if the angular momentum is high enough, but the story of binary stars is more complicated than that, because there are different types. Table 4-4 summarizes current thinking on the origins of binary, multiple, and planetary systems based on astronomical evidence.

TABLE 4-4 Possible Origins of Binary, Multiple, and Planetary Systems

Very closely spaced systems ($\ll$ solar system)	Closely spaced systems ($\approx$ solar system)		Widely spaced systems ($>$ solar system)
	Stellar companions	Planetary companions	
Fission of nearly formed, rapidly rotating protostar Inward tidal evolution of initially closely spaced pair formed by subfragmenting protostar (circular orbits produced by tidal forces)	Subfragmentation of protostar (due to higher-than-average angular momentum?) and subsequent separate evolution by gravitational collapse (orbit types may vary)	Aggregation of dust component of nebula by gravitational collapse and/or collisional aggregation (circular, coplanar orbits)	Capture of adjacent star in original star cluster (elliptical, noncoplanar orbits)

The widely separated pairs of binary stars, with orbits larger than the scale of the solar system, are probably formed by **capture,** a process whereby one star picks up a companion during gravitational interactions during a close encounter in the presence of a nearby third star or a resisting gaseous medium. Probably this happens just after the star-forming process while the stars are crowded in their initial star cluster. Arny and Weissman (1973) found that half the protostars in a typical cluster undergo collisions or close encounters before the cluster disperses.

Similarly, Hills (1977) found a high probability that if one already formed binary interacts with a single star, the single star may sometimes steal one member of the binary as its own companion. Hills found that 1 in 1,000 of the field stars of the solar neighborhood have undergone this process, whereas most binaries with wide separations in open clusters have had one or more such "exchange collisions." Thus, stars engage both in picking up singles and mate swapping. Both processes could drastically affect any chances for growth of planetary progeny.

Another category of binaries, very closely spaced pairs only a few stellar radii apart (or actually in contact in some cases), are probably formed by **fission,** a process whereby a rapidly rotating star splits into two similarly sized objects late in the star's formative process (Figure 4-4). Some of these close pairs may also result from tidal evolution of more widely spaced pairs. If the satellite star revolves faster than the primary star rotates, it would tend to pull ahead of the tidal bulge, contrary to the Earth-moon case (compare with Figure 3-11); hence, the tidal bulge would pull backward on the satellite, causing its orbit to spiral closer to the primary and producing a very close pair.

The large group of remaining binaries, with orbits about the size of planetary orbits, may form from **subfragmentation** of the collapsing protostar, a process in which the protostar divides into several separate condensations, each of which then separately evolves into a star, producing a co-orbiting system (Bodenheimer, Ruzmaikina, and Mathieu, 1993). Lucy (1977) found such a result in pioneering computer simulation of a rotating, axially symmetric, contracting, optically dense protostar, in which the initial protostar divided into a triple system whose components contained 60% of the original total mass, the rest of the mass being ejected. This same process may occasionally produce subcondensations with planetary mass.

Thus, the last class of multiple star systems—those in which the collapsing protostar breaks into several smaller collapsing objects—could possibly include some systems with small companions in the planetary mass range. However, the next chapter gives evidence that planetary bodies in our solar system formed not simply by gravitational collapse of part of a protostar but by still another process: the aggregation of the dust grains during their collisions.

Calculations suggest that planets like those of our system would not form in most systems with two or more stars because the complicated perturbations caused by the stars prevent the surrounding dust grains from aggregating (Hartmann, 1978; Greenberg and others, 1978). Thus we also conclude that systems of planets are less likely to accompany known binary and multiple stars than to accompany the other, seemingly single, stars. The stars most likely to have planets are probably the solitary stars of about 0.1 to 2 M_{sun}.

SUMMARY

In the complex interstellar environment of gas and dust, stars continue to form today, giving us examples of what the solar system may have looked like during planet formation. These examples confirm the nebular hypothesis: Circumstellar cocoon nebulae of gas and dust hide the stars during their early formation but may spawn planets and then dissipate. Such cocoon-shrouded new stars have been observed as infrared stars. Combining our discussions of various types, we see that the cocoon nebula is a thick disk that initially hides the stars. In later stages, such as the T Tauri

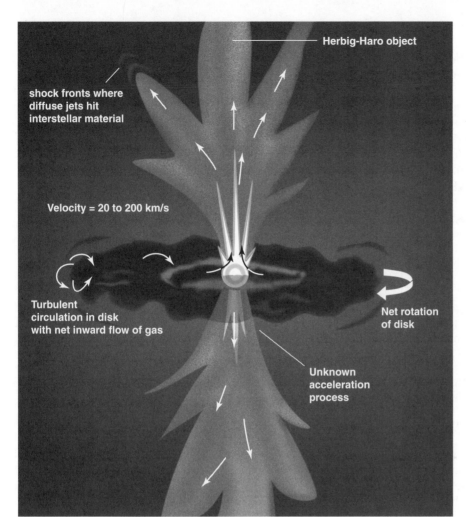

shock fronts where
diffuse jets hit
interstellar material

Herbig-Haro object

Velocity = 20 to 200 km/s

Turbulent
circulation in disk
with net inward flow of gas

Net rotation
of disk

Unknown
acceleration
process

Figure 4-11. Schematic summary of astrophysical observations of cocoon nebulae. The dust disk partly obscures the central star. It has turbulent motions but net inflow toward the star. Material is accelerated and ejected by bipolar outflow, creating Herbig-Haro glowing nebulae at a distance "above" and "below" the system.

stage, the nebula thins and disappears, part of it flowing inward due to drag in the disk and part being blown outward, initially by bipolar outflow. The cocoon then dissipates on a timescale of the order 10–20 My.

The cocoon nebulae themselves are observed to contain silicate dust and icy grains, exactly resembling the conditions under which planets are believed to have formed. Some relevant features are shown in Figure 4-11, which should be compared to Figures 4-9 and 4-7. After 100 My or so, the star may look normal from a distance but still be surrounded by a gas-poor disk of dust grains giving off far-infrared thermal radiation, like that of Beta Pictoris. Even our own 4,500-My-old system would look something like a weak Beta Pictoris system from a distance, because sensitive detectors could pick up the disk-shaped swarm of dust generated by asteroids and comets of our planetary system.

As a cocoon nebula clears, the coarser debris may remain behind to continue aggregating into larger bodies. Based on such an evolutionary theory, planets should form around at least some other stars. Brown dwarf-sized objects and probably Jupiter-sized planets have been detected around at least a few other stars.

It is uncertain whether planets *normally* form from this dust and ice, though that is what happened in the case of the solar system. A complication is that most stars form with binary or multiple stellar companions, and the relation of binary star formation to planet formation remains unclear. Some planet-sized masses may have formed in some binary or multiple systems, but planetary systems like ours seem more likely to be found near seemingly single, solar-type stars.

big bang	Bok globules	T Tauri star
supernova	cocoon nebula	FU Orionis star
open star cluster	nebular hypothesis	Beta Pictoris
molecular cloud	R Mon	binary star
nebula	infrared star	multiple star
interstellar material	proplyd	double star
infrared wavelengths	Herbig-Haro objects	visual binary
interstellar dust	H-R diagram	spectroscopic binary
absorption bands near	main sequence	astrometric binary
10 and 20 μm	post-main-sequence star	astrometric observation
gravitational collapse	pre-main-sequence star	brown dwarf
protostar	evolutionary track	capture
virial (Jeans) theorem	Hayashi track	fission
protocluster	Henyey track	subfragmentation

PROBLEMS

1. Give some lines of evidence that star formation is continuing at the present time in our part of the galaxy and that we can observe in the sky objects that exemplify the early solar system.

2. How are the Pleiades, Hyades, and other features in the Orion region of the winter sky relevant to star formation? Sketch this part of the sky and identify the Pleiades, Hyades, Orion belt and sword, and the location of R Mon.

3. List four properties of astronomical objects that can be measured by spectroscopic techniques and describe how each measurement is made.

4. A distant star has a spectrum exactly like that of the sun and has no Doppler shift. In addition, infrared observations show emission of infrared radiation peaking near 3 μm. Superimposed on this infrared spectrum are broad absorption bands at 10 μm and 20 μm. Also superimposed are some absorption bands arising in a circumstellar nebular cloud that are slightly blue-shifted from their normal positions. What physical conclusions can you draw about this star? Can you prove from this information whether the star is a newly formed star, a normal, "middle-aged" main sequence star, or a star near the end of its main sequence lifetime?

5. Two infrared stars are observed. Star A has its radiation maximized at wavelength 3 μm, and star B at 6 μm. (a) Compare whatever physical properties of A and B that you can. (b) The two stars are found to have the same luminosity (total Joules radiated per second). What further comparison of physical properties can you now make?

6. How do discoveries of complex organic molecules and grains of rock-forming minerals in interstellar space and in circumstellar clouds support, but not prove, the idea that life might exist on planets outside our planetary system?

7. If a shock wave (high-pressure wave that at least temporarily compresses gas in its path) rushes out from a supernova explosion through neighboring gas clouds and compresses some of the clouds, how might this help initiate star formation in the region near the supernova?

8. Given that supernovas occur in massive stars only about 10^8 y or less after such stars form, would a supernova explosion like that considered in Problem 7 be likely to have gas clouds and a potentially star-forming environment around it, or would it be more likely to occur isolated in deep space? Why?

9. Ejected material from a supernova is likely to contain some radioactive dust grains and gas that decay to distinctive stable isotopes. These stable isotopes are less abundant in ordinary interstellar gas and dust. How might a planetary system blasted during its formation by a nearby supernova be distinguished from a system of the same age never exposed to a nearby supernova?

10. Why are T Tauri stars thought to mark a later stage of formation than infrared stars?

11. (a) What evidence, if any, is there that microscopic rock material exists in the universe beyond the solar system? (b) What about rocks of, say, 1-m dimension? (c) What about "rocks" of 1,000-km dimension, that is, planets?

12. (a) Does the existence of binary and multiple star systems help prove that planetary companions may exist around other stars? Why or why not? (b) Do reports of astrometric binary companions of Jupiter mass prove that planets exist near other stars? (c) If planetary material is prevented by perturbations from aggregating into planets in all binary or multiple star systems, what percentage of other stars could still have planets, according to recent observational estimates?

13. Cite evidence, based on observations of suspected young stars, that during the formation of the solar system, the sun may have been much more active in terms of flares, strong solar winds, and similar instabilities than it is now.

14. According to the thermodynamics discussed here, most dust grains vaporize (evaporate into gaseous form) at 2,000 K. Suppose the gas in a collapsing protostar, a few astronomical units from the center, reaches 3,000 K and then forms a stable, circumstellar disk that slowly cools, so that the same gas remains in a relatively dense state. The gas cools to, say, 1,000 K. What would you expect the history of the grains and condensable elements in this nebula to be?

15. A T Tauri star has its maximum radiation at 0.5 μm. What are the probable sources of these two sets of radiation, and what are their temperatures?

16. (a) If the emission lines near 0.5 μm in the T Tauri star of Problem 15 are blue-shifted up to 0.0005 μm, and the circumstellar absorption lines are blue-shifted by 0.0005 μm relative to the star, what are the motions of the circumstellar material? (b) What would you conclude if a red shift of 0.0001 μm was also seen faintly in some of the stronger absorption lines?

17. Two optically thick infrared stars are measured to be at the same distance from the sun, and both have infrared radiation peaking at 2.0 μm. However, one is 2.56×10^2 times brighter in total luminosity than the other. What is the main difference between them, assuming that they are both blackbodies?

18. Two optically thick infrared stars are at the same distance. Star A has peak radiation at 2.0 μm, and B at 4.0 μm. Star A is 16 times as bright as B. What physical conclusions can you draw about these stars if they both radiate as blackbodies?

19. Once the collapse of a protostar gets well under way, the inequality of the virial theorem becomes an approximate equality, so that

$GM\mu M_{sunH} \approx 5kTR$

Assuming that the cloud remains approximately isothermal (having equal temperature throughout), calculate a representative internal temperature for the protosun (a) when it reaches the size of the solar system (use Pluto's orbit) and (b) when it reaches its present size. (c) Compare result (a) with the reported temperature for infrared stars and comment on its significance.

20. The present internal (central) temperature of the sun is a few million Kelvin, a temperature sufficient to initiate nuclear reactions that produce the heat and radiant energy to maintain the sun at its present size. Nuclear reactions started in the center of the sun when the center reached that temperature. In view of Problem 19(b), explain why the sun stopped contracting when it reached its present size.

21. A condensation of gas and dust with Jupiter's mass forms within a cocoon nebula whose temperature is 500 K. It has an orbit equal in size to Jupiter's orbit and a radius 0.1 that size. (a) What density must the condensation have for it to collapse into a Jupiter-sized planet rather than dissipating if no tidal or other forces act to break up the condensation before it can collapse? (b) Suppose the mean density in this nebula is 10^{-6} kg/m³. Is such a nebula likely to produce planets of Jupiter's size or *smaller* by purely gravitational collapse?

22. The condensation in Problem 21 is acted on by the central star, which has 1 M_{sun} and is 5.2 AU away. Estimate the density that the condensation must have to keep it from being torn apart by tidal forces from the sun. (*Hint:* The condensation would have to be outside Roche's limit.)

23. A distant asteroid has a blackbody spectrum with radiation peaking at 30-μm wavelength. What is its temperature?

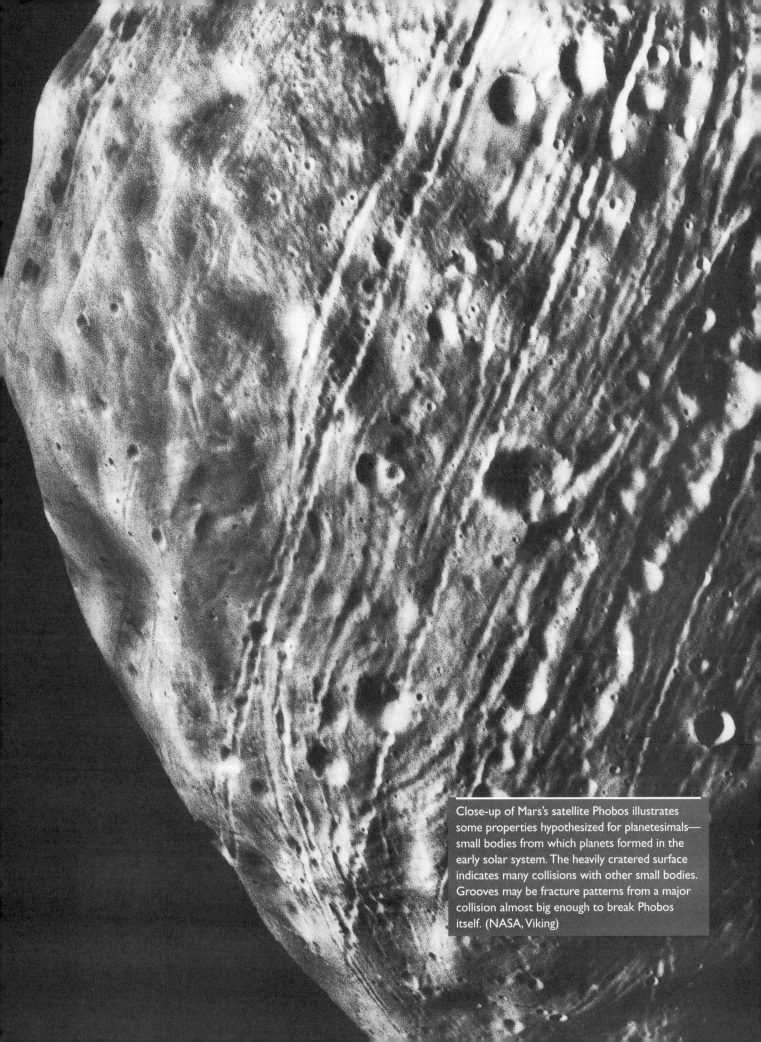

Close-up of Mars's satellite Phobos illustrates some properties hypothesized for planetesimals— small bodies from which planets formed in the early solar system. The heavily cratered surface indicates many collisions with other small bodies. Grooves may be fracture patterns from a major collision almost big enough to break Phobos itself. (NASA, Viking)

CHAPTER FIVE

The Formation of Planets and Satellites

Chapter 4 emphasized astrophysical evidence about formative conditions among other stars. In this chapter we emphasize evidence derived from our own star and planet system—evidence that provides more detail on how planets form. In the case of our own system, we can address specific questions of physical processes, chemistry, and timescales.

Date and Duration of Solar System Formation

The age of the solar system is about 4.6 Gy. To be more accurate, meteorites indicate the formation of the first solid planetesimals (the refractory rich inclusions such as found in the meteorite Allende) around 4,559 ± 5 My ago (Taylor, 1992); it is interesting to give this number in terms of millions of years, because meteorites allow us to resolve many events of that era with an accuracy of a few My. The number is verified from studies of additional rocks from three main planetary sources: other meteorites (probably representing fragments of asteroids), the moon, and Earth. We also have rocks from Mars, but they are less useful in studying primordial events because we have fewer of them and less information about their geologic context.

The derived ages come from **radiometric dating,** the determination of a rock's age from the radioactive and radiogenic isotopes in the rock. **Radioactive isotopes (parent isotopes)** are unstable and eventually decay at a known rate into **radiogenic isotopes (daughter isotopes).** Because the radioactive decay process proceeds continually, the ratio of daughter-to-parent isotopes in a given mineral continually increases and becomes a "clock" measuring the age of the mineral. The actual age determination involves a sequence of highly complicated and precise chemical measurements.

Just what do we mean by the "age" of the mineral? Usually this refers to the elapsed time since the mineral solidified from some previous molten or gaseous material. Prior to solidification, when all atoms are in gaseous or liquid form, the parent or daughter atoms (for example, a daughter element that exists in gaseous form) may escape from the system. Solidification traps both the parent and daughter atoms in the crystal structure of the mineral grain, thus allowing the radiometric clock to start running.

Consider an idealized example in which crystals of some mineral attract a certain radioactive parent element but not the daughter. As these crystals actually solidify, they incorporate a certain allotment of parent but no daughter atoms, and the radiometric clock starts running. A rock age measured by using this principle is called a **solidification age,** because it indicates when the mineral and rock material solidified into its present form. The actual techniques of measurement are complicated by the need to measure other isotopes as well, giving estimates of the numbers of parents and daughters (not necessarily zero) actually present in the original rocks.

Even more complex measurements of various isotopes allow estimates of the time when the initial isotopic mixture of a rock's parent material was established—that is, the date for formation of the original planetary material. This is called the **formation age** of the material. Because a given rock might have formed in a recent lava flow or during some other event on the planet, solidification ages may be younger than formation ages for the same material. However, many ancient meteorites have equal solidification and formation ages, showing that they have been little altered since the formation of the solar system. Other types of ages can also be measured, and these are discussed in the next chapter.

Formation Age of the Solar System

Table 5-1 shows ages related to the formation of three types of planetary material. All indicate that planetary matter formed about 4560 My ago. The sun itself is also believed to have formed at this time. This is consistent with the astronomical data indicating that planet-spawning cocoon nebulae last only about 10 My or so after the star itself forms. Notice that the table pins down the entire planet-forming process, from the first solid rocky materials to the formation of the Earth and moon, to the brief period of not more than about 50 to 100 My—only about 1% or 2% of the age of the whole system! Planets formed fast. Meteorites give the best determinations because they are abundant and little modified since they formed.

TABLE 5-1   Formation Dates of Some Planetary Bodies

Sample	Date of origin (My)	Notes
Meteorites:[a] Most primitive available samples		
Earliest solid inclusions in meteorites	4,559 ± 5	Average of 6 refractory inclusions in Allende meteorite, cited by Taylor (1992)
Early, never-melted meteorites	4,526 ± 30	Average of 8 dates on 7 samples of chondrites, by 3 different isotopic methods, cited by Taylor (1992)
Solidification of once-melted basaltic meteorites	4,539 ± 4	Average of dates on 5 achondrites by 3 different isotopic methods cited by Taylor (1992)
Moon: Rare samples of ancient, little-altered crustal rock		
72417, dunite crustal differentiate (probable fragment of earliest solidified crust)	4,550 ± 100	Papanastassiou and Wasserburg (1976). (Albee, Gancarz, and Chodos, 1973, got 4.6 ± 0.07.)
60025, lower limit on lunar age (anorthosite crust fragment)	4,440 ± 20	Carlson and Lugmair (1988).
Estimated lunar age from various samples	4,500 ± 50	Range 4530–4550 applies *if* moon had chondritic Rb/Sr ratio. Range 4440–4510 plausible if moon had low Rb/Sr. Carlson and Lugmair (1988).
Earth: Old rock sample		
Amîtsoq 3590 My-old gneiss sample from Greenland	4,500 ± 0.05	Average of U-Th-Pb ages by Gancarz and Wasserburg (1977), based on reconstructions of original chemistry of parent material. This 4500 result interpreted as the approximate age of the Earth.

[a]Meteorite types are discussed in Chapter 6.

The terrestrial samples require a special comment. Although both the moon and meteorites give samples with solidification ages of 4400 to 4500 My no known rocks on Earth solidified earlier than about 3600 My. Older rocks have been eroded, and so heavily modified by chemical and geologic processes that formation ages are hard to estimate. However, Gancarz and Wasserburg (1977) analyzed 3600 My-old specimens from Greenland, in which the initial isotopic chemistry had been little altered. As a result, they could derive formation ages for Earth of 4470 or 4530 My, depending on the assumptions made about the isotopic history of the material.

Did a Nearby Supernova Explode Just Before the Solar System Formed?

In 1960, J. H. Reynolds discovered that many 4.6-Gy-old meteorites contain radiogenic isotopes that must have come from highly unstable radioactive parents that last only a few millions of years. In particular, he found xenon 129, a xenon isotope that forms by decay of radioactive iodine 129, which has a half-life of only 0.017 Gy, or 17 My. In any radioactive decay process, the **half-life** is the time during which half the atoms present at a given moment will decay. For example, if a billion iodine 129 atoms were initially present, one-half billion would be left after 17 My, one-fourth billion after 34 My, one-eighth billion after 51 My, and so on.

Clearly, because of this relatively rapid decay rate, the original unstable iodine 129 could not have been created too long before the meteorites were created, or else there would have had to be fantastically large amounts of initial iodine 129 for any to be left and trapped in the meteorite. What created the iodine 129?

Similar questions have been prompted by other radiogenic material in meteorites. For example, magnesium 26 has been found that formed by decay of radioactive aluminum 26, which has a half-life of only 0.72 My! This finding suggests that the aluminum 26 must have been created only a few million years before being incorporated into meteorites. What created the aluminum 26?

Growing evidence indicates that the unstable radioactive elements are the type created by nuclear reactions inside stars. Cameron and Truran (1977) wrote a famous paper arguing that a supernova went off near the sun just before it formed, spraying out fresh radioactive isotopes that impregnated the solar nebula. It became widely believed that a supernova happened near the solar nebula, but Cameron (1985) retracted the model, arguing instead that the radioactive nuclei were created in ordinary novas, which are smaller explosions involving binary stars. One line of evidence is that even the ordinary interstellar gas seems enriched in Al-26. Lee and others (1997) re-argued that all the short-lived radioactive isotopes came from a single nearby supernova explosion just before the solar system formed. In any case, the evidence shows that the

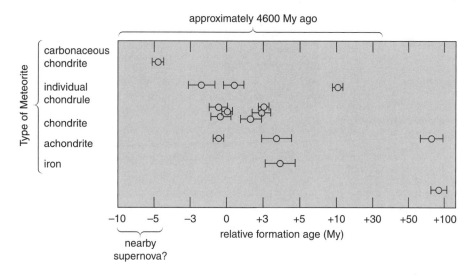

Figure 5-1. The first 110 My of solar system history. Formation ages of Earth and of the several meteorites (identified by type; see also Chapter 6) are not absolute but relative to the formation of a well-studied chondrite meteorite named Bjorböle, whose age is arbitrarily designated as zero. The results show that most meteorites formed during a 20-My period. Dates are interpreted from decay of radioactive iodine 129, apparently produced by a nearby supernova a few million years before the solar system formed. (Adapted from Podosek, 1970; Wasson, 1974; Pepin and Phinney, 1975)

radioactive isotopes came from material formed inside stars 100 to 400 My before the planetesimals formed, with some late additions no more than 3 My before the planetesimals formed (Wasserburg, 1985).

These conclusions dovetail nicely with Chapter 4, which shows that the primordial sun probably resided in a cluster, close to many stars, including unstable ones that could have exploded as novas and/or supernovas. Aside from being dramatic, the hypothetical creation of short-lived radioisotopes just before the formation of the solar system gives us the opportunity for another type of dating that reveals the *duration* of the planet-forming process as well as its date.

Formation Interval and Duration of the Formative Process

When the short-lived radioactive isotope iodine 129 decays, the inert xenon 129 gas that is formed escapes unless the parent and daughter materials have already been incorporated in a mineral crystal. Suppose a supernova impregnated the solar system material with a certain amount of iodine 129. A meteorite that formed early would have had a certain amount of this iodine, which would have decayed rapidly, leaving xenon 129 trapped in the rock. A meteorite that formed later would have had less initial iodine 129 because the iodine 129 would have been decaying all along; thus, the second meteorite would now have less xenon 129. Therefore, laboratory measurements of the radiogenic xenon now found in a meteorite gives a measure of the time when one meteorite formed relative to another, a measure called the **relative formation age.**

Measurements of this type, together with certain assumptions, lead to estimates of the **formation interval**— the time between the nucleosynthesis (element-forming) event (a supernova?) and the inclusion of the short-lived isotopes into planetary material. Estimates for the forma-

tion interval are a few million years but are model dependent. Rather than reporting model-dependent formation intervals for each meteorite, the current practice is to reference all formation ages to that of the well-studied meteorite Bjorböle, as shown in Figure 5-1 (Wasson, 1974). Figure 5-1 shows that most meteorites formed within about 20 My of each other. Using a similar technique, Pepin and Phinney (1975) found that Earth formed (that is, was big and solid enough to retain its xenon) 96 ± 12 My after Bjorböle. These results are all consistent with the conclusion of Table 5-1, that the ages of meteorites and planets scatter over a range not larger than about 50 to 100 My.

Solar System Characteristics to Be Explained by a Successful Theory

In constructing a theory of the events of 4.6 Gy ago, we have to choose carefully from among the observational clues. For example, Kepler's laws did not need to be explained by a theory of origin because they are required by Newton's laws of gravity. On the other hand, the circularity and coplanar nature of the planetary orbits do need to be explained because the laws of gravity do not require such conditions.

Nobel laureate Hannes Alfvén, who spent years researching the solar system's origin, once said, "To trace the origin of the solar system is archaeology, not physics." He meant that our ignorance of the initial conditions forces us to work backward through time, reasoning from whatever clues we can find. The most important clues are *facts about the solar system that have no obvious explanation from present-day conditions or known physical laws.* Table 5-2 gives a list of such facts and we refer to them in the following discussion, in which we explain them, one by one.

TABLE 5-2 Characteristics of the Solar System to Be Explained by a Theory of Origin

1. All the planets' orbits lie roughly in a single plane.

2. The sun's rotational equator lies nearly in this plane.

3. The planets and the sun all revolve in the same west-to-east direction, called *prograde* or *direct* revolution.

4. Planetary orbits are nearly circular.

5. Planets have much more angular momentum (a measure of orbital speed, size, and mass) than does the sun, and the sun spins slower than expected if it spun off the solar nebula during its collapse. (Failure to account for this was the great flaw of early evolutionary theories.)

6. Some meteorites contain inclusions of minerals that formed from grains condensed at higher temperatures than were other meteorite materials; these inclusions contain materials with unique isotopic abundances.

7. Planets differ in composition, roughly correlating with distance from the sun. Mercury is dense and metal rich; other terrestrial planets are less metal rich. The giant, hydrogen-rich planets are in the outer solar system.

8. Meteorites differ in detailed chemical and geologic properties from all known terrestrial and lunar rocks.

9. The distances between the planets mostly obey the simple Bode's rule.

10. All closely studied planets and satellites show impact craters indicating collisions with interplanetary bodies whose diameters ranged up to 100 km.

11. Except for Venus, Uranus, and Pluto, all planets rotate prograde with obliquities of less than 29°.

12. Most planets and asteroids rotate with similar periods of about 5 to 10 h, except in cases where obvious tidal forces have slowed them (as is the case with Earth).

13. As a group, comets' orbits define a large, spherical swarm around the solar system.

14. Major planet-satellite systems resemble the solar system on a smaller scale.

The Solar Nebula: The Nebular Hypothesis Confirmed

Because of a tradition in planetary science that arose before astronomers discovered cocoon nebulae, the sun's cocoon nebula is called the **solar nebula.** The following terms describe objects *in* the solar nebula, to which we refer in this chapter:

Grains or **dust grains:** microscopic particles such as silicates and ices that condensed from the solar nebula or interstellar grains that became trapped in the solar nebula.

Planetesimals: bodies from submillimeter size up to hundreds of kilometers in diameter (even up to 1000-km-scale bodies in some instances) that formed during the planet-forming process in the solar nebula.

Protoplanet: (1) any large precursor of a planet, usually referring to an extended mass, including any large atmospheric cloud that may have later dissipated, and usually implying a body as massive as or more massive than the planet today; (2) in theories involving gravitational collapse, the large gravitationally unstable units from the time they first gain an identity by initiating collapse to the time they reach planetary size; initially they may extend across an appreciable part of the orbital circumference of the planet in question.

Origin and Mass of the Solar Nebula

The solar nebula was the sun's cocoon nebula. It consisted of hot gas and dust extending at least 40 AU and probably several hundred AU from the sun by the time the sun began to take on its identity as a central condensation in the gravitationally contracting, prestellar cloud. Stages 2 and 3 of Figure 4-4 showed the development of such a nebula as the star collapses rapidly in the nebula's center. Thus, the solar nebula is not believed to have been material that was ever inside the sun or thrown off its incandescent surface.

Some early theorists incorrectly assumed that the mass of the solar nebula was equal to the total mass of the planets. However, the mass of the nebula must have been many times the planetary mass, as the planets are composed of only residual elements from a nebula originally of cosmic composition. For example, the terrestrial planets are composed mostly of silicon and iron. Table 4-1 showed that only 0.0006 of the cosmic material is silicon, and thus 1,700 grams of interstellar gas would be needed to obtain even one gram of silicon.

Most of the solar nebula was not condensable material but consisted of gases such as hydrogen, helium, the inert gases, water vapor, and so on. These materials are called **volatiles,** the elements and compounds that, under the given conditions, tend not to form solids. Volatiles have a high vapor pressure, meaning that if they are incorporated into solid or liquid material and find themselves in a vacuum environment, they will readily escape from the solid, atom by atom, and build up the surrounding gas pressure.

A minimum mass for the solar nebula can be calculated if we estimate the total amount of missing volatiles that must be restored to each planet to produce a protoplanet of cosmic composition, as shown in Table 5-3. Because silicates and iron constitute only a small fraction of cosmic material, the mass of Earth must be multiplied by a large factor to restore the lost mass. Interestingly,

TABLE 5-3   Masses of Nebular Matter Required to Form Planets

Planet	Assumed major components	Present mass (kg)	Estimated ratio required mass: present mass	Estimated nebular mass (kg)
Mercury	Silicates, iron	3×10^{23}	410	1×10^{26}
Venus	Silicates, iron	5×10^{24}	380	2×10^{27}
Earth	Silicates, iron	6×10^{24}	380	2×10^{27}
Mars	Silicates, iron	6×10^{23}	370	2×10^{26}
Asteroids	Silicates, iron (ices?)	$\sim 1 \times 10^{21}$	250	3×10^{23}
Jupiter	Hydrogen, ices	2×10^{27}	10	2×10^{28}
Saturn	Hydrogen, ices	6×10^{26}	16	1×10^{28}
Uranus	Methane, ammonia, water, ices	9×10^{25}	67	6×10^{27}
Neptune	Methane, ammonia, water, ices	1×10^{26}	64	6×10^{27}
Pluto	?	$7 \times 10^{23}?$	75?	$5 \times 10^{25}?$
Comets	Methane, ammonia, water, ices	$>1 \times 10^{27}?$	5	$>5 \times 10^{27}?$
			Total nebular mass:	$>5.1 \times 10^{28a}$

Source: Adapted from Kuiper (1956a), Cameron (1962, 1975), Hoyle (1963), and Whipple (1964).
[a]Equivalent to $>0.03 M_{\odot}$.

the masses contributing to the various planets (the last column in the table) were much more nearly equal than the final masses of the planets themselves. The table shows that the minimum material involved in planet formation approached 0.03 M_{sun}. Of course, the nebula might have been still more massive, as we do not know how much extra interplanetary gas and dust may have been blown away without contributing to the planets. As early as 1966, the Russian theoretician Victor S. Safronov (1966) reviewed solar nebula physics and inferred a mass of 0.15 M_{sun}.

This figure is consistent with estimates of the mass of cocoon nebulae around young stars, which are in the range of one or a few tenths of the stellar mass. T Tauri stars apparently blow away up to 0.4 M_{sun} of material, and this very strong, outflowing stellar wind also carries off angular momentum that slows the star's rotation from the fast rates observed for T Tauri stars to the slow rate observed for the sun (Kuhi 1978). Magnetic fields during the loss of a 0.15-M_{sun} nebula could have slowed the sun to its present rotation rate (Schwartz and Schubart, 1969). More massive nebulae, up to 1-2 M_{sun}, have been suggested from theoretical models of star formation but are more difficult to defend on observational grounds.

Composition and Shape of the Solar Nebula

On the basis of both meteorite chenistry and the above conclusion that the nebula was a solar remnant, researchers have long argued that the solar nebula had approximately the same composition as the sun, as indicated in Table 4-1.

However, as noted above, variations in isotopic composition among planets and meteorites indicate that the nebula composition was not uniform and that some interstellar (supernova-derived?) material was injected into the nebula, perhaps primarily the outer nebula.

As noted in the last chapter, Laplace deduced as early as 1796 that the solar nebula was disk-shaped and rotated in the plane of the sun's equator. What are the mechanics that control the disk shape? Consider a system of particles (gas atoms or dust grains) orbiting almost randomly around the newly formed sun. If the orbits are random, many will intersect, causing collisions. The net effect of these collisions will be to average out the different directions of motions until most particles are moving in one plane defined by the original net angular momentum of the system. The particles would also come to move in nearly circular orbits. (If a particle tried to remain on an elliptical orbit, it would cross other orbits, be hit, and be deflected toward the common direction of motion.) This is the same reason that particles near Jupiter, Saturn, and Uranus have formed flat ring systems and that the dust and gas complexes of the Milky Way and most other galaxies have formed disks. It also accounts for characteristics 1–4 in Table 5-2, the list of characteristics to be explained by a theory of solar system origin.

Theoretical studies of the collapse process (Tscharnuter, 1978) show that the material "above" and "below" the sun plunges in rapidly toward the equatorial plane, so that a disk shape is attained probably within a million years or so. These results agree with observations mentioned in Chapter 4, favoring a disk shape for some observed cocoon nebulae.

As shown in Figure 5-2, the nebula finally adopts a quasi-equilibrium shape when all forces on the particles

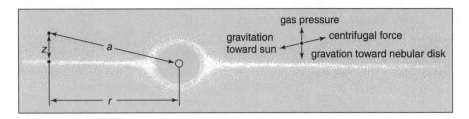

Figure 5-2. Cross section of the solar nebula, showing balanced forces on a particle in the nebula. The disk shape is indicated. Density and pressure are greater in the disk than out of it and increase toward the sun. The outer regions are thicker and less dense than inner regions. Dimensions a, r and z are employed in the optional mathematical discussion.

are balanced—primarily gravity (directed inward) and gas pressure and centrifugal force (outward). Mathematical models using the **hydrostatic principle**—which assumes balance among these forces—were pioneered by Von Weizsacker (see Chandrasekhar, 1946) and Kuiper (1951). Interestingly, the stronger gravitational forces toward the plane in the central regions near the sun make that central region thinner and denser, whereas the outer portions are thicker; therefore, instead of resembling a convex lens, the dust disk has some resemblance to a concave lens with diffuse edges. The disk is only a quasi-equilibrium configuration because, as the nebula cools, some of the dust

aggregates into planets and some dust and gas eventually blow away.

Density and Pressure in the Solar Nebula

For given temperatures, such as those observed in infrared cocoon nebulae, the density and pressure distributions in the solar nebula can be found from the principles described above. The mean density, of course, is just the total mass divided by the total volume. If we take the mass

---

Mathematical Notes on the Shape of the Solar Nebula

The structure of any gaseous atmosphere or nebula in equilibrium can be analyzed by noting that all the inward forces are equal to the outward forces. In its simplest form, this analysis involves only two equations, the perfect gas law, which merely states that the pressure, temperature, and density are controlled by the fact that the substance is a gas:

(1) $\quad P = \dfrac{\rho k T}{\mu M_H}$

and the hydrostatic equation, which states that the pressure in the gas responds to a gravitational compressive force:

(2) $\quad P = -\rho g r$

In these equations,

$\quad P$ = pressure
$\quad T$ = temperature
$\quad \rho$ = density
$\quad \mu$ = mean gas molecular weight
$\quad g$ = gravitational acceleration at the point in question
$\quad r$ = distance from sun

Other symbols are constants defined in Table 2-2.

We could use these equations to analyze the pressure and density changes along the ecliptic plane in the direction toward and away from the sun. However, the result shows that ρ does not change too much in this direction, and it is more interesting to analyze the structure perpendicular to the ecliptic plane, as shown in Figure 5-2.

For the upward (perpendicular to ecliptic) forces to equal the downward forces, the gas pressure must balance the hydrostatic pressure. The downward force is the z component of gravity:

(3) $\quad g_z = g\dfrac{z}{a} = \dfrac{GM_\odot}{a^2}\dfrac{z}{a} = \dfrac{GM_\odot z}{a^3}$

Since the variables depend on one another, we set up a differential equation for the change in pressure dP over a small distance equating dP from the perfect gas law (1) with dP from the hydrostatic equation (2):

of 0.2 M_{sun}, mentioned above, and spread it across a 0.5-AU-thick nebula the size of Pluto's orbit, the mean density of the gas is only of the order 10^{-7} kg/m³. Similarly, if we distribute the minimal 380 M_{earth} of cosmic material necessary to make Earth (see Table 5-3) in a toroid 0.1 AU thick around Earth's orbit, we have a mean gas density again around 10^{-7} kg/m³. Dynamical studies show that the density was greater nearer the midplane of the nebula and also nearer the sun (see mathematical notes on the shape of the solar nebula).

For a hydrogen-rich gas at 800 K and with density of, say, 10^{-6} kg/m³, the gas pressure would be of the order 10 N/m³, which is 0.1 mbar, or about 10^{-4} normal atmospheric pressure.

Cooling of the Solar Nebula

The last chapter showed how a contracting gas cloud about to form a star gets warmer as it contracts. This is the source of the heat that warmed the inner solar nebula to temperatures probably exceeding 2,000 K (Cameron and Pine, 1973). However, once the nebula assumed a disk shape rotating approximately in hydrostatic equilibrium, its contraction was greatly slowed and it eventually cooled by means of infrared radiation. Temperatures dropped toward values of around 800 K, as determined for infrared cocoon stars, and eventually reached values as low as a few hundred Kelvin.

Magnetic Effects

Our discussion so far has assumed that the solar nebula evolved as a neutral gas with some admixture of dust particles. Evolution of such a nebula has been theoretically analyzed in terms of motions controlled by gravity plus additional small forces such as gas pressure or radiation pressure.

However, a different situation exists if there was sufficient energy to keep the nebular gas ionized, as is the solar corona and interplanetary gas today. Whereas a neutral gas can move freely in any direction, a highly ionized gas, or **plasma**, cannot move freely with respect to a magnetic field. As shown in Figure 5-3, a plasma can move freely *along* magnetic field lines, but not *across* them. (**Magnetic field lines** are imaginary lines stretching through space from the north to the south pole of a magnetic source, such as the sun.)

If a magnetic field moves through a plasma, the plasma tends to be dragged along with the field lines; this

---

(4) $\quad \dfrac{kT}{\mu M_H} dp = -\rho g_z dz$

which assumes that T is about constant at the given distance from the sun r.* Substituting for g_z,

(5) $\quad \dfrac{kT}{\mu M_H} dp = -\rho \dfrac{GM_\odot z}{a^3} dz$

giving

(6) $\quad \dfrac{dp}{\rho} = -\dfrac{\mu M_H GM_\odot}{a^3 kT} z dz$

Integrating this gives

(7) $\quad \ln\left(\dfrac{\rho}{\rho_o}\right) = -\dfrac{\mu M_H GM_\odot z^2}{2kTr^3}$

where ρ_o = density in the ecliptic plane, and z is assumed to be not too far out of the plane, so that $a \approx r$, being constant. This gives the density structure

(8) $\quad \rho = \rho_o \exp - \dfrac{\mu M_H GM_\odot z^2}{2kTr^3}$

This says two interesting things. First, the density decreases above and below the ecliptic plane as z increases. Second, the greater the distance r from the sun, the thicker the disk, as illustrated in Figure 5-2. This can be seen by looking at the exponent. Let us define scale height H as the distance "above" the ecliptic over which the density drops by a factor e. By examining the exponent, one sees that

(9) $\quad H = \sqrt{\dfrac{2kTr^3}{\mu M_H GM_\odot}} = 1.11(10^{-8}) T^{1/2} r^{3/2}$

The numerical solution is for a hydrogen-dominated nebula. Thus, the scale height increases with r. At 1 AU from the sun in a 1000-K gas, this gives a scale height of about 0.14 AU; at 10 AU, it gives a scale height of about 4 AU.

---

*Incidentally, Safronov (1972, p. 25) gives a similar derivation and this is his equation 2.

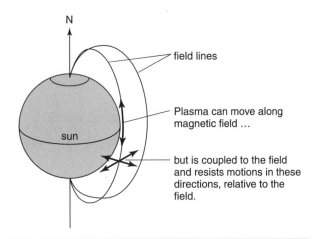

N

field lines

Plasma can move along
magnetic field ...

sun

but is coupled to the field
and resists motions in these
directions, relative to the
field.

Figure 5-3. In the magnetic field of the sun (represented
by two magnetic field lines), neutral gas can move in any direction,
but ionized gases (plasma) are deflected if they do not move
along field lines. Rotation of the sun would rotate the field lines
and cause the plasma to be dragged along in the rotational
direction.

is called **magnetic coupling** to the field. If the nebular gas
were neutral, therefore, its motions would be analyzed by
hydrodynamics (the science of fluid and gas motions),
but if the gas were a plasma, it would be dragged around
by the rotating solar magnetic field (rotation of the sun
and field can be visualized in Figure 5-3), and its analysis
would require **magnetohydrodynamics** (the science of
plasma motions).

Was the Solar Nebula Highly Ionized?

Most theorists have developed models with neutral nebular gas (because they are easier?); a few, however, have developed plasma models. Nobel laureate Hannes Alfvén
(1954) and later Alfvén and Arrhenius (1976) championed the view that the nebula was largely ionized by its
energy of infall, and its motions were controlled by magnetohydrodynamics.

A new reason for proposing at least partial ionization
of the solar nebula was suggested in 1978. Consolmagno
and Jokopii (1978) noted that if the solar nebula really
contained the proposed amounts of radioactive aluminum
26 or potassium 40 from a supernova, the energetic sub-
atomic particles shot out of these atoms during radioactive decay would have ionized neighboring atoms. Their
calculations for reasonable models involving aluminum
26 content indicate that ionization would be partial but
sufficient to couple the nebular gas to ambient magnetic
fields. These calculations suggest that further magnetohydrodynamic studies will be required for precise understanding of the solar nebula's evolution.

Magnetic Braking of the Sun's Rotation

Although opinions on the importance of ionization in the
solar nebula vary, most researchers agree on one exceedingly important magnetohydrodynamic effect. The ionized atmosphere, or corona, of the primeval sun merged
with the inner nebula, so that at least the inner nebula had
a large ion content. As the sun's magnetic field turned
through gas at some distance from the sun, the gas would
have tended to be dragged along with the field, as can be
visualized by imagining the sun and field turning in Figure 5-3.

Initially, as a result of the contraction process, the
sun should have been spinning much faster than it does
today, as pointed out by Laplace. The big problem with
Laplace's original nebular hypothesis is that it did not explain why the sun slowed from this initial fast spin rate.
Now, however, we see that the sun's magnetic field would
have tended to grip the nebula. The nebula in turn would
have caused a drag on the magnetic field, twisting the
sun's field and slowing the sun's spin, just as a spinning
tennis ball dropped in a swimming pool will rapidly come
to rest because of the drag of the water on the fibers in
the ball's surface. Thus, magnetohydrodynamics explains
the fifth important characteristic of the solar system in
Table 5-2.

As the sun slowed down, its angular momentum was
transferred to the nebular gas. Much of the angular momentum was eventually carried out of the system, as the
gas rushed away from the sun in an expanding solar wind.
The role of the observed bipolar outflow of gas in this
whole picture is still not fully understood. Nonetheless,
the transfer of angular momentum from the star to the gas
explains not only the sun's spin rate but also the fact that
T Tauri stars seem to be rotating much faster than the
main sequence stars into which they evolve. Apparently
the T Tauri stars will also be slowed by magnetic braking,
which occurs between the T Tauri stage (as the nebula begins to clear) and the final shedding of the nebula.

First Planetary Material:
Evolution of Dust in the Solar Nebula

The next steps in planet formation involved the condensation of dust as the nebular gases cooled.

The Condensation Process

Suggestions were made as early as the 1940s and 1950s
that solid microscopic mineral grains would condense out
of such a nebula. Pioneering work in this area was done by

the physical chemist and Nobel laureate Harold C. Urey (1952), who discussed the condensation of grains. He showed from meteorite chemistries that the meteorites and planets must have formed from individual solid grains at temperatures as low as a few hundred Kelvin rather than being directly formed from gravitationally bound, incandescent masses of gas as had once been supposed. Urey was virtually the only physical chemist working on planetary science at the time, and a decade passed before his ideas were fully absorbed and a new generation of chemistry-oriented planetary scientists began to make great strides.

The condensation process is simple enough. If a hot gas contains a condensable substance and starts cooling, eventually a condensation temperature will be reached at which the condensable substances change from gaseous form to liquid or solid form—the latter being microscopic dust grains. This is a familiar phenomenon: we know that raindrops or snowflakes appear when water vapor condenses in air that is lifted to cool regions of the atmosphere.

In the 1960s, Wood (1963), Lord (1965), Larimer (1967), Larimer and Anders (1967), and Blander and Katz (1967) began to sketch out the sequence of mineral types that must have condensed in the cooling solar nebula. In the 1970s, planetary chemists such as Lewis (1972a, 1972b, 1974), Grossman (1973, 1977), and Goettel and Barshay (1978) brought this work to fruition with models that explained in surprisingly simple terms the gross chemical and mineralogical properties of the solar system. Actual experiments have confirmed the condensation of microscopic smoke particles, such as Si_2O_3, from silicon monoxide (SiO) vapor (Day and Donn, 1978).

The published models determine the concentrations of different components at a series of temperatures as the nebula cooled and as various gases became depleted by condensation of solid grains. The results are seen in Figure 5-4, which shows the history of condensation as the various elements change from gaseous to solid form. The figure shows only which gases disappeared, not what minerals resulted.

The composition of the mineral dust depends on the cooling rate and the degree to which the early grains remain available to interact with the later material. Fortunately, however, the sequence of events is fairly predictable. Figure 5-5 gives an example, based on Lewis's (1972a, 1972b) calculations and his assumption that the dust and gas remain in **chemical equilibrium** with each other—that is, that the minerals present are in equilibrium with the gas and other minerals for the given temperature and pressure. At this point, or within about a million years, the scene in the solar nebula would have resembled Figure 5-6, with solid bodies beginning to aggregate out of the dust grains, and the sun barely visible through the thinning dust. The following sections describe these events in more detail, starting with the initial high-temperature conditions and then following the condensates as the temperature dropped.

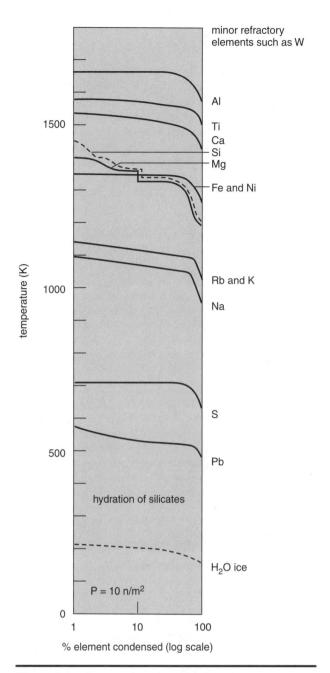

Figure 5-4. Condensation of elements from gas in the solar nebula as temperature drops, calculated by Grossman (1975) at $P = 0.1$ mbar. Graph does not indicate resulting mineral forms, except to note that water vapor begins to hydrate the silicate minerals below about 500 K and freezes out as ice at about 200 K (Lewis, 1972a). Curves for silicon and water, materials of special planetary relevance, are dashed.

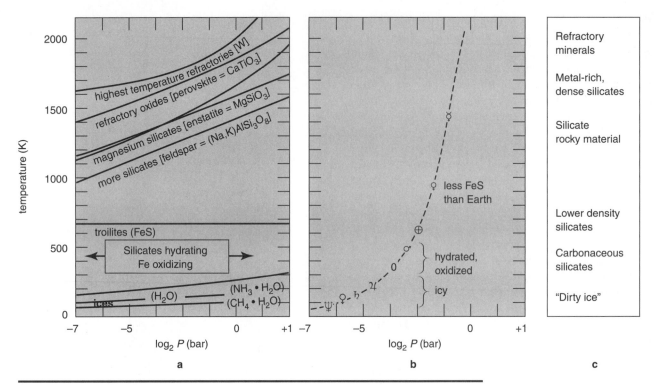

Figure 5-5. (**a**) The types of minerals condensing as the temperature declines, for different pressures. Curves indicate locations for various mineral groups, and brackets give most prominent examples. (**b**) The dashed curve gives a plausible adiabatic distribution of temperature and pressure in the solar nebula at a given moment, with the relative positions of the planets given. If the gas were suddenly blown away, the mineral grains remaining would have compositions given in part (a), which approximately agree with observed planetary properties. (**c**) The bulk composition of solid material that would dominate at various temperatures. (Adapted from Lewis, 1972a, 1972b)

Condensation of the High-Temperature Refractories

Refractory compounds are the last to melt during a heating process and the first to condense during cooling, at around 1,500 K. Extreme refractories include relatively rare elements such as tungsten, osmium, and zirconium, which appeared among the earliest crystals condensed. More abundant amounts of refractories appeared as oxides of calcium, titanium, and aluminum.

Proof of the early condensation of refractories came in 1969 from nuggets of refractory material found in a primitive type of meteorite called a carbonaceous chondrite (see Chapter 6 for more details on meteorite types). **Carbonaceous chondrites** are carbon-rich meteorites that, after being formed, were never strongly heated, thus preserving mineralogy from the days of their formation 4.6 Gy ago. On February 8, 1969, several tons of such a meteorite fell at Allende, Mexico. Samples were rushed to numerous labs, which were just gearing up to analyze the first lunar samples, due later that year. As visible in Figure 5-7, about 5% to 10% of the Allende meteorite consists of centimeter-scale, irregular inclusions of whitish minerals, which turned out to be calcium-, titanium-, and aluminum-rich minerals such as spinel ($MgAl_2O_4$) and perovskite

($CaTiO_3$). These **Allende inclusions** are believed to be actual pieces of the first material condensed in the solar nebula. They contain little iron and they condensed at around 1,625 to 1,125 K as indicated by their mineral content (Grossman, 1973; Haggarty, 1978). Apparently the Allende meteorite itself formed at a relatively late date but trapped a range of much earlier refractory particles that were floating around in the nebula. This accounts well for item 6 in Table 5-2.

A problem with the inclusions is explaining how material that solidified at high temperature got trapped in the carbonaceous material, which formed at quite low temperature. Some models in the 1990s suggest that the refractory nuggets, after forming in the inner nebula near the sun, got caught in the bipolar outflow and were shot "upward" out of the inner disk, only to fall back into colder outer regions where carbonaceous material dominated.

Major Condensates: Nickel-Iron and Silicates

At about 1,400 K more abundant materials began to condense (see Figures 5-4 and 5-5). Iron and a fair amount of nickel condensed to form an alloy, and magnesium

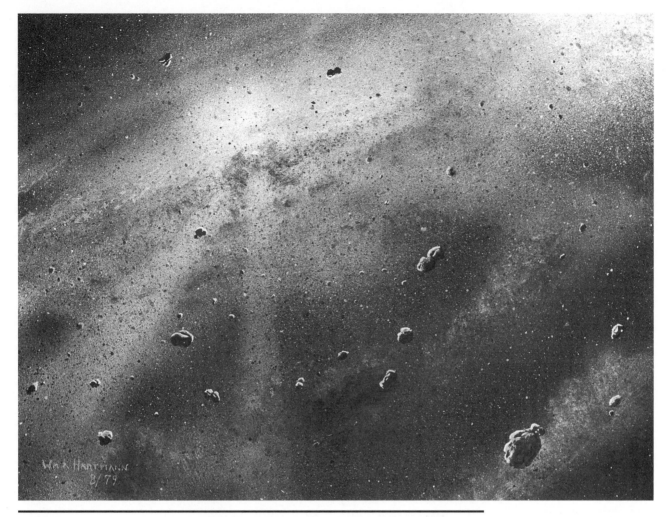

Figure 5-6. A scene within the solar nebula within a few million years of its formation. Mineral dust grains are aggregating into asteroid-sized planetesimals. (Painting by the author)

silicates (very common rock-forming minerals on Earth) condensed as rocky particles. Because these materials are more abundant than the refractories, the dust existing at nebular temperatures around 1,300 to 1,400 K became mostly a mixture of nickel-iron alloy grains and rocky particles. One of the most common minerals that condensed at this temperature was **enstatite** ($MgSiO_3$).

As the temperature continued to drop, sodium and potassium also formed silicate minerals, especially the feldspars. **Feldspars** are familiar aluminum-silicate minerals making up as much as 60% of Earth's crust and having formulas such as $NaAlSi_3O_8$, $KAlSi_3O_8$.

Also, as the temperature dropped, chemical reactions probably occurred between the gas and the dust. In particular, the iron metal began to be oxidized to other mineral forms. Some iron reacted with hydrogen sulfide gas (H_2S) to form the mineral **troilite** (iron sulfide, FeS). Other iron reacted with oxygen to form ferrous oxide (FeO), which in turn reacted with enstatite to form the important min-

eral **olivine** (a mixture of Fe_2SiO_4 and Mg_2SiO_4). These reactions were relatively complete by the time temperatures reached 500 K, when the dust probably consisted of olivine, feldspars, and other silicate and oxidized iron minerals.

The types of primitive 4.6-Gy-old meteorites that were little altered by heating actually confirm these chemical models. One type of meteorite, the enstatite chondrite, consists of nearly pure enstatite and 19% to 25% by weight of nickel-iron, as predicted for the dust in the 1,300-K nebula (Mason, 1962, p. 71; Wasson, 1974, p. 27). These enstatite-rich objects may be objects whose formation ended at about 1,300 K; they amount to only about 2% of the meteorite population.

Most of the remaining meteorites, called ordinary chondrites and amounting to about 78% of meteorite falls, are rich in olivine, contain troilite, and have about 25% by weight of nickel-iron in the form of pure nickel-iron alloy and oxidized iron minerals. They seem to reflect the basic

Figure 5-7. A fragment of the Allende carbonaceous chondrite meteorite, showing the prominent whitish inclusions of refractory minerals, incorporated as the meteorite accreted. (Photo courtesy of R. S. Clarke, Smithsonian Institution)

process of condensing silicate minerals in our part of the solar system; they have not been drastically altered (by melting, for example) since their formation.

The Carbonaceous Condensates

In more distant parts of the nebula, where the temperature dropped still lower, massive amounts of so-called carbonaceous material formed; it is characterized by carbon in the form of black graphitelike material mixed with organic molecules and ordinary silicate minerals. An important type of meteorite, called a carbonaceous chondrite (see Chapter 6), consists of this material (see Figure 5-7, which shows the black matrix material). Smith and Buseck (1981) found the black matrix material to be rich in poorly crystallized graphite with grain sizes of only 1 to 100 nm. The carbon and organic molecules are not organized in well-defined mineral crystals but in poorly organized organic matter sometimes called **kerogens.** (This name came from the petroleum industry and refers to sludges of organic rich matter left after petroleum products are distilled from oil shales.) Partly because the carbon and carbon compounds are not found in well-defined mineral crystals, they are hard to study; their exact mode of formation is somewhat uncertain. Chemical models suggest that the carbon grains and organics may not condense in large amounts directly from the nebula. Instead, some cosmochemists have suggested that the carbon and kerogens are products of reactions in other carbon-bearing material. For example, carbon dioxide ice (CO_2) and methane ice (CH_4), when hit by cosmic rays, may produce carbon and organics. In other words, the carbonaceous material may have been synthesized in icy grains and planetesimals rather than condensed directly.

Telescopic observations reveal what appears to be this type of dark material—in various shades from black to dark reddish-brown—on the surfaces of all asteroids and comets from the outermost asteroid belt outward to the Kuiper belt. Theoretical calculations by Consolmagno and Lewis (1977) suggest that planetesimals in Jupiter's region might have had initial compositions of about half carbonaceous material and half ice. Temperatures in the region inside the middle of the asteroid belt may never have dropped low enough to have allowed production of abundant carbonaceous minerals.

More Major Condensates: The Ices

Below 500 K, water played an increasingly important role. Between 500 and 200 K, water vapor in the gas reacted with some of the dust grains to form complicated hydrated minerals, such as serpentine ($Mg_6Si_4O_{10}[OH]_8$, formed when olivine reacts with water vapor), tremolite ($Ca_2Mg_5Si_8O_{22}[OH]_2$), and talc ($Mg_3Si_4O_{10}[OH]_2$).

In colder regions at around 200 K, many minerals were hydrated and water began to appear in the form of H_2O ice crystals. In even colder regions, in the outermost parts of the solar system, ammonia (NH_3) and methane (CH_4) also condensed as ices. Probably they were mixed with water, forming hydrated ices such as ammonia hydrate ($NH_3 \cdot H_2O$) and methane hydrate ($CH_4 \cdot 8H_2O$) (Lewis, 1972a).

These ices were so abundant that they swamped the rocky materials that had condensed earlier in these regions. Note that the elements involved, H, O, C, and N, are numbers 1, 3, 4, and 5 in abundance, respectively (see Table 4-1). (Number 2 is helium, an inert gas that does not readily form compounds.) There was more than enough hydrogen to combine with all available O, C, and N, and the amount of the O, C, and N was about 8 times that of the Si, S, Fe, and Mg of the silicate minerals. Thus, planetesimals formed in the coldest regions were mixtures of ice somewhat dirtied by trapped dust grains.

Where were ices stable in the early solar system? As long as a significant fraction of the planetary material existed as subcentimeter dust, the nebular cloud was opaque and shielded from direct sunlight but may have been warmed by infrared radiation within the cloud. Once the dust grains clumped together into meter-dimension bodies, the nebula began to clear (Hartmann, 1970) and sunlight sublimed any unshielded ice crystals in the inner regions. Under such conditions, as at present, H_2O ice is stable only beyond the asteroid belt (Watson, Murray, and Brown, 1963; Urey, 1952).

Observations support these theoretical considerations. The black matrix material of the most primitive, carbonaceous chondrites such as Allende consists of low-temperature hydrated minerals containing as much as 22% bulk water content. The water is chemically bound in the minerals (not in the form of ice) but can be easily driven off by mild heating to a few hundred degrees C. This might represent the form of carbonaceous materials formed in outer-belt asteroids that also included some ice; the minerals may have become hydrated when these bodies heated, melting the ice and allowing moisture to percolate through the body.

From Jupiter's orbit outward, ice-covered satellites, such as Europa, Enceladus, Rhea, Titania, and Saturn's ring particles, are common. Cometary nuclei, from the same region, are ice-rich chunks containing black carbonaceous grains.

Gross Compositions of the Planets

Lewis (1972b) pointed out an interesting consequence of the condensation theory. Recall from the last chapter that during the T-Tauri stage, roughly 10 My after the star forms, the nebular gas is swept away, leaving the dust behind. At a distance from the sun corresponding to each planetary orbit, the dust would have the bulk composition of condensates at temperatures corresponding to that distance. Refractory and metal-rich dust would lie close to the sun, near Mercury's orbit. More silica-rich, lower-density dust would be farther from the sun, in the terrestrial planet zone. Low-density ices and associated carbonaceous materials would lie farther from the sun.

These ideas can be quantified by theory. Given a temperature at one point, solar nebula models give the temperature at other points, which fall along an **adiabatic curve,** or temperature-pressure relation, affected only by the gas and the sunlight. Figure 5-5b shows such a curve and the relative positions of planets on it. The result is an amazingly good first-order match to planetary compositions, indicated by Figure 5-5c. It shows that if the gas cleared at the moment when the Mercury material was in the refractory- and metal-rich part of the diagram, then Venus and Earth would be silica rich with lower densities, as observed. Mars would have a still lower density, as observed, being rich in oxidized iron compounds (the rust-like minerals that color Mars's surface red). The outer planets would be rich in ices and have low densities, also as observed.

The first-order explanation of planet compositions accounts for item 7 in Table 5-2. Because meteorites are mostly fragments of interiors or bodies formed in the region of the asteroid belt and elsewhere, it also accounts for item 8 in the table.

Chemical Complexities

Nature is rarely as simple as a satisfying model. The solar nebula is no exception. Although temperature no doubt declined with increasing distance from the sun, producing different minerals in different zones, there may have been some turbulence in the nebula that moved material toward or away from the sun and also "up" and "down" out of the midplane to regions of lower pressure and different temperature. One example would be the redistribution of nuggets of refractory material to produce the calcium-rich inclusions in carbonaceous meteorites, as mentioned earlier. Also, giant planets' gravities may have widely scattered the volatile-rich carbonaceous planetesimals that formed beyond the outer asteroid belt. Numerous theorists require an admixture of carbonaceous material to explain the composition of Mars and other bodies (Pepin, 1991; McSween, 1989). Such mixing would complicate Lewis's simple picture. A current research problem is whether planet compositions are seriously affected by such mixing.

On the other hand, the total bulk mixing must have been limited because each zone has preserved a unique composition, as noted above. Even within the asteroid belt, such distinct zones exist, from light-toned rocky asteroids in the inner belt to black carbonaceous ones in the outer belt. Wood (1985) notes that inclusions have different properties in different meteorite types, showing that even such small bodies did not mix throughout nebula before being caught up in larger planetesimals; they stayed in their own zones. Finally, a crucial discovery (Clayton, Grossman, and Mayeda, 1973) is that materials from different zones, such as Earth/moon rocks, rocky meteorites from the inner asteroid belt, and carbonaceous meteorites from more distant regions, have distinctly different ratios of oxygen isotopes O^{16}/O^{18}, showing that the nebular gas did not mix uniformly throughout the nebula.

The rate of accretion versus the rate of nebular cooling is also critical to planet makeup. At one extreme, if accretion happened faster than cooling, early refractories would accrete into metal-rich bodies before later silicate and carbonaceous dust disappeared. Planets might have started with at least partially formed iron cores (Slattery, 1978). This is often called the **heterogeneous accretion model.** Alternatively, in the **chemical equilibrium model,** accretion was slower than cooling, and the dust remained dispersed through the nebular gas, reacting with it as new components condensed. Most theorists lean toward some variant of the latter model, with the iron cores forming by the melting or partial melting of the planet interior, almost as fast as the planet formed.

Further Evolution of the Solar Nebula Dust

The dust grains must have aggregated. If they hadn't, Earth wouldn't be here and we wouldn't be here, and the dust would have been blown outward into interstellar space or perhaps would have remained behind as a set of rings like Saturn's or as a belt of miniasteroids. The question is, how did the dust grains aggregate?

This issue aggravated researchers for years. Kerridge and Vedder (1972) designed an experiment with silicate particles hitting each other at speeds of 1.5 to 9.5 km/s (typical of collisions in today's asteroid belt) to test whether any sticking or impact welding occurred. They found none; the particles shattered. Other workers hypothesized electrostatic processes or sticky coatings of organic molecules to explain aggregation.

The dynamics of these particles as they moved around the sun suggest a different approach. Collisions and gas drag would quickly damp them into nearly circular orbits. If the orbits were precisely circular, there would be virtually no high-speed collisions because the orbits would not intersect.

At least three effects would cause collisions at low speeds. According to Kepler's laws, particles orbiting closer to the sun move faster than those further away, so that inner particles would catch up to outer particles; this effect is called Keplerian shear. Second, gas drag effects between particles and the nebular gas would cause particles of different sizes to move at different speeds relative to the gas. For example, among bodies 0.1 mm to 10 m across, Weidenschilling (1980) described drag effects produced by the nebular gas, giving relative velocities of only a few centimeters to a few meters per second. Greenberg and others (1978b) suggest collisions as slow as 0.01 to 1 cm/s for the initial microscopic dust particles, about 1 μm across. Third, particles passing near each other would perturb each other's orbits gravitationally by amounts proportional to the masses of the particles. Thus, orbits would become eccentric and collisions would occur. However, even if two particles in Earth's orbit developed eccentricities equaling Earth's current eccentricity ($e = 0.017$), their collision would occur at only about 500 m/s, only a tenth the speed of modern asteroid collisions (see formulation by Safronov, 1972, p. 69).

Weidenschilling, Wetherill, and their co-workers, from the 1970s onward, made extensive studies of the collisional evolution of such systems of particles. For example, Hartmann (1978) fired rocky bodies together and measured the collision speed marking the transition between when they simply bounce apart and when they shatter; the transition speed is only 2 m/s for weak objects like carbonaceous chondrites with the consistency of dirt clods, but closer to 40 m/s for intact rock. The early dust grains nestled up to each other like snowflakes falling out of a snowstorm. As with snowflakes, the grains probably clustered together into fluffy clumps—the first aggregated planetary matter. Forces tending to hold them together could have been **electrostatic forces** such as Van de Waals forces (Weidenschilling, 1980). Actual evidence of the proposed fluffy structure comes from microscopic meteoritic particles collected in space and in the uppermost atmosphere.

Further direct evidence that planets grew from interactions of innumerable small particles comes from the general properties of the solar system. Regularities such as nearly coplanar, circular orbits, prograde revolution, and the usual prograde rotation all suggest that final planetary properties derived from statistical averaging of properties of countless constituent particles. This was first empha-

sized as early as the 1940s by the Russian planetary researcher O. Y. Schmidt and his students, even while Western researchers pursued other types of theories involving semi-independent, massive protoplanets. The Russian followers of Schmidt made many of the pioneering studies of the dynamical evolution of swarms of particles orbiting around the sun or planets. These studies were little noticed in the West until V. S. Safronov published a summary of the subject, which reached the West in 1972 as the product of an Israeli translation program. This book has had a major impact on subsequent investigations of the planets' early evolution.

As the dust aggregations grew larger into full-fledged planetesimals, they continued to perturb each other during near collisions. They were now large enough not to be affected by gas drag; they were decoupled from the gas and they approached each other faster than before. Calculations by Safronov (1972) and Greenberg and others (1978a) demonstrated the important rule of thumb that the approach velocities of the orbiting planetesimals tend to become equal to the escape velocities of the larger, but not the largest, bodies in the swarm, because these are large and abundant enough to have strong, perturbing gravitational effects. The large bodies become so-called "gravitational spoons," stirring the smaller bodies in the cosmic mixing bowl of the solar nebula.

By numerical coincidence, the escape velocity (V_{esc}) of a planetesimal in meters per second is about numerically equal to the radius of the planetesimal in kilometers.* For example, a 1-km-radius planetesimal would have an escape velocity around 1 m/s. Mars's satellite Phobos, at mean radius 12 km, has $V_{esc} \approx 12$ m/s, the speed of a fast baseball pitch. So the Safronov rule of thumb is that if the planetesimals have grown to a meter in size, they interact at speeds around 10^{-3} m/s; if they've reached 1 km, around 1 m/s; and if 10 km or so, around 10 m/s.

Collisional Accretion of Subkilometer Planetesimals?

Combining Safronov-descended orbital considerations with the mechanical measurements of the collision mechanics Stuart Weidenschilling, George Wetherill, Glen Stewart, and other researchers have created numerical models of the evolution of innumerable small bodies orbiting the sun; these models show the rapid emergence of asteroid-sized and moon-sized planetesimals from initial smaller bodies, as seen in Figures 5-6 and 5-8 (Spaute and others, 1992).

The larger objects in a planetesimal swarm would be hit by smaller objects that rebound too slowly to escape, fall back onto the surface, and add to the mass of the larger body. Furthermore, the granular layers thus accumulated

---

*This is exactly true if the density of the planetesimal is 1,790 kg/m^3, typical for low-density rock-ice mixtures.

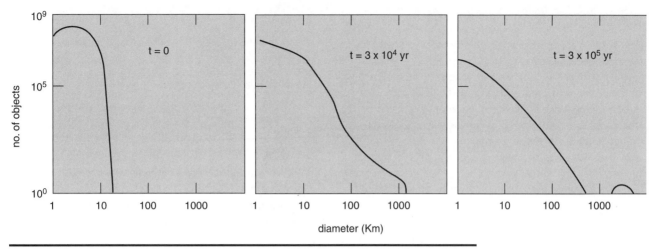

Figure 5-8. Simulation of planetesimal growth in the solar nebula. The computer model, developed at the Planetary Science Institute, follows collisional accretion of bodies orbiting the sun using realistic orbital and collisional mechanics. In the first step, the size distribution in the zone of the terrestrial planets is assumed to be peaked toward bodies with diameters of a few kilometers, perhaps formed by gravitational collapse. After 30,000 years, larger bodies are emerging. After 300,000 years, runaway growth has produced a dominant body some 2,000 km across, and collisional fragmentation has broken smaller bodies, producing a tail of small fragments. (After Spaute and others, 1992)

would slowly rebound and aid the efficiency of this process, making growth even more rapid.

This method of planetesimal growth, with bodies striking each other one at a time in random collisions and often sticking together, is called **collisional accretion.** The computer models indicate that primitive dust grains or aggregates of grains could rapidly accumulate into bodies hundreds of kilometers across by collisional processes alone. The timescales for this stage of growth are very short, probably less than a million years. Because the larger bodies, with their stronger gravities, could grow faster than smaller bodies, there was a tendency toward *runaway growth,* in which large bodies emerged rapidly and swept up the smaller debris.

Gravitational Collapse of Kilometer-Scale Planetesimals?

Collisional accretion is not the only way to get planetesimals of kilometer scale or even larger. Another way is by **gravitational collapse,** which was illustrated in Figure 4-3. Gravitational collapse might seem a promising way to make a planet in one fell swoop, without the tedious business of making it one collision at a time, rock upon rock. You might simply imagine a whole piece of the solar nebula becoming dense enough to become gravitationally unstable and then collapsing at once into a planet-sized object.

There are two things wrong with this scenario. First, the nebular gas and dust mix is unlikely ever to have become dense enough to make bodies as small as Earth or the parent bodies of meteorites by one-stage collapse of the gas and dust mixture (see Figure 4-3). Second, geochemists have long realized (Urey, 1952) that Earth did not form this way, because Earth lacks a full cosmic complement of the inert gases such as neon, argon, krypton, and xenon. Current data show that in Earth, Venus, Mars, the moon, and meteorites, these gases have no more than 10^{-7} to 10^{-10} of their normal cosmic abundances (Taylor, 1975; Kaula, 1968, p. 377). The gases are heavy and chemically inert, so that if they had ever been gravitationally trapped in Earth's material, they would not have floated off into space (as most of the light hydrogen has done) or have been chemically tied into mineral compounds hidden inside Earth. Therefore, if the terrestrial planets had formed by gravitational collapse of part of the solar nebula, neon and the other inert gases would have been brought along with everything else in their normal cosmic complement and would be prominent in the atmosphere today.

Thus, terrestrial planets did not form by simple gravitational collapse of the gas and dust solar nebula.

But what about the dust alone? Important studies by Gurevich and Lebedinskii (1950), Safronov (1972), and Goldreich and Ward (1973)* showed that the freshly condensed dust grains and aggregates, while orbiting around

---

*The Cal Tech researchers Goldreich and Ward began their work without knowledge of the earlier Russian results and independently derived a quite similar model. At about the time of publication, they became aware of Safronov's (1972) book and credited the Russian work in a note added at the end of their paper—an interesting example of the independent confirmation of scientific ideas.

If two bodies are approaching each other from a great distance apart at a speed V_{app}, at what speed V_{imp} will their actual impact occur? Suppose a small body or meteorite m is colliding with a much larger body M. (This is the usual situation, since a wide distribution of sizes makes the collision of two equal-sized bodies relatively rare.) As they approach, the small body accelerates due to the gravity of M (and vice versa, to a negligible extent, since m is much smaller). The collision thus occurs at $V_{imp} > V_{app}$. It is easy to derive V_{imp} in terms of other knowns by the conservation of energy:

$$E_{initial} = E_{final}$$

Total energy E is kinetic energy plus gravitational potential energy. Thus,

$$\frac{1}{2}mV_{app}^2 - \frac{GMm}{\infty} = \frac{1}{2}mV_{imp}^2 - \frac{GMm}{R^2}$$

After cancelling m's, we have

$$V_{imp}^2 = V_{app}^2 + \frac{2GM}{R}$$

$$V_{imp}^2 = V_{app}^2 + V_{esc}^2$$

In other words, the square of the final impact velocity is just the sum of the squares of the initial approach velocity and the escape velocity of the larger body. If the approach is so slow that $V_{app} \ll V_{esc}$, the meteorite will fall in at escape velocity. If the approach is so fast that $V_{app} \gg V_{esc}$, then the meteorite will slam in at its approach velocity without ever "feeling" the gravity of the larger body.

---

the sun, also tended to settle through the gas toward the ecliptic plane, being attracted not only by the sun but also by the disk-shaped nebula. **Stoke's law,** which gives the maximum velocity at which a particle can move through a gas under the influence of a given force, indicates that the formation of a relatively thin, dense layer of dust in the ecliptic plane was rapid. Safronov (1972, pp. 26–27) estimates that micrometer- and submicrometer-scale particles settled to the plane in only about 10^3 to 10^5 y; larger particles fell faster. Goldreich and Ward (1973) found that early condensates such as iron may have grown to a few centimeters just by further chemical condensation on their surfaces during their fall to the plane (a source for Allende inclusions?).

Once the dust layer formed, the density of material in the dust layer was considerably more than the density in the earlier nebular gas. Gravitational instability may have occurred within this dust layer, forming planetesimals with diameters of a few kilometers. Aggregation velocities were very low, so that individual grains, aggregations of grains, and centimeter-scale chunks of early condensates gently accumulated into kilometer-scale bodies.

Goldreich and Ward (1973) concluded that several stages of gravitational collapse occurred. The first formed planetesimals with diameters up to about 2 km, directly as a result of collapse of the primeval dust. This layer of kilometer-scale bodies was also gravitationally unstable and aggregated into a second generation of 10-km-scale planetesimals—clusters of first-generation bodies that resembled asteroids. For these reasons, many of the computer simulations of planetesimal growth, such as Figure 5-8, start with objects a few kilometers in diameter. From the point of view of an observer riding on an early dust grain, however, these processes were all mixed up. This observer would have just seen grains and clumps of material getting closer and closer, with collisions, electrostatic attraction, and gravitational attraction all mixed up.

What matters is that the dust aggregated into asteroid-like planetesimals and that aggregation continued toward even larger sizes.

Collisional Accretion of Full-Scale Terrestrial Planets

In the final stages of growth to planetary size, collisional accretion was probably the dominant mechanism, at least among terrestrial planets and the asteroids (Greenberg and others, 1978; Wetherill, 1981; Spaute and others, 1992). The runaway growth of the larger bodies probably helped ensure that one large planet dominated each zone of the solar system.

Evidence supporting this theory is found in the current asteroid belt. The largest asteroid, Ceres, is about 1,000 km across, the next three bodies are about 500 km across, and a host of smaller fragments exists. The size distribution is similar to that calculated by the models as seen in Figure 5-8 (right), where a 2,000 km body has pulled away from the others in that zone of the solar system after only 0.3 My.

The asteroid belt is thus a fossil tableau of the early colliding planetesimals. Today, in the belt, collision velocities are high enough to break up many of the smaller colliding pairs. As seen in Figure 5-9, striking testimony of the collisional history of planetesimals is offered by certain asteroids and satellites with craters representing impacts almost big enough to have shattered them.

Growth to the full-fledged planetary state apparently slowed as the largest planetary bodies in each zone began to sweep up the remaining debris, leaving one large body dominating most zones. Isaacman and Sagan, as early as

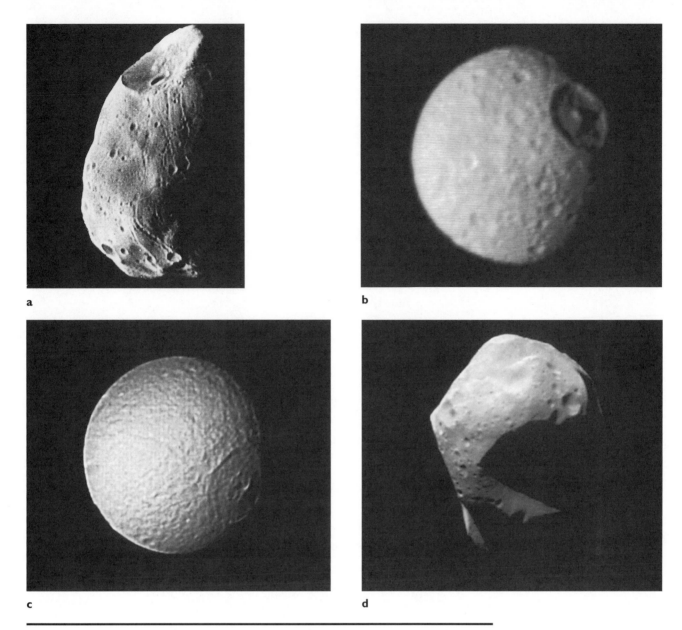

Figure 5-9. Many small bodies of the solar system have suffered impacts nearly big enough to disrupt them. (**a**) Mars's satellite Phobos has a crater Stickney (top) from which radiate fracturelike grooves. The crater diameter is 37% of the 27-km-long axis of the satellite. (**b**) Saturn's 394-km moon, Mimas, has a crater about 34% as big as the satellite. (**c**) The nearby 1,048-km moon, Tethys, has a similar crater about 40% as large as the satellite. (**d**) Asteroid 253 Mathilda, which is 59 × 47 km, has two large craters. The one on the left edge is about 62% of the asteroid's mean diameter. (NASA Viking, Voyager, and NEAR photos)

1977, simulated this growth in a simplified computer model. Instead of using specific experimental collision data, they merely assumed that each collision resulted in particles sticking together. The result was a variety of planetary systems, as seen in Figure 5-10. These included many systems resembling the solar system, with some massive Jupiter-like planets, some Earth-sized planets, and often some leftover asteroidlike bodies. The competition to sweep up the planetesimals leads to a spacing of the large planets in Bode's-rule-like zones, helping to explain item 9

in Table 5-2. Isaacman and Sagan stressed that similar results came from a variety of starting conditions, supporting the idea that planet systems have formed elsewhere.

Models of the final stages of accretion have been made by George Wetherill (1990). He suggests that runaway growth of the largest bodies produced a host of Mars-sized planetesimals in the terrestrial zone, spaced about 0.01 to 0.02 AU apart, within as little as 100,000 y. Further collisional accretion among these eventually produced the terrestrial planets.

THE FORMATION OF PLANETS AND SATELLITES

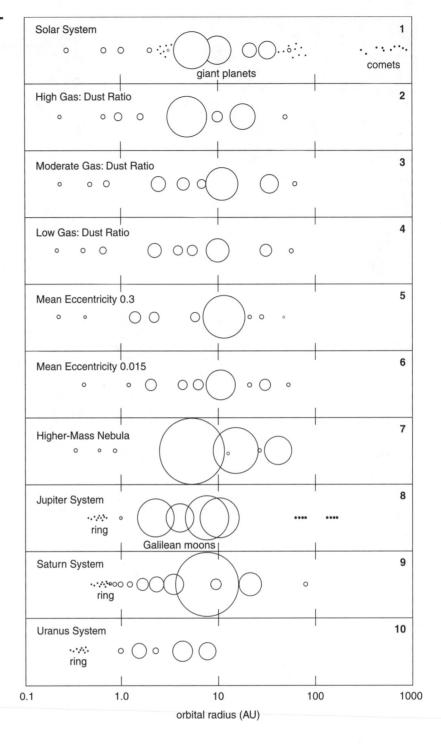

Figure 5-10. Schematic representation of the distribution of bodies in various real and theoretical planetary systems. Orbit scale is given at bottom, and circles represent relative sizes of planets (on a scale different from the orbit scale). Plots 1–7 are after Isaacman and Sagan (1977); plots 2–7 are systems computed by a theoretical model of planetesimals' collisional accretion. Plots 8–10 show satellite systems of three giant planets by a similar schematic method, with one of the inner satellites arbitrarily chosen to represent "Earth." All systems show a tendency to form several bodies, with "giant planets" occupying intermediate orbits.

Forming the Giant Planets

In the outer solar system, ices supplemented the planetesimal mass supply, and still larger bodies formed. According to "traditional models" from the 1980s and 1990s, the process combined collisional accretion and gravitational collapse. Once rocky/ice planetary bodies grew to 10 to 20 M_{Earth}, they had such strong gravity that they began to pull in large amounts of gas directly from the solar nebula,

forming a giant planet within 10^7 years (Mizuno, 1980; Kortenkamp and others, 2001). In this scenario, the bulk composition of Jupiter might have an intermediate content of inert gases, though the atmosphere, added last, might have the solar complement of inert gases.

More recent work (e.g., Kortenkamp and others) revived an earlier idea that giant planets formed directly and rapidly by gravitational collapse of the solar nebula gas. In this view, Jupiter-scale masses could form all at once (in timescales as short as 10^2 years!), before the terrestrial planets grew by conventional accretion. This idea might help explain properties of observed extra-solar planets. The choice between forming giant planets by accretion followed by gravitational collapse, and direct one-step gravitational collapse, is an ongoing area of active research.

Sweep-Up of the Last Planetesimals

Eventually, planet growth reached the stage at which the planets had most of their present mass and were sweeping up the last planetesimals. Each collision with a planetesimal left a crater on the planet's surface, accounting for item 10 in Table 5-2.

Visible craters on the terrestrial planets date back at least 4 Gy and record the sweep-up of interplanetary bodies ranging in diameter up to roughly 200 km. The size distribution of the craters is similar to that of present-day asteroids, and the number of craters corresponds to roughly half the number of asteroids (Hartmann, 1977).

From these statistics of planetesimal sizes, even larger impacts can be inferred for the period before 4 Gy ago. Safronov (1966) pointed out that these large impacts affected planetary obliquities. An important feature of the accretion theory is that statistical averaging of the impact effects of innumerable small planetesimals during planet growth tended to produce planets rotating prograde with periods of about 5 to 20 h (a few times slower than the limiting case of rotational instability, in which material would be thrown off the equator) and with rotation axes nearly perpendicular to the ecliptic (Harris, 1977).

A growing planet was only the largest planetesimal in its local swarm, and eventually it had to interact with the 2nd-, 3rd-, and 4th- . . . nth-largest planetesimals. These largest impacts were statistical flukes, and some of them may have been major. Their effects did not necessarily average out. For example, Safronov concluded that the odd obliquity of Uranus came from the impact of a mass about 0.05 that of Uranus itself. Hartmann and Davis (1975) and Wetherill (1976, 1990) calculated that the second-largest bodies in the terrestrial zone grew as large as some 4,000 to 6,000 km across. Hartmann and Vail (1986) concluded that statistics of planetary obliquities and inclinations were consistent with secondary late impactors, typically growing to 0.3% to 20% of the mass of the planet itself (about 15% to 60% of the planet's radius). The details of the sweep-up process thus may account for items 11 and 12 in Table 5-2.

The outer planets were surrounded by many icy planetesimals. Some of them were left stranded in the present region of Neptune; they form what is called the Kuiper belt, analogous to the asteroid belt that was left between Mars and Jupiter. During near-miss encounters between the planets and planetesimals, the strong gravity of the giant planets was adequate to throw many other icy bodies nearly out of the solar system. Hence, they accumulated in a swarm with enormous semimajor axes, spending most of their time on the outskirts of the solar system. This accounts for item 13 on our list.

Ring Systems as Clues to Planetary Origin

Chapter 2 described the discoveries of ring systems around the giant planets. Chapter 3 described some of the dynamical processes controlling the structure of these ring systems. What of the **origins of ring systems?** Are ring systems primordial structures left over from planet formation, or are they more recent or transient?

Although Saturn's ring particles are fairly pure and bright-colored pebbles of water ice, the ring particles of the system of Jupiter, Uranus, and Neptune are much darker in color. The infrared spectra of those systems are inconsistent with clean ice but match that of dark soil (e.g., Neugebauer and others, 1981). The particles may be rocky or carbonaceous.

The differences between Saturn and the other three systems may arise from the source and recent history of the ring particles. Saturn's rings may have formed from the breakup of a modest-sized ice moonlet, hit by an asteroid or comet. The fragments would be relatively pure ice from the interior of the body. The other ring systems seem to consist of dark particles knocked off the dark surfaces of tiny moonlets by more steady-state micrometeorite erosion. The surface layers of such moonlets may have dark material because they are ice-depleted by micro-meteoroid impacts that preferentially vaporize the ices.

To take an example, the dense, outer edge of Jupiter's ring (Figure 3-14) lies just at or slightly inside the orbit of the innermost known moon, Metis, and the ring particles are believed to be μm-scale particles knocked off Metis by impacts. Particles can easily be knocked off Metis into Jupiter's ring; velocities of only about 4 to 10 m/s are needed (Burns, 1980). The albedos of Metis and neighboring satellites are believed to be 0.04 to 0.10, implying coloring by dark, carbon-rich soil. Jewitt and Danielson (1981) calculated a lifetime of 250,000 y for a 5-μm diameter particle to spiral from the satellite orbit into Jupiter. The Jupiter ring is thus analogous to a river, which may always be visible but consists of different water at different times. Just as a river may occasionally flood, a major impact on Metis might send a mass of particles through the ring; the ring might have had ancient episodes of greater prominence. The current total material in the ring would make a ball only 30 m across, and the estimated total material cycled through the ring since the solar system's origin would make a satellite only a few kilometers across (Jewitt and Goldreich, 1980; Burns, 1980, 1990).

Similarly, in the case of Uranus's rings, Goldreich and Tremaine (1979) found orbital decay of typical ring particles by Poynting-Robertson and collisional effects in only 10^7 to 10^8 y unless the rings are stabilized by small unseen satellites of diameter around 20 km. Further, Uranus's ϵ, or outer, ring is apparently slightly elliptical or inclined to the other rings, suggesting that the Uranus system may be a collisional product whose fragments have not had time to collapse to a circular disk (Millis and Wasserman, 1978; Nicholson and others, 1978).

In other words, ring systems may be relatively changeable and transient structures whose appearance today depends on their bombardment history during the last part of solar system history.

Whether such a model has relevance to Saturn's rings is problematical. Because the Poynting-Robertson lifetime increases with particle radius and decreases with the solar flux, the larger particles in Saturn's rings—of centimeter to kilometer scale—would have lifetimes longer than the solar system, although there might still be a population of μm-scale dust particles spiraling inward after being eroded off the larger particles.

In summary, ring particles have probably been provided by the breakup or erosion of other bodies, but the process operates on two different scales. In the Saturn system, a body of moderate size may have fragmented and provided the ring particles. It may have been a satellite near the Roche limit that was hit by a passing interplanetary body. Alternatively, it may have been a body that found itself inside the Roche limit and broke up spontaneously because of tidal stresses—as Comet Shoemaker-Levy 9 did when it passed too close to Jupiter. This body might have been a passing comet (Wetherill, 1976) or a satellite that spiraled in under the influence of tides, gas drag, or growth of the planet (Harris, 1978; recall that under Kepler's laws, as a planet's mass grows, its satellites' orbital radii shrink).

In the Jupiter, Uranus, and Neptune systems, the ring system may consist of small particles eroded at a more steady rate by micrometeorite impacts on the surfaces of nearby moons.

Origin of Satellites

For several reasons, it is difficult to trace the origins of satellites by any direct knowledge of their orbital histories. We cannot trace the past history of satellite orbits by contemporary celestial mechanics because theory and observational parameters are only accurate enough to trace the orbits through roughly the last 100 My—a small percentage of solar system history (Kuiper, 1956b). Thus, we cannot rigorously calculate the initial orbital properties of satellites.

Second, some regularities of satellite orbits are associated with resonances between pairs of satellites (such as Saturn's), not with initial conditions. Enceladus is in a 1:2 resonance with Dione, Mimas 1:2 with Tethys, and Titan 3:4 with Hyperion. These satellites may have "fallen into" these stable positions during the tidal evolution of their orbits (Goldreich, 1965; Dermott, 1968; Greenberg, 1977; Peale, 1978).

Third, tidal transfer of angular momentum from planets to satellites tends to force prograde satellites outward toward escape, as with the moon (Chapter 3). By the same mechanism, retrograde satellites and satellites inside the synchronous point are drawn in toward planets. (Can you confirm this from Figure 3-11?) Many more retrograde satellites may have existed long ago. Similarly, Triton will approach and crash into Neptune. Once thought imminent (within 100 My; McCord, 1968), this crash is more likely some 3,600 My in the future (Chyba and Nicholson, 1987). Phobos is expected to crash into Mars in as little as 50 My (Burns, 1990)!

Without the ability to track the satellites backward in time, we have to use other clues about composition and general orbital distributions to surmise satellite origins.

Major Prograde Satellites: Miniature Solar Systems

Satellites consist of three basic types: "normal" systems of giant planets; captured satellites; and satellite systems of the Earth-moon type, which may have formed in catastrophic collisions. The "normal" types are the families of large satellites orbiting Jupiter, Saturn, and Uranus. They all orbit in the prograde direction with low inclinations, and as shown in Figure 5-10, they seem to be miniature analogues of the solar system itself. They are symbolized by Figure 5-11.

A general consensus is that these moons formed out of miniature, disk-shaped cocoon nebulae surrounding the giant planets as they formed. This idea would explain item 14 in Table 5-2. For example, the smooth decrease in density and rock:ice ratio as one moves outward among the Galilean satellites mimics the decreasing density and metal:rock:ice ratio in the solar system, suggesting similar sequences of condensation with higher temperatures toward the central "star." The dynamics of giant planet gravitational collapse and the high luminosity of the primordial giant planets kept the central regions of their "miniature solar nebulae" hottest. Chemical studies of the condensation sequence in such nebulae have helped explain satellite compositions. For example, Figure 5-12 shows a diagram of condensation in the Jovian nebula, adapted from Consolmagno (1981). Based on an assumed model of the nebula, Figure 5-12 shows that Io and Europa, in the warm region near Jupiter, would have accumulated mostly from hydrated rocky material (perhaps like carbonaceous chondrites; later internal heating could have driven the water out to form the observed ice on Europa's surface). Ganymede's and Callisto's materials, condensing in a colder region, would have been about

Figure 5-11. The satellite Io, seen against the immense backdrop of Jupiter's clouds. Io is almost exactly the size of our moon, and a quarter the size of Earth, but only 2.5% the size of Jupiter. This size ratio is roughly the same as that for Neptune relative to the sun. (NASA, Voyager 1, March 2, 1979)

one-half water ice, mixed with rocky dust and ammonia ice. Once condensed, the dust and ice grains would aggregate into planetesimals and grow by collisional accretion into major satellites (Harris, 1978; Burns, 1978).

If this scenario is correct, the larger moons, generally in the midst of the systems—such as Ganymede, Callisto, and Titan—would be the equivalent of giant planets. Smaller, inner moons, such as Amalthea, Io, Mimas, and Miranda, would be equivalent to terrestrial planets. Fragmentation among the innermost moons, at the edge of rings, is favored because of the high impact rate of asteroids and comets being attracted to the giant planets themselves.

As mentioned above, satellite orbits can evolve because of tidal forces. A major satellite, once formed, was not necessarily safe from orbital change or even destruction. Tidal interactions may have been quite effective in removing primeval moons, either by making them spiral into the planet or by making them spiral out, escaping into solar orbit, only to collide later with the planet or be ejected from the solar system during a near miss with the planet (Burns, 1973; Ward and Reid, 1973). As discussed in Chapter 3, observed orbital resonances between the largest moon and other smaller moons might have been established as the larger moons' orbits changed under tidal forces. Harris (1978) noted that surviving satellites may be the last-formed ones, composed partly of late-arriving planetesimals with somewhat different chemistries from those that formed the parent planet. Thus, because of the orbital and chemical evolutionary complexities of satellites, discovering their origin by direct rock sampling remains difficult.

Captured Satellites

A second class of satellites include relatively small moons, mostly outermost satellites, that appear to have originated as interplanetary bodies, later captured into orbits around certain planets, as suggested by Kuiper (1956b). This group would include both moons of Mars, eight moons of Jupiter (Figure 5-13), Phoebe going around Saturn, and Nereid going around Neptune, and perhaps even the large moon Triton, also going around Neptune. Of that group of 13, six are moving in retrograde or backward orbits. The other seven are prograde. Except for the Martian case, all the inclinations are moderately high, some 17° to 33° out of the planet's equatorial plane.

The captured moons (excepting large, ice-covered Triton) are all very dark objects, consistent with all known

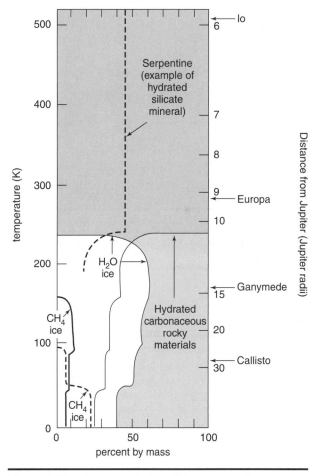

Figure 5-12. A chemical analysis of the condensation sequence among the Galilean satellites, based on an assumed model of the "miniature solar nebula" around primordial Jupiter. The format is a modification of that in Figure 5-4, showing the bulk composition of material condensing at different temperatures (left) and hence different distances from Jupiter (right). For example, materials condensed at more than about 240 K were entirely rocky (about 45% being the strongly hydrated mineral serpentine). At around 200 K, the percent of rock (shaded area) drops to 40% and the rest is water ice. Smaller amounts of ammonia ice and methane ice appear at lower temperatures. (Adapted from Consolmagno, 1981)

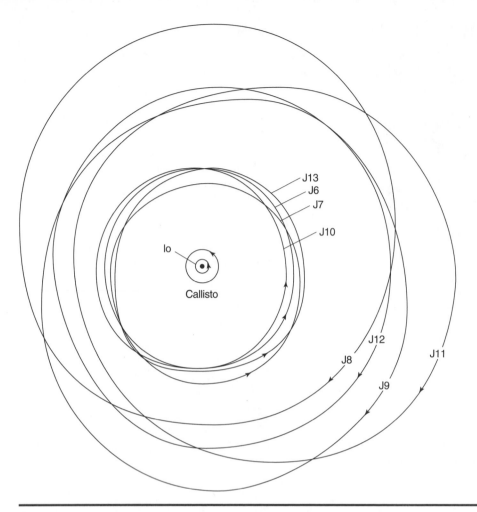

Figure 5-13. Schematic diagram of orbits of the eight largest outer Jupiter satellites, J6 through J13. Callisto is the outermost Galilean satellite. Satellites beyond Callisto fall into two groups: an inner, prograde group with nearly coinciding orbits, and an outer, retrograde group with similar orbits.

asteroids and comets of the outer solar system, believed to be carbonaceous-rich bodies. Figure 5-14 shows the dark color dramatically, in the case of Phobos. The Soviet PHOBOS-2 probe confirmed a Phobos density of 1,950 kg/m³, which is in the range measured for volatile-rich carbonaceous chondrite meteorites. These relations suggest that these moons are carbonaceous asteroids perturbed into planet-crossing orbits in the early days of the solar system; as the giant planets accreted massive atmospheres and grew rapidly from one dozen Earth-masses to many dozens of Earth-masses, there must have been a brief epoch when they scattered thousands of carbonaceous bodies from nearby positions or resonances onto orbits passing throughout the solar system (Hartmann, 1987, 1990).

There are several possible capture mechanisms for these bodies:

1. The massive extended atmospheres of the early planets could have captured passing bodies and perturbed them eventually into low-inclination orbits around the planet. Phobos and Deimos seem especially good candidates for this mechanism, as their carbonaceous-chondrite-like compositions suggest an origin distant from Mars (cf. Chapter 7; Hunten, 1979; Burns, 1978; Veverka and others, 1978; Pollack and others, 1979).

2. New work by Kortenkamp (2004, in press) indicates that gas drag in the solar nebula brings planetesimals into resonances with giant planets, facilitating capture.

3. The close approach of a passing large planetesimal to an existing satellite might have altered the satellite orbit and led to the capture of the planetesimal. The Neptune system, with Triton's weird orbit, suggests this. This theory looks especially attractive if we remember that the largest planetesimals kicked out of the neighborhood of Jupiter and Saturn may have been enormous—possibly larger than Earth. They might easily have disturbed outer satellite systems before colliding with planets.

4. Collision with a satellite might have thrown out debris that could have reaggregated into a new satellite. Inner moonlets and ring debris of giant planets are candidates for this process.

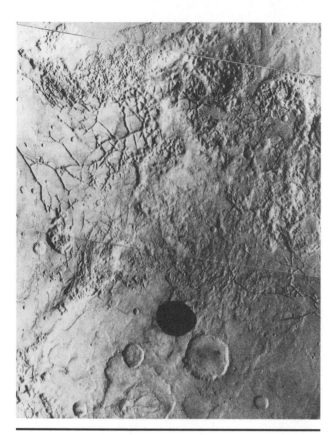

Figure 5-14. Satellite Phobos (dark object) photographed against the surface of Mars, showing its small size relative to features of the fractured terrain in Mars's equatorial region. Because Phobos is covered with carbonaceous-chondrite-like material with only one-fourth the albedo of Mars, its surface is very underexposed in this image, which was exposed for Mars and is correctly rendered as virtually black in this image. See page 98 for a more detailed image of Phobos. (NASA, Viking Orbiter 1, 1977)

In this context, it is interesting to examine Neptune's largest moon, Triton. Triton is strange in several ways. It moves around Neptune in a retrograde direction, opposite to that of most other satellites. This suggests that it may have originated as an interplanetary body and been captured into its present orbit around Neptune. At a diameter of 2,705 km, it is about 18% bigger than Pluto, which orbits around the sun in the same region (its orbit crosses Neptune's). Triton and Pluto may be two surviving examples of a whole group of bodies that formed near Neptune's orbit, with Triton getting captured and Pluto being left behind. Others may have crashed into Neptune long ago.

The most spectacular case of captured satellites is the outermost set of Jovian satellites. The eight largest are shown in Figure 5-13 and are clustered in two sets of orbits; the orbits in each group are very close and similar. Russian scientists (Aitekeeva, 1968; Bronshten, 1968) suggested that each group might represent a capture event, with the satellite broken up by a collision or drag forces. Pollack and others (1979) suggested that the original bodies were captured by interaction with a massive gaseous envelope around proto-Jupiter. Kortenkamp's gas

drag and resonance effects (item 2 above) are likely to be involved.

The eight captured satellites shown in Figure 5-13 have been known for some decades. Their region, so distant from Jupiter, was not effectively searched for additional satellites by passing spacecraft, but sensitive telescopic searches since around 2000 have netted many captured moons, both prograde and retrograde, around all four giant planets. A fifth member of Jupiter's prograde group was found, and another 21 of the outermost retrograde group were also found, not to mention another prograde moon, Themisto, between Callisto and the inner captured group.

An obvious question is whether members of each group of captured moons all have the same spectral type, as consistent with each group coming from a single broken body; or whether each group has a distribution of different spectral types consistent with their being a grabbag of captured Trojan asteroids, or captured comets, or captured objects from some other pool. Unfortunately, the objects are faint and hard to observe. Spectrophotometric studies suggest the eight largest ones among Jupiter's captured moons all have dark surfaces, consistent with their being captured Trojans or comets, but different observers have disagreed on the details of the spectra (Luu, 1991; Tholen and Zellner, 1984). Better observations are needed to use spectra for this important test of origin.

Nonetheless, the general possibility of capture by Jupiter has been verified observationally! Calculation of the past orbital evolution of several comets revealed seven that spent periods of a few months to a few years as captured satellites of Jupiter, mostly within this century (Carusi and Valsecchi, 1980). Comet Shoemaker-Levy 9, which split apart as it looped around Jupiter and then fell into the planet, is also in this category. These observations prove that some satellites could be former interplanetary bodies that were captured by the planet. These recent captures involve three-body interaction between the planet, the sun, and the comet. Ancient captures, at the beginning of the solar system, may have involved drag forces in the very extended primordial atmospheres that were being concentrated from the nebular gases and were facilitated by the very high flux of planetesimals being scattered around the solar system at that time by gravitational encounters with giant planets as they finished growing.

Origin of Earth's Moon

Earth's satellite is different from the others discussed so far. For example, it has an unusually large ratio of satellite mass to planet mass (0.012, which is around 60 times greater than for the largest satellites mentioned above). One of the main problems in explaining lunar origin centers is the difference between the bulk composition of the moon and Earth. The mean densities are 3,340 and 5,250 kg/m^3, respectively. Earth's is higher because of the larger amount of iron and nickel concentrated in its core. From the densities alone, we see that the moon has little iron.

Lunar samples confirm that the moon is deficient in **siderophile elements**—iron, nickel, and other elements with chemical affinity for iron and nickel. Also, the moon is much richer than Earth in refractory elements and is depleted in volatile elements and compounds, such as lead and water. These findings suggest that the moon's material was once very hot and that the volatile elements vaporized and disappeared into space during this heating. A problem, however, is to explain why the moon lacks the iron and siderophiles that Earth has.

Prior to the Apollo mission, there were three general theories about the moon's origin: (1) the **fission theory,** asserting that the moon was spun off of Earth's outer layers, due to Earth's rapid initial rotation. (This was suggested by a descendant of Charles Darwin, George Darwin, who studied tidal forces in 1898); (2) the **binary accretion theory,** asserting that it accreted as an independent body in orbit around Earth at the same time the Earth formed; and (3) the **capture theory,** asserting that it accreted elsewhere in the solar system (where there was little iron) and was captured later by Earth. Researchers hoped that the Apollo flights would help establish one of these theories, but they did not.

The fission theory was abandoned because the angular momentum of the Earth-moon system did not seem adequate, nor was there an adequate energy source to spin off the moon spontaneously. The binary accretion theory was abandoned because there seemed no way to create an iron-poor moon from the same material that produced the sister-object, an iron-rich Earth. The capture theory was abandoned not only because it seemed difficult for Earth to capture a large object (atmospheric drag would not work) but also because Apollo samples showed that the moon has virtually identical proportions of the various oxygen isotopes as Earth, whereas all material from other parts of the solar system (Mars rocks and various classes of meteorites) have different values. Thus, the moon is unlikely to have formed far away.

Study of samples brought back by Apollo astronauts and Soviet Luna probes has shown that the moon has a bulk composition close to that of Earth's mantle or upper mantle (Wanke and Dreibus, 1986). A very important step in acceptance of this idea comes from work on oxygen isotope ratios. As mentioned above, an important discovery has been that different zones of the solar system have different O^{16}/O^{18} ratios, but lunar rocks have oxygen isotope ratios virtually identical to those of Earth rocks (Clayton and others, 1984). This finding appears to prove that the Earth and moon both formed from material at about the same solar distance.

To resolve these problems, Hartmann and Davis (1975) developed a new theory, the **giant impact theory.** We pointed out that the second-largest, third-largest, and other planetesimals in Earth's neighborhood may have grown to very large size before hitting Earth. The fate of the largest subordinate planetesimal could have a strong but stochastic effect on the final planetary configuration. The most likely fate would be a close encounter that would throw it to another region of the solar system, possibly to a Jupiter encounter that could throw it out of the system entirely. But a few planets might be hit by large planetesimals. A tangential impact could change the spin rate or obliquity, and so on. We suggested that a very large planetesimal hit Earth after the core formed, and blew hot mantle material into orbit where it lost its volatiles and aggregated into the moon (Figure 5-15). Our paper suggested second-largest planetesimals 1,000 to 6,000 km across. Independently, Cameron and Ward (1976) developed almost the same picture from angular momentum calculations, concluding that Earth was hit by a Mars-sized body (6,800 km across). Adding support, Wetherill's accretion models (1976) predicted second-largest bodies 4,000 to 6,000 km across.

Subsequent computer models of the impact have supported the concept that a fraction of the hot mantle debris could go into orbit, rapidly forming a ring around Earth, where the moon would aggregate close to Earth and then move to its present distance. A number of other studies support the concept, following the aggregation of the ring into multiple moonlets (Canup and Esposito, 1996) and eventually a single large moon (see also the collection edited by Hartmann and others, 1986). Currently, geochemical studies of the mantle are seeking evidence of whether such an impact left distinctive chemical traces in the mantle mineralogy. For example, computer models predict transient high temperatures in the mantle that would never have existed without the impact, and these might leave distinctive traces in the mineral chemistry.

Other Satellites of Catastrophic Origin: Charon, and Selected Asteroids?

Pluto's satellite, Charon, is half as big as Pluto itself—a size ratio even larger than that between the moon and Earth. Pluto has a density close to 2,000 kg/m^3, indicating that it is composed of about 70% rock and 30% ice. Charon's density is similar but may be a bit lower; Charon may be more ice-rich. At the time of Charon's discovery in 1978, the satellite seemed very mysterious. How did a "planet" only 2/3 the size of our moon get a satellite 1/3 the size of our moon? On the other hand, anyone who wanted to regard Pluto as the largest interplanetary body faced the problem that no other interplanetary body was known to have a satellite. The idea of a fission origin had been suggested for Charon, but seems ruled out by the available angular momentum data on the system (McKinnon, 1989).

Today, Charon's origin may be more understandable, and the link to asteroids seems stronger. Ida, one of the three asteroids visited by spacecraft by 1997, was found to have a small satellite, and in 1997, lightcurve data indicated that asteroid Dionysius has a satellite about half its own size. Hartmann (1979) suggested that asteroid satellites and bilobed compound asteroids might arise by interaction of neighboring fragments in large collisions, and computer modeling of collisions has supported this (Durda, 1996). McKinnon (1984, 1989) suggested that

a

b

Figure 5-15. The giant impact that triggered the formation of the moon. (**a**) One-half hour after the giant collision that led to formation of the moon. A Mars-scale planetesimal has hit Earth somewhat tangentially and is shearing off mantle materials from both bodies in a blaze of incandescent material. (**b**) Five hours after the impact a streamer of debris is spreading out from Earth. Only a small fraction of this will go into orbit to make the moon. (Painting by author, based on computer reconstructions by Jay Melosh, Willy Benz, A. Cameron, and others)

Charon originated through catastrophic impact analogous to the giant impact that may have formed the moon, and this is currently the leading hypothesis.

Thus, our moon, Pluto's Charon, and certain asteroid pairs may testify to formation of a class of satellites through catastrophic impact events.

SUMMARY

In most current thinking, the solar nebula is pictured as a cloud left after the sun's formation, analogous to other cocoon nebulas, and impregnated by material from a nearby supernova or supernovas. The cloud was heated by its gravitational contraction during solar formation but then cooled. During its cooling, solid microscopic grains formed by condensation. Refractory minerals, followed by metals, ordinary silicates, carbonaceous materials, and then ices, condensed as the temperature fell. The local composition at any time and place was controlled largely by the ambient temperature.

The dust settled toward the ecliptic plane and aggregated into planetesimals, probably largely by gravitational collapse. Collisions among the earliest planetesimals (of kilometer scale or less) led to their accretion into larger bodies. Some collisions led to fragmentation, with fragments later accreted by other bodies. The asteroids probably never completed the growth process.

The largest bodies in each zone of the solar system grew fastest, so that one body eventually came to dominate each zone (except the asteroid belt). The dominant body—the incipient planet—accreted the remaining small planetesimals or scattered them out of their original zones to be swept up by neighboring planets. The giant planets probably grew to sizes of 10 to 15 M_{Earth}, and then rapidly attracted massive hydrogen-rich atmospheres from the nebula, bringing them up to their present masses of 14 to 317 M_{Earth}.

Giant planets probably developed miniature, disk-like solar nebulae, in which satellites accreted. The extended atmospheres of primeval planets also apparently captured some planetesimals, adding additional outer satellites. Earth's moon probably formed when a giant planetesimal blew mantle material into orbit; Pluto's moon may have formed in an analogous way on a smaller scale.

CONCEPTS

radiometric dating	hydrostatic principle	adiabatic curve
radioactive isotopes	plasma	heterogeneous accretion
parent isotopes	magnetic field lines	model
radiogenic isotopes	magnetic coupling	chemical equilibrium
daughter isotopes	hydrodynamics	model
solidification age	magnetohydrodynamics	electrostatic forces
formation age	chemical equilibrium	collisional accretion
half-life	refractory compounds	gravitational collapse
relative formation age	carbonaceous chondrite	Stoke's law
formation interval	Allende inclusions	origin of ring systems
solar nebula	enstatite	siderophile elements
grains	feldspars	fission theory
planetesimals	troilite	binary accretion theory
protoplanets	olivine	capture theory
volatiles	kerogen	giant impact theory

PROBLEMS

1. (a) What is the date of solar system origin? (b) How long did the origin process take to create planet-sized bodies?

2. Describe how answers to Problem 1 can be determined by measurements.

3. What is the evidence for a supernova explosion near the contracting solar nebula shortly before the solar system formed?

4. (a) Describe the solar nebula. (b) Summarize evidence that it existed.

5. Assume that you observed the sun and solar nebula from a nearby star during their formation; compare your observations with modern observations of T Tauri stars and related objects.

6. Calculations indicate that if a sun-sized star formed from a contracting, rotating, interstellar cloud, it would end up spinning much faster than the sun unless slowed in some way. What theory has been advanced to explain why the sun does not spin much faster?

7. (a) You are traveling through space and come to a star of normal solarlike composition but with planets composed of refractory silicate minerals rich in aluminum, titanium, and calcium, and containing no water or ice. What can you conclude about formation conditions? (b) Compare these planets with our own moon and comment on formation conditions of the moon.

8. Citing their distance from the sun, temperature history, and gravity, give two reasons why the giant planets have so much more volatile and icy material and mass than do terrestrial planets.

9. What are the Allende inclusions, and why are they important in interpreting early solar system conditions?

10. (a) By what factor are inert gases depleted in the terrestrial planets? (b) Why does this enormous depletion indicate that terrestrial planets formed by accretion of small, independent planetesimals rather than by a single gravitational collapse of a whole planetary mass from the solar nebula?

11. (a) Describe some theories of the moon's origin and some problems associated with each. (b) How old are typical lunar rocks? (c) What circumstances prevent us from getting many lunar rocks 4.5 Gy old that might reveal clues about the moon's origin?

12. (a) Describe the satellite system of Jupiter. (b) What circumstances indicate that the outer Jupiter satellites are fundamentally different from the Galilean satellites and might have formed by capture?

13. Describe ring systems in the solar system and indicate some possible modes of origin for them.

14. (a) If a planetesimal has an effective cross-sectional area of πR^2 and is sweeping through a cloud of smaller planetesimals of fixed size with velocity show that the number of collisions per second will be

$$\frac{dn}{dt} = \frac{\pi R^2 V \rho_N}{m}$$

where ρ_N = the spatial density (kg/m³) of particles in the surrounding nebular cloud, and m = the mass of each small particle. (b) Show that if each collision results in the target mass sticking to (or embedding in) the planetesimal, the planetesimal will gain mass at a rate of

$$\frac{dM}{dt} = \pi R^2 V \rho_N$$

where M = planetesimal mass.

15. Under the assumptions of problem 14 and assuming that R is the actual geometric radius of the body, show that an expression for the time to grow to radius R is

$$T = \frac{4R}{V}\frac{\rho_P}{\rho_N}$$

where ρ_P = density of the planetesimal itself. Assume while solving the differential equation that $_N$ and V remain constant during the growth process (that is, the nebula acts as a very large reservoir for material to be accreted rather than used up).

16. (a) Assuming the equation in problem 15 holds and that the reasonable value for density ρ_N of accretible material in the nebula is 10^{-7} kg/m³, estimate the timescale to accrete a body 1,000 km across in the solar nebula. Assume a reasonable ρ_P. (b) How does the timescale change if the accretion process is inefficient, so that not every collision results in mass gain? (c) Is the calculated result reasonable compared with the measured formation interval?

17. (a) What is the escape velocity of a 1,000-km-diameter planetesimal if its density is 3,000 kg/m³? (b) A small meteoroid at a great distance from this planetesimal is approaching it directly at 1 km/s. At what speed will it impact the surface of the planetesimal?

18. If the density of gas and dust in the solar nebula had been 10^{-3} kg/m³ in some region and the temperature had been 500 K, what would have been the smallest body that could have formed directly by gravitational collapse?

One important consequence of meteorites is that they have hit virtually all known planetary surfaces, causing impact craters. The low-angle lighting on this portion of the lunar surface indicates that meteorites of many sizes have fallen and created a surface nearly covered with craters. (NASA, photo by Apollo astronauts orbiting the moon)

CHAPTER SIX

Meteorites and Meteoritics

Between the planets, bits of interplanetary flotsam pursue their orbits around the sun. They range in size from dust particles to substantial worldlets. Among the terrestrial planets, the largest are about 30 km across. Among the giant planets, the largest are at least 300 to 500 km across, or 2000 km if you count Pluto. In the asteroid belt between the terrestrial and giant planets, the largest object is Ceres, about 1000 km across.

All these interplanetary objects—dust particles, asteroids, and comets—are collectively known as the **meteoritic complex** from the Greek term *meteöron*, meaning a celestial phenomenon or airborne object.* Some of these objects collide with Earth, surviving the passage through the atmosphere and hitting the ground. They are called *meteorites*. They are only fragments of once-larger bodies, which are usually referred to as their *parent bodies*.

Meteorites, and the whole meteoritic complex, shed light on the past. Some meteorites come from parent bodies that were never much heated nor chemically altered after the days of planet formation. They give better clues about planetary origin than most planetary rock samples, which have experienced severe geologic processing.

---

*Until the 1900s, this term was used to refer to rainbows, lightning, snowflakes, and so on, as in "the meteor of the ocean air shall sweep the clouds no more" (Oliver Wendell Holmes). Hence, the discipline of meteorology refers to the atmosphere—not meteors.

Fate of Planetesimals

To understand the history of meteorites, consider the five possible destinies awaiting planetesimals left over after planet formation.

Ejection from the Solar System

Most of the early planetesimals remaining after planetary accretion were perturbed by the planets' gravity until they made a close encounter with one of the planets. Usually, such encounters sent planetesimals into the outermost solar system, but some experienced such strong perturbations that they were kicked into interstellar space on hyperbolic orbits (Öpik, 1966; Arnold, 1965; Wetherill, 1976, 1977a, 1977b).

Collision with Planets

Many planetesimals actually collided with a planet, with each crash causing a crater. This explains the densely cratered surfaces we see on most worlds studied so far at close range, such as the moon, Mercury, Deimos, and Phobos as well as asteroids Gaspra, Ida, and Mathilde, and satellites Callisto, Ganymede, Mimas, Rhea, Dione, and Miranda. Examples are shown in Figure 6-1. Even geologically active worlds like Earth, Venus, and Europa, with their younger surfaces, have some impact craters. As noted above, planets reached essentially their current sizes in the first 50–100 My, and the remaining planetesimals were mostly swept up in 500 My. A remaining steady supply of interplanetary bodies, which continue to hit planets, is supplied from the asteroid belt, the Kuiper comet belt, and the Oort comet cloud.

Capture into Satellite Orbits or Resonant Orbits

As mentioned in the last chapter, primeval planets may have had extended atmospheres or satellites whose interactions with the approaching planetesimals allowed capture of the planetesimal into a satellite orbit around the planet (Hunten, 1979). A variant of this fate would be capture into a dynamically stable configuration, such as the L_4 and L_5 Lagrangian points in Jupiter's orbit, where Trojan asteroids are now located.

Fragmentation

Suppose a planetesimal gets perturbed by Jupiter from the asteroid belt onto an Earth-crossing orbit. With a typical eccentricity of 0.67, it would speed by Earth's orbit at about 10 km/s. In contrast, original planetesimals moving in or near Earth's orbit with eccentricity 0.05 would pass each other at only about 0.7 km/s. (Note that velocities of

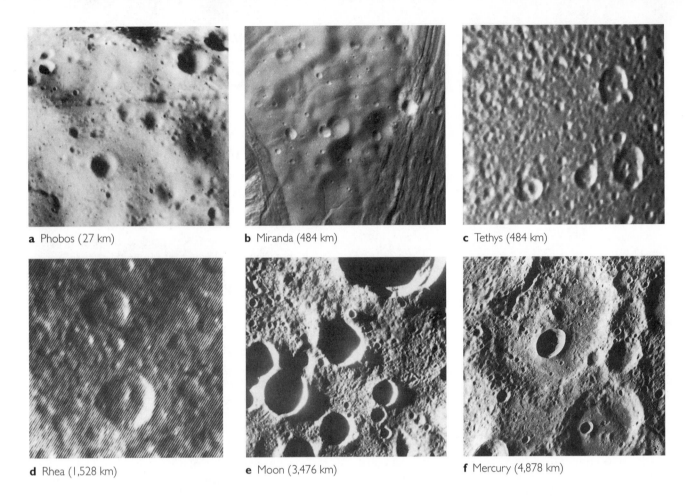

a Phobos (27 km) **b** Miranda (484 km) **c** Tethys (484 km)

d Rhea (1,528 km) **e** Moon (3,476 km) **f** Mercury (4,878 km)

only about 0.04 km/s are necessary for rocks to fragment each other instead of merely bouncing apart.) Thus, as planetesimals got scattered around the solar system by the larger planets, encounter velocities between them increased to the point that fragmentation rather than accretion dominated (Kaula and Bigeleisen, 1975; Wetherill, 1977a). An additional planetesimal fate, then, was to be smashed to bits. The fragments, which were smaller than the original, could of course undergo ejection, collision with planets, or further fragmentation.

This effect probably explains why accretion stopped in the asteroid belt before a planet grew there. Calculations indicate that even by the time Jupiter reached only 0.1 percent of its present mass, it scattered bodies through the asteroid belt at relative speeds as high as 10 to 15 km/s (Ip, 1978), causing much damage. As of today, the mean collision speed between pairs of asteroids in the belt is 5 km/s, causing catastrophic fragmentation when similar-sized asteroids collide.

Preservation until Today

Planetesimals and fragments that escaped the first four fates have lasted until today. There are few places in the solar system where bodies could be "stored" with much hope of lasting this long. Most bodies, even if placed on circular orbits midway between two adjacent planets, will

get perturbed onto planet-crossing orbits and swept up within a few hundred My or less. This is beautifully shown by calculations of Jack Wisdom, as seen in Figure 6-2. In a computer model, Wisdom placed hundreds of imaginary particles in orbits from Jupiter outward; none survived, except those beyond about 43 AU, which is just where the Kuiper belt objects are actually observed. The main asteroid belt between Mars and Jupiter, and the Kuiper Belt are thus two reservoirs that have survived. Wetherill (1977a) describes how some planetesimals that originated among terrestrial planets may have been perturbed into the inner edge of the asteroid belt and preserved there.

Meteorites as "Free Samples" of Planetesimals

The preserved planetesimals and their fragments are especially interesting because some of them fall on Earth and give us "free samples" of ancient materials from the beginning of the solar system. Orbits of these bodies in space are typically slightly inclined to the plane of the solar system and therefore pass through the ecliptic plane at two points, called **nodes.** If either node happens to lie very

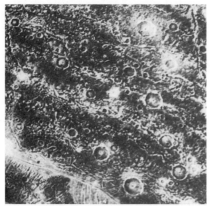

g Ganymede (5,262 km)

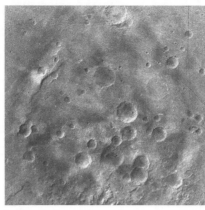

h Mars (6,786 km)

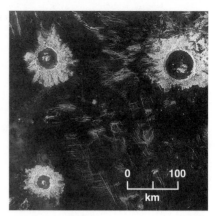

i Venus (12,104 km)

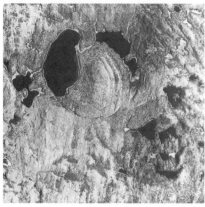

j Earth (12,756 km)

Figure 6-1. Impact craters throughout the solar system are evidence that countless planetesimals have crashed into worlds of all sizes. The diameter of each world is given. The arrangement by size shows that larger worlds (Mars, Venus, Earth) have fewer or more eroded craters. The crater shown on Earth is the 4-km, 450-My-old Brent crater in Ontario, Canada; it has been filled by glacial erosion. Scale varies from photo to photo (although the Phobos and Earth photos cover similar areas). (Photos **a–g** and **i,** NASA; **h,** Soviet Mars 5 orbiter photo, courtesy Vernadsky Institute, Moscow; **j,** Earth Physics Branch, Department of Energy, Mines, and Resources, Ottawa)

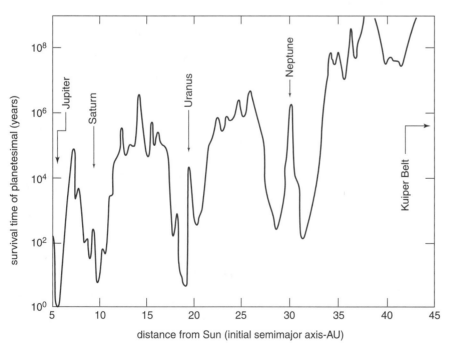

Figure 6-2. Calculated lifetimes of interplanetary bodies placed among the four giant planets. The graph shows interplanetary bodies in this part of the solar system cannot be stable for the life of the system, but are swept up or ejected by the giant in a few hundred million years or less. If a fifth planet had tried to grow in this space, it would have been swept up or ejected. The only stable region is beyond about 40 AU— where the Kuiper belt of leftover planetesimals actually exists. Spikes at the positions of the four planets represent Lagrangian point resonances. (After calculations by M. J. Holman and Jack Wisdom)

close to Earth's orbit, sooner or later the planetesimal and the Earth will arrive at this point at the same time. For a typical planetesimal on such an orbit, an encounter will occur typically within about 20 My. The most common result of such an encounter is a near miss that gravitationally alters the orbit of the object. After several near misses, however, the object may eventually hit the planet.

These objects come in all sizes. The larger the size, the rarer they are. Larger ones can be spectacular. On April 10, 1972, a house-sized interplanetary body approached Earth from behind at 10 km/s on a 15°-inclined orbit that reached as far as 2.3 AU from the sun—in the asteroid belt. It entered the atmosphere at 15 km/s, traveling south to north across the western United States. Friction with the atmosphere heated its outer skin to incandescence, and it became visible as a brilliant glowing object over Utah, moving north over Idaho and Montana. Here it dipped as low as 50 to 60 km in the atmosphere, at which point most such bodies would have fallen into the ground with explosive energy. But like a stone slicing into water and skipping back out again, this body skipped off the atmosphere, ascending at a shallow angle over Alberta, Canada, and exiting into space once again on a different orbit.

Sky and Telescope magazine over the next few months published 11 still photos and 72 movie frames (a small sample of the total movie footage) of the object. A classified Air Force satellite detected it from above with an infrared radiometer designed to look for heat emitted from missiles.* All these observations permitted analysis of the orbit and the object, which was found to be a rocky body with a diameter between a few meters and 80 m (Jacchia, 1974). This, then, was a rare case of a large interplanetary body that encountered the atmosphere and barely escaped plummeting into the ground. The miss was fortunate, as a hit by this object would have caused a (nonnuclear) explosion of about the magnitude of the Hiroshima bomb, probably somewhere in Alberta. More recently, military satellites have detected smaller meteoroid explosions high in the atmosphere over the Pacific and other sparsely inhabited locations.

Meteoritics and Some Associated Definitions

The study of the interplanetary samples is called **meteoritics.** Meteoritics is a rich science attempting to determine where these samples came from and how they relate to either the primordial or present-day planetary bodies. Many results in this field are published in the international journal, *Meteoritics and Planetary Science.* A number of useful terms have evolved in this field, including the following:

meteoroid: Any small, extraterrestrial, solid body floating in space. The term is usually used for bodies smaller than a kilometer—and frequently millimeters in size. Larger objects are usually classed as either asteroids or comets.

meteor: A heated, glowing meteoroid moving through the atmosphere, having not yet hit the ground. A meteoroid may enter the atmosphere at 10 to 70 km/s, and friction between it and the atmosphere heats its surface and ionizes atmospheric molecules. Airless planets such as the moon would have no meteors, for an atmosphere is needed to heat the meteoroids and make them visible. "Shooting star" is a folk name for a meteor.

meteorite: A body from space that hits the ground. There are three broad classes: stones, irons, and stony-irons. We cannot assume that the meteors we see on an average night are physically related to the meteorites we see in a museum. Indeed, we will describe contrary evidence. Meteorites are generally named after a city or geographical landmark near where they fell or were found.

fireball: A very bright meteor. Fireballs are infrequent and can be much brighter than the full moon, in rare cases rivaling the sun.

fall: A meteorite witnessed during descent. Some 94% of falls are rocks, but a few are metallic iron alloy objects. Thus, most meteoroids in space are believed to be rocks.

find: A meteorite not seen to fall, but found on or in the ground. Many finds may have fallen thousands of years ago. The people in locales of some ancient large falls recount legends or place names associated with the sky, as at Campo del Cielo, Argentina. Such legends may have persisted since the time of the fall. Most finds are irons, as these resist weathering longer than rock material.

parent body: Meteorites are fragments of larger objects in space, which are called parent bodies. The interiors of these hypothetical parent bodies sustained the pressures and temperatures suitable for forming the minerals found in meteorites.

---

*This large body of solid observational reports contrasts strongly with the absence of good observations of UFO cases allegedly involving alien spacecraft. One might infer that if passages of such spacecraft through our airspace occur as frequently as proposed by a few UFO researchers, we ought to obtain better observational evidence within a few years, resulting from the increasing numbers of amateur photographers and surveillance devices.

History of Meteorite Studies

Early Chinese, Greek, and Roman writings describe the fall of "stones from the sky." Often these celestial stones

Figure 6-3. The Winona stone meteorite buried in a crypt by prehistoric Indians in northern Arizona. (Museum of Northern Arizona, Anthropological Collections; photo by Gayle Hartmann and the author)

were collected and enshrined as sacred objects. The famous black stone in the Kaaba, the sacred shrine of Islam in Mecca, is apparently a meteorite that was enshrined before Muhammad conquered the city around A.D. 600 (Sagan, 1975). An iron meteorite's devastating impact on an island in the Baltic Sea around 400–800 B.C. left 9 craters up to 110 m across, created legends, and wiped out agriculture for 100 years (Veski and others, 2001). Cold-hammered artifacts of meteoritic iron have been found at prehistoric mound-builder sites in Ohio (Carr and Sears, 1985), and at Canadian Eskimo sites from A.D. 800 (McCartney and Kimberlin, 1988). Figure 6-3 shows meteorite fragments buried around A.D. 600 in a stone-slab crypt by Pueblo Indians near Winona, Arizona (Heineman and Brady, 1929).

During the Middle Ages in Europe, fallen meteorites were likely to be brought by incredulous peasants to local priests. Thus a number of meteorites were preserved in European village churches.

In 1751, a widely witnessed fall occurred in Zagreb. The bishop of Zagreb carefully collected and protected pieces of the iron meteorites and sent them, along with a tabulation of witness testimony, to the Austrian emperor. These were passed on to the Vienna museum but were not treated as important. However, by the 1790s, philosophers and scientists were aware of many claims of stones falling from the sky. Most eminent scientists were skeptical. The breakthrough came in 1794 when a German lawyer and physicist, E. F. F. Chladni, published a study of some alleged meteorites, one of which was found just after a fireball had been sighted. Chladni concluded that these meteorites really did fall from the sky, and he correctly inferred that they were extraterrestrial objects that were

heated by falling through Earth's atmosphere. Chladni even postulated that they might be fragments of a broken planet—an idea that set the stage for early theories about asteroids,* the first of which was discovered seven years later. Chladni's ideas were widely rejected, not because they were ill-conceived, for he had been able to collect good evidence, but because his contemporaries were simply loath to accept the idea that extraterrestrial stones could fall from the sky. Gradually the evidence became more compelling. A 1796 paper accepted the celestial origin but argued that the stones had been swept up by tornadoes, only to fall again to Earth (Wood, 1968).

The last holdouts were members of the austere French Academy of Sciences, who were considered the scientific leaders of the day. These gentlemen argued that stones could not fall out of the sky and argued that peasant witnesses could not be expected to make reliable observations. In retrospect, it appears that the academy's reluctance was generated in part by subtle social pressures. In the years following the French Revolution, there was a philosophic wave of antipathy to religious authorities, who were often the caretakers of the mysterious sky-stones.

As if in answer, a shower of fragments from a large meteorite pelted the French town of L'Aigle on April 26, 1803. The academy sent the noted physicist Biot to investigate. His exhaustive report finally convinced the scientific world that stones do fall from the sky and that Chladni had been right.** A century of further work revealed that meteorites were unlike any terrestrial stones and had very complex histories, as we see in this chapter.

In 1891, a naval expedition on the H.M.S. *Challenger* dredged up seafloor sediments containing numerous submillimeter spherules. Spherules such as these have been proven to be extraterrestrial and probably formed as molten droplets sprayed off meteorites burning in the atmosphere. They are probably the most common form of extraterrestrial material on Earth (Ganapathy, Brownlee, and Hodge, 1978). By the 1940s, the collection of known meteorites was considerably expanded by the pioneer meteoriticist H. H. Nininger (1952), who toured midwest America and obtained falls and finds from farmers and other residents.

In 1969, Japanese geophysicists discovered Antarctic ice fields littered with meteorites. These often lie in clusters at the mouths of glaciers, where the melting glacier drops its load of stones that have fallen on its surface. Japanese and American teams have collected thousands

---

*This accounts for the once-popular theory that asteroids were fragments of a single planet that exploded for unknown reasons. Modern work shows this is incorrect.

**The knowledge only slowly filtered to the fledgling United States. According to an apocryphal story, on hearing two Yale professors report a meteorite fall, the Virginia naturalist (and politician) Thomas Jefferson remarked, "It is easier to believe that Yankee professors would lie than that stones would fall from heaven."

of new meteorite samples. Many are processed through NASA's lunar rock curatorial facility and preserved in sterile surroundings.

Recovery of a new meteorite specimen is a rare privilege and should be reported to the Smithsonian Institution in Washington or to a nearby major university, as each meteorite has its own peculiarities and might provide some major insight into the history and environment of the solar system.

Phenomena of Meteorite Falls

Modern data show that larger meteoroids often explode high in the atmosphere; many hand-sized specimens are fragments dropped from such explosive events. Such events are more common than was once thought. For example, on October 9, 1996, a fireball exploded over El Paso, Texas, with energy equivalent to about a kiloton of TNT, rattling windows and making a dust trail photographed by many observers. The event was reported in the media within hours (a scorched field was mistakenly attributed to it); but a few days later when scientists had gathered interesting data on the reality of the event, there was virtually no follow-up report because it was considered "yesterday's news." Because of the media's fixation in our culture on news of the moment, the public gets very poor information on this type of phenomenon. Current estimates indicate that a meteorite explosion of this size happens every few days somewhere on Earth, but it is rarely over a city where it is well observed.

Air Force sensors picked up a still larger meteorite explosion over the south Pacific in the mid-1990s, but again this event was only poorly reported and scarcely noticed by the public.

Sound

Large meteorites produce energetic shock waves when they enter the atmosphere at supersonic velocities. These "sonic booms" have long been known, being described in terms of the contemporary culture: "like distant guns at sea" (England in 1795) and "like a horse and a carriage clattering over a bridge" (America in 1913). Peculiar sounds reported from fireballs brighter than the full moon have been suspected to have an origin involving electromagnetic radiation at very low frequencies (Keay, 1980), but they are poorly understood.

Brightness

The larger the object, the brighter it appears. For example, when a 70-ton iron meteorite broke up and fell at Sikhote-

Alin, Siberia, in 1947, it dazzled the eyes of observers and cast shadows even though the sun was shining (Krinov, 1966).

Train

Many large meteorites leave a dusty **train** made up of debris blown off during their atmospheric passage. Meteors and meteorites seen at night may leave a faintly luminous train visible from a few seconds to several minutes. This trace is caused by photochemical reactions among high-altitude air molecules disturbed by the meteor's passage.

Temperature

A meteorite's passage through the atmosphere is so swift that only its outer surface is heated; the meteorite's interior remains cold. The widespread belief that a meteorite may remain too hot to touch for some hours after its fall is a fallacy because the outer hot layer, about 1 mm thick, cools rapidly. A thin **fusion crust,** usually black, is produced on meteorite surfaces by the atmospheric heating of the thin surface layer. An example of the black fusion crust is shown in Figure 6-4.

Velocity

A typical meteorite strikes the atmosphere at about 10 to 20 km/s. Meteoroids with the highest velocities (up to 70 km/s) are the ones most likely to fragment in the air. Those smaller than about 1000 t lose a substantial part of their initial velocity due to atmospheric drag and fall like rocks dropped from an airplane. Larger meteorites can strike the ground with most of their initial velocity; impacts at supersonic speeds cause explosive craters, typically larger than 50–100 m across; subsonic meteorites do not make explosions, but merely dig small holes in the ground.

Impact Rates: Meteoroid Flux

Various studies estimate that the total **meteoroid flux,** or mass of meteorites that falls onto Earth, is about 10^7 to 10^9 kg/y (Dodd, 1981); the mass flux onto the moon is about 4×10^6 kg/y (Wetherill, 1976; Hartmann, 1980). The terrestrial rate equals hundreds of tons per day! Because the number of particles increases with decreasing size, microscopic particles, called **micrometeorites,** are the most abundant. Objects large enough to hit the ground subsonically and form modest craters hit Earth roughly annually, and giant explosions happen once or twice per century. Six out of seven occur over oceans or poles, so they have mostly gone unreported in history.

Figure 6-4. A black fusion crust on a freshly fallen meteorite is well shown by the Lost City meteorite, an H-type chondrite. The meteorite was detected by a network of cameras designed to detect meteoritic events and was recovered in Oklahoma 6 days after it fell in 1970. The fusion crust is caused by heating during the fall through the atmosphere. (NASA)

Present-Day Impact Rate

Figure 6-5 shows the numbers of objects of different sizes presently hitting Earth. The curve is based on spacecraft measures of tiny meteoroids in space and on counts of craters formed on Earth and the moon (Morrison, Chapman, and Slovic, 1994). The scale at the top shows the sizes of craters produced if the objects strike surfaces at 10 to 20 km/s, speeds typical of planetary impacts (see Chapter 9 for further discussion of crater formation).

Primeval Impact Rates—
The Early Intense Bombardment

As early as 1952, Urey postulated an intense bombardment of planets during the close of planet formation because the newly formed planets were efficient at sweeping up debris. Planetesimals on orbits similar to planets' orbits had half-lives as short as 15 to 20 My, according to dynamical calculations (Öpik, 1963, 1966; Arnold, 1965; Wetherill, 1977a, 1977b).

As shown in Figure 6-6, lunar data confirm an intense early bombardment. Dates determined for surface rock layers at different landing sites, combined with crater counts at those sites, give impact rates in different time intervals. The box-shaped data points measured in this way clearly show the decline from the early intense bombardment some 4 Gy ago to the more constant value experienced in the last 2 Gy or so.

Lunar data give little evidence about conditions before the oldest surfaces were formed, 4 Gy ago. However, the impact rate necessary to accrete the moon and Earth is off the scale to the upper left (Hartmann, 1980).

One interpretation (curve A) is that the cratering rate simply declined smoothly, with impacts pulverizing most lunar rocks formed before 4 Gy ago. A decline of this sort would result if the Earth and moon simply swept up a population of nearby leftover planetesimals, as shown by the "W" points in Figure 6-6, based on sweep-up calculations by Wetherill (1977a). However, some lunar rock analysts believe that the sharp cutoff in numbers of lunar rocks older than 4.0 Gy indicates an intense burst of cratering at that time, destroying older rocks (Tera, Papanastassiou, and Wasserburg, 1974). Such bursts (curve B) could come about if large planetesimals from the outer solar system broke up in the inner solar system (by tidal stresses during close encounters with terrestrial planets?), supplying bursts of meteoroidal fragments (Wetherill, 1975, 1977a, 1977b).

Impact Rates on Other Planets

Impact rates vary from planet to planet. Spacecraft have measured the flux of micrometeoroids at various planets; however, the fluxes of larger objects, such as asteroid fragments ejected from the asteroid belt, and comets, are uncertain. Table 6-1 on page 135 lists estimated cratering rates for various planets, assuming a mixture of cometary and asteroidal meteoroids. Comparable data for the outer solar system are even less certain, because the number of interplanetary bodies there is unknown. Estimated cratering rates for inner moons, like Io, Mimas, and Miranda, are much higher than rates on outermost moons because of gravitational focusing toward the giant planets.

Because the planets formed at about the same time and jointly swept up planetesimals during early solar system

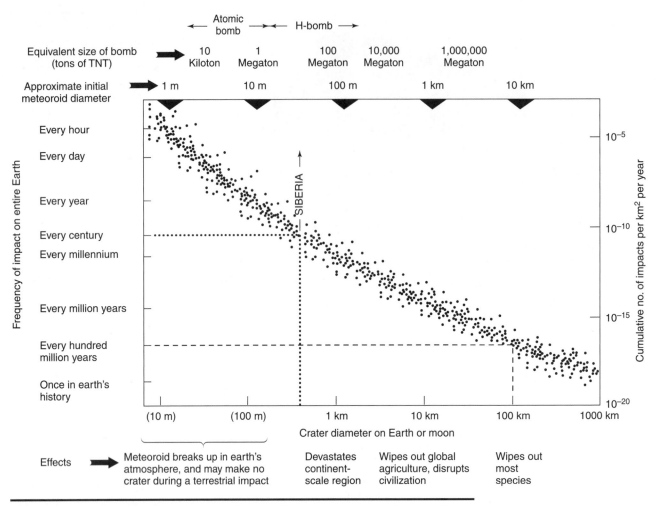

Figure 6-5. The rate of meteoroid impacts at the top of Earth's atmosphere as a function of meteoroid size. Bottom scale gives crater size for typical impact velocity of about 15 km/s. Top scale gives meteorite size and energy in terms of tons of TNT. The dotted line shows the scale of the Siberian meteoritic explosion of 1908 (see end of chapter). Dashed line shows scale of impact 65 My ago that wiped out dinosaurs and other species. (After Hartmann and Impey, 1994, and 1993 data from E. Shoemaker, C. Chapman, D. Morrison, G. Neukum, and others)

history, the rate of cratering as a function of time is believed to resemble Figure 6-6 for all planets, with only the absolute rate varying from planet to planet.

A Meteorite Classification System

Now we turn from impact effects to the meteorites themselves. In the last chapter, we saw that chemical properties of certain primitive meteorites fit beautifully with the condensation theory of the origin of planetary matter. Other meteorites have been **metamorphosed,** or altered, by heat, pressure, mechanical shock, or other effects. Many have been gently heated so as to alter their minerals slightly. Others have been completely melted, allowing heavy ma-

terial such as metal to separate from light materials. These materials then resolidified into lavalike rocks or metal.

Any process that segregates chemicals into different proportions from their original mixture is called **differentiation.** Meteorites have been classified primarily by their degree of differentiation; the classification scheme helps us to understand meteorites' clues about early conditions in the solar system and about the interiors of other planetary bodies. Table 6-2 summarizes the current classifications, of which beginning students should learn at least the "broad classification." We give a brief overview first and then discuss the various classes in more detail.

Meteorites are broadly divided into **stony meteorites, iron meteorites** (pure metal, essentially nickel-iron alloy), and **stony-iron meteorites** (mixed iron and stony portions). Stony meteorites, usually called simply "stones," are by far the most common (95%) and include the ones that are least differentiated. Among the stones, the most

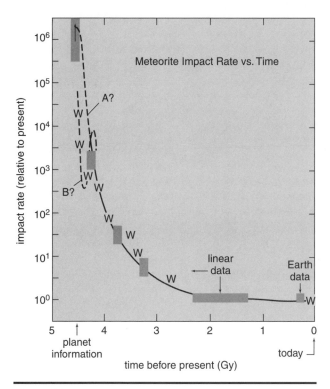

Meteorite Impact Rate vs. Time

Figure 6-6. The history of impact rates in the Earth-moon system, as derived from terrestrial and lunar data (adapted from Hartmann and others, 1980). Sizes of boxes indicate approximate precision of available data. Two interpretations of earliest cratering, A and B, are discussed in text. Points labeled W mark calculated impact rates at different times, based on an assumed initial swarm of planetesimals with a = 0.9 AU, e = 0.27, i = 5.7°. (from Wetherill, 1977a)

TABLE 6-1

Rates of 2-km Crater Production of Various Planets Relative to Earth

Planet	Minimum likely value	Most likely value	Maximum likely value
Mercury	0.5	1.3	3.3
Venus	0.5	0.7	1.4
Earth	1.0	1.0	1.0
Moon	0.5	0.7	0.9
Mars	0.6	1.3	2.6
Outer planets	?	?	?

Source: Data adapted from Hartmann and others, 1981.

Note: These values are not quite the same as relative meteoroid fluxes, since a crater on one planet may have been formed by a somewhat different-sized meteorite than the same-sized crater on another planet due to an impact velocity difference. Estimates depend on the assumed distribution of Mars-crossing asteroids, Earth-crossing asteroids, short-period comets, long-period comets, and so on.

numerous are **chondrites,** a class named because they contain BB-sized spherules called **chondrules,** after a Greek term for seedgrains. The mysterious chondrules are discussed in more detail later in the chapter. Chondrites make up about 80% of all meteorites falling on Earth.

A smaller class of stones are **achondrites,** which have been heated, melted, and/or shattered enough to destroy the chondrule structures by melting or chemical alterations. Some achondrites resemble volcanic lavas of the moon and Earth. These imply heating and melting processes within at least some parent bodies.

Another, smaller class is of special interest. These are the **carbonaceous chondrites,** stony meteorites colored black by a matrix of carbon-rich material. Most have been heated so little that they retain water (bound in the minerals) and other **volatile elements and compounds**—the ones that would be easily driven off by moderate heating. These are the meteorites most representative of the primitive material that condensed at low temperature from the solar nebula.

Many meteorites, among all classes, are *breccias,* or rocks formed of broken fragments cemented together. These fragments are often from different meteorite types—for example, chips of relatively unheated carbonaceous

chondrites interspersed among broken pieces of once-melted achondrites! Many other meteorites are intensely fractured. Breccias and fractures suggest violent collisions in the history of meteorites (Bunch and Rajan, 1988).

All these facts are clues that help explain the histories of planetesimals. Early planetesimals were made of primordial condensates, as outlined in the last chapter. Some of them may have survived with little alteration to produce meteorites like carbonaceous chondrites and some ordinary chondrites. For example, lab experiments by Lauretta and others (1997) simulated formation of various types of iron-nickel sulfide grains that resemble grains in a carbonaceous chondrite; the latter probably condensed directly from the solar nebula. Others were partly melted by heating, probably caused by radioactivity or by electric currents induced by magnetic fields in the passing solar wind.* During melting, chemical differentiation occurred. Because of their high density, metals such as iron and nickel, and elements with a chemical affinity for iron, called **siderophile elements,** drained toward planetesimal centers, forming a nickel-iron core region. Lighter, silica-associated elements, called **lithophile elements,** floated toward the surface layers, forming stony mantles. Some of this lavalike material erupted into some planetesimal surfaces. When these planetesimals were blown apart during collisions, fragments were ejected from different regions—metal objects from the cores, mixed objects from core-mantle interfaces, lavalike rocks from upper mantles and surfaces, and primitive rocks from the least-heated regions. In a first-order way, this explains meteorite types such as irons, stony-irons, achondrites, and chondrites. During

---

*Heating processes and evolution of interiors is discussed in more detail in Chapter 9.

TABLE 6-2 Classification of Meteorites

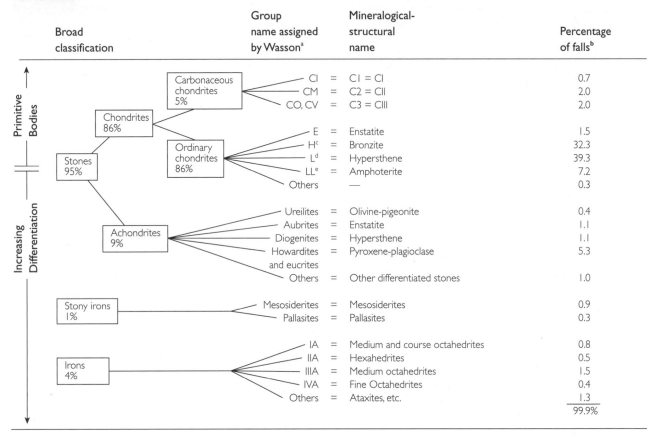

Broad classification	Group name assigned by Wasson[a]	Mineralogical-structural name	Percentage of falls[b]
Primitive Bodies ↑			
Stones 95% → Chondrites 86% → Carbonaceous chondrites 5%	CI =	CI = CI	0.7
	CM =	C2 = CII	2.0
	CO, CV =	C3 = CIII	2.0
Ordinary chondrites 86%	E =	Enstatite	1.5
	H[c] =	Bronzite	32.3
	L[d] =	Hypersthene	39.3
	LL[e] =	Amphoterite	7.2
	Others =	—	0.3
Achondrites 9%	Ureilites =	Olivine-pigeonite	0.4
	Aubrites =	Enstatite	1.1
	Diogenites =	Hypersthene	1.1
	Howardites and eucrites =	Pyroxene-plagioclase	5.3
	Others =	Other differentiated stones	1.0
Stony irons 1%	Mesosiderites =	Mesosiderites	0.9
	Pallasites =	Pallasites	0.3
Irons 4%	IA =	Medium and course octahedrites	0.8
	IIA =	Hexahedrites	0.5
	IIIA =	Medium octahedrites	1.5
	IVA =	Fine Octahedrites	0.4
	Others =	Ataxites, etc.	1.3
			99.9%

Increasing Differentiation ↓

[a] From Wasson (1974).
[b] From Scott (1978) and Dodd (1981).
[c] H = high iron content.
[d] L = low iron content.
[e] LL = very low iron content.

collisions, fragments from different parent bodies, or from different regions of one parent body, were intimately mixed and buried in ejected rubble on planetesimals, where they were welded together, only to be released again during a later collision. This explains the complex breccias found in many meteorites. With these complexities in mind, and with Table 6-2 as a guide, we turn now to more specific properties and types of meteorites.

Chondrules

Chondrules are glassy spherules (Figure 6-7) embedded in meteorites. They are composed principally of moderately high temperature minerals such as olivine, $(Mg, Fe)_2SiO_4$, and pyroxenes such as enstatite, $Mg_2(Si_2O_6)$.* They range in size from about 0.5 to 5 mm, and their structure shows that they formed during rapid cooling of droplets of molten material. Chondrules can make up 70% of a chondrite's total mass, in which case the meteorite consists of a mass of chondrules held together by a dark matrix, as shown in Figure 6-7a. The matrix typically consists of submicrometer-scale original condensate grains as well as fragments of broken chondrules, as can be seen in the magnified view in Figure 6-8.

How chondrules formed is hotly debated. There has hardly been much advance since the English mineralogist Sorby in 1877 invented the process of making thin sections of rocks for study through the microscope and then first described the chondrules. Sorby inferred they formed from "melted . . . glassy globules, like drops of a fiery rain." A number of chemical indicators suggest that the initial cooling of the chondrules occurred very fast, with temperatures dropping from 1500 K to solidification within minutes. One bit of evidence is the glassy structure of some chondrules. Properties of metal particles in the chondrules (Figure 6-8) suggest that the final stages of cooling to a few hundred Kelvin was much slower, requiring 10^4 to

*See the discussion in the preceding chapter of mineral condensation, but note that minerals and rocks are discussed in more detail in Chapter 9.

a

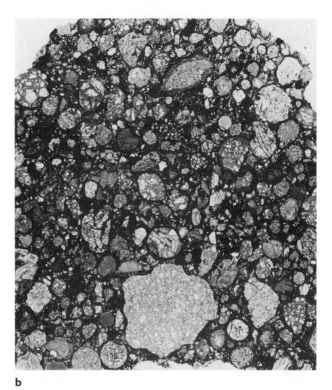

b

Figure 6-7. Meteorites with chondrules. (**a**) A 3-cm portion of the chondritic meteorite Björbole. Protruding from the surface are chondrules of various sizes, including a large one at lower left. (**b**) A cross section of the highly chondritic meteorite Chainpur, which consists mostly of chondrules embedded in a black matrix. (**a:** Center for Meteorite Studies, Arizona State University; **b:** Courtesy Laurel Wilkening, University of Arizona)

10^7 y (Wood, 1968). What process in the earliest history of the solar system could have produced "drops of a fiery rain"?

One group of theories states that chondrules formed directly out of the solar nebula gas, requiring gas pressures higher than normally estimated for the nebula (Suess, 1949; Wood, 1963; Wood and McSween, 1977). These theories suggest that chondrules condensed during disturbances such as shock waves that boosted local pressures in the nebula to unusually high values. Weidenschilling (1997) suggested the shock waves might have arisen as bodies were accelerated through the gas by Jupiter resonances. Wood and McSween criticize "a tendency to oversimplify [chondrules], to see all of them as quenched droplets of silicate melt." They emphasize a wide variety of forms and composition, presumably requiring different gas pressures and gas:dust ratios in different regions of the nebula. Whipple (1966) and others have suggested that lightning discharges or other electromagnetic effects in

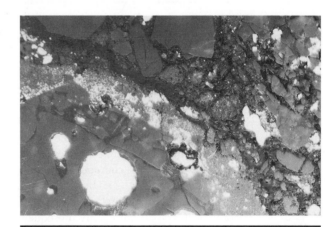

Figure 6-8. This microscopic view of the Chainpur chondrite shows the interface between a chondrule (rounded body, lower left half) and the dark matrix, which consists of irregular fragments of chondrules and fine grains (original condensates from the solar nebula?) a few micrometers across. The white, circular body in the chondrule is a 50-μm metal spherule. (Courtesy Laurel Wilkening, University of Arizona)

the nebula could have produced fused droplets resembling chondrules.

A more widespread idea is that chondrules formed during impacts on surfaces of planetesimals (Fredriksson, 1963; Ringwood, 1966; Dodd, 1971), which may have been meter-sized or smaller (Kieffer, 1975). The idea of "fiery droplets" spraying out of impact craters is buttressed by the finding of glassy spherules around at least one explosion crater of Earth (Hartmann and Wilkening, 1981) and chondrule-like spherules in lunar soil (Figure 6-9). But lunar soil has less than 1% chondrule-like objects and is composed mostly of angular fragments, so the problem is to explain how so many chondrules accumulated on a surface, somewhere, to be aggregated into chondritic rocks. Impact velocities at least 2.5 km/s are needed to melt rock, whereas the escape velocity of a 500-km planetesimal is only about 0.5 km/s, indicating that chondrules would probably initially be blown clear off all but the largest planetesimals (Wasson, 1974). Vast numbers of chondrules may have accumulated in space, only to reaccumulate later on planetesimals. Dodd (1965) showed that elongated chondrules often lie with their long axes parallel, typically the result of an orderly deposition process, perhaps during chondrules' reaccumulation on planetesimals' surfaces. Because meteorite data show that many planetesimals melted very early, an idea that might deserve further work is chondrule formation from droplets splashed out of impacts into asteroid-like planetesimals that were *already* molten inside.

Carbonaceous Chondrites

If we consider meteorites in order of their temperature of formation and the amount of heating they have experienced, carbonaceous chondrites might be listed first. They are the lowest temperature condensates, and probably formed in a reasonably cold part of the solar system where even ices may have been initially present—the outer asteroid belt or even beyond. There are several indications that they formed at low temperature and have not been highly heated, compressed, or otherwise altered: (1) They have unusually high substances of volatile compounds, such as water. (2) They have low densities. (3) They are rich in organic compounds that condensed at relatively low temperatures and would be driven off during any subsequent heating (Anders, 1963). (4) They contain the heavier elements in nearly their original cosmic proportions. (5) They apparently were never heated above about 500 K.

The black color of carbonaceous chondrites (abbreviated CC's) comes from finely dispersed black, opaque carbon and magnetite in their fine-grained matrix material

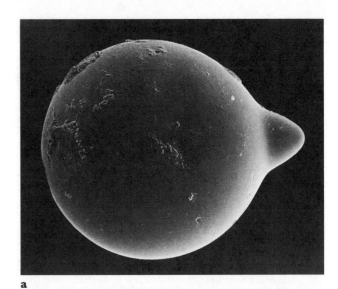

a

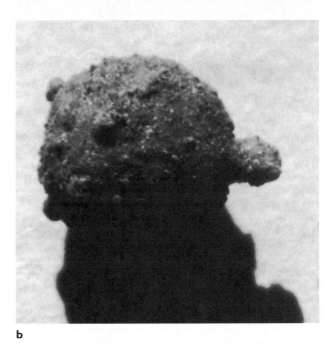

b

Figure 6-9. Glassy spherules possibly related to chondrule formation. (**a**) Green glass spherule from lunar soil. Such spherules are probably splashed from lunar impact craters. Diameter, about 0.2 μm. (NASA, courtesy David S. McKay) (**b**) Glassy spherules, welded during flight, from ejecta of experimental crater formed on Earth by explosion of 500 tons of TNT. Possibly due to moisture in original soil, interior texture is much more frothy glass than found in chondrules. Diameter about 3.5 mm. (Photo by author)

between the chondrules. For some years the nature of this matrix was a mystery, but detailed studies have revealed carbon in the form of poorly crystallized graphite (Smith and Buseck, 1981), fine-grained clay minerals, organic polymers, magnetite, iron sulfide (Wilkening, 1978), and even microscopic interstellar diamonds and other mineral grains trapped in the matrix during its formation (Lewis and others, 1979).

Carbonaceus chondrites were divided into three types by Wiik (1956) on the basis of chemical analysis—a classification later improved by Wasson (cf. Wasson, 1974). The groups are shown in Table 6-2 with the group names assigned by earlier authors. Type CI CC's have compositions closest to solar composition. They have high volatile content, including some 8% to 22% water content in the form of water of hydration—water bound in their minerals. Their density is only 2200 kg/m^3 compared with about 3600 kg/m^3 for other kinds of chondrites. Their minerals formed at low temperatures. Thus it appears that CI CC's have never been subjected to high heat or high pressure (that is, they have not been buried deep inside planet-sized bodies). However, they virtually all show brecciation (Bunch and Rajan, 1988, p. 147), suggesting they have been subjected to impact processing. They may have formed in a layer of surface soil as shallow as 100 to 300 m (Heymann, 1978), and seem to represent primitive material from the outer part of the solar nebula.

Some CC's have been exposed to small amounts of liquid water in their histories (DuFresne and Anders, 1963; Zolensky and McSween, 1988), and microscopic veins running through CI CC's contain carbonate and sulfate deposits (the type of deposits forming bathtub rings and caliche layers, left after evaporation of mineral-bearing water), apparently having accumulated over several generations of chemical activity as water seeped through the material (Richardson, 1978). The parent bodies must have been ice rich (i.e., comet nuclei), with the water liberated during mild heating. The present CI CC's may be the solid residue left after the ice disappeared.

CI CC's have no chondrules—the nomenclature is misleading! Perhaps their region of formation or their initial parent bodies were so ice rich that the chondrule forming process could not produce silicate spherules. Alternatively, CI's may represent planetesimal material from a prechondrule-forming era.

The remaining three classes of CC's have chondrules and progressively less water. CM's and CV's have typically 2% to 16% bound water in their minerals. CO's average around 1% bound water. Breccias are common among CM's and CV's, but rare among CO's.

Table 6-2 lists CC's as being only about 5% of meteorite falls, with the CI's being only about 1%. These numbers must be smaller than the actual percentages of CC meteoroids in near-Earth space, for several reasons. First, CC's are weak, so they probably break up in the atmosphere more easily than other types, thus being under-represented on the ground. Second, the dominant type of meteorite in the lunar soil is CI material (Ganapathy and others, 1970; Taylor, 1975). Third, direct photographic studies of meteors and meteorites show that a substantial fraction of small, weak bodies burn up in the air, and these may be the "missing" CC particles that accumulate in lunar soil.

As we see in the next chapter, most asteroidal and cometary bodies beyond the middle of the belt, starting at about 2.5 AU, are black bodies with spectra similar to those of carbonaceous chondrites. Researchers believe that CC's must come from that material, but it is unclear if they come from the outer asteroid belt or from comets.

Carbonaceous chondrites get their name from their relatively high abundance of carbon. The carbon matter is mostly not in the form of well-crystallized minerals but in more amorphous structures including graphite grains silicon carbide, and amorphous mixtures of organic molecules. Organic compounds—complex molecules involving the C-H bond—make up about 1% or less of the meteorite, as discovered in 1834 by Berzelius. An exciting area of research has been the relation of these organic materials to the origin of life. By the 1960s, various specific organic compounds had been identified in meteorites, but there was controversy over whether any of them had biological origin. The situation gained still more interest when Kvenvolden and others (1970) conclusively identified in the freshly fallen Murchison (Australia) carbonaceous chondrite amino acids that had not formed on Earth. Amino acids are the class of molecules that join to form proteins and are thus an important building block of life. Muchison's amino acids were proven to be extraterrestrial because they had different proportions of "right- and left-handed" molecular structure than do terrestrial amino acids.*

There is no evidence that any living material ever evolved in meteorite parent bodies, although infalling CC's delivered organic material to planets, where biochemistry went further, at least in the case of Earth. Some experiments suggest that the organic material compounds formed when cosmic rays zapped ice mixtures containing methane ice (CH_4) (Thompson and others, 1987; Strazzula and Johnson, 1991; Bohn and others, 1994). In particular, Thompson and others (1987) and Khare and others (1993) showed that methane/water ice mixtures, irradiated by cosmic-ray-like atomic particles, produced complex organic compounds that match spectra of comets and outer solar system asteroids. Other experiments (see Wasson, 1974, for review) show that the types of organic compounds found in CC-like material can be formed by heating initial mixtures of carbon monoxide (CO), water (H_2O), and ammonia (NH_3) with catalysts to about 1000 K and then allowing extended baking at temperatures around 500 K—presumably in the CC parent bodies. These ingredients and ovens were undoubtedly common in the great kitchen of the solar nebula.

---

*The chains of H, H_3N^+, COO^-, and other molecules can join in two different mirror-symmetrical ways to form amino acids. Nearly all living organisms on Earth formed from and replicate the "left-handed" type. This type dominates on Earth but not in space. See Pellegrino and Stoff (1979) for a nontechnical review.

Chondrites

The rest of the chondrites, sometimes called ordinary chondrites, are the most numerous meteorites. They have bulk chemical compositions similar to CC's, and, like CC's, they have not been melted. However, they do not have the carbon and water-bearing matrix, and they have been chemically or geologically processed slightly more than CC's. The spectra of asteroids in the inner half of the asteroid belt (from 2 to about 2.5 AU) indicate that they have rock-forming minerals more similar to ordinary chondrites than to CC's. Also, proportions of oxygen isotopes in ordinary chondrites and rocks from Earth, the moon, and Mars are all quite similar, but they are different from those in CC's. For these reasons, ordinary chondrites are believed to be representative of the ancient planetesimals formed among the terrestrial planets and inner asteroid belt out to 2.5 AU. This contrasts with the CC's, which seem to have formed beyond 2.5 AU.

The chemistry of iron in chondrites is one indicator of the amount of chemical processing. Iron may either be strongly bound to oxygen atoms in silicate rock-forming minerals (**oxidized**) or exist as pure metal (**reduced**). It is found in both states in chondrites. The geochemist G. T. Prior (1920) estimated that all chondrites have about the same iron content, and he developed the so-called **Prior's rules,** which state that as the amount of oxidized iron decreases, the amount of reduced iron correspondingly increases. This relationship indicates some chemical reactions in the material but not enough geologic activity to melt the rock or drain molten iron away from the system. As an example, metallic iron could be oxidized (rusted, in fact) by interacting with water vapor or with moisture inside heated, ice-rich planetesimals.

In a now classic review, Urey and Craig (1953) revised Prior's result somewhat by showing that different groups of chondrites have different total iron contents. As shown in Table 6-2, these came to be called the high-iron **H group** (Figure 6-4), the low-iron **L group,** and the very low iron **LL group.** These discontinuities in total iron might be evidence of chondrite origin in several environments, such as different parent bodies or different nebular regions.

One small group of chondrites, the enstatite chondrites, were noted in the last chapter as having a composition quite close to that for early-condensed material formed at around 1,300 K. Some well-preserved enstatite chondrites may therefore exemplify condensation products in an early, high-temperature nebular regime in what late became the terrestrial planet zone. Dodd (1981) notes that others are metamorphosed and chemically related to certain achondrites and irons. He therefore hints that these groups might represent the interior core-to-surface structure of the original enstatite parent body.

Another subgroup of chondrites is called the unequilibrated chondrites. As with carbonaceous chondrites, much of their iron is oxidized, perhaps by interaction with water or steam, but the iron and other minerals are in mixtures that could have not survived if they had ever been heated to high temperatures. Unequilibrated chondrites may also be primitive objects accumulated at lower temperatures than was true of enstatite chondrites.

Although no chondrites were melted by heating in the parent bodies, minor heating did affect some of them. This gives us an important tool for estimating the sizes of the parent bodies. Cooling rates can be calculated from properties of individual crystals in the metal particles of chondrites (see Wood, 1968, pp. 29–39, for an introductory discussion). If a rock were heated deep inside a parent body, the overlying material would insulate it and make it cool slowly. The closer it was to the surface, the faster it would cool. Results (Wasson, 1974, p. 185) show that the chondritic material was typically insulated by overlying material about 2 to 50 km thick. Therefore, the parent bodies must have had diameters at least 4 to 100 km. They could have been that far down inside still larger bodies, up to hundreds of km. These sizes are characteristic of actual asteroids, which fits the theory that chondrites are fragments of asteroids.

Figure 6-10 shows one of many examples of a chondrite that includes an intact, angular fragment of another meteorite (a CC in this case), showing that deposits of chondrules and rock chips were in intimate contact in at least some parent bodies. This finding supports the theory that different types of parent bodies collided, and many chondrites formed in the layers of mixed fragmental soil in the outer kilometers of planetesimals, similar to the fragmental soil layer on the moon.

Gas-Rich Chondrites

Further evidence for chondrite formation in soil-like layers on the surfaces of parent bodies came in the 1970s, when meteorite researchers realized that many brecciated chondrites consist of grains that were once exposed to cosmic rays and solar wind gases. Microscopic tracks can be seen in these crystals where cosmic atomic particles penetrated them. (This phenomenon was clarified by lunar studies that showed the same effect in lunar surface soils whereas buried soils were shielded from such effects.) Detailed studies of meteorites and lunar soils show episodes of burial, re-excavation, and reburial from impacts. For example, a gas retention age measured by Bogard and Husain (1977) showed that the fragments now cemented together in the Plainview chondrite were blasted out of an impact site 3.7 Gy ago. As the fragments themselves contain solar gases, they must have lain together on some surface before being cemented together. The solar gases accumulated post-impact, because the impact would have driven out the loosely bound solar gases. Thus, the data tell a complete story: the material was broken loose in a collision 3.7 Gy ago; instead of being dispersed in space, the pieces were thrown together onto a surface where they

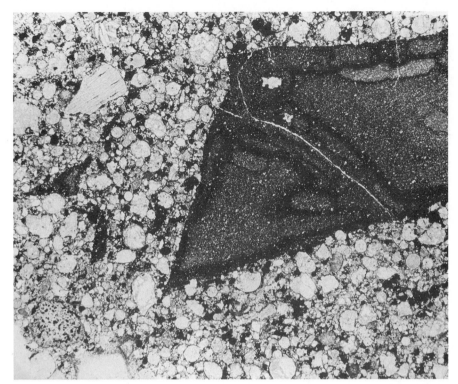

Figure 6-10. A meteorite breccia. An angular carbonaceous chondrite fragment (dark mass) lies inside the chondrite Sharps, illustrating mixture of rock chips from different parent bodies. The white "fish" in the upper corner of the dark mass is a white refractory inclusion, similar to the famous inclusions in Allende (cf. Figure 5-7). Dark rim on the carbonaceous chondrite fragment is indicative of some heating at the time the entire mass was forming into a chondritic rock. (Courtesy Laurel Wilkening, Lunar and Planetary Laboratory, University of Arizona)

were irradiated by solar wind, and then buried and cemented together into a single rock. In another example, two different pieces of the Djermaia gas-rich chondrite breccia showed different sequences of burial and exposure in the upper meter or so of a soil-like layer on some parent body (Lorin and Pellas, 1979). Layers of fragmental debris on the surfaces of cratered planetesimals seem the only plausible setting for such mini-dramas of the past. Modeling of the evolution of such layers has been published by Housen and others (1979).

Achondrites

Achondrites are stony meteorites without chondrules. The crystal structure of achondrites is coarser than that of chondrites. Coarse crystal structure is usually an indication of slow cooling in insulated surroundings—for example, it is found in underground intrusions of solidified magma. Of all the meteorites, achondrites are closest to terrestrial rocks. In particular, they are similar to **igneous rocks** (rocks produced from molten material). Some even resemble basaltic lavas and are called **basaltic achondrites.**

These characteristics indicate that the achondrites were produced when some kind of parent material melted and differentiated. One question that naturally arises is whether chondrites were the parent material. The answer is—probably (Kerridge and Matthews, 1988; McSween, 1987; Wasson, 1985). As early as 1962, Mason empha-

sized that if chondritic material melted, it could cool and recrystallize into mineral groups similar to those observed among the achondrites. The chondrules would indeed be destroyed if chondrites had melted, and chondrites do show a range of chondrule states—from perfect preservation to nearly obliterated, recrystallized examples. Also, the iron content has been severely reduced in many achondrites, as we would expect if the heavy iron had drained away from the silicate phase during the melting and differentiation. Furthermore, most achondrites are breccias (Wasson, 1974; Bunch and Rajan, 1988); this finding is consistent with the ideas that achondrites formed as lava flows on surfaces of parent bodies, and then were chewed up by impacts, in the same way that the lunar surface layers have been pulverized into dust and rock fragments.

The suggestion that achondrites were produced by melting chondrites is qualified, because the parent material of the achondrites probably did not have precisely the same composition as that of the chondrites. For example, certain isotope ratios are different.

Certain types of achondrites stand out as prime examples of lavalike basaltic igneous rock, formed from solidification of molten material. Among these are the **eucrites,** a type of lavalike rock that sometimes displays the bubbly texture of lava flows on Earth—called vesicular texture. They are believed to be pieces of basaltlike lava flows extruded onto the surface of some parent body in the form of lava billions of years ago (Sears and others, 1997). The parent body seems to be asteroid 4 Vesta, as its spectrum shows it to be the only large asteroid with a eucritelike lava surface (McCord and others, 1970; Binzel

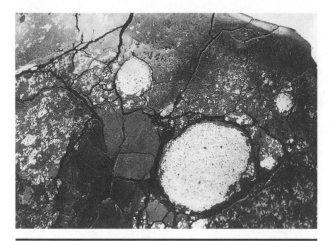

Figure 6-11. Cut and polished section of a stony-iron meteorite from the Bondac Peninsula, Philippines. The dark, fractured matrix is silicate rock, possibly fractured in collisions. The bright nodules are nickel-iron inclusions a few centimeters across. (Photo by author)

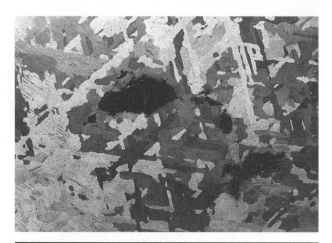

Figure 6-12. Cut and polished section of the Campo del Cielo (Argentina) iron meteorite. Intersecting metal crystals can be studied to reveal environmental conditions inside the parent body. The dark inclusions are stony silicate bodies, the largest being about 4.5 cm across. This pattern of metal crystals is called the Widmanstätten pattern, after its discoverer. (Photo by author)

and Xu, 1993). The idea that fragments of Vesta reached Earth was supported when researchers recognized that Vesta fragments might reach a nearby resonance and be perturbed onto Earth-crossing orbits (Migliorini and others, 1997), and also when Cruikshank and others (1991) discovered that three small Earth-approaching asteroids have spectra indicating basaltic-eucritic composition. Finally, in 1997, Hubble Space Telescope photos revealed a fuzzy image of a probable 460-km crater on one side of 458×578 km Vesta, from which large fragments might have been expelled (Thomas and others, 1997).

Stony Irons

Stony irons (Figure 6-11) are divided into two main classes, pallasites and mesosiderites, depending on the minerals in the silicate stony part. The pallasite silicates are principally olivine, a dense mineral believed common in planetary mantles. The pallasites may thus have formed where mantle materials abutted the surface of an iron core or the surfaces of iron blobs near the core. In this vein, Stoffler (1997) emphasized that modern studies of meteorites are an increasingly important tool in understanding Earth's mantle structure.

Mesosiderites' silicates are mainly plagioclase and pyroxene minerals and thus resemble crustal lavas rather than deeper mantle rocks. How the more crustlike rocks of mesosiderites came into contact with iron is more problematic. Suggestions include sinking of solid basaltic slabs into a melted interior (Greenberg and Chapman, 1984) or intimate mixing during collisions.

Irons

Iron meteorites are the most useful in telling us about the size, number, and history of the meteorite parent bodies. They are classed according to nickel content, crystal structure, and related parameters.

Widmanstätten Pattern: Size of Parent Bodies

In many irons, the two major minerals, kamacite and taenite, occur in crystals that interlock in an interesting geometry called the **Widmanstätten pattern*** shown in Figure 6-12. The pattern is prominent only after the meteorite is cut, polished smooth, and then etched with acid. Large crystals mean long cooling times, suggesting that these materials were buried deep inside parent bodies, where they would have been insulated against rapid cooling. Measures of the structures of the crystals suggest locations near the centers of bodies about 30 to 400 km across (Wood, 1968; Wasson, 1974; Scott, 1977).

Irons were first classified according to the Widmanstätten pattern, but later workers such as Wasson (1974) reclassified them according to chemical properties. When abundances of certain specific elements such as gallium, germanium, and nickel are considered, certain groups of meteorites are found to have nearly the same

---

*Named after Count Alois von Widmanstätten, director of the Imperial Porcelain works in Vienna and discoverer of the pattern in a meteorite in 1808.

amounts of these elements, giving rise first to the concept of four major **gallium-germanium groups** and later to Wasson's groups Ia, IB, IIA, and so on. Each of these groups may have come from a separate parent body, suggesting perhaps one-half dozen or dozen parent bodies supplying most of our iron meteorites.

Neumann Bands: Evidence for Collisions

Neumann bands are fine striations visible in a polished and etched surface of most hexahedrites and some other irons. They are caused by an alteration of the crystal structure within the kamacite crystals, known as *twinning* (not to be confused with the interlocking Widmanstätten arrangement of the crystals themselves). Uhlig (1955) showed that Neumann bands are the result of strong shocks applied to the meteorites when they were at low temperatures, not above 900 K and probably below 600 K. Again, the interpretation of this finding is that after the interiors of parent bodies had solidified through cooling, they were subjected to violent impacts, probably collisions with other parent bodies.

Brecciated Meteorites: Proof of Collisional Mixing

The brecciated character of many stony meteorites offers considerable further insight into their history (Wilkening, 1977). Three categories are recognized: **monomict breccias,** in which all fragments come from the same meteorite type; **polymict breccias,** in which fragments come from very different objects or environments (for example, the chondrite and carbonaceous chondrite breccia in Figure 6-10); and **genomict breccias,** a category coined by Wasson (1974), in which the fragments have similar basic elemental compositions but show evidence of different geologic processing. Breccias, especially the polymict type, show that parent bodies went through complicated collisions and fragmentations that mixed pieces of one or more parent bodies. Monomict breccias might have arisen when one region of a parent body was broken and mixed (during impact crater formation?); polymict breccias, when two different parent bodies contributed their fragments (Wahl, 1952); and genomict breccias, when different parts of the same parent body (such as core and mantle fragments) were mixed by an explosive collision followed by reaggregation of the fragments into a new body (Davis and others, 1979; Hartmann, 1979). All these processes testify to the wide range of collisional energies that affected the parent bodies—from simple cratering of the surface to completely shattering the whole body.

Meteorite Ages

The last chapter described how formation ages, formation intervals, and solidification ages can be measured from planetary rock materials. Table 6-3 lists some age relations that can be determined; they are discussed below.

An important generalization to remember is that the combined isotopic and age studies indicate that meteoritic material generally went from a condition of dispersed dust to incorporation within solid bodies, between about 4560 and 4571 My ago. Within about 10 to 20 My of the beginning of the solar system, most of the melting, partial melting, and/or differentiation within the asteroidlike solid bodies had occurred, and the parent bodies were cooling.

Gas retention ages of meteorites measure the interval from the most recent degassing event of the rocks to the present. The "degassing event" almost certainly refers to the last major impact. The method utilizes gaseous daughter elements, such as argon, which accumulate in the crystal-lattice interstices. Any major collision among parent bodies can cause tiny fractures and heating, which allow the gas to escape from the crystal. New gas then starts to collect, and measuring this gas dates the time since the shock event. If the rock was never shocked strongly enough to cause heating, the gas retention age may date the original solidification of the rock.

Gas retention ages are concentrated at about 4000 to 4500 My ago, but perhaps a third are dotted through more recent time. These almost certainly reflect a long history of collisions among parent bodies, with most major collisions occurring early in the 4.0–4.5 Gy period, which also marked the intense early cratering of the planets.

Cosmic ray exposure ages, or CRE ages, measure the duration of the meteorite's exposure to space. Cosmic rays penetrate only about a meter into solid material. Thus, the CRE age is the interval from the meteorite's first exposure to space (or when it was first within 1 m of the surface of a larger body) until its impact on Earth. Some meteorites had a multiphase exposure history, first being just under the surface of the parent body, then perhaps being buried deeper if ejecta were thrown on top of them, and then being exposed directly to space when the parent body fragmented.

Exposure ages are measured either by utilizing radioactive and stable isotopes created by nuclear reactions with cosmic rays, which penetrate roughly 1 m into the parent body, or by counting microscopic particle tracks left in the material by the passage of cosmic rays through it. The former method is widely applied to meteorites, and both methods have been used on lunar samples.

TABLE 6-3 **Types of Meteorite Ages**

Type of age and interval measured	Examples of isotopes involved	Meteorite type	Results
Formation age (establishing of isotopic identity)	$^{87}Sr-^{86}Sr$	Stones	4.55–4.65 Gy
Formation interval (nucleosynthesis to gas retention)	$^{129}I-^{129}Xe$	Stones and irons	c. 20 My
Solidification age (solidification to present)	$^{87}Rb-^{87}Sr$ $^{235}U-^{207}Pb$ $^{238}U-^{206}Pb$ $^{232}Th-^{208}Pb$ $^{40}Ar-^{39}Ar$	Stones and irons	Mostly 4.4–1.6 Gy. A few lavalike meteorites solidified later. Nakhlites (a type of basaltic achondrite) solidified about 1.3 Gy ago.
Gas retention age (onset of retention to present)	$^{40}K-^{40}Ar$	Stones (irons contain little K)	Often 4.4–4.6 Gy. Some younger, suggested to be due to degassing during shock events.
Cosmic ray exposure age (onset of exposure to present)	$^{3}He^{a}$	Stones	0.1–100 My (young due to erosive impacts in space?)
		Irons	1–1000 My
Date of fall (end of cosmic ray exposure to present)	^{14}C ^{26}Al	Stones Irons	c. 0–10 000 y c. 0–50 000 y

[a]Produced by cosmic rays.

Mathematical Notes on the Rubidium-Strontium System of Rock Age Measurement

Suppose a rock crystallizes from a uniform molten mass, as shown in the top of Figure 6-A. As it solidifies, different mineral crystals will form, with different affinities for the two chemicals rubidium and strontium. For simplicity, imagine the rock composed of three minerals, A, B, and C, each having increasing initial amounts of rubidium. Initially there is relatively little strontium 87, as that is the daughter element produced by radioactive decay of rubidium 87, with a half-life of 72 Gy. A chemist is now about to measure the strontium 87 and rubidium 87 in each mineral crystal, A, B, and C, but he will express his results normalized to the strontium 86 isotope. That is, rubidium 87 will be expressed not in terms of numbers of atoms, but as the ratio rubidium 87:strontium 86 and strontium 87 as the ratio strontium 87:strontium 86. One convenience of this is that the initial strontium 87:strontium 86 ratio will be the same in all crystals; this is because the crystals grow according to only the chemical nature of the strontium (the outer electron structure of the atom being all that the crystal "sees"), rather than the isotopic nature (determined by the nucleus of the atom). Thus each crystal incorporates the same initial proportions of strontium 87 and strontium 86. This initial abundance is marked in Figure 6-A, along with the initial compositions of the three crystals.

As soon as the rock forms, the rubidium 87 decays, of course. For each atom of rubidium 87 lost, the amount of strontium 87 increases by one atom, which is trapped in the rock. Therefore the compositions of minerals A, B, and C will each evolve to the upper left on the 45°-angle dashed lines, producing points A′, B′, and C′ by the time a chemist makes his measurement a billion years or so after the rock forms. The composition measured at other times would fall along other lines. The line is called an *isochron* (*iso* = same, *chron* = time), as all its points represent the rock at a specific age. The slope of the isochron is a measure of the age.

Note that all the isochrons marking different ages would go through the same point on the ordinate. This point, marked BABI in Figure 6-A, denotes the initial isotopic composition ratio of the strontium, strontium 87:strontium 86, at the beginning of the solar system's history. The peculiar acronym comes from "basaltic achondrite, best initial," meaning the best estimate of the initial strontium ratio, based on studies of a basaltic achondrite, particularly studies by Gerald Wasserburg and his colleagues at Cal Tech.

As the rubidium-strontium dating system was first worked out, planetary chemists assumed that the BABI value was a universal standard for all material in the solar system.

Cosmic ray exposure age measurements show that stones have typically been exposed to space for about 5–50 My, whereas irons have been exposed much longer, typically 100–1000 My. This is the amount of time it has taken the meteoroids to travel from the point where they were meter scale bodies (small enough so their insides were not shielded from cosmic rays) to Earth.

Cosmic Ray Exposure Ages and the Yarkovsky Effect

Why are cosmic ray exposure ages not 4.5 Gy, and why are irons' ages longer than those of stones? The answer to the first question is that the CRE age measures the time since the last collision ejected a body small enough to be penetrated by cosmic rays. The answer to the second question involves an interesting interplay between collisions, erosion, and dynamic orbital effects (review Chapter 3). Remember that meteorites appear to come from the asteroid belt. Ejection velocities from collisions in the belt are too slow to get meteoroid fragments out of the belt onto Earth-crossing orbits; they are probably delivered onto Earth-crossing orbits by resonances with Jupiter. Dynamical studies show that if meteoroids reach resonances, they get kicked onto Earth-crossing orbits very fast, within a few My. Therefore, the meteoroids must not be created in the resonances but must drift for 10 s or 100 s of My before they reach the resonance. This drift is probably due to the Yarkovsky effect, which changes the orbital size of bodies in the size range of 0.1 to 10 s of meters.

Why do stones have lower CRE ages than irons? For one thing, they drift faster by the Yarkovsky effect. In addition, they are weaker than irons, so that micrometeorites continually erode them and larger impacts may shatter them. Thus, a plausible scenario is that when meter-scale irons are ejected by catastrophic breakups, they undergo Yarkovsky drift in the asteroid belt for 100 My or more until they hit a resonance and then are perturbed onto Earth-crossing orbits in a few My, and then hit Earth a few My later; but when meter-scale stones are ejected, they have a high probability of destruction within 50 to 100 My, so they must start near enough to a resonance that they can reach it and be sent Earthward in tens of My.

Clustering of Ages: Evidence for Specific Collisions

Certain groups of chemically related meteorites have about the same gas retention or cosmic ray exposure age, implying that they all broke loose from some parent body

However, lower values were found in the Angra dos Reis anomalous achondrite meteorite (the so-called ADOR value), an inclusion in the Allende carbonaceous chondrite meteorite. Later studies suggest that there was an evolution of strontium isotope compositions, perhaps different in different parts of the early solar system, culminating in a metamorphosing event that mixed and homogenized the strontium isotopes about 74 My after the time when the BABI value applied (see summary by Wasson, 1974). The clustering of rubidium-strontium formation ages, together with the quick evolution of the original strontium value, gives striking evidence of the rapidity of the planet-forming events that happened in the first 2% of the solar system's history.

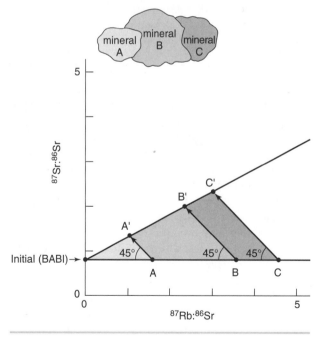

Figure 6-A. The rubidium-strontium dating system, explained in the text.

at once. For example, studies of H chondrites reveal strong concentrations of CRE ages at about 6 and 20 My ago and show that each age group contains H chondrites belonging to various petrologic subtypes (Crabb and Schultz, 1981). These results indicate that two major collisions released fragments from parent bodies which already had mixtures of fragments. Similarly, a certain group of L chondrites, recognizable by blackening due to mineral changes induced by a strong mechanical shock (impact?) have the same gas retention age of approximately 450–500 My (Goplan and Wetherill, 1971). Swedish investigators found fossil evidence of a few-My shower of L chondrites (embedded in limestone) at that time—marking the initial sweep-up of fragments of that collision (Schmitz and others, 1997; Kerr, 2001). Table 6-4 shows that major collisions created many of our meteorite types in the last few percent of solar system history, even though the first-generation meteorite parent bodies date back to the formation of the solar system itself.

Date of Fall

Fall dates measure the interval from the time the meteorite ceased being shielded from cosmic rays (its arrival on Earth) to the present. Stones are not identifiable much longer than about 10,000 y on most parts of Earth's surface because they are subject to weathering and erosion. However, some Antarctic stony meteorites fell as long as 1.5 My ago, a dating that is possible because of the quality of deep-freeze preservation in the Antarctic ice sheet.

Tektites

Tektites are small, distinctive, and rare glass objects rich in silica. Their dimensions are typically a few cm. They are often found in fields broadly strewn over hundreds of km. Their origin was a mystery for many years, but in 1963, Chapman and Larson (1963) used wind tunnel experiments to show that the tektites had either entered or reentered the atmosphere at hypersonic speeds, causing them to melt and reshape. Their cosmic ray exposure ages are very different from those of meteorites—most important in that tektites have been exposed in space less than 300 y, perhaps only minutes, which is much too short for asteroidal or cometary origin. This discovery restricts their place of origin to the Earth-moon neighborhood.

The high silica content of tektites (about 73% SiO_2) suggests that they come from Earth, where silica-rich sediments abound, and not from the moon (as once suggested), because lunar igneous rocks are not silica rich. The current interpretation is thus that tektites are bits of fused silicate blasted out of terrestrial meteorite impact craters. They apparently flew outward in huge expanding

TABLE 6-4	Dates of Meteoroid Collisions

Meteorite type	Collision dates (years before present)[a]
H chondrites	4.5×10^6
	2×10^7
Aubrite achondrites	4×10^7
L chondrites	5×10^8
Hexahedrite irons	2×10^8?
Medium octahedrite irons	6×10^8

Note: See Wood (1968), and Wasson (1974) for discussions of the data. The collisions and suspected collisions greater than 10^8 y ago were probably major collisions that fragmented large objects, possibly original parent bodies. More recent collisions may have fragmented only pieces of parent bodies. Taylor and Heymann (1969) have shown that many meteorites give evidence of other major collisions scattered in time between 0.6 and 4.5 Gy ago.

[a] Dates determined by clustered ages. All ages determined from cosmic ray exposure except for L chondrites, the age of which was determined from gas retention.

clouds of debris, solidified, and fell back through the atmosphere (Faul, 1966; Taylor and Kaye, 1969).

Each "strewn field" of tektites is thus related in age and location to an ancient impact crater. The best example is the Czechoslovakian tektite group called moldavites (after the Moldau River). They are 14.8 My old, exactly matching the age of the nearby 24-km-diameter Rieskessel crater in Germany (see Table 10-4).

Where Do Meteorites Come From? — Orbits and Origins

We've already given several lines of evidence that most meteorites are fragments of asteroids. For example, the cooling times indicate that before they crashed to Earth, many meteorites were inside bodies a few hundred kilometers across—the size of asteroids.

There are some further lines of evidence. The orbits of Earth-approaching meteoroids have half-lives against Earth-collision of only a few My. Therefore, virtually all meteorites would have been swept away long ago unless the supply was constantly being replenished by a reservoir somewhere in the solar system. Orbits of meteorites, determined from photos of them as they fall, directly suggest the asteroid belt as the reservoir. Although only a few meteorites have well-measured orbits, they all have aphelia in or near the asteroid belt, as shown in Figure 6-13.

Furthermore, many stony meteorites come from a soil layer like the moon's. Their cosmic ray exposure ages and the amount of inert gases their surfaces have absorbed from the solar wind are only around 0.01% to 1% of the

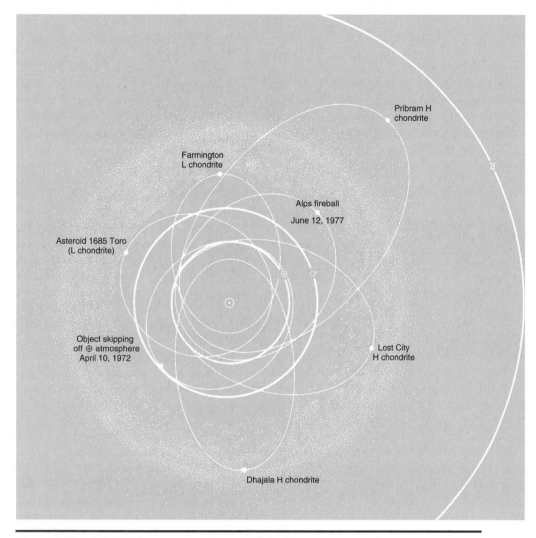

Figure 6-13. Orbits of four meteorites and three related objects. Most have aphelia in a region from the inner edge of the asteroid belt to somewhat beyond the belt, which is shown stippled. Drawing shows orbit sizes, perihelia, and aphelia to scale, but orientations of semimajor axes are chosen for clarity. (Data from Wasson, 1974; Ballabh and others, 1978; Jacchia, 1974; Clepecha, 1981)

lunar values. As pointed out by geochemist Edward Anders in 1978, such data are consistent with a meteorite origin among asteroids.

All these clues point strongly toward the asteroid belt as the reservoir producing most meteorites. The larger asteroids have the sizes inferred for meteorite parent bodies, and calculations show that they undergo numerous collisions. Ejected fragments would not fly off fast enough to intercept Earth directly but could undergo a two-stage process: (1) The collision would eject fragments onto orbits in the belt. (2) Small fragments, in the 0.1 to 10 m diameter range, would undergo a drift in their orbits by the Yarkovsky effect, until they hit a resonance. (3) Resulting perturbations would then send the fragments onto orbits that intercept Earth (Farinella and others, 1997). The resulting orbits would resemble those shown in Figure 6-13.

A seeming clincher to the argument is that belt asteroids and Apollos have spectra indicating various compo-

sition classes close to carbonaceous, stony, stony-iron, and iron meteorite classes. Indeed, the second-largest belt asteroid, 550-km Vesta, matches eucrites and perhaps one or two other subtypes of basaltic achondrite (McCord and others, 1970), whereas Earth-crossing asteroid Toro matches the most common meteorite type, L chondrites (Chapman, McCord, and Pieters, 1973). We probably have pieces of Vesta and Toro in our museums.

In spite of these strong arguments, some problems remain. First, comets may contribute some meteorites. Stony material inside the icy matrix of comets may be weak carbonaceous-chondrite-like material. Thus, a large flux of cometary CC's may enter the top of the atmosphere but disintegrate before reaching the ground. In general, there is still some uncertainty about how much CC material comes from among the black asteroids in the outer half of the belt and how much comes from cometary sources beyond the belt.

An additional complexity is the possibility that the current distribution of meteorite types falling on Earth is not representative of the average over geologic time, perhaps because the meteorite types falling in the last million years are strongly dependent on the last few major asteroid collisions, as shown in Figure 6-14. This means that our meteorite sample depends on the "luck of the draw" among recent large colliding asteroid pairs that may have sent fragments our way. The meteorite record may bear some evidence of type variations resulting from this effect.

Meteorites from the Moon and Mars

An irony followed the massive American Apollo program and Soviet Luna program to bring samples back from the moon: Lunar samples were found here on Earth among meteorites collected in Antarctica! Antarctic ice fields have produced major meteorite collections because meteorites falling on ice are trapped and later released when glaciers melt or erode. From 1982 to 1991, 12 lunar rocks were found there and elsewhere. These samples don't negate the value of Apollo and Luna, because without lunar flights we would not have known how to recognize them; in any case only the Apollo and Luna samples have known geologic context on the moon. The Antarctic lunar rocks add a random sampling from other lunar locations (including the far side?) not visited by landers. They also

prove that impacts can blast intact rock samples off a body as big as the moon (escape velocity 2.4 km/s).

These discoveries in turn led to recognition of even more exotic celestial rocks: rocks from Mars! The rocks are mostly lavalike meteorite types called shergotites, nakhlites, and chassignites. Initially these were called **SNC meteorites,** or SNC's (affectionately dubbed "snicks"); some writers simply call them martian meteorites (MM's) to avoid limiting the classes. Most examples are basaltic and about 1.3 By old—much younger than other meteorites. Oxygen isotope ratios proved they weren't from Earth or the moon; asteroids should have cooled too early to have produced such young lavas.

As early as 1979, Nyquist and others (1979) and Wasson and Wetherill (1979) suggested Mars as a possible source, but others scoffed that intact rocks couldn't be blasted off Mars, with its 5 km/s escape velocity. (This was based on an erroneous "golf ball" conception of the process, as if the rock were blasted off with an instantaneous impulse; in a large cratering event it is cushioned by the massive ejecta cloud.) Soon the SNC's were found to contain nitrogen and noble gases matching the ratios and isotopic composition of gases found on Mars by the Viking lander (Bogard and Johnson, 1983). Also, the 1.3 By age matched estimated ages of widespread basaltic lava flows on Mars, based on crater counts (Hartmann and others, 1981). Gladman and others (1996) simulated trajectories of particles knocked off Mars with speeds about 1 km/s faster than escape velocity and found that although the largest fraction (38%) are destroyed by falling into the sun,

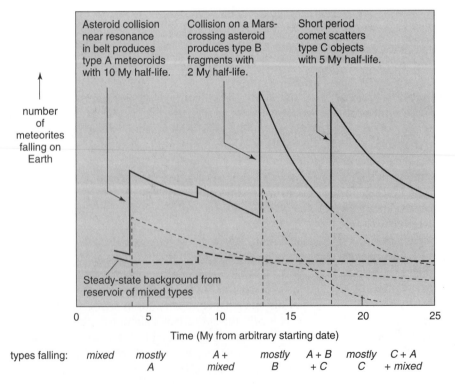

number of meteorites falling on Earth

Asteroid collision near resonance in belt produces type A meteoroids with 10 My half-life.

Collision on a Mars-crossing asteroid produces type B fragments with 2 My half-life.

Short period comet scatters type C objects with 5 My half-life.

Steady-state background from reservoir of mixed types

Time (My from arbitrary starting date)

| types falling: | mixed | mostly A | A + mixed | mostly B | A + B + C | mostly C | C + A + mixed |

Figure 6-14. Schematic imaginary history of meteorite types falling during a 25-My interval on Earth. Various events (top; subject to random chance) provide bursts of fragments of different compositional types. These have different half-lives against sweep-up by Earth, depending on their orbits. Thin dashed lines show individual major collisions. Solid line (top) shows net additive result of all collisions and fragments. The net result is episodic bombardment by different groups.

about 8% eventually impact Earth—establishing that delivery is possible once objects escape from Mars.

Thus, most researchers now believe we have a handful of Martian rocks from unknown Martian sites (see Pepin and Carr, 1991, for review of evidence). By 1997, about a dozen had been recognized. The next step is to visit Mars and compare the chemistry and geology of the Mars meteorites with ground truth at different "known" sites!

Siberia, 1908: A Grand Meteoritic Explosion

Meteoritic bodies hit Earth more frequently than most people recognize. Upper-atmosphere blasts comparable to the scale of atom bombs (but not involving nuclear energy) occur every century or so. One example happened at 7:17 AM on the morning of June 30, 1908, in the skies over Siberia. It was caused by the explosion of a large meteorite—which we could also call "a tiny asteroid fragment"—at an altitude roughly six kilometers in the atmosphere. This was the largest meteorite explosion recorded in history, and it offers a good chance to review some of our knowledge of meteoritics. Two broad questions are these: What happened and what caused it?

No eyewitness pictures of the event are available. However, Russian scientists collected eyewitness accounts of the event (Krinov, 1966). I have used these—together with modern knowledge of impacts—to reconstruct the event's appearance (Figure 6-15).

Seismic vibrations were recorded by sensitive instruments as much as 1,000 km (600 mi) away. At 500 km (300 mi), observers reported "deafening bangs" and a fiery cloud on the horizon. At about 400 km (240 miles), witnesses in the town of Kirensk and nearby towns at the same distance saw the object (Figure 6-15a) as

> a ball of fire . . . coming down obliquely. A few minutes later [we heard a] separate deafening crash like peals of thunder . . . followed by eight loud bangs like gunshots.

Some of the closest reporting witnesses were at a trading station about 60 km (40 miles) from ground zero (Figure 6-15b). One of these witnesses said:

> I was sitting on the porch of the house at the trading station, looking north. Suddenly in the north . . . the sky was split in two, and high above the forest the whole northern part of the sky appeared covered with fire. I felt a great heat, as if my shirt had caught fire. . . . At that moment there was a bang in the sky, and a mighty crash. . . . I was thrown twenty feet from the porch and lost consciousness for a moment. . . . The crash was followed by a noise like stones falling from the sky, or guns firing. The earth trembled. . . . At the moment when the sky opened, a hot wind, as if from a

cannon, blew past the huts from the north. It damaged the onion plants. Later, we found that many panes in the windows had been blown out and the iron hasp in the barn door had been broken.

A second witness said

> I saw the sky in the north open to the ground and fire pour out. The fire was brighter than the sun. We were terrified, but the sky closed again and immediately afterward, bangs like gunshots were heard. We thought stones were falling . . . . I ran with my head down and covered, because I was afraid stones [might] fall on it.

Because the object blew up in the atmosphere instead of hitting the ground, it left no crater. The effect on the ground was limited to a broad forest area. At ground zero, tree branches were stripped, leaving bare trunks standing up like poles, but at distances from roughly 5 out to 15 kilometers, the trees were blown over, lying with tops pointed away from the blasts. No one was known to have been this close to the blast. The closest humans were probably herders camped in tents roughly 30 km from ground zero. They related:

> Early in the morning when everyone was asleep in the tent, it was blown up in the air along with its occupants. Some lost consciousness. When they regained consciousness, they heard a great deal of noise and saw the forest burning around them, much of it devastated. The ground shook and incredibly prolonged roaring was heard. Everything round about was shrouded in smoke and fog from burning, falling trees. Eventually the noise died away and the wind dropped, but the forest went on burning. Many reindeer rushed away and were lost.

One older man at about this distance was reportedly blown about 40 feet into a tree, suffering a compound fracture of his arm; he soon died. Hundreds of the herders' reindeer, in the general area around ground zero, were killed. Many campsites and storage huts scattered in the area were destroyed (Figure 6-15c).

What was the nature of the object itself?* Because the meteorite did not strike the ground or make a crater, early researchers thought the object might be a weak, icy

---

*Pseudoscientific literature popularized occult explanations of the Tunguska explosion, including the idea that it was a nuclear explosion caused by the crash of an alien spaceship. Among the alleged evidence was a report of radioactivity at the impact site and elsewhere, suggesting a nuclear rather than a chemical explosion. As reviewed by Oberg (1977), the "increased radioactivity" during the year of the blast was actually within normal statistical fluctuation and hence not considered significant. Indeed, physicist Willard Libby pointed out that the levels of radioactivity in his measurements rule out the possibility of a nuclear blast as large as 10 megatons. Furthermore, the identification of chondritic or carbonaceous chondrite dust in the trees and soil at the site appears to clinch the identification of the Tunguska object as some kind of interplanetary debris.

A thought-provoking novel about the consequences of a larger comet impact on Earth is *Lucifer's Hammer* (Niven and Pournelle, 1977). TV and Hollywood discovered the big-budget film possibilities in the 1990s.

Figure 6-15. The Siberian fireball and explosion of 1908. First three views are reconstructions, based on eyewitness descriptions. (**a**) View one second before the explosion, from 400 km, where the fireball was described as a brilliant fiery mass leaving a trail. (**b**) The moment of the explosion, from 60 km, where the explosion knocked one witness off a porch, and broke windows and a barn door hinge. (**c**) Five seconds after the explosion, 10 km from ground zero. From ground zero to a distance of 10–15 km, the explosion killed raindeer, blew over trees, and started fires in densely forested regions. (**d**) Devastation of the forest, showing trees blown over, photographed about three decades later. (**a–c:** Paintings by author; **d:** E. L. Krinov, 1966)

fragment of a comet, which vaporized explosively in the air, and left no residue on the ground. However, modern planetary scientists disagree. They can model meteorite explosions produced by different bodies hitting the atmosphere. In 1993, researchers Chris Chyba and others studied the Siberian explosion and concluded that a weak stone meteorite had exploded in the atmosphere. Calculations by Svetsov (1996) suggest that a large stony meteorite would blow itself to dust and bits smaller than 10 cm, reducing the chance of finding fragments. The indication of a stony meteoroid was supported when current Italian and Russian researchers reportedly found tiny chondritic stone particles embedded in the trees at the collision site (Gallant, 1994). The diameter of the object has been estimated at 30 to 60 meters, and the energy released has been estimated at 10 to 15 megatons of TNT, in the range of H-bombs.

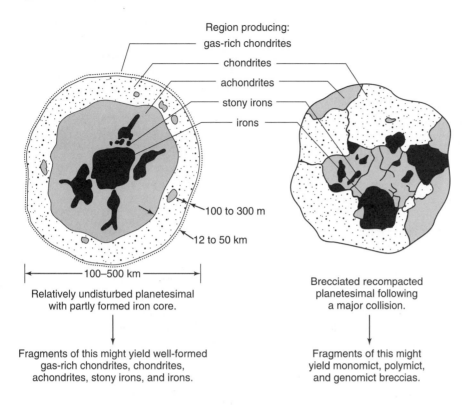

Region producing:
gas-rich chondrites
chondrites
achondrites
stony irons
irons

100 to 300 m
12 to 50 km
100–500 km

Relatively undisturbed planetesimal
with partly formed iron core.

Brecciated recompacted
planetesimal following
a major collision.

Fragments of this might yield well-formed
gas-rich chondrites, chondrites,
achondrites, stony irons, and irons.

Fragments of this might
yield monomict, polymict,
and genomict breccias.

Figure 6-16. Left diagram shows layered structure and dimensions of a hypothetical parent body moderately affected by internal heating and melting. Other parent body (right) shattered and reaggregated into breccias during large-scale collisions. Still later fragmentation of these two bodies would yield many types of observed meteorites.

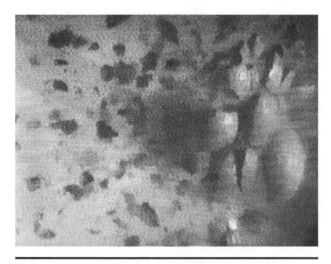

Figure 6-17. Laboratory collisions of rocky objects at asteroidal speeds, in Japan, the United States, and Italy, have been used to study breakup of asteroids and production of meteorites. Here a basalt sphere is shattered by the impact of a smaller projectile from the left at 3.2 km/sec, similar to actual collision speeds in the asteroid belt. Studies of fragments clarify size distribution and other properties of meteorites and asteroids. (Courtesy Akiko Nakamura, Kyoto University, Japan)

Figure 6-18. An asteroid collision ejects meteoritic fragments. Larger asteroid is light-colored rocky type from inner half of asteroid belt; smaller is black carbonaceous type from outer half of the belt. Fragments may be polymict breccias. (Painting by author)

SUMMARY

Accretion and chemical fractionation processes in the solar nebula about 4567 My ago, within a 10–20 My period, produced planetesimals with compositions different from the nebular composition and varying with distance from the sun. Such planetesimals ranged in composition from the nearly primordial CI carbonaceous chondrites in the colder, outer solar system to a range of ordinary chondrites inside of 2.5 AU. This material was incorporated into parent bodies, or planetesimals of asteroidal dimension, up to at least 200 km in diameter and probably larger. Based on petrologic and chemical data, we can conclude that the parent bodies, whose fragments are our meteorites, were probably asteroids in the asteroid belt. (However, some carbonaceous chondrite material may be associated with comets.)

Some parent bodies were heated around 4,560 My ago, perhaps by short-lived radioactive isotopes, magnetic induction heating, or other mechanisms. The interiors of some bodies got hot enough to destroy chondritic textures, produce achondrites, or melt and produce irons, as shown in Figure 6-16. Other parent bodies (smaller ones?) may have remained cold, preserving chondrite textures and volatiles as in the case of carbonaceous chondrites.

From the beginning, collisions among parent bodies affected their structures. Such collisions have been reproduced by studies in laboratories (Figure 6-17). Small-scale collision layers created fragmental layers on the surfaces of parent bodies, which later compacted into brecciated chondrites and achondrites. Large-scale collisions (Figure 6-18) may have shattered entire parent bodies, sometimes scattering the fragments and sometimes allowing them to resettle into confused, brecciated masses. Sometimes, fragments from both colliding objects reaccumulated into polymict breccias. Many fragments hit planets and moons, producing craters.

As Figure 6-16 shows, the collisions produced small fragments of widely varying character, many of which became our meteorites. Some fragments may have been ejected directly into resonances in the belt, where they ejected from the belt onto Earth-approaching orbits; other fragments in the meter-diameter size range may have drifted from their points of origin to resonances by Yarkovsky effects. Some specific major collisions can be dated by clusters of cosmic ray exposure ages.

In general, meteorites confirm the planetesimal theory of planet formation, indicate conditions inside certain ancient planetary bodies, explain craters, and imply material resources, such as relatively pure metals, available in space for future space explorers.

CONCEPTS

meteoritic complex
nodes
meteoritics
meteoroid
meteor
meteorite
fireball
fall
find
parent body
train
fusion crust
meteoroid flux
micrometeorite
metamorphosed material
differentiation
stony meteorite
iron meteorite
stony-iron meteorite
Neumann bands
monomict breccias
polymict breccias
genomict breccias

chondrite
chondrule
achondrite
carbonaceous chondrite (CC)
volatile elements and compounds
siderophile element
lithophile element
oxidized
reduced
Prior's rules
H group
L group
LL group
igneous rock
basaltic achondrites
eucrites
Widmanstätten pattern
gallium-germanium groups
gas retention age
cosmic ray exposure (CRE) age
tektite
SNC meteorites

PROBLEMS

1. Describe some different groups of small bodies in the solar system. Comment on any relationships, such as meteorite relationships with Apollo asteroids. (These relationships will be dealt with further in later chapters.)

2. (a) Assuming that manned space vehicles continue to evolve, what materials do meteorites suggest might be available in interplanetary space that would be useful in space exploration? (b) Depending on the destination, why might it cost less energy (hence money) to haul raw materials from a near-Earth meteoroid (approaching Earth at, say, 5 km/s) than to haul the materials from Earth's surface?

3. (a) How do we know that the meteoroid flux was hundreds or thousands of times greater 4 Gy ago than it is today? (b) Relate this finding to theories of planet formation.

4. If you saw a meteorite fall, describe some phenomena you might expect to experience and state what you might do with the recovered samples.

5. How many meteorites heavier than 100 kg fall into the atmosphere in a year over an area of a million square kilometers (about the area of Texas plus New Mexico)? Would the full 100 kg reach the ground?

6. Give some lines of evidence that most multikilometer craters on planetary bodies are formed by meteorite impacts rather than, for example, volcanic action.

7. (a) Here is a theory of meteorite origin: Two large parent bodies formed, each 500 km across. Their interiors melted, and each formed an iron core surrounded by stony-iron material overlaid by an achondritic layer and surface layers of chondritic materials. The parent bodies then collided, and their fragments were shot onto Earth-crossing orbits, becoming meteorites and Apollo asteroids. Indicate several lines of evidence showing that this theory is too simple to explain meteorites. (b) How would a similar theory with at least 10 such parent bodies undergoing multiple collisions and partial reaccumulations be more realistic?

8. Describe evidence that meteorites formed inside parent bodies that were about 30 to 400 kilometers across.

9. (a) Must a meteorite have melted 1.5 Gy ago in order to have a gas retention age of 1.5 Gy? (b) What are such young gas retention ages interpreted to mean? (c) As a number of meteorites have gas retention ages less than 2 Gy, why are the nakhlites, with solidification ages of 1.2 Gy, considered so unusual?

10. Why are cosmic ray exposure ages not equal to gas retention ages of meteorites?

11. List some lines of evidence that meteorites and their parent bodies underwent major high-energy collisions.

12. Why are the Antarctic meteorites especially valuable to planetary scientists?

13. How do we know that tektites did not come from Mars, Venus, or some more distant location that could explain their special chemical properties?

14. Convert the velocity equation from Chapter 3 into the form

$$v^2 = k\left(\frac{2}{r} - \frac{1}{a}\right)$$

where r and a are given in astronomical units and v is given in kilometers per second. (Identify k.)

15. (a) Using the result in problem 14 (or the original general form in Chapter 3), calculate the orbital velocity of Earth and the orbital velocity at 1 AU of a prograde meteoroid with its aphelion in the middle of the asteroid belt at 2.8 AU and perihelion at 0.9 AU. (b) Assuming that the meteoroid orbit lies in the ecliptic plane, at what speed does it approach Earth and from what direction? (c) Compare this with quoted meteorite approach velocities. (d) Assuming that the same meteoroid has an inclination to the ecliptic of 30, at what relative speed would it approach Earth? (*Hint:* Think of the problem as the subtraction of two vectors.) (e) Modest inclinations are typical of meteoroids. Is the result in (d) more typical of meteorite velocities than the result in (b)?

16. (a) How old, in half-lives, is a meteorite whose rubidium-strontium isochron lies at 45° to both axes in Figure 6-14? (b) How old in years? (c) Would this result be likely in a real meteorite?

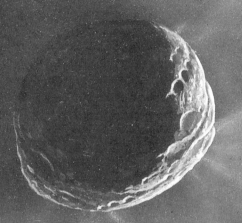

"Asteroid" 2060 Chiron surprised astronomers in 1988 when it developed cometary activity, emitting clouds of dust and gas. It is shown here in one of its periodic close encounters with Saturn. Saturn's rings are backlighted by the sun. (Joint painting by author and Ron Miller)

CHAPTER SEVEN

Interplanetary Worldlets: Asteroids and Comets

Figure 7-1. Comets are a common spectacle in the sky. This 1965 view of Comet Ikeya-Seki shows a typical example of the appearance of a bright comet. (Photo by Steven M. Larson, 20s exposure at f1.9, Tri-X film)

Meteorites, discussed in the last chapter, are mostly samples from interplanetary worldlets. The more we've learned about those worldlets, the more complicated they have made the solar system—even as they have given us more clues about its history. The solar system of a few hundred years ago was simple: five planets (or six counting Earth) and some comets. Asteroids were unknown.

Comets: From Omens to Interplanetary Bodies

Comets are ice-rich interplanetary bodies that give off gas and dust when warmed by the sun. In ancient times, comets were not recognized as astronomical bodies in the solar system but were regarded as mysterious evil omens (Figure 7-1). The appearance of Halley's comet in A.D. 66 was said to have heralded the destruction of Jerusalem in A.D. 70. Five circuits later it was said to mark the defeat of Attila the Hun in 451. In 1066 it presided over the Norman conquest of England. In 1456, its appearance coincided with a threatened invasion of Europe by the Turks, who had already taken Constantinople three years before. Pope Calixtus III ordered prayers for deliverance "from the devil, the Turk, and the comet."

Seneca, the Roman contemporary of Jesus, wrote that "someday there will arise a man who will demonstrate in what regions of the heavens the comets take their way." That man was Tycho Brahe, who in 1577 deduced from the slow movement of comets and their lack of parallactic shift, when observed from different locations on Earth, that they are more distant than the moon, and are not in Earth's atmosphere, as some people had supposed.

In 1704, the English astronomer Edmond Halley applied Newton's law of gravity and some newly developed methods of computing orbits, and discovered that comets travel on *long, elliptical orbits* around the sun and that certain comets reappear many times. Calculating the orbits of 24 well-recorded comets, Halley found that four comets (seen in 1456, 1531, 1607, and 1682) had the same orbit and appeared approximately 75 y apart. Halley correctly inferred that these appearances were by *a single comet* and that the slight irregularities in periodicity were caused by gravitational disturbances from the planets, especially Jupiter. Halley predicted that the comet would return about 1758. It did so on Christmas night of that year. It was named Halley's comet. The discovery that comets are visitors on ordinary elliptical orbits with predictable motions helped dispel the superstition that comets are evil omens.

At present, astronomers have cataloged nearly 2,000 comets and shown that all of them travel on elliptical orbits around the sun. Table 7-1 lists some notable examples, along with their orbital particulars. They are divided into two groups, as the table shows. Long-period comets travel on near-parabolic orbits with aphelia often tens of thousands of AU from the sun. They come from a distant, spheroidal swarm of comets surrounding the solar system like bees around a hive; this swarm is called the **Oort cloud** after its discoverer, the Dutch astronomer Jan Oort. Oort (1950, 1963) used orbit statistics to show that these comets drop in from a swarm extending out to about

TABLE 7-1 Orbits of Selected Comets

Comet	a (AU)	e	i	P	Date of next or last perihelion	q[a] (AU)	Q[b] (AU)	Remarks[c]
Short Period								
Encke	2.2	0.85	12°	3.3	12/80, 3/84	0.34	4.2	D = 0.6−3.5 km
Schwassmann-Wachmann II	3.4	0.39	4°	6.4	3/81, 8/87	2.1	4.8	
Grigg-Skjellerup	3.0	0.67	21°	5.1	5/82	0.99	6.0	
d'Arrest	3.4	0.62	20°	6.4	9/82	1.3	5.5	
Swift-Tuttle	~24	0.960	114°	~120	~1982	0.96	~47	Perseid meteor source
Tempel II	3.1	0.55	12°	5.3	5/83, 9/88	1.4	4.8	
Arend-Rigaux	3.5	0.60	18°	6.8	11/84	1.4	4.9	D = 0.5−3 km
Giacobini-Zinner	3.5	0.71	32°	6.6	9/85	1.0	6.1	Draconid meteor source
Halley	17.9	0.97	162°	76	~2/86	0.59	28.4	
Schwassmann-Wachmann I	6.1	0.04	10°	15	10/89	5.8	6.7	D = 6−38 km; sporadic outbursts
Long Period								
Donati	~157	0.996	117°	~2000	1858	0.58	~313	
Humason	~204	0.990	153°	~2900	1962	2.13	~400	
Morehouse	Large	1.00	140°	Large	1908	0.95	Large	
Burnham	Large	1.00	160°	Large	1960	0.50	Large	
Kohoutek	Large	1.00	14°	Large	1973	0.14	Large	

[a] q = perihelion.
[b] Q = aphelion.
[c] D = diameter.

50,000 AU from the solar system. The observation that no long-period comets travel at faster than parabolic velocity proves that they are gravitationally bound to the solar system. Further, isotopic ratios of C, N, and S measured among several comets show solar system values, not interstellar values (Jewitt and others, 1997). Thus, these objects are part of our system, like asteroids, and not from interstellar space.

Short-period comets have aphelia among the planets or somewhat beyond Pluto. They started as long-period comets but were perturbed into short-period orbits by close approaches to planets as they dipped into the solar system.

The Comet Discovery Process

Roughly a dozen comets are observed each year, but only about one per year reaches naked-eye visibility or prominence. About half are new discoveries. The other half are known comets on return trips around the sun, which are known as recovered comets. Most of the recoveries are first detected on astronomical photographs taken specifically to locate them. Many new comets are found by amateur astronomers who make a hobby of scanning the skies with modest telescopes, checking any suspicious fuzzy objects against the catalogs of known nebulas (which can easily be mistaken for comets in small telescopes) and looking for slow movement against the starry background.

When a newly discovered comet is confirmed, an announcement is sent to astronomers around the world by the International Astronomical Union, and the comet is assigned a designation consisting of the year and a letter in order of discovery or recovery: 1973a, 1973b, and so on. Comets are also popularly named after their discoverers, such as Comet Burnham and Comet Shoemaker-Levy 9 (the 9th found by the Shoemaker-Levy team). After a year or two, when observations of the comets have been collected and reliable orbits determined, they are assigned new, permanent designations with Roman numerals for the order in which they passed their perihelion points: 1999 I, 1999 II, and so on.

Asteroids: Discovering a New Class of Solar System Bodies

The popularization of Bode's rule about 1772, followed by its confirmation through the discovery of Uranus in 1781, led to widespread belief that there ought to be a "missing" planet at 2.8 AU. In 1800, six German astronomers, under the leadership of Johann Schröter, determined to search out the missing planet. Before these "celestial police" (as they were nicknamed) could succeed, the Italian astronomer Giuseppe Piazzi announced discovery of an unknown body as he made routine stellar observations at the observatory at Palermo, Sicily, on New Year's Day, 1801. The new object moved with respect to the stars and Piazzi called it a new planet, Ceres. By the fall of that year, the famous astronomer Gauss had derived the first general method for determining orbits from observations of celestial bodies, and he computed Ceres's orbit—at 2.8 AU, just as predicted by Bode's rule.

Because Ceres seemed too small to be the sought-for planet, the German astronomers continued their survey program. In March 1802, one of them, Olbers, discovered a second body, which he named Pallas. Pallas was even smaller and fainter than Ceres but was located at about the same distance. This raised the possibility that a normal planet had once occupied the predicted position but had somehow fragmented, producing many small bodies. Working from this hypothesis, the searchers continued looking. Juno was discovered in 1804 and Vesta in 1807. No more had been found by 1815 and the search stopped until a Prussian amateur named Hencke began a one-man program in 1830 and eventually discovered Astraea. Other discoveries followed. The bodies came to be called **asteroids** after their faint, starlike images; today we know they are rocky and interplanetary bodies. By 1890, the total known was 300. In 1891, the German astronomer Max Wolf began the first photographic patrol, detecting asteroids by their trails on long-exposure plates guided on the stars, as shown in Figure 7-2. This mechanization

greatly increased the rate of discovery. Thousands are known now. Table 7-2 lists some examples, along with their orbital and other properties.

Even though more asteroids were found in the mid-1800s, the solar system still looked pretty simple. The planets were the eight large objects going around the sun. Of 13 satellites that had been found by 1840, all were intermediate in size and moving in direct, or prograde, orbits around their respective planets. The asteroids, or "minor planets," were small bodies grouped neatly between Mars and Jupiter—fitting the prediction by Titius and Bode that this niche would be filled by some sort of planetary body. The comets seemed different from the other classes; each had a bright fuzzy head and an enormous, diffuse tail. These four classes of objects seemed distinct and had their own properties.

Small Bodies: Confusion Among Types

Some problematic objects soon appeared and messed up this neat picture. Two small moons the size of asteroids were found around Mars in 1877, and Phoebe was discovered in 1898 moving at the outskirts of Saturn's gravitational sphere of influence in a retrograde orbit! Conceivably, such bodies could be captured asteroids. Were they really more related to asteroids than to the big moons?

Planetary astronomers eventually had to acknowledge that the solar system was less tidy than they had thought. To take satellites as an example, nine more satellites had been discovered by 1960, of which four moved in

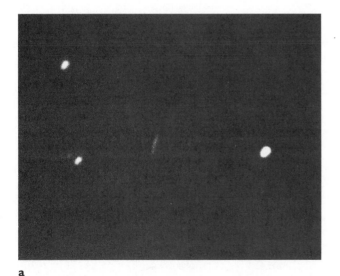
a

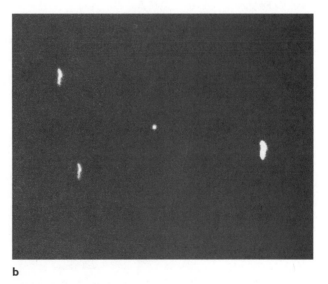
b

Figure 7-2. Two images of asteroid 887 Alinda. (**a**) The telescope was guiding on the stars, and the asteroid appears as a streak due to its own orbital motion during the time exposure. Asteroids are often discovered by such trails on stellar photos. (**b**) The telescope was guiding on the asteroid, causing the star images to be trailed. (New Mexico State University Observatory)

INTERPLANETARY WORLDLETS: ASTEROIDS AND COMETS

TABLE 7-2   Properties of Asteroids

Number and name	a (AU)	e	i	Diameter (km)	Rotation period (h)	Geo-metric albedo	Spec-trum[a] type	Est. composition	q[b] (AU)	Q[c] (AU)
Inner solar system										
Aten type (a < 1 AU; Q ⩾ 0.983 AU; Earth-crossers interior to Earth's orbit)										
2100 Ra-Shalom	0.83	0.44	16°	3.4	19.8	—	C	Carbonaceous	0.47	1.20
2062 Aten	0.97	0.18	19°	1.0	—	—	S	Silicates	0.79	1.14
2340 Hathor	0.84	0.45	6°	0.2	—	—	—	—	0.46	1.22
Apollo type (a ⩾ 1 AU; q ⩽ 1.017 AU; Earth-crossers exterior to Earth's orbit)										
1866 Sisyphus	1.89	0.53	41°	10	—	—	—	—	0.87	2.92
1685 Toro	1.37	0.44	9°	4.8	10.2	0.03	S	L chondrite	0.77	1.96
1974 MA	1.78	0.76	38°	6?	—	—	—	—	0.42	3.13
1973 NA	2.39	0.63	68°	6	—	—	—	—	0.88	3.91
1863 Antonius	2.26	0.61	18°	3	4.0	—	S?	Silicates?	0.89	3.63
1947 XC	2.25	0.63	1°	3?	—	—	—	—	0.83	3.67
1864 Daedalus	1.46	0.62	22°	3.4	8.6	—	SQ	Silicates	0.56	2.36
1620 Geographos	1.24	0.34	13°	2.0	5.2	—	S	Silicates	0.83	1.66
Amor type (a > 1 AU; 1.017 < q < 1.4 AU; Mars-crossers)										
1036 Ganymede	2.66	0.54	26°	30?	10.3	0.17	S	Silicates	1.22	4.10
433 Eros	1.46	0.22	11°	7×19×30	5.3	0.18	S	Silicates	1.13	1.78
1474 Beira	2.73	0.49	27°	11?	—	—	—	—	1.39	4.07
1974 UB	2.12	0.36	36°	8?	—	—	—	—	1.36	2.89
1975 AD	2.37	0.38	20°	8?	—	—	—	—	1.48	3.26
1139 Atami	1.95	0.26	13°	7?	—	—	S	Silicates	1.45	2.44
1963 RH	2.38	0.38	21°	7?	—	—	—	—	1.48	3.28
1580 Betulia	2.20	0.49	52°	6	6.1	—	C	Carbonaceous	1.12	3.27
1627 Ivar	1.86	0.40	8°	6?	4.8	—	S	Silicates	1.12	2.60
Inner Edge of Belt (a < 2.5 AU)										
4 Vesta	2.36	0.09	7°	500	5.3	0.24	V	Eucrite basaltic achondrite	2.15	2.57
7 Iris	2.39	0.23	6°	204	7.1	0.16	S	Olivine, nickel-iron	1.84	2.93
6 Hebe	2.43	0.20	15°	192	7.3	0.16	S	Silicates, nickel-iron	1.94	2.92
19 Fortuna	2.44	0.16	2°	190	7.4	0.03	C	Carbonaceous	2.05	2.83
8 Flora	2.20	0.16	6°	153	13.6	0.12	S	Pyroxenes, nickel-iron	1.85	2.55
11 Parthenope	2.45	0.10	5°	150	10.7	0.12	S	Silicates, nickel-iron	2.20	2.70
18 Melpomene	2.30	0.22	10°	150	11.6	0.14	S	Silicates, nickel-iron	1.79	2.81
20 Massalia	2.41	0.14	1°	131	8.1	0.16	S	Silicates, nickel-iron	2.07	2.75
12 Victoria	2.34	0.22	8°	126	8.7	0.11	S	Silicates, nickel-iron	1.83	2.85
21 Lutetia	2.43	0.16	3°	115	6.1	0.09	M	Metallic	2.04	2.82
Mid Belt (2.5 < a < 3.1 AU)										
1 Ceres	2.77	0.08	11°	914	9.1	0.05	C	Carbonaceous	2.55	2.99
2 Pallas	2.77	0.23	35°	522	7.9	0.08	B	Altered carbonaceous?	2.13	3.41
704 Interamnia	3.06	0.16	17°	334	8.7	0.04	C	Carbonaceous	2.57	3.55
15 Eunomia	2.64	0.19	12°	272	6.1	0.17	S	Silicate, nickel-iron	2.14	3.14
16 Psyche	2.92	0.14	3°	264	4.3	0.09	M	Nickel-iron, enstatite?	2.51	3.32
3 Juno	2.67	0.26	13°	244	7.2	0.15	S	Olivene, pyroxene, nickel-iron	1.98	3.36
324 Bamberga	2.69	0.34	11°	242	8	0.03	C	Carbonaceous	1.78	3.60
451 Patientia	3.06	0.07	15°	230	7.1	0.03	C	Carbonaceous	2.82	3.30
13 Egeria	2.58	0.09	16°	214	7.0	0.03	C	Carbonaceous	2.35	2.81
45 Eugenia	2.72	0.08	7°	214	5.7	0.03	C	Carbonaceous	2.50	2.94

TABLE 7-2 Properties of Asteroids (*continued*)

Number and name	*a* (AU)	*e*	*i*	Diameter (km)	Rotation period (h)	Geo-metric albedo	Spec-trum[a] type	Est. composition	*q*[b] (AU)	*Q*[c] (AU)
Outer Edge of Belt (3.1 < *a* < 4.1 AU)										
10 Hygiea	3.15	0.10	4°	443	18	0.04	C	Carbonaceous	2.84	3.46
511 Davida	3.18	0.17	16°	336	5.2	0.03	C	Carbonaceous	2.65	3.73
87 Sylvia	3.48	0.10	11°	272	5.2	0.04	P	Silicate?	3.13	3.83
31 Euphrosyne	3.15	0.22	26°	248?	5.5	0.07	C	Carbonaceous	2.46	3.84
65 Cybele	3.43	0.12	4°	246	6.1	0.02	C	Carbonaceous	3.02	3.84
107 Camilla	3.49	0.07	10°	236	4.6	0.04	C	Carbonaceous	3.25	3.73
24 Themis	3.14	0.12	1°	228	8.4	—	C	Carbonaceous	2.76	3.52
94 Aurora	3.15	0.09	8°	212	7.2	0.03	C	Carbonaceous	2.84	3.47
702 Alauda	3.19	0.03	21°	202	8.4	0.06	C	Carbonaceous	3.09	3.29
165 Loreley	3.13	0.08	11°	202	7.0	0.07	C	Carbonaceous	2.88	3.38
Extreme Outer Belt (4.1 < *a* < 5.1)										
279 Thule	4.26	0.03	2°	60?	—	0.03	D	Reddish, carbonaceous	4.12	4.40
Outer Solar System (*a* > 5.1 AU)										
Trojans[d] (Lagrangian clouds in Jupiter's orbit; 5.1 < *a* < 5.3 AU)										
624 Hektor (P)	5.15	0.02	18°	150 × 300	6.9	0.02	D	Reddish, carbonaceous	5.0	5.2
911 Agememnon (P)	5.19	0.07	22°	148?	7?	0.04	D	Reddish, carbonaceous	4.8	5.5
617 Patroclus (F)	5.21	0.14	22°	140	—	0.02	P	Carbonaceous	4.5	5.9
1437 Diomedes (P)	5.08	0.05	21°	130?	18?	0.03	C	Carbonaceous	4.8	5.3
1172 Aneas (F)	5.17	0.10	17°	125	—	0.02	P	Carbonaceous	4.4	5.9
588 Achilles (P)	5.17	0.15	10°	118?	—	0.03	D	Reddish, carbonaceous	4.4	6.0
1143 Odysseus (P)	5.21	0.09	3°	118?	—	0.04	D	Reddish, carbonaceous	4.7	5.7
659 Nestor (P)	5.26	0.11	4°	102?	—	0.04	C?	Carbonaceous?	4.7	5.8
1208 Troilus (F)	5.17	0.09	34°	98?	—	0.04	C?	Carbonaceous?	4.7	5.6
1583 Antilochus (P)	5.28	0.05	28°	98?	—	0.05	D	Reddish, carbonaceous	5.0	5.6
944 Hidalgo	5.82	0.66	42°	39	10.1	—	D	Reddish, carbonaceous	2.00	9.64
Centaurs (q beyond Jupiter but < ~25 AU)										
2060 Chiron	13.7	0.38	6°	300	5.9	0.08?	C	Carbonaceous + ices	8.5	19.0
5145 Pholus	20.4	0.57	25°	175?	—	—	super-D	Carbonaceous + ices + organics	8.8	32.0
1993 HA$_2$	24.9	0.52	16°	100?	—	—	D	Carbonaceous + ice	11.9	37.7
1995 GO	—	—	—	?	—	—	super-D	Carbonaceous + ices + organics	6.6	21.6
Largest Kuiper Belt Objects (Region of Pluto)										
50,000 Quaoar	43.2	0.04	43°	~1200	—	0.12	D?	Carbonaceous + ices?	41.7	44.7
28,978 Ixion	39.4	0.24	20°	~1065	—	0.09	D?	Carbonaceous + ices?	29.8	49.0
20,000 Varuna	43.2	0.05	17°	~ 900	—	~0.07	D?	Carbonaceous + ices?	41.0	45.5
Pluto	39.5	0.25	17°	2300	153.3	—	—	Methane-rich ices	29.7	49.3

Source: Various references have been used to construct the table, including Asteroids II (1989, University of Arizona Press); Wetherill (1979); Chapman, Williams, and Hartmann (1978); Kowall (1988); Beatty and Chaikin (1990); Binzel and Sauter (1992).

Note: Diameters from various sources. Question marks indicate estimates; other values measured with fair precision.

[a] Spectrum lists the type classification; U indicates an unusual spectrum; B is a variant of class C; SQ is probably a variant of S.

[b] *q* = perihelion distance.

[c] *Q* = aphelion distance.

[d] (P) or (F) indicates preceding or following Trojan cloud.

the retrograde direction, and six were less than 100 km across. A number of these moons are now accepted as captured interplanetary bodies.

To make matters worse, asteroids began popping up in unexpected parts of the solar system. In 1898, a German astronomer discovered asteroid 433 Eros, whose perihelion lies inside the orbit of Mars. In 1932, asteroid 1862 Apollo was found passing inside the Earth's orbit. Others of this type came to be called **Earth-approaching asteroids.** They are asteroids that have been perturbed into the inner solar system and whose orbits come close to Earth's orbit. They are subdivided as shown in Table 7-2 into groups named after prototype asteroids. Amors come inside Mars' orbit but don't cross Earth's orbit. Apollos cross Earth's orbit. Atens, which are rare, travel entirely inside Earth's orbit. Nearly 200 Earth-approachers were cataloged by 1993, and asteroid survey statistics by 2001 indicated 1227 ± 170 objects with perihelia inside 1.3 AU larger than 1 km (Stuart, 2001). The biggest are 1036 Ganymede at roughly 30 km diameter (it comes within about 0.2 AU of Earth's orbit) and 433 Eros at roughly 10×30 km (it comes within about 0.1 AU of Earth's orbit). Apollos, or Earth-approaching asteroids that actually cross Earth's orbit, are the only ones that have a chance of hitting us in the foreseeable future. The biggest of these is 1627 Ivar at about 8 km diameter. It would cause a global disaster if it actually hit Earth—but that is unlikely to happen within the next million years. Many small ones have been seen, including some that are only tens of meters across. Some of these may hit Earth within a million years. One such body passed by in 1991 at half the distance to the moon!

Asteroid 624 Hektor, spotted in 1907, turned out to be the first of still another distinct group, moving in the same orbit as Jupiter but 60° ahead of the planet. Numerous other asteroids were later found sharing Jupiter's orbit, averaging either 60° ahead of Jupiter or 60° behind the planet, though they may wander along Jupiter's orbit 20° or so from those positions. The asteroids in these two swarms are called **Trojan asteroids** (they are named after characters in Homer's Trojan war epics). The two semi-stable positions are named Lagrangian points after Joseph Lagrange, the French dynamicist who discovered the gravitational effects that define them. A number are listed in Table 7-2.

The discovery of numerous asteroid groups outside the belt, plus short-lived comets with similar orbits, led to questions of which objects came from the asteroid belt and which ones might be dormant or extinct comets. What did astronomers really mean by this distinction anyway?

Surprisingly, astronomers did not seriously begin to compare asteroids and comets or to study their relationships until the late 20th century. Instead, an artificial sense of distinction was encouraged by their appearance, semantics, and the methods of studying them. Comets give off gas and their spectra show emission lines; they came to be studied by astronomers who were gas spectroscopists.

Asteroids' spectra show faint absorption features of the solid minerals on their surface; they came to be studied by astronomers who were interested in solid surfaces, and by petrologists who had tried to infer their properties from meteorites. Although comets and asteroids are both solid bodies, some kilometers in diameter, they ended up being studied by completely different communities of scientists. At a typical planetary science meeting, the "comet session" would occur on one day and be attended by one group of researchers whereas the "asteroid session" would occur on another day and be attended by a largely different group!

A more cosmopolitan way of looking at things is to consider the whole range of planetesimals—from comets to asteroids; they are all planetesimals formed in different parts of the solar system, with a range of different ice contents. The classic asteroids in the belt are dominated by rocky, metallic, and carbonaceous compositions and don't show comet activity, but other objects are more ambiguous in nature.

Can Some Asteroids Be Comets and Vice Versa?

The tidy semantic distinction between asteroids and comets began to break down when telescopic observers realized that as comets become inactive (i.e., when they recede from the sun), they can become telescopically indistinguishable from asteroids. For example, "asteroid" number 4015, "discovered" in 1979, turned out to be the same object as periodic "comet" Wilson-Harrington, discovered in 1949, except that it had lost its coma and tail.

Another example of the blurred distinctions came with Charles Kowal's 1977 discovery of an object cataloged as asteroid 2060 Chiron, moving in a cometlike orbit between Saturn and Uranus. Chiron was cataloged as the then-most-distant known "asteroid," and its colors were soon found to be similar to those of other blackish asteroids and comet nuclei of the outer solar system. Yet in 1988, as Chiron's orbit brought it closer to the sun, Dave Tholen, Dale Cruikshank, and I discovered that Chiron unexpectedly doubled in brightness, apparently due to cometary activity; in a few months, the cloud of dust from this activity had expanded to the point that observers Karen Meech and Michael Belton directly detected the fuzzy coma around "comet Chiron." "Asteroid" Chiron was now an active comet! By 1991, S. J. Bus and colleagues had spectroscopically detected CN molecules emitted from "comet Chiron"; they interpret the cometary activity as driven by sublimation of CO_2 ice. Water ice has also been found on the surface (Luu and others, 2000). With a diameter of about 250 km, it is the largest known comet—15 times larger than the nucleus of Halley's Comet. Even its orbit is not long for this system. Many comets interact with giant planets as they work their way into the inner solar system from these regions, and Chiron fits this pattern, as it comes close to Saturn every 10,000 years or so. Such interactions will cause its orbit to evolve rapidly. If one of these passes throws it into the inner solar system, it would become the biggest and brightest comet in history!

In view of such cases, how do we know whether a given interplanetary body cataloged as an asteroid is merely an inactive comet? The physical definition of a comet clearly involves ices, which sublime to supply the gas blowing off the nucleus to form the tail; this was established by Fred Whipple in the 1950s. By the 1970s, spectroscopic observations indicated that asteroids' surfaces are dominated by stony, metallic, or sooty carbonaceous materials. Until the 1980s, observers expected that bare comet nuclei would look like Whipple's "dirty snowballs"—white with a smattering of dust. But telescopic observations in the 1980s and the Giotto probe flyby of Halley's comet's nucleus revealed that comet nuclei are as black or dark brown as carbonaceous asteroids! Their ices are darkened by admixtures of dust—perhaps needing only a few percent of dust to blacken the ice.

Many outer solar system asteroids with dark, carbonaceous surfaces might thus contain ices trapped inside. Consistent with this idea, surface soils of many large, black asteroids in the outer asteroid belt contain minerals with chemically bound water of hydration; and many carbonaceous meteorites have carbonates and deposits, apparently deposited when the interiors were heated enough that liquid water percolated through them, reacting with the minerals. Probably these bodies originally contained ice, which later melted and seeped through internal fractures. Interestingly, Trojans generally don't show the hydration feature, indicating that they never heated enough to melt the ice and create moisture in contact with the carbonaceous minerals.

Thus—and this is the important point—instead of thinking of the Victorian, semantic distinction between two distinct types of bodies, we have to learn to think in terms of a continuum between less ice-rich bodies and more ice-rich bodies, depending on the original location and conditions of origin.

Expanding the Inventory: Centaurs and Kuiper Belt Objects

By the 1990s, the solar system inventory of small bodies had expanded even further. Chiron turned out to be just the first known of a group of bodies moving in elliptical orbits that cross the orbits of giant planets, in the general region from Saturn past Uranus out to Neptune (roughly 20 to 50 AU). These bodies came to be called *Centaurs,* as they were named after the man/horse beasts of mythology (see Table 7-2). One well-known example is 5145 Pholus,

which turned out to be the reddest asteroid ever observed, with a visual color about as red as Mars. Other Centaurs of similar extreme redness were later found.

Beyond the Centaurs, observers such as Jewitt and Luu used sensitive detectors to discover another group of still more distant objects, generally in the region of Neptune and Pluto and beyond, starting in 1992. Well over two dozen such objects were found in the next three years. Most are in the estimated size range of 100 to 300 km. From the statistics, observers David Jewitt and Jane Luu estimated that 35,000 objects in this size range might orbit the sun between 30 and 50 AU. These objects are usually called the *Kuiper belt objects.* The Dutch-American astronomer Gerard Kuiper, in the 1950s, had hypothesized that such objects should be found in this region, being the debris from the fringes of the primeval solar nebula, a region where there was not enough material to form planets. (Sometimes this group is called the Edgeworth-Kuiper belt, adding the name of another midcentury astronomer who also wrote about such objects.) Examples are listed in Table 7-2.

The Kuiper belt is believed to extend from roughly 40 to 400 AU. Figure 6-2, in the previous chapter, showed that interplanetary bodies in that zone can survive from the days of solar system origin, whereas Centaurs, moving among the giant planets, are essentially on temporary orbits. Unlike the Oort cloud, the Kuiper belt is flattened and is part of the disk-shaped planetary system; most of the Kuiper belt objects have orbital inclinations less than 8°. The longevity of the Kuiper belt and its flattened shape indicate that it is a surviving remnant of the planetesimals formed in the solar nebula disk.

Table 7-2 emphasizes the similarity of orbits between known Kuiper belt objects and Pluto. Observers have pointed out that some of these objects are actually in the same 3:2 mean-motion resonance with Neptune that Pluto is in. Jewitt and his co-workers use the term *Plutinos* to indicate Kuiper belt objects that have become trapped in the same resonance as Pluto. Jewitt, Luu, and Chen (1996) found that out of a sample of 32 known Kuiper belt objects, 12 were Plutinos. From this point of view, Pluto is just the largest known Kuiper belt object, and the largest Plutino.

Some short-period comets (Table 7-1) have characteristics similar to those of Centaurs except that they have interacted with planets and are on orbits whose aphelia are within the planetary system. Comet Schwassmann-Wachmann I, for example, is on a nearly circular orbit just beyond Jupiter's orbit. Remember that orbits like these are not stable over the history of the solar system. Furthermore, a typical comet in such an orbit can make only about a thousand trips around the sun before running out of ice. Thus, Centaurs and short-period comets have not been in these orbits since the beginning but were perturbed into these orbits by close encounters with planets.

Centaurs, short-period comets, Kuiper belt objects, and Oort cloud objects are probably physically similar:

dark objects containing ice colored by carbonaceous dirt. However, most don't come close enough to the sun to become active comets. We can see only the Centaurs and Kuiper belt objects and none of the Oort cloud objects (until they drop into the inner solar system). Even though the Centaurs and the Kuiper belt objects are almost certainly inactive comet nuclei, they are named and numbered according to asteroid traditions because they look like asteroids. Like the first Centaur, 2060 Chiron, most Centaurs and Kuiper belt objects would turn into comets if they came close enough to the sun.

Here we seem to have a hodgepodge solar system of small bodies, a main asteroid belt, two Trojan swarms, Centaurs, the Kuiper belt, the Oort cloud, and scattered interplanetary comets and asteroids. Some bodies are prograde, some retrograde. There are rocky bodies, dark carbonaceous bodies, and icy bodies cataloged as asteroids, which later turn into comets. How can we make more sense of this welter of bodies?

Compositional Trends: The Taxonomic Classes

One way to make sense of the interplanetary hodgepodge is to look at more detailed information on compositions of the interplanetary bodies. These data come from spectra of asteroids and comets (preferably inactive comets so that we can see the solid nucleus). Today, as a result of the work of the current generation of astronomers, we have a clearer sense of the continuum of compositional types among planetesimals, ranging from rocky/metallic "asteroids" to ice-dominated "comets," or more precisely, "comet nuclei." The specific type probably depends mostly on the temperature where these objects formed, which depends in turn on the distance from the sun where they formed.

Spectroscopic Properties and the Taxonomic Classes

The various astronomical measurements that can reveal interplanetary bodies' properties—such as polarimetry, radar, and spectroscopy—are called techniques of **remote sensing.** The strongest technique is **infrared reflectance spectrophotometry,** the study of infrared absorption bands produced when sunlight passes through the outer micrometers of certain minerals in the rocks before reflecting back to the observer. Important examples are the **pyroxene absorption bands** near 1 and 2 μm, which are indicative of pyroxene, a silicate mineral family important in basaltic and some meteoritic rocks. Its formula is XY (Si_2O_6) where X and Y may be the same or different elements. An example is enstatite, Mg_2 (Si_2O_6), mentioned in Chapter 5. The pyroxene band appears in Figure 7-3,

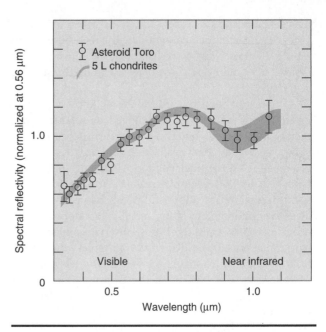

Figure 7-3. Reflectance spectrum of Apollo asteroid 1685 Toro, from blue to near infrared, showing close match to L-type chondrite meteorites, whose minerals are predominantly iron-magnesium pyroxenes and olivine. Other meteorite types lack the absorption near 1.0 μm or have absorptions at different wavelengths. Carbonaceous chondrites and C-type asteroids, for example, roughly match the left third of this spectrum but are nearly flat in the right two-thirds. (After Chapman and others, 1973)

which displays the close match between the spectrum of the 12-km Earth-approaching asteroid 1685 Toro and the most common type of meteorite, the L chondrites. As mentioned in the last chapter, many L chondrites may actually be pieces knocked off Toro or the same parent body as Toro. Binzel and others (1993) reported an even closer match to an ordinary chondrite for 7-km asteroid 3628 Boznemcova, which lies near the 3:1 Jupiter resonance in the belt, where its fragments might be perturbed to Earth.

Another important group of absorption bands are those of ices, such as H_2O and CH_4, and of water of hydration. These bands typically fall in the near infrared part of the spectrum, for example 1-4 μm.

Remote sensing tools have been applied to many hundreds of asteroids and dormant comets. As a result, these bodies were grouped in different spectral classes, signifying different compositional types (Bowell and others, 1978; Tedesco and Gradie, 1981; Tholen and Barucci, 1989). The major recognized types are listed in Table 7-3. These so-called **taxonomic classes** are defined purely in terms of spectral features but are believed to be loosely related to the various meteorite types because comparisons of the various classes of asteroids' spectra and meteorites' spectra show broad similarities. It seems likely that taxonomic class M, for example, represents metal-rich bodies like the iron or stony-iron meteorites. The black C's, P's, and D's are likely to be generally similar to carbonaceous meteorites. In detail, however, there is much argument

TABLE 7-3 Selected Asteroid Compositional Types (in Order Outward from the Sun)

Taxonomic Class	Albedo (%)	Composition (meteorite type)	Location[a]
E	>23	Enstatite chondrite?	Concentrated near inner edge of belt
S	7–23	Stony material; possibly chondrites?	Major class in inner to central belt
M	7–20	Stony-iron or iron	Central belt
V	38	Basalt	Vesta and a few Earth-crossers
A	~25	Olivine rich; pallasite	Main belt (rare)
C	2–7	Carbonaceous chondrite	Dominant type in outer belt beyond about 2.7 AU
P	2–7	Possibly carbonaceous; spectrum resembles M but albedo is lower	Outer and extreme outer belt
D	2–7	Dark reddish-black organics including kerogens and carbonaceous material?	Extreme outer belt and Trojans; about two-thirds of Trojans are type D.
Z (provisional)	4–10?	Organics? Extremely red	Centaurs and Kuiper belt

[a]After Gradie and Tedesco (1982).

over whether specific asteroid classes really have the composition of specific meteorite types.

Do the S-class Asteroids Equal Chondritic Meteorites?

An enormous controversy has raged, curiously, about one of the most common taxonomic asteroid classes and the most common meteorite class. Most of the inner belt asteroids are class S, and 81% of Earth's recovered meteorite falls are ordinary chondrites. It might seem, therefore, that class S objects are chondrites. One camp of researchers defends that position. In detail, however, class S asteroid spectra and chondrite spectra show some differences. As mentioned earlier, only a few obscure asteroids, like 3628 Boznemcova near the 3:1 belt resonance, match chondrite spectra very well (Binzel and others, 1993). The first camp argues that most S's are chondrites, but that processes in space, such as cosmic ray exposure, production of impact glasses, and pulverization by micrometeorites, cause "space weathering," which alters the spectra. Space weathering has clearly been observed on the moon, where the surface soils have different spectra from those of the parent rocks.

In support of this, the NEAR spacecraft revealed a chondritic composition for major elements of asteroid 433 Eros, along with evidence for space weathering, but apparently did not rule out a weakly differentiated basaltic achondrite composition (Sullivan and others, 2002).

Another camp argues that most S spectra do not match undifferentiated chondritic material but do match differentiated material. In this view, most S's would be differentiated asteroid mantle materials. For example, spectral observations by D. Tholen, D. Cruikshank, and the author (unpublished) seem to show a fairly smooth transition from the S class to the olivine-rich A class, which is differentiated, olivine-rich material, probably from asteroid mantles. If many common S's are not chondrites, then the high abundance of chondrite falls on Earth would be a statistical fluke of delivery mechanisms: a breakup of a chondrite parent body near a resonance might have placed many chondrites on Earth-crossing orbits during recent geologic time. For example, argon gas retention dates of the L subgroup of ordinary chondrites indicate that a major collision produced them around 500 My ago. Breakup of a chondrite parent body at that time increased the infall rate of meteoritic material by a factor of about ten for a few million years (Schmitz and others, 1997).

Resolving the possible equivalence of S's and ordinary chondrites is important, not only to find out where our most common meteorites come from but also to establish what resources may be in space; if humans are ever to consider using the iron, ice, platinum-group metals, and other resources of asteroids, they need to be able to recognize the mineral suite of each taxonomic type before flying to a given asteroid.

The Small-Body Zones of the Solar System

Cataloging of the taxonomic types has revealed a striking zonal structure among asteroids, as indicated in Table 7-3, according to which the various asteroid types have preferred locations at different distances from the sun. This zonal distribution was discovered by Gradie and Tedesco in 1982. The rare, high-albedo E type (possibly related to enstatite chondrites) and the very common S type (possibly related to ordinary chondrites) are, for example, mostly on the inner edge of the belt. Examples of the unusual E-type asteroids are 44 Nysa and 434 Hungaria (albedos around 45%!) whose orbits are well placed to produce Earth-impacting meteorites (Zellner and others, 1977). S-type asteroids continue into the mid belt, where M-types (possibly stony-irons?) also peak in abundance.

The most abundant asteroid type is probably the C-type of the outer belt. The C's are thought to be carbonaceous chondrite material, whose strikingly low albedos of 2% to 7% are attributed to opaque minerals (graphite, other carbon-based minerals, and perhaps magnetite) disseminated through their matrix. Remember that carbonaceous chondrites include low-temperature condensates. They contain chemically bound water along with fine-grained clay minerals like montmorillonite, and organic polymers, magnetite, and iron sulfide. The relation between C-type asteroids and carbonaceous chondrites was strengthened when Lebofsky and his co-workers (1977, 1982), found spectral evidence for chemically bound water and montmorillonite clay minerals in the surface material of the largest C-type asteroid, 1 Ceres.

Beyond the main asteroid belt, the Trojan asteroids are composed also of very dark asteroids with 2% to 7% albedos, but with redder colors. There are two classes: P's have slightly red color, and D's have strongly red color. The reddish color is generally believed to be caused by an admixture of organic materials. As early as 1980, Gradie and Veverka predicted that organic materials—molecules defined by C-H bonds—would readily form at large solar distances and cause redder colors at large distances. Observations have supported this prediction and have revealed a rough color trend that seems to involve reddish organic compounds. They may contain ice in their interior, although ice in the surface soils has apparently long ago been vaporized by impacts and exposure to the sun.

The D's (originally called RD for "reddish, dark") asteroids were first noted and defined by Degewij and van Houten (1979). D's constitute about two-thirds of the Trojans. Gradie and Veverka (1980) reproduced D spectra, roughly, with mixtures of coal-tar residues, magnetite, carbon black, and montmorillonite. A principal coloring material in the coal-tar residues is the organic compounds called kerogens, also found to some extent in carbonaceous chondrites. The brilliant "Tagish Lake" fireball over Canada in 2000, caused by a 4–6 m meteoroid, produced the first meteorites to match the D-class spectrum (Brown and others, 2000; Hiroi and others, 2001). These carbonaceous chondrites contained a unique mix of 4–5% carbon by weight, carbonates, and 100 parts/million of organic compounds including aromatic hydrocarbons (Pizzarello and others, 2001).

Comet nuclei fall in the general realm of C, P, and D taxonomic types. 2060 Chiron, for example, was a C when it developed its cometary activity. Most comets are a bit redder; Halley's was more D-like (Cruikshank, Hartmann, and Tholen, 1985).

Centaur "asteroids" (of which Chiron is one) are mostly P or D in coloration, or even redder. One of the earliest Centaurs measured, 5145 Pholus, is the reddest interplanetary body yet measured; it would appear visually about as red as Mars.

An important fact in interpreting the dark C, P, and D classes of the outer solar system is that only a few percent admixture of carbon minerals mixed with pulverized ice can lower the albedo from the pure ice value around 70% to a value around 5% (Clark, 1980). This observation suggests the likelihood that many blackish C-, P-, or D-type asteroids of the outermost belt or Trojan regions may contain large amounts of ice in their interior. The presence of the ice is disguised by the opaque carbonaceous minerals. Trojans may have had ice removed from their surfaces by aeons of afternoon heating, when peak temperatures were high enough to sublime the surface ices. In summary, the C-, P-, and D-class "asteroids" beyond the belt are likely to include a great many objects that are icy and could be called inactive comet nuclei.

Looking at things in this way, we can roughly divide the solar system into a sequence of what we might call **primordial taxonomic zones,** depending on the compositions of the solid bodies smaller than the giant planets. (Giant planets don't count because their bulk compositions were modified when their enormous gravity allowed them to accrete massive, hydrogen-based atmospheres and mantles from the hydrogen-rich nebular gas.) This system of primordial zones is shown in Figure 7-4a. Figure 7-4a is a more detailed view of the condensation sequence discussed in Chapter 5. The zones are controlled by the materials that condensed at different distances, and the interplanetary bodies preserve a record of this.

Thus, we see that the inner solar system, from the sun to the mid-asteroid belt at about 2.5 AU, is a zone of rocky and metallic bodies. As we move away from the sun, the dominant compositions change. At about 2.5 to 2.7 AU, a remarkable change occurs: The outer half of the belt is dominated by black objects composed of soot-like carbonaceous materials—carbon, various carbon-rich compounds, plus ordinary rock-forming minerals. I refer to this position at about 2.5 to 2.7 AU as the "soot line"; it marks the point beyond which black carbonaceous compounds formed by some uncertain means. (Whether any of the carbonaceous material was a direct condensate in the solar nebula, or required modification of carbon-rich material like CH_4 ice, is an open question.) The black, carbonaceous materials were added to the silicate/metallic materials, and they dominate the colors of the aggregate solid material.

Somewhere nearby, perhaps around 3 to 4 AU, is what we could call the "frost line"; it marks the point beyond which the temperature was low enough for water to freeze into ice. The ice was added to the silicate/metallic/

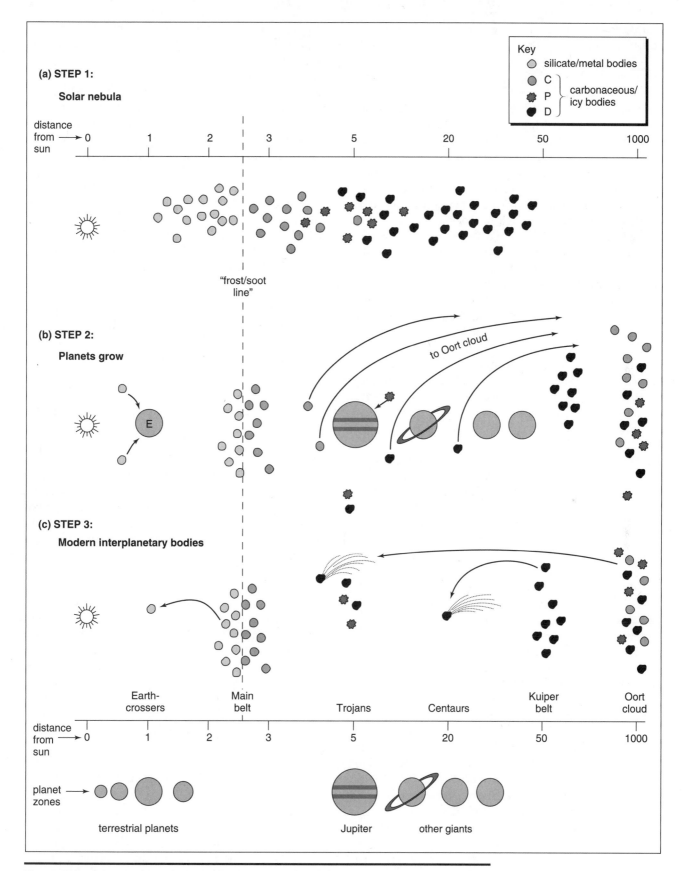

Figure 7-4. Schematic history of interplanetary bodies and distribution of taxonomic classes (not to scale). (**a**) Initial composition zones among planetesimals. (**b**) Perhaps 50 My later, planets are forming and the interplanetary bodies are being depleted by collisions with planets and ejection into the Oort cloud. The main asteroid belt and Kuiper belt are left stranded. (**c**) In the present system, a population of Earth-crossers, Centaurs, and active comets is maintained by processes of perturbation.

carbonaceous mix. At large enough distances it dominated the total composition. Beyond Saturn, other volatiles such as carbon dioxide, methane, and ammonia would freeze into additional ices. The colors of bodies beyond the frost line are not white but still black or dark brown, because they are colored by the carbonaceous material, like snow on an urban curb in February, after too much ash has been spread on the streets.

Another trend that can be seen from spectral observations involves the history of the frozen water. In the outer asteroid belt, among black carbonaceous asteroids, water of hydration is often found. The ice apparently melted in the past, percolating through the rock and creating hydrated minerals that have been detected spectroscopically. Carbonaceous meteorites, probably from this region, actually show deposits of carbonates in their fractures, left by water that evaporated and left dissolved minerals behind—similar to the carbonate "bathtub ring" that forms in our homes. Farther from the sun, among more distant asteroids, the water feature is not found; there may be more ice there, but the temperature has always been so low that the ice never melted and the water was never allowed to be mobile. Ice always acted like a solid. In spite of being in contact with water ice, the silicate minerals are dry, not hydrated.

The zonal trends continue into the outer solar system, where progressively redder colors seem to be the norm. This phenomenon suggests greater production of organic compounds on these bodies. The mode of production remains uncertain but probably involved reactions among initial carbonaceous materials, or very low temperature methane ice (CH_4) and other materials bearing the C-H bond. These reactions may be catalyzed by high-energy cosmic rays (which are not "rays" but energetic atomic particles). For example, cosmic ray particles may break molecules into fragments called free radicals, which often support formation of colored compounds. Such processes have been demonstrated in the laboratory by the Italian researcher G. Strazzula and others. Similarly, McKay and Borucki (1997) reported synthesizing molecules such as C_2H_6, HCN, CH_3NH_2, and other organic materials by simulating impacts into CH_4-containing ices. Methane-related ices would be stable only at the greatest solar distances, and so they would come into play only in distant objects that had never looped into the inner solar system. In support of these ideas, University of Arizona astronomer Robert H. Brown and colleagues reported in 1997, for example, that the reddish Kuiper belt object, 1993 SC, has a spectrum revealing the presence of hydrocarbon ices, such as methane and C_2H_6.

If the red colors are produced by reactions between cosmic ray particles and methane ice, these objects might tend to lose their reddish colors when they are perturbed into orbits closer to the sun. The reason is that cosmic rays penetrate only a few meters into solid material, so the reddish organics would be concentrated in the surface layers. Because these ices are easily sublimated by solar heating,

the red colored carbonaceous dust might be blown off the surfaces by cometary activity. Deeper, more neutral dust would finally be exposed. 2060 Chiron offers an interesting example. It has migrated inward from the Kuiper belt and may have been redder originally. Its present neutral gray color may be the result of its red layers having been blown off during its sporadic, perihelion cometary activity. Other Centaurs and Kuiper belt objects, which have not had as much solar exposure, might retain redder colors.

In summary, differences in color among Centaurs and Trojans (ranging from neutral C's to redder than normal D's) might thus depend on factors such as orbital history, availability of the relatively unstable ices in the surface soils, impact history that might vaporize ices, and even the stochastic presence of large, recent impacts that might expose fresher material.

How Asteroids and Comets Reached Their Present Locations

The key to making sense of the interplanetary hodge-podge is understanding their dynamical evolution—the changes in their orbits, which left the asteroids and comets in the positions and compositional zones we see today. The mere existence of the zones today indicates not only that compositional zones existed but also that the original populations have not been thoroughly mixed. However, at the same time, it is clear that certain groups of objects have been moved from their original positions. For example, many bodies that formed among the giant planets were thrown into the Oort cloud. Figure 7-4 b and c illustrate the evolution as currently visualized. As the largest planetesimals grew into planets, they swept up all the planetesimals in the intervening spaces (as also indicated by Figure 6-2); in fact, that is what left the main planets spaced as they are. This process was aggravated among the giant planets by the fact that they accreted massive atmospheres, swelling their masses from around 10 M_{Earth} to 100–300 M_{Earth} in the cases of Saturn and Jupiter. In spite of their smaller masses, Uranus and Neptune were especially effective at perturbing objects into the Oort cloud because they are on the gravitational fringes of the solar system. All four giants thus became very effective at scattering interplanetary bodies into the Oort cloud, as shown in Figure 7-4b. Thus, the color trends discussed above do not necessarily continue into the Oort swarm, because the Oort swarm was populated by a grab-bag mixture of different types of comets from the regions of the four giant planets. Perhaps some of them lacked enough methane-related ices to produce red colors. Thus, we cannot assume that there is a single type of "pristine, Oort-Cloud comet." Furthermore, some of the bodies might have been hundreds of kilometers across or even bigger—much bigger than typically observed comet nuclei (1 to 50 km).

At the same time, Jupiter's scattering of planetesimals helped to define the structure of the asteroid belt itself—not only the Kirkwood gaps but the outer edges of the belt as well (Liou and Malhotra, 1997).

To a lesser extent, the Earth and terrestrial planets kicked some bodies onto new orbits. Calculations of the orbital evolution of early planetesimals in the inner solar system show that many were thrown by Earth outward onto orbits in the inner fringe of the belt. Thus, the inner fringe may be enriched in (stony?) volatile-poor asteroids that formed in the inner solar system. These might explain, for example, the E-type (enstatite chondrite?) objects on the inner edge of the belt; they are the only significant group of meteorites with oxygen isotope ratios near the terrestrial/lunar values (Clayton, Mayeda, and Rubin, 1984). Even the Martian SNC's have a greater difference from terrestrial/lunar values (e.g., Taylor, 1988). Enstatite chondrites thus may originally have formed in Earth's zone. Further calculations show that some asteroids thrown from Earth's zone into the inner belt would have been thrown rapidly back onto Earth-crossing orbits, whereas others would have been preserved until today (Wetherill, 1977), consistent with the observations.

Jupiter perturbations and its scattering of planetesimals around the solar system also rapidly pumped up the mean collision velocities in the asteroid belt, causing accretion to end and catastrophic fragmentations to begin. Asteroid 1 Ceres, which had started to pull ahead of the others, was left as the largest asteroid at D = 1000 km. A few other bodies in the 200 to 500 km range were left, but a number were probably destroyed. 4 Vesta was left as the only large body with an intact basaltic surface layer.

Thus, a hundred My after planet formation ended, the asteroid belt was established between Mars and Jupiter, other interplanetary zones were cleared, the Kuiper belt of residual objects was left in the nebular disk beyond Pluto where there was insufficient material to make a planet, and the Oort cloud had been populated by ejected bodies.

As passing stars perturbed the icy bodies in the Oort cloud, some occasionally were directed in random-inclination, prograde or retrograde orbits back into the inner solar system and became active comets. Many of these were one-shot comets, dipping into the solar system on virtually parabolic orbits and returning to another near-eternity in the Oort deep freeze. For some, however, the story was not so simple. If they encountered giant planets, especially Jupiter, their orbits were altered once again. Early work by Oort (1963) and Öpik (1963, 1966) showed that over 96% of such Jupiter encounters would lead to re-ejection from the solar system in 1 to 100 My, but a tiny percentage might evolve into short-period comets. More recent calculations have shown how additional comets may evolve stepwise into the inner solar system, one planet at a time. Everhart (1977) started with a group of comets assumed to have a mean distance from

the sun of 50,000 AU, an inclination of 6°, and perihelia in the range of 30 to 34 AU, close enough to interact with Neptune. He found that about 0.15% of those interacting with Neptune were perturbed inward, with about 0.01% eventually ending up as short-period comets. Of such comets interacting initially with Uranus, Saturn, or Jupiter (near perihelion), he found 0.03%, 0.14%, and 0.77%, respectively, ending up as short-period comets.

The discovery of the Kuiper belt in recent years led to another source for short-period comets. Planetary perturbations cause some Kuiper belt objects to drop into the region of the giant planets, and again a process of encounters leads a fraction of those to become short-period comets, mostly with prograde revolutions, as shown in Figure 7-4c. This fits with observations by Jane Luu, Dave Jewitt, and S. F. Green and colleagues, suggesting that the reddest colors exist among Kuiper belt objects and Centaurs, with less red colors occurring among comets.

In an analogous process, Jupiter perturbations may kick asteroids from the main belt into the region of the terrestrial planets, as also shown in Figure 7-4c. In this sense, Earth approachers are to the main belt as Centaurs are to the Kuiper belt. This same process, Jupiter perturbation, clears the Kirkwood gaps in the asteroid belt, described in Chapter 3.

Trojan Asteroid Histories

One question has not been settled: Were the Trojans locally formed planetesimals that were hung up primordially in the Lagrangian points, or are they a hodgepodge of Centaurs and other cometary bodies captured into the Lagrangian points? A first step in answering this question would be to compare the distribution of colors and spectral classes of the Trojans with those of the Centaurs and Kuiper belt objects. So far, the Trojans seem less red, on average, than the Centaurs and Kuiper belt objects; among dozens of well-observed Trojans, none is as red as the reddest Centaurs. This might seem to suggest that Trojans did not originate as captured Centaurs and Kuiper belt objects; but if Centaurs grow less red after exposure to the sun, perhaps Trojans are captured Centaurs after all—Centaurs that have lost their red color. Clearly, more data are needed to give a firm answer to the history of Trojans and their relation to Centaurs.

Relating Small Moons to Asteroids and Comets

This framework leads to interesting questions about the smaller moonlets of the solar system—bodies in the size range of modest asteroids. Do they show color and composition similarity to one or more groups of asteroids and comets, or are they totally distinct? There is a special class of these bodies for which this question is especially important: the outermost moons of Jupiter, Saturn, and Neptune, and the two moonlets of Mars. The orbital properties of these moons suggest that they may have been captured into their orbits from interplanetary space. For example, about half these moons move in "backward" or retrograde directions around their planets, compared to the other moons—a situation that would be expected only if they were captured in random encounters between the planet and passing bodies. In other words, they were originally asteroids or comets.

If they were really captured, one of the first questions we could ask is what color/spectral classes they belong to. This might clarify where they originated before they were captured. Were they errant asteroids thrown out of the main asteroid belt? Are they more like Trojans? Or could they be drawn from the Centaur population as those bodies were perturbed from the Kuiper belt? Or, as a group, are they different from any of those populations?

Interestingly enough, the observations of these small, faint moons show that they all have very low reflectivity, meaning that they match the sample of objects that originated outside 2.5 AU, not inside that distance. Beyond that, oddly, the results on color are controversial. Phoebe, the retrograde moon of Saturn, is pretty clearly a neutral black C object, similar in color (and size) to 2060 Chiron. Only the brightest few of Jupiter's eight captured moons have been studied in this way, and the results of different observers disagree. Some observers reported them as C's, but others report some of them as more reddish-brown, similar to Trojans. A similar situation is the embarrassing case of Phobos and Deimos, the closest moons beyond our own, where reported colors are all over the map; one reason is that it is hard to separate their light from the red glare of Mars. Around 1980, they had been confidently reported as being neutral black C-like; the data included measurements from the Viking spacecraft. Then Hubble Space Telescope data indicated that the colors match taxonomic type D. But observations from the Mars Pathfinder lander gave a color less red than a D. The color is reddish, but perhaps more like class P, which is between C and D in color (Smith and others, 1997). The sketchy spectral data available for six or seven of these twelve probably captured moons suggest that they do not match a sample from either the inner or outer asteroid belt. They may match a sample drawn from the Trojans or the Centaurs, which means they may have originated as comets. The densities (about 1,900 to 1,950 kg/m³) measured by Viking and PHOBOS-2 are consistent with carbonaceous composition but could also represent rubble pile structure of some inherently denser material.

In 1997, two similar small, outer objects were discovered with the Hubble Space Telescope on the outskirts of Uranus's system, beyond Oberon. They may also be captured moons, the first known in the Uranus system. First reports indicate highly eccentric, possibly retrograde orbits, but the orbits have yet to be confirmed. However, they are so faint that no spectral taxonomic classes have yet been determined for them.

Evidence of Collisional Fragmentation: Hirayama Families

The previous two chapters indicated that collision and fragmentation were fundamental, defining processes in creating the solar system we know today. Planetary properties indicate origin by collisional accretion. Meteorite breccias testify about ancient smash-ups. In the asteroid belt, we see direct evidence of the inferred breakups.

In 1918, the Japanese astronomer K. Hirayama published the first of a series of papers pointing out that various asteroids clustered into groups with similar orbital elements. The groupings are called **Hirayama families.**

Each Hirayama family contains the fragments of an asteroid that suffered a collision and fragmented long ago. When any orbiting body breaks up with a random distribution of fragment velocities, the net result will be for the pieces to spread out along the orbit of the original body, and this is what is seen in Hirayama families. When the original asteroid collided with another (smaller) asteroid, it was blown apart and the pieces separated. Brouwer (1951) showed that the differences in velocity among family members are about 0.1 to 1.0 km/s—values that must approximate the original ejection velocities. Such speeds are only a small fraction of orbital velocity; hence, all the fragments remain in approximately the initial orbit. The speeds are also consistent with average ejection velocities of fragments from laboratory impacts at about 2 to 10 km/s (Gault, Shoemaker, and Moore, 1963). Although we have spoken of "fragments," some studies suggest family members may not all be coherent splinters. Michel and others (2001) made computer simulations suggesting that collision debris re-aggregate into rubbly bodies that are the family members.

The family involving the 160-km asteroid 8 Flora is especially interesting. It has a bimodal distribution of semimajor axes among its fragments, possibly reflecting the two colliding bodies. The members would combine to make a spherical parent body estimated at 165-km diameter. But Flora itself, if added to the second-largest body (78-km asteroid 34 Ariadne), would require a long axis of 240 km for the parent. Also, Flora is spherical and rotates slowly. These properties suggest that Flora itself is not a fragment. Tedesco (1979c) suggested that proto-Flora was a binary asteroid and that only the satellite was broken in the collision.

The spectra of some families show uniform composition among all members, strongly supporting the theory of origin by collisional breakup of a parent body, but other families show mixed compositions (Chapman, 1979). Perhaps their fragments came from different regions in the interior of two colliding bodies that were already nonhomogeneous.

Brouwer (1951) described 9 families and 19 additional "groups" identified by similar orbits. The number of asteroids in the families ranges from 9 to 62, and in the groups from 4 to 11. Arnold (1970) reviewed Brouwer's and Hirayama's data, confirmed the existence of families 1 to 9, and added a number of new families. As many as half of all asteroids have been associated with 100 proposed Hirayama families, but only about a dozen families are populous and well established.

It is intriguing to think that many Hirayama families may be drifting debris from the titanic collisions that have sporadically marked solar system history.

Physical Nature of Asteroids and Comets

Amazing strides were made in the 1980s and 1990s in understanding the physical nature and surface properties of asteroids and comets. The first close-up image of an interplanetary body was that of Halley's comet by the European probe, Giotto, in 1986. In 1989, Steven Ostro began to bounce radar off Earth-approaching asteroids with enough precision to show irregular shapes and craterlike depressions. The first close-up images of asteroids were made in 1991 and 1993, as the Galileo probe passed 951 Gaspra and 243 Ida, on its way to Jupiter. The NEAR (Near-Earth Asteroid Rendezvous) spacecraft photographed asteroid 253 Mathilde at close range (Figure 5-9d) and then orbited asteroid 433 Eros, landing on it in 2001 (Sullivan and others, 2002). The Stardust spacecraft photographed comet Wild 2 at close range in 2003. All the asteroids were lumpy and cratered, having undergone a history of intense collision and cratering. All the asteroids' surfaces are densely crowded—perhaps saturated—with craters; the bodies themselves are shaped by the larger craters, with whole sides seemingly lopped off and turned into cavities. Mathilde surprised researchers by having a crater 62% of its mean diameter; some thought this was larger than an asteroid could sustain, but Davis (1998) calculates that the impactor needed to form the crater was smaller than the size needed to blow Mathilde apart.

The comet close-up photos (Figure 7-6) are less sharp; the environment was hazy with dust. The photos clearly show distinct jets of gas and dust streaming off specific regions (fissures?) on the surface, and they show irregular, peanut shapes. Halley and Wild 2 are 15 and 3 km long, respectively. Many craters can be seen on Wild 2, with unusually flat floors. They may be impact scars modified by gas venting. Surface characteristics may be controlled as much by venting as by impact.

Closeup images of Phobos and Eros have been obtained, showing rocks, regolith soils, and gooves (Figure 7-7). Phobos may be a captured asteroid of taxonomic class D. Its largest crater shows semi-radial swarms of grooves that may mark deep fractures caused by the explosion. Curiously, these grooves have rows of craterlets with raised rims; I have suggested that they may record venting of water vapor released from interior ice during or after the impact (Hartmann, 1980), and Fanale and Salvail (1990) have argued that the Russian PHOBOS-2 results indicated OH molecules escaping from interior ice. If this is correct, Phobos might be an example of a body that once had substantial cometary activity and the grooves may be analogous to the vents on the nucleus of Halley's comet.

Asteroids' Moons and Compound Shapes

During the 1970s and 80s, when asteroids passed in front of stars, astronomers observing the event occasionally recorded not only the few-second dimming of the star as the asteroid passed by, but also a secondary dimming event. These secondary events were rarely if ever confirmed, but they led to controversy over whether some asteroids might have satellite bodies—a possibility unexpected by most theorists.

In 1993, the second close-up photo of an asteroid, that of 52-km asteroid 243 Ida by the Galileo spacecraft, revealed a 1-km moon orbiting the asteroid at a distance of roughly 100 km from Ida's center (not shown in the close-up view of Figure 7-5). By 2002, precision Earth-based telescopic imaging revealed satellites of seven more main belt asteroids, one Trojan, and seven out of 500 known Kuiper Belt objects (Merline and others, 2002). Also, radar images show 800-m wide Earth-approacher 2000 DP107 as having a 300 m satellite (Margot and others, 2002). The existence of such moons not only explains the occultation results but also an older discovery: Earth, the moon, and Mars show numerous pairs of adjacent impact craters of the same age. Clearwater Lakes, a pair of adjacent craters 32 and 22 km across in Canada, are the best example. Such crater pairs formed when an asteroid and its nearby satellite hit at the same time. Analyses of these crater statistics suggest that at least 10 percent of asteroids have sizable moonlets moving around them; the fraction may be maximized among Earth approachers that split as a result of tidal forces during close encounters (Bottke and Melosh, 1996).

A related discovery came from a new technique that creates images of nearby asteroids by means of radar signals bounced off them from large radio telescopes as they pass close to Earth. Results have astonished astronomers. In 1989, when the 1×1.5-km asteroid 4769 Castalia passed by Earth, Ostro obtained a fuzzy radar image that showed a two-lobed dumbbell shape rotating end over end. In 1992, when the 3×5-km asteroid 4179 Toutatis passed

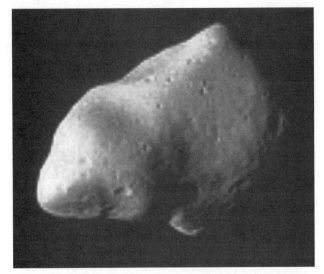

a

b

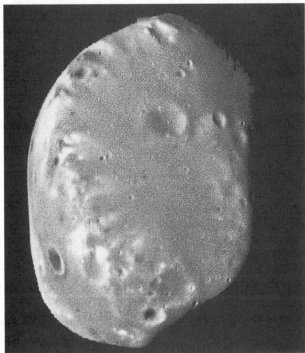

c

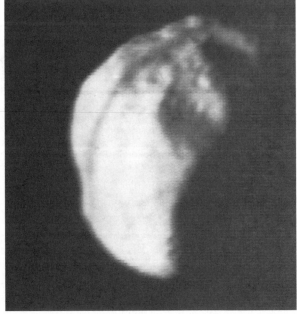

d

Figure 7-5. Irregular-shaped small bodies. This picture compares the first two close-ups of asteroids (returned by the Galileo spacecraft on its way to Jupiter in 1991 and 1993) with similar small moons. (**a**) Asteroid 951 Gaspra, which is about 12 × 16 km in size. It is believed to be a fragment of a larger asteroid, and the number of impact craters on its surface suggests that it may have broken off its parent body a few hundred million years ago. (**b**) Asteroid 243 Ida is larger, about 52 km long. It has more impact craters, and therefore is believed to be older. (**c**) Mars satellite Deimos is 10 × 16 km, and appears to have "softer" surface crater profiles, possibly due to "sandblasting" by dust impact in the Mars environment. (**d**) One of Saturn's inner satellites, S11 Epimetheus (about 100 × 140 km) also shows craters and an irregular shape. In this unusual view, the shadow of Saturn's edge-on rings was moving across the moon. (**a** and **b:** NASA, courtesy Michael J. Belton; **c** and **d:** NASA Viking and Voyager photos)

a

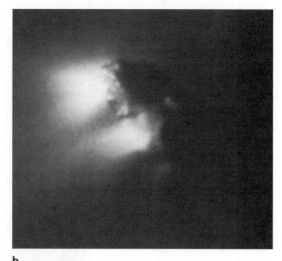

b

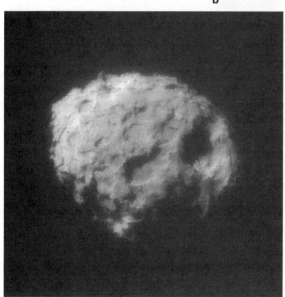

c

Figure 7-6. The two comet nuclei imaged at closest range are Halley's Comet (in 1986 by European Giotto probe) and Comet Wild 2 (in 2004 by American Stardust probe). (**a**) Distant view shows dusty haze around 15-km-long Halley nucleus, with bright, sunlit jets of gas and dust. (**b**) Closest view of Halley nucleus shows jets and irregular surface, with dark side silhouetted against the softly glowing coma. (**c**) View of 5-km-wide Wild 2 nucleus revealed rough surface with crater-like depressions—possibly impact craters modified by venting through the impact site. Longer exposure showed gas jets like those of Halley. (Giotto images courtesy Harold Reisema and Alan Delamere, Ball Aerospace: copyright 1986 MPAE; Wild 2 image: NASA)

Earth, radar images again showed a crudely double-lobed, lumpy shape (Figure 7-8). Such images, together with other techniques, show that some asteroids must be loosely bonded clumps of two or three large fragments. These compound asteroids are within reach of our spaceships; it would be a strange experience for an almost weightless astronaut to float among the rocks of an asteroid built like two rounded mountains, just touching each other!

The compound asteroids and asteroid satellites may be part of the same phenomenon: asteroid fragmentation events. Dynamical studies of the evolution of orbiting pairs shows that they are unlikely to last over solar system history; tidal forces lead to rapid orbit evolution on timescales of 10^5 y (Binzel and van Flandern, 1979). Satellites outside the synchronous point evolve outward and at about 100 primary diameters, they escape the primary's sphere of influence. Satellites inside the synchronous point drift inward until they land on the primary. Thus, satellite pairs require an ongoing creation mechanism.

Hartmann (1979a, 1979b) suggested that some of these originate when one asteroid hits another and blows it apart in a titanic collision; as a jumble of fragments races outward from the explosion, adjacent fragments may bump into each other, fall together as pairs and make dumbbell-shaped compound objects or go into orbit around each other. Models by Durda (1996) and Durda et al. (2004) support this. Thus, the seemingly disconnected discoveries of asteroid moons and compound shapes may both be clues to the violent histories of small bodies.

Aside from immediate creation in collision, another source for compound shapes might involve the orbital evolution of satellite pairs. If a satellite evolves inward and lands on a primary, it would create a compound asteroid.

Some support for these ideas comes from Trojan asteroid 624 Hektor, which may be two similar-sized lobes (Hartmann and Cruikshank, 1978; Weidenschilling, 1980), or from 4179 Toutatis (D = 3 × 5 km) which seems to consist of two main lobes (Figure 7-8).

Figure 7-7. Close-ups. (**a**) Landscape on Phobos offers a possible analog of an asteroid surface. Photo was made from 120 km away by Viking Orbiter. Grooves near horizon may be surface expressions of regolith covered fractures caused by the largest collision. The area is about 2 km across and the smallest craters are about 15 m across. (NASA Viking orbiter photos, slightly blurred by spacecraft motion) (**b**) Regolith soils on asteroid 433 Eros. Area is 33 m across. (NASA, NEAR photo).

Absolute Magnitudes, Sizes, Albedos, and Masses of Interplanetary Bodies

For most interplanetary bodies, the property simplest to measure from Earth is brightness. The apparent brightness depends on the distance from both the sun and the Earth, the reflectivity, and in the case of a comet, the presence of a surrounding halo of dust or gas. The total area of reflecting dust in a comet halo is difficult to know, so the following techniques are applied primarily to asteroids or inactive comets.

A standardized way to express the brightness is as follows: The **absolute magnitude** of a solar system body is its brightness at zero-phase angle (sun-object-observer) when the object is 1 AU from the sun and 1 AU from the observer. The astronomer's stellar magnitude system is used, in which the brightest stars are around magnitude 0 and the faintest visible with the naked eye are around 6. Depending on whether visual (yellow) wavelengths or blue wavelengths are used, the absolute magnitude is des-

ignated $V(0,1)$ or $B(0,1)$, where the "0" implies zero phase angle, and the "1" implies the distance of 1 AU. Sometimes $V(0,1)$ is simply designated g.

Measurement of the visible brightness (reflected light) and the thermal infrared brightness (re-radiated thermal radiation) allows calculation of the **albedo,** or percentage of sunlight absorbed. The albedo can then be used to calculate the area needed to reflect the observed sunlight, and hence the diameter. These diameters are calibrated by occasional occultations, when the time needed for the asteroid to pass in front of a star gives a direct measurement of a chord drawn across the body. If enough ground stations observe the asteroid, the shape and diameter can be measured.

Masses are much harder to measure. Masses of a few large objects, such as 1 Ceres, 2 Pallas, and 4 Vesta have been estimated from their barely measurable perturbations on other asteroids. Recent spacecraft flights past asteroids have allowed measurement of the gravitational effect on the spacecraft; hence, the mass.

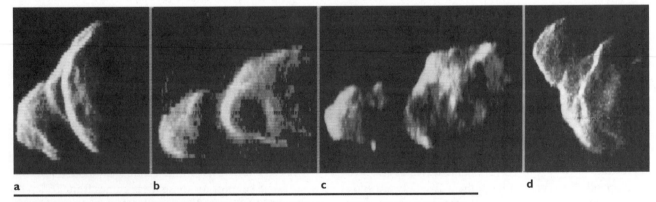

a   b   c   d

Figure 7-8. Radar images of asteroid 4179 Toutatis, made when it passed close to Earth in 1992, at a distance only about six times that of the Moon. The images show several sides of the asteroid as it rotated, and indicate a lumpy shape composed of two main bodies. (NASA radar image from Jet Propulsion Laboratory, courtesy Steven Ostro)

TABLE 7-4 Asteroid Densities

Asteroid	Density kg/m³	Taxonomic Class	Notes
1 Ceres	2400 ± 200	C	Largest asteroid; data from Millis and others, 1987; Yeomans and others, 1997
2 Pallas	2800 ± 500	B (subclass of C)	Millis and Elliot, 1979; Yeomans and others, 1997
4 Vesta	3300 ± 500	V	Millis and Dunham, 1989; Yeomans and others, 1997
22 Kalliope	2370 ± 400	M	Margot and Brown, 2003; metal excluded
243 Ida	2000 to 3100	S	Spacecraft measure of mass; Chapman, 1996
253 Mathilde	1300 ± 200	C	Spacecraft measure of mass; Yeomans and others, 1997
433 Eros	2670 ± 30	S	NEAR; Sullivan and others, 2002
Phobos	ca. 1900	P or D?	Spacecraft measure of mass; Veverka and Farquhar, 1997
Deimos	ca. 1800	P or D?	Veverka and Farquhar, 1997
Statistical average for class C asteroids	1200 ± 100	C	E. M. Standish, cited by Yeomans and others, 1997

Densities of Interplanetary Bodies

Once we have a size and mass for a body, we can compute its density. Densities give considerable evidence about asteroids' structure and history. Do asteroids have densities that match proposed corresponding meteorite types? Table 7-4 shows some results. The first three crude measurements from ground-based data suggest that the large asteroids have densities in keeping with the corresponding meteorite types. Carbonaceous chondrite rock densities range 2200 to 3600 kg/m³ for CI types at the low end and CV/CO types at the high end. Basalt-like achondrites, possible analogs of Vesta or its surface, range 3200 to 3400 kg/m³. A satellite of 22 Kalliope permitted a density measurement. The result rules out nickel-iron composition, which had been considered likely for M class objects. Radar results indicate some M's are iron and others are stone (Magri and others, 2001). As for S-class Ida, the issue is whether it can be an ordinary chondrite. Ordinary chondrites have densities 3400 to 3800 kg/m³. Ida's density is around 70% of this. Therefore, for Ida to be a chondrite, it probably is highly fractured inside, having a rubble-pile structure, with considerable empty pore space. Rubble piles—rather than solid bodies—were long ago suggested as a likely structure for asteroids because of their history of collisions and fragmentation events (Davis and Chapman, 1977).

The interpretation of density reduction by rubble pile structure in smaller asteroids was supported by the flyby of 253 Mathilde in 1997. C-class Mathilde's density is 50% to 70% of the density of moderate-to-low-density carbonaceous chondrites. Similarly, Phobos, Deimos, and average C's are underdense. An open question is whether water ice might contribute to the low density in C-, P-, or D-class objects.

When the Ice Sublimes: Phenomena of Active Comets

The word *comet* comes from the Greek word *kome* ("hair"). The name describes the long wispy **tail** for which comets are famous. Orientals knew comets by the equally descriptive term *broom stars*. The tail is composed of gases released by the ice as it is warmed by the sun and sublimed, and by dust blown outward with the gas.

Comet tails extend away from the sun, as shown in Figure 7-9. During the part of its orbit when a comet moves away from the sun, the tail leads; picturing the tail as always trailing behind the comet is inaccurate.

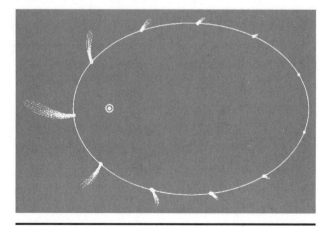

Figure 7-9. Orbit of a typical comet, showing development of the tail as the comet approaches and recedes from the sun.

In addition to the tail, a comet has a **head,** composed of the **coma,** a bright, diffuse region immediately surrounding the **starlike nucleus** (Figure 7-10). Among the less spectacular comets, the inner coma is all that can be seen with the naked eye from Earth. With a large telescope, the true nucleus of even a spectacular comet appears as no more than a starlike point. Nonetheless, these ice chunks a few kilometers across produce glowing gas comas that expand to sizes bigger than planets, and dusty tails that span an astronomical unit.

Comets approaching the sun from the outer solar system usually begin to show activity around the orbits of Saturn or Jupiter, depending on what ices are present: CH_4 and CO_2 ices, for example, sublime at larger distance than H_2O ice. Comets are usually discovered when they are faint and far away, approaching the sun at a distance of several astronomical units. In general, they brighten and their tails develop and lengthen as they approach the sun. There is a tendency for their comas to reach a maximum diameter at roughly 1.5 to 2.0 AU from the sun, because fewer gases are released at greater distances, whereas the molecules in the coma are readily disassociated by solar ultraviolet radiation when the comet is nearer to the sun. When a comet swings close to the sun, it is very bright but often is not visible from Earth because it is lost in the glare of the nearby sun. However, recent satellites designed for solar studies have picked up numerous "sun-grazing" comets as they pass near the sun or even crash into the sun.

Tails and Comas: Appearance and Composition

As shown in Figure 7-10, comets have two types of tails, Type I and Type II. **Type I tails** are essentially straight, often structured by fine, linear streamers or "rays." Their spectra show emission lines indicating a composition of mostly ionized gas. As a mnemonic aid, it is helpful to think of the *I* in Type I: This Roman numeral is linear and can stand for ionized. The principal light-emitting molecules are CO^+, N^+_2, and CO^+_2. Additional neutral atoms and molecules that have been identified include hydrogen, hydroxide (OH), oxygen, carbon monoxide, and carbon, all at typical emission rates of 1 to 65×10^{28} molecules/second at 1 AU from the sun.

Type II tails are usually broad, diffuse, and gently curved. Their spectra show only the reflected spectrum of the sun. This reflection indicates that instead of being composed of gas atoms and molecules that absorb or give off their own characteristic light, they are composed of grains of dust that merely reflect (scatter) sunlight. They may show a delicate rayed structure, seen in Figure 7-11, caused by individual dust jets from the nucleus.

Some comets have Type I tails, some have Type II, and some show both. The appearance of tails is affected not only by type but also by aspect angle. Occasionally a comet displays a seemingly anomalous tail that appears to point toward the sun; Comet Arend-Roland in 1957 was a widely publicized example. This is only a projection effect. As we look along the line from Earth to the comet, most of the tail may be on the antisolar side, but a portion of the curved tail may appear in projection against the sky to be on the sunward side. In the case of Arend-Roland, the "solar-directed tail" was part of a Type II tail seen in projection. It appeared as a narrow spike only as Earth passed through the comet's orbit plane. This narrowness indicated that the dusty Type II tail was highly confined to the plane of the comet's orbit.

The tiny dust grains that compose the Type II tails are repelled from the sun by radiation pressure (see Chapter 3), and each particle follows an orbit determined by gravity and radiation. According to Kepler's second law, the particles farther from the sun revolve more slowly and so they lag behind, giving the Type II tail its characteristic curvature. But this law does not explain all the properties of Type II tails, not to mention Type I tails. Observations of comets showed that knots of material in Type I tails were accelerated away from the sun faster than could be explained by radiation pressure. This observation led Biermann (1951) to the discovery of the solar wind streaming out from the sun. Magnetic interactions couple the ions of the Type I tails to the outward-expanding ionized gas, or **plasma,** of the solar wind.

Composition of Comet Solid Materials

The next step beyond identifying gases in comets' tails is to derive from the spectra the relative amount of each gas and the production rate at which the gas is being emitted by the comet nucleus. From such data, Delsemme (1977) was able to show that water (H_2O) ice must be the dominant constituent of most nuclei. Methane (CH_4) and ammonia (NH_3) ices are also probably present in most comet nuclei.

More specific measurements of composition were made when the European Giotto probe and the two Soviet Vega probes flew close to Halley's comet in 1986. These probes, together with Earth-based data, give Halley's ice composition by numbers of molecules (Brandt, 1990):

water ice (H_2O)	80%
carbon monoxide (CO)	10%
carbon dioxide (CO_2)	3.5%
formaldehyde polymer ($H_2CO)n$ and/or other organic compounds	few %

These numbers confirmed the abundant water ice in the nucleus. The Halley probes were also able to analyze dust particles being blown off the comet. The particles were found to be rich in C, H, O, and N, and were called **CHON particles.** They differ from familiar terrestrial dust, which is richer in silicon, iron, other metals, and other oxides. The latter type of silicate dust constituted a second type of particle measured in Halley, and a third, most abundant type, seemed to be a mixture of the two, rather similar to carbonaceous chondrite material. The CHON particles indicate that the nucleus soils are rich in

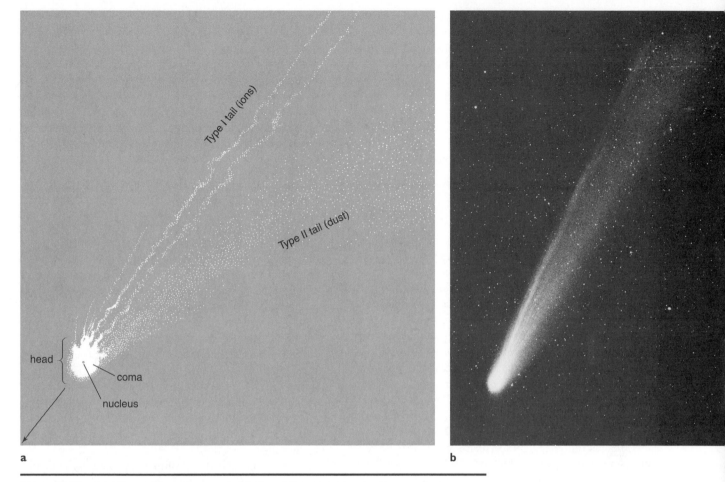

a

b

Figure 7-10. (**a**) Nomenclature of a comet. (**b**) Comet Kohoutek, showing the same features, with minimal separation of Type I and Type II tails. (NASA, Lunar and Planetary Laboratory, University of Arizona)

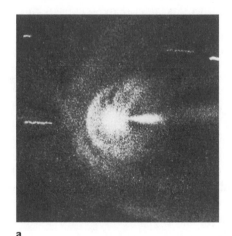

a

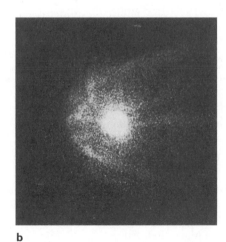

b

Figure 7-11. Spiral and fountainlike jet forms are seen near the nucleus of Halley's comet as jets of gas and dust are emitted from the rotating nucleus and blown back (to the right) by the solar wind. These photos were made on different days as the comet passed near Earth in 1910, but the jets were revealed only in 1984 when astronomers Z. Sekanina and S. M. Larson used modern image processing techniques to enhance low contrast detail. Streaks are star images, trailed by long exposure. (Courtesy S. M. Larson, Lunar and Planetary Laboratory, University of Arizona)

carbon and probably organic compounds (Sagdeev and others, 1986), though no life is expected to have evolved there.

In summary, a typical comet nucleus is now believed to be mostly ice by composition (frozen H_2O and other volatiles) but impregnated with black sooty carbonaceous

and CHON particles. Just a small percentage of the sooty material, scattered through the comet ice, could make even the "fresh ice surfaces" appear very black. As the ice sublimes at the surface of a comet, during passes near the sun, the ice/soil ratio in the surface layers probably decreases, leaving an even darker surface regolith of carbonaceous soil. Gases from subliming ice may build up pressure and eventually break through in certain areas, explaining the sporadic eruptions of comets and the well-defined jets photographed on Halley's surface.

Rotation and Shapes of Interplanetary Bodies

Many asteroids, as well as the few bare comet nuclei that have been observed, vary in brightness periodically, as found by measuring the **light curve,** or brightness as a function of time. This variation can have two causes: **1.** The body is elongated in shape and is tumbling end over end. In this case the true rotation period would include two maxima in the light curve. **2.** The body has dark and light markings. Most likely, one hemisphere would on average be darker than the other (as with the moon), and the true rotation period could include only one cycle of the light curve.

Observations show that the first case is much more common: asteroids and comet nuclei are often elongated. Elongated shapes have been confirmed by direct imagery (Figures 7-5 and 7-8). Much earlier, in 1931, Apollo asteroid 433 Eros came so close to Earth (23 million kilometers) that astronomers could visually see the 7 × 19 × 30 km elongated object tumble end over end every 5.3 h. Elongated objects have more symmetric light curves than spotted objects do, and searches for spectrophotometric differences from one side of an asteroid to another during rotation have generally been negative or inconclusive.

Another way to distinguish elongation from spottedness is to compare light curves of reflected solar and emitted thermal infrared radiation. If an asteroid is elongated, both curves will have simultaneous maxima; but if it is spotted, the dark side will be warmer and will emit maximum thermal infrared radiation at the same moment that the reflected visual light curve is at minimum due to the low albedo. This test was used to prove that the peculiar Trojan, 624 Hektor, is unusually elongated (Hartmann and Cruikshank, 1980).

Many elongated asteroids vary in brightness by a factor of as much as 3 or more. Light-curve statistics reveal fewer kilometer-scale elongated asteroids in the belt than among similar-sized Mars-crossers, possibly due to the higher rate of collision and erosion in the belt. Also, Trojan asteroids and comets seem to have a higher frequency of elongated shapes than do belt asteroids (French, Kramer, and Vargass, 1986; Hartmann and others, 1988); this might be due to processes of ice sublimation or processes of collision (Hartmann and Tholen, 1990).

Photometry indicates not only rotation, shapes, and markings but also the obliquity of asteroids. To envision this, suppose an irregular asteroid can be seen pole-on from Earth during one apparition; it would have zero am-plitude. Some months later we would see it at some quite different aspect angle, because of our relative motion. The light variations due to rotation would now become appreciable. By studying variations in the light curve with aspect angle, we can derive the position of the asteroid's axis of rotation. Taylor (1979) reviews pole positions published for 17 asteroids with obliquities ranging from 30° to 90°. The high, seemingly random obliquities are consistent with spins imparted during major collisions.

Feasibility of Irregular Shape as a Function of Size

If Earth, moon, and other planets are almost exactly spherical, why are small bodies irregular in shape? Asteroids and comets are in a sense large chunks of rock, metal, ice, or mixtures; obviously, rocks and icebergs can take a variety of shapes. Perhaps, then, the question should be rephrased: *Why are the larger planets round?*

It is a question of strength of material. The larger the asteroid or planet, the greater is the pressure at the center. If the central pressure exceeds the strength of the rocky material, the material will deform, either by plastic flow or by fracture, as a result of failure of the normal elastic properties of the solid rock. Higher internal temperatures also favor deformation toward an equipotential shape. In the case of an ideal nonrotating body, this shape will be perfectly spherical.

As calculated in the accompanying optional mathematical notes, the diameter at which the central pressure begins to exceed the strength of the planetary material is around 400 to 700 km. As plotted in Figure 7-12, there is a strikingly sharp cutoff in irregular shapes at diameters around 400 to 600 km. Bodies smaller than about 400 km often are noticeably irregular; worlds larger than 600 km are not.*

Centrifugal Force Versus Gravity

Because the asteroids spin rapidly, with periods as short as a few hours, a frequent question has been whether they spin so fast that any loose material would be thrown off their equators by centrifugal force. The answer is no. An asteroid would have to rotate in less than about 2 h for centrifugal force to exceed gravity. Even the fastest-rotating asteroid, the Apollo object Icarus, has a period of 2.25 h, so loose debris could exist on its surface. Polarimetric data support this conclusion by indicating a dusty surface on many asteroids.

Asteroid Rotations as a Clue to Planet Evolution

The planets' rotation rates, with periods ranging from 0.4 d for Jupiter to 243 d for Venus, might appear at first

---

*See Chapter 8 for more detailed discussion of planetary oblateness caused by rotation.

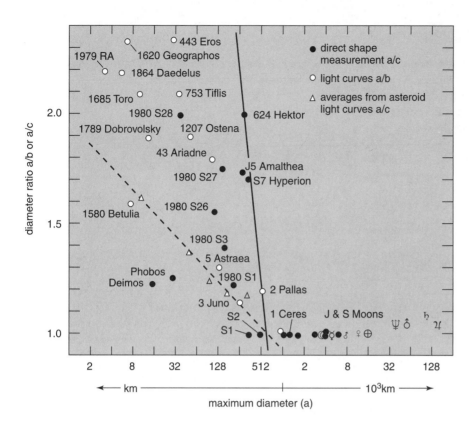

Figure 7-12. Irregular shapes in planetary bodies, as a function of size. Closed circles and planetary symbols show direct measure of a/c, where a and c are maximum and minimum diameters. Open symbols are based on asteroid light curves, which reveal the lower ratio a/b, where b is the intermediate diameter. Both the trend of the a/c and average a/b ratios indicate that the maximum diameter for irregular bodies is about 600 km. Larger planetary bodies have spheroidal shapes because planetary interiors are not strong enough to support irregularities of such large scale.

glance to be random. However, if we add data on asteroid rotations, a remarkable trend emerges: Though the masses of asteroids and planets range over 15 orders of magnitude, the rotation periods are relatively constant. Of a sample of 137 planetary bodies, 83% rotate with periods between 4 and 16 h.

The general relation appears to apply to planets as well. (Recall that Mercury's rotation is tidally locked by a resonance with the sun, and Venus's is probably similarly influenced by solar tides or resonance with Earth, so they do not count.) Earth and Pluto have been slowed by tidal interaction with their relatively large satellites. Restoring the Earth-moon and Pluto-Charon systems to initial, faster rates based on conservation of angular momentum, we find that they fit the above rule.

Why shouldn't planetary bodies rotate with widely scattered periods, such as 1 h, 100 h, or 3 mo? A clue comes from considering centrifugal force versus gravity, as mentioned above. The collisional processes appear to have spun planetesimals up toward high rotation speeds of a few hours, but they could not have grown if they had reached rotations faster than about 2 h, because they would then throw material off their equators. Weidenschilling (1981) considers the further complexity that a weak asteroid spinning up toward this rate will begin to deform, first into a Maclaurin spheroid (a figure with an equatorial bulge), then into a Jacobi ellipsoid (an elongated, football-like figure), and finally into a body that fissions into two parts. The elongation inhibits spin-up, so that the periods of weak asteroids do not get shorter than

about 4 h. Note that if an asteroid once melted and then solidified into strong igneous rock or metal and if it escaped later weakening by impact-induced fractures, it could spin faster without deforming. Weidenschilling (1981) suggests that 321 Florentina is the best candidate for such solid-body rotation; its period of 2.87 h is the shortest known in the main belt.

The cumulative effects of collisional spinup but increased loss rate of ejecta from fast-spinning bodies and the tendency to elongate were probably compensating influences that left nearly all asteroids with periods between 4 and 16 h (Harris, 1977). Long-period exceptions, such as 128 Nemesis and 393 Lampetia, may have been struck and slowed down, late in their careers.

Number and Size Distributions of Asteroids and Comets

Statistical knowledge of numbers of bodies in the asteroid belt are good enough that we have interesting information about the total numbers and size distribution of bodies in that region. To estimate the "total number" of asteroids, we must specify a minimum size, as the smaller the size, the more asteroids there are. Only 3 are bigger than 500 km across, but 12 are bigger than 250 km, and roughly 200 exceed 100 km. Among Trojan asteroids, there may be about

Mathematical Notes on Maximum Rotation Rates

It is reasonable to suppose that the fastest rotation rate that could be acquired by an object during its formation by accretion would be the rate at which loose material would be thrown off its surface by centrifugal force. Setting gravity equal to centrifugal force, and expressing the velocity in terms of rotation period, we have

$$\frac{GMm}{R^2} = \frac{mv^2}{R} = \frac{4\pi^2 R^2 m}{P_c^2 R}$$

where P_c is the critical rotation period at which material will be thrown off the equator. This reduces to

$$P_c = \sqrt{\frac{3\pi}{G\rho}} = \frac{104}{\sqrt{\rho}} \text{ hours}$$

For material with densities ranging from 2200 to 3800 kg/m³ (typical of carbonaceous chondrites and chondrites; Wasson, 1974), this formulation gives critical periods from 1.7 to 2.2 h.

Mathematical Notes on Maximum Sizes of Irregular Bodies

In Chapter 8 we will derive the following equation for the pressure of overlying material, measured at some distance r out from the center of a planet with uniform mean density $\bar{\rho}$ and with radius R:

$$P_r = \frac{2\pi G\bar{\rho}^2}{3}(R^2 - r^2)$$

Now consider a planetesimal in whose central region the strength of the material, is greater than the pressure exerted by the overlying material. (Note that strength S and pressure P_r have the same units, newtons/meter².) As long as S exceeds P_r in the central region, the planetesimal will be strong enough to retain any irregular shape it might acquire, say, by collisional fragmentation. But if the planetesimal grows to a size such that P_r begins to exceed then the central material will be crushed and the planet can begin to assume the spheroidal shape of hydrostatic equilibrium. If we set P_r equal to S for material 20% of the way out from the center (an arbitrary choice of distance), the equation reduces to the following condition for the largest planetesimal radius R_{Max} that can sustain a substantially irregular shape:

$$S = 1.34(10^{-10})\bar{\rho}^2 R_{Max}^2$$

(assuming SI units)

We might expect the inner regions of a reasonably large planetesimal to be made either of moderately strong rock (like chondritic meteorites) or of iron (like iron meteorites). For these two cases we have the following conditions:

(1) For a *rocky planetesimal,* strengths of chondritic stone meteorites range from about 6×10 to 4×10^8 N/m² (Wasson, 1974), and corresponding densities range from 2500 to 3800 kg/m³, respectively. Using a representative strength and density of 2×10^8 N/m² and 3500 kg/m³ we find

$$R_{Max} \simeq 349 \text{ km}$$

(2) For an *iron-cored planetesimal,* using an iron meteorite's strength of 4×10^8 N/m² (Wasson, 1974) and the iron density of about 7870 kg/m³, we find

$$R_{Max} \simeq 220 \text{ km}$$

These calculations suggest that the largest irregular planetesimals might be expected to lie in the *diameter* range of 440 to 700 km. This prediction, though based on a simplified analysis, is in excellent agreement with the empirical data in Figure 7-4, where the largest irregular asteroids and satellites show a well-defined upper diameter limit in the range of about 360 to 600 km.

20 that exceed 100 km in diameter. Thousands of belt asteroids have been cataloged down to sizes of tens of km. The relation between number and size is called the **size distribution** (or **mass distribution** if mass instead of size is the variable). The estimated size distribution is shown in Figure 7-13.

Mass Distribution as a Clue to Collisional History

Figure 7-13 contains clues about the history of asteroids. In the first place, the trend roughly resembles a power law of the form $N = K D^{-2}$, where N is the number in a given diameter (D) interval on the logarithmic scale and K is a proportionality constant. As shown in Figure 7-13, this type of size distribution is produced when rocks smash against each other in nature (Hartmann, 1969b), thus supporting the idea that asteroids have undergone mutual fragmentation.

Ceres is somewhat larger than expected from the trend predicted by the power law and may have begun to pull away from the rest during its growth by runaway accretion. It is nearly twice as large as the second-biggest asteroid. At about 1.4×10^{21} kg, it contains nearly half the mass of all asteroids. The total asteroid mass, around 3×10^{21} kg, is only about 4% the mass of the moon.

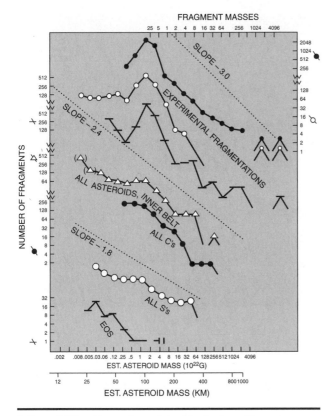

Figure 7-13. Comparison of size distribution of impact-shattered rock fragments and asteroids, showing general slopes. Dotted lines are reference power laws with indicated exponents b, in equation of N = kD^{-b} (see text). Top three curves show fragments of artificial aggregate targets of rock with masses on upper scale. Bottom four curves summarize actual asteroids in the inner half of the asteroid belt (a<2.8 AU), with masses on lower scale, including members of Eos Hirayama family, which may be fragments of a single asteroid collision event.

Numbers and Size Distribution of Comets

Numbers and size distribution of comets and bodies in the Kuiper belt are much more sketchy, because these are too far away to observe well. From the statistics of discovery orbits and discovery rates for comets, Oort and others were able to estimate populations as high as 10^{11} objects of normal cometary size in the Oort cloud reservoir. As mentioned earlier, Jewitt and Luu estimated 35,000 objects in the 100 to 300 km size range in the Kuiper belt, based on discovery statistics of those objects. Davis and Farinella (1997) concluded from population numbers that many Kuiper belt objects have undergone collisions, and that the size distribution is controlled by collisional fragmentation, like that of main-belt asteroids. If the size distribution is asteroidal, the numbers down to 1 km-sized objects (like ordinary comet nuclei) could be in the range of 10^8 to 10^9.

Fragmentation and Outbursts of Comets

Comet behavior during a period of activity gives interesting clues about their structure. They are fragile and their gas outflow is not uniform. Some comets, for example, suddenly flare up in brightness, suggesting sudden releases of gas. Comet Humason is known for its peculiar variability in brightness and the unusual intensity of CO$^+$ emission lines in its spectrum. Such changes have been seen as much as 6 AU from the sun, considerably farther than the limiting distance for such activity levels of most comets (about 3 AU).

Some comets return earlier and some later than predicted. As early as 1836, Bessel attributed this variation to accelerations produced by the loss of mass from the nucleus. Whipple (1950) showed how the jet action of pockets of escaping gas in a rotating nucleus could either accelerate or decelerate a comet. Indeed, thin, curved streamers of luminous material emanating from many comet nuclei in the inner coma strongly suggest jets of material erupting from the nuclei surfaces (Figure 7-11).

Most catastrophic of all is the fragmentation or complete disruption of comets, shown in Figure 7-14 on page 180. Comet Biela split in 1846 and was seen as a pair of comets when it returned in 1852, but was never seen again. Certain comets, known as sun-grazers, pass less than about 0.007 AU from the sun at perihelion. Many of these break apart at this time. The breakup is probably caused by heating, gas jetting, and spinning of the nucleus. Sun-grazer 1882 II broke into several pieces during perihelion passage, and the 1965 sun-grazer Ikeya-Seki also broke into two pieces that separated at about 12 m/s (Pohn, 1966). Separation velocities of only around 20 m/s have been found for a dozen other disintegrating sun-grazers, suggesting an upper mass limit of 10^{13} kg and a

Chapman and Davis (1975) asserted that the initial belt could have been several times more massive than it is today. This massive belt would have rapidly ground itself down to the present belt, with the smallest dust fragments being removed by Poynting-Robertson and radiation forces. The "bump" in the mass distribution at diameters around 100 km might involve a difference in strength among different asteroid types, such as a group of 100-km iron cores exposed by fragmentation, but others have viewed it as a remnant of the initial mass distribution (Davis and others, 1979; Ryan, Hartmann, and Davis, 1991).

In spite of differences in interpreting the detailed shape of the size distribution, analysts agree that the general form is probably the result of collisional evolution among a swarm of an uncertain initial number. Typical encounter velocities in the belt are 5 km/s, quite adequate to fragment many asteroids (Piotrowski, 1953).

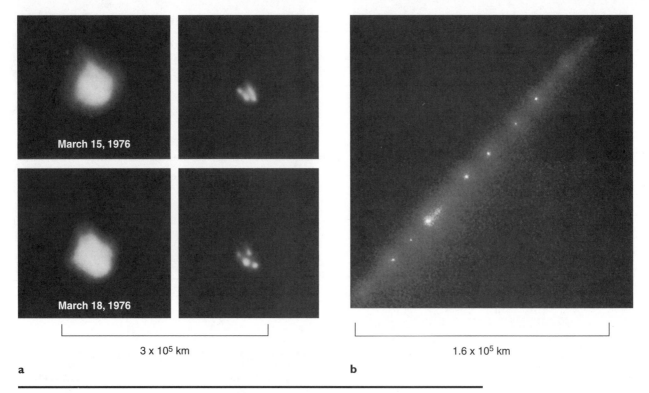

March 15, 1976

March 18, 1976

3 x 10^5 km

1.6 x 10^5 km

a   b

Figure 7-14. Fragmentation of comets. (**a**) Breakup of Comet West (1975n) into four pieces in March, 1976. Left-hand pictures show longer exposures dominated by coma. Right-hand pictures show shorter pictures revealing four fragments separating at speeds of about 1 to 5 m/s. One piece was seen for only two weeks; others were followed for more than six months. (**b**) After a close pass by Jupiter, Comet Shoemaker-Levy 9 broke into a string of fragments. A year after this image was made, the fragments all crashed into Jupiter. (**a:** Lunar and Planetary Laboratory Mt. Lemmon 1.5 m telescope photo, University of Arizona, courtesy S. M. Larson. **b:** Hubble Space Telescope image by H. Weaver and T. Smith, Space Telescope Science Institute and NASA)

diameter <1.3 km for those nuclei (Sekanina, 1977; Brandt and Chapman, 1981). The motions of the fragments are often affected by strong gas jetting as they break apart, perhaps from sublimation of gas off freshly exposed icy surfaces.

Not all comet disruptions are caused by solar heating in sun-grazers. Strangely, Comet Wirtanen (1957 VI) broke into two pieces near the orbit of Jupiter for no apparent reason. Other comets break up during close approaches to giant planets, because of strong tidal forces. Comet Shoemaker-Levy 9 was a famous example that broke into a string of fragments (Figure 7-14b) when it encountered Jupiter in 1992. It looped around the planet, and crashed into it in 1994 (Figure 2-21) giving astronomers their first look at a large-scale collision between a planetary body and an interplanetary body.

All these observations support the view that comets are extremely weak. Some may be aggregations of primordial planetesimals, only weakly bound by their gravity. Most have presumably not been heated enough to melt or form into cohesive homogeneous blocks of ice.

Meteors and Cometary Meteor Showers

Meteors, or "shooting stars," are familiar to anyone who has watched the night sky for more than a few minutes. They are small particles that heat up from friction with air molecules when they plunge into the atmosphere, just as a reentering spacecraft heats up and creates a spectacular fireball. As early as 1861, Kirkwood (of Kirkwood gap fame) argued that meteors are debris of comets, and this conclusion has been confirmed through orbital studies.

Velocities and Orbits

Trajectories derived from photographic and radar observations of meteors show that they travel in orbits around the sun, similar to those of comets. Some of these orbits are directly related to specific comets, and most reach farther into the outer solar system than do typical meteorite orbits. Photographic data on 2587 meteors and radar data

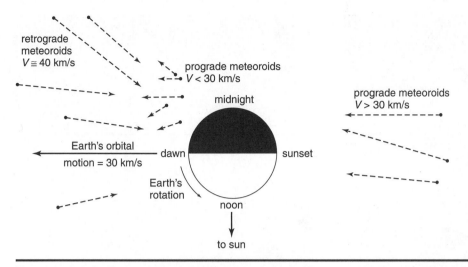

retrograde
meteoroids
$V \cong 40$ km/s

prograde meteoroids
$V < 30$ km/s

midnight

prograde meteoroids
$V > 30$ km/s

Earth's orbital
motion = 30 km/s

dawn

sunset

Earth's
rotation

noon

to sun

Figure 7-15. Preferential arrival times for meteors. Because Earth is moving left through the meteor swarm at 30 km/s, the majority of meteors are intercepted on the dawn side of Earth, and the best time to see them is a few hours before dawn. Retrograde meteors are strongly concentrated on the dawn side. All velocities are heliocentric.

TABLE 7-5 Dates of Prominent Meteor Showers

Shower name (constellation)	Date of maximum activity[a]	Associated comet
Lyrid	April 21, morning	1861 I
Perseid	August 12, morning	Swift-Tuttle
Draconid	October 10, evening[b]	Giacobini-Zinner
Orionid	October 21, morning	Halley
Taurid	November 7, midnight	Encke
Leonid	November 16, morning	1866 I
Geminid	December 12, morning	?

[a]Showers can appear several days before and after the peak activity on the listed date. Observations are best when the constellation in question is high above the horizon, usually just before dawn.

[b]The Draconids are now weak because their orbits have been disturbed by the gravity of planets, but future perturbations may again strengthen the shower.

on some 39,000 indicate no orbits from beyond the solar system (Millman, 1979).

Meteor Showers and Comets

On a typical night, most meteors are sporadic and quasi-randomly directed on downward paths across the sky at a rate of roughly eight per hour. The number seen usually increases toward dawn, for the reason diagrammed in Figure 7-15. However, on the nights of about the same dates each year, one can see a **meteor shower,** a large number of meteors falling within a few hours or few days and appearing to come from the same direction in space.

Showers are named for the constellations that lie in the direction from which they come. One of the most famous

showers is the Perseid shower, which comes in mid-August each year. The **radiant,** or direction of its source, lies in the constellation Perseus. Some other well-known meteor showers are listed in Table 7-5. During a shower the rate may rise to 60 meteors per hour (typical of the Perseids) or more. The most spectacular shower of recent times occurred on November 17, 1966, when the observers counting Leonid meteors in the predawn hours in the western United States noted a dramatic rise in the rate that continued until they could no longer count fast enough. For about half an hour meteors fell like snowflakes in a blizzard, and the estimated rate exceeded 2000/min! A time exposure of part of this shower is shown in Figure 7-16. A similar shower of Leonid meteors fell in 1833.

How can we explain such observations? Obviously the earth is encountering a cluster of particles called a

a

b

Figure 7-16. Rare intense displays of the Leonid meteor shower. (**a**) The shower of November 17, 1966. The rate of meteors visible to the naked eye was estimated to exceed 2000/min. This exposure of a few minutes' duration was made with a 35-mm camera. Meteors are the longer streaks. (D. R. McLean) (**b**) The intense Leonid shower of November 12, 1833, drawn by a contemporary artist. The meteors were described as "falling from the sky like snowflakes." (NASA)

meteoroid swarm. In the case of the Leonid swarm, the duration of the shower indicated that the inner core was some 32,000 km across (McLean, 1967). Similarly, Davies and Lovell (1955) determined a diameter of 100,000 km for the inner core of the Draconid (Giacobinid) shower. In 1866, the Italian astronomer G. V. Schiaparelli (later more famous as the discoverer of the "canals" on Mars) showed that the Perseid meteoroids were clustered along the orbit of Comet Swift-Tuttle. Within a few years, observers discovered that several shower meteoroids lay in swarms coinciding with comet orbits. The inference was unmistakable: Meteors and meteor swarms are associated with comets.

Their orbits are understandable in terms of celestial mechanics. Should a group of meteoroids become detached at low speed from a comet, they would initially lie in a swarm moving in the same orbit. But perturbation theory shows that the swarm would soon stretch out along the same orbit so that eventually the meteoroids would be more or less smoothly distributed all the way around the ellipse. If Earth's orbit intersects the cometary orbit and in-

tercepts a relatively young, undispersed cluster of meteoroids, a spectacular shower like that of November 17, 1966, ensues. Older swarms are more dispersed.

Physical Nature of Cometary Meteoroids

Carefully calibrated photographs of meteors show how they decelerate as they hit the atmosphere. From this deceleration we can calculate the drag due to atmospheric friction and hence estimate the density of the particles. Most cometary meteoroids are very low-density, fragile objects. In fact, no known cometary *shower* meteor has been big enough or dense enough to reach the ground. An investigator of meteoroids has compared some of them to bits of burnt newspaper or the balls of fluff that collect under beds!

Various studies (Jacchia, Verniani, and Briggs, 1967; Verniani, 1969) also showed that different showers (that is, different comets) produce meteoroids of different character. Short-period comets, which have been exposed to the sun more often than long-period comets, tend to pro-

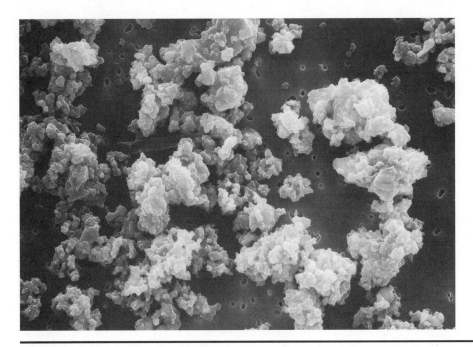

Figure 7-17. A cluster of microscopic chondritic meteorite particles (probably carbonaceous chondritic) fragmented into hundreds of pieces when it hit a collector plate on a U-2 aircraft at an altitude of about 20 km. It is spread across 100 μm, but if reconstituted it would be a fluffy object about 10 μm across. Its fragile, low-density structure as clumps of 0.5-μm scale particles shows clearly. Numerous such particles have been collected. (Electron microscope photo courtesy D. E. Brownlee, University of Washington)

duce higher-density, stronger particles. There are exceptions to this tendency; the lowest density meteors are the Draconids, associated with periodic Comet Giacobini-Zinner. The average densities of photographed meteors range from less than 10 kg/m^3 (Draconids) to 1060 kg/m^3 (Geminids, comet association unknown but possibly associated with a prehistoric, burnt-out comet).

Collections of microscopic interplanetary meteoroid fragments have been made by rockets and high-altitude aircraft. Micrometer-sized particles of this type are strewn off larger meteors in the atmosphere or may even enter the stratosphere without being severely destroyed. Particles are typically some tens of micrometers across but composed of aggregates of micrometer- and submicrometer-scale dust grains, as shown in Figure 7-17 (Brownlee, Rajan, and Tomandl, 1977). The composition of these particles most closely resembles C1-type carbonaceous chondrite material. Some of them may be similar to the CHON particles found in Halley's comet.

In addition to the dominant carbonaceous chondritic dust aggregates, other denser particles have been found in micrometeoroid collections. Nickel-bearing iron sulfide and crystals of olivine and pyroxene have been reported (Brownlee, Rajan, and Tomandl, 1977). Among the microscopic particles netted in one collection experiment was a possible Perseid with mass of 5×10^{-10} kg, density of 7700 kg/m^3, and a magnetic composition presumably mostly iron. Geminid micrometeoroids were found to have densities of 2000 to 2600 kg/m^3, and an apparently glassy, not crystalline, structure.

Meteoroid Ejection Velocities

The weak structure of most suspected cometary particles indicates that they are gently separated from their parent material. This is consistent with the idea of dust emitted from evaporating ice in a comet. From statistics of Draconids, Davies and Turski (1962) found that these meteoroids separated from their parent comet with velocities less than 10 m/s, again indicating that the particles are not ejected with any great force. Note that these velocities are of the same order as the separation of comet fragments during breakups.

Zodiacal Light

With microscopic dust grains being spewed out of comets in the form of Type II tails, as well as being blasted off asteroids in numerous large and small collisions, it is hardly

a   b

Figure 7-18. Sequence of wide-angle (72° × 52°) views of winter sunset from Hawaii, showing emergence and setting of the zodiacal light. (**a**) Sunset. (**b**) 1 h after sunset, with zodiacal light extending about 50° above horizon. (**c**) 1½ h after sunset. (**d**) 2 h after sunset. (Photos by author; altitude 4 km; Mauna Kea Observatory)

surprising that interplanetary space contains much fine dust.

You can satisfy yourself that this is true by observing the dust with your naked eye. It can be seen in the form of the **zodiacal light,** a faint reflection of sunlight off dust grains concentrated in the ecliptic plane. The zodiacal light is visible shortly after sunset and before sunrise, as seen Figure 7-18. In the evening, just after the stars have come out, the zodiacal light is a faint band of light, widest at the western horizon, and tapering upward to a width of only 5° or 10°. Seen from mid-northern latitudes, it slants upward to the left, following the ecliptic. In other words, it lies along the plane of the solar system.

The zodiacal light is about as bright as the Milky Way and under good conditions can be traced at least 45° from the sunset point. At a point 180° from the sun is a faintly glowing patch, not as bright as the zodiacal light, called the **gegenschein** (German for "counterglow"). It is sunlight reflecting off particles at the antisolar point.*

Zodiacal light is seen only from sites with dark skies. An observer must be several miles from dense population and several hundred meters from street lights. Dark adaptation, averted vision, and a technique of scanning—swinging the head back and forth—may be necessary at first to notice the zodiacal light.

Measurements of the brightness, distribution, color, and polarization of the zodiacal light have helped to clarify the nature of the interplanetary dust. The particles responsible for most of the zodiacal light are tiny dust grains about 1 μm in diameter. The total mass of this dust is about 2.5×10^{16} kg (Whipple, 1967), a mass less than that of a single moderate-sized asteroid. Most of this mass is concentrated among the smallest particles.

The shape of the zodiacal dust system can be partly inferred from the narrow width of the zodiacal light; our

---

*This phenomenon of brightening at the antisolar point can be seen around the shadow of your own head on dusty, leafy, or other complex surfaces, particularly if your shadow is cast over a long distance. It is also often seen around the shadow of an airplane in which the observer is riding. In all cases, it lies 180° from the sun. This general phenomenon, partly caused by the absence of visible shadows at the antisolar point, is called *heilgenschein* (German for "holy glow")

c

d

cross-sectional view shows that the system is flattened toward the plane of the solar system. A total width of the zodiacal light of about 30° near the sun would imply its extension only about 0.2 AU above or below the ecliptic plane, for example.

Spacecraft have measured the radial distribution of the meteoroid dust outward through the solar system both by optical sensors and by impacts on the vehicles themselves. The largest (asteroidal) bodies are concentrated in the asteroid belt, but subcentimeter pieces are redistributed by Poynting-Robertson and the radiation effect. The smallest pieces and the zodiacal glow are concentrated near the sun. The outer luminous atmosphere of the sun, called the **corona,** grades into the interplanetary medium and has two components, the K-corona and the F-corona. The **K-corona** is the gaseous portion, and the **F-corona** is the dust, mixed with the gas. This dust can be traced to within 4 solar radii of the sun by its infrared radiation.

For the zodiacal particles larger than 1 μm, the mean lifetime in the solar system is only about 10^5 y (Whipple, 1967), and still smaller particles are blown away (as in comet trails) in about a year (Biermann, 1967). Whipple concludes that the zodiacal dust cloud is completely replenished in about 2×10^5 y, requiring about 2.5×10^{11} kg/y, or 8 metric t/s in the form of particles ranging from many microscopic dust grains to a few boulders. Most of this material comes from comets, particularly in the inner solar system, where the comets erode as their dust-laden ices sublimate (Millman, 1979); additional particles come from the asteroids colliding in the asteroid belt.

Pristine Bodies

As more missions to asteroids and comets are planned, including sample return missions, how can we choose the bodies that might give us the most pristine samples, telling us the most about the history of the ancient solar system? The word *pristine* is used by some authors to refer to lack of mechanical processing of primordial condensates (processes such as collisional disruption) and by other authors to refer to low-temperature environments that would preserve primordial chemicals. However, all planetesimals probably underwent at least some collisional processing and cratering; that was inherent in the accretional growth process. Thus there is a real question whether any objects are really "pristine."

Hypothetical pristine objects have been called the Rosetta Stones of the solar system because they would—in theory—preserve ancient clues to be read by us. However, Dale Cruikshank (private communication) has noted that a host of objects have been called "the Rosetta Stone of the solar system"—from the moon to various asteroids and comets! The pristine Rosetta Stone gets shunted to more remote regions as we discover unexpected degrees of processing in nearby objects.

Comets are often billed as the most pristine objects in the solar system, having been long in deep freeze. However, Oort cloud comets originated much closer to the sun than they are now and were once warmer. Furthermore, comet activity itself alters and blows off the pristine surface layers, and even a gradual reduction in orbit radius exposes a body to stronger sunlight and can cause loss of the most low-temperature ices. Even Centaurs and Trojans may have lost ice. Thus, neither Centaurs nor any active comets are as pristine as Kuiper belt objects, which have been in their original locations, and in deep freeze, ever since they formed. A pristine Kuiper belt object might reveal the original accretional structure of planetesimals, formed by low-velocity collisions (Weidenschilling, 1997). However, as noted earlier, Davis and Farinella (1997) calculated that even the Kuiper belt objects underwent collisional evolution and some disruptions.

Exploring Asteroids

The inner solar system has so many Earth-approaching asteroids that many of them are easier to reach (in terms of total fuel or energy expenditure) than Mars, and many are easier to reach and return from than either Mars or the Moon (because their low gravity does not require spaceships to carry much fuel to land and take off). Partly because of the threat of collisions of asteroids and Earth, these bodies will see increasing interest and visitation in coming decades. It is interesting to discuss them in further detail.

More than 100 Apollos are now known to have orbits that come in far enough to cross Earth's orbit. Helin and Shoemaker (1979) estimate a population of 700 ± 300 larger than about 1 km diameter. The largest has an estimated diameter of 8 km (object 1978 SB).

Surveys are rapidly increasing the numbers of known Apollos, and the public consciousness of the fact that they can hit Earth. Twelve Apollos were known before 1960, 15 by 1970, and 60 by 1990. The high current rate of discovery is due mainly to various search programs started in recent years by Carolyn Shoemaker, the late Eugene Shoemaker, Eleanor Helin, and Tom Gehrels.

Visiting Asteroids

An Apollo asteroid might be the next planetary body to be explored by humans, as suggested in Figure 7-19. Such a trip would be scientifically interesting because Apollos are probably sources of meteorites. In fact, an expedition might merely discover a huge example of a meteorite type already in our museums! Nevertheless, the structure and surface of such a body (unaltered by atmospheric entry) would be of interest, and any major differences from meteorites would be thought provoking indeed. Further, an Apollo's materials might be of use in constructing large space habitats.

Another reason for interest in Apollos is the asteroid threat. Most Apollos will end their days by hitting earth many million years in the future. A 100-m object capable of making a 1-km crater hits Earth every 3000 y or so (70% falling into oceans and not leaving permanent craters). Such impacts would devastate areas comparable to moderate-sized cities, send out shock waves causing injury many kilometers away, and throw dust into the high atmosphere, coloring sunsets around the world for years.

One example of a close approach by an Apollo capable of making a 10-km crater (!) was the approach of 1566 Icarus on June 14, 1968. It passed only 6 million kilometers away, only about 16 times the distance to the moon. Though this approach was predicted well in advance, the incident touched off a number of scare articles in tabloids by sensationalistic "reporters" who allegedly did not trust the assurances of astronomers that Earth would be spared.

The Asteroid Threat Versus the Asteroid Opportunity

The media have focused on the threat of asteroid impacts, but the opportunities offered by asteroids may be more important to human beings in the next century. Because asteroid research has revealed the likelihood of large objects containing nickel-iron alloy and other rare elements, scientists have come to realize that asteroids may have enormous economic value (Gaffey and McCord, 1977). Unlike lunar igneous rocks, in which many elements melted and blended into forms difficult to retrieve, some asteroids (judging from meteorites) contain elements in the form of primitive condensates, and others are made of nearly pure nickel-iron alloy. Given that we humans are digging ever deeper into Earth to retrieve and smelt lower-grade ores, and dumping the waste products into our biosphere, asteroids may come to offer a cleaner and more attractive source of at least some resources. Even those with minimal metal content may contain water, hydrogen from the solar wind, and building materials useful to future astronauts in space.

Consider a 1-km-diameter asteroid, of which dozens approach Earth's orbit (not to mention the thousands still in the belt). If it contains typical chondritic stony material, it has around 7% to 17% by weight of free nickel-iron alloy, whose market value at 1982 rates would be on the order of $100 to $300 billion. If such an asteroid had a substantial regolith layer, the iron flecks in that layer

Figure 7-19. The future of asteroid research? Dozens of small Apollo asteroids approach Earth and may become targets for future exploration. They may offer not only interesting samples of ancient material but also metals and minerals of economic interest. In this imaginary view, an exploration vessel approaches an asteroid about 100 m long. Bright objects in the lower right are Earth and the moon, about 0.1 AU away. (Painting by author)

could conceivably be harvested with simple magnetic rakes! Gaffey and McCord (1977) discuss the case of a similar nickel-iron asteroid, whose value would be closer to $4 trillion!

If we allow human capability in space to increase, we will inevitably reach the time when the cost of going after these resources will be less than their value. Some writers have raised the specter of humanity despoiling the solar system, in the same manner that overindustrialization is beginning to despoil Earth's environment, but to this writer the prospects seem just the opposite. With a careful balance of research and exploitation, we could learn from and process materials in space in a way that would begin to take the pressure off Earth's ecosystem. A transition from Earth-based manufacturing to interplanetary manufacturing could eventually reduce pollution and ravaging of Earth by an Earth-based society bent on ripping the last dwindling resources from the land.

SUMMARY

Asteroids and comets represent two ends of a continuum of ancient planetesimals: the silicate-metal-rich end and the ice-rich end. Asteroids and perhaps comets have undergone extensive collisional evolution; they are heavily cratered and most smaller asteroids in the main belt are fragments of once-larger parent bodies. Meteorites also show that many but not all rocky asteroid parent bodies have undergone heating and melting.

The zonal distribution of taxonomic types reveals traces of the original composition gradient in the solar nebula, but perturbations have cleared most of interplanetary space and thrown many objects into the surrounding Oort cloud.

CONCEPTS

comet	Type I tail
Oort cloud	Type II tail
asteroid	plasma
Earth-approaching asteroid	CHON particles
Trojan asteroid	light curve
remote sensing	size distribution
infrared reflectance spectrophotometry	mass distribution
proxene absorption bands	meteors
taxonomic classes	meteor shower
primordial taxonomic zones	radiant
Hirayama families	meteoroid swarm
absolute magnitude	zodiacal light
albedo	gegenschein
tail	corona
head	K-corona
coma	F-corona
nucleus	

PROBLEMS

1. (a) How do you think a comet should be distinguished from an asteroid observationally? (b) By physical properties of the body itself? (c) By origin?

2. Assume that you had free access to satellite and planet surfaces and to equipment monitoring interplanetary bodies. Suggest some ways to determine whether impact craters on the moons and planets are predominantly from cometary, stony asteroid, or metallic asteroid origin.

3. If you are scanning the skies with a small telescope and find a faint, fuzzy object not marked on your star map as a nebula, what could you do to test whether it might be a newly discovered comet? (Assume no spectroscopic equipment is available.)

4. If Jupiter were magically moved a few percent closer to the sun, describe how the Kirkwood gaps' structure would change.

5. Suppose members of a certain Hirayama family were created by breakup of a large asteroid by high-speed collision with a much smaller one. Why might the family members have different compositions (and hence spectral properties) even though mostly originating in a single parent body?

6. (a) Describe three major spectral classes of asteroids and their likely composition. (b) How does 4 Vesta differ from these, and what might its composition be? (c) How does 1685 Toro differ, and what might its composition be?

7. Name some important gases seen in comets. How do they suggest that comets contain—in the form of ices—water, methane, and ammonia, which are common in other primitive solar system regimes, such as the atmosphere of Jupiter?

8. Describe some possible solutions to the paradox that most belt asteroids are C type, whereas only a small fraction of meteorites are carbonaceous chondrites.

9. (a) How do shapes, statistics, and dynamics of comet orbits imply a vast number of unseen comets in the Oort cloud beyond the planetary system? (b) For every short-period comet we see, roughly how many comets made it from the Oort cloud in as far as Neptune's orbit, only to be ejected again into the Oort cloud or out of the solar system entirely?

10. Give some lines of evidence that meteors are associated with comets and are not simply examples of small ordinary meteorites.

11. Give some lines of evidence that many asteroids melted very early in their histories but that the melting was not smoothly correlated with asteroid size.

12. Give lines of evidence that asteroids as a group have undergone many collisions.

13. Give evidence that comets contain some carbonaceous chondrite material. How might this help solve the mystery that the dominant meteoritic component in the lunar soil seems to be carbonaceous chondritic, while only 5% of meteorites reaching Earth's surface have this composition.

14. What change of velocity would be required to move an asteroid fragment from a circular orbit on the inner edge of the belt (say 2.0 AU) to an Earth-crossing orbit? (Note: Most analysts believe this change is too high to occur in a single acceleration during a collision or fragmentation—an argument against direct delivery of belt asteroids to Earth.)

15. What change in velocity would be required to move an asteroid fragment from the innermost belt at 1.8 AU onto a Mars-crossing orbit?

16. Asteroid 1566 has an unusual orbit that reaches perihelion only 0.19 AU from the sun. It has a visual albedo of about 0.18. Estimate the maximum likely surface temperatures to be found during perihelion passage.

17. An asteroid 100 km across is suspected of having a satellite 10 km across and 500 km away. (a) What revolution period would the satellite have? (b) Which piece of information is irrelevant? (c) What other numerical datum is needed in part (a)? Defend the value you assumed. (d) Observers monitoring the asteroid's rotational light curve claim to see an event confirming the satellite's existence. What might their evidence look like and what might the event be? (Note: There are four different types of such events!) (e) Describe changes that would be expected in the light curve evidence over a period of a few months after the event was detected.

18. Confirm the derivation of the relation between strength S and maximum R_{max} of an irregular body given in the math notes in this chapter.

19. (a) Confirm the derivation in the mathematical notes on maximum rotation rates. (b) You are captain of a research vessel approaching a coherent, rocky, 4-km-diameter Apollo asteroid rotating with a period of 1.5 h. Describe a strategy for landing your ship or crew on this asteroid in order to carry out surface investigations.

20. (a) Show that the maximum possible encounter velocity between Earth and a member of the solar system is about 72 km/s. (b) Why can a meteor observed to hit Earth at this speed be associated with long-period comets and not any known asteroids?

21. A water-icy body with albedo 70% and emissivity 0.2 falls toward the inner solar system. (a) What is its surface temperature at Jupiter's orbit? (b) At what distance from the sun does its temperature reach $0°C = 273$ K?

22. The same body as in problem 10 is thinly mantled by carbonaceous dust that reduces the visual albedo to 3%, while leaving the emissivity at 0.2. Answer problem 10(a) and 10(b) for this body.

23. (a) A particle of radius 1 cm breaks loose from a short-period comet having a nearly circular orbit at semimajor axis 3 AU. What will be its orbital history and roughly how long will it last in the solar system? Answer the above for a particle with radius (b) 100 μm (0.01 cm); (c) 0.1 μm.

The smooth, fractured ice surface of Jupiter's satellite Europa. The network of dark lines marks fractures that break an otherwise relatively featureless global ice plain. The ice pack may average 100–200 km thick is believed to be floating on a deep, interior global ocean, similar to the ice pack in Earth's arctic regions. (NASA, Galileo spacecraft photo, 1997)

CHAPTER EIGHT

Planetary Interiors

Chapters 4 and 5 described the origin of the planetary system and Chapters 6 and 7 described the various surviving representatives of the planets' building blocks. Thus, the stage is set for a direct confrontation with the structure of the planets themselves—the topic of this chapter.

Planetary interiors is the field that attempts to learn not only the compositions, pressures, temperatures, densities, and dynamical activity inside planets today but also how these conditions have changed through time. The first portion of this chapter is devoted to determining the current conditions inside planets. We can't drill to significant depths in planets, so the conditions have to be deduced through a careful combination of observation and theory. Meteorites and asteroids offer especially important clues. Nature has sliced some asteroids open for us and sent us samples in the form of meteorites that suggest layering. In planets, the dense materials such as metals tend to drain toward the center whereas light materials float. The densest, innermost region is called the **core;** its surroundings, a broad zone of intermediate density, are called the **mantle;** and the thin, low-density "scum" on the surface is called the **crust.**

Four Basic Observations and a Basic Concept

Imagine that we are approaching a planet in a spacecraft. How can we diagnose its internal properties? Does it have an iron core? Is the interior cold? Warm? Partly melted? Is the planet cooling and contracting? Heating and expanding?

Before answering these questions, we should determine four basic properties: the mass, diameter, mean density, and surface rock properties. The mass can be found by measuring the planet's gravitational influence on our spacecraft (or on a natural satellite, or even on a distant planet) and applying Kepler's and Newton's laws. Measuring the diameter is easy once we have a high-resolution image of the disk.

The **mean density,** which is simply the mass divided by the volume, immediately tells us something about the planet's gross composition. For instance, a mean density of 1000 kg/m^3 is typical of ices; values of 2800 to 3900 are typical of volcanic rocks and stony meteorites; values of 5000 to 6000, of iron-rich minerals, and values around 7900, of iron to meteorites (see Table 7-4 and accompanying text).

We can certainly do better than to characterize the planet by its mean density if we add the fourth observation, obtained by sending down a lander and sampling the surface rocks. If the "planet" we are approaching is a small one, say a 1-km asteroid, we will probably find that the surface rocks have a density value equaling the mean density of the whole object, perhaps 3500 kg/m^3, which is typical of chondrites. Then we could reasonably conclude that the asteroid is relatively uniform in composition, being composed of the surface material throughout.

Consider a somewhat bigger world, the moon. Surface rocks typically have a density around 2800 kg/m^3, with a mean density of 3300 kg/m^3. Now we must conclude that the internal material is somewhat denser than the crustal material. Perhaps a small iron core is hidden in the center, or perhaps the higher central pressure simply compresses the rock material to higher density.

In the case of Earth, we have a striking discrepancy of 2800 kg/m^3 for typical surface rocks and 5500 for the mean density. The reason turns out to be twofold: The mantle has high-density, iron-rich rock, and there is a large iron core.

A basic concept evolving from this discussion is that we can use the surface materials as an observable boundary condition and then reason our way into the interior by learning as much as possible about how materials behave under the higher pressures of the interior.

Theoretical Techniques for Calculating Interior Conditions

The Equation of State

The pressure (P), density (ρ), and temperature (T) in any material are called the **state variables** because they determine the state of the material. For instance, if the pressure gets too high or the temperature too low, molten material may change to the solid rather than the liquid state. As we go down into a planet, P increases, causing material to compress and reach higher ρ's. For gases, there is a single, simple equation that relates the state variables (called the universal gas law), but for the rocks of a planet, the relation is much more complex. This complex relation can be measured to some extent in the laboratory. Any such relation between P, ρ, and T is called an **equation of state** for the given material, and it tells what density is attained for a specified pressure and temperature.

Computer Models of Planets

Given the equation of state of a planet's inner material, it is easy to see how a theoretical model of the planet's interior can be made. The **hydrostatic equation**

$$dP = \rho \, g \, dz$$

says that the increase in pressure dP as we descend through a layer is equal to the density ρ of the material in that layer times the gravitational acceleration g times the layer thickness dz (with z increasing downward).

A computer program can be devised that divides a planet into a thousand layers (or some other large number of layers) and then starts iterating its program down from the surface values. At the surface, the pressure is zero (we can neglect the atmospheric pressure in most cases). We know by direct measurement the density of rocks at the surface, at least in the case of Earth and the moon, and we can measure or compute the gravitational acceleration g. The computer starts with these figures and multiplies them by the layer thickness dz to get the pressure at the bottom of the first layer. Then the computer inserts that pressure into the equation of state to compute a new density for the rocks at the bottom of the first layer. Computing the new gravitational acceleration that applies at that depth, it then performs a new multiplication to get the pressure at the bottom of the second layer, and so on.

There are two obvious problems in this sort of scheme. For one thing, the equation of state involves the temperature, and we have not yet supplied the temperature to the computer. Usually a theoretical model of the temperature profile in the planet must be supplied, based on measured temperature gradients at the surface and estimates of interior heat sources. Fortunately, however, the density of a rock does not change very much as the temperature changes (unlike the density of gases). A second problem is that the equation of state must be accurate, and no one yet knows precisely the composition of any planet's interior, so the equation of state has to be guessed at.

In spite of these difficulties, a number of observational facts guide us in constructing our models. One is that we know the total mass of each planet. Therefore, when the computer reaches the bottom of the last layer (that is, the center), it must have "used up" precisely the mass of the planet. If mass is left over, the densities computed were systematically too low; if the allotted mass is used up in the 999th layer or sooner, the densities were

too high (and the model would predict a hollow cavity at the center that a sudden collapse would immediately fill in a real planet). To solve this problem, an adjustment is made to the equation of state, and the computer program is run again. A number of other similar checks can also be made—we discuss this later. (Alternatively, the computer program can be designed to start by guessing central conditions and then work outward. It can then check to see whether the correct radius and mass have been reached at the level where the pressure equals zero—that is, the surface.)

Plastic Flow Inside Planets

In Chapter 7, we noted that planetary bodies smaller than a few hundred kilometers across are often irregular in shape. Worlds larger than 1000 km are round because they obey the hydrostatic equation, because the pressures and temperatures in their interiors are great enough to cause failure of the normal elastic properties of rock. Under these conditions the material deforms to meet the equilibrium shape under the influences of gravity and rotational forces.

This shape is called the **figure** of the planet, and because of centrifugal force, the figure of an isolated, fluid, rotating planet is not a sphere but a flattened shape called an **oblate spheroid.** Additional effects, such as tides, interior processes, and irregular mass distribution, cause small departures from the oblate spheroid shape. An example is Earth's slightly pear shape, noted by the first artificial satellites in 1957–1958.

Figure 8-1 dramatizes the remarkable ability of planetary material to deform plastically. This fluidlike deformation of rock is called **plastic flow,** or **rheidity,** and is the property that allows planets to deform into spheres or spheroidal shapes. A **rheid** is a substance that has brittle properties on a short time scale but fluid ones over long times. Pitch or asphalt is a famous example; if you strike it with a hammer it chips, but if you place a cannonball on the surface of a large vat of pitch, it will eventually sink out of sight.

These properties are believed to arise at least in part from small cracks or imperfections in the crystal lattices of the material. A sudden force may fracture the crystal entirely, but a prolonged gentle stress may cause imperfections (such as "holes" due to missing atoms in the lattice) to propagate through the material, so that on a macroscopic scale the material appears to flow. It is thus rheidity, as much as melting or crushing, that allows planetary rock to adjust to the hydrostatic equation, bringing planets into round shapes.

For this reason, a planet with a dense crustal layer overlying a lower-density layer is unlikely because such a combination is unstable; given enough time, the dense

Figure 8-1. Folded chalky strata at Lulworth Cave on the south coast of England. Examples of folded strata around the world show that rocks subjected to stresses over long periods of time can deform plastically, thus allowing the principle of hydrostatic equilibrium to control planetary configurations. Note figures (top) for scale. (Photo by author)

material is likely to flow downward while the lighter material rises, just as the cannonball sinks in the pitch. To the first approximation, then, planets are likely to have their densest materials in the center and lightest materials near the surface, even if they have never been completely molten. Local dense concentrations, such as lava-filled impact basins, may exist on planetary surfaces, however.

Changes of State

Minerals, Rocks, and Ices

Planetary samples from Earth, the moon, and meteorites show that terrestrial planets are composed principally of silicate and metallic compounds. Such compounds solidify into different forms known as **minerals.** Assemblages of different minerals make up **rocks.** These traditional geologic definitions need to be broadened for application to outer planets and satellites, where much of the solid bulk is ice. Thus, *minerals* may also refer to frozen water (H_2O),

methane (CH_4), or ammonia (NH_3), and the geologic role of "rocks" may be played by both ices and silicates. Nonetheless, we usually use the term *rocky* to mean largely silicate. Chapter 9 gives a more complete review of various mineral and rock types.

To deduce the conditions inside planets, we must have equations of state that describe not only relations among pressure, density, and temperature for each rock type, but also any *changes* in the state of the constituent minerals. The two most important **changes of state,** melting and solid state phase changes, are described next.

Melting

Measurements in drill holes show that temperature increases with depth inside Earth. The same probably applies to all other planets, since planetary interiors contain primordial heat and radioactive atoms that produce heat as they decay. If the interior temperature is high enough, rocks or metallic materials could be molten. In the case of Earth, for example, seismic evidence indicates that the **outer core** is liquid, probably liquid iron. On the other hand, the high pressures at planetary centers may

Two expressions can be combined in a rather simple exercise to derive the pressure inside planets and to indicate the somewhat secondary role of temperature as a state variable. Let r be the radius to a point in the planet. Then in any thin layer of thickness the pressure P changes by

$$dP = -\rho g \, dr$$

from the hydrostatic equation. The minus sign arises because P increases as r decreases.

If we want to integrate this to obtain P at different r's, we note that both the density ρ and the gravitational acceleration g are functions of r. However, ρ is almost a constant, changing by less than a factor of 2 in many planetary objects. The more important change is in g, which is given by our second expression

$$g = \frac{GM}{r^2}$$

from Newton's gravitational law. The mass of the planet M has to be written in terms of r and is given by

$$M = \frac{4}{3}\pi r^3 \bar{\rho}_r$$

where $\bar{\rho}$ is the mean density out to point r. Substituting all this back into the hydrostatic equation, we have

$$dP = -\rho \frac{4\pi G \bar{\rho}_r}{3} r \, dr$$

To simplify matters, let us assume that the density really is some constant (which is quite true in the outer layers and less true throughout), so that

$$\rho = \bar{\rho} = \bar{\rho}_r$$

where $\bar{\rho}$ equals the mean density for the whole planet. Now we can integrate:

$$\int_{P_{surf}}^{P_r} dP = -\frac{4\pi G \bar{\rho}^2}{3} \int_R^r r \, dr$$

P_{surf} is the pressure at $r = R =$ the surface, which is assumed to be zero (the pressure of the atmosphere is relatively small). Thus,

$$P_r = \frac{2\pi G \bar{\rho}^2}{3} (R^2 - r^2)$$

$$= 1.4 \, (10^{-10}) \, \bar{\rho}^2 (R^2 - r^2)$$

in SI units.

This expression gives a central pressure of about 46 kbar (4.6×10^9 N/m^2) in the moon, which is about right. Because of our cavalier assumptions about ρ, this expression yields an answer that is about a factor of 2 too low in Earth, where it gives a central pressure of 1700 kbar, compared to more sophisticated estimates of about 3700 kbar.

Our expression can be improved by keeping the two densities ρ and $\bar{\rho}_r$ separated in the integration, but it has much illustrative value even in this form. For instance, we find that you have to go down 1000 km in the moon to get to a pressure of 40 kbar (4×10^9 N/m^2), while this is reached at only about 100 km down in Earth.

Note also that we derived a planet's pressure structure without any reference to temperature T. So where is T needed in understanding interiors? One place is in understanding the true value of ρ as a function of r. As pointed out in the text, this utilizes the equation of state, and ρ values in each layer are adjusted according to T. Also, as we will see in the text, T controls the dynamics by determining whether rock is molten, in convective motion, and so on.

---

compress material that would otherwise be molten, forcing it back into the solid state. The innermost part of Earth's core, called the **inner core,** has been found to be solid.

Any equation of state used in modeling planets must thus take into account possible sudden transitions between solid and liquid states at various points in the interior. In other words, given the planet's composition, we must know at just what pressure, density, and temperature the material will melt so that the computer can be instructed to change to the liquid state if the critical conditions are encountered while "building" the planetary model.

Solid State Phase Changes

A different and more vexing problem arises because rock-forming chemical compounds may exist in different solid forms at different pressures and temperatures. That is, if

the temperature and pressure increase, a mineral with a certain density and crystal structure may change phase to become another mineral with the same composition but denser crystal structure. An example is carbon's existence as graphite and diamond. Other examples occur in meteorites. Around 1970, predicted high-pressure forms of olivine and pyroxene (named ringwoodite and majorite after two well known Australian planetary researchers) were found in meteorites, and in 1997, Japanese and American teams found even higher pressure forms that are the first discovery of minerals long hypothesized to exist in Earth's lower mantle, as pointed out by Stöffler (1997). These minerals were not created by sustained high central pressures in the meteorite parent body, but rather by transient high shock pressures during parent body collisions, but they allow scientists to study the proposed mineralogy of the mantle.

The existence of multiple phases for various minerals obviously affects our efforts to determine the structure of a planet. They could cause a systematic layering—a structure consisting of shells of different mineralogy with discontinuities at the shell interfaces. Thus, a planet could have a complicated internal structure even if it were chemically uniform.

Some of the layered structure in Earth is almost certainly the result of phase transformations. The best example is a transition of the mineral olivine, $(Mg,Fe)_2SiO_4$, a mineral common in Earth's mantle, to a denser form called spinel polymorphs (including ringwoodite), at pressures around 140 kbar.* This **olivine-spinel transition** causes a 10% jump in density at a depth of about 400 km in Earth, where the pressure is about 135 kbar. Other phase transitions may also be important.**

Laboratory Experiments Versus Theory of State Changes

How can we determine accurate equations of state describing the phase changes among planetary materials? If we could simulate conditions inside planets, we could directly measure how various rocks behave. Many tests of this kind have been done, but the highest pressures obtained in *most* laboratory studies are around 20 to 40 kbar, which equals the pressure at depths of only about 63 to 125 km inside Earth, less than 2% of the way to the center (Cole, 1978). The high-pressure shock-induced mineral forms found in meteorites correspond to pressures above 600 kbar (lower mantle pressures) (Stöffler, 1997).

Higher sustained pressures have been reached in laboratories—exceeding 1700 kbar, a state that corresponds to pressures part way into Earth's core (*Geotimes,* August 1980, p. 25). However, because temperatures are also high at this depth, and because it is hard to produce high temperature and pressure at the same time in the lab, we lack accurate simulations of rocks at such depths. Better experimental data should be forthcoming. Still higher transient pressures can be reached in the lab during instantaneous shock or explosion tests, but these give limited insight

into equilibrium conditions in a planet. The pressure at Earth's center, around 3700 kbar, is well beyond any value sustained in the lab.

Pressure Ionization and Metallic States in Giant Planets

In still larger planets, the pressures increase to values beyond about 1 Mbar, where the electron swarms of the individual atoms are squashed together and the atomic structure breaks down. (This 1 Mbar value was recently revised downward from a long-accepted figure of 3 Mbar, meaning that it happens at shallower depths in giant planets than had been thought.) Because ordinary chemical properties depend on the interactions of the electron swarms, the matter loses these properties and is said to exhibit **pressure ionization**—a high-pressure state different from those of familiar solids, liquids, or gases. Many or all of the electrons may be freed to move around (as happens in a metal), while remaining ions attempt to locate themselves in some sort of crystal lattice. As the state of the electrons is similar to their state in a metal, an element in this pressure-ionized form is sometimes said to be in a metallic state,* even if it is not a metal under ordinary pressure. In the interiors of the giant planets, for example, we encounter **metallic hydrogen,** a high-pressure form of hydrogen.

Because the crystal lattices have broken down in metallic matter, many of the complexities of minerals' chemical interactions are absent, and the equations of state are easier to predict than for rocks under intermediate pressure. Also favorable is the fact that hydrogen, a major constituent in the giant planets, is the simplest element and relatively easy to treat theoretically. Much of the early work on pressure-ionized materials in planets was done by the geophysicist W. C. DeMarcus (1958), whose thesis in 1951 first detailed the structure of the hydrogen planets.

A Density-Mass Diagram For Planets

Many of the principles just discussed are clearly displayed on a simple diagram that plots two of the most basic measurements that can be made on planets: density and mass. This is shown in Figure 8-2. Suppose we picked some

---

**Note that 1 bar = the pressure exerted by the atmosphere at sea level = 1.0×10^5 N/m² = 1.0×10^5 pascals = 1.0×10^6 dynes/cm².

**Perhaps the reader has already realized that the study of planetary interiors is laced with uncertainty and difficulties. If not, the following note by the geophysicist F. Birch may make the situation clearer:

Unwary readers should take warning that ordinary language undergoes modification to a high-pressure form when applied to the interior of Earth. A few samples of equivalents follow:

High-pressure Form	Ordinary Meaning
certain	dubious
undoubtedly	perhaps
positive proof	vague suggestion
unanswerable argument	trivial objection
pure iron	uncertain mixture of all the elements

---

*This state is also sometimes referred to as a degenerate state, though this description is not quite correct. Degeneracy in physics involves the role of quantum statistics in determining the condition of the electrons.

Figure 8-2. Densities of solar system bodies plotted against their masses. Satellites are entered by number (S1 = Mimas). Entries along the left edge show densities of common planetary materials in solid, unfractured form. Dotted lines are schematic lines of constant composition. Upward curvature at right is due to matter compressing to higher density and then turning degenerate (shapes based with Wildt, 1972; Lupo and Lewis, 1979). Obviously, planetary bodies comprise a wide range of composition. Those closer to the top of the diagram typically involved heat sources that excluded accumulation of volatiles.

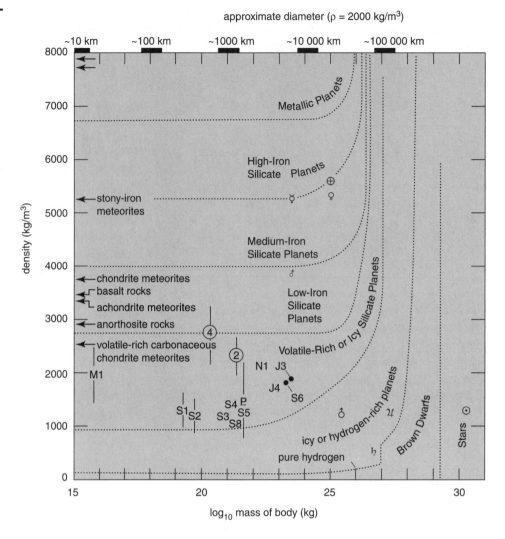

standard material, such as a volcanic rock, and began constructing worlds of it. A small planetesimal would have just the density of the uncompressed rock, around 3,300 kg/m³. As we get into worlds in the range of 10^{22} to 10^{25} kg (typically 1,000 to 10,000 km across) the rock substantially compresses and may change phase, increasing the density as shown by the dotted lines. At somewhere around 10^{26} or 10^{27} kg, the central material begins to be pressure ionized and the density shoots up.

The planets, satellites, and asteroids do not lie along a single such dotted line. They scatter widely up and down the diagram, which means that they must be made of different materials. The diagram labels the different general types of materials that fall in the different regions. Bodies near the top have been exposed to heat sources that drove away the volatiles and left high-temperature (usually denser) condensates, as in the case of the inner planets and Io. On the other hand, bodies near the bottom are either cold, icy worlds, such as Saturn's moons, or the giant planets that have gravitationally trapped low-density volatiles such as hydrogen and ices.

Differentiation

As mentioned in Chapter 6, **differentiation** is any process by which initially homogeneous material gets divided into masses of different chemical composition and physical properties. Figure 8-2 shows that strong differentiation has occurred among the planets: The initially uniform solar nebula has produced planets of remarkably different compositions.

Differentiation is also common inside planets. For example, if a rock mass melts and forms a homogeneous **magma** (molten rock), different minerals may crystallize at different temperatures as the mass cools. As they form, the heavy minerals may sink and light ones rise to the surface, just as happens in the smelting of metals. Thus, the final mass may be differentiated into regions with mineral mixes of different compositions.

Complete melting is not necessary for differentiation; if even partial melting occurs, fluids can move from one region to another, altering compositions. For example, if

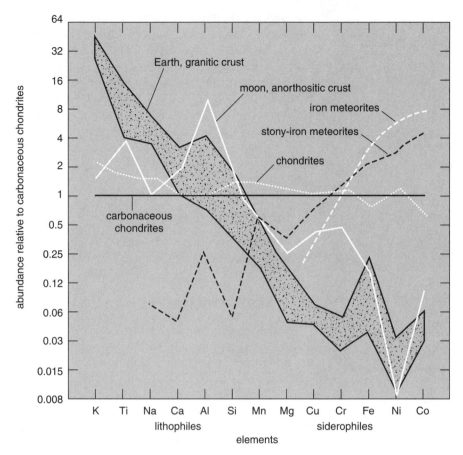

Figure 8-3. Differentiation in various planetary bodies. Carbonaceous chondrites are believed closest to the primitive material in terms of heavy element abundances. Chondrites have been only slightly modified. In the moon's anorthositic "crust," and more especially in Earth's granitic crust, lithophile elements have been strongly concentrated and siderophiles depleted. Conversely, in stony-iron and especially iron meteorites, which probably represent planetesimal interiors, lithophiles have been strongly depleted and siderophiles enhanced. The width of the Earth crust band shows the typical range of variation. (Data from Fairbridge, 1972; Taylor, 1975; Wasson, 1974; and other sources)

you heated a mixture of ice and rock, the ice would melt first; the rock would not melt but would settle out of the slushy mixture. Europa's smooth, virtually pure ice surface may have formed in this way. Similarly, even in a layer of melting rock, some mineral crystals melt at a lower temperature than others. **Partial melting** is the condition in which some mineral components have melted and others have not, depending on whether their melting temperature is above or below that of the material. During partial melting, differentiation can occur not only by gravitational settling of the heavy solid minerals through the molten material, but also by solution of some minerals into the hot fluid component, which may then flow and carry them away from their original locations.

In studying meteorites (Chapter 6), we saw that certain groups of elements have a chemical affinity for each other and cluster together during differentiation processes. The most important examples are **lithophiles,** the elements that chemically bond into low-density silicate minerals and hence concentrate near the surface, and **siderophiles,** the elements that chemically bond into higher-density iron minerals and hence tend to concentrate in the lower mantle or core of a planet.

Figure 8-3 shows the chemical differentiation that has occurred in planets. It plots abundances of some of the heavy elements, ranked from strongly lithophilic to strongly siderophilic properties. The carbonaceous chondrites have solarlike abundances of these elements and have not been strongly differentiated; they establish a reference line. Stony-iron and iron meteorites are, of course, extremely enriched in siderophiles and depleted in lithophiles. They seem to be samples from well inside a melted parent body, where iron and siderophiles have collected near the center. On the other hand, the surface rocks of the moon are enriched in lithophiles and depleted in siderophiles; the moon must have once been at least partially molten, allowing the lighter materials to rise and carry lithophiles toward the surface. Rocks from Earth's granitic crust are even more lithophilic. Probably Earth's differentiation is continuing even today as molten materials flow and volcanic magmas erupt, allowing light materials to accumulate on the surface.

Additional Observational Checks on Planetary Models

Given a model of a planet computed according to the scheme outlined earlier, we could use a variety of direct observations to check its accuracy. We've already described four basic measurements: (1) the planet's total

TABLE 8-1 Values of Moment of Inertia Coefficient k for Various Bodies

Body	Moment of inertia coefficient k
Idealized cases	
Hollow sphere	0.667
Homogeneous sphere	0.400
Sphere with core having half the total radius and twice the mantle density	0.347
Mass concentrated entirely at center	0.000
Actual cases	
Moon	0.40
Mars	0.37
Earth	0.33
Neptune	0.29
Jupiter	0.26
Uranus	0.23
Saturn	0.20
Sun	0.06

Source: Data from Cole (1978); Mars value 0.3654 ± 0.001 from Reasenberg (1977).

mass, (2) its diameter, (3) its mean density, and (4) the composition of the surface rocks. Now we discuss more observational checks: (5) the planet's moment of inertia, (6) its geometric oblateness, (7) the form of its gravitational field, (8) its rotation rate, (9) its heat flow, (10) the composition of neighboring planets and meteorites, (11) the form of its magnetic field, (12) drilling and direct sampling, and, most important, (13) seismic properties.

Moment of Inertia

The **moment of inertia** of an object is a measure of the degree of concentration of mass toward the object's center. This is important in planets, as we expect the heavier materials to have concentrated toward the center. For example, if a heavy mass M is whirling around on the end of a long (massless) wire of length R, the moment of inertia of this configuration is MR^2. In planets and other objects of geometrically regular shape, the moment of inertia is said to be kMR^2, where k is called the coefficient of the moment of inertia. For spheroidal objects, k must lie between 0.0 and 1.0, and the numerical value tells how concentrated the mass is. Examples are shown in Table 8-1.

The moment of inertia of a spinning object can be determined by its response to an external torque, which causes it to wobble. For instance, a spinning top wobbles when gravity tries to pull it down. Torques on planets are exerted by the sun and large satellites. For example, the

sun and moon produce Earth's precessional wobble, which takes 26,000 y and indicates a moment of inertia coefficient of 0.33. As shown in Table 8-1, this value indicates considerably denser material in Earth's core than in its outer regions. In the case of Mars, an improved moment of inertia from the 1997 Pathfinder mission constrained the Martian iron core to a smaller fraction of the planet's mass than Earth's core (Golombek and others, 1997; Folkner and others, 1997). Table 8-1 also shows that the giant planets have dense cores and vast, low-density envelopes.

Geometric Oblateness

A rotating planet is an oblate spheroid. Its equatorial diameter a exceeds its polar diameter b. The oblateness, or flattening, is a measure of this difference:

$$\textbf{geometric oblateness} = \epsilon = \frac{a - b}{a}$$

Oblateness is a useful quantity because it reflects the planet's internal properties, such as mass distribution or departure from hydrostatic equilibrium. It is possible to calculate the value of ϵ that will characterize a planet in equilibrium, given the planet's mass size, rotation period, and its mass distribution. We can calculate values that would correspond to idealized cases, such as a rocky world with an iron core half as big as the body itself, and then list the observed values to see how the planets compare.

Gravitational Field and Dynamical Ellipticity

If a planet departs from spherical symmetry—by being oblate, having an asymmetric mass distribution, or in some other way—its gravitational field will not be spherically symmetrical. A nearby satellite (either natural or artificial) will respond to the asymmetries by precession of its orbit and by departing slightly from a true Keplerian orbit. By measuring the satellite's motions, therefore, it is possible to estimate the departure from spherical symmetry of the mass inside the planet. These departures are often given in terms of a series of numerical coefficients known as J values, which describe greater degrees of complexity than the mere central concentration of mass described by the moment on inertia coefficient k.

The simplest way to estimate the planetary structure from gravity measures is to assume that the planet's mass is distributed in smooth concentric, equipotential shells that have various departures from sphericity, and then to calculate the planetary figure that would produce the observed J values. The amount of flattening derived in this way is called the **dynamical oblateness.** If it equals the observed geometric oblateness, the concentric shell assumption is probably fairly accurate.

If not, it may indicate "lumps" of higher density scattered here and there. A famous example was found in the lunar gravity field, where measures of space vehicle motions near the moon led to the discovery of **mascons,** or mass concentrations in the lunar surface layers (Muller and Sjogren, 1968). These turned out to be large lava flows that fill ancient impact basins, forming rock sheets denser than their surroundings. Since the moon's outer layers formed, they have been too rigid to allow the mascons to achieve hydrostatic equilibrium by sinking into the moon's interior.

Another type of inconsistency between the geometric figure and mass distribution has been found in the case of the moon and Mars. In both bodies, the geometric center is offset from the center of mass. This condition shows that the mass is not distributed symmetrically around the geometric center and that full hydrostatic equilibrium has not been attained.

In the moon, the center of mass is about 2 km closer to Earth than the center of the figure. This asymmetry arises because the low-density crust is several kilometers thinner on the Earth-facing side than on the far side. The thinner condition may have resulted when the crust was blasted away during the formation of the largest impact basins on the Earth-facing side. In Mars, the center of mass lies north of the center of the figure. This is associated with the southern cratered highlands, standing about 4 km higher than the northern volcanic plains, a fact that may also be associated with an asymmetry in crustal thickness (Wu, 1978).

Rotation Rate

The rotation period must be known in order to interpret the moment of inertia, the geometric oblateness, and the dynamical oblateness. Telescopic measures, radar, spectroscopic Doppler shifts, and radio results, together with photos and magnetic data from the Voyager probes, have been used to confirm planetary rotation rates for all planetary bodies.

Surface Heat Flow and Temperature Gradient

The rate at which heat escapes from the interior of a planet and the rate at which temperature increases as we go down into the planet (the **temperature gradient**) are both important parameters for judging the interior conditions. They can be measured in the following way on a planet's surface. First, a hole is drilled and then a thermal probe is lowered into the hole. The hole must be deep enough (>10 m?) to get away from **diurnal** (day and night) surface changes of temperature. The probe is several meters long and has a thermometer to measure the temperature at each end. Next, we must wait till the temperatures in the hole reach equilibrium values, because the drilling of the hole created heat by friction. After some hours, the frictional heat will have been dissipated and we

TABLE 8-2	Heat Flows in the Solar System	
Object	Measured heat flow ($J/m^2 \cdot s$)	Energy generated per kilogram per sec ($J/kg \cdot s$)
Sun	6.6×10^7	2×10^{-4}
Earth	0.055	4.7×10^{-12}
Moon	0.03	1.5×10^{-11}
Jupiter	7.6	2.4×10^{-10}
Saturn	2.8	2.2×10^{-10}
Uranus	<0.18	$<2 \times 10^{-11}$
Neptune	0.28	2×10^{-11}

Note: Giant planet data from Hubbard, 1980 preprint, *Intrinsic Luminosities of the Jovian Planets.*

can measure the environmental temperatures in the hole. The difference between the temperature at the top and bottom of the probe divided by the length of the probe gives the temperature gradient.

The surface heat flow (that is, the amount of energy coming through a square centimeter of the surface in 1 s [J/m^2 s]) is directly proportional to the temperature gradient and to the conductivity of the rock. If we take a sample of rock from the drill hole and measure its thermal conductivity in the laboratory, we can multiply the conductivity by the temperature gradient and get the heat flow coming out the region of the hole. Such experiments have been done all over Earth. In the outer planets, heat flow is sufficiently large to be measured by infrared detectors. Values are summarized in Table 8-2.

Suppose that on a certain planetary surface we measured a certain rate of heat flow from the interior. Furthermore, suppose that on measuring the radioactivity of the surface rocks, we found that a layer of those rocks only 30 km thick would provide sufficient heat from radioactive decay to account for the measured heat flow. We would have to conclude that the radioactive elements were strongly concentrated near the surface. This is just the situation on Earth's continents. Heat flow on the terrestrial continents and the flat ocean floor averages about 0.04 to 0.08 J/m^2s.

Over certain volcanic areas, especially the mid-ocean ridges, heat flow ranges 0.08 to 0.32 J/m^2s. Ascending currents inside Earth are bringing hot material upward in these regions. Because radioactive elements are lithophilic, they ascend with the lithophilic chemicals and end up concentrated in the lithophilic surface rocks, especially in the 30-km-thick continental blocks.

Another example is the moon. Before Apollo, many researchers thought the moon might have a chondritic composition, with a sufficiently low content of radioactive elements to give a low heat flow. However, measurements at the Apollo 15 and 17 sites gave unexpectedly high values, around .03 J/m^2 s, approaching Earth's continental values. Because the lunar surface rocks are not as differentiated as Earth's, they don't have as much radioactivity as the very lithophilic continental rocks, and the radioactivity levels measured in the lunar surface rocks must extend to deep levels in order to provide the observed heat flow. These findings support the assumption that the moon's bulk composition resembles the general composition and radioactivity of Earth's upper mantle, as noted in the discussion of lunar origin in Chapter 5.

Composition of Neighboring Planets and Meteorites

An additional clue about planets' interiors comes from the run of compositions as a function of solar distance, controlled by the condensation sequence (see Chapter 5). Also, the composition of meteorites, particularly irons and stony-irons, confirms the belief that nickel-iron cores formed inside asteroids, as hypothesized for Earth.

Magnetic Field

The presence of a strong planetary magnetic field probably indicates the presence of a fluid core in the planet, as we will see in more detail later in this chapter. Even if the planet lacks a strong field, magnetic observations near the planet can reveal secrets of its interior. In this case, the magnetic field associated with the solar wind moves through the planet, as the planet's lack of magnetic field causes no deflection. Deflections in the magnetic field around the planet are controlled by the conductivity inside the planet, and a dry, poorly conducting rock would produce a different result from that produced by a conductive interior.

Deep Drilling and Direct Sampling

The deepest drill holes penetrate about 6 km into Earth, only 0.1% of the way to the center, so one might conclude that drilling could hardly reveal any important properties of planetary interiors. This is not so, however, because important layering effects occur within a few kilometers of the surface. The base of Earth's crust, the surface layer of relatively low-density rock, is only about 5 km below the ocean floor.

A national effort to drill through the crust and directly sample the underlying mantle, the **Mohole Project,** was launched in the 1960s but was eventually canceled in a political-scientific wrangle. However, that project was replaced in the 1970s by the building of the extremely successful research ship **Glomar Challenger,** which has sailed all oceans and drilled as much as 1.8 km into the seafloor, including 0.6 km into the basaltic layer at the base of the crust under the ocean sediments. For example, its finding that the oldest known seafloor rocks are only 160 My old proves that the present-day ocean configura-

tions are very young features of a very active terrestrial crust (Nierenberg, 1978).

As for the mantle material, some free samples may be provided by volcanoes. Certain deep-rooted volcanoes erupt nodules of dense rock relatively rich in iron-bearing minerals such as **olivine** (the mixture of Fe_2SiO_4 and Mg_2SiO_4 mentioned in Chapters 5 and 6). Note that olivine is made of the four most abundant elements in Earth: iron, oxygen, silicon, and magnesium, in that order.

Especially important volcanic vents of this type are the **kimberlite pipes** in South Africa, Wyoming, and elsewhere. These are root structures of volcanoes exposed by erosion (Cox, 1978). They contain kimberlite, a dense rock rich in olivine, and even denser nodules composed of olivine and **pyroxene** (a mixture of $MgSiO_3$ and $FeSiO_3$). Kimberlite pipes also contain diamond, a high-pressure form of carbon, and hence have received much (not wholly academic) attention, which has illuminated mantle composition. On other planets, volcanoes and deep meteorite craters are areas where mantle rocks might be exposed. Confirming this, Pieters (1982, 1984) and Pinet, Chevrel, and Martin (1993) applied spectroscopic techniques to the moon and discovered olivine-rich rock, unusual on the moon, outcropping in the central peak of the crater Copernicus; these findings support the view that central peaks are upthrusts of deep rock created during impact crater formation.

Seismic Properties

Of all observations, seismic observations give the most useful experimental data about Earth's interior. During the first Apollo flights in 1969, the first seismometers were deployed on another planetary body, opening the way to planetary seismology. Seismology permits us to measure positions of layers inside a planet, determine their densities, and look for special structures such as cores or mass concentrations nearer the surface. Such measures sketch the geometry of the planet's interior and limit possible compositions. Seismology is considered in more detail in the next section.

Seismology and Earthquakes

The Seismometer

An earthquake is a series of vibrations, swayings, or sudden jiggles of the ground. The jiggles can be so small as to be scarcely detectable; rarely, they can knock over buildings, as shown in Figure 8-4. Imagine that we wanted to record such jiggles; how could we do it? Suppose we suspended a 100-kg weight on the end of a 10 m wire in a quiet building and attached a scribe to the bottom of the weight so that it just touched paper on a tabletop. If the

Figure 8-4. Destructive effects of the San Francisco earthquake of 1906, caused by movement along a major active fault that passes near the downtown center. (Courtesy Steven M. Larson)

building suffered a sudden horizontal jolt, the scribe on the bottom of the pendulum would mark a line on the paper. This would be a record of the earthquake; the length of the line would measure the violence of the jolt.

This is basically the principle of the **seismometer,** except that the pendulum must be designed so that its own natural swinging frequency will not be confused with the

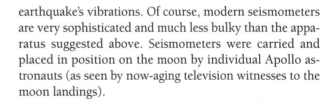

Figure 8-5. Pressure (*P*) and shear (*S*) seismic waves.

earthquake's vibrations. Of course, modern seismometers are very sophisticated and much less bulky than the apparatus suggested above. Seismometers were carried and placed in position on the moon by individual Apollo astronauts (as seen by now-aging television witnesses to the moon landings).

Wave Types and Early Seismic Observations

The vibrations felt during an earthquake are waves propagating along Earth's surface. The nineteenth-century physicists Poisson (1829) and Rayleigh (1887) predicted theoretically the main types of seismic waves that could be propagated by earthquakes. The most important of these are the **P waves** and **S waves** predicted by Poisson (Figure 8-5). P waves are *pressure* waves, such as sound

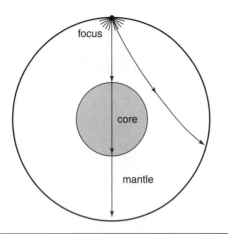

Figure 8-6. Paths of seismic waves through Earth. Waves that arrive on the far side of Earth from an earthquake focus are altered by the core, a fact that led to the discovery and size measurement of the core.

waves, in which the motion of individual particles is along the direction of the wave's motion (as a mnemonic aid, think of P waves as *push-pull* action). S waves are shear waves, in which the motion of individual particles is at right angles to the wave's motion; an example of such motion occurs in a wave along a rope or garden hose.

Because P and S waves have different velocities, they reach the seismic observer at different times. If the observer is located close to the **focus,** the earthquake's point of origin, the interval between the P and S arrival times will be short, but if he is farther away, it will be longer. The time interval between the P and S waves is thus a measure of the distance to the focus.

A minimum of three observing stations in a triangular array is needed to locate an earthquake. Each observes the **P-S arrival time interval** and thus computes the radial distance to the focus. On a map, a circle of this radius can be drawn around the station to indicate possible locations. If three stations provide three circles, there will usually be only one mutual intersection. This marks the earthquake's **epicenter,** or surface point directly above the focus. Thus an array of three simultaneously operating seismometers on any planet is far more valuable than a single seismometer.

English geologist R. D. Oldham (1900) was the first to distinguish the P and S waves in seismic records, and he first constructed useful tables of travel times for the two waves. Seismologists quickly discovered that they could even observe earthquakes that occurred on the other side of the globe. In 1906, Oldham found that waves arrived at the anticenter (180° from the earthquake) later than would be predicted by his travel time curves. This finding meant that the center of Earth contained a low-velocity region, as shown in Figure 8-6. This was the first evidence of a distinct core.

Another famous discovery was soon made by the Yugoslavian seismologist Mohorovičić (1909). In studying shallow earthquakes, he found that there were two sets of P waves and two sets of S waves. One set of each type trav-

eled at a high velocity and was traceable over large distances; the second set traveled at a low velocity and was best observed near the epicenter. Apparently the shallow rocks transmitted waves more slowly, as shown by Figure 8-7. The interpretation was that a discontinuity in density and wave velocity—hence, rock type—occurred at a depth of some 30 km. This is called the **Mohorovičić discontinuity** (sometimes called the "Moho" for short), and its discovery was the first proof of a distinct crust overlying the interior of Earth. Seismic and geological properties showed that the crust was composed of a lower-density rock type than the underlying mantle. Seismology thus first defined the three major compositional regimes in Earth: core, mantle, and crust. Various minor layerings also appear in the mantle.

In 1926, the German-American geologist Beno Gutenberg pointed out an upper mantle layer in which seismic wave velocities were lower than normal. This zone, which lies about 100 to 350 km deep, was for many years called the **low-velocity zone.** There was no evidence for major rock density change in this zone, and it was considered poorly defined and unimportant. However, in recent years it has been identified as a weak layer of partially melted material within the upper mantle and is now regarded as very important in interpreting both the dynamics of the upper mantle and the surface geological features of the crust, as we shall see. In this zone some of the lower-temperature minerals have melted, but higher-temperature minerals remain suspended as solid crystals, rather like a slushy mixture of water, ice, and soil. This layer is now called the **asthenosphere** (from a Greek word for "weak").

In 1937, a Danish seismologist, I. Lehmann, detected structure inside the core. P waves traveled faster in the interior part of the core than in the outer part. Studies of the outer part of the core revealed that it did not transmit any S waves, indicating that it has zero rigidity. Therefore, the outer core must be liquid, while the inner core is solid.

Origin of Earthquakes

What causes earthquakes? Earth is a dynamic body, even though its evolution is very slow. To drive an evolving system, energy is required, and in the case of Earth this energy is thermal energy (heat). Earth continually tries to

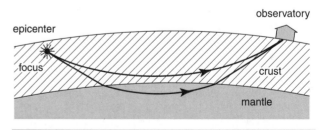

Figure 8-7. Discovery of the Mohorovičić discontinuity by observation of two sets of seismic waves with different velocities coming from a single, shallow earthquake.

TABLE 8-3 Destructive Earthquakes and Volcanic Eruptions

Date	Location	Event[a]	Log energy (J)	Deaths (approx.)[b]
Aug. 24, 79	Pompeii, Italy	V	—	15,000
Feb. 2, 1556	Shensi, China	E	—	800,000?
1737	Calcutta, India	E	—	300,000
Nov. 1, 1755	Lisbon, Portugal	E	~17.5	60,000
1815	Sumbawa Island, Indonesia	V	—	75,000
Mar. 26, 1872	Owens Valley, California	E	≥16.8	~36
Aug. 26, 1883	Krakatau, Indonesia	V	—	37,000
May 8, 1902	St.-Pierre, Martinique	V	—	29,000
Apr. 18, 1906	San Francisco, California	E	16.7	600
Dec. 16, 1920	Kansu, China	E	17.1	~180,000
Sep. 1, 1923	Kwantō, Japan	E	16.8	~143,000
Feb. 29, 1960	Agadir, Morocco	E	13.0	~12,000
May 31, 1970	Peru	E	16.0	~70,000
Feb. 9, 1971	San Fernando, California	E	~14.0	62
Feb. 4, 1976	Guatemala City, Guatemala	E	—	22,000
July 28, 1976	T'angshan, China	E	—	700,000
May 18, 1980	Mt. St. Helens	V	16.7?	65
Aug. 6, 1945	Hiroshima	Atom bomb	13.9	66,794
1969–1970	Moon	Total lunar seismic activity	~ 7	—

[a] V = volcanic eruption; E = earthquake.

[b] Various sources; see Tazieff (1964) and 1978 reference book *Disaster!* prepared by editors of Encyclopaedia Britannica. Figures include all related destruction. Mt. St. Helens energy from *Volcano News*, July 1980.

readjust its temperature distribution: Radioactivity produces new heat, while the interior as a whole tries to cool.

Earth might be compared to a house cooling at night; occasionally it creaks or pops. Because of contraction, expansion, or slow movements of materials, stresses build up in the solid outer part of Earth. The rock may stretch elastically, but if the stress is too great, the rock may fail, like a twig bent too far. The sudden splitting and slippage of the rock to a new position of less stress is an earthquake. This process is called faulting, and any rock fracture along which there has been an offset motion is called a **fault.** Many of California's earthquakes are caused by slippage along the San Andreas Fault.

A less important source of earthquakes is volcanic activity at Earth's surface. The spewing forth of billions of tons of lava during a volcanic outburst may be locally accompanied by vibrations and shocks that can literally shake the ground to pieces.

A typical large earthquake may release 10^{17} J of energy. The annual release of earthquake energy is estimated to be about 5×10^{17} J, most coming from a very few large quakes. Table 8-3 lists some of history's most energetic earthquakes and volcanic eruptions.

Not all earthquakes are disastrous. The lower the energy, the more common is the earthquake. "Garden variety" earthquakes cause a slight vibration of the ground or a rattling of windows and walls. In addition, seismographs detect vibrations that range in energy all the way down to tremors produced locally by wind, traffic, and so on.

Studies that utilize the seismic signals from naturally occurring earthquakes are called **passive seismology.** In order to better analyze the probing seismic signals, it is necessary to know the exact time of the initial shock and the exact location of the focus; for these purposes, man-made explosions are used as sources. Such studies are called **active seismology.**

Earthquake Distribution: Plates, Asthenosphere, and Lithosphere

Earthquakes are not distributed randomly, either geographically or in terms of depth. Figure 8-8 shows the geographic distribution of shallow earthquakes, deep earthquakes, and volcanoes. This map is full of clues concerning relations between Earth's surface layers and interior, but geophysicists

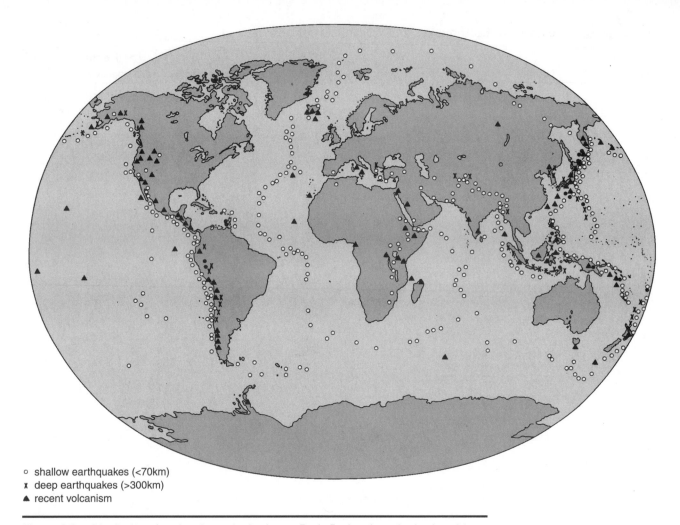

○ shallow earthquakes (<70km)

✕ deep earthquakes (>300km)

▲ recent volcanism

Figure 8-8. Distribution of earthquakes and volcanism on Earth. Earthquake and volcanic activity occur along the margins of plates where the plates abut and move relative to each other. Earthquake zones tend to slope downward under continental margins of plates, as shown by the pattern of shallow versus deep earthquakes.

needed many years to understand these clues. The map shows that Earth's surface is divided into **plates,** or coherent blocks bounded by zones of earthquakes and volcanism. Some of the plates contain continents, but others are huge segments of the seafloor. Earthquakes are concentrated at the margins of plates.

A great deal of geological as well as seismological evidence has revealed the fundamental importance of the plates and their linkage with the asthenosphere, or weak, partly melted zone mentioned earlier. The combination of heat flow and fluidity within the asthenosphere produces slow currents that drag along the overlying layers. These layers are solid and more brittle than the asthenosphere and are called the **lithosphere** (from the Greek root for "rock"). The 100-km-thick lithosphere has probably been broken into plates because of movements in the asthenosphere. Earthquakes tend to occur near plate margins, where collisions between plates occur. The motions of plates are very slow—only a few centimeters per year—but these movements are sufficient to build up great

stresses in the rock, which eventually are released as faults and accompanying earthquakes.

The depth distribution of earthquakes, shown in Figure 8-9, reveals more information. Earthquakes are most concentrated in the 100-km-thick lithosphere because it is continually being cracked by the asthenospheric currents. The number of earthquakes drops sharply as we proceed down through the more plastic asthenosphere—from about 100 km to 350 km in depth. Other concentrations may mark alternating fluid layers with strong, brittle layers. The numbers drop to near zero at depths below about 700 km, implying that the lower mantle below 700 km is too plastic—too near hydrostatic equilibrium—to build up stresses that cause earthquakes.

"Moonquakes" and "Marsquakes"

Similar concepts apply to other planets. Small planets cool more rapidly than large planets and thus have developed thicker lithospheres. This phenomenon is clearly shown

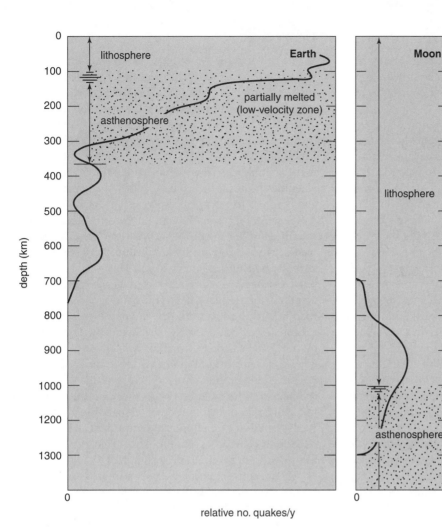

Figure 8-9. Depth distributions of "earthquakes" on Earth and the moon. Earthquakes tend to occur along the bottom edge of the solid lithosphere and the top of the more plastic asthenosphere, where currents build up stresses in the rocks. Moonquakes number nearly 3,000/y but are much less energetic than larger earthquakes, so that the two scales are not directly comparable.

in the moon, which is seismically much quieter than Earth. During the first year of lunar observations, lunar quakes rated only 1 to 2 on the well-known Richter scale, contrasted with ratings of 5 to 8 for major terrestrial quakes. The total seismic energy released annually in the moon is estimated at about 10^7 J (Latham, 1971), contrasted with 5×10^{17} J in Earth.

As shown in Figure 8-9, the quakes inside the moon occur in a band around 800 to 1000 km in depth, a level that is interpreted as the base of the lunar lithosphere. Various geophysical measures, such as seismic data and estimated temperature profiles, suggest a partially molten asthenosphere below that level (Taylor, 1975).

The situation on Mars appears similar, according to results from one relatively crude seismometer placed by the Viking 2 lander (a duplicate seismometer on the Viking 1 lander failed to be deployed). Analysis indicated that Mars is probably less seismically active than Earth and perhaps close to the moon in its level of activity.

Thermal Histories of Planets

Definition of the Problem

Most of this chapter so far has dealt with ways of determining the **statics of planets**—that is, descriptions of their interiors in their present-day states. But planets are evolving systems. We must therefore also come to grips with the **dynamics of planets**—that is, the evolution of their internal properties through time.

The heating or cooling of a planet is one of the driving factors in its evolution. The problem in the study of thermal histories is to discover the temperature and related properties (motions, liquid or solid state, and so on) of planetary interiors as a function of time. This is a more grandiose goal than modeling the present-day interiors.

Theory of Heat Transport

Heat is a form of energy. The atoms and molecules of a hot body move faster and hit each other harder than those of a cold body. **Temperature** is a measure of this thermal energy.

Why are the interiors of planets hot? Some of the heat is original; the material from which the planet formed had a certain temperature, and more heat was generated as the planetesimals crashed into planets and transferred their kinetic energy into the growing world. Other heat is added to the planet over the course of time by radioactivity, whereby atoms emit small particles. These particles speed out into the surrounding atoms, hitting them and thereby adding to the total thermal energy.

Heat moves naturally only from hot regions to cold regions—never from cold to hot. **Heat transport,** the process of transferring the heat and equalizing temperatures, can occur in three ways. First is **conduction,** the process by which the atoms or molecules in a substance strike each other, preferentially transferring energy from the fast-moving atoms in the hot region to the slow-moving atoms in the cold region. Second is **radiative transfer,** the process by which high-energy atoms emit radiation that travels through the surrounding material and is absorbed by low-energy atoms in a cooler region. Finally, **convection** is the process by which an entire hot region—because it has expanded to a lower density than a cold region and is therefore lighter—rises upward as a unit through the colder material while nearby colder material sinks. An example of this is the hot smoke from a burnt match rising through the air; an example with geological relevance can be seen by heating a shallow pan of cooking oil, which will set up a cellular pattern of convection currents reminiscent of the ascending and descending currents postulated by the theory of plate tectonics.

To apply these concepts to the problem of planetary thermal histories, we must have a great deal of quantitative information. First, we must have an adequate mathematical theory describing each of the three modes of heat transport. Also, we need to know how much heat is being created at each point by radioactivity. Because of their lithophilic chemistry, the main radioactive elements are concentrated in the crust, but we need to estimate their abundance throughout the planet.

Next, we must determine which mode of heat transport is most important in the given planet. If heat transport is by conduction, we need to know the rock conductivity at different depths; if by radiative transport, we need the transparency of the rock to the kind of radiation involved; if by convection, we need to know the mobility of the material. These properties are estimated from lab measurements. Furthermore, at different points and different times inside the planet, different modes of heat transport may dominate.

Convection

To understand the importance of convection, consider some parcel of the fluid or plastic material inside a planet (or, for that matter, in a planetary atmosphere). Suppose some random perturbation starts this material moving upward. As it ascends to a new level, it immediately adjusts its pressure to the surrounding pressure, which is lower. (*Immediately* means at the speed of sound if the medium is truly fluid, as this is the speed at which pressure inequalities transmit themselves.) Thus, the parcel expands and becomes less dense as it rises.

Any heat transfer between the parcel and its new surroundings takes much longer than the pressure adjustment; therefore, the pressure adjustment takes place with virtually no external energy input. This is called an **adiabatic change.** An adiabatic change creates a new density and temperature in a parcel, and if the new density is less than that of the surroundings, the parcel will be buoyed up and will continue to rise automatically. The resulting motion is called convective motion.

Steep temperature gradients favor convection. That is, convection proceeds only if there is a large temperature difference in the material over a short distance. Convection can begin only if the actual temperature gradient exceeds the **adiabatic temperature gradient,** which is that experienced by an ascending parcel that rises adiabatically—without external energy input.

Thus, convection may develop in an asthenosphere because the material is plastic and the temperature drops rapidly from the heat of the interior to the cold of the surface in a short distance. If the temperature gradient is great enough, and if the material is fluid enough, convection currents are sure to arise, creating **convective cells** in which hot material rises, spreads outward, and descends along the edge of the cell. If the lithosphere is thin enough, the convection currents may drag on it strongly enough to disrupt it.

Initial Thermal State of Planets

A model of the thermal evolution of a planet gives the calculated temperatures at all points in its interior as a function of time. Once the theory of heat transport is available, calculating such a model requires only that we specify the materials composing the planet and, of course, the equations of state of those materials.

Also, we need to specify conditions throughout the planet at some specific time. Therefore, modelers pay special attention to initial conditions and usually calculate forward from the initial state, using the present conditions as a check.

Following the chemical equilibrium theory discussed in Chapter 5, early modelers assumed that each planet formed as a nearly homogeneous ball, without a core or crust. As an alternative, the inhomogeneous accretion theory of Chapter 5 predicts that the iron-rich early condensates accrete first, forming a ready-made iron core that was overlaid by later silicate condensates (Slattery, DeCampli, and Cameron, 1980).

In any case, geochemical evidence indicates that planets (or at least their surface layers) were partly molten by the time they finished forming (Fricker, Reynolds, and Summers, 1974). Lunar rock chemistries and ages seem to require that the moon was initially (4500 My ago) covered by a "**magma ocean**"—a molten layer at least several hundred kilometers deep—whereas meteorite parent bodies apparently also melted and resolidified 4500 My ago.* Recent dating of 4400 My-old Australian zircons indicates continental crust had formed on Earth by that time (Wilde and others, 2001). So most models begin at 4500 My ago with the surface layers already molten to great depths.

Several heat sources are possible for this initial melting. One or more may have been effective:

1. The impact of the planetesimals during accretion provided heat. If significant growth occurred in only a few thousand years or less, this heat may have been trapped and may have helped to melt planets, and encouraged iron to drain toward the center even as planets formed.

2. Radioactive elements found today, such as uranium, potassium, and thorium, produced more heat at the beginning, as shown in Figure 8-10, but not enough to melt planets in the first few million years.

3. Short-lived isotopes (dashed line in Figure 8-10), such as aluminum 26 and iodine 129, described in Chapter 5, were so abundant that they produced enough heat to melt planets a few million years after they formed.

4. Sonett and others (1970) and Herbert and Sonett (1979) pointed out that strong solar winds during the T Tauri phase of the sun could have induced electric currents in the outer parts of planetesimals or planets as the solar wind carried solar magnetic fields past the bodies. These currents could produce so-called **electromagnetic induction heating** in the outer parts of the planets—possibly enough to cause melting in bodies in certain size ranges and solar distances. This theory had the remarkable success of showing how asteroids of certain sizes and distances could have been melted whereas others were unmelted, as shown in Figure 8-11, on page 208. For instance, 4 Vesta's basaltlike surface indicates that it was melted whereas larger and smaller neighbors were not. However, many of the parameters in this theory are poorly measured, and results are therefore highly model dependent. As T Tauri conditions rapidly died out, the theory would explain only why the planets were strongly heated at the beginning.

---

*Note: This doesn't mean that planets grew from incandescent material, as thought half a century ago. Urey (1952) proved from meteorite chemistries that planets formed from cold dust, typically at only a few hundred K. They apparently then melted (at least partially) after reaching dimensions around 500 to 1,000 km.

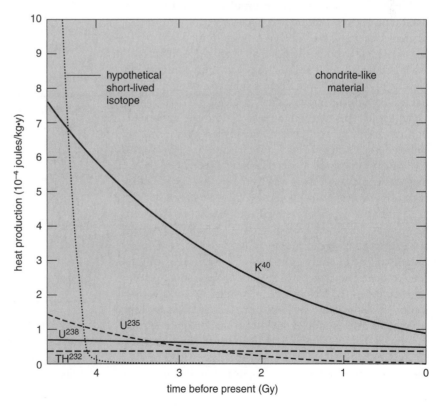

Figure 8-10. Relative heat production in chondritic material as a function of time for important radioactive elements, showing that planetary materials produced several times more heat during the planet-forming period than they do today. Thorium has such a long half-life that it produced little more then than now. On the other hand, nucleosynthesis events may have produced short-lived isotopes that were the dominant heat source during the first few million years (dashed line). (After data of Kaula, 1968, pp. 110–111)

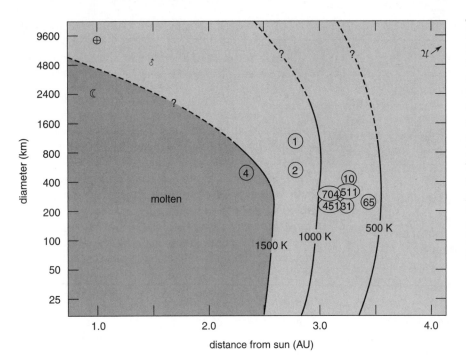

Figure 8-11. Temperatures reached in carbonaceous chondritic planetary bodies of different sizes as a function of distance from the sun, under an assumed model of electromagnetic heating by electric currents induced by early, strong solar winds. The model is consistent with observations of all asteroids larger than 300 km: 4 Vesta has a basaltic surface that was apparently once molten; the surface materials of 1 Ceres and 2 Pallas are chondritic and were never molten; and the surfaces of more distant asteroids resemble lower-temperature carbonaceous chondrites. (After Herbert and Sonett, 1979)

Principles of Crust and Lithosphere Formation

Any initial magma ocean cooled rapidly and solidified not only because the heat sources declined rapidly but also because the ocean was at the surface, where its heat radiated into space. However, in the interior, which was insulated by the overlying layers, heat accumulated from the long-lived radioactive uranium and potassium, trapped in the planet's silicates. This heat began to melt the deep interiors of planetary bodies larger than perhaps 1000 km, allowing differentiation and sinking of the iron component. Settling of the metals into a core produced still more internal heat, acting like gravitational contraction. Thus, while a magma ocean rapidly disappeared from the surface, a molten or partly molten asthenospheric zone grew at some depth.

During the asthenosphere phase, the lower-density magmas could work their way to the surface, either during normal volcanic eruptions (see Chapter 11) or during large impacts, which penetrated into magma-rich layers.

As you recall, the lithosphere-asthenosphere boundary is defined by *physical state* (solid versus liquid); the crust-mantle boundary, however, is defined by *chemical composition*. The crust, which may involve some of the same layers as the lithosphere, is merely the surface collection of distinctively low-density materials. Now we can see why **feldspars,** one of the lowest-density types of silicate minerals ($2600 \ kg/m^3$), are the most common minerals in Earth's and the moon's crust: In silicate-rich planets, the crusts are likely to be feldspar-rich rocks because

feldspars float to the top of magma oceans or subsurface melted zones. Examples of feldspar-rich rocks are **basalts,** a commonly erupted type of volcanic lava, and **gabbros,** which are similar to basalts but have coarser crystals and form when magma solidifies at some depth instead of on the surface. Much of Earth's crust is basalt, and the lunar crust is anorthositic gabbro, a type of gabbro composed almost entirely of feldspar. Furthermore, Viking landers discovered basaltic boulders and soil on Mars (see the next chapter for more details).

The lithosphere thickness is defined by the physical state of solidity, which in turn depends on the chemistry, as chemistry determines the melting point. For dry silicates (as in the moon), this is about 1100 K, which may be fairly deep in the planet. But with a small percentage of water (as in Earth), this amount is reduced to about 900 K, which may occur at a shallower depth, producing a thinner lithosphere (Warner and Morrison, 1978). Similarly, for a water-rich icy body (such as a satellite in the outer solar system), the melting temperature may be only 200 to 300 K, requiring minimal heating to produce a "molten" zone.

Heat Balance of Planets

If a planet were in equilibrium with its surroundings, it would radiate just as much heat as it receives from the sun. If it radiated less than it received, it would be heating up; if more, cooling down. We have just pointed out, however, that in addition to sunlight, a planet contributes a certain amount of heat of its own from radioactivity and possibly from early initial heat.

In the case of Earth, the solar heat flux coming in (that is, the solar constant) is about 20,000 times the terrestrial heat flux going out. Thus, at Earth's surface, sunlight is the dominant energy source in maintaining the temperature. In the case of the giant planets, Kuiper (1952) speculated that the Jupiter atmosphere was heated primarily from internal sources, not sunlight, because of the spectacular cloud activity. The internal heat can be measured by infrared detectors on telescopes or spacecraft. Öpik (1962) and Taylor (1965) made the unexpected discovery that Jupiter and Saturn radiate *more* heat than they receive from the sun! Some approximate current figures are given below (Hubbard, 1990; Voyager team press releases; Stone and Miner, 1989):

	heat generated (watts)	heat emitted
	10^{12} kg of total mass	heat absorbed
carb. chondrite	3	—
Earth	4	1.00005
Jupiter	210	2.5
Saturn	150	2.3
Uranus	~10	~1.1?
Neptune	20	2.7

The first two lines show the heat generated from radioactive isotopes in chondritic meteoritic material, and show that Earth generates about this much heat per kilogram. The next lines show that at least three of the giant planets have additional, stronger heat sources. Their heat is believed to be the last stages of the long, slow gravitational contraction that brought these planets from their initial protoplanetary size down to their present size.

Magnetism of Planets

A compass is a magnetized needle, and if a magnetic field is present, the needle aligns itself in what we call the direction of the field. Arbitrarily, we define a "north-seeking" end of the needle, and then say that end points toward the north direction of the magnetic field. The stronger the field, the stronger will be the tendency of the needle to align itself. Magnetic fields, then, have both strength and direction.

To make matters conceptually easier, physicists often speak of imaginary "**lines of force,**" which are lines paralleling the direction of the field (the direction in which the compass needle comes to rest). This makes it possible to represent a magnetic field in a drawing such as the inset in Figure 8-12 on page 210, showing the field around a bar magnet. As shown by Figure 8-12, a planet's field resembles a bar magnet's field in having two well-defined poles;

such a field is called a **dipole field.** The north and south magnetic poles are defined simply by the direction in which a north-seeking magnet would point when placed near the planet. (By contrast, the north rotational pole is defined as the pole on the north side of the ecliptic plane.)

The magnetic poles do not necessarily coincide with the poles of rotation. In Earth's case, they depart by a slowly changing angle, which is now about 11 . This explains why surveyors and hikers must apply a **magnetic correction** to directions determined with a compass. Furthermore, there are local, complicated non-dipole components of the field about one-tenth as strong as Earth's main field.

Magnetic Interactions with the Solar Wind

Figure 8-12 shows the interaction between a planetary magnetic field, like Earth's, and the field being carried along by the ions of the solar wind. There are several characteristic features. The **magnetosphere** is the volume of space in which the magnetic field is essentially that of the planet. This is bordered by a **magnetopause,** or region where the planet's field is in equilibrium with the impinging solar wind field. Because the solar wind rushes by the planet at hypersonic speed, a **bow shock** builds up where the solar wind strikes the planetary field, and the planetary field is said to be encased in a **magnetosheath,** as shown in Figure 8-12.

Changes in solar activity alter the solar wind and the configuration of these boundaries, whereas the interior state of the planet controls the strength of the planetary field itself. Spacecraft magnetic measurements to map the magnetic field near a planet are thus important in deducing the planet's properties. Interestingly, Voyager measurements showed that the solar wind drags Jupiter's magnetic field into a tail at least 3 AU long and possibly extending far enough to interact with Saturn (Scarf and others, 1981)!

Changes in a Planet's Magnetic Field

One might suppose that a characteristic as major as a planetary magnetic field would be relatively permanent. On the contrary, measurements show that Earth's field is continually changing strength, drifting so as to change its pole positions and even reversing direction! About 250 My ago, Earth's north magnetic pole was located in the present north Pacific. Measurements show that the present field is fluctuating at a rate of about 0.1%/y, so that major changes in strength are expected in about 1000 y. Measurements suggest that Earth's field in 500 B.C. was 50% stronger than it is now (Kaula, 1968).

Most mystifying, the direction of the field has reversed itself at irregular intervals. No one yet knows why this happens, although evidence for reversed magnetism has been known since 1906. In the last few million years these **magnetic field reversals** have occurred every few hundred thousand years, but there have been intervals as

Figure 8-12. Schematic cross section of the magnetic field around a planet with an intrinsic magnetic field. The magnetic field of the passing solar wind is disturbed as it hits the magnetic field of the planet, forming a bow shock analogous to that formed in the wave pattern of lake waters moving past a speedboat. Spacecraft-borne magnetometers can map field configurations near planets and reveal internal planetary properties. The inset illustrates the resemblance between the dipole field of a bar magnet and an inner planetary field.

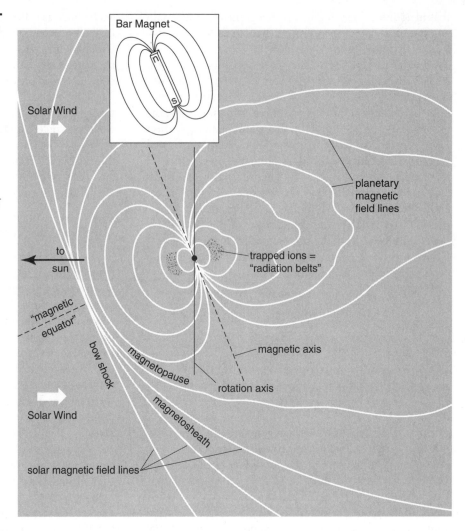

long as 50 My without a shift. The process of reversal takes less than 10,000 y, and some believe that the field may reach nearly zero strength for a short time during this interval. There has been speculation that without a strong field to deflect cosmic ray particles, the incidence of cosmic rays at the surface goes up during the times of reversal, increasing biological mutation rates and affecting the course of biological evolution.

Paleomagnetism

How are past magnetic fields determined? This subject is called **paleomagnetics.** Whenever a rock crystallizes from molten lava in the presence of a magnetic field, the magnetic elements in the rock, like compass needles, are frozen into position aligned with the field. This happens not at the freezing (that is, melting) point, but at a lower temperature, called the Curie temperature. As long as the rock remains below the **Curie temperature** and fixed in position, it retains a record of the strength and direction of that original field.

Geologists locating an ancient, undisturbed outcrop of such rock can carefully remove a sample, noting its orig-

inal orientation, take it back to the lab, and measure the slight magnetism of the rock itself, which is called **remanent magnetism** (not *remnant!*). This gives a measure of the direction and strength of the ancient magnetic field.

Origin of Planetary Magnetic Fields

Paleomagnetic studies suggest that Earth's magnetic field is not a frozen remanent field derived from some strong field in the early solar system, because a remanent field would have decayed in around 10^4 y (Cole, 1978). Further, the field cannot be due to the presence of some giant, permanent dipole magnet such as a magnetized iron core because (1) the interior temperature is far above the Curie temperature and (2) the field is too variable. These observations show that planetary magnetic fields must arise from some dynamic process in the interior.

The **dynamo theory** is the most widely accepted explanation of planetary magnetic fields. In this theory, if a conductive fluid (such as a molten iron or metallic hydrogen core) moves (by means of convection currents or planetary rotation) through an external magnetic field (such as that of the solar wind or a random initial plane-

TABLE 8-4 Surface Planetary Magnetic Fields

Object	Magnetic field (nT)[a]	Approximate inclination of dipole to rotation axis
Sun	200 000[b]	6°
Mercury	220	<10°
Venus	<30	?
Earth	30 500	11°
Moon		
3.3 Gy ago	2 000	—
Today	10	—
Mars	40	—
Jupiter	420 000	9.°5
Saturn	20 000	<1°
Uranus	23 000	58°.6
Neptune	~100 000	46°.8

Note: Field values are variable from place to place, typically by about a factor of 2. Values from Taylor (1982); Van Allen, 1990; Hubbard, 1990.

[a]The tesla is the preferred SI unit for magnetic field strength, though values have traditionally been given in gammas or gauss until quite recently. $1 \text{ nT} = 10^{-9}$ tesla = 1 gamma = 10^{-5} gauss. For comparison, the interplanetary field in the inner solar system is typically a few (~3) nanoteslas.

[b]Highly variable, up to 10^8 nT in sunspots.

tary field), electrical currents are induced that create their own local magnetic fields. The presence of these fields causes a feedback effect that further alters the motions of the fluid medium. The effect is the same as that in a dynamo motor. The upshot is that a partitioning of energy would occur between the kinetic energy of fluid motion and the magnetic energy of the magnetic field being created. An initially small field may thus be converted to a strong field at the expense of some of the motion.

Theoretical analysis shows that production of a strong field with good dipole symmetry requires two factors: a large volume of conductive fluid (that is, a large core region) and a rapid enough rotation to create a single symmetric pattern of motion (Cole, 1978). Internal irregularities, such as growth and decay of convective cells because of radioactive heating, could cause magnetic field variations such as polar wandering and field reversals. The patterns in Earth's field suggest that the liquid outer core is convecting with a set of perhaps 10 to 20 convection cells.

Strength of Planetary and Solar Magnetic Fields

Table 8-4 summarizes available data on magnetic fields of various bodies in the solar system. The sun has a magnetic field in some ways similar to Earth's except with a shorter time scale. The sun's magnetic field reverses every 11 y, in step with a cyclic behavior in the number of sunspots. Every 22 y the sun goes through one complete magnetic cycle and two sunspot cycles.

Spacecraft measures near Venus, the moon, and Mars indicate that none of these bodies have magnetic fields comparable to Earth's. This finding supports the theoretical ideas explained earlier, as Venus rotates very slowly, and the moon and Mars are so small that they have probably cooled to a point that they no longer have large enough molten cores to generate such a field. Mercury's substantial field suggests a large, convecting iron core, an idea supported by the planet's high density. The strong fields of Jupiter and Saturn are believed to relate to their rapid rotations and the presence of metallic hydrogen inner mantles.

Synthesis: Interiors of Specific Worlds

With this theoretical background, we can now describe what is known about some specific cases, starting with the best-known one: Earth.

Earth: Core-Mantle-Crust Evolution

The core of Earth is believed to be about 80% to 90% iron. Its material is an iron-nickel alloy with an admixture of up to 20% sulfur, cobalt, and other minor materials. Iron meteorites (92% iron, 7% nickel, plus FeS) support this. No volcanoes tap the core, but iron-nickel nodules erupted from deep volcanic sources may be core-like material that was trapped in the mantle. Though the inner core is hottest, the high pressures there make it solid whereas the outer core stays liquid, as shown in Figure 8-13. Circulation at an estimated 0.02 cm/s in the outer core helps produce Earth's magnetic field.

As noted in Chapter 5, tungsten isotope studies and other studies show that the core must have formed early, probably within the first 40 to 50 My (Halliday and others, 1996). Some analysts have suggested that the core formed intact when iron accreted directly from the solar nebula and before silicates condensed (Slattery and others, 1980; Stevenson, 1981); most, however, believe it formed somewhat later during the accretionary heating and partial melting of Earth's interior. Since remanent magnetism appears in 3.5-Gy-old rocks, we can infer that the core-generated magnetic field was up and running within the first 1000 My. Similarly, Basu and others (1981) find from chemical abundances in 3.8-Gy-old East Indian rocks that at least some parts of the mantle had already begun to differentiate from the core and crust by that early era.

The mantle is siderophile-rich silicate rock, which probably has layers of different mineral phases and different rock types, such as the layering caused by the transition

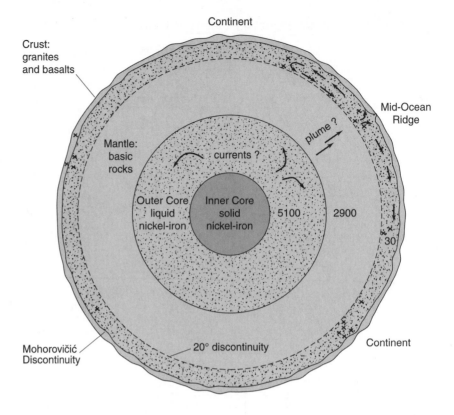

Figure 8-13. Structure of Earth's interior as revealed by seismology, with compositions inferred from additional evidence. Earthquake positions are indicated by Xs. Stippling indicates molten material in the outer core and partially molten material in the asthenosphere. Numbers indicate depth in kilometers.

from the olivine to the spinel phase about 400 km down. As recognized by the 1960s, the heat flow through the top of the mantle under continents is substantially less than the heat flow out of the mantle under the seas; hence, there must be lateral as well as vertical variations in mantle structure. Aside from the olivine-rich rock types, pyroxene-rich and other rock types are suspected to exist in the upper mantle; all these are denser and more iron rich than common rocks on the surface, but they are sometimes found in material erupted from deep volcanoes (see also the discussion of rock types in the next chapter).

The crust is the outermost rock skin—the lowest-density rocks that have accumulated on the surface after aeons of recycling mantle materials through volcanic and erosive processes. In the sea floors, the crust is thinnest—about 5 to 10 km of basaltic rock. Basaltic crustal layers now appear common among the terrestrial planets. In the continents, the crust is composed of thicker blocks—20 to 60 km thick—consisting primarily of granitic rock (more silica rich than basalts; see the discussion of rock types in Chapter 9). These blocks overlie a more basaltic layer 5 to 10 km thick and resembling the ocean floor crust.

From geochemical abundance patterns in crustal rocks, Jacobsen and Wasserburg (1979) found evidence that after the period of rapid cycling and differentiation of mantle and crustal materials in the first 1,000 My, the present crust began to form at the end of that period, and that today's continental crustal rocks have a *mean* age of 1.8 to 1.5 Gy. They concluded that the rate of continental crustal rock accumulation in the last 0.5 Gy is less than the earlier rate, but different workers' models of the crustal thickness as a function of time vary substantially.

Earth: Unraveling the Dynamics of the Interior

For centuries, naturalists pondered features of Earth's surface. How do mountain ranges form? Why do certain regions sink and get invaded by shallow seas? Why are some areas shaken by earthquakes while other areas remain quiet?

One fact recognized early was that the continental masses have a lower density than do the heavier mantle rocks. Furthermore, the crust is thickest under high mountain ranges, which thus have low-density roots supporting their weight. An analogue is an iceberg, which has a large volume of ice below sea level to support the small volume that protrudes.

These facts show that the continental masses, which rise above the mean crustal surface, are not piled on a rigid substrate, but rather are very nearly an **isostatic equilibrium**—that is, they are floating in equilibrium on the denser mantle. Another observation is that the central parts of continents, the so-called **continental shields,** are ancient, flat-lying regions not recently disturbed by mountain building. They contain the intact oldest rocks, about 3.5 to 3.7 Gy old. An example is the Canadian Shield, surrounding Hudson's Bay.

Major mountain ranges, as shown in Figure 8-14, are the youngest, most active, and most distinctive features of Earth. They are vast crumpled crustal masses (Figures 8-1 and 8-15, p. 215) usually lying around the outer edges of continental shields. Many of them formed in recent episodes such as the "Laramide revolution" about 70 My ago, which saw much of the warping and thrusting that led to the Rocky Mountains.

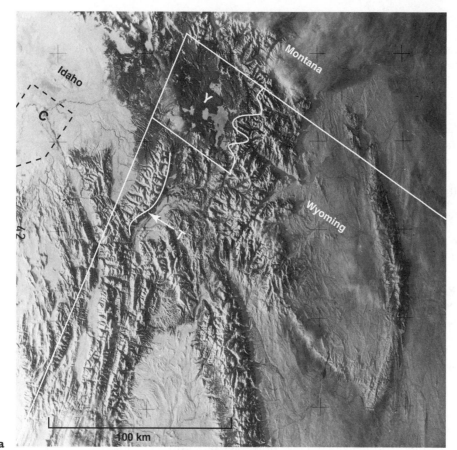

Figure 8-14. Major folded mountain chains are unique to Earth in the solar system. (**a**) A portion of the Rocky Mountains photographed with a handheld camera by a Skylab astronaut. Y = Yellowstone Park, a crustal hot spot with active geysers and fumaroles; C = an area of recent cinder cones and lava flows near Craters of the Moon National Monument; T = a large fault that exposed the east face of the Grand Teton Mountains. Arrow marks position and direction of view in part (**b**). (NASA) (**b**) East face of the Grand Tetons (Middle Peak) showing cliffs exposed by faulting and a massive dike of basaltic rock. This basaltic rock apparently intruded as magma into a large fracture when the original rock mass was underground, prior to faulting. The dike, partly covered by snow, runs several hundred meters from the summit to a point where it disappears behind loose talus at the cliff base. (Photo by author)

Classic geologic studies of continents did not solve the problem of the origin of Earth's dynamic processes, because the continents are so contorted and active that the geological messages are lost in the chaos. Historically, it was the 71% of Earth's surface that is hidden under the sea that has provided the key to the dynamic evolution of Earth's crust.

As early as 1620, Francis Bacon noted that the east and west coastlines of the Atlantic could fit like jigsaw puzzle pieces. He made the remarkable suggestion that the Americas may once have been in contact with Europe and Africa. By the late 1800s, paleontologists discussed similarities of 300-My-old fossil plants from opposite shores of the Atlantic, and the famous Austrian geologist Eduard Suess noted jigsaw puzzle fits of ancient geological formations across the Southern Hemisphere. Suess hypothesized a primeval southern continent, which he named Gondwanaland (from a key geological province in India). It was the parent of the present southern continents, which took shape as Gondwanaland broke up.

This idea became the theory of **continental drift,** that continents broke apart and drifted. In 1908 the American geologist F. B. Taylor discussed possible mechanisms to keep the continental masses moving. Continental drift was then championed by the German meteorologist Alfred L. Wegener in 1922. He pointed to cross-oceanic fits of geological structures and fossil flora and fauna on two sides of the ocean, indicating that coastlines had actually been in contact as recently as a few hundred million years ago. In the 1920s and 1930s, South African geologists added maps of distinctive glacial deposits in South America, Africa, Australia, India, and Madagascar, indicating that these areas were once closer together and grouped near Earth's southern rotational pole. Also in the 1930s, geophysicist F. A. Vening Meinesz's gravity measurements led him to suggest that convection currents in the upper mantle (that is, asthenosphere) drag the floating continental blocks along, causing drift.

Still, the continental drift theory remained extremely controversial, and as recently as the early 1960s, most geologists simply did not believe it. Aside from an inbred belief in *terra firma,* many argued that drag forces on moving continental blocks would be enormous and that no energy sources in Earth were sufficient to push the continents along.

Direct mapping of the ocean floor led to a revolution in geological thinking, in which the theory of continental drift was somewhat modified and finally accepted. In the late 1940s, oceanic expeditions discovered that the ocean floor was not simply a submerged continental landscape; instead, it was marked by major fractures, and shallow earthquakes concentrated along winding **oceanic ridge** systems (see Figure 8-8), where fresh lavas were being erupted and sediments were rarely older than 100 My. This finding showed that the ocean floors are young. At the same time, geophysicists recognized that heat flow at the oceanic ridges was several times higher than elsewhere in the ocean. There could now be little doubt that magma is ascending from the mantle beneath oceanic ridges, erupting, and forming new seafloor, while old seafloor is being dragged away from the ridges in both directions by subcrustal currents (Figure 8-16). Iceland, which is an exposed bit of the mid-Atlantic ridge, dramatically shows the evidence of being split by **rift valleys** created by horizontal stretching (Figure 8-17).

A pivotal meeting of the Geological Society of America in San Francisco in 1966 was the final turning point in acceptance for the combination of these ideas into the theory of **plate tectonics,** the theory that Earth's lithosphere is broken into several plates that move relative to each other (Hurley, 1968). This theory explained many hitherto puzzling features of Earth and emphasized that today's geography represents the configuration of the plates occurring in just the last few My. As summarized in Figure 8-18 on page 216, most present continents were associated, 300 My ago, in a protocontinental land mass known as **Pangaea** (pan-GEE-a). Two major portions were Gondwanaland in the south and **Laurasia** in the north. Evidence includes glacial striations marking south polar glaciation in Gondwanaland, and jigsaw fits of different-aged provinces (X's and dots in Figure 8-18) in South America and Africa. Pangaea was rifted apart when spreading started in sites such as the modern Mid-Atlantic Ridge. The Atlantic's age is generally less than 100 My— 2% of the age of Earth! Similar spreading is still rifting modern continents—for example, in East Africa and along the Sea of California.

Plate tectonic theory explained mountain belts as areas where plates were colliding and crumpling their leading edges. The types of mountains produced depend on the collision speed of the continental and seafloor units, which range from about 4 to 10 cm/y. Among the features produced are **trenches,** the lowest suboceanic spots on Earth's surface, which are created when the seafloor is dragged downward as it descends under a continent (Figure 8-19, p. 217). The fact that deep earthquakes are landward of shallow quakes (Figure 8-8) is also explained by the angle of descent of the seafloor crust under the continental crust (Menard, 1969; Dewey and Bird, 1970; Marsh, 1979).

The theory of plate tectonics views Earth's lithosphere as analogous to a thin scum of wax floating on heated water. The water is not boiling but is stirred by convection currents that break the scum into units, or **plates,** that move independently. On Earth, the plates are large units of the lithosphere. Some of the plates are continental masses, but others are large seafloor units. Rifting of plates occurs along mid-oceanic ridges and other related faults.

What drives the tectonic motions? Probably the lithospheric plates are driven by convection currents in the asthenosphere and/or larger-scale currents in the mantle, which deforms by plastic flow over long periods of time. In Earth, the adiabatic temperature gradient is about 0.2 K/km, whereas the observed temperature gradient in

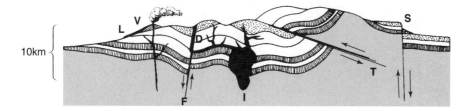

Figure 8-15. Schematic cross section of a terrestrial mountain belt, showing folds and tectonic structures. L = lava flow; V = volcano; F = normal fault; D = dike intruded along the fault; I = intensive body; T = thrust fault; S = scarp (cliff) formed along fresh normal fault.

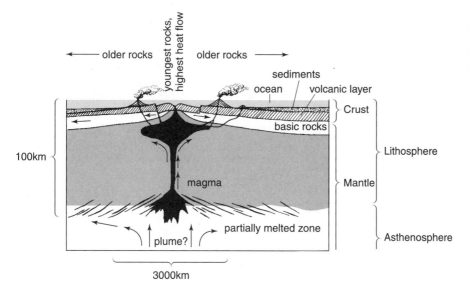

Figure 8-16. Schematic cross section of a terrestrial oceanic ridge. The central rift, high heat flow, and the increasing age of rocks outward from the ridge are accounted for by ascending, outwardmoving magma currents. Note exaggeration of vertical scale.

Figure 8-17. Example of rift faulting. The Mid-Atlantic Ridge is exposed in Iceland, which is being split apart by parallel fissures. This view looks across one such fissure, about 11 m wide, that exposes layers of basaltic lava formed by recent eruptions along the ridge. (Photo by author)

the crust is about 3 to 30 K/km. Thus, the observed gradient is super-adiabatic, favoring convection. Although rocks seem to have a great deal of strength, we have already seen that if they are stressed long enough, they can deform by flowing. Thus, it appears theoretically likely that slow convection currents exist in the relatively fluid asthenosphere. Seafloor spreading rates suggest that the currents have velocities of a few centimeters per year.

Some long-lived regions of high heat flow on Earth's surface, such as the "hot spot" under the Hawaiian islands, are thought to mark columns of hot or molten material ascending from fairly deep in the mantle. These are called **mantle plumes.** In the case of the Hawaiian Islands, a lithosphere plate has apparently slid to the northwest across the top of a plume. The plume produced magmas that occasionally erupted to form volcanoes.

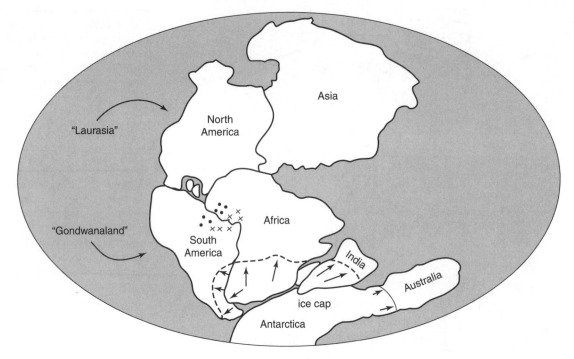

a 300 million years ago

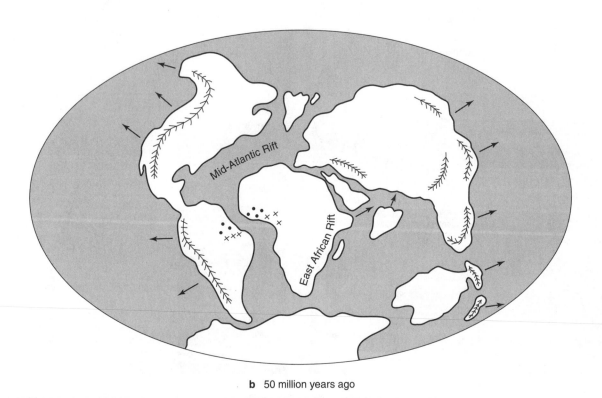

b 50 million years ago

Figure 8-18. (**a**) Configuration of continental masses roughly 300 My ago at the beginning of the breakup of Pangaea. See text for further discussion. (**b**) Configuration roughly 50 My ago as continental masses were drifting toward their present positions. Mountain masses are shown along leading continental edges, at collision zones between plates.

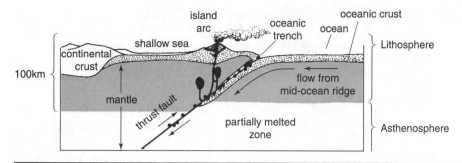

Figure 8-19. Cross section of the collision zone between continental and oceanic plates. The oceanic lithospheric plate, moving outward from the mid-oceanic ridge, slides under the lighter continental plate. Blobs of heated oceanic crustal material rise as magma and produce island arc volcanism. Dots indicate earthquake foci. (After Marsh, 1979)

Instead of piling up into one enormous volcano (as happened on Mars, for example), each intermittent eruption formed a volcanic mass—an island—on the part of the crustal plate occupying the "hot spot" over the plume at that time. As the plate slid by, various islands formed, with the islands at the northwest end being oldest. The youngest island (Hawaii itself) is at the southeast end, and a subterranean volcano lies off the southeast coast of that island, where a new island will someday be born.

Researchers are uncertain about the long-term history of the plate motions. Did convection become more vigorous in the last few hundred million years and thus cause a sudden breakup of a previously intact Pangaea? Or was Pangaea just a temporary conglomeration of earlier continental blocks that had crashed together? Some 3.7- and 2.9-Gy-old belts of "greenstone" granitic crust in Canada and Australia have been interpreted as evidence of ancient plate collisions (Windley, 1976), and other eroded mountain chains have been interpreted as island arc or mountain zones at the sites of ancient plate collisions. Ben-Avraham (1981) describes bits of old continents embedded inside today's continents and seafloors. Thus, there were probably some plate motions long before the "recent" continental breakup 300 My ago.

Going back even further, several researchers have suggested that the primeval intense cratering of Earth, 4,500 to 4,000 My ago, may have blasted holes in the crust at large impact sites and piled up crustal ejecta in other regions. This activity may have led to the first well-defined continental blocks (Goodwin, 1976; Frey, 1977). As mentioned earlier, ancient zircons indicate crustal rock formed by 4400 My ago (Wilde and others, 2001).

Venus: Earth's Enigmatic Sister

Because Venus is about the size of Earth and located close by, we expect it to have a roughly similar radiogenic heating history, with similar bulk thermal and insulating properties, and a similar thermal evolution (Figure 8-20). The theory of the condensation sequence suggests less volatiles and sulfur in Venus than in Earth, implying that

Venus's core might have less iron sulfide. Because of the planet's smaller size and lower internal pressure, Venus's inner solid core may be smaller than Earth's. These facts, as well as Venus's slow rotation, may help explain the weakness of Venus's magnetic field (Taylor, 1992, p. 215; Fuller and Cisowski, 1987).

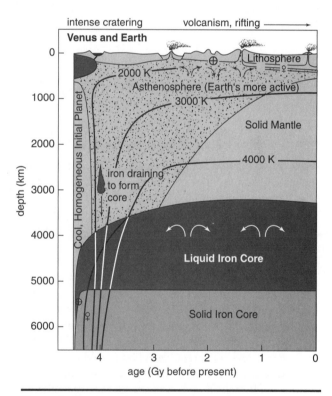

Figure 8-20. Schematic guide to calculated thermal histories of Venus and Earth. Curves show thermal profiles at any given moment in time. For example, the initial condition includes assumed near-surface melting (black). Stippled area indicates partial melting. During accretion, heating melted Earth's interior, causing metal to drain to the center and form the core. Venus's core is estimated to have formed somewhat later. Venus's lithosphere may be thicker than Earth's due to its lower water content (see text), thus making its surface geology less active. (Adapted from calculations by Hsui and Toksöz, 1977; Warner and Morrison, 1978)

a

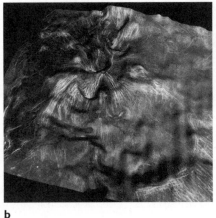

b

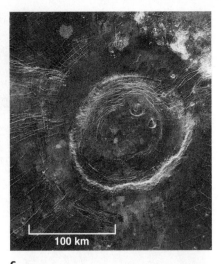

c

Figure 8-21. Three examples of coronas in two regions of Venus. Each region is shown in a near-vertical radar view and a computer-constructed oblique view showing exaggerated relief (using radar altimetry data). Some scientists believe these coronas show an evolutionary sequence. Rachel corona (upper left of [**a**] and [**b**]) may represent an early stage. It is an uplifted volcanic complex, radially faulted, with collapse in the center. Bhumidevi corona ([**c**] and [**d**]) may be an intermediate case, where the whole central area has collapsed, leaving a raised ring with circular and radial fractures. Rebecca corona (lower right of [**a**] and [**b**]) may be an old example, where relief almost totally relaxed back into the rolling terrain. (NASA Magellan radar images, courtesy of Daniel M. Janes, Cornell University, and Jet Propulsion Laboratory)

d

Although Venus was expected to have an internal history similar to Earth's, the surface geology shows some dramatic differences. Most important, there is little evidence for plate tectonics, large crustal motions, or continental-scale compressional mountain ranges formed by plate collisions. As described in Chapter 2, most of the surface is rolling volcanic plains, broken by a few continentlike areas and a few tall volcanic peaks. Unlike Earth's mountains, the tall Venusian volcanoes are believed to be too high to be supported isostatically by low-density roots, because Venus's lithosphere is warm and plastic (Phillips, Grimm, and Malin, 1991). The nature of the support, age, and stability of the volcanoes is still at issue (Kaula, 1995).

Venus's surface is mostly volcanic and basaltic, similar to Earth's oceanic crust. Lava flows are shown in Figure 2-7. Soviet landers Venera 8-10 and 13-14, and VEGA 1 and 2 made rock composition tests of varying quality on the plains near the equator, some closer to uplands than others. Generally, Veneras 9, 10, 13, and 14 and VEGA 1 and 2 found roughly basaltic compositions. Venera 8 found a more granitic composition. (See Chapter 10 for more detail.) The general feeling is that some differentiation has occurred but not as much as on Earth's surface.

Various workers have tried to interpret Venus's interior from these facts, but the literature suggests that we need more information, such as seismic probes, to make

much progress. For example, many workers believe that because of the compositional zoning of the solar system, Venus got less water than Earth to begin with (see commentary by Grinspoon, 1987). Using this logic, Warner and Morrison (1978) concluded that a lack of water in mantle rocks would raise their melting temperature, creating a lithosphere twice as thick as Earth's. On the other hand, Kaula (1990) and Phillips and others (1991) conclude that for likely compositions, the hot surface would keep the mantle warm and create a lithosphere only about 100 km thick, similar to or thinner than Earth's. Phillips and Hansen (1998) subsequently proposed that the early lithosphere was thin and deformed by plumes, but that it recently thickened, about 1 Gy ago.

If the volcanic peaks are too high to be supported by the lithosphere, they must be recent, temporary lava accumulations, supported by convective flow. Kaula (1995) cited estimates of a few 100 My for the time needed for high peaks, like Maxwell Montes, to slump and subside if they were not actively supported. This assumption implies that active volcanic upwelling and eruption is building the mountains.

Plumes may cause uplift as they hit the underside of the lithosphere. High volcanic mountains may mark long-lived plumes. Shorter-lived plumes may explain coronas (Figure 8-21). Coronas are often filled with lava. Belts of chaotic, folded, and faulted mountains around some of the high areas may be the result of compression caused by slumping of material off uplifted areas over ascending plumes. On the other hand (showing the perversity of Venus studies to date), some workers assert that the same features are located over areas where convecting mantle material flows *downward*.

Widespread fractures show that Venus's crust and lithosphere are stressed, even if not as mobile as Earth's. Kaula (1990) and Phillips and others (1991) agree that while subduction drags segments of Earth's crust back into the mantle for further differentiation, Venus lacks subduction; hence, instead of a few very differentiated, granitic continental blocks as occur on Earth, it has accumulated a relatively uniform global crust without much plate motion, composed of mildly differentiated basalts. More proof will come from future measurements.

The Moon: A Low-Density World That Cooled Rapidly

The moon's interior lacks the clear structure of Earth's interior because most of the moon resembles Earth's upper mantle rocks and has only enough iron to form a small metallic core. Seismic evidence is inconclusive, but the lunar moment of inertia and the small, ancient lunar magnetic field suggest a small core that formed 3500 to 4500 My ago. A core of nearly pure iron may have a radius of 300 to 400 km; a core rich in iron sulfide could reach 500 to 600 km (Solomon, 1979).

Seismic records indicate that the lunar crust is about 60 km thick and composed of anorthosites and anorthositic gabbro, which is made up mostly of low-density feldspar minerals. As shown in Figure 8-22, seismic wave speeds in layers shallower than 25 km suggest that the near-surface rocks are intensely fractured by impacts. From the 60-km base of the crust to the core, at about 1300 km down, the rock probably resembles a dense gabbro rock, rich in olivine and similar to Earth's mantle rocks.

These data suggest that the moon started out with an Earth-mantlelike composition and then differentiated during very early melting or partial melting. However, it never underwent such efficient churning and differentiation as did Earth.

Lunar researchers quickly recognized that the 4400–4100 Gy-old lunar crustal rocks are the low-density feldspars that would be expected to float to the surface if the outer layers of the moon had been molten initially. The initial magma ocean may have been several hundred kilometers deep, as shown in the calculated thermal history of Figure 8-23. Consistent with the magma ocean concept, Warren and Wasson (1979) made a special search among lunar samples for "pristine" crustal fragments and found eight that indicated an initial lunar crust composed of feldspar-rich gabbroic rocks: namely, anorthosites, troctolites, and norites (see Chapter 9 for further discussion of rock types).

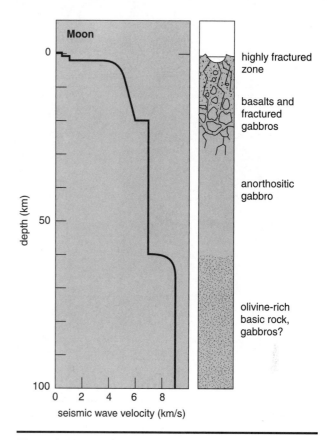

Figure 8-22. Seismic velocity profile of the moon, measured by Apollo equipment. The interpretation is shown at right. (After data from Taylor, 1975)

If the moon formed as the result of a giant impact on the nearly complete Earth, it probably assembled itself from orbiting debris very rapidly, in millenia or less; this extremely fast accretion would have caused strong surface heating and made magma ocean production unusually efficient on the moon. The magma ocean quickly cooled as lunar accretion ended, forming an impact-shattered, rocky skin that thickened until the ocean disappeared, after perhaps 400 My or less. At the same time, radioactive atoms inside the moon released heat that accumulated until partial melting occurred, probably about 400 km down, around 4000 My ago. So the magma ocean gave way to a deeper source of molten material, which is believed to have been the source of the mare lavas, with their typical ages of 3800 to 3200 My, as shown in Figure 8-23.

Because the moon is quite small, the ratio of heat-radiating surface to heat-producing volume is large, and the moon cooled much faster than Earth. The lithosphere thus thickened to its present 1000-km depth, too thick to allow plate tectonic activity (cf. Figure 8-23). The primeval surface features such as craters and lava plains were thus preserved and not disrupted by mountain formation or rift fractures. The lunar surface lacks obvious compressional or extensional faults and thus did not experience major expansion or contraction. This lessened activity constrains thermal models and precludes an

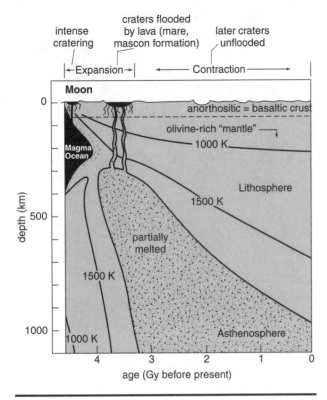

intense cratering
Expansion → ← Contraction →

craters flooded
by lava (mare,
mascon formation)

later craters
unflooded

Figure 8-23. Schematic guide to the calculated thermal history of the outer part of the moon. As the magma ocean cooled, radioactivity in the interior caused temporary partial melting at greater depths. Due to the moon's small size, rapid cooling produced a thick lithosphere (see text). (Adapted from calculations by Solomon and Head, 1979; Hsui and Toksöz, 1977; Herbert, Drake, and Sonett, 1978; and others)

initially molten moon that cooled and contracted after formation (Solomon, 1979).

Although the paleomagnetic studies indicate a weak magnetic field when the lunar maria formed as early as 3800 Gy ago, the thermal models together with geological data suggest that the moon's interior may not have melted much before that. Thus, the lunar core may not have formed long before that time. Possibly the lunar magnetic field never got very strong because the core was not large or hot enough to develop the vigorous convective currents needed to drive the electromagnetic dynamo.

Mercury: Moonlike, but not Entirely Moonlike

Mercury is moonlike in its cratered appearance and lack of atmosphere, but it has a much greater density. Seismic data and moment of inertia figures are lacking, but four lines of evidence help us hypothesize about the interior (Solomon, 1976; Nelson, 1997): (1) The high density indicates a large iron content of about 60% ± 10%, much higher than Earth's and the moon's. This density may result from a giant impact that blasted away some mantle

silicates. (2) The magnetic field suggests that the iron has differentiated into a hot, convecting core. (3) Radioactivity consistent with the condensation theory would also require that Mercury has melted and differentiated. (4) Surface features indicate extensive volcanic flooding and a type of compressional faulting not seen on the moon. These activities suggest planetary contraction.

Figure 8-24 shows a theoretical model of the thermal history of Mercury. The time of core formation depends strongly on the assumed conditions. The draining of such large amounts of iron to the center released substantial potential energy in the form of heat, causing expansion and also adding to the radiogenic heat that produced the initial internal melting. Thus, the whole mantle may have became molten (Solomon, 1976).

Contraction began as the planet cooled after core formation, and the models predict a decrease in radius of around 2 km. This would compress the surface and cause **thrust faults**, a type of fault in which one rock unit slides up over another. As shown in Figure 8-25 on page 221, such faults have actually been discovered and indicate a contraction of about 2 km (Strom, Trask, and Guest, 1975). Thus the calculation and the observation agree.

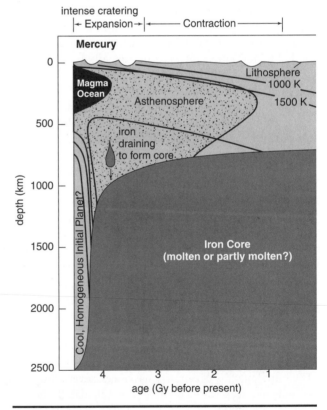

Figure 8-24. Schematic guide to the calculated thermal history of Mercury. This model is less constrained by observations than lunar and terrestrial models. A massive core (about 75% of the planetary radius) must have formed relatively early during the intense accretionary cratering. At the same time, the entire mantle was melting. Due to rapid cooling, the lithosphere has reached great thickness. (After Hsui and Toksöz, 1977; Solomon, 1976)

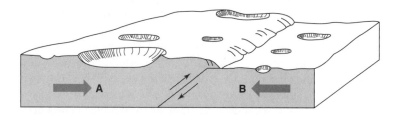

a

b

Figure 8-25. (**a**) Cross section of a large thrust fault produced by compression of the planetary lithosphere (open arrows) accompanying substantial planetary contraction, as is believed to have occurred on Mercury. (**b**) A portion of the "Discovery Scarp" on Mercury, a feature believed to have been created by compressive thrust faulting, described in part (**a**), during planetary contraction after core formation. The portion of the fault shown is about 360 km long, and the largest crater through which the fault passes (upper center) is about 55 km across. (NASA)

According to some models, the asthenosphere has disappeared entirely, so that Mercury's lithosphere reaches hundreds of kilometers down to the iron core, a condition that would be consistent with the absence of recent eruptions or plate activity. Most analysts believe that the generally high temperatures of Mercury have helped keep the core molten, thus aiding production of the magnetic field.

Mars: The Intermediate Case

Mars is intermediate in size between the small terrestrial worlds (the moon and Mercury) and the large ones (Venus and Earth). It appears to be intermediate in surface and internal properties, too, thus shedding important light on the relationships among the planets. One hemisphere is dominated by ancient, eroded impact craters and the other by sparsely cratered lava flows, relatively recent volcanoes, and huge rift valleys.

Mars Global Surveyor data indicate that Mars has an iron core with radius 1520 to 1840 km, about 45% to 54% the radius of the planet (Yoder et al., 2003). This core size and density implies it is rich in low-density contaminants, such as iron sulfide and magnetite. Mars Global Surveyor measurements also showed that the ancient parts of the crust are magnetized while the younger parts are not, implying an early dipole magnetic field, but that it shut down in the first few hundred My. Martian meteorites as well as crater counts on volcanoes suggest that some Martian volcanism has occurred in the last 20% of the planet's history, and that some volcanoes are possibly still intermittently active (Figure 8-26, right). Discussion of the evolution of Mars' interior will be deferred to Chapter 13, which treats Mars as a case study in planetary science.

Icy Satellites of Outer Planets: Overview

The American geochemist John Lewis began studies of the interiors of the outer plants' satellites in 1971. From their low mean densities and from condensation chemistry, he predicted that they consist primarily of ices and silicates. Lewis (1971a, 1971b) suggested that with even minimal heating, such satellites would form rocky or "muddy" cores and icy surface crusts. The mantles, composed mostly of

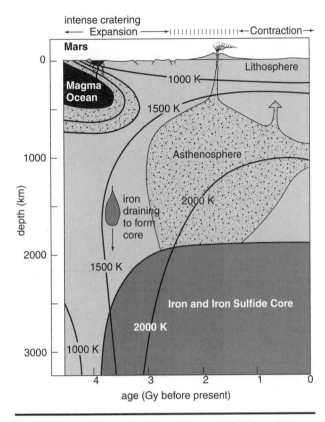

Figure 8-26. Schematic guide to the calculated thermal history of Mars. Calculations are model dependent but suggest early melting of iron, which worked its way to the center. Silicates melted later at a higher temperature, producing an asthenosphere and a volcanic era. This era may have been short-lived, possibly having already terminated. (Adapted from calculations by Hsui and Toksöz, 1977; Solomon, 1979; Toksöz and Hsui, 1978)

watery material, might be either frozen or liquid, depending on the thermal state.

Early spectral evidence generally supported this picture, revealing absorption bands of water frost on major satellites such as Europa, Ganymede, Rhea, Dione, Tethys, and one side of Iapetus; and close-up studies by the Voyager and Galileo spacecrafts have spectacularly confirmed the view that some of these satellites have icy crusts but warm and/or melted interiors.

Early thermal history calculations for such worlds (Consolmagno and Lewis, 1977) showed that the degree of melting, in the absence of outside influences, would be strongly dependent on the satellite's size and composition. A typical bulk composition for a Jupiter satellite condensing at 160 K would be about half C1 carbonaceous chondrite and half water ice by mass, a composition that would lead to substantial internal melting at diameters larger than 2000 km and some melting at diameters of 1300 to 2000 km. The presence of ammonia ices would reduce the melting point. These models suggested cores in such worlds consisting of hydrous silicates, an extensive asthenosphere of ammonia-rich liquid water, and a thin "lithosphere" of ices (if we may apply that term with its Greek root meaning "rock").

A new twist was added in 1979 when California researchers S. J. Peale, P. M. Cassen, and R. J. Reynolds, and also Charles Yoder, established the effect of tidal heating in Io, as explained in Chapter 3. The presence or absence of tidal heating thus turns out to be the dominant effect in establishing the character of satellite interiors. Io, Europa, and Ganymede are locked in a three-way resonance relation that forces eccentricity of their orbits and uses tidal heating with maximum effect on Io. The Voyager flybys of Jupiter in 1979 confirmed tidal heating and volcanoes on Io. The two Voyagers, followed by the Galileo Jupiter orbiter in the mid 1990s, began an era of intense geologic exploration of the Galilean satellites. The general history of this exploration was a transformation of these worlds from imagined dormant iceballs to worlds of unexpected geologic variety and even activity. To see the pattern, let's consider the Galilean moons from outermost to innermost.

Callisto

Callisto, the outermost and second biggest of Jupiter's four large moons, is lowest in density. It is too far from Jupiter to have significant tidal heating and its surface is a relatively unbroken panorama of craters with virtually no large-scale fracture systems or young, crater-poor resurfaced units (Figure 8-27). These facts suggest that Callisto has the highest ice content and least active internal structure of Jupiter's large satellites. Schubert, Stevenson, and

Figure 8-27. Callisto, seen in a composite view utilizing an ultraviolet image that enhances contrast variations in surface materials. The dark surface soil is believed to be depleted in ice due to micrometeorite "sandblasting," but the craters eject brighter, more ice-rich underlying icy material. (NASA, Voyager 2)

Ellsworth (1981) proposed a never-melted, virtually un-differentiated interior, though earlier models call for differentiation with a rocky core. Initial Galileo data showed no magnetic field, probable lack of any core, and an estimated composition of 40% compressed ice and 60% rocky material, with solar elemental abundances including iron (Anderson and others, 1997; Khurana and others, 1997). The data suggest that the undifferentiated model is close to the truth (see Figure 8-28).

Ganymede

The largest Jupiter satellite is the opposite; it is highly differentiated. Ganymede surprised scientists on the Galileo mission by having a magnetic field, suggesting an active interior with a distinct core structure. Current models call for a differentiated interior with a lunar-size rocky core,

containing an unexpected metallic inner core that may be the source of the magnetic field. The rocky/metallic core is overlaid by a massive icy mantle (Figure 8-28).

Initial calculations suggested that Ganymede is far enough from Jupiter that it probably had only modest tidal heating. However, Tittemore (1990) showed that as the Ganymede, Europa, and Io orbits evolved into their current three-way Laplace resonance, they may have had high-eccentricity episodes that could have produced tidal heating in Ganymede (Stevenson, 1996). Among Ganymede's broad, cratered, old regions, there are swaths of fractures (Figures 8-29, 8-30, p. 224), which indicate large stresses and even incipient plate tectonic activity. Ganymede's fractures may have been formed by the jostling of icy plates, one against another. The bright, ridged bands may have formed as water erupted or fresh ice squeezed up between plate margins. Figure 8-30a even shows an offset fault

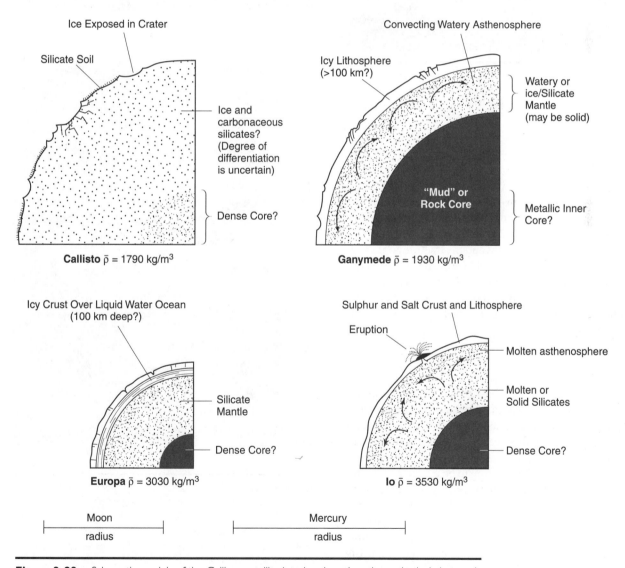

Figure 8-28. Schematic models of the Galilean satellite interiors based on thermal calculations and data from Voyager and Galileo probes. (Adapted from data reviews by Torrence Johnson, plus recent references mentioned in text)

Figure 8-29. Global view of Ganymede showing old, dark, cratered crust and bright, highly fractured regions. Freshest impact craters, in general, are the brightest. The surface is mostly ice. (NASA, Galileo spacecraft)

a

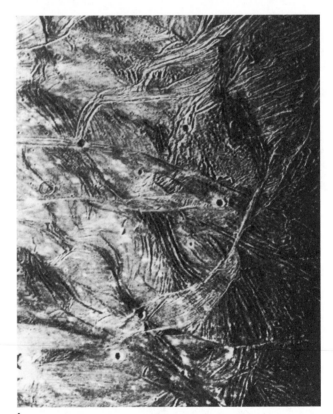

b

Figure 8-30. Fracture systems on Ganymede. (**a**) Bright, 150-km-wide band of fractures crossing an older Callisto-like cratered terrain. Note offset in broad fracture band by a narrower fracture band in lower left. (**b**) Close-up of fracture patterns. The frame width is about 580 km, and the smallest features are about 3 km across. (NASA, Voyager)

similar to those that occur where Earth's plates are offset. The chaotic fracture systems shown in Figure 8-30b are not unlike the fracture patterns observed in the ice pack "plate" that floats on Earth's Arctic Ocean. Impact craters superimposed on both the older regions and the fracture zones suggest all these features are fairly old (probably 3 to 4.4 Gy), so that Ganymede's geological activity or "plate jostling" may have been mostly confined to a short period of maximum heating, when a thin ice "lithosphere" overlay a watery asthenosphere.

Europa

The two Voyager probes established that Europa's interior had been heated enough to resurface the moon with smooth, young, bright ice plains, seen in Figure 2-22 and the opening image of this chapter. Initial dynamical calculations suggest that Europa is only marginally close enough to Jupiter for tidal heating to be effective, but later calculations indicated that the three-way resonance with Io and Ganymede may have forced tidal heating that melted the interior. This heating has allowed water to erupt and resurface the moon with young ice plains. Promptly after the Voyager data were received, scientists proposed a largely rocky interior, overlain by an ice-water layer 75 to 150 km deep (Smith and others, 1979). A liquid

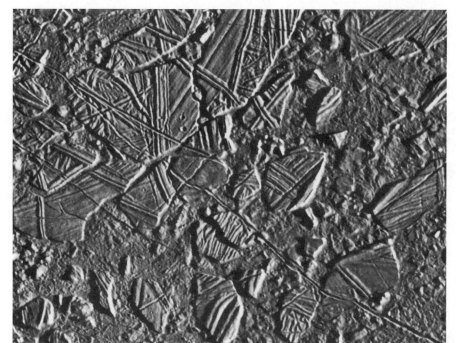

Figure 8-31. The Europa ice pack breaking into massive icebergs. Close study of this image reveals the history of this region. An initial thicker section of the ice crust (upper left) was intensely fractured (criss-crossing pattern of fracture lines). In some places, the fractured pieces broke off (along diagonal fracture line, upper center to lower left). Pieces of the old crust can be seen drifting away. As fast as pieces broke off, new solid surface formed, with a hummock texture. At least one subsequent fracture has broken both the old and the new surface (thin fracture line, upper left to lower right corner). The scarcity of impact centers indicates that the surface is much younger than most planetary surfaces. (NASA, Galileo spacecraft)

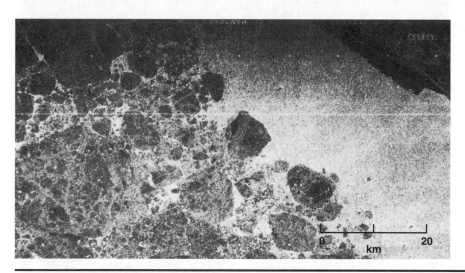

Figure 8-32. Fractured ice pack (left) in Arctic Ocean displays fracture patterns similar to those seen on Europa, which may also involve icy slabs floating on a watery substrate. Dark zone in the upper right is the open sea. Compare with chapter opening image, lower right, and broken ice blocks in Figure 8-31. (NASA, Seasat satellite radar image)

water ocean under a thin ice crust seemed likely. Arthur C. Clarke promptly suggested the suitability of such an ocean for life forms in his novel *2010*.

The Galileo probe, which orbited Jupiter in the mid-1990s, affirmed an ocean estimated to be 100 to 200 km deep (Carr and others, 1998). Impact crater central peaks show that the icy crust must average >3 km thick (Turtle and Pierazzo, 2001); it may reach tens of km thickness (Proktor and Pappalardo, 2000) but local thin spots (<1 km?) are possible. The crust has a jigsaw-puzzle surface of shattered ice blocks, resembling Arctic ocean ice flows (Figures 8-31 to 8-33). The sparse numbers of craters, plus estimates of the rate of impact by comets and asteroids (Shoemaker, 1996), suggest older surfaces are 10 to 100 My old, but that intervening surfaces may be as young as a few My. These observations imply that the resurfacing is an ongoing, contemporary process and that liquid water has access to the surface at some points. The Galileo specrometer found that the dusky brownish markings show spectra of hydrated salts such as magnesium sulfate and sodium carbonate. This suggests the dusky markings along fractures are evaporites from water

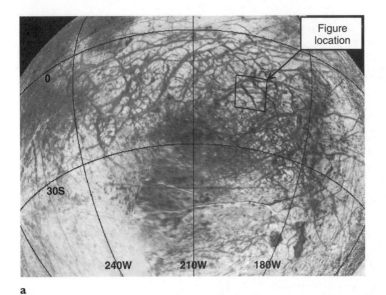
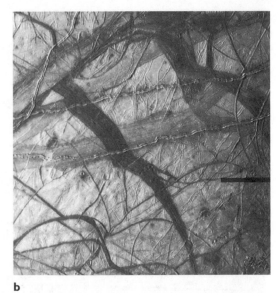

a

b

c

observed
configuration

reconstruction

50 km

Figure 8-33. Solving the Europa jigsaw puzzle. (**a**) Global view of Europa showing location of detailed figure. (**b**) Close-up view of present-day fractured system. The darker bands mark fractures that have separated the brighter platelike masses of ice. Contrast is enhanced in Figures **a** and **b**. (**c**) Reconstruction of ice plates in Figure **b**. Fracture spaces have been filled in with black (left) and then the plates have been reconstructed into the original surface (right). This shows dramatically how Europa's ice crust is constantly fracturing and ice plates separating into new positions, with fresh ice filling in the fractures. (NASA, Galileo images courtesy Robert Sullivan, Cornell University; see Sullivan and others, in press)

eruptions, and that the ocean water is salty, with dissolved minerals (McCord and others, 1998).

Galileo gravity and magnetic measurements suggest that the silicate rocky interior contains a denser, sulfide-rich metal core (Kargel, 1998). Pappalardo and others (1998) suggest that circular deformation features in the surface are evidence of solid-state rheid convection in the ice crust itself.

Io

In contrast to all other moons of outer planets, little or no ice is present in Io—a conclusion supported both by the high density of 3530 kg/m³, the surface composition of sulfur-rich, non-icy compounds, and the active volcanism (Figure 2-23, Chapter 2 opening figure). Even primordial Io may have been deficient in ice because of the high temperatures in the proto-Jupiter nebula at Io's distance (see Figure 5-12). Consolmagno (1981) estimated that Io started with approximately C2 or C3 carbonaceous chondrite composition. Whatever ice may have been incorporated into Io initially has been heavily depleted by the heating and volcanic emission of the volatiles. Proposed models of Io's interior have a thin sulfur crust overlying a silicate mantle. In one such model, the sulfur crust has its own solid surface layer, or "lithosphere," perhaps only a kilometer thick, riding on a fluid asthenosphere of molten sulfur and sulfur dioxide. This in turn rides on a solid silicate lithosphere that overlies the deep, molten silicate mantle (Smith, 1979).

Many volcanoes on Io (Figure 8-34, p. 228) have colors matching those of molten sulfur eruptions (black at high temperature, cooling to redder and yellowish colors further from the vent). However, Io's measured lava temperatures are far above the 715 K melting point for sulfur in a vacuum. Many Io lavas have temperatures as high as 1700 to 2000 K. This suggests silicate lavas (melting temperatures 1300 to 1450 K) (Johnson and others, 1988; Johnson, 2000). Veeder and others (1994) established a global heat flow of around 2.5 W/m² or more—high enough to confirm tidal heating as the only viable mechanism. Schubert and others (1981) calculated that convection in the solid interior carries most of the internal heat, so that only a 100-km-thick asthenosphere is molten in their model. Underlying it is a solid, thick mantle and core.

Heating and eruptions exhausted water from the crustal layers, leaving a crust rich in sulfur and oxidized iron. Galileo orbiter data allowed refinement of interior models. Analyzing Io's gravitational effect on Galileo during its first Io encounter, Anderson and others (1996) reached the "inescapable conclusion" that Io has a large metallic core that ranges from 20% of its mass (FeS-rich core) to 10% of its mass (pure iron core). Core radii in these two cases would be 52% and 36%, respectively.

Satellites of Saturn

Saturn's moons (except for huge, haze-covered Titan) are smaller than the Galilean satellites. They have icy surfaces and icy interiors. The moons Mimas, Enceladus, Tethys, Dione, Rhea, and Iapetus have densities about 1200 to 1400 kg/m³. Even giant Titan has a density of only 1880 kg/m³. These densities are so low that they imply large portions of ice in the interior. For example, models of Rhea and Titan by Lupo and Lewis (1979) called for about 66% ice by mass and small rocky cores with about half the radius of each moon.

What is the state of internal heating in such ice-balls? Peale, Cassen, and Reynolds (1980), who successfully predicted the tidal heating and volcanism of Io, concluded that "tidal heating is not an important contributor to the thermal history of any Saturnian satellite." The small sizes of most of these icy moons also suggested that internal radiogenic heat would be inadequate to produce geologically active worlds. For example, in models of icy satellites by Consolmagno and Lewis (1978) and Lupo and Lewis (1979), bodies smaller than 1000 km in diameter never melted; bodies smaller than 1500 km had some early melting but refroze. Even in bodies as big as 3000 km across, endogenic (internally derived, nontidal) heating was inadequate to drive surface eruptions of water. According to these models, Saturnian moons other than Titan, which are all less than 1500 km, should show old, cratered surfaces with minimal tectonic features. Most of them do.

But this is not the whole story. Yoder (1979) made a remarkable prediction. He noted that of all Saturn's satellites, the Enceladus-Dione pair* are in a tidal resonance that provided a good chance for tidal heating. Yoder wrote, "It is possible that originally [the eccentricity of Enceladus] was pumped up to some critical eccentricity which caused catastrophic fracturing. . . . The Voyager 2 flyby of Enceladus may reveal another curious satellite whose past and present state is controlled by tidal friction."

As shown dramatically by Figure 2-29, the Voyager 2 photos did reveal that 500-km Enceladus is a "curious satellite." Unlike its neighbors, it is only moderately cratered, and the cratered regions are broken by swaths of smoother plains. The surface material is the brightest in the solar system, with nearly 100% reflectivity. The appearance of the plains suggests that water erupted and formed fresh ice flows on the surface—a surprise for such a small body. From crater counts, Smith and others (1982) concluded that the younger plains units are less than 1 Gy old. Combining this result with the observed presence of debris in a thin circum-Saturn ring spread along Enceladus's orbit, some theorists have suggested that Enceladus may be geologically active even today, occasionally erupting

---

*Note that the resonance associated with tidal heating does not have to be with the neighboring satellite.

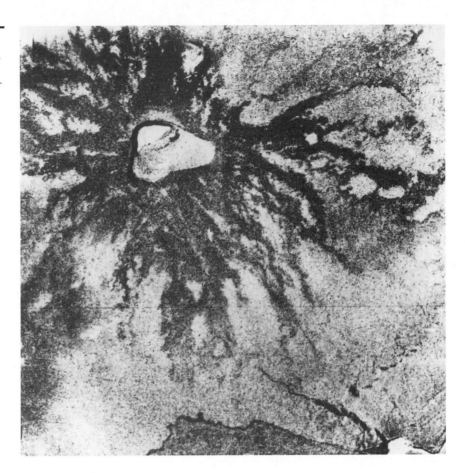

Figure 8-34. A volcanic caldera on Io, showing spidery pattern of radiating flows. The flows may be sulfur or sulfur compounds. Active calderas with molten sulfur are black; this one may be inactive and coated inside with whitish sulfur dioxide frost. (NASA, Galileo)

debris off its surface! Uncertainty remains, however, about whether the past tidal heating rate has been adequate to melt an H_2O ice body (Dermott, Malhotra, and Murray, 1988). Some alternate suggestions are that an admixture of ammonia ice (NH_3) lowers the melting point or that impacts have created localized fracture zones where the tidal heating process was maximized (Smith and others, 1982).

Even some other moderate-size satellites show features suggesting internal activity. Cratered terrains with different crater densities, hence different ages, have been found on several of the moons. As shown in Figure 8-35a and b on page 229, 1048-km Tethys has a globe-girdling rift or trough. Named Ithaca Chasma, it is an estimated 3 to 5 km deep and 100 km wide; it extends around three-fourths of the circumference. Calculations suggest that if Tethys had once been melted with only a thin icy lithosphere, the subsequent expansion during the freezing of the water could have produced a crack with the area of Ithaca Chasma (Smith and others, 1982). As shown in Figure 8-35c, Uranus's 1580-km moon, Titania, shows similar major riftlike canyons. Why the tectonic forces produce one major rift instead of a network of smaller fractures is unknown.

Miranda

Another surprise came in the Uranian system when the smallest of the five large moons, 470-km Miranda, turned out to be by far the most intensely fractured (Figures 2-32c and 8-36, p. 230). Why should such a small moon have so much deformation? The first theory, proposed by the Voyager team, noted that moons close to giant planets have a much higher impact rate than more distant moons because the planet attracts asteroids and comets toward itself. Miranda might have been fragmented and reassembled one or more times, resulting in a chunky, irregular structure. Heat released as gravitational contraction led to a more spherical shape and could have caused geologic activity (Smith and others, 1986).

What about tidal heating as a source of tectonic energy? Originally, it was thought too weak to do the job on such a small moon. However, Dermott, Malhotra, and Murray (1988) and Marcialis and Greenberg (1987) analyzed Miranda's dynamics and concluded it may have been temporarily trapped in resonances in the past, and forced into higher eccentricity orbits and/or chaotic rotation that may have led to heating. Thus, the heat sources remain uncertain, but tidal heating now seems likely.

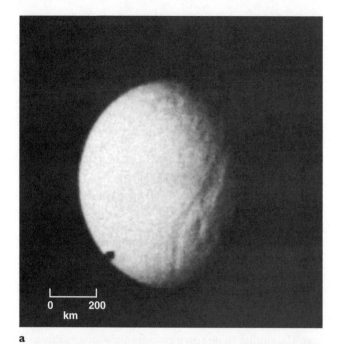

a

b

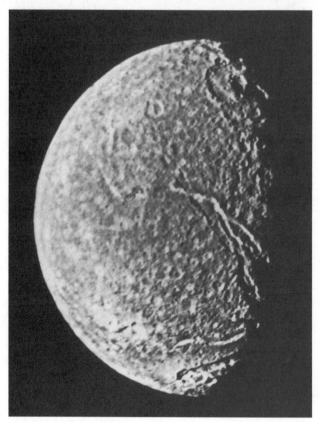

c

Figure 8-35. Fracture systems on icy moons. (**a**) and (**b**) show two views of Saturn's moon, Tethys, showing different parts of the riftlike trough, Ithaca Chasma. The portion shown in (**a**) is at the extreme left of (**b**). Crater in upper left and craters in lower left center can also be seen in Figure 2-28c, lower right. (**c**) Uranus's largest moon, Titania, also showing riftlike troughs. (**a**: NASA, Voyager 1; **b** and **c**: Voyager 2)

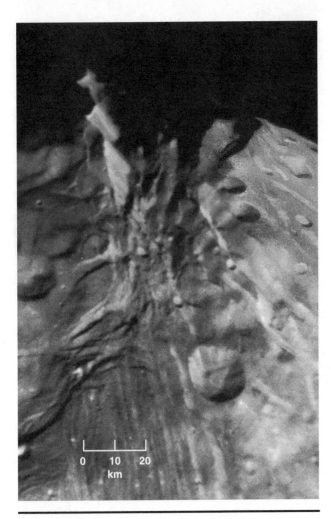

Figure 8-36. One of the largest sheer cliffs in the solar system is a fractured offset on Uranus's moon, Miranda. Seen at a steep oblique angle, the cliff follows fracture lines and extends at top onto the dark night side. It has an estimated height of 5 km (16,000 feet) and slopes down to the right at an angle close to 45°. (NASA Voyager 2)

The Giant Planets

In the early 1900s, many astronomers thought the giant planets were gaseous throughout. However, the English geophysicist Jeffreys (1924) calculated that such objects would have cooled to a nongaseous state; he suggested that Jupiter was a small, solid planet surrounded by a vast hydrogen-rich atmosphere. Wildt (1938) computed a more realistic Jupiter model that had an inner core of mean density 5500 kg/m³ (the same as for Earth as a whole), an inner mantle of ice, and an outer mantle of frozen hydrogen. Jupiter was recognized as having such a strong gravitational field that it could retain all its original complement of even the lightest gas, hydrogen. Because hydrogen is by far the most abundant material in the universe, attention focused on it as a major constituent of Jupiter.

DeMarcus (1958) then applied theoretical equations of state for the pressure-ionized metallic states of hydrogen and helium to calculate models of Jupiter and Saturn. He found that Jupiter consists of at least 78% by weight of hydrogen, distributed in various pressure-modified forms. The exact distributions depend on the initial and later thermal states (Podolak, 1978; Hubbard, 1981, 1990) and on the equation of state of the material. Hubbard (1970, 1990) applied the measures of Jupiter's heat flow to conclude that most of the interior consists of convecting liquid metallic hydrogen, as shown in Figure 8-37; this model supplied the heat flow observed in the thermal infrared spectrum and coincidentally accounted for the strong magnetic field. Such models also give a cooling rate consistent with the type of evolutionary track shown in Figure 4-8 (Hubbard, 1977).

More detailed calculations of Jupiter's interior conditions are limited by uncertainties about the equation of state of the mix of hydrogen, helium, and trace compounds. In the atmosphere, hydrogen forms a molecular gas, H_2, similar to our own gaseous atmosphere. Pressure, density, and temperature increase toward the lower atmosphere, following an adiabatic relation. If the temperature were cool enough, an ocean surface of liquid hydrogen would appear at a depth where the pressure exceeded 13 bars, perhaps 50 to 100 km below the clouds. But the temperature is too high, and the gas simply gets denser and denser, turning into a mush resembling a hot liquid at depths where the pressure exceeds 0.1 Mbar, without a well-defined "ocean surface." A major question was the pressure at which the hydrogen electron structure breaks down and yields liquid metallic hydrogen. In 1996, this was measured in the laboratory for the first time, giving a pressure of 0.9–1.4 Mbar at which the hydrogen becomes liquid metallic hydrogen, with protons surrounded by loose electrons. This result gives the depth to the top of the conducting liquid hydrogen as about 8500 km below the clouds, about half the depth estimated earlier. The central pressure is a whopping 45 Mbar.

The disposition of silicates and metals in Jupiter is uncertain. Most analysts conclude that Jupiter has lost some volatiles in spite of its strong gravity, so that the relative abundance of heavy elements is perhaps 3 times that in solar material, or the equivalent of roughly 15 M_{Earth} of silicates and metals. A few models suggest that these heavy elements could be dispersed throughout the deep interior; more likely, differentiation has allowed them to accumulate in the center along with some helium. Probably buried deep in the gaseous and liquid hydrogen bodies of Jupiter and of other giants are cores that look strangely like huge terrestrial planets (Figure 8-37). A typical Jupiter model calls for a density of 1000 kg/m³ at a depth 14,000 km below the visible clouds, and a dense core with a mass of 14 M_{Earth} compressed to 22,000 kg/m³ and a radius of 1.5 R_{Earth}, with a temperature around 20,000 K at the core-mantle boundary (e.g., Klepeis and others, 1991).

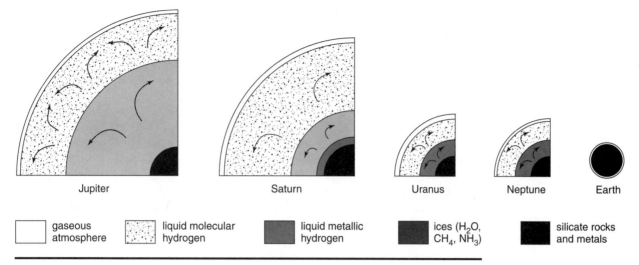

| | gaseous atmosphere | | liquid molecular hydrogen | | liquid metallic hydrogen | | ices (H_2O, CH_4, NH_3) | | silicate rocks and metals |

Figure 8-37. Schematic models of the giant planets. Some variations can be found, depending on model assumptions. (Adapted from calculations by Podolak and Cameron, 1974; Hubbard, 1981, 1990; and others)

The situation in Saturn is similar but with lower pressures, temperatures, and densities. There is probably a more complete differentiation of heavy elements into a core, perhaps due to less initial heating and less convective stirring (Podolak, 1978). A massive fluid hydrogen ocean with an ill-defined surface overlies a liquid metallic hydrogen region, containing a core with a mass of 17 M_{Earth} compressed to 10,000 kg/m^3 and having a radius of 2.1 R_{Earth}, with a temperature around 10,000 K at the core-mantle boundary (Klepeis and others, 1991). The central pressure is estimated at 10 Mbar. Ices form on the core surface, but the outer portions of Saturn are so hydrogen-rich and distended that the mean density of the planet is only 690 kg/m^3—the lowest of any planet, and low enough to float in water!

Uranus and Neptune are much denser than theoretical pure hydrogen planets of the same mass, as shown in Figure 8-2. They are richer in heavy elements than Jupiter or Saturn. Of the following two factors, one or both may have been involved: (1) They accreted smaller rocky cores than Jupiter or Saturn, and thus did not gravitationally pull in such large H/He atmospheres from the nebula. (2) Accretion was slower at Uranus and Neptune because of their slower orbital velocities, and the nebula had partially cleared by the time large rocky cores formed, resulting in less atmosphere captured. Because Uranus and Neptune are so much smaller than Jupiter or Saturn, the pressure at the base of the H-rich layers does not exceed 4 Mbar; there may be no metallic hydrogen layer (Hubbard, 1981, 1990).

Typical models of Uranus and Neptune call for rocky cores ranging from 4 to 15 M_{Earth}, with central temperatures around 7000 K and convecting icy mantles (see Figure 8-37). Pollack (1985) infers that all giants have rocky cores with masses within a factor of 2 to 15 M_{Earth}.

SUMMARY

The concept of differentiation is helpful in understanding planetary interiors. Differentiation includes any process that arranges elements in different abundance patterns than they had originally, whether in the nebular gas, inside a planet as a whole, or within a single geological setting, such as a volcanic magma chamber. Figure 8-2 shows that differentiation in the solar nebula led to planets with different initial compositions. Planets near the sun are composed of dense, iron-rich refractories, and outer planets of low-density volatiles.

Further differentiation occurred inside the planets because of heating. The smallest bodies, such as some asteroids, radiated away this heat before enough accumulated to allow even partial melting. But larger planets were better insulated by their own bulk, which prevented heat from escaping. They melted or partially melted inside, and their dense materials drained to the center to form cores. In the terrestrial worlds, iron cores formed whereas lower-density, more silica-rich materials floated to the surface to form crusts of feldspar-rich rock such as basalt, gabbro, anorthosite, and, on Earth, granite, the most silica rich of all. Icy satellites formed rocky cores and icy crusts. Tidal heating in Io and Ganymede created enough differentiation to produce iron cores. Giant planets formed rocky cores overlaid by icy layers and liquid hydrogen seas.

Understanding heat sources is crucial to understanding planetary interiors. More work in this area will be fruitful to distinguish effects of initial accretional/impact heating, short-lived radioactive isotopes, long-lived radioactive isotopes, tidal heating, and magnetic induction heating.

Cooling of surfaces led to the formation of solid, ever-deepening lithospheres, which overlay molten or partly molten asthenospheres. The thickness of the lithosphere, especially during maximum heating, had a crucial influence on the development of surface features. Small planets tended to cool fast and form thick lithospheres. Large planets and worlds heated by tides acquired only thin lithospheres.

Planets that developed thick lithospheres preserved intensely cratered surfaces. If the lithosphere was a moderately thick, planet-wide, single "plate," its surface character was influenced by whether the planet underwent significant expansion or contraction during its thermal history. Planets with moderately thin lithospheres developed more modified surfaces, with more fracturing and volcanic activity, and hence fewer surviving craters. Worlds with thin lithospheres and convecting asthenospheres (Earth and Europa being prime examples) had their lithospheres broken into plates, with resulting major tectonic shifting and deformation of the surface.

planetary interior
core
mantle
crust
mean density
state variables
equation of state
hydrostatic equation
figure
oblate spheroid
plastic flow
rheidity
rheid
mineral
rock
change of state
outer core (of Earth)
inner core (of Earth)
olivine-spinel transition
pressure ionization
metallic hydrogen
differentiation
magma
partial melting
lithophiles
siderophiles
moment of inertia
geometric oblateness
dynamical oblateness
mascon
temperature gradient
diurnal
Mohole Project
Glomar Challenger
olivine
kimberlite pipes
pyroxene
seismometer
P waves
S waves
focus
P-S arrival time interval
epicenter
Mohorovičić discontinuity
low-velocity zone
asthenosphere

fault
passive seismology
active seismology
plates
lithosphere
statics of planets
dynamics of planets
heat
temperature
heat transport
conduction
radiative transfer
convection
adiabatic change
adiabatic temperature gradient
convective cell
magma ocean
electromagnetic induction heating
feldspars
basalts
gabbros
lines of force
dipole field
magnetic correction
magnetosphere
magnetopause
bow shock
magnetosheath
magnetic field reversal
paleomagnetics
Curie temperature
remanent magnetism
dynamo theory
isostatic equilibrium
continental shield
continental drift
oceanic ridge
rift valley
plate tectonics
Pangaea
Laurasia
trench
plate
mantle plumes
thrust fault

1. Why is understanding a planet's interior helpful in understanding its surface?

2. (a) Does rock have to be molten or partly molten to deform like a fluid over long time periods? (b) Would such flow be reduced or enhanced if all the crystal lattices in all minerals were "perfect," with no microfractures of missing atoms? Why? (c) Why is flow enhanced if the rock is heated, even if not completely melted? (d) Does a rock with remanent magnetism have to be melted in order to destroy the remanent magnetism? Why or why not?

3. The highest lab pressures are around 40 kbar, a pressure reached near a depth of 100 km in Earth, but at a depth of nearly 1000 km in the moon. Explain why this means that knowledge of lunar and Martian mantles could be better than knowledge of Earth's mantle if we had equally accurate information on the chemical composition of the rocks from the two bodies.

4. (a) Compare current seismic knowledge of the mantles of Earth, the moon, and Mars. (b) When better Martian data are available, do you think Mars will be found to display more, less, or the same amount of internal seismic activity as the moon?

5. How does a crust-mantle interface differ from a lithosphere-asthenosphere interface? Define each.

6. (a) What is pressure-ionized matter? (b) In which planets is it important?

7. Explain why the soil in your backyard is not chondritic, even though Earth was made from chondritelike planetesimals and chondrites are the dominant type of "soil" falling out of the sky.

8. Spacecraft are placed in orbit around two identical-looking planets with no natural satellites. The first planet exhibits perfect Keplerian motion. The second exhibits nearly Keplerian motion, but with minor accelerations and decelerations. What do you infer about the interiors of the two planets?

9. Because the moon and Earth may have formed from very similar material, why doesn't the lunar crust have the same composition as the terrestrial crust?

10. Pretend that a sheet of paper is a map. Place several dots labeled A, B, C, and so on at random to represent seismometers. Mark an X at random to represent the epicenter of an earthquake. (a) How do observers at A determine how far the earthquake was from them? (b) With only their own data, can these observers locate the epicenter? Mark the points where their data indicate that the earthquake could be located. (c) Show that a minimum of three stations are needed to locate the epicenter precisely.

11. Explain why, if the giant impact theory of lunar origin is correct, the nature of the moon proves that Earth's core must have formed within the first 100 My of Earth's existence, or perhaps even as Earth was forming.

12. A satellite orbits at a distance of 100,000 km from a planet of 22,000-km radius. The satellite has a mass of 10^{12} kg, a circular orbit, and a period of 1 d.

(a) Find the velocity of the satellite.

(b) Find the mass of the planet.

(c) Find the mean density of the planet.

(d) Which solar system planet does this most resemble?

(e) Describe briefly some plausible characteristics of this planet's interior. Which piece of information is irrelevant (assuming that we can see at first glance that the planet is much larger than the satellite)?

13. At what depth in the ocean does the pressure equal Venus's surface pressure of 90 bars?

14. Make a diagram showing the pressure as a function of depth inside the outer 3000 km of the moon, Mars, Earth, and Jupiter. Contrast the depths at which the pressure reaches the maximum laboratory pressure of roughly 40 kbar.

15. The specific heat c of basaltic rock is the amount of energy involved in a temperature change of 1 K for a cubic meter of rock; it is $c = 2.4$ J/m^3 K. Suppose a magma ocean starts at time $t = 0$ with mean temperature 1800 K and cools with no significant input of additional heat from below or above. The ocean is 3000 km deep and has a surface crust that radiates at a mean temperature of $\overline{T} = 900$ K. Would you expect a significant cooling in 1 y? 10^3 y? 10^9 y? (a) Make an estimate of the time scale for cooling of the mean ocean temperature by 400 K. (b) Give some physical complications that require a more exact calculation in order to derive the complete cooling history from an initial molten state to today's conditions.

An astronaut's sampling of a large boulder on mountain slopes at the edge of the moon's Sea of Serenity symbolizes the exploration of planetary surfaces to learn about early surface-forming processes. Foreground soil is powdery dust and rock fragments created by meteorite bombardment of the lunar surface. (NASA, Apollo 17)

CHAPTER NINE

Planetary Surfaces 1: Petrology, Primitive Surfaces, and Cratering

Surface environments of other planets are among the most intriguing topics in planetary science. What is it like to stand on another planet? What processes formed those alien landscapes? Our formal goals in studying planetary surfaces are to describe the present-day conditions (such as rock types, surface structures, and temperature) and to understand the evolution of the surface by processes such as impact, volcanism, and erosion.

In the last few chapters, we have confronted (and sidestepped) the details of rock chemistry. Now it is time for a more detailed review of minerals and rock types. As the Reverend John Fleming noted as early as 1813, "He who has the boldness to build a theory of the earth without a knowledge of the natural history of rocks will daily meet with facts to puzzle and mortify him" (quoted by Geikie, 1905).

Petrology

Petrology is a study of rocks—of their description, classification, evolution, and origin. This vast subject is reviewed in many geology textbooks. The introduction here is intended only as a guide for students with little geological background.

Minerals

Minerals are solid inorganic substances that compose solid planetary material. They are characterized by forms determined by their constituent elements or compounds, and they are defined by composition and structure. The crystals visible in rocks are usually individual minerals. **Rocks** are merely assemblages of different minerals, usually in the form of crystals, but sometimes in glassy or amorphous forms that may occur under conditions such as rapid cooling.* Many minerals are defined by simple, fixed atomic composition (such as SiO_2, which is called silica or quartz), but they do not all have such simple formulas. For example, olivine has the compositional formula $(Mg,Fe)_2SiO_4$, which means that ions of either magnesium (Mg) or iron (Fe) may *substitute* in the crystal lattice. The governing factor in substitution is often the size of the atom or ion; the situation is analogous to a structure made of packed Ping-Pong balls in whose interstices BB's, but not golf balls, can fit. The number of substituting ions is, of course, constrained by the requirement of electrical neutrality. Many minerals thus cannot be defined by one fixed composition, but must be defined in terms of ranges of composition.

Although they are defined chemically, minerals can be visually identified in the field by a series of semiquantitative tests of such properties as hardness (on a 0–10 scale called Moh's scale), streak (color of the powdered form produced when a sample is rubbed against a small, hard tile, or "streak plate"), shape, luster, and density.

In typical rock specimens, the mineral crystals range from easily visible to microscopic. In the latter case, the rock seems homogeneous but is really a mass of the fine mineral crystals. Microscopic examination is often required to identify the various minerals present.

Minerals and rocks are sometimes loosely classified according to composition, ranging from **acidic** or **siliceous** (high silica content) through **basic** or **mafic** (high content of heavy elements such as iron or of elements with chemical affinity for iron) to **ultrabasic** or **ultramafic** (very rich in heavy elements and low in silica). A very rough rule of thumb among terrestrial planet rocks is that the darker or denser the rock type, the more basic it is. In a planet that has been at least partially molten, crustal minerals and rocks tend to be siliceous, erupted lavas tend to be basic, and deep-seated minerals and rocks tend to be ultrabasic.

The vast majority of rocks on terrestrial planets are composed mostly of a few important minerals. The most important minerals include the following:

Feldspars: $(K, Na, Ca)AlSi_3O_8$; density 2600 to 2800 kg/m^3. Feldspars are among the most important groups of minerals in planetary sciences and by

---

*Glasses are noncrystalline; that is, the atoms and molecules have not been able to arrange themselves in the characteristic orderly, geometric manner of a crystal. The atoms may occur in tangled molecular chains or mixed, incomplete crystal lattices. The difference between a crystal and a glass, as Russian-American physicist George Gamow once remarked, is like the difference between a carefully built brick wall and frozen caviar.

far the most common mineral in terrestrial surface rocks, making up about 60% of the crustal minerals (Mason, 1966) and an even higher fraction on the moon. Feldspars are silicates of aluminum (Al), with a variable admixture of potassium (K), sodium (Na), or calcium (Ca). The formula shows how K, Na, or Ca may substitute for each other in the aluminum silicate lattice, providing a range of compositions. Note that feldspars contain a majority of the most abundant elements in Earth's crust (and probably other terrestrial planet lithospheres): O, Si, Al, Fe, Ca, Na, K, and Mg. One important property of feldspars is that they are among the lowest-density silicate minerals; this is why they rise to the surface or crustal strata during any episode of melting and recrystallization on a planet. The density of about 2700 kg/m^3 contrasts with densities of 3300 to 4300 kg/m^3 for other common silicate minerals, such as olivine, spinel, and garnet, and values such as 5200 kg/m^3 for the iron-rich mineral magnetite. A consequence is that feldspars tend to float in magmas and accumulate closer to planetary surfaces than do the denser minerals. Because feldspars are so common, they are intricately subdivided. **Orthoclase** feldspars are the K-rich group, tending to occur in lighter-colored and more silica-rich rocks. **Plagioclase** feldspars are the Na- and Ca-rich group, tending to occur in darker-colored, less silica-rich rocks, although they are nonetheless common in most igneous rocks. The plagioclase feldspars are still further subdivided according to the Na/Ca ratio. The sodium-rich end of the spectrum is a mineral called albite (sometimes abbreviated Al), and the Ca-rich end is anorthite (sometimes abbreviated An). We do not emphasize these subtypes, though their names are as likely to be encountered in rock analyses as the general term *feldspar.*

Quartz: A form of **silica** or SiO_2; density 2600 kg/m^3. In a cooling magma, minerals solidify in a certain sequence, and if a cooling magma has silicon atoms left after the feldspars have formed, silica is likely to solidify. The most common form is the familiar mineral quartz. Because of its low density, quartz, like feldspars, is likely to float in magma and accumulate in surface rocks. If there is very active geological processing and differentiation, quartz may become very concentrated at the surface, because its density is even less than that of many feldspars. These conditions are especially prevalent on Earth, where surface rocks are often richer in quartz crystals than are surface rocks sampled on the moon and Mars.

Pyroxenes: A group of Mg, Fe, Ca, Na, Al, and Ti silicates, including specific minerals such as augite (probably the most common), enstatite, and hypersthene. The density is high, ranging from 2800 to 3700 kg/m^3. These minerals are common in meteorite as well as basic planetary igneous rocks. They constitute roughly 10% of Earth's crustal minerals.

Amphiboles: A group of Mg, Fe, and Ca silicates with different crystal structure from the pyroxenes. The densities are only slightly less than those of pyroxenes. The amphiboles make up perhaps 7% of Earth's crustal minerals, being common in basic igneous rocks.

Micas: K, Al, and Mg silicates, with intermediate densities ranging from 2760 to 3200 kg/m^3. They are common in igneous rocks and compose roughly 4% of Earth's crustal minerals.

Beyond about 4 AU from the sun (the outermost fringe of the asteroid belt), H_2O ice constitutes as much as 60% by mass of the material that condensed in the solar nebula. Hence, H_2O and other ices play the role of major "rock"-forming minerals on worlds in the outer solar system.

Water ice: H_2O; density 932 kg/m^3 at 100 K and 1 bar (917 at 273 K, 1 bar). Water ice dominates the surfaces of many distant moons, such as Europa, Ganymede, Enceladus, Rhea, Titania, Charon, and Saturn ring particles (Cruikshank and Morrison, 1990), and the polar caps of Earth and Mars. Water ice appears to be the main constituent in most comets.

Carbon dioxide ice: CO_2; density 1560 kg/m^3 (216 K, 5 bar). This ice sublimes at 194 K at 1 bar pressure, and is thus unstable up to about 10 AU from the sun. It is an important volatile in powering cometary activity from 4 to 10 AU.

Ammonia ice: NH_3; density 817 kg/m^3 (194 K, 1 bar). Ammonia ice has a low melting point—195 K at 1 bar—and may be a partial constituent of some outer planet satellites. Because it melts at a lower temperature than does H_2O ice, it could play an important geological role as a fluid or gas in slightly heated bodies.

Methane ice: CH_4; density 415 kg/m^3 (109 K, 1 bar). Methane ice is very volatile because of its low melting temperature. It should be stable in sunlight only on the most distant bodies. It is present on Pluto (discovered by Cruikshank and Silvaggio, 1980) and Triton (discovered by Cruikshank and Silvaggio, 1979), where it is probably mixed with frozen N_2 ice (Stone and Miner, 1989). Methane is among the first gases to sublime as a comet approaches the sun from the Oort cloud.

Some other types of minerals are commonly encountered in planetological literature:

Olivine: $(Mg,Fe)_2(SiO_4)$; density 3300 to 4400 kg/m^3. Because of its high density, olivine sinks to the lowest parts of magma volumes. It is observed to be important in materials formed at great depth, such as Earth's mantle, some meteorites, and planetary lavas. Olivine is believed to be a major component of all terrestrial planet mantles. It is usually a greenish crystal and is often found as inclusions in lavas. Because the Mg/Fe ratio varies, olivine has a variety of subtypes. A-class asteroids are very rich in olivine.

Iron oxides: Important examples: **magnetite** (Fe_3O_4) and **maghemite** (an unusual form of Fe_2O_3), which are magnetic and likely candidates for the 1% to 7% of loose Martian soil that adhered to magnets on Viking and Pathfinder landers (Hargraves and others, 1977; Hviid and others, 1997). Other nonmagnetic forms are **hematite** (the more common form of He_2O_3), **goethite** (pronounced GER-tite; $HFeO_2$), and **limonite** ($FeO\ nH_2O$). These are oxidized iron minerals, cousins of ordinary rust, and have reddish colors ranging from yellow to brown. They are believed to be the minerals that redden the soil of Mars, where they may occur as fine coatings on dust grains of other compositions. Limonite is the hydrated form, of variable composition, caused by alteration of other iron minerals, often by intermittent exposure to water.

Troilite: FeS; density 4600 kg/m^3. An accessory mineral in most meteorites, troilite is typically 5% to 6% by weight of chondrites. Iron sulfide is frequently mentioned as a contaminant mineral in iron cores of planetary bodies, altering the chemical properties of the cores.

Graphite (C) and **carbonaceous minerals:** Density 2200 kg/m^3. Various forms of carbon-based minerals are low-temperature condensates that appear to play a very important role in the outer solar system, creating the dark, blackish (albedo 2% to 8%) surfaces common among comet nuclei and "asteroids" located there, as described in Chapter 7. Organic compounds, probably reddish in most cases, are common among these materials, causing brown and red colorations. The formative process for the carbonaceous minerals in general and the organics specifically is an open question; it may involve cosmic ray interaction with associated methane ices.

Clay minerals: Example: **montmorillonite,** $(Al,Mg)_8$ $(Si_4O_{10})_3(OH)_{10}\ 12H_2O$. Clay minerals are essentially hydrous aluminum silicates. They play two important roles in the solar system. First, they are major minerals in the erosion products of Earth and Mars. Montmorillonite and **nontronite** are considered to compose roughly 20% and 50%, respectively, of the fine dust of Mars; they are frequent products of weathering of volcanic lavas (Toulmin and others, 1977). Second, they are major minerals in the carbonaceous matrix that condensed among C-, P-, and D-class interplanetary bodies. Spectra suggest the presence of montmorillonite on Ceres, the largest asteroid. The open crystal structure of clay minerals allows them to absorb and hold large amounts of chemically bound water.

Rocks and Rock Types

Rocks are traditionally divided into three groups. **Igneous rocks** are rocks formed directly from the cooling of molten magma. **Sedimentary rocks** are formed by deposition and cementing of small particles, either of biogenic or nonbiogenic origin. **Metamorphic rocks** are rocks originally formed in either of the above modes but changed to a new rock form by high pressure, high temperature, or the addition of new chemicals.

Igneous Rocks. Igneous rocks have a variety of compositions and textures, depending on the composition and cooling history of their parent magmas. Terrestrial magmas typically have temperatures of 1300 K and may solidify underground. If they work their way upward and extrude onto the surface, they are called **lavas.** Rock formed on the surface from cooling lava is also loosely called lava or lava rock.

Igneous rocks are the most important rock type in planetary surface studies. Some 95% of Earth's outer few kilometers is composed of igneous rocks. The high percentage of sediments we see around us is misleading, because sediments and sedimentary rocks form on the surface, accounting for much area but little depth. The moon's rocks are all igneous. Lavas abound in the dark plains, but in the uplands, which are older regions, intense cratering has shattered most igneous rocks. There they have recemented into breccias, or rocks formed of welded angular fragments. Many Martian and Venusian rocks and soils are also igneous, as determined from their appearance and soil analysis by several landers, though sedimentary layers also exist on Mars.

If we broaden our horizons to include the predominantly nonsilicate (that is, icy) crusts of the outer solar system, we can anticipate that ices play the role of igneous rocks on these worlds. "Volcanism" on icy worlds may occur when ice melts and erupts watery "magma" that resurfaces the crust with icy "lava"-like plains.

Igneous rocks are subdivided into compositional classes and subclasses, as shown in Figure 9-1. Arranged down the figure are compositions ranging from silica rich to silica poor.

The classifications shown in Figure 9-1 reflect the eruptive history of the magma. If the magma remains underground, it is called **intrusive.** There it is well insulated, cools slowly, has lots of time for crystals to grow, and hence produces coarse-grained, or **phaneritic,** rock. Such rocks may be exposed on the surface by later erosion. If the magma erupts into the surface, it is called **extrusive;** it cools rapidly and produces fine-grained, or **aphanitic,**

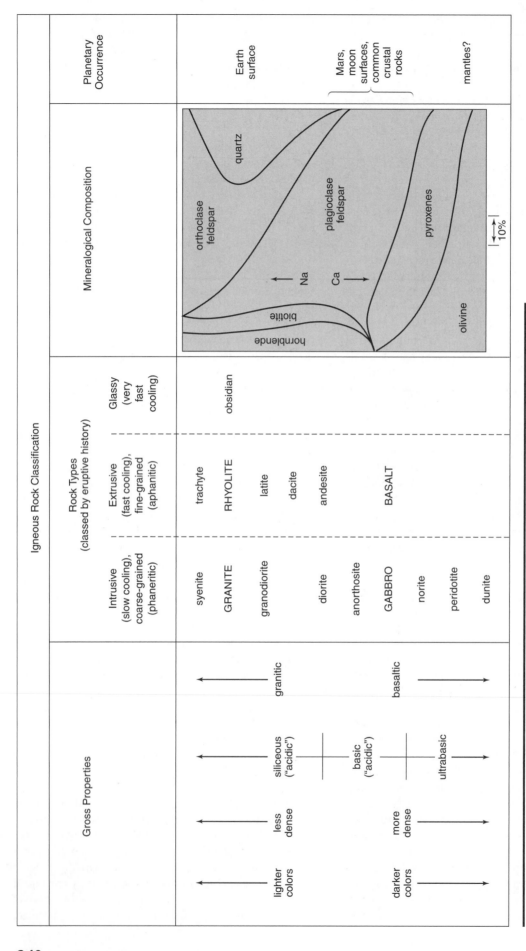

Figure 9-1. Classification scheme for igneous rocks. Crude classifications are at left. Middle columns show rock names associated with different mineralogical makeups diagrammed in the box at right. The four most important names are capitalized. As shown at the far right, the vertical arrangement tends to match that in the planetary crusts and upper mantles.

rock. Extrusive materials may cool so fast that no crystals grow because the molecules do not achieve a crystal lattice structure; in this case the material is called **glass.**

In planetary science, the most important group of igneous rocks are the **basaltic rocks.** Space exploration has revealed such prevalence of basaltic rocks on other worlds that the meteoriticist, Robert Dodd (1981, p. 262) was moved to comment, "It appears that basaltic volcanism [is] a *leitmotif* of planetary geology, and has been such for 4.55 billion years." The general category of basaltic rocks includes **gabbros** (the coarse-grained, intrusive type) and **basalts** themselves (fine-grained, extrusive). Basaltic rocks are common because they form from the feldspars and other materials common in terrestrial planet crusts, through repeated melting and differentiation processes. Basaltic rocks cover the dark lunar plains and much of Mars, Venus, and asteroid Vesta. The lunar uplands have many gabbros, some erupted basalts, and a gabbroic rock made largely of plagioclase feldspar, called **anorthosite,** an anorthositic gabbro.

The next most important igneous rocks are **granitic rocks.** These are quartz-rich rocks formed by repeated melting and differentiation processes that concentrate quartz and feldspars. Granitic rocks include **granite** itself (coarse-grained, intrusive) and **rhyolite** (fine-grained, extrusive). On Earth, granites are common. Rhyolite flows, the result of eruptions of granitic magma, are increasingly being recognized as common.

The importance of basaltic and granitic rocks on Earth is shown by the distribution of silica contents in random rock samples. There are two peaks: a basaltic group averaging 53% SiO_2 and a granitic group averaging 73%. Thus, basaltic and granitic rocks, with their distinct colorations, often appear as distinctive outcrops in terrestrial landscapes, as shown in Figure 9-2.

Granitic rocks seem rare on other terrestrial planets because the planets apparently have not undergone enough cycles of differentiation to produce them. Venus is a possible exception. One out of seven Soviet landers, Venera 8, measured radioactivity levels comparable to a granitic rock called granodiorite; the other landing sites yielded chemistry typical of basalts (see Chapter 8). Searches for granites on Mars, among the rocks sampled

Figure 9-2. Contrast of basaltic and granitic rocks on Earth is prominent in this Mexican range, where basalt lavas broke through a light-toned granitic ridge. The dark basaltic cap has eroded and produced basaltic talus spreading down the granitic hillside. (Pinacate volcanics, Sonora; photo by author)

by the Sojourner rover in 1997, were unsuccessful; only basalts and andesites were found.

The kind of rock produced by a given magma depends on the magma's composition and what happens during the cooling. As the temperature drops, minerals form and react chemically with each other and with the remaining magma. Therefore, the composition of the resulting rocks depends on the rate of the cooling process and on the degree of **fractionation**—the differentiation process by which minerals crystallizing early separate from the rest of the magma. Bowen (1928) expressed this concept in his **reaction series,** the sequence of minerals that crystallize from a given melt as the temperature drops.

The significance of the reaction series can be seen better by considering the example of a melt of basaltic composition. Among the first minerals to crystallize are olivine and calcium-rich feldspars. An eruption at that point might produce basaltic lavas studded with olivine crystals. If there is no eruption and no fractionation, these crystals remain suspended in the melt and eventually react chemically with the remaining hot fluid magma. The olivine is converted into pyroxenes. If this mass erupts, it produces normal basalt, and if it cools in place instead of erupting, it produces an underground mass of gabbro.

Fractionation, however, is likely to separate the heavy crystals from the light ones, and this changes the scenario. The olivines are denser than the melt—the molten material—and tend to sink to the bottom of the melted region, if there is enough time. This sinking stops the reaction of the olivine with most of the melt, so that the melt itself evolves differently. Amphiboles and more potassium-rich feldspars may form. Feldspars, too, may fractionate by floating to the top of the molten material and forming feldspar-rich rocks there. If the fractionation continues, the last stages of crystallization will produce quartz, often in a watery solution of hydrothermal minerals (minerals affected by the presence of hot water and associated dissolved material). Eruption onto the surface during any stage of the proceedings might produce lavas with specialized compositions.

Conversely, if igneous rocks are heated, the first stages of melting may produce a hydrothermal mineral solution (which might drain off), then melting of quartz, and last, melting of certain basic minerals.

Sedimentary Rocks. **Sedimentation** is essentially a planetary skin effect, the interaction of the atmosphere and hydrosphere with the crust of the planet. Chemical and mechanical **weathering** constantly attacks rocks exposed at the surface of a planet with an atmosphere. Chemical weathering involves chemical reactions that break down the rocks; mechanical weathering involves nonchemical breakdown—for example, by abrasion from blowing sand or moving water. The rock particles are transported away in suspension and in solution to form new rocks of different compositions and textures in other places.

The most important type of sedimentary rock on Earth is **shale,** a rock composed of very fine grains derived from consolidated beds of clays, silts, or mud. Shales make up an estimated 4% of the upper 6 km of Earth's crust (Mason, 1966). **Sandstone,** a coarser-grained rock, is composed of cemented sand grains (often of quartz but sometimes of other minerals). On Earth (and perhaps Mars), shale and sandstone are usually composed of rounded grains because the grains have been abraded by being tumbled in wind or water.

If rock fragments are violently broken and recemented, they will be highly irregular and angular; a rock made of such fragments is called a **breccia.** Breccias are important in planetary science. As mentioned earlier, many meteorites consist of breccias produced by mixing and compacting impact-produced fragments. Similarly, 85% of the moon is ancient highlands where virtually all rocks are breccias composed of gabbroic and basaltic fragments blasted out by meteorites. Similar conditions probably exist on Mercury and perhaps some asteroids. On Earth, breccias are usually classed as sedimentary because they usually involve fluvial transport and cementation. The impact breccias of other worlds, however, are usually not called sedimentary. (It is misleading to say that the moon is 85% covered by deep beds of sedimentary rocks!)

Another type of sedimentary rock important on Earth, Mars, and perhaps elsewhere is a type known as **evaporates**—solid materials left behind when solutions evaporate. When water flows on the surface or in subsurface layers, it dissolves (or "leaches") salts (such as NaCl), carbonates and sulfate minerals; then, when the water ponds and evaporates, the water molecules speed off into the air, but the salts, carbonates, and sulfates are left behind to form evaporites. For example, gypsum ($CaSO_4$ $2H_2O$) is a major mineral that is left behind as moisture evaporates from soil. Salt (NaCl, sometimes called rock salt) is another. Calcium carbonate "bathtub rings" left in your tub are another example; calcium carbonate "caliche" deposits are also common in nature. The presence or absence of such materials on Mars is of considerable interest as an indicator of past conditions (see also Chapter 13, about Mars).

Significant geochemical differentiation can occur during the formation of sedimentary rocks. Certain minerals, especially quartz, are hard—resistant to solution during low-temperature weathering processes—and are dropped as fragments; sandstones may result. Other minerals are broken down; the aluminosilicates give rise to clays, muds, and shale; ferrous iron (+2 charge) is oxidized to the ferric state (+3 charge, usually producing red mineral colors), which may produce hydroxide precipitates abundant enough to form ore deposits; calcium minerals produce calcium carbonate solutions, which may yield limestones. Certain elements that remain in solution are carried preferentially from the land to ocean basins; sodium, the best example, is constantly being added to oceans.

Metamorphic Rocks. Metamorphism is the net effect of all processes (short of complete remelting) acting to alter and recrystallize solid rock material beneath a planet's surface. These processes are a result of changes in pressure, temperature, and chemical environments. *A specific suite of minerals* is stable only for a certain set of environmental conditions. If these conditions change, the minerals may change to a new mix. One example is a pressure increase as more sediments are added above, which may cause a structural rearrangement of minerals along parallel bands. Another example is a temperature increase due to local geothermal heating. Probably the most important example is the presence of chemically active volatiles, which may carry off or introduce new material to the system, a process called **metasomatism.**

Some common metamorphic rocks are listed in Table 9-1. Pressure plays an important role in producing the banding and cleavage characteristic of the first three entries.

A Survey of Planetary Rocks

Many principles of planetary geochemistry become clearer if we review the compositions of planetary samples collected to date. The most basic way to look at rock compositions is to look at elemental abundances, as shown in Table 9-2. However, analysts commonly report compositions in terms of oxides, as if the available elements were all combined into compounds with oxygen. Because most of the minerals in most rocks are likely to be oxides, this type of table comes close to indicating the mineral makeup of the rocks, especially in the case of igneous rocks. This method of presentation is shown in Table 9-3 on page 245. But in the real rock, many of the atoms may be bound up in minerals more complex than the simple oxides indicated in Table 9-3. We examine information to be gleaned from both modes of presentation.

Table 9-2 starts with the most primitive solar system materials, such as the solar gas from which the planets

ultimately derive. Note the tremendous amount of differentiation implied by going from the primeval gas to the composition of solid matter. Even as we go from carbonaceous and ordinary chondrite meteorites to planetary rocks, we see increasing differentiation, a point also made graphically in the last chapter (Figure 8-3). The material of the terrestrial and lunar mantles, judged from deep-seated samples, is depleted in siderophiles and correspondingly enriched in lithophiles, such as silicon (Si), magnesium (Mg), and aluminum (Al), relative to meteorites. The process that produced crusts, such as floating of low-density minerals and chemical affinity of the lithophiles for silica-rich minerals, produced complex patterns in crustal rocks. Crustal rocks are generally enriched in oxygen (O) and silicon (Si) and depleted in iron (Fe). The moon's crust of anorthositic rocks is also enriched in aluminum (Al). Earth's crust, as a result of more complete differentiation, is particularly rich in silicon. The planetary basalts, which are postcrustal lavas erupting from the upper mantles of the planets, are remarkably similar.

More data are available in Table 9-3, and some of the same patterns can be seen. Primitive meteorites are richer in iron than are planetary mantle rocks; they represent planetary material before the iron differentiated into cores. The crustal rocks, such as lunar highland materials and Earth's crust, are even more depleted in dense iron-rich minerals. Note the trends toward extreme silica contents with greater differentiation. Earth's continents are the most silica rich, while the oceanic crust resembles the lunar crust and basaltic materials.

The numerous basaltic samples in the table show a range of silica contents. They may also show anomalous enrichments of one mineral or another. Many lunar mare basalts are unusually high in titanium (Ti), whereas the older, so-called **KREEP basalts,** often found in the lunar uplands, are enriched in potassium (*K*), rare-earth elements (*REE*), and phosphorus (*P*). Most Martian meteorite rocks are basaltic and andesitic (see Figure 9-1). The Viking and Pathfinder landers found Martian soils that were interpreted as derived from basalt, but with an excess of sulfur compounds and related materials left as evaporites. Only Mars Pathfinder analyzed the chemistry of surface rocks on Mars. It landed in an area where a mixture

TABLE 9-1	Common Metamorphic Rocks	
Metamorphic rock	Common parent rock	Description
Gneiss	Granite	Marked parallel bands, coarse grained
Schist, phyllite	Shale, granite	Banded, fine grained
Slate	Shale	Fine grained, splits into thin sheets
Quartzite	Sandstone	Massive quartz rock, breaks through quartz grains
Marble	Limestone	Familiar in architecture

TABLE 9-2 Elemental Abundances in Cosmic Materials (percent weight)

Material	H (V)	He (V)	O (L)	Fe (S)	Si (L)	Mg (L)	S	C	Ca (L)	Ni (S)	Al (L)	Na (L)	Ti	K	Total	Primary reference
Primitive Materials																
⊙ ("cosmic" abundance)	78	20	1	0	0	0	0	0	0	0	0	0	0	0	99	Gibson (1973), p. 72
Comets	7	?	76	?	6	?	?	3	?	?	?	?	?	?	97	Delsemme (1977), p. 9
CI carbonaceous chondrite	0	0	46	18	11	10	6	3	1	1	1	1	1	1	100	Wasson (1974), p. 78
Meteorites (avg.)[b]	0	0	33	29	17	14	2	0	1	2	1	1	0	0	100	Mason (1962), p. 151
Planetary Interiors: Ultrabasic Rocks																
Iron meteorites	0	0	0	91	0	0	0	0	0	9	0	0	0	0	100	Wyllie (1971), p. 97
⊕, core (est.)	0	0	0	85	5?	3	4	0?	0?	3	0?	0?	0?	0?	100	Wyllie (1971), p. 104
⊕, mantle (est.)	0	0	44	10	23	19	0?	0?	2	0?	2	1	0	0	101	Wyllie (1971), p. 104
☾, bulk (est.)	0	0	42	8	20	18	0	0	5	0	4	0	0	0	97	Taylor (1975), p. 316
Crustal Materials																
☾, highland soils	0	0	45	5	21	4	0	0	11	0	13	0	0	0	99	Taylor (1975), p. 64
⊕, crust (avg.)	0	0	47	5	28	2	0	0	4	0	8	3	0	3	100	Mason (1958), p. 44
Basaltic Materials																
☾, mare basalt	0	0	40	11	21	6	0	0	8	0	8	0	6	0	100	Lofgren and others (1981), p. 239
☾, mare soils	0	0	43	11	21	5	0	0	8	0	8	0	2	0	98	Taylor (1975), p. 64
♂, Chryse soil	0	0	42?	13	21	5	3	0	4	0	3	0	1	0	92	Clark and others (1977), p. 4588
♂, Utopia soil	0	0	42?	14	20	?	3	0	4	0	?	0	1	0	84	Clark and others (1977), p. 4588
⊕, basalts	0	0	43	9	23	5	0	0	8	0	8	2	1	1	99	Lofgren and others (1981), p. 14ff
Other Material																
♂, duricrust	0	0	42?	13	21	5	4	?	4	?	3	?	1	0	93	Clark and others (1977), p. 4588

[a] (V) = volatile (easily lost by heating); (S) = siderophile (tending to associate with iron minerals); (L) = lithophile (tending to associate with silica minerals).

[b] Essentially ordinary chondrites.

of rocks had apparently been washed down from the Martian highlands in floods; five measurements of rocks and one of a possible cemented aggregate rock gave basic compositions, generally of a basaltic and andesitic nature (Golombek and others, 1997; Rieder and others, 1997). The possible andesites would imply greater crustal differentiation than basalts alone (see Figure 9-1). Of great interest would be comparably accurate data on Venusian and Mercurian rocks.

Among the nonigneous rocks, Earth's sediments represent an averaging of different rock types, and the limestones are noteworthy for having a biological origin that (presumably!) never occurred on other solar system worlds. From a rather detached viewpoint, the sea creatures whose shells produced limestone represent an extreme geochemical differentiation favoring calcium!

Early Evolution of Lithosphere Materials

Now we have the tools to discuss in more detail the formation of the primeval surfaces of planets. We have emphasized that feldspar crystals, because of their early appearance during solidification, their low density, and their abundance, tended to float during melting (for example, in primordial magma oceans), thus accumulating

TABLE 9-3   Compositions of Planetary Rocks (percent weight)

Rock	SiO$_2$	Total iron[a]	MgO	Al$_2$O$_3$	CaO	Na$_2$O	TiO$_2$	K$_2$O	H$_2$O	Total	Primary reference
Primitive Materials											
Carbonaceous chondrites	28	27	19	2	2	1	0	0	13	92	Mason (1962), p. 74
Enstatite chondrites	38	32	22	2	1	1	0	0	0	96	Mason (1962), p. 74
Hypersthene chondrites	40	27	25	2	2	1	0	0	0	97	Mason (1962), p. 74
Ultrabasic (Mantlelike) Igneous Rocks											
⊕, deep-source nodules	43	9	43	2	1	0	0	0	0	98	Wyllie (1971), p. 112
⊕, dunitic peridotite, Switzerland	42	8	42	2	2	0	0	0	4	100	Rittmann (1962), p. 105
⊕, mantle (est.)	44	8	40	3	2	0	0	0	?	100	Wyllie (1971), p. 114
☾, bulk silicate portion	43	9	28	11	9	0	0	0	0	100	Warren and Wasson (1979), p. 2054
"Crustal" Igneous Rocks											
☾, breccias, Descartes highlands	44	3	3	30	17	1	0	0	0	98	Taylor (1975), p. 216
☾, anorthosites, highlands	44	1	1	35	19	1	0	0	0	98	Taylor (1975), p. 234
☾, gabbroic anorthosites, highlands	44	3	3	31	17	0	0	0	0	98	Taylor (1975), p. 234
☾, highland rocks and soils (avg.)	45	6	7	25	15	0	1	0	0	99	Taylor (1975), p. 81
⊕, oceanic crust (avg.)	48	10	7	16	12	3	2	1	1	100	Wyllie (1971), p. 154
⊕, crustal composition (avg.)	57	8	5	16	8	3	1	2	1	101	Wyllie (1971), pp. 153, 155
Basaltic Materials											
☾, high titanium mare basalts, Tranquillitatis, and Serenitatis	39	20	8	9	10	0	12	0	0	98	Taylor (1975), p. 136
☾, mare basalts, Imbrium, Procellarium	45	20	9	9	10	0	3	0	0	96	Taylor (1975), p. 136
☾, KREEP basalt, highlands	49	8	10	18	12	0	1	1	0	99	Taylor (1975), p. 228
♂, Chryse soil[b]	45	18	8	6	6	0	1	0	2?	86	Toulmin and others (1977), p. 4629
♂, Utopia soil[b]	43	20	?	?	5	0	1	0	2?	71	Toulmin and others (1977), p. 4629
⊕, olivine basalt, Mauna Loa, Hawaii	46	13	24	6	6	2	2	0	1	100	Rittmann (1962), p. 105
♂, meteorite Zagam, (basaltic)	50	19	10	6	10	1	1	0	0	97	Rieder and others (1997)
♂, "Wedge rock" (Pathfinder basalt)	52	15	5	10	7	3	1	1	0	94	Rieder and others (1997)
⊕, basalt, Kilauea volcano, Hawaii	50	11	7	14	12	2	3	1	0	100	Rittmann (1962), p. 105
⊕, plateau flood basalts (avg.)	49	13	7	14	9	3	2	1	2	100	Rittmann (1962), p. 105
⊕, oceanic basalt layer (avg.)	50	9	7	17	12	3	2	0	1	101	Wyllie (1971), p. 153
⊕, continental basalt layer (avg.)	58	8	4	16	6	3	1	3	1	100	Wyllie (1971), p. 153
Siliceous Igneous Rocks											
♂, "Shark rock" (Pathfinder andesite)	61	12	3	10	8	2	1	1	0	98	Rieder and others (1997)
⊕, continental granitic layer	64	5	2	15	4	3	1	3	2	99	Wyllie (1971), p. 153
⊕, dacite, Krakatoa 1883 explosion	68	4	1	15	3	6	1	2	1	101	Rittmann (1962), p. 105
⊕, rhyolite, New Zealand	72	2	0	12	1	3	0	4	4	98	Rittmann (1962), p. 105
⊕, rhyolite, Sicily	75	1	0	13	1	4	0	5	1	100	Rittmann (1962), p. 105
Nonigneous Rocks											
♂, duricrust pebbles[c]	44	18	9	6	5	0	1	0	3	86	Toulmin and others (1977), p. 4629
⊕, sediments, all types (avg.)	58	5	3	13	6	1	1	3	3	93	Mason (1958), p. 147
⊕, shales (avg.)	58	6	2	15	3	1	1	3	5	94	Mason (1958), p. 147
⊕, sandstones (avg.)	78	1	1	5	6	0	0	1	2	94	Mason (1958), p. 147
⊕, limestone (avg.)[d]	5	1	8	1	43	0	0	0	1	59	Mason (1958), p. 147
Tektite											
Australite (avg.)	73	5	2	12	4	1	1	2	0	100	Taylor (1975), p. 81

[a] Iron may exist in various states of reduction and oxidation. Primarily, these are Fe, FeO, Fe$_2$O$_3$, and FeS. For example, iron is all in the form of FeO in carbonaceous chondrites, and mostly in the form of Fe metal in enstatite chondrites. In the basaltic samples, it is almost always FeO, except in the Martian soil samples, where it is more oxidized and listed as Fe$_2$O$_3$.

[b] The Martian soil is believed to be derived mainly from weathered basaltic material similar to the boulders scattered at the two Viking landing sites. It has an additional component of weathering products, such as 7% SO$_3$ and 1% Cl.

[c] These are fragments of crumbly material believed to be evaporites. They have a high content of sulfur and other weathering products: 10% SO$_3$ and 1% Cl.

[d] Limestone has a very high carbon content, listed by Mason as 42% CO$_2$. The terrestrial sediments listed above this entry also have 3% to 5% CO$_2$.

near surfaces. This explains why the crustal layers of Earth, the moon, and Mars appear to be anorthositic, gabbroic, and basaltic in composition, as these are the rock types associated with abundant feldspars. On Earth, of course, the initial crust has been nearly obliterated by geological activity, and even more extreme differentiates, such as granites and sandstones, have concentrated at Earth's surface due to tectonic and surficial recycling of the material.

The moon's crust is a special case. If the moon really aggregated in Earth orbit from debris after a giant impact, the rate of its accretion was much faster than that of the planets. (The duration of accretion scales more or less as the orbital time period, which is of the order of a year for terrestrial planets, but less than a day for the primordial moon.) The fast accretion produced an extreme impact heating of the surface layers, producing a hot, deep, magma ocean. As the lunar magma ocean solidified, the Ca-rich feldspars (that is, plagioclase feldspars) fractionated from the magma so efficiently that many lunar highland samples are classed as anorthosites (nearly pure feldspar rocks, defined by very high plagioclase feldspar content). Related, specialized rock types also commonly described among primitive crustal samples from the lunar highlands are **norites** (hypersthene-plagioclase-rich gabbros) and **troctolites** (olivine-plagioclase-rich gabbroic rocks).

Some researchers have attempted theoretical extensions of these principles to characterize other planets' primitive crusts. Geochemist J. Wood (1977) emphasized that feldspars crystallize from magma readily at pressures less than 12 kbar but are replaced by denser minerals at pressures greater than 12 kbar. This means that in any planet, feldspars would not float upward from below depths where the pressure equals 12 kbar, thus limiting the thickness of feldspar-rich crusts. Morrison and Warner (1978) carried this a step further and calculated the following maximum thicknesses for feldspar-rich crusts that extend to the 12-kbar level in terrestrial planets:

Earth	40 km
Venus	45 km
Mars	105 km
Mercury	108 km
Moon	250 km

These depths would be reached only if there were enough feldspathic material to fill such zones.

Did the primitive Earth therefore have a 40-km anorthositic crust? Anorthosite masses are found in some of the primitive shield areas of Earth, but they are thought to be products of fractional crystallization of large, localized magma masses rather than relics of a primeval crust.

Aside from the thinness of the feldspathic layer available for Earth-sized planets, Warren and Wasson (1979) raise another reason for concluding that "no substantial anorthositic crust is to be expected on Earth-sized plan-

ets": Much of the magma ocean crystallization would have taken place on the bottom of the magma layer, which may have been beneath the 12-kbar level on the larger planets. The denser minerals crystallizing there in place of plagioclase (namely garnets and spinels) would rob the remaining magma of much of the calcium and aluminum needed to make the plagioclase feldspars before the feldspar-rich crust could form on the top of the ocean. With Earth's surface materials thus seriously depleted in plagioclase relative to the moon's crust, a more granitic primitive composition would be favored. On the other hand, if Earth and the other terrestrial planets aggregated slowly enough, deep magma oceans may not have existed. The crustal compositions of Venus, Mars, and Mercury will provide an interesting test of all these ideas.

If initial surface crusts had formed without external effects, primitive worlds like the moon ought to display nicely stratified sequences of accumulated crustal rocks, with numerous 4500-My-old **Genesis rocks** lying around, waiting to tell us their stories of how it was when the planets had just finished forming. Lunar astronauts were trained to look for such rocks. They weren't there, at least not in any great number. Primeval rocks were mostly pulverized by the intense meteoritic bombardment associated with the sweep-up of the planetesimals.

Regolith and Megaregolith

Prior to the 1960s, many scientists and artists pictured the moon's surface as incredibly rough, craggy, and fissured. Yet the first close-up photos, obtained as Ranger 7 crashed into the moon in 1964, revealed not crags, but a smooth, rolling topography, with only scattered rocks. The explanation came about the same time. McCracken and Dubin (1964) pointed out that the amount of meteoritic material that has accumulated on mare surfaces in the last few billion years amounts to a layer a few centimeters deep. As each meteorite dislodges several hundred times its own mass (Gault, Heitowit, and Moore, 1964), the meteorites would have created a layer of rubble a few meters deep. Summarizing a 1963 conference on the lunar surface, Salisbury and Smalley (1964) wrote: "It is concluded that the lunar surface is covered with a layer of rubble of highly variable thickness and block size. The rubble in turn is mantled with a layer of highly porous dust." A more accurate description could hardly have been written, as seen in Figure 9-3.

Following the first Ranger photos, the continuous layer of rubble and fine dust came to be called the **regolith** (Greek for "rocky layer"). Shoemaker (1965) considered more carefully the statistics of post-mare impact craters and the amount of debris thrown out by each one, and he

a

b

Figure 9-3. Views of the lunar regolith. (**a**) Regolith surface texture. The Surveyor spacecraft bounced before coming to rest, leaving a footprint in the flourlike soil. (**b**) Regolith cross section in the wall of Hadley Rille. Among the loose boulders on the slope, layered outcrops of rock can be seen some tens of meters below the surface. These layers are probably intact portions of successive, bedded lava flows. (NASA)

confirmed depth estimates of about 10 m for the regolith in the maria.

The first lunar astronauts remarked on the dust kicked up by their rocket exhaust during their landings and by themselves during their hikes. Their photos of a valley hillside (Figure 9-3b) show a rare glimpse of the regolith and underlying rock beds in cross section. The great amount of fine dust shows that most surface rocks have been pulverized by impacts, but the scattered boulders and subsurface rock layers indicate that bigger impacts penetrate rock layers and throw out intact rocks.

Apollo data show that in the lunar maria, the regolith is typically about 2 to 20 m deep, grading into broken or coherent lava flow layers below. The regolith is quite variable, and depths as much as 36 m were estimated at some points near the Apollo 17 site in Mare Serenitatis (Taylor, 1975; Short, 1975).

The ancient lunar uplands are much more heavily cratered than the lunar maria. Hence, they are believed to contain a much deeper regolith than the maria. Short and Forman (1972) and Hartmann (1973) independently used crater statistics and other data to estimate that the

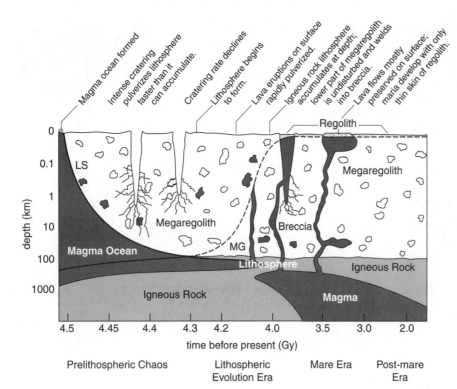

Figure 9-4. Schematic cross section of the moon's surface layers, showing evolution during early lunar history. See text for further description.

densely cratered highlands contain a roughly 2-km layer of pulverized material, which was termed the **megaregolith.** Still earlier craters, now obliterated by later craters, may have created even greater amounts of debris, possibly tens of kilometers deep. However, the deeper layers of debris were probably bonded into coherent breccia rock rather than remaining in the form of loose regolith.

The term *regolith* has been extended to dust layers or other planetary surfaces created by meteoritic bombardment. One controversy about asteroids is whether the low gravity has resulted in regolith being blown off, producing bare rock surfaces, or whether asteroids are so fragmented that they are essentially megaregolith "rubble piles" all the way through. Low bulk densities measured for at least asteroids favor fragmental structures for them.

Primeval Lithospheric Evolution: Rock Formation Versus Rock Destruction

Some consequences of early intense cratering for the moon's primeval lithosphere structure can be seen in Figure 9-4, which shows depth cross sections through time. The curves involved are based on quantitative calculations, but are model dependent (Solomon and Longhi, 1977; Herbert and others, 1977; Hartmann, 1980a). The solid curve LS shows the rate of lithosphere solidification resulting from cooling of the magma ocean. It represents

the tendency to create coherent igneous rock. The dashed curve MG shows the amount of megaregolith generation produced by cratering in any interval of about 100 My. It thus represents the tendency to destroy coherent igneous rock by grinding it into dust and debris.

These two processes—magma solidification and cratering—competed. Initially (the first few hundred My), cratering was so intense that impacts frequently smashed through any primitive lithosphere, splashing magma about and completely pulverizing the earliest rock materials. Eventually (at about 4.3 Gy on Figure 9-4), curves LS and MG crossed so that rapid pulverization no longer occurred at the level where new rock was forming. Therefore, a true lithosphere, intact except for the sites of the deepest late impacts, began to evolve at that depth, overlaid by megaregolith. The rubble just above this level, which had been rapidly churned before, was now less disturbed because the cratering rate had declined, and impacts rarely penetrated to this depth. With its long exposure to moderate pressure from the overlying material and heat from below, this material may have welded into the breccias so commonly found in the uplands. Note that any lava flows erupting at this time (such as the one indicated at about 4.12 Gy) would have been rapidly eroded by cratering. A new era began later, around 3.9 or 4 Gy ago, when lava flows such as the 20- to 50-m-thick flows measured in Mare Imbrium could partially survive, with only a thin layer of regolith on the top. The lava flows of this era became the familiar lunar maria.

This scenario of early evolution divides lunar history into four phases. In the first phase, marked **prelitho-**

spheric chaos in Figure 9-4, no substantial layers of igneous lithosphere formed above the magma ocean. In the second phase, the **lithospheric evolution era** (perhaps 4.3 to 4.0 Gy ago), intact lithosphere formed at depth but no flows were well preserved on the surface. In the **mare era** (4.0 to 3.0 Gy), flows erupted from the deep-seated magma and surface structures were preserved. The moon, being small, cooled quickly enough that much of its history involved a fourth phase, the **post-mare era,** when eruptions had virtually ceased.

If some other planets lacked magma oceans initially, cratering would have simply pulverized their surface layers until a decline in cratering allowed preservation of surface features and undisturbed welding of lower layers into coherent breccias. Also, the larger planets maintained hot mantles and eruptive activity longer, so that their "postmare" periods would have been shorter or nonexistent. Various cratering rates in different parts of the solar system competed with various thermal histories of different-sized planets and led to different patterns of surface evolution.

It is clear from Figure 9-4 that cratering is an exceedingly important effect in the development of planetary surfaces. On some small, rapidly cooling planets, eruptive activity never occurred, and the surface and landscapes were entirely produced by cratering effects. Because of the importance of cratering, most of the rest of this chapter is devoted to consideration of cratering effects.

Discovery of Meteorite Impact Craters

Galileo first turned his telescope to the moon on November 30, 1609, and within a few months he announced that circular pits pockmarked the moon's surface (Whitaker, 1978). Galileo used the Greek word **crater** ("cup") to describe these. Until the middle of this century, a debate raged about the origin of craters. Some scientists thought they were impact features, but others thought they were volcanic or formed by giant gas bubbles that rose through a molten primeval moon and broke on the surface.

Not until this century were meteorite craters, complete with meteorite fragments, firmly identified on Earth. Further geological fieldwork revealed terrestrial **astroblemes** (from Greek words for "star wounds"), or eroded circular structures that have turned out to be remnants of ancient, eroded meteorite craters (see Table 9-4 and Figure 9-5, p. 251). These results favored the meteorite theory of planetary craters. Further evidence came when astrophysicist Ralph Baldwin (1949) showed that the properties of lunar craters matched those expected for impact explosions. Next, fieldwork and space probes of the 1960s revealed compelling similarities between known meteorite craters on Earth (Table 9-4) and the ubiquitous craters on other planetary bodies. These similarities included hummocky rim forms and patterns of ejected rubble. Furthermore, fresh craters on various planets have size distributions consistent with size distributions of the interplanetary meteoroids. So the majority of craters larger than a few kilometers across on planetary bodies are now believed to have been formed directly by impact of interplanetary bodies.

Craters are interesting in several regards. They are fascinating as weird landscapes and scenes of ancient cataclysms.* Their structures reveal subsurface properties, and their numbers reveal the age of the surface: the greater the number of craters that have accumulated, the longer the surface has been exposed.

Mechanics of Impact Crater Formation

Because of the combination of orbital speed and planetary gravity, meteoroids typically strike planets at 10 or more kilometers per second. If the planet has an atmosphere, the smaller meteoroids are slowed and do not form craters.

The kinetic energy of high-speed meteorites is converted on impact into thermal, acoustic, and mechanical energy that distorts, fractures, and rejects rocks. The result is like an explosion centered a few meteorite diameters below the ground, and meteorite impact craters are thus somewhat like bomb craters. When the meteorite enters the ground, it is usually **hypersonic,** meaning that it is moving faster than the local speed of sound, because seismic waves (sound waves in rock or soil) typically move at 1 to 4 km/s. A hypersonic meteorite thus enters the rock faster than the energy can propagate outward as seismic waves. A **shock wave,** or highly compressed zone in front of a supersonic body, thus builds up around the impact point. It carries a high density of energy and matter and builds up around the impact point, resembling a shock wave around the front of a supersonic aircraft. The explosion is the spreading of this shock wave, which compresses the rock and initially makes it deform almost like a fluid around the impact site. The rock bends backward, upward, and outward, excavating a volume of material much larger than the meteorite itself. (In a subsonic collision, the rock simply is mechanically shattered, but no such explosion occurs, and so a true explosion crater is not created.)

A relation exists between the kinetic energy of the hypersonic meteorite and the size of the crater, though the crater size and shape also depend to a lesser extent on

---

*A meteoriticist friend tells of visiting Meteor Crater, Arizona. As he stood on the rim, looking at the contorted strata, he imagined the thunderous explosion of the impact, the fiery burst of ejecta shooting upward, and the shock wave racing out across northern Arizona, devastating life throughout the region in a matter of minutes. His mood was broken when a woman approached and remarked with disappointment, "Oh, my! It's nowhere near as big as the Grand Canyon!" Knowledge, coupled with imagination, aids landscape appreciation.

TABLE 9-4	Selected Meteorite Impact Craters on Earth		
Name	Location	Est. original diameter (km)	Est. age (My)
Sudbury	Ontario	140	1840 ± 150
Vredefort Ring	South Africa	140	1970 ± 100
Popigai	Russia	100	39 ± 9
Lake Manicouagan	Quebec	100	212 ± 2
Acraman	Australia	90 (or 160?)	~600
Puchezh-Katunki	Russia	80	183 ± 3
Siljan	Sweden	52	368 ± 1
Kara	Russia	50	57 ± 9
Charlevoix	Quebec	46	360 ± 25
Araguainha Dome	Brazil	40	<250
Carswell Lake	Saskatchewan	37	117 ± 8
Clearwater Lakes	Quebec	32, 22[a]	290 ± 20
Manson	Iowa	32	<70
Slate Island	Ontario	30	350
Lake Mistassini	Labrador	28	38 ± 4
Teague	Australia	28	1,685 ± 5
Rieskessel	Germany	24	14.8 ± 0.7
Gosses Bluff	Australia	22	142 ± 5
Wells Creek	Tennessee	14	200 ± 100
Sierra Madera	Texas	13	<100
Deep Bay	Saskatchewan	12	<100 ± 50
Bosumtwi	Ghana	10.5	1.3 ± 0.2
Kentland	Indiana	9	~300
Serpent Mound	Ohio	6.4	~300
Decaturville	Missouri	5.6	320
Crooked Creek	Missouri	5.6	320 ± 80
Brent	Ontario	3.8	450 ± 30
Flynn Creek	Tennessee	3.6	360 ± 20
Steinheim	Germany	3.4	14.8 ± 0.7
New Quebec	Quebec	3.2	<5
Meteor Crater[b]	Arizona	1.1	0.05

[a] Two craters lie almost tangent to each other. They may indicate fracture of a meteorite into two pieces in the atmosphere or impact by a body and its satellite.

[b] Meteorites are known at this location and at many lesser sites in various locations, ranging down to clusters of pits caused as a meteor broke up in the atmosphere. The smallest example is an 11-m single pit at Haviland, Kansas.

Source: Based mainly on Grieve, R., and four others (1988): "An Astronaut's Guide to Terrestrial Impact Craters" (Houston: LPI Tech. Report 88–03).

Note: Table lists probable meteorite impact craters, with known evidence of mineral alteration due to shock wave metamorphism. List is complete down to 28-km diameter, with selected smaller examples.

momentum of the meteorite and the nature of the surface materials. Because the explosion center is below the ground, strata that initially lay flat are heaved upward and outward, bent back, and even overturned in big flaps like petals of a giant flower opening. The rim is built partly from this upthrust rock and partly from excavated debris dumped on the crater edge.

An issue in planetary science is the size of meteorite that produced any specific-sized crater on surfaces of different worlds, with their different surface materials, gravi-

Figure 9-5. An ancient, eroded impact crater on Earth: Brent crater, Ontario, Canada. The circular structure has been partly filled by glacial and other erosive activity, leaving two lakes (black) in the original interior. The crater's diameter is 4 km and it has been dated at 450 ± 30 My. Other larger and older craters are known in the Canadian shield, an old and stable region of Earth's crust. (Aerial photo, Earth Physics Branch of the Department of Energy, Mines, and Resources, Ottawa, Canada)

ties, and so on. Valuable mathematical tools have been developed to scale—from lab-sized impacts to large craters in nature, in order to relate meteorite size and crater size (Holsapple and Schmidt, 1982; Schmidt and Housen, 1987; Chapman and McKinnon, 1986; and especially the book by Melosh, 1989). These relations allow scientists to predict the crater size and shape that would result if a given meteorite hit a given world at a given speed. General results are shown in Figure 9-6.

In detail, crater diameter is found not to depend directly on energy but on a combination of energy, surface gravity, projectile size, and target strength. An interesting consequence is the dividing of cratering phenomena into two regimes. The *strength regime* applies to smaller energies and craters, in which the final crater shape is found to depend on the strength of the target materials. The *gravity regime* applies at larger energies and sizes. Imagine the bowl-shaped *transient* cavity blasted out an instant after an impact explosion. In the gravity regime, the cavity is large; the weight of the steep wall is too great for the rock layers to support the initial shape and slumping occurs, so that the final shape is controlled ultimately by gravity. In the strength regime, such as a bullet fired into rock, the

crater is simply chipped out of the rock. The transition size depends on the gravity and target materials, but Melosh (1989, p. 120) gives a transition from strength to gravity regimes at about a 220 m diameter for craters in lunar basaltic lava flows. Photos show that craters below this size typically have more ragged shapes and broken bedrock fragments littering the floor whereas at larger sizes, craters begin to assume more symmetric, bowl shapes. At still larger sizes, generally 10 km or so, more complex forms appear, as clarified below.

As a rule of thumb (see Figure 9-6), the ratio of crater diameter to meteorite diameter for most cosmic impacts ranges from about 20 for 100-m craters, to about 10 for 1–10 km craters, to around 5 for large, 100-km craters.

Features of Impact Craters

Craters formed by meteorite impacts are often called **primary impact craters.** Figures 9-7, page 252, and 9-8, page 253, show some common features. The rim has a hummocky structure grading outward into a thinner layer of debris (Figure 9-7b). All this externally deposited material is called the **ejecta blanket.** Discrete ejected blocks

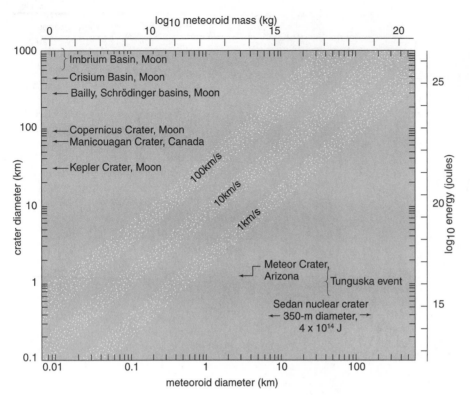

Figure 9-6. The diameter of an impact crater as a function of meteorite size for three different impact velocities. Scale at right gives the estimated energy required to make each crater. Documented craters on Earth and the moon are shown. The assumed meteorite density is 3 g/cm³. (Crater energy and diameter data from Baldwin, 1963; Wasson, 1974, p. 145; Vortman, 1977)

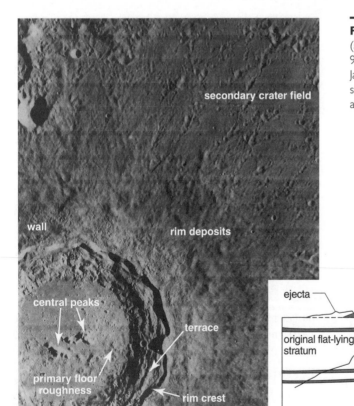

Figure 9-7. Features of impact craters. (**a**) Vertical view of the relatively fresh 90-km lunar crater Copernicus. (Courtesy James Head, Brown University) (**b**) Cross section of a typical large crater showing additional features (see text).

a

b

c

Figure 9-8. Features of impact craters. (**a**) View along inner wall of Meteor Crater, Arizona, showing upended strata. Original beds (as found nearby) lie flat. (Photo by author) (**b**) View toward the lunar crater Copernicus (see Figure 9-7a) on the horizon. Ejecta has been thrown over the Carpathian Mountains (the Imbrium rim, middle distance), littering mare surface with secondary craters and bright rays. (NASA; National Space Science Data Center) (**c**) View of the moon, centered directly above the young crater Tycho, showing the Tycho rays, which can be traced more than 1000 km. Note that many rays are tangent, not radial, and begin outside a dark nimbus around the crater's rim. This photo was made by projecting an Earth-based photo on a globe and rephotographing the projected image from above Tycho. (Lunar and Planetary Laboratory, University of Arizona)

Figure 9-9. Test explosion of 91 metric tons of TNT illustrates features of crater formation (see text). This 1972 test in Colorado produced a crater 9 m across and 7 m deep with secondary impact craters up to 110 m away and ejecta as much as 201 m away. (Photo by author)

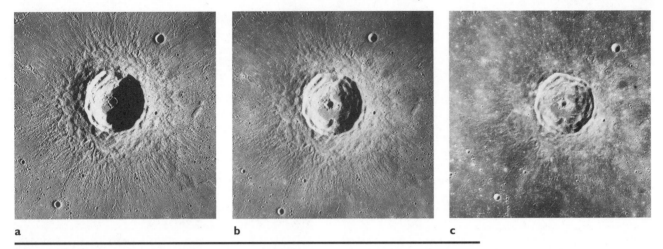

Figure 9-10. Three views of the young, 35-km lunar crater Timocharis under different lighting. Low lighting exaggerates relief; high lighting brings out ray material and bright crater walls. (NASA, National Space Science Data Center)

Figure 9-11. Surface views of impact craters. (**a**) North Ray crater, near Apollo 16 site in the lunar uplands. Crater is about 900 m across and about 50 My old according to rock sample dates. (NASA) (**b**) Interior of Meteor Crater, Arizona, about 1100 m across and 20,000 y old. (Wide-angle photo by author)

Figure 9-12. Central mounds in two small craters. (**a**) Terrestrial test explosion crater about 30 m across, formed by 109 t of explosive. Central mound may be related to a resistant caliche layer below dusty sediments in which the crater was formed. (Miser's Bluff, Arizona; photo by author) (**b**) Lunar crater adjacent to Apollo 11 site. Central mound may be related to a resistant lava layer below surface regolith. (NASA)

Figure 9-13. Lunar crater about 150 m across in Oceanus Procellarum shows terraced rim of ejecta blanket of boulders averaging about a meter in size. These features may indicate that the crater intruded a layer of intact lava below the regolith surface. (NASA, Orbiter 3)

or clumps can fall back to the surface and form **secondary impact craters,** well displayed in Figures 9-7b and 9-8b. Powdered and melted material resolidified as glassy beads is thrown out at very high speed (as much as 1 km/s or more) and leaves long, bright, linear deposits called **rays** (Figure 9-8c). The rays radiate from the primary crater, often with secondary craters clustered along them.

Drilling reveals other features. The crater floor is typically a lens-shaped mass of breccia, rubble, and small amounts of lava produced by melting during the impact. This is ejecta that has fallen back into the crater's original cavity. Below this is highly fractured **bedrock** (rock not moved out of its original position). The fractures typically penetrate about 3 times deeper than the depth of the crater itself.

Some idea of the origin of crater features can be gleaned from Figure 9-9 on page 253, showing the explosion of 91 t of TNT in a cratering experiment. The turbulent cloud around the base of the fireball is expanding, soil-laden gas called a **base surge,** which deposits some of the hummocky, dunelike ejecta around the crater rim. The high-speed jets angling upward may be analogous to the spurts of material that created ray systems.

Visibility of crater features depends strongly on the lighting angle, as shown by Figure 9-10 on page 254. Rays are prominent under high light but disappear under low light; but low light brings out relief and allows rim and mountain heights to be measured from the lengths of shadows.

Simple Craters, Complex Craters, and Multiring Basins

As mentioned above, the structural features of craters change with increasing size. On the Earth and moon, fresh craters about 1 km across tend to have smooth, bowl-shaped interiors and are called **simple craters.** Examples are seen in Figure 9-11. At larger sizes, the floor flattens, although resistant layers at the crater floor level can cause central mounds or other uneven structures on the floor (Figures 9-12 and 9-13). At still larger sizes, the floor develops a full-fledged **central peak,** or mountain mass, such as seen in Figures 9-7 and 9-10. Central peaks are probably formed by a rebound phenomenon like the rebound of a droplet in a coffee cup. Terraces may also appear on the inner walls, apparently due to slumping of the rim inward. Craters with such features are called **complex craters.**

The transition from simple to complex occurs at smaller sizes on larger planets because of their larger gravity. For example, lunar central peaks are most common in young craters larger than 60 km in diameter; Martian central peaks, in craters larger than 10 to 30 km; and Earth's, in craters above 1 to 3 km. The difference relates to the wall height that can be sustained without slumping, given the gravity of the planet.

Still other features occur at larger crater diameters. For example, at about 100 to 300 km on the moon, Mars, and Mercury, a rare transitional shape develops, in which the central peak broadens and turns into a ring of hills, or **peak ring,** a term coined by Hartmann and Wood (1971), as shown in Figure 9-14 on page 256.

Finally, the largest impact features are huge systems of concentric rings, called **multiring basins** (or just **basins**). Figure 9-15 on page 257 shows spectacular examples on three worlds. Some of the rings, especially the inner ones, may be only roughly defined circles of hills, sometimes partly flooded by lavas that have covered the basin's inner floor. Other rings, often outer rings such as the Apennine arc around the lunar Imbrium basin, have a well-defined crest resembling the rim of an ordinary crater. These outer rimlike rings may be the true rim of the original impact crater, with other rings being rebound or slump features. Curiously, well-developed rings are often spaced at intervals of about √2 times the inner ring radius (that is, 1.0, 1.4, 2, 2.8 . . . radii from the center). Multiring basins are the largest individual geological structures in the solar system. Some examples on the moon had been recognized before Apollo. Concentric, multiring lunar structures were described by Baldwin (1949, 1963), but they were more clearly recognized as a class after a number of others were discovered by "rectified" photography (projecting ordinary lunar photos on a globe and then photographing selected regions from "overhead." The great Orientale system (Figure 9-15a) was discovered in this way by Hartmann and Kuiper (1962) prior to lunar mapping by spacecraft.

Discovery of additional multiring basin systems on Mercury, Mars, Callisto, and Ganymede spurred interest in understanding their formative processes and the roles of giant impacts in establishing crustal heterogeneity on the planets. They appear to provide fracture systems allowing lava to gain surface access (Hartmann and Wood, 1971), and their ring spacings may be indicators of subsurface

a

b

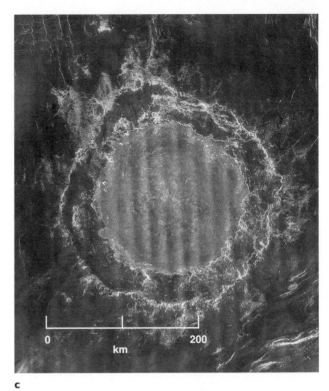

c

Figure 9-14. Craters with peak rings, a transitional form between central peaks and multiple concentric rings. (**a**) Lunar crater Schrödinger, about 320 km across. (**b**) Mercurian craters Ahmad Baba (top) and Strindberg (200-km diameter, bottom). (**c**) Venusian crater Mead, about 280 km across, is the largest on Venus and shows $\sqrt{2}$ ratio between its outer ring and inner peak ring diameters, as opposed to 2:1 ratios in above cases. (NASA)

layering (Wilhelms, Hodges, and Pike, 1977). In addition, they are the centers of vast systems of radiating valleys and ridges that suggest profound fractures radiating from the impact sites (Figures 9-15a and b). Multiring systems on the icy satellites often appear more like shatter patterns in glass (Figure 9-15c), an appearance that may be related to the thinness of the brittle lithosphere of solid ice.

Utilizing Impact Craters to Learn About Planets

Impact craters give several types of evidence about planets. They excavate and expose material. Their central peaks expose material originally about one-tenth of the crater diameter below the surface. For example, Pieters (1982) found that the central peak of the lunar crater Copernicus has a spectrum not seen in other lunar areas, and he interpreted it as olivine-rich, upthrust material from a 10-km-deep crustal layer. As indicated in Figure 9-16 on page 258, the structure of an ejecta blanket may indicate properties of the material at the impact site. Figure 9-16 shows one of many Martian ejecta blankets with lobate structure (lobe-shaped sheets extending from the crater). These are

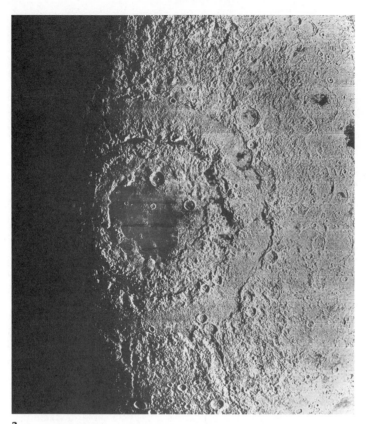

a

Figure 9-15. Multiring basins on three worlds. The systems nearly match, with outer rings being about 1300 km to 1800 km in diameter, or capable of stretching from the Great Lakes to the Gulf of Mexico. (**a**) Orientale Basin, a bulls-eye on the limb of the moon, discovered in 1962, has a well-defined outer ring at 930 km diameter (and a fainter 1300 km surrounding ring not seen as well here). (**b**) Caloris Basin on Mercury, discovered in 1974 by Mariner 10, has dimensions similar to the previous example. (**c**) A view of Callisto shows a large multiring basin with closer spacing of rings. The bright center may mark flooding of the initial impact basin by fresh ice. The outer rings span about 1800 km diameter. (NASA images)

b

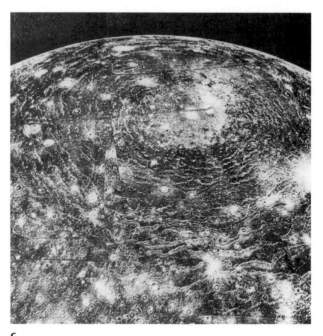

c

Figure 9-16. Lobate ejecta around the 25-km Martian crater Arandas. This pattern, unique to Mars, is interpreted as resulting from impact into water- or ice-bearing soil. Patterned ground to the right is also possibly related to the freeze-thaw cycle of ice in the soil. (NASA, Viking 1)

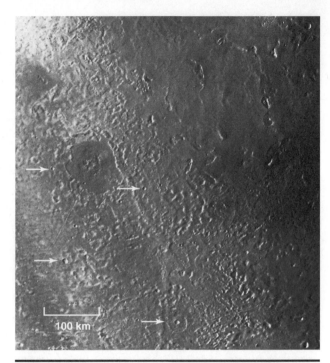

Figure 9-17. Lack of abundant impact craters is a sign of a youthful surface, recently resurfaced. This part of Neptune's icy satellite, Triton, shows rolling hills similar to "cantaloupe terrain" found in other areas. Very small impact craters (arrows) are sparsely scattered through this region—similar in number to those of lunar maria. In upper right are darker, smooth plains with still fewer impact craters. Narrow linear fractures and smooth "ice lake" (left center) can also be seen. (NASA Voyager 2)

unknown on the moon and Mercury. Such craters have been termed **rampart craters;** they are attributed to impacts into soils containing large amounts of ice or water, which formed muddy ejecta flows with entrained gas (Carr and others, 1977). Rampart craters are the most common craters in many areas of Mars, though other areas contain primarily the lunar type of crater, probably caused by impacts into drier soil.

Craters as Tools for Dating Surfaces

The most important use of craters has been to date planetary surfaces. On an erosion-free surface, such as a lava flow on an airless world, meteorite craters accumulate with time. Thus two types of dating are possible.

The simpler but less informative type of dating measures the **relative age** of geological provinces by merely measuring their crater densities (craters per square kilometer). On a given planet, a less-cratered province is younger than a more-cratered province. Relatively uncratered surfaces, as found on Europa, Io, Triton, and parts of other satellites, are thus recognized at once as being unexpectedly young (Figure 9-17). **Stratigraphic relationships**—that is, young formations overlapping older formations—assist in establishing the relative ages of different units on a single world.

The more exact type of dating determines the **absolute age** in years. This can be estimated if the number of craters/km^2 is counted and the rate of formation of craters is known. Formation rates are indicated in Figure 6-5 and Table 6-1. Unfortunately, crater formation rates outside the Earth-moon system remain quite uncertain—errors in age may be a factor of $\pm$ 2 or 3. This sounds like a large error, but in reading a world's history it is essential to sort out whether stratigraphic units date from primordial times a few 1000 My ago, or only a few 100 My years old (essentially current geological times, indicating ongoing activity). Absolute ages could be improved by better knowledge

of the asteroid/comet impact rate, or by datable rock samples from certain geological provinces on other planets; these data could then be used to calibrate the relationship between crater density and relative age for each planet.

Stratigraphic Studies of the Earth-Moon System

Systematic work on crater densities was begun when U.S. Geological Survey researchers Shoemaker and Hackman (1962) laid out the beginnings of a stratigraphic system for the moon. Just as terrestrial geologists developed a system of names such as Cambrian and Jurassic (often from geographic locales) to designate periods whose relative ages were established from fossils, Shoemaker and Hackman chose names such as Imbrian and Copernican (from lunar locales) to designate periods whose relative ages were established from crater counts and overlap relations.

In the 1960s, Geological Survey geologists preparing for the Apollo missions mapped much of the moon's stratigraphy in this way (Mutch, 1970). The Apollo missions then gave absolute calibration for certain geological provinces, allowing the time scale based on lunar crater counts to be calibrated globally. Absolute dates can now be estimated for all major lunar features.

Table 9-5 compares the stratigraphic systems built up for the Earth-moon system in this way. Earth's stratigraphic column has a hierarchy of divisions including broad eras and finer periods (and even finer subdivisions).

Note that fine detail appears in the terrestrial system only in the last 10% of solar system history (the only interval with a good fossil record and abundant rock samples), whereas fine detail appears only for the earlier history of the moon because that was the period of active volcanism and intense cratering. The important point is that the two

TABLE 9-5 Stratigraphic Systems Used for Earth and the Moon

Absolute Time Scale (Gy before present)	Earth		Moon	
	Era	Period (or system)	Period (or system)	Events[a]
0 ----------------				
	Cenozoic			Mammals Modern continents forming
	Mesozoic	Cretaceous		
		Jurassic		Dinosaurs
		Triassic		Ferns, conifers
	Paleozoic	Permian		
		Carboniferous		
Note expansion of scale (0 to1 Gy)		Devonian		Fish
		Silurian		Early land plants
		Ordovician		
		Cambrian		Earliest well-formed fossils (trilobites, etc.)
1 ----------------	Precambrian or Proterozoic	Late	Copernican	Sporadic large craters forming on Earth and the moon
		Middle		Oxygen increasing in Earth's atmosphere
2 ----------------		Early	---------- ?[b] ----------	Soft life forms in Earth's seas; stromatolite fossils forming at seacoasts
			Eratosthenian	Decline of mare lava flooding on the moon
3 ----------------				Crustal and atmospheric evolution on Earth
	Archean	Archean	----------------------	Early microscopic fossils on Earth (poor record)
				Mare flooding on the moon
			Imbrian	Oldest terrestrial rocks
4 ----------------			----------------------	
			Nectarian	"Modern" basins forming on the moon Intense cratering
5 ----------------			Pre-Nectarian	Magma ocean and earliest features (obliterated by subsequent cratering)
				Planet formation

[a] Events are on Earth unless specified otherwise.

[b] Date of this division is uncertain.

Note: The Quarternary (0–3) and Tertiary (3–62 My) periods are too brief to show within the Cenozoic Era.

a

b

c

Figure 9-18. Examples of surfaces with cratering near saturation equilibrium. (**a**) Lunar upland surface around Bailly basin (left center). (Lunar Orbiter photo, NASA) (**b**) A portion of the cratered uplands of Saturn's satellite Rhea. (NASA) (**c**) An imaginary surface generated in a computer simulation that creates realistic craters and ejecta blankets on an initially craterless surface. Note rolling terrain created by old craters that have been partly eroded and blanketed by younger craters. (Robert Gaskell, Jet Propulsion Laboratory)

systems dovetail and rough out the entire history of the Earth-moon system, providing a complete story that could not have been obtained from either body alone.

Crater Saturation Equilibrium

Before looking at actual data on crater counts and absolute ages, we need to consider that there is an upper limit to how many craters can accumulate on a surface. Consider a fresh lava flow on a planet. As time passes, the surface will accumulate craters. Eventually a point is reached where craters are so crowded that new impacts destroy old craters as fast as they make new ones. This condition is called **crater saturation equilibrium;** the observable crater density can't rise much beyond this level. Figure 9-18 shows some examples of surfaces that are saturated, and

Figure 9-19 shows crater counts for several such surfaces. The saturation crater density is roughly 30 times that found among the multikilometer craters on the lunar maria. As shown in Figure 9-18c, realistic computer simulations confirm that crater densities saturate and stop rising at about this same value (Hartmann and Gaskell, 1997). It might seem that on these surfaces more small craters could be added, but in a realistic impactor population, adding small craters leads statistically to a large impact that obliterates many small features, so the density never rises much above these values, and the general character stabilizes.

A surface that is nearly saturated with craters must be about 3900-4400 My old, because only surfaces dating back to the era of early intense bombardment acquired enough hits to approach saturation. We can't tell from photos whether a given saturated surface has been exposed long enough to absorb enough hits barely to achieve saturation or long enough to absorb many times more craters (which would have been obliterated). Surfaces anywhere in the solar system younger than 3500 My should have less than the crater densities shown in Figures 9-18 and 9-19.

Saturation equilibrium cratering is also important because it leads to the formation of deep megaregoliths on planets, as described in our previous discussion of megaregolith formation (see Figure 9-4). Although cratering has covered only about 3% of the lunar maria with craters larger than 4 km and has made a regolith in those areas about 10 m deep, cratering has covered about 100% of the highlands with such craters and thus stirred the surface to a depth of at least 1 or 2 km.

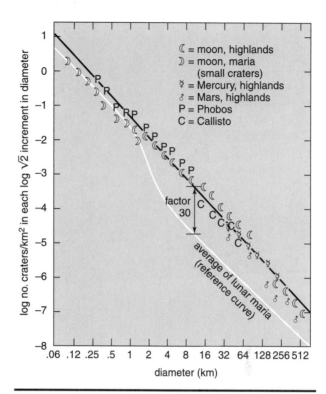

Figure 9-19. Crater counts on some of the most heavily cratered surfaces of the solar system mark a fairly well-defined band in the crater-diameter distribution diagram, with about 30 times the crater density on the lunar maria. This band may approach the maximum number of visible craters that can be crowded on a planet, given the size spectrum of impacting meteorites, and it marks surfaces that have developed megaregoliths.

Crater Counts and Isochrons

We can estimate the mean effective crater production rate for the last 3,500 My on different planets, using Apollo lunar data and Table 6-1. Combining this with the saturation equilibrium level reached during the intense early bombardment about 3,900–4,400 My ago, we can prepare diagrams of crater diameter distribution, complete with isochrons, or lines of equal age, showing crater densities that would be reached in 10^8 y, 10^9 y, and so on. These diagrams are shown in Figure 9-20 on pages 262–263. The isochrons are shown as stippled bands to indicate the uncertainties involved in the predictions. Figure 9-20 compares observed numbers of craters with predicted numbers for surfaces of various ages on 11 planetary bodies. The small, airless worlds have retained the most craters of various sizes. Large, geologically active worlds have younger surfaces and fewer craters. These diagrams contain key data for understanding the evolution of planetary surfaces in the solar system.

Crater Retention Ages

What age is actually measured when we count primary impact craters? If the surface is a sparsely cratered, thick

lava flow on an airless world, the number of craters measures the time since the flow's origin. But if the flow is on a planet such as Mars, where dust blows into craters and other erosion occurs, small craters may last only (say) 1 My and large ones, 100 My. In these cases, the number of craters of a certain diameter indicates not the age of the surface but the length of time a structure with that dimension can withstand erosion. Thus, ages measured by crater counts have been termed **crater retention ages** (Hartmann, 1966). With care, these ages can be used not only to learn about ages of certain features but also to interpret rates of erosive activity. For example, note in Figure 9-20f that in the Martian polar sediments, 30-km craters have accumulated over about 10^9 y, but the crater curve bends over at a shallow slope, indicating that the sedimentation rate is so high that 1-km craters have lasted only around 10^8 y before being obliterated.

The D_L Method of Dating

Soderblom and Lebofsky (1972) developed a different technique. They noted that if all primaries of a given size began life as bowls of a certain shape, subsequent impacts would have required a fixed amount of time to erode the

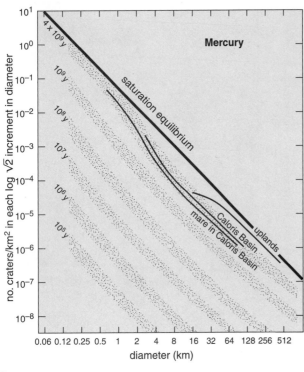

a

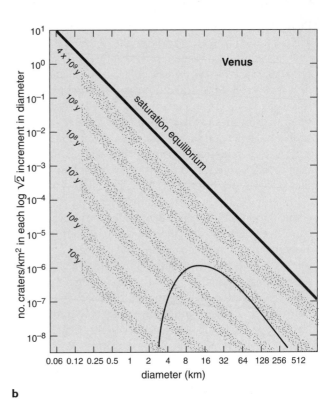

b

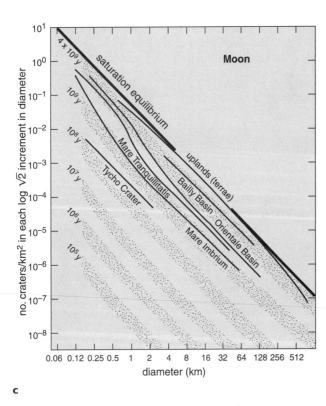

c

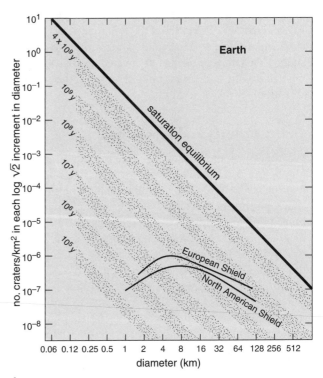

d

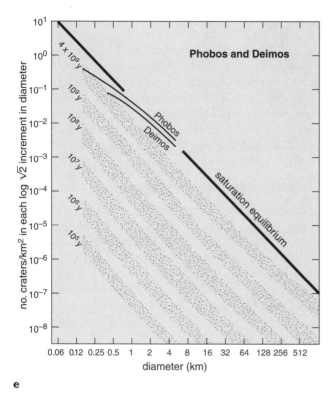

e

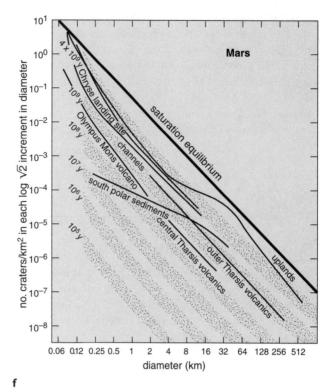

f

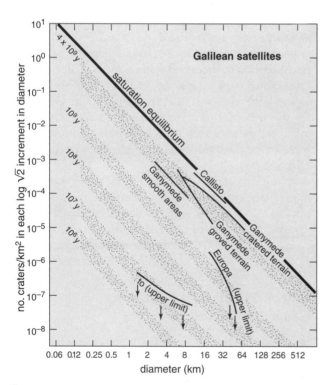

g

Figure 9-20. Crater counts and isochrons for 11 planetary bodies. Solid lines give schematic observed diameter distribution of craters. Dotted isochrons are estimates of numbers of primary impact craters expected for surfaces that have preserved all craters since the times of formation listed at left. Data show that Phobos, Deimos, the moon, Callisto, Ganymede, and Mercury are relatively primitive bodies with surfaces that are a few billion years old in most regions. Mars has many old surfaces but also some extensive younger volcanics, perhaps a few hundred million years old, and evidence of erosive loss of smaller craters in some regions. Venus and Earth have regions that have preserved large craters for timescales around 800 My. Smaller craters have been eroded on Earth. Venus's thick atmosphere prevents formation of craters <3 km in diameter. Europa's surface (based on the detection of three 20-km craters) is perhaps a few 100 My old. Io is being resurfaced, probably in less than 1 My, by volcanic eruptions.

bowls to a degraded state where their interior walls had a shallow angle of, say 1°. A parameter called D_L ("D sub L") was defined as the diameter of a crater eroded to this state of degradation. D_L was then used as a measure of the age of the surface. A fair correlation exists between Apollo age data, crater counts, and D_L ages for various surfaces. Because D_L can be found for quite small craters (100 m or so), it is a useful criterion for dating local surfaces.

Microeffects on Airless Surfaces

Micrometeorite Effects

We have been discussing effects caused by the impact of *macro*scopic meteorites. Still more abundant are *micro*meteorites. Their integrated smoothing or eroding effect is called **sandblasting.** The coarser plowing effect by somewhat larger (up to centimeter-scale) meteorites is known as **gardening. Pitting,** or **microcratering,** is the production of microscopic or subcentimeter-scale craters by impacts of individual micrometeorites on rock surfaces. Most lunar boulders, for example, have an array of microcraters (often called "zap pits"), whose number density depends on the length of time the rock was exposed on the surface. A given rock may be exposed for a while and then turned under by gardening; it may even go through several cycles of exposure.

On icy satellites like Callisto and Ganymede, the most heavily cratered surfaces are generally darkest. This may indicate a process on two-component ice-stone worlds whereby microcratering preferentially sublimes the icy component while the stony material remains behind. Long exposure to cratering would thus tend to build up a dark, dust-enriched regolith on the surface of these worlds, even if the interior is ice-rich (Hartmann, 1980b). The very slow sublimation of the ice component because of weak solar heating at 5 AU would help produce the same effect. New craters would punch through the dark layer and expose fresher, brighter surface with less dirt, as well seen on Callisto.

Consistent with this idea, the highest albedo in the solar system (exceeding 90%) belongs to Saturn's sparsely cratered Enceladus, which appears to be extensively resurfaced by fresh ice. More cratered Saturnian moons, such as Rhea and Mimas, are darker, with albedos around 60%. Within the Uranian system, the highest albedos belong to the two moons with the most evidence of resurfacing: Miranda and Ariel (though the Uranian system albedos are 20% to 40% less than those in the Saturn system, for unknown reasons) (Smith and others, 1986).

Clark (1980) and others emphasized an important related effect. Intimate mixtures of ice and carbonaceous dust tend to show the dark coloration of the dust, even if the dust amounts to only a percentage or so of the mixture, because light penetrates through the ice grains until it is absorbed by the dust. Thus, dark outer solar system surfaces such as Callisto (albedo 19%) or C-type and D-type asteroids (albedo 4%) could contain large percentages of ice in their surface regoliths even though they are very dark. Clark (1980) suggests 30% to 90% ice by weight in Callisto. He finds that a frost layer on such dark materials would produce an obvious ice spectrum only if the frost thickness exceeds about 0.6 mm.

Glasses and Iron Coatings

Glass is produced during impacts when molten ejecta is cooled so fast that it does not have time to form a crystal lattice structure. Lunar craters up to 10 cm across are commonly lined with glassy material, and larger craters have also splashed glassy blobs or sprayed out fine beads of glass (Figure 6-9) that are mixed with the nearby soil. As a result, most lunar soil has a component of microscopic glassy beads, and the amount increases with the age of the soil. Soils near a young crater at the Apollo 14 site have 10% glass; mare lava soils, 20% to 40%; upland soils, 40% to 85% (Short, 1975).

The glass beads have different colors and brightness, depending on the minerals in the parent rock that melted to produce them. The beads also tend to darken with age as volatile elements are lost during shock events from subsequent gardening. Further, small impacts produce a micro-coating of vaporized iron on surfaces of soil grains, first pointed out by Hapke and others (1975). As later recognized, the iron, as well as the glass, suppresses some of the spectral absorption bands of some minerals, and thus may alter the spectrum of regolith-covered bodies, compared to the expected spectrum of the bedrock (Clark and others, 2002).

Both small and large impacts send shock waves, with momentary pressures up to several hundred kilobars, through nearby rocks and soils. Resultant **shock damage** can include alterations of the crystal shapes of certain crystals, partly melted fragments, or microfractures.

Albedo Effects and Sputtering Caused by Irradiation

In addition to micrometeorites, atomic particles such as protons and electrons strike airless surfaces. These come directly from the solar wind and may also be trapped in the Van Allen belts formed in magnetic fields of planets, where close satellites orbit. Hapke's (1965) experiments showed that such irradiation can redden and darken rocky soils, accounting for the reddish tinge and low albedo of Mercury and the moon. Astronauts, however, found that the uppermost lunar surface is usually a bit lighter than the soil a few centimeters below. Glasses may play a role, but researchers also suspect an additional bleaching effect from radiation.

Sputtering is microscopic erosion caused when low-energy ions strike a solid surface. (High-energy particles penetrate farther and produce other effects.) One result, in addition to erosion, is a loose cementing of granular materials. Principal sputtering agents in the planetary environment are solar protons and alpha particles in the energy range of 1 to 20 keV.

The moon and other small bodies lose mass to the sputtering process because some rock particles are blasted off into space. KenKnight, Rosenberg, and Wehner (1967) estimated that exposure of the lunar surface for 4500 My resulted in the loss of a layer a few centimeters thick.

Compositional Effects

Impact experiments show that each meteorite dislodges 100 to 1000 times more local material than its own mass; therefore there will be an admixture of only a small fraction by mass of meteoritic material in planetary regolith (Gault, Heitowit, and Moore, 1964). Therefore, the external meteorite component gets diluted in planetary soils. Lunar mare soil samples have only about 1.5% to 2% meteoritic (mainly Type 1 carbonaceous chondrite) material, indicating an infall rate of about 4 $g/cm^2/Gy$ on the moon in the last few Gy (Ganapathy and others, 1970). Addi-

tional cratering churns soils to greater depth, so that the highland soils have not accumulated greater concentrations of meteorite material (Taylor, 1975).

Compositional effects also arise on the atomic level. The solar wind strikes airless surfaces and implants atoms in the soil. The lunar soil has about 0.005% to 0.01% by weight of hydrogen from this source. As the moon lacks water, this hydrogen may someday be an important resource for astronauts. It is easily removed by mild heating and is thereby available for combination with oxygen to convert 0.05% by weight of lunar soil into water (Williams, 1976; Arnold, 1977). Similarly, around 0.01% of the lunar soil is a mixture of rare gases such as helium, neon, and argon, also trapped from the solar wind.

Another change on the atomic level is the production of rare or short-lived isotopes, such as helium 3 or aluminum 26, by **spallation,** the splitting of heavier atoms by impacts with atomic particles. Concentrations of such materials give ways of dating surface exposures of lunar soils. For example, material from the 64-m South Ray crater visited by Apollo 16 astronauts turned out to have been lying on the surface for only 2 My, making this crater one of the youngest sampled. Most randomly sampled lunar soils have roughly 400 My exposure or residence time at the surface, but with much variation (Taylor, 1982).

Figure 9-21. Saturn's moon Iapetus has dark surface material on its leading hemisphere (lower-left side in this view from above the north pole) and bright material on its trailing hemisphere (upper right). Boundary of two materials is ragged and well defined. (NASA, Voyager 2)

Production of Carbon and Colored Organics

As noted in Chapters 5 and 7, cosmic ray or solar wind protons, hitting methane ice on outer solar system surfaces, can break carbon bonds, create free radicals, and alter chemistry. Colored organic compounds may be built up, as demonstrated in lab experiments. Histories of organic materials on surfaces may be thus complex, and as noted in Chapter 7, some of the red colorations of Trojan and Kuiper belt objects may be caused by this process.

The Dark Side of Iapetus: A Case Study

As discovered telescopically, the leading side of Iapetus has an albedo of only 5%, whereas the trailing side is at least 6 times brighter, with albedo 28% to 60% (see Figures 2-26 and 9-21 on page 265). The bright trailing side is primarily water ice.

    Beginning with Cruikshank and others (1983), various observers showed that the dark side spectrum is similar to the reddish black D asteroids and matches a mixture of organic compounds from carbonaceous meteorites, ice, and hydrated silicates (Bell and others, 1985; Vilas and others, 1996). But why is this material all on one face of Iapetus? As first pointed out by Soter (1974), Enceladus is hit by dust knocked off Phoebe, the next moon outward. Phoebe is a retrograde-orbiting black object of spectral class C, probably captured. When meteorites blast dust off its surface, the dust spirals inward toward Saturn, hitting the leading side of prograde Iapetus at unusually high speed. A buildup of only a few percentages of dark dust would darken an icy regolith on Iapetus. This would explain everything except that the Phoebe dust is neutral black C material, whereas the Iapetus dark side has the reddish-black taxonomic type, D. The reddish color on Iapetus may come from a synthesis of colored organic compounds resulting from cosmic ray impacts, utilizing fresh methane-bearing Iapetus ice exposed during the Phoebe dust impacts (Strazzula, 1986; Wilson and Sagan, 1995). The irregular outlines of the white/black borders (Figure 9-21) may result from later cratering or endogenic events (Smith and others, 1982).

Figure 9-22. Albedo variations stand out on the impact-eroded surface of lumpy-surfaced Deimos at nearly full-phase lighting. Some bright areas are clearly associated with the younger impact craters, but others may mark "downslope" migration of soil particles, exposing fresher material. (NASA, Viking Orbiter)

Space Weathering

All the effects discussed above, such as glass production, sputtering, cosmic ray damage, and so on, are lumped under the term **space weathering.** Space weathering microscopically alters the surfaces of minerals, and suppresses spectroscopic features such as mineral absorption bands. It explains why many asteroid spectra don't quite match spectra of proposed equivalent meteorite types, as discussed in Chapter 7 (Clark and others, 2002). These effects are not fully understood, but it is clear empirically that on some small bodies, at least, fresh craters expose unaltered native material that is different from the space-weathered surface soils (Figure 9-22).

SUMMARY   Two subjects dominate this chapter—petrology and cratering. Surface petrology divides naturally into silicate surfaces in the inner solar system and carbonaceous/icy mixtures in the outer solar system. A first-order understanding of silicate surfaces comes from a simple differentiation model in which low-density feldspar minerals accumulate in anorthositic crusts as magma oceans cool. Eruptions of mantle melts and partial melts have produced basaltic volcanism on the surface of terrestrial bodies, though Mercury's volcanism may be somewhat nonbasaltic.

    Beyond 2.7 AU most surfaces are dark because of the carbonaceous component, unless thermal activity has produced water eruptions, bright ice flows, or bright frost condensates from escaping vapors. Water ice and frost are important in the Jupiter-Uranus region, but more volatile ices, including methane (CH_4), nitrogen (N_2), and carbon dioxide (CO_2), are important at still greater distances, and the CH_4 may be important in creating organic compounds through cosmic ray bombardment.

All surfaces have been battered by impact craters, which give a clue to their age and level of geologic activity. Primitive bodies—the ones that have little modification by internal activity—are heavily cratered, often saturated by craters during the intense early bombardment that accompanied sweep-up of planet-forming debris. The competition between rock formation and rock pulverization was initially won by cratering, producing deep megaregoliths on the most primitive surfaces. Later eruptions produced lava plains where later cratering accumulated to different degrees. Crater counts thus produce useful information about the age, erosional state, and subsurface character of those surfaces.

Of course, not all planets were dominated by primitive, heavily cratered surfaces. On large, less primitive worlds, lava flows continued to erupt because thermal histories were more protracted. Volcanic edifices built up. Tectonic effects may have faulted or folded the surface, and winds or water may have stripped materials from some regions and deposited them in sedimentary blankets elsewhere. The next chapter discusses those modifying effects in greater detail.

CONCEPTS

petrology	phaneritic
minerals	extrusive
rocks	aphanitic
acidic	glass
siliceous	basaltic rock
basic	gabbros
mafic	basalts
ultrabasic	anorthosite
ultramafic	granitic rock
feldspars	granite
orthoclase	rhyolite
plagioclase	fractionation
quartz	reaction series
silica	sedimentation
pyroxenes	weathering
amphiboles	shale
micas	sandstone
water ice	breccia
carbon dioxide ice	evaporites
ammonia ice	metasomatism
methane ice	KREEP basalts
olivine	norites
iron oxides	troctolites
magnetite	Genesis rocks
maghemite	regolith
hematite	megaregolith
goethite	prelithospheric chaos
limonite	lithospheric evolution era
troilite	mare era
graphite	post-mare era
carbonaceous minerals	crater
clay minerals	astrobleme
montmorillonite	hypersonic
nontronite	shock wave
igneous rock	primary impact crater
sedimentary rock	ejecta blanket
metamorphic rock	secondary impact crater
lava	ray
intrusive	bedrock

base surge
simple crater
central peak
complex crater
peak ring
multiring basin
rampart crater
relative age
stratigraphic relationships
absolute age
crater saturation equilibrium

crater retention age
D_L
sandblasting
gardening
pitting
microcratering
shock damage
sputtering
spallation
space weathering

PROBLEMS

1. Why did the discovery that the lunar highlands are a relatively uniform mass of anorthositic gabbros and anorthosites imply that a lunar magma ocean had once existed?

2. Why are the lunar highlands heavily cratered and the lunar maria only sparsely cratered?

3. Compare surface rock types that you might expect on an imaginary, 5000-km-diameter planet of bulk composition like Earth given the following conditions:
(a) The planet had never melted.
(b) The planet had a magma ocean with very little convection or vigorous stirring.
(c) The planet had a magma ocean in which there had been vigorous convection and efficient differentiation of siliceous and basic minerals.
(d) The events in (c) had occurred but were followed by extensive volcanism that tapped magmas in the planet's upper mantle.

4. How would discovery of wide regions of granite on Venus imply a history different from that implied by discovery of a wholly basaltic or gabbroic crust?

5. Why is Mars red?

6. Why might evaporite deposits be common on some outer solar system satellites if they had near-surface heat sources such as tidal heating or radiogenic heating? What surface material is most common on these bodies in the absence of such heating?

7. Why are feldspar-rich crusts probably thinner on the larger terrestrial planets than on the moon?

8. If the lunar (or other) magma oceans solidified in the first 100 My or so of planetary history, why are 4.5-Gy-old Genesis rocks so rare on these worlds?

9. Compare the appearance of the surface topography (especially the presence of heavily cratered highlands) on two moonlike worlds (for example, the moon and Mercury) if magma ocean activity or near-surface melting lasted only until 4.4 Gy ago on one but until 3.8 Gy ago on the other.

10. A 1-km-diameter, long-period comet nucleus in retrograde orbit and an equal-sized Apollo asteroid in direct orbit hit the moon. (a) Will they make the same-sized crater? (b) If not, which makes the bigger crater? (c) Roughly what is the diameter of the two craters?

11. You are an astronaut traversing a 10-km-diameter crater on Mercury. The crater sits in a 100-m-deep flow of basaltic lava, overlaying upland material of unknown composition X. You begin on the lava, cross the ejecta blanket, descend into the crater; and climb the central peak. Describe some of the rocky or soil-like materials you find during this traverse.

12. A volcano on Mars is found to have one percent as many craters of all sizes as the average lunar maria. What is its absolute age?

13. Consider a typical Apollo asteroid of diameter 2 km approaching the moon on an orbit with aphelion in the outer asteroid belt and perihelion at Earth's orbit.
(a) At what velocity does it approach the Earth-moon system?
(b) At what velocity does it hit the moon?
(c) What energy, in joules, is expended during the collision?
(d) What diameter crater does it make?

14. How do the answers in problem 13 vary if the object falls into Earth instead of the moon?

15. Suppose you were going to dig a 10-km-diameter hemispherical crater in such a way as to throw the material out of the hole (shovelful by shovelful or all at once) about 10 crater diameters upward into the air, allowing the material to disperse outside the rim. (a) Calculate the potential energy consumed in this operation. (b) Compare this figure with the kinetic energy indicated to be necessary to create a 10-km-diameter crater and comment on the similarity or difference.

16. Suppose an incoming comet nucleus strikes the moon and ejects 500 times its own mass from the crater that is formed. The fastest 0.2% of this ejecta is moving faster than 2.4 km/s. (a) Does the moon gain or lose net mass during this impact? (b) What would be the answer for the same circumstances during an impact on Mercury?

A midday view across the plains of Mars, near the mouth of Ares Vallis, a dry river channel. This view shows scattered rocks in the foreground (believed to be flood debris washed out of the Martian uplands by ancient water flow in Ares Vallis). On the horizon is a hill about 1.0 km west of the lander. The rocks are mostly dark basalts, but the tops of many rocks are bright, possibly due to weathering products. (NASA, 1997 Mars Pathfinder mission)

CHAPTER TEN

Planetary Surfaces 2: Volcanism and Endogenic Processes of Surface Evolution

Planetary surfaces are not all cratered, rock-strewn primordial crusts. A glance at the moon reveals not a uniformly cratered globe but rather the prominent features of "the man in the moon"—dark lava plains produced by volcanism that obliterated many cratered areas. Numerous icy satellites show massive fractures. Larger worlds are still more geologically active. Mars and Earth, for example, show faulted terrains, vast dune fields, polar ice caps, canyons, and river valleys.

This chapter deals with the many processes beyond crust formation and cratering that have shaped planetary landscapes. The first part of the chapter deals with volcanism and related processes. Middle portions of the chapter deal with erosional effects on larger worlds, whose atmospheres and hydrospheres have modified their surfaces. Chemistries and temperatures of surface are also discussed. The last part of the chapter attempts to draw all these diverse processes together by summarizing the surface conditions of the planets.

An organizing principle underlies the division between this and the last chapter. Volcanism, tectonics, and other processes discussed here are **endogenic processes**—those that arise from within the planet and its atmosphere. Most processes discussed in the last chapter, such as cratering and solar wind sputtering, are **exogenic processes**—those acting on the planet from the outside.

Volcanism and Tectonics

Volcanism includes all processes by which material—gas, liquid, or solid—is expelled from the planet's interior. **Tectonics** includes all processes causing movement or distortion of planetary surfaces. Tectonic processes are driven by endogenic forces, and produce folded mountains, faults, rifts, and so on. Volcanism and tectonic activity often occur together but can also occur independently.

Rocks Produced by Volcanism

The last chapter described the basic rock material that formed primeval crusts, not so much by conventional volcanic processes as by basic processes of magma-ocean solidification near the surface of the planet. Volcanism can be thought of as eruptions that bring new rock materials up from depth and extrude them onto the surface. The most important rock produced by volcanism is basalt (refer back to Figure 9-1), an extrusive rock rich in plagioclase feldspars but having a composition range that includes minerals such as pyroxenes and olivine.

Basalts are the main rock in the volcanic parks of the American West and Hawaii, the lunar maria, among the boulders on Mars, and at most sites visited on Venus (based on Venera soil chemistry analyses). Geochemical studies of lunar basalts indicate that they erupted from magma sources perhaps 300 km down, probably produced around 4,000 to 3,200 My ago by radiogenic heating.

In the lunar uplands, we find a more complex story. The last chapter showed that the lunar uplands are mostly low-density rocks, primarily anorthosites (nearly pure feldspar), anorthositic gabbros (coarse-grained equivalent of feldspar-rich basalt), and some norites and troctolites (gabbros with special mineralogy). These rocks, called **cumulates,** are formed by mineral assemblages that more or less floated to the top of the magma ocean. But erupted noncumulate rocks are found among them. Most noteworthy are the KREEP basalts (Chapter 9), which are older than mare basalts. Their chemistry suggests that they were erupted from shallower magmas, probably the last magmas in the magma ocean, at depths around 60 km at the base of the anorthositic highland crust (Taylor, 1975).

Basaltic lava flows can take many forms, depending on their rate of flow and viscosity, which in turn depend on the angle of slope, the gas content, temperature, and so on. The two most common forms have Hawaiian names. **Aa** (AH-ah) is rough and clinkery (Figure 10-1) and is formed when solid blocks on the top of the flow are broken and upended by the flow's motion. (A mnemonic device: Think of the pointed top of the *A* as representing the pointy texture). **Pahoehoe** (pa-HOY-hoy) is smooth (Figure 10-2) and sometimes has a glassy surface produced by rapid cooling. Often these varieties are mixed

Figure 10-1. Flow of aa basalt lava. Solidifying crust was broken into rough clinkers by the motion of flow behind it, creating an intricate surface structure. (Pinacate volcanics, Sonora, Mexico; photo by author and Joyce Rehm)

Figure 10-2. Tongue of solidified pahoehoe basalt lava lapping onto a road. In the distance, the road is cut by an aa flow. (Kilauea, Hawaii; photo by author)

and overlapping in adjacent flows or even different parts of the same flow.

Other volcanic rock types occur, especially on Earth, and are arranged by mineral content, especially silica content, as diagrammed in Figure 9-1. Eruptions of rhyolite, a light-colored, siliceous rock, may be even more voluminous on Earth than eruptions of basalt, though they are not as easily recognized, because they are more readily eroded. Many rhyolite deposits were emplaced not in the form of massive lava flows but as clouds of pulverized, red-hot rock fragments suspended in hot gases. These are known as **fluidized eruptions.** There is no flow of liquid, but the cloud of gas and suspended particles is dense enough to flow along the ground like a swirling flash flood or like the base surge in Figure 9-9. The cloud itself is called a **nuée ardente** (glowing cloud). A nuée ardente from Mt. Pelée, in Martinique, raced across the city of St. Pierre at 150 m/s on May 8, 1902, carrying along large boulders, leveling buildings, and killing 29,000 inhabitants. There are terrestrial rhyolitic deposits less than 1 My old and covering 26,000 km² (Rittmann, 1962).

A number of other volcanic products are common and of planetological interest. Very fine volcanic fragments—of millimeter scale—are called **ash.** Volcanic fragments

about 10 mm in size are called **cinders** (Figure 10-3). **Tuff** is a type of rock formed by consolidation of ash particles, either by deposition into water or by coalescence directly from the air. Tuff is sometimes classified as sedimentary because of its depositional character. **Welded tuff,** or **ignimbrite,** forms when hot, glassy ash particles weld to each other during deposition. Ignimbrites can form vast, blanketing deposits, usually of siliceous composition, as a result of nuée ardente eruptions.

Structures Produced by Volcanism

If we photograph, fly over, or land on a planet, we see a great number of surface features. To an untrained observer, these may be bewildering in their variety. For example, on the worlds seen so far, features include mountain ranges, impact craters, landslides, lava plains, floating ice crusts, shifting dune fields, fault scarps, canyon systems, volcanic mountains, geysers, and fracture systems. The study of the structure and origin of landforms is a part of geology known as **structural geology.**

Figure 10-3. Ash and cinders blown out of the background cinder cone over a period of weeks blanketed and destroyed most of a forest at this site at Kilauea, Hawaii, 20 y prior to this photo. Foreground drainage craters drained into fissures in underlying lava flows. (Photo by author)

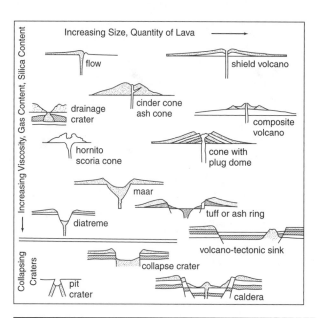

Figure 10-4. Schematic diagram of some volcanic landforms arranged by quantity and quality of lava. (See text)

Figure 10-5. Terrestrial cinder cones and associated lava flow near Flagstaff, Arizona. In middle distance is a diatremelike crater. On the horizon are the San Francisco peaks, an eroded shield volcano dominating this volcanic region. (Infrared aerial photo by author)

On Earth, structural geologists deal mostly with folded mountain belts and strata. In planetary science, structural geological problems deal mostly with volcanic and tectonic landforms. Most planetary geologists believe that a thorough knowledge of all volcanic and tectonic structures of Earth would be the best preparation for interpreting other planetary surface features.

In this field it is important to learn the terminology associated with these structures, which is discussed below. Many of the volcanic landforms are sketched in Figure 10-4, where they may have either positive or negative relief.

Cinder cone (Figures 10-5 and 10-6): A conical hill typically 50 to 300 m in height that is built up around a vent that ejects cinders, ash, and boulders. Smaller cones built as viscous lava spatters and welds into a hill around the vents are called **spatter cones, scoria cones,** and **hornitos.**

Shield volcano (Figures 10-7, 10-8, and 10-9): A gently sloping volcanic mountain built by fluid, basic lava flowing out from a central vent. Shield volcanoes may grow to enormous size. The largest known single mountain of any kind in the solar system is the vast Olympus Mons shield volcano on Mars (Figures 10-7 and 10-8). Its base is big enough to cover most of Missouri, and it rises about 24 km above the surrounding lava plains. Among the largest shields on Earth are the Pacific volcanic islands. Mauna Loa, on Hawaii (Figure 10-9) rises about 9 km from its ocean floor base and may be the largest single mountain on Earth. By comparison, Everest rises a bit over 8 km above sea level and less above its own mountainous base.

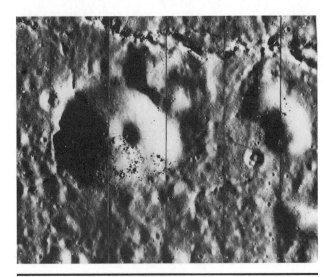

Figure 10-6. A possible lunar cinder cone. Several cones and pit craters on the floor of Copernicus suggest that the floor has been filled by volcanic eruptions younger than most lunar maria. Lighting from the right in this vertical view reveals a cone about 1 km across with a summit crater and numerous boulders on its slopes. At right is a C-shaped remnant of a similar feature. Most lunar eruptions apparently produced lavas that were too fluid to create cones and instead extruded into extensive sheetlike flows. (NASA, Viking Orbiter)

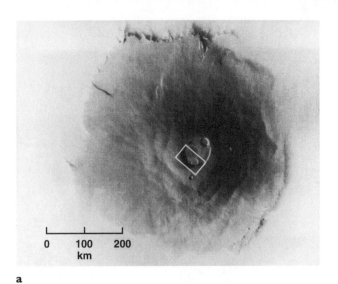

a

Figure 10-7. One of the largest shield volcanoes in the solar system is Olympus Mons, on Mars. (**a**) (above) General orbital view; boxes show positions of close-ups. (**b**) (below) Oblique view of summit calderas. Note the small, lobate landslide on the rim at top, the slumped and faulted terrain at lower left, and the wrinkle ridges on the lava flows on the central caldera floor. (**c**) (next page, top left) Lava flows on the flanks. A prominent fissure follows a ridge crest, resembling the terrestrial feature in Figure 10-9b. At upper center, a fingerlike flow runs downhill toward the upper right, parallel to the fissure. The absence of impact craters testifies to the youth of this surface, which may be only several hundred million years old. Region is about 43 × 55 km. (NASA; (**a**), Viking; (**b** and **c**), Mariner 9)

b

c

Composite volcano: A large volcanic mountain with an irregular profile, built by eruptions of viscous and fluid lavas. Often it resembles a shield surmounted by a large cinder cone.

Crater: A term used (unfortunately) for almost any hole in the ground, whether caused by meteorite impact, volcanic explosion, or collapse. The terms listed below specify different types of volcanic craters.

Caldera (Figures 10-10, 10-11, and 10-12): A volcanic crater, usually 1 km or more across, caused primarily by collapse. It may or may not have a raised rim caused by eruptive activity, and it often surmounts a volcanic mountain. As lava erupts and flows downhill, the overlying rock layers often sag or collapse to make up the vacated volume, producing a summit collapse crater. As shown in

Mathematical Notes on Volcanic Eruptions

Consider the hydrostatic pressure on a magma chamber below rock layers of density ρ_R at depth z. Suppose this is just equal to the pressure exerted by the weight of the magma of lower density ρ_M extending up to the top of a volcano of height h. This would be the greatest height that such a volcano could reach, as no pressure is available to push the magma any higher.

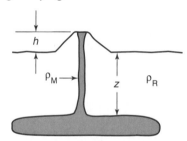

Thus, the pressure P at the depth of the magma chamber is

$$P = \rho_M g(z + h) = \rho_R g z$$

where g = gravitational acceleration.
Then,

$$h = \frac{\rho_R - \rho_M}{\rho_M} z \simeq \frac{\Delta\rho}{\rho_M} z$$

where

$\Delta\rho$ = difference in density between magma and surrounding rock.

Example

If lunar magma has few volatiles and is thus only (say) 40 kg/m³ less dense than lunar rock, then the maximum height h of a volcano originating from a 150-km-deep magma zone would be only about $(40/3000)150 = 2$ km, which is considerably

less than the relief of large shield volcanoes on Earth (5 to 10 km) or Mars (up to 24 km). This effect might account for the paucity of tall lunar volcanic structures.

Example

Eaton and Murata (1960) analyzed volcanism at Kilauea, Hawaii, as shown, using a measured magma density of 2770 kg/m³. Thus, adding the pressures contributed by each layer,

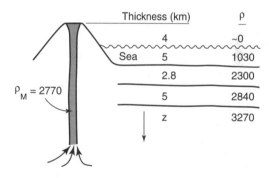

$$2770\,(16.8 + z) = 1030(5) + 2300(2.8) + 2840(5)$$
$$+\ 3270z$$
$$500z = 20746$$
$$z = 41$$

Total depth for equilibrium = 57 km below summit

They concluded that this is, in fact, the source depth of Kilauea magmas, as "great swarms of tiny to moderate, sharp earthquakes, totaling several thousand during the few days they last, emanate from depths between 45 and 60 km beneath the summit of Kilauea. These are the deepest quakes . . . in Hawaii, . . . 4 times deeper than the Mohorovičić discontinuity under Kilauea."

Figure 10-8. An eroded shield volcano on Mars. Apollinaris Patera is a shield volcano that is more cratered and older than Olympus Mons and many other volcanoes of the Tharsis region. Like Olympus Mons, it is surrounded by a scarp at its base, but the scarp is buried by lava at the right. The flanks are cut by a filigree of fine channels, probably from lava flows. The central caldera is about 100 km wide. (Viking mosaic, NASA)

a

b

Figure 10-9. Features of the terrestrial shield volcano Mauna Loa, in Hawaii, for comparison with Figures 10-7 and 10-8. (**a**) Collapsed calderas mark snow-covered summit, with a profile of the Mauna Kea shield volcano in background. Pits are smaller but similar in form to the summit calderas of Olympus Mons. Striated slopes resemble the texture shown in Figure 10-7c. (U.S. Geological Survey, D. W. Peterson) (**b**) Downhill view of a fissure on a ridge built by eruptions along a rift zone radial to the summit. The fissure has been widened by posteruption collapses along its edges. (Photo by author)

Figure 10-10. Road cut through basalt flows reveals subsurface cavity typical of such flows. (Kilauea, Hawaii; photo by author)

Figure 10-11. Caldera about 1.6 km across, with eroded rim of tuff beds. Floor is filled with eroded tuff and windblown dust, reminiscent of Martian craters. (MacDougal Crater, Pinacate volcanics, Sonora, Mexico; infrared aerial photo by author)

Figure 10-12. Inner rim of a pit crater is a vertical wall with a succession of basalt flows that build the surrounding countryside. Wall reveals a feeder dike (left of center) along which some lava may have erupted. The flat rim indicates little eruptive activity from the crater itself. (Mauna Loa summit, Hawaii; photo by author)

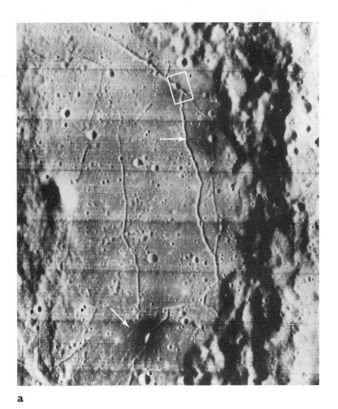

a

b

Figure 10-13. Medium- and high-resolution views of dark halo craters (arrows) on the floor of the lunar crater Alphonsus. Boxed example is shown in more detail at right. (NASA, Ranger 9)

Figure 10-10, underground cavities and **lava tubes** up to tens of meters across and several kilometers long are often left when lava drains or erupts, and the roofs of these can also collapse to form small craters about 10 to 100 m across, usually called **pit craters** (Figure 10-12).

Diatreme: A volcanic crater caused primarily by eruptive and explosive activity, often associated with gas-charged eruptions producing ash, that creates a gently raised rim (Figure 10-5, middle distance). A **maar** is a similar crater associated with groundwater that gains access to the magma chamber and causes an explosion. A diatreme often has a lake in its interior. Related to diatremes may be some of the lunar pits called **dark halo craters** (Figure 10-13), which are rounded craters with low rims forming dark aprons of debris that mantle the surround topography. Some are located on fissures and may be eruptive sites. They have not been visited by astronauts. Features between diatremes and cinder cones are tuff rings (nearly synonymous terms: *ash rings, ash cones,* and *tuff cones*), which have fairly high rims and floors that may be barely depressed. Diamond Head and nearby craters in Honolulu, Hawaii (Figure 10-14) are examples.

Volcano-tectonic sink: A depression larger and more irregular than a caldera, sometimes exceeding 30 km in diameter.

What Causes Volcanism?

How can we explain the rise of dense, molten rock from the interiors of Earth, Io, Mars, the moon, and other planets? What makes the planet spew forth with dazzling fire fountains, sluggish flows, and other forms? The basic cause is that magma is slightly less dense than most of the surrounding solid rock of the upper mantle. This means that the magma tends to rise just as oil blobs rise in water. Imagine a magma chamber connected by a crack to the surface. In the magma chamber, many kilometers below the surface, the weight of the overlying rock layers pressing on the magma exceeds the weight of (less dense) magma in the crack. Therefore, the magma gets pushed to the top of the crack and beyond, onto the surface. Combining this scenario with measurements of Hawaiian lava densities, Eaton and Murata (1960) calculated a source region for Hawaiian magma 60 km below the surface. This figure agrees with seismic evidence that magma originates at that level.

Of course, other factors may come into play. If tectonic forces squeeze the magma source region, they add to the pressure driving the magma upward. An analog for this would be a plastic bottle filled with water and buried in sand with the bottleneck flush with the surface. Pressure

Figure 10-14. Hawaiian tuff ring, eroded and breached at one end by the sea. For scale, note swimmers and beach at right. (Hanauma Bay, Honolulu; wide angle photo by author)

or weight on the sand would "erupt" a flow of water. Another factor is volatile content. As magma rises close to the surface, the pressure is less, and dissolved gases may begin to form bubbles. The bubbling gas may froth up the magma and create enough pressure near the vent to shoot out fountains of flowing material. An analog for this would be agitated soda pop, rather than water, in the plastic bottle. Sudden release of the bottle cap (representing the sudden access of magma to the surface) would produce fountaining. If the amount of gas is modest, it may produce vesicular lava, but if the amount is large, it may disrupt the magma into cinders and ash.

Even the geometry of the event may affect the type of eruption (Wilson and Head, 1981). For example, a narrow constriction could produce high pressure and a high fountain, whereas a large opening or long rift might produce a gentler flow. The shape of the vent often changes during the eruption, owing to erosion of the side walls by the lava, plus tectonic motions that may open new vents. As indicated by Figure 10-15, actual eruptions may be very complex, since the magma conduits may be linked by many interconnected passages, with eruptions breaking out and ending in various places. Early opening of a vent may yield a gentle eruption, whereas the buildup of pressure could produce an explosive eruption that blows away overlying crust.

Structures Produced by Tectonic Activity

Many planetary surfaces are subject to **tectonism,** or formation of landforms by compressive or stretching forces.

Figure 10-16 reviews some types of **faults** or fractures along which offset motion has occurred.

Thrust faults and **strike-slip faults** predominate where compressive forces have squeezed an area.

Normal faults and **grabens** (A German-derived term; sometimes *graben* is also used as the plural form) are produced where an area is stretched, as in rift zones. Grabens are illustrated in Figures 8-17, 10-7b (lower left), 10-16, and 10-17.

Lineament is a more general term for any linear structural feature, typically hundreds of meters to kilometers in length. It may be a fault, a graben, or a linear array of pits or cinder cones lying on some hidden underground fracture. Many lineaments are associated with volcanism, but others are not. The latter include radial structures around multiring impact basins, probably involving impact-generated fractures.

Volcanic and Tectonic Landforms of the Moon and Planets

The terminology of terrestrial volcanic landforms, developed through years of geological fieldwork, and many terms mentioned above have been extended to other worlds. At the same time, an additional terminology of lunar landforms developed independently through telescopic observations. Features were given picturesque names simply to describe their appearance, not their origin. Many of these names have been retained and transferred

a

b

c

Figure 10-15. Three different styles of eruptions. (**a**) Fire fountain eruption of Kilauea Iki, Hawaii, in 1959. The spray of hot ash and cinders reached as high as 600 m, one of the highest such fountains ever recorded. Note debris on the road. Posteruption view of the devastated forest is shown in Figure 10-3. (U.S Geological Survey, J. P. Eaton) (**b**) Molten lava splashing a few meters into the air, with twisted pahoehoe in foreground. (Boone Morrison) (**c**) A non-magmatic eruption sprays hot water containing dissolved minerals. The rim is built around the vent by the deposition of evaporite minerals left by the eruptions. (Castle Geyser, Yellowstone National Park; photo by author)

Figure 10-16. Examples of fault structures.

normal fault

thrust fault

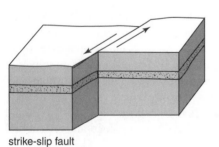

strike-slip fault

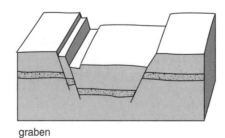

graben

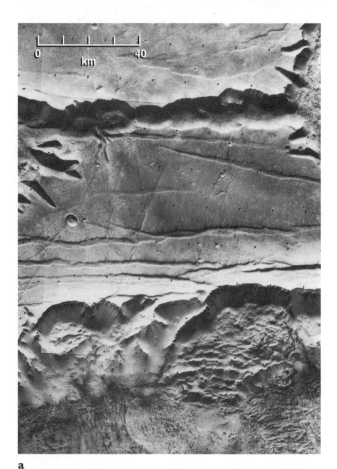

a

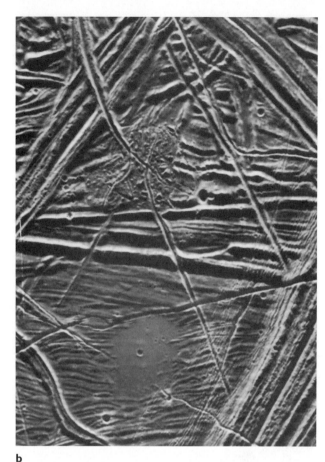

b

Figure 10-17. Grabens and intense tectonic faulting. (**a**) A family of grabens (center) paralleling the Valles Marineris rift valley (bottom) on Mars. These and other lineaments run roughly radial to the Tharsis volcanic complex. The north edge of Valles Marineris is seen at bottom; hummocky terrain at foot of scalloped cliffs (lower left and lower right) marks massive landslides where the cliff collapsed into the valley. (NASA, Viking Orbiter 1 mosaic) (**b**) Intense fracturing of Jupiter's moon, Europa. Fine systems of parallel faults, of different ages, cut the surface, which is probably a thin ice crust. At bottom center is an unusual smooth "pond," probably marking where water erupted and formed a patch of new surface. (NASA, Galileo)

to features on other planets, giving us a set of terms that are noncommittal about mechanisms of formation. Most of these features have not been examined at close range by astronauts or spacecraft, but most are believed to be volcanic or tectonic in origin.

Mare: As noted in Chapter 2, this term originally applied to any dark area on a planet (dark areas were once thought to be seas). Though Martian maria turned out to be associated with windblown dust deposits, the term is now generally applied to lava-covered plains (see Figure 10-18a, right side) that are dark because of the dark, basic rock. As correctly argued in 1949 by astrophysicist Ralph Baldwin, lunar maria were formed by very fluid lavas. (This was opposed to other incorrect theories—for example, that the maria were deep deposits of fluid dust.) Apollo rocks indicate unusually high eruption temperatures of 1400 to 1600 K, increasing their fluidity (Epstein

and Taylor, 1970). The flatness of mare lava fields and the lack of large volcanic mountains is apparently due to the high fluidity of the lavas.

Wrinkle ridge (Figure 10-18): A crinkly type of ridge, often a few kilometers across and 100 km or more in length and usually found in mare lava plains. These ridges often arc around edges of mare lavas in circular basins and may be compressional ridges formed during the final stages of lava filling and sagging in these basins. Some may be sources of lava flows. They are not recognized on Earth, possibly being disguised by erosion and deposition.

Dome: A roughly circular hill with a conical or rounded profile and sometimes a summit pit. Domes may be volcanoes. They often appear in mare areas. Particularly rough ones can be seen in Figure 10-19a; the lunar cone in Figure 10-6 might also be called a dome.

a

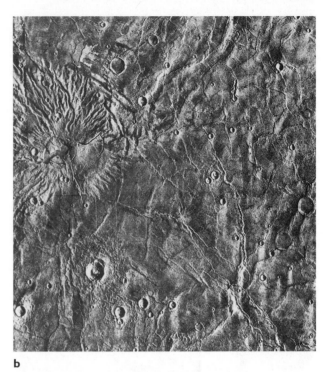

b

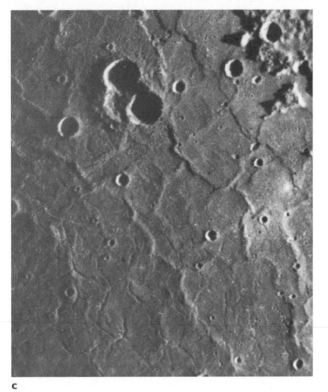

c

Figure 10-18.   Wrinkle ridges on three planets. (**a**) Southwestern part of Mare Serenitatis on the moon. Some of the ridges and other features lie radial to the Imbrium basin (upper left on horizontal). Rilles cut the older terrain on the mare "shelf," lower left. Foreground with 150 km. (**b**) Tyrrhena Patera volcano (upper left, cut by rilles) and nearby volcanic plains, laced by wrinkle ridges. The ring features, as along right margin, have an uncertain origin; they might be partly buried impact craters or volcanic features. (**c**) Marelike surface in the Caloris impact basin. Same scale as in part **b**; width about 320 km. (NASA)

Rilles (Figures 10-19, 10-20, and 10-21): Long valleys of two main types. A **sinuous rille** is a winding valley superficially resembling a channel cut by a river or lava flow. A **linear rille** is straight sided and more like a graben. These types, which can grade into each other, may involve four origin mechanisms, operating singly or jointly: (1) collapse of lava tubes; (2) buildup of levees of spattered lava along the sides of the still-molten central part of a flow, thus defining a channel for the flow; (3) draining of the central molten part of a flow away from its borders or levees, thus leaving a depressed channel (terrestrial flows do not seem efficient at eroding channels out of underlying rock); and (4) graben formation. The Apollo 15 astronauts' brief visit to the sinuous Hadley Rille (Figure 10-21) did not allow enough fieldwork to establish its origin.

Similar but longer rilles, with the surprising appearance of river channels, have been found on Venus, apparently cut either by very hot silicate lavas, or by carbonatite lavas (formed from molten carbonates and salts). The latter, if confirmed, would imply that Venus had early oceans

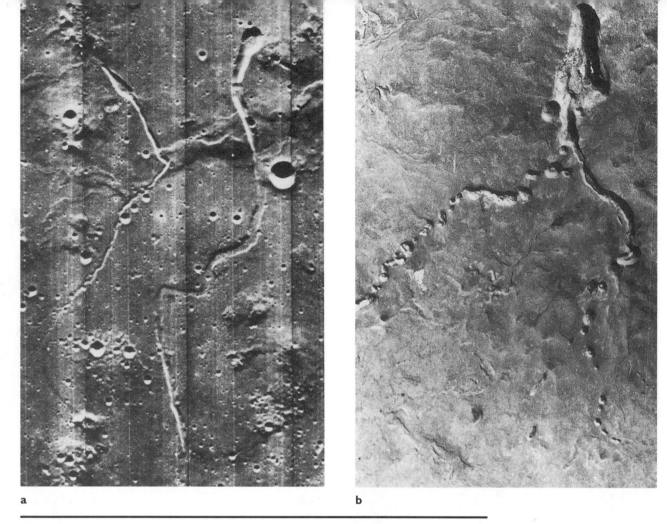

a

b

Figure 10-19. Comparison of lunar rilles with a terrestrial lava flow channel and associated pits. (**a**) Lunar sinuous rilles among Marius Hills dome field. (**b**) Icelandic flow channel and collapse pits. Such comparisons are instructive, though one must be careful of scale differences. As in this case, lunar examples are often larger in scale. (NASA)

Figure 10-20. Terrestrial lava flow channel possibly analogous to a sinuous rille. Channel was opened partly by a lava tube collapse (especially around the "head," right middle distance); lavas associated with the background cinder cone also flowed into and down the channel. (Mauna Loa, Hawaii; photo by author)

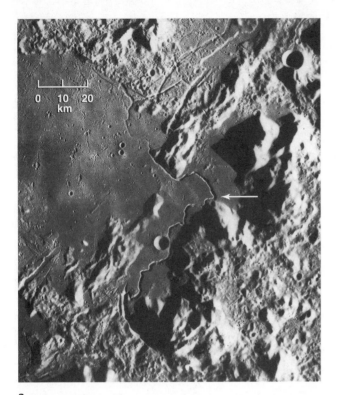

a

Figure 10-21. Two views of Hadley Rille, a lunar sinuous rille. (**a**) Vertical view shows Imbrium basin's Apennine Mountain rim (right) and the Mare Imbrium edge (left). Rille originates in the depression in the mountain shadows of bottom. Arrow shows position and direction of view in **b**. (**b**) Ground view, looking north from the sharpest bend at the base of the prominent Apennine peak. See also the close-up of the rille wall, Figure 9-3b. Rille width about 1.3 km. (NASA, Apollo 15)

b

Figure 10-22. Davey crater chain on the moon, photographed from orbit with a camera handheld by Apollo 14 astronauts. Alignment of such craters suggests an internal eruptive mechanism along a subsurface fissure. (NASA)

that evaporated and left carbonate deposits as a source for the molten carbonate lavas (Kargel, 1997).

Crater chain (Figure 10-22): An alignment of craters, usually not more than a few kilometers in diameter, but often many tens of kilometers in length. Some crater chains are associated with rilles, as in Figures 10-18 and 10-22. Flat rims suggest collapse pits, whereas raised rims suggest eruptive vents along fissures.

Large-Scale and Global Lineament Systems

Many lineaments, such as rilles, faults, and crater chains, appear not just as local features but also in vast parallel or radiating swarms. Some radiate from large impact basins, but others involve uncertain global forces.

Tectonic Patterns Associated with Impact Basins

Because we have treated impact structures and volcanic structures separately, we have only hinted at the complex landforms that develop as volcanism and tectonic activity modify the primeval, impact-dominated landscapes of planets. The large impacts that formed multiring basins on many worlds 4,400 to 3,900 My ago created patterns of concentric and radial fractures in the lithospheres around the basins. These were immediately masked by radially striated ejecta blankets. Later, the underlying fractures helped

a   b

Figure 10-23. Latticelike grid systems of lineaments, apparently formed by tectonic development and lava flooding along radial and concentric fractures around impact basins. (**a**) Ridges primarily radial to the Imbrium basin. These form the Haemus Mountains, which are the rim of the Serenitatis basin (left edge), which predates Imbrium. Width about 200 km. (Earth-based telescopic photo; Hale Observatories) (**b**) Radial and concentric ridge pattern in the rim of Caloris basin, Mercury. Basin floor is at bottom. Width about 450 km. (NASA, Mariner 10)

control the distribution of volcanic eruptions, as is clear in the arc-shaped lava deposits at the feet of the bullseye rings of the Orientale Basin on the moon (Figure 9-15a). Rims of impact basins and nearby preexisting craters are often broken along radial and concentric fractures, producing the latticelike landforms of Figure 10-23. Even postmare faulting in the lava plains was probably affected by the basin-created fractures in the underlying lithosphere, as suggested by Figure 10-24.

Figure 10-25 shows the cross section of such a giant impact basin, modified by lava flooding and tectonic activity along its concentric and radial fractures. Sagging of the central basin as lava extrudes and weighs down the surface creates additional tectonic modification, including concentric rilles (grabens) and wrinkle ridges in the mare surface (compare with Figure 10-16).

Mass Movements

Even if a planetary landscape had negligible impacts, no volcanism, no major fracturing, and no atmosphere to transport debris, it would change. Regolith or soil can move in response to earthquakes or gravity, even under seemingly insignificant stress. Various kinds of mass motions of this type are described in standard textbooks on physical geology. Of special interest are motions such as **slumping,** whereby material on gentle slopes (considerably less than the angle of repose) migrates downhill. The motion can be discontinuous, as in landslides, or a series of small slumping movements, or virtually continuous **creep.** On Earth, this process is aided by water, not acting as a lubricant under the flow but as a destroyer of the surface tension created by the normally small amounts of moisture between grains of the soil.

On the moon, an interesting difference between the hilly uplands and mare plains may be due to creep (Milton, 1967). As shown in Figure 10-26, upland hills, where they contact the maria, often have a rounded "toe" at their bases; and the slopes have fewer craters than do the maria, even though the upland surfaces are believed to be older. Both these effects may be due to a slow creep of the regolith downslope, perhaps during seismic tremors caused by impacts or moonquakes. Boulders on the slope are exposed by this process, and craters at the bottom of the

a

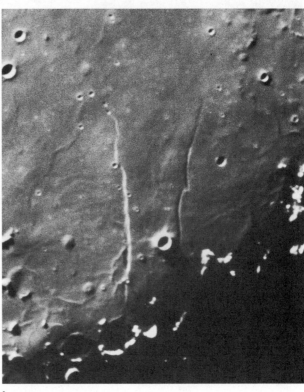

b

Figure 10-24. Two similar examples of lunar normal faults with parallel rilles, both lying roughly radial to the Imbrium basin and both with downthrown side (left) toward the interior of a mare-filled basin. Each is roughly 130 km long. They may be caused by sagging along premare Imbrium faults. (**a**) The Straight Wall in Mare Nubium. (Earth-based photo: University of Arizona) (**b**) The Cauchy Fault in Mare Tranquillitatis. Note adjacent domes, one with a summit pit. To the right of the fault is a rille. (Earth-based photo; Yerkes Observatory)

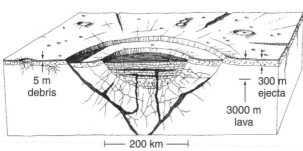

Figure 10-25. Schematic cross section of a multiring basin system, showing fracture patterns and later flooding by lavas.

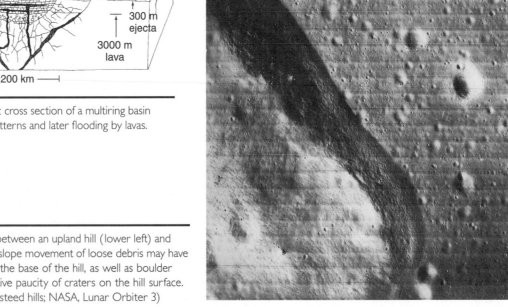

Figure 10-26. Contact between an upland hill (lower left) and mare surface (right). Downslope movement of loose debris may have caused the convex "toe" at the base of the hill, as well as boulder outcrops, texture, and relative paucity of craters on the hill surface. Picture width 4.3 km. (Flamsteed hills; NASA, Lunar Orbiter 3)

slope are sometimes invaded and filled in. Tracks in the dust on lunar hillsides show the paths followed by some of these boulders that have rolled downhill. The lower part of Figure 10-17a shows a more dramatic downslope movement—the collapse of the wall of Valles Marineris in enormous Martian landslides.

Atmospheric Effects on Planetary Surfaces

In addition to cratering from space and volcano-tectonic forces from below the surface, many worlds show effects of atmospheric processes.

Windblown Deposits on Mars

Mars and arid parts of Earth have developed astonishingly similar landforms as the result of wind action, as shown in

Figure 10-27. In fact, research on Martian surface features has emphasized that traditional geology texts underrate the importance of wind action on Earth. The study of geology evolved in European climes; arid regions were often neglected because they were economically unimportant (until the oil discoveries of this century). Yet many of Earth's equatorial regions, like those of Mars, are arid and dominated by wind-formed landscapes.

Even before the detailed mapping by Mariner 9 and landings by the Vikings and Pathfinder probes, researchers recognized that the thin Martian winds could carry Martian dust grains smaller than about 0.3 mm whereas the wind would pick up larger grains, enabling them to hop erratically over the surface—a process called **saltation.** This phenomenon is familiar to anyone who has lain on a windy, sandy beach (see Sagan and Pollack, 1969). From Earth, large-scale Martian dust storms have been seen, proving the existence of Martian dust transport. Saltation can play a striking role in erosion, as shown in Figure 10-28. Because the saltating grains stay within a meter or so of the

a

b

c

Figure 10-27. The closest analogs of Martian landscapes on Earth occur in the most arid regions. (**a**) In this coastal desert region of Peru, rainfall is so rare that landforms are controlled mostly by wind, which has stripped fine material and left a "pavement" of various-sized boulders and gravels. Rare (every few decades?) storms cause runoff that erodes channels, which resemble those on Mars. (Photo by author) (**b**) Scattered basalt boulders in a once-glaciated area on the arid summit of Mauna Kea volcano, Hawaii. (Photo by author) (**c**) Scattered basaltic boulders and dust at Viking 1 landing site in Chryse Planitia, Mars. (NASA)

Figure 10-28. "Mushroom Rock," a wind-eroded outcrop of basalt in the dusty wastes of Death Valley. Saltating sand grains have cut into the lower parts of the rock and will eventually sever the upper part. (Photo by author)

ground, they undercut rock outcrops, wearing fresh supplies of dust from rock surfaces while old debris weathers into fine, iron-oxide red dust. Some of the dust sticks to the Martian rocks by processes not entirely clear. Pathfinder scientists in 1997 were surprised to find windblown dust plastered onto the windward sides of some rocks. They had to place the composition-measuring probes of the Sojourner rover against the clean, leeward side to get a composition measurement less contaminated by dust (Rieder and others, 1997).

Most patchy markings seen on Mars through Earth-based telescopes appear that way probably because of thin dust deposits, either lighter or darker than the background (Veverka, Thomas, and Greeley, 1977). The Martian winds are seasonal, so the markings often change seasonally and annually, as seen in Figure 10-29. For example, new dust storms often start in the summer in the southern hemisphere, sweeping new masses of dust to other regions of the planet. This explains the seasonal variations in the dark markings, interpreted incorrectly by Lowell and other early observers as evidence of growing vegetation.

One form taken by the thin dust deposits is a pattern of long **wind streaks** downward from craters and other

obstacles, as seen in Figure 10-30. As the wind sweeps across the Martian plains, dust is often deposited on the leeward sides of craters. It may be darker or lighter than the background. In some regions, streaks from recent dust storms overlap earlier, fainter streaks in other directions, created by earlier storms in other seasons when prevailing wind directions were different. As shown in Figures 10-30b and 10-31, similar effects can be found on Earth. Wind tunnel studies with simulated Martian conditions produce similar streaks on the leeward side of scale model craters (Greeley and others, 1974). Spacecraft monitoring of streaks and associated dust storms shows that the global Martian wind patterns roughly repeat from year to year.

Localized "drifts" around the Viking 1 lander were found to lie in the same orientation as the larger-scale streaks in that region observed from orbit (Greeley, Papson, and Veverka, 1978).

Dune fields, or large masses of wind-sculpted dunes, are common on both Mars and Earth. Winds pile fine materials into ripple patterns on a vast range of scales, from centimeter-size waves to graceful ridges kilometers apart (Figure 10-32). When Martian dunes were discovered during the Mariner 9 orbital mapping program in 1971–1972, many planetary geologists turned to the classic work by

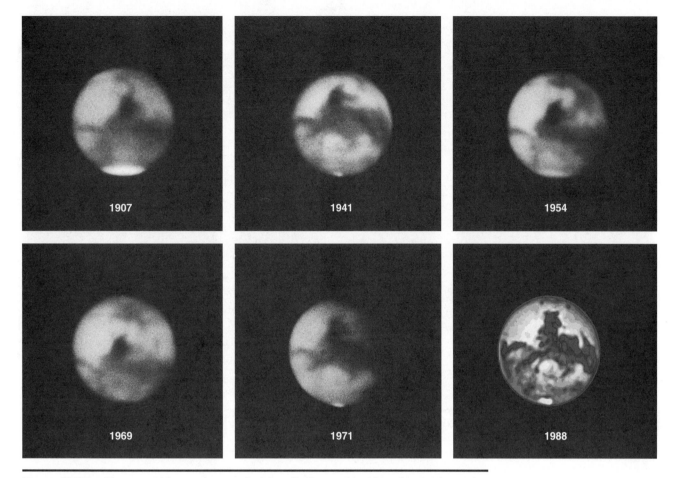

Figure 10-29. Changes on Mars as photographed from Earth over 66 y. Note changing shape of "tail" extending east from Syrtis Major, especially between 1941 and 1954, and brightening of Hellas bright region (bottom) between 1969 and 1971. Changes probably involve variable dust deposits. (Lowell Observatory; 1988 view is from France with a CCD imaging system)

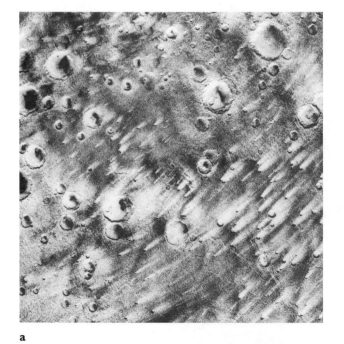

Figure 10-30. Wind streaks on Mars and Earth. (**a**) (left) Intermediate-size craters in this region have downstream streaks; dark deposits in larger craters are probably dune fields. (NASA) (**b**) (above) Light-toned streak left in the wake of a bush on darker background material, Williams Wash, western Arizona. Similar streaks appear in the background. (Photo by author)

Figure 10-31. Cinder cone and associated basaltic lava flow (extending toward camera), covered by a thin deposit of light, windblown dust. The flow structure can be seen, but the albedo is lighted. Dark dust derived from a cinder cone is deposited beyond the cone in a wind streak pattern. (Pinacate volcanics, Sonora, Mexico; aerial photo by author)

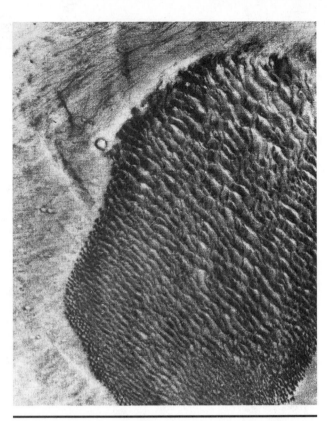

Figure 10-32. Dark dune field forms a 35 × 55 km patch on the floor of a Martian crater. Many Martian craters have deposits of dunes or other dust depositions in their floors. (NASA, Mariner 9)

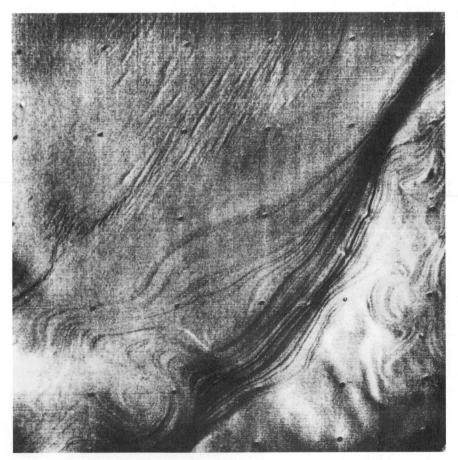

Figure 10-33. Orbital view of stratified terrain about 15° from the south pole of Mars. Width 40 km. Individual strata are about 20 to 50 m thick. Lack of craters suggests that the deposits are very young, perhaps still forming. Striations (top) may be due to wind erosion. (NASA, Mariner 9)

Bagnold (1941), who made the first serious studies of dunes while stationed in the African desert in the 1930s. On Mars, because crater rims disturb the wind flow, many of the most remarkable dune fields are formed in crater bottoms. The most spectacular dune field in the solar system rings the north polar cap of Mars (Cutts and others, 1976). Its area exceeds that of the Sahara and Arabian deserts combined.

In the polar regions of Mars, thick deposits of sediments have accumulated. These are revealed in orbital photos by the contourlike edges of the stratified beds, as seen in Figure 10-33. Researchers believe that the annual condensation of carbon dioxide and water frost deposits in the polar regions (see next chapter) plays a role in the dust deposition (Squyres, 1979; Cutts, Blasius, and Roberts, 1979). Probably the volatiles condense on dust particles and then carry them to the surface. Annual deposits of dust and ice may form, with the dust left behind during the spring "thaw" (or more properly, sublimation). Variations in dust storm intensity from year to year and in the polar climate from one era to another may aid in creating layered deposits of different thicknesses and strengths. The "individual" layers visible in orbital photos like Figure 10-33 might reflect not annual cycles, but larger cycles on million-year time scales. Crater counts in the Martian polar sediments are very low (Figure 9-20f), suggesting that resurfacing occurs at the 1-km scale in 100 My or less.

River Channels on Mars

The whole question of resurfacing, erosion, and transport on Mars is complicated by the existence of dry river channels and evidence that the ancient climate on Mars was different from the present climate, and more conducive to liquid water—at least during some episodes. This problem is taken up again in Chapter 13, an epilog about planetary science as applied to the modern exploration of Mars.

Geochemical Cycles

Terrestrial geologists have defined the **geochemical cycle** as a schematic description of the complete history of rock material after it is injected into the planetary surface environment from deep magma sources. The concept can usefully be applied to other planets.

Each planet has its own geochemical cycle. On a small, airless, primitive world, the possible histories of rock material are limited. Lava on the surface may be shock damaged, pulverized, reburied, and/or welded into coherent breccia before remelting occurs. The cycle is slow because of general geological inactivity. On planets with atmospheres and active interiors, such as Earth, Venus, and Mars, the cycle may be faster and more complex.

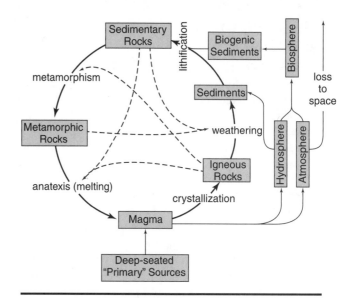

Figure 10-34. The terrestrial geochemical cycle.

Figure 10-34 shows the cycle known in most detail—that for Earth. Magmas may produce atmospheric, hydrospheric, and rock products. The rock materials may break down into sediments, be metamorphosed, or be remelted. Most of the continental materials have probably been through the cycle several times. The cycles for Venus and Mars are understood less well, but several important elements may be sketched.

Carbon Dioxide, Carbonates, and the Urey Reaction

Harold Urey, in his pioneering 1952 book on the geochemistry of the planets, pointed out the importance of the reaction between CO_2 gas and the silicate minerals to form carbonate minerals and quartz. An important example is

$$MgSiO_3 \; + \; CO_2 \; \rightarrow \; MgCO_3 \; + \; SiO_2$$

enstatite   carbon dioxide   magnesite   quartz

High temperatures drive this reaction to the left, whereas the presence of water drives it to the right. This and similar reactions are sometimes called the **Urey reaction.**

The great importance of this type of reaction is that it allows us to understand the difference between the surface environments of the two similar-size planets, Earth and Venus. On Earth, the CO_2 from volcanic gases dissolved in oceans and made a weak carbonic acid, and the weak carbonic acid reacted with the silicate rocks to form carbonate rocks. Thus, most of the CO_2 on Earth is tied up in carbonate rocks and in the oceans. On Venus, because there are no oceans, the carbon dioxide could not follow this route into solution and into rocks, and thus remained in the atmosphere, building up to a massive CO_2 atmosphere. As we see in the next chapter, the total CO_2 inventories on Earth and Venus, taking the carbonate rocks into account, are the same!

The Chemistry of the Venusian Surface

In 1967, Connes and others made the surprising discovery that hydrogen chloride (HCl) and hydrogen fluoride (HF) are trace constituents of the carbon-dioxide-dominated Venus atmosphere. Later, Sill (1973) and Young (1973) showed that the droplets forming the Venus clouds are sulfuric acid. Clearly, Venus has an unearthly chemistry!

Chemical analyses soon showed that the hot (750 K!) atmosphere of Venus would react with the rocky surface to affect the surface chemistry. Lewis (1968) attempted to explain the atmospheric chemistry in this way and suggested that even mercury and lead could be driven out of the crustal rocks and into the atmosphere, with lead possibly condensing out at the poles. The Russian geochemists Khodakovsky and others (1979) made more detailed analyses of the chemistry of rocks containing silicon (Si), titanium (Ti), aluminum (Al), iron (Fe), manganese (Mn), magnesium (Mg), calcium (Ca), and other elements in the presence of CO_2, CO, and H_2O at several altitudes and temperatures on Venus. Consistent with the above discussion of the Urey reaction, they found that the CO_2 would remain in the atmosphere because of a lack of water to dissolve it. Any water may have been driven into the rocks by oxidizing iron minerals,

$$2FeO + H_2O \rightarrow Fe_2O_3 + H_2$$

the same reaction that may have used up some of the water on Mars and produced the rust-red, oxidized iron minerals there.

Hunten (1973) considered a different reaction that would use up water:

$$CH_2 + 2H_2O \rightarrow CO_2 + 3H_2$$

The idea here is that Venus may have accreted from hydrocarbon-rich planetesimals, with about 25% as much water as Earth's planetesimals. The available water was then consumed in oxidizing the hydrocarbons (represented as CH_2 in the equation) and perhaps iron minerals, as in the earlier equation. The hydrogen from these reactions, being very light, escaped readily into space and left the carbon-dioxide-rich atmosphere behind. On Earth, water was left over after the hydrocarbons were oxidized. Grinspoon (1997) gives an entertaining book-length introduction to the problems of the Venusian environments evolution, which is also discussed further in the next chapter.

Chemistry and Red Color on the Surface of Mars

One important process on Mars has been the oxidation of iron minerals. There is much evidence that Mars had liquid water flowing on its surface in the past, either in an early wet period or in intermittent episodes. We return to this evidence in the epilog chapter on Mars. The ancient water on Mars is probably associated with the red color. Everyone is familiar with the expanses of red-colored deposit in the deserts of the U.S. Southwest, and geologists have found that similar deposits of sediments, called **red beds,** are common in many parts of Earth's stratigraphic sequence. Studies show that fresh terrestrial sediments derived in high humidity regions tend to be brownish and blackish and involve hydrated iron oxides, often in amorphous or poorly crystallized states. Arid regions with intermittent supplies of moisture produce the red iron oxide minerals. A fluctuating water table, alternately wetting and drying the sediments, thus encourages production of red beds (Van Houten, 1973). The same process produces rust on iron tools exposed to the weather. Generally, the red color is associated with oxidation of iron atoms in the minerals from the $Fe++$ state to the $Fe+++$ state. Thus, an arid Mars with postulated episodes of a moist climate would be ideal for producing the observed red Martian dust.

Understanding of the red Martian mineralogy came gradually. Early theorists, such as Percival Lowell and Gustav Arrhenius, correctly associated the red colors with the red oxidized minerals of terrestrial red deserts, but incorrectly looked for mechanisms to produce deep layers of oxidized iron, such as concentration of rusted iron meteorites on the surface. By the 1960s, spectra and polarimetry had focused the discussion on oxidized iron minerals, such as limonite, which usually occurs as a mixture of hydrated iron oxides. Binder and Cruikshank (1966) correctly emphasized that the iron oxides are unlikely to exist as pure soil layers but rather are likely to exist as oxidized skins and stains on the surface rocks and silicate dust grains, as they do in terrestrial deserts. Binder and Cruikshank (1966) showed that the spectrum of Mars was a close match to the spectra of Arizona desert rocks (especially basalts) stained with limonite and related iron oxide minerals. Other workers quite early concurred that oxidized basalts are the best fit to the red spectrum of Mars (Adams, 1968; McCord and Adams, 1969). The Viking landers in 1976 made the first close-up photos of the red soils (presumably dust grains well coated with iron oxides) and a range of rock colors, from natural-gray basalts to red-stained basalts. The Pathfinder lander in 1997 showed a number of gray rocks, coated on the windward side with red dust or stains. Presumably, most of the basalts on Mars, if broken open, would have gray interiors.

The primary minerals now discussed as sources of the red stains on Mars, and presumably produced during fluvial periods, are

Limonite:	$FeO(OH) \cdot nH_2O$	orange-brown
Goethite:	$HFeO_2$	yellow-brown
Hematite:	Fe_2O_3 (alpha form)	reddish brown
Maghemite:	Fe_2O_3 (gamma form)	reddish

Maghemite was favored as the reddish, slightly magnetic airborne dust component picked up by magnets on the Pathfinder lander in 1997 (Hviid and others, 1997).

Figure 10-35. Evaporite beds: "The Devil's Golf Course," Death Valley. The deposits are 95% pure salt plus silt. They are formed by evaporation of water draining into the plain, which is the floor of a large graben. (Photo by author)

Evaporite Minerals on Mars

If there was once abundant liquid water pooled on Mars, it must have contained dissolved mineral salts, and when the water evaporated, these would have been left behind on the surface in deposits of **evaporates** (see previous chapter). Figure 10-35 shows a spectacular terrestrial example: dry lake beds, or **playas,** often colored whitish by salts and carbonate deposits. Could playas, with distinct evaporite deposits, be found on Mars, proving existence of ancient lakes or seas? The search is already proceeding in current Mars missions.

As of this writing, the best evidence has been found not in large-scale features, but in Martian soils, which are rich in sulfates, a common evaporite component. Typical Martian loose dust has about 100 times the sulfur content of average materials in Earth's crust (Clark and others, 1977; Rieder and others, 1997). Evaporite salts, sulfates,

and carbonates tend to bond surface dust into crumbly or flaky layers. Supporting the idea that water has played a role in the Martian mineral chemistry, such flaky layers were found on Mars first at the Viking 2 site (Figure 10-36). They were given the name **duricrust.** The duricrust was found to be 20% to 50% richer in sulfur than the loose dust (Mutch and others, 1977).

Temperatures of Planetary Surfaces

If planetary surfaces consisted only of solid rock and had no atmospheres, their temperatures would be fairly simply determined by the amount of solar radiation striking them. At dawn, the surface would warm rapidly until it radiated energy at the same rate as it absorbed energy. The

Figure 10-36. Furrow across middle foreground (left center to lower right) of this Martian landscape in Utopia Planitia may be part of a larger patterned ground system associated with underground permafrost. Flaky texture of the duri-crust is visible between rocks in the lower right corner where sampling was conducted. (NASA, Viking 2)

temperature would peak soon after noon and then drop at night as the surface radiated its energy into space.

In reality, various factors complicate this. One is that dark areas absorb more heat than light areas. Another is that most planets have a dusty regolith that acts as a good insulator, preventing much heat from being transmitted more than a few centimeters into the ground during the daylight period. The subsurface thus shows much less temperature variation during a day than the surface does. Similarly, dusty areas thus store less heat than bedrock. At night, bedrock remains slightly warmer than dust or sand because of its stored heat. These principles allow orbiting infrared radiometers to make temperature measurements that allow interpretation of the surface materials.

As we see in the next chapter, atmospheres greatly alter the surface temperature by absorbing sunlight or the reradiated planetary thermal radiation at some wavelengths and not at others.

Table 10-1 summarizes some observational and theoretical data on planetary temperatures. The measured day and night temperatures in the table are either direct measurements or estimates closely based on measurements of thermal properties. The calculated average temperatures are based on the simplified theoretical model given in the accompanying mathematical notes. Scanning the table, we see that the dominant effect in determining temperatures is simply the distance from the sun. The considerable albedo differences among Saturn's moons, for instance, make only about a 30 K difference in temperature. Outstanding anomalies are the 730 K (855° F) surface temperatures of Venus, caused by its special atmospheric properties (see Chapter 11), and the erupting hot spots of Io, ultimately produced by Jupiter's tidal heating effects.

Summary: Synthesizing Planetary Surface Processes

To summarize this and the preceding chapter, we review surface environments planet by planet. It is an exciting challenge to attempt to draw together all this information so that we can produce comprehensive pictures of the surfaces of other planets. We ask not simply for predictions of numerical measurements to be made by instruments, but for an idea of the landscape, history, and "feel" of each planet as it may be experienced by the first human explorers. Table 10-2 summarizes some of this information, with the planets entered in order of size so that the trend can be seen from smaller bodies with primitive materials to larger bodies with more complex, differentiated surfaces—a trend made more complicated by the additional changes from silicate mineralogies in the inner solar system to carbonaceous and ice mineralogies in the outer solar system.

Phobos, Deimos, Amalthea, and Other Small Bodies

Close-up photos of Phobos and Deimos (Figures 2-17, Chapter 5 opening photo, 6-1, 5-9a, 7-5c, 7-7, 9-22) and mid-range photos of Amalthea, Hyperion (Figure 2-27), S11 (Figure 7-5d), and other small Saturn moons reveal heavy (saturated?) cratering on all these bodies. The cratering itself may control the shapes, first by creating some of these bodies as fragments, and later by blowing chunks off one side or another in super-cratering events. A good

TABLE 10-1 Planetary Temperatures, Measured and Calculated

Planet/Layer	Measured temperature (K)		Calculated mean (or typical) temp. (K)	Bond albedo (assumed or measured)	Effective emissivity (assumed)	References and notes
	Day	Night				
Mercury						
Soil surface	700 max.	100	452	0.10	0.95	Mariner 10; Strom (1979).
Venus						
Cloud tops, 65-km altitude		215–240	261	0.76	0.65	Pioneer Venus IR; Taylor (1979).
Surface air	721–731	732	—	—	—	Pioneer Venus 4 landers; Russian landers; Seiff and others (1979). Strong greenhouse effect.
Earth						
Surface air	277–310	260–283	281	0.39	0.65	
Moon						
Soil surface	380 max.	100 min.	280	0.11	0.95	Taylor (1975). Subsurface, 210 K; soil = excellent insulator.
Mars						
Surface air	240 max.	190 min.	—	—	—	Viking landers; Hess and others (1977).
Soil surface	250–280	150–180	—	—	—	Viking orbiter IR; Kieffer (1977).
Vesta						
Soil surface	—	—	175	0.24	0.95	—
Ceres						
Soil surface	—	—	171	0.05	0.95	—
Jupiter						
Cloud tops or haze	120–150	—	120	0.44	0.65	Voyagers, Hanel and others (1979a, 1979b). P = 150–800 mbar.
Io						
Surface	125–135	—	106	0.56	0.8	Voyagers; Hanel and others (1979a, 1979b).
Warm spots	270–310?	—	—	—	—	Voyagers; Hanel and others (1979a, 1979b). 180-K spots cover 5%.
Eruptions	385–2000	—	—	—	—	Voyagers; Hanel and others (1979a, 1979b); Witteborn, Bregman, and Pollak (1979). Molten sulfur at 385 K.
Europa						
Surface	125 max.	85	103	0.58	0.9	Voyagers; Hanel and others (1979a, 1979b).
Ganymede						
Surface	145	85	107	0.48	0.95	Voyagers; Hanel and others (1979a, 1979b).
Callisto						
Surface	153 max.	79 min.	122	0.13	0.95	Voyagers; Hanel and others (1979a, 1979b).
Small moons						
Surface	140–170	—	126	0.03	0.95	Amalthea, quoted by Morrison (1977, p. 283).
Saturn						
1-bar level	120–160	—	88	0.46	0.65	Orton (1979).
Icy moons						
Surface	67–93	—	70	0.7	0.9	Hanel and others (1982).
Titan						
Cloud tops	90	—	97	0.21	0.65	—
Surface	82–136	—	—	—	—	Morrison (1977, p. 281).
Iapetus						
Trailing	—	—	85	0.35	0.9	—
Iapetus and Phoebe						
Leading	≥110	—	94	0.05	0.9	Hanel and others (1982).
Uranus						
1-bar level	78	—	59	0.56	0.65	Hanel and others (1986).
Icy moons						
Surface	84–86	—	62	0.27	0.9	Hanel and others (1986).
Neptune						
1-bar level	69	—	48	0.51	0.65	Conrath and others (1989).
Icy moons						
Surface	38	—	33	0.85	0.9	Conrath and others (1989).
Rocky moons						
Surface	—	—	52	0.1	0.9	—
Pluto and Charon						
Surface	—	—	37	0.6	0.9	Cruikshank and Silvaggio (1980).

TABLE 10-2 Dominant Surface Material on Selected Bodies

Object	Approx. diameter (km)	Surface material
Saturn ring bodies	<0.1	H_2O ice
Deimos	12	Dark, reddish carbonaceous silicates?
Phobos	22	Dark, reddish carbonaceous silicates?
J7 Elara	80	Dark, carbonaceous silicates (+ H_2O ice?)
J6 Himalia	180	Dark, carbonaceous silicates (+ H_2O ice?)
S9 Phoebe	220	Dark, carbonaceous silicates (+ H_2O ice?)
J5 Amalthea	270 × 155	Dark, reddish (carbonaceous and sulfur?) soil
S2 Enceladus	500	H_2O ice (virtually pure)
1 Ceres	914	Dark carbonaceous hydrated silicates (+ NH_3-bearing silicates or H_2O ice?)
S3 Tethys	1048	H_2O ice
S4 Dione	1120	H_2O ice
U1 Ariel	1160	H_2O ice + dust?
P1 Charon	1190	H_2O ice
S8 Iapetus	1440	Dark, carbonaceous silicates (leading side); mostly H_2O ice (trailing side)
U4 Oberon	1520	H_2O ice + dust?
S5 Rhea	1530	H_2O ice
U3 Titania	1580	H_2O ice + dust?
Pluto	2300	CH_4 ice
N1 Triton	2700	N_2 ice + Ch_4 ice
J2 Europa	3130	H_2O ice + salts
Moon	3476	Basaltic soil and rock
J1 Io	3630	Sulfur compounds
J4 Callisto	4840	Dark (carbonaceous?) soil + H_2O ice
Mercury	4878	Basaltic but not ultrabasic soil (based on spectra)
S6 Titan	5150	? (obscured by clouds, differentiated?)
J3 Ganymede	5280	H_2O ice and (carbonaceous?) soil
Mars	6787	Basaltic, chemically weathered soil and rock
Venus	12,104	Basaltic and some granitic soil and rocks
Earth	12,756	H_2O liquid; basaltic and granitic soil; rock
Neptune	49,528	H_2 liquid?
Uranus	51,118	H_2 liquid?
Saturn	120,660	H_2 liquid? (no real surface)
Jupiter	142,800	H_2 liquid? (no real surface)

Note: Table lists bodies for which observations or inferences are available; see also tables in Ch. 7 for related asteroidal data.

Source: Data based on Cruikshank (1979; private communication, 1981); Degewij (private communication, 1980); Strom (1979); Veverka, Thomas, and Duxbury (1978); Burns (1977); King and four others (1992).

example of irregular shape is seen in Jupiter's inner moon, Amalthea, which has a potato shape and is sculpted by large craters (Figure 10-37).

In spite of the similarities of cratered surfaces and irregular shapes, these small moons show interesting variety in surface texture. Even Phobos and Deimos differ at a scale of 10 to 100 m. The terrain on Deimos seems smooth and blanketed, with boulders protruding here and there; the craters on Phobos seem more rugged, and the terrain is cut by the striking grooves best seen in Figure 5-7a. Part of the difference may be that Phobos was struck more recently by a large impact. If the giant Stickney impact on Phobos was a relatively late addition, it may have changed the surface character of Phobos as well as creating the swarms of surface fracture-grooves.

Soter (1971) pointed out that although meteorites easily knock debris off Phobos and Deimos, this dust cannot readily escape from the gravitational field of Mars. Instead, it goes into Martian orbit, spreading along the orbits of Phobos and Deimos. The velocity required to knock debris off these satellites is only around 10 m/s, but the ve-

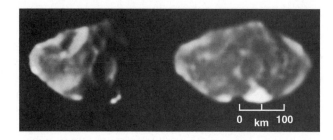

Figure 10-37. Jupiter's reddish, inner moon Amalthea was revealed by Voyager and Galileo photos to be highly elongated and heavily cratered. These views show the same face under side and full lighting. (NASA, Galileo, Courtesy Elizabeth Alvarez)

locity to make the debris escape Mars entirely is about 960 m/s for Phobos and 500 m/s for Deimos. Thus, impacts, especially on Phobos, would eject debris, forming "dust belts" around Mars. These would eventually be swept up by the satellites themselves. Each new crater thus provides

Mathematical Notes on Planetary Surface Temperatures

Any surface heated by radiation approaches an equilibrium temperature at which the amount of radiation received equals the amount radiated for any given area over any given time. Therefore, the equilibrium temperature of a planetary surface can be calculated by equating the flux of incoming solar radiation with the flux of outgoing thermal radiation. The Stefan-Boltzmann radiation law states that for a blackbody

$$E = \sigma T^4$$

where

$E =$ energy flux (J/m²·s) radiated by body of temperature T

The Stefan-Boltzmann constant σ was defined in Table 2-2. We also defined the solar constant $F_\odot$ there, which is the flux of energy (J/m²·s) received by a surface 1 AU from the sun. Because the flux of sunlight declines as the inverse square of the distance from the sun the flux of sunlight striking any planet is $F_\odot/a^2$, where a is given in astronomical units. Thus, for any blackbody, we would have the equilibrium condition, where incoming flux equals outgoing flux:

$$\frac{F_\odot}{a^2} = \sigma T^4$$

But a planet isn't a blackbody. The amount of incoming radiation actually absorbed is $(1 - A)$ times the total, where A is the Bond albedo, defined in Chapter 2. Further, for a spherical planet with zero obliquity, only the equator receives full illumination; a factor $\cos \phi$ needs to be inserted to

correct for other latitudes. (For planets with high obliquity, the situation is still more complex.) Finally, if the planet rotates with moderate speed, the surface doesn't come into full thermal equilibrium; it heats up during the day and cools at night. A given zone of latitude absorbs sunlight across its diameter (as seen from the sun) but reradiates around its full circumference, so another correction factor of diameter/circumference, or $1/\pi$, is added.

Similarly, the right side of the equation needs a correction for a nonblackbody. A gray body radiates e times as much flux as a blackbody of the same temperature, where e is a property of the given material, a fraction called the emissivity. The emissivity for brick, glass, and rough dull metal is around 0.8–0.9, but for shiny metals it can be as low as 0.02–0.1.

With these corrections, our estimate of temperature T at latitude ϕ on a rotating planet is given by the equation

$$\frac{1 - A}{\pi} \cos \phi \, \frac{F_\odot}{a^2} = e \sigma T^4$$

If we insert numerical values and assume we are interested in a "typical" latitude of 30°, this equation reduces to

$$T = \sqrt[4]{\frac{1 - A}{ea^2}} \; 285 \text{ K}$$

where the Bond albedo A, emissivity e, and solar distance a are known or can be estimated. Often only the geometric albedo is measured, but the Bond albedo is usually not very different (see Chapter 2 for clarification of these terms).

a source of sandblasting that erodes the satellite and yet ultimately adds to the regolith. Such a regolith, at least 1 mm deep, has been inferred on Phobos by photometric and infrared thermal observations (Pollack, 1977); consistent with this observation, PHOBOS-2 spectra showed that fresh craters exposed deep material that was different from the (mature regolith?) surface. The "Soter effect"—sandblasting of satellites by their own debris—applies to all satellites but more so to satellites close to planets, where it is harder for satellite debris to be blown out of the system entirely. This effect may cause a systematic difference between these satellites and asteroids.

Craters lying along the Phobos grooves offer a puzzle. They are too well aligned to mark secondary impacts, and they can't mark simple drainage of regolith into cracks because they have raised rims. Hartmann (1980) suggested that Phobos originally contained volatiles that vented through the fracture-grooves, piling up regolith around the vents. Fanale and Salvail (1990) calculated that Phobos could still contain ices and might be leaking H_2O molecules. Independently, Russian investigators on the PHOBOS-2 mission reported detection of H_2O molecules at about the flux predicted by Fanale and Salvail. Phobos may thus be one of many volatile-rich, carbonaceous interplanetary bodies (P or D type?) scattered through the solar system as giant planets grew, and in this case captured by Mars (Hartmann, 1990).

Another small moon, Jupiter's Amalthea (155 × 270 km), is unusually red and may be coated with sulfurous material that was blown off Io (by its volcanoes or by impacts) and spiraled inward toward Jupiter, similar to the contamination of Iapetus by Phoebe dust, described in Chapter 9 (see Figure 9-21).

The Intermediate-Size Satellites of Saturn

Mimas (390-km diameter), Enceladus (500), Tethys (1048), Dione (1120), and Rhea (1530) have cratered surfaces of clean, bright water ice, with albedos 60% to 95% (Cruikshank, 1979; Smith and others, 1982). Though mostly larger than Miranda, they show some related features. Several have fracture systems.

As discussed in Chapters 2 and 8, Enceladus is especially important because it appears to offer a "missing link" between heavily cratered, inactive worlds (Rhea, Dione, Callisto) and resurfaced or fractured, more active worlds (Europa, Ganymede, Miranda). The ages of Enceladus's young plains have been estimated at less than 1 Gy by crater counts (Smith and others, 1982). Tidal heating may have produced volcanic eruptions of watery "magma" (see Chapter 9). Some heating and eruptive mechanism seems to have been at work: the E ring of Saturn—a weak ring whose density peaks near the orbit of Enceladus—suggests that eruptions have thrown material off Enceladus into orbit around Saturn.

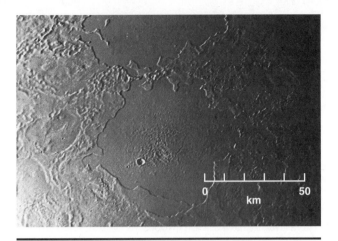

Figure 10-38. "Ice lake" of uncertain origin on Neptune's largest satellite, Triton, may mark an ancient, degraded impact or tectonic feature. Deposits around it seem clearly layered (see overlapping layers at bottom center).

Voyager scientists estimated the cratering rate in the Saturn system to be so high that the inner moons, especially Mimas, may have been fragmented and reassembled several times. Heavy cratering on Mimas and Tethys yielded one crater on each body almost large enough to have shattered the body. Similar satellite breakups might have contributed to the plethora of small inner moonlets and ring material.

Triton and Pluto

Even before Voyager got to Neptune, scientists expected a strange surface on its moon Triton. Cruikshank and Silvaggio discovered atmospheric methane gas and surface methane ice in 1979. Color changes suggested atmospheric hazes; Cruikshank and others (1989) later suggested nitrogen frost or even seas of liquid nitrogen. Voyager found no seas but an extraordinary world (Figures 2-33 and 2-34), and current data suggest the surface ices are mainly N_2 mixed with CH_4 and CO, with some areas of CO_2 ice (Cruikshank and others, 1993). A tenuous atmosphere consists mainly of nitrogen. Alternate illumination of the two poles by the sun (82 years each due to Triton's high orbit inclination) leads to seasonal transfer of frost deposits from one hemisphere to the other. The south cap had been illuminated for 30 years when Voyager arrived, but still had a frost layer of N_2 ice, with albedo >80% (Figure 2-34). The young, nearly uncratered surface is mostly N_2 and CH_4 ice with some carbonaceous dirt; there is no spectral evidence of H_2O ice. Unfamiliar geologic units include cantaloupe-textured terrain (Figures 2-34 and 9-17) and smooth "ice lakes" (Figure 10-38).

Most surprising of all were at least two active vents of some kind, marked by plumes of dark smoke rising some 8 km vertically, then streaking off parallel to the surface in upper-atmosphere winds (Figure 10-39). Numerous parallel dark surface streaks mark windblown deposits of debris from such vents. Researchers presume that the vents are driven by gas, probably N_2 jetting from below the surface (Smith and others, 1989). Some of the required heat probably comes from its being kept molten for about 1 Gy by tidal forces after capture by Neptune (Goldreich and others, 1989). Such a capture event would also explain Triton's odd, inclined, retrograde, circular orbit.

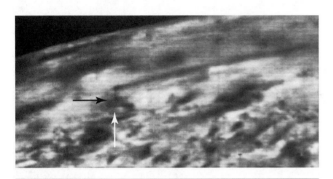

Figure 10-39. One of Triton's dark volcanic smoke plumes drifts as a long dark streak from left to right across image. Plume rises vertically from a vent on the surface (arrows). Vent details were not resolved, but elevation of plume was confirmed by stereoscopic pairs of images. Horizon at top. (NASA Voyager 2)

Numerous researchers have commented on the similarity of Triton and Pluto, whose diameters are 2,700 and 2,300 km, respectively. They both have densities close to 2,000–2,100 kg/m^3, tenuous atmospheres of comparable surface pressure, and atmospheres and surface frosts involving N_2, CH_4, and CO. Carbon dioxide appears less abundant on Pluto than on Triton (Owen and others, 1993). (Some of Pluto's CH_4 surface frost probably involves a transfer of volatile CH_4 from its moon Charon, which curiously shows primarily H_2O on its surface—see Chapter 11.)

As remarked above, Triton may have been an interplanetary body that was captured by Neptune. Thus, extending our earlier comments, instead of viewing Pluto as a planet and Triton as a moon, it may be more fruitful to visualize *both* as only among the largest bodies of the inner Kuiper cloud.

Io

At still larger size, the most volcanically active world in the solar system is 3,630-km Io, with its dark volcanic vents continuously spewing elegant, umbrella-shaped plumes as much as 280 km into space (Figure 10-40). During 6 ½ days of monitoring in 1979, Voyager 1 found eight erupting volcanoes. Four months later, Voyager 2 found at least six still erupting and one or two possible new eruptions.

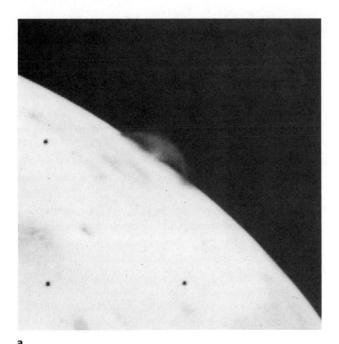

a

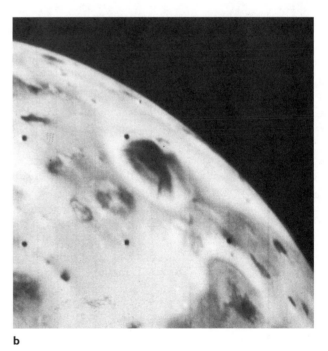

b

Figure 10-40. An erupting volcanic plume on Io. The plume is about 70 km high and 250 km wide. It was seen for 18 h by Voyager 1 and was still active 4 mo later, as revealed by Voyager 2. (**a**) Plume silhouetted on horizon. (**b**) Plume seen obliquely against Io's disk. Note distinct arcs of ejecta, indicating complex jetting at vent. (NASA, Voyager 1)

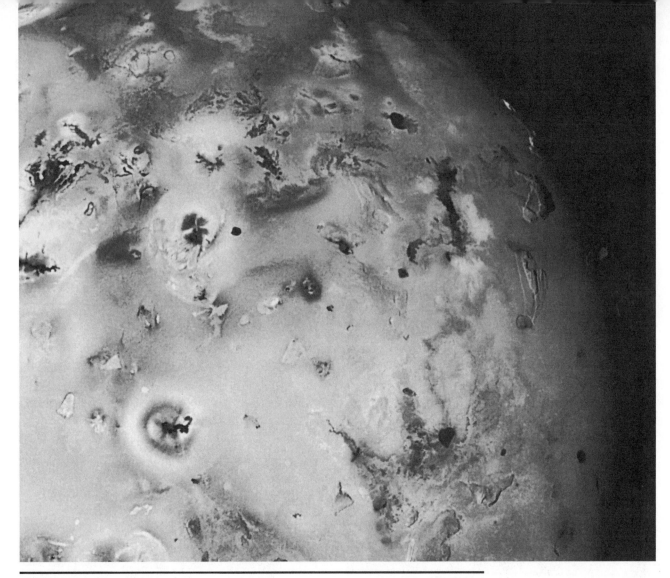

Figure 10-41. The mottled surface of Io. The surface is relatively flat, although mountain masses can be seen along the terminator (right). Small black spots are volcanic calderas, possibly containing molten sulfur. Dark flows emanate from some of them. Vertical view through an erupting volcanic plume produces the doughnut-shaped feature surrounding a dark caldera, in lower left center. (NASA, Galileo spacecraft view)

Based on the plume heights, the volcanoes eject material at 0.5 to 1.0 km/s. Voyager data suggested 270 to 310 K for material in the plumes, with the material at the vent probably much hotter (Hanel and others, 1979a, 1979b). Mauna Kea Observatory measurements in the 1980s gave 900 K for some hot spots, and Galileo spacecraft measurements in the 1990s measured typical vent temperatures of 420 to 700 K (Carlson and others, 1996, Belton and others, 1996). The temperature of melting sulfur in a vacuum is 715 K (Johnson and others, 1988), and the detection of hot spots hotter than this, together with color properties, led early workers to infer that Io "lavas" were sulfur (Sagan, 1979).

Io volcanoes usually display a blackish central caldera about 100 km across. This is often surrounded by a lighter patchy region, sometimes crossed by radial, reddish or yellowish flows that presumably run downhill from the caldera, as seen in Figures 8-33 and 10-41. Plume deposits are often reddish-brown and background areas are often whitish. The color properties match sulfur. Molten sulfur is black and has the fluidity of basaltic lavas. As it cools, it becomes red. This would account for black calderas with red, lavalike flows (Figure 10-41). At temperatures around 400 K, the color is yellow-orange. This could account for the wider flows and deposits of yellowish materials beyond immediate caldera rims (Belton and others, 1997). White deposits may be SO_2 frosts.

In spite of this work, evidence now supports silicates as a major component of at least some Io lavas, as mentioned in Chapter 8. The Galileo probe detected Io lavas

Figure 10-42. Imaginary view of phenomena of Io. Jupiter subtends an angle of nearly 20°. The sun is coming out from behind Jupiter following an eclipse. On the horizon is a relatively small plume from an erupting volcano. (Painting by author and Ron Miller)

with temperatures 1700 to 2000 K [basaltic lava typically melts at 1300–1450 K (Johnson, 2000)].

Detection of Io volcanic eruptions from Earth has an interesting history. About a year before the Voyager flybys, several observers reported a dramatic flare-up of the 3- to 5-μm thermal radiation from Io, which lasted a few hours (Witteborn, Bregman, and Pollack, 1979). The explanation was unknown, though a 600-K hot spot roughly 50 km in diameter was proposed. A similar event was observed from Earth between the two Voyager flybys (Sinton, 1979). This time, the geometry permitted approximate location of the source on Io, and Voyager 2 confirmed the formation of a new caldera surrounded by new deposits at this site, which were not present four months earlier. Subsequent Earth-based observations have allowed monitoring of Io's volcanic outbursts (Johnson and others, 1988).

Most of Io's topography is subdued, with isolated mountains that may be uplifted blocks (Figure 10-41). Stratified features may be lava masses; such strata in polar areas have led to suggestions of deposits of SO_2, condensing from volcanic gases. The polar regions are some tens of Kelvin colder than Io's equatorial regions. Voyager 1 found an SO_2 cloud with partial pressure of 0.0001 mbar near a volcanic plume, which could condense in the colder polar regions. Just as Mars has polar strata involving CO_2 and dust precipitation, Io may have polar strata involving SO_2. Modeling of the crustal structure and the roles of molten silicates versus sulfur compounds contin-

ues. Figure 10-42 characterizes some of the phenomena of Io as they might appear to a surface visitor.

Europa, Ganymede, and Callisto

Jupiter's satellites, and presumably others, were clearly exposed to heavy cratering between their formative era and today, since Callisto and parts of Ganymede are nearly saturated with craters. The less-cratered regions appear to have been resurfaced too recently to have accumulated the full complement of craters. The question is, how were they resurfaced? Figure 10-43 suggests the interesting possibilities. Among these bodies, the darkest surfaces are most cratered and light surfaces usually less cratered, with the bright surface of Europa being relatively pure H_2O ice according to spectral data. These facts suggest that surface melting or eruption of a watery "lava" allowed the silicate dirt to sink, with subsequent freezing creating a clean ice surface. What, then, was the heat source?

As explained in Chapter 8, the energy driving the resurfacing events was probably tidal heating involving orbital resonances among Europa, Ganymede, and Io. Spectra for all three bodies, taken by the Galileo space probe, show water ice and water frost spectral features and regional differences on each body that suggest frosts with different grain sizes, deposited atop more solid ices (Carlson and others, 1997). In addition, Galileo spectra of Callisto and Ganymede show evidence of hydrated minerals, possibly including hydroxides (OH-bearing minerals),

Figure 10-43. Ancient Europa? The sparsely cratered ice plains of Europa indicate that watery magmas erupted in the past and flooded large regions. In this imaginary reconstruction, a geyser (background) accompanies massive water eruptions on Europa, creating temporary surf and a thin transient atmosphere. Details of such hypothetical events remain to be studied. (Painting by Michael Carroll)

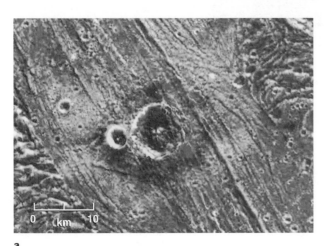

a

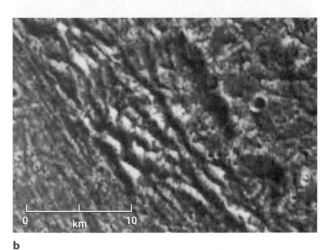

b

Figure 10-44. Tectonic destruction of craters on Ganymede. (**a**) The largest crater in this view, with a dark ejecta blanket, appears younger than some of the fractures, but other fractures appear to cut through the right side of the crater. (**b**) Extremely intense fracturing of this region of Ganymede has all but obliterated two old impact craters, slicing them laterally into many segments, as if cut by an egg slicer. (NASA, Galileo, courtesy Elizabeth Alvarez)

and also a spectral feature attributed to ice clathrates. Clathrates are ices in which one type of ice is mixed in another; one candidate for Callisto and Ganymede is CO_2 ice mixed into the ice structure of the frozen H_2O.

Continuing puzzles revealed by the Galileo images and spectra are the mode of eruption, fracturing, and fresh ice emplacement on Europa and Ganymede. Some of Ganymede's bright, sparsely cratered regions, which had been thought to be fresh ice, were revealed in photos to be older ice that was simply fractured to an extraordinarily intense degree, deforming old craters (Figure 10-44). On Europa, dark reddish-brown lines once thought to be sharply defined fractures were found to be diffuse, with sharper bright lines running down their centers; the bright

lines may represent fractures with fresh ice deposits, and the brownish bands bordering them may be deposits of debris erupted from the fractures (Chapter 8 opening image, right side).

On Ganymede and Callisto, bright, smooth, circular patches are presumably icy scars of old, filled-in craters. They were given the unusual name **palimpsest** by the Voyager team; the term refers to a reused writing surface on which the original writing has been erased. Figure 10-45 shows examples. Also, the systems of parallel arcuate rings look like remains of multiring basin systems flattened by some process. In both cases, the flattening process may involve filling by water, plus isostatic filling of depression by the glacier-like flow of the viscous ice

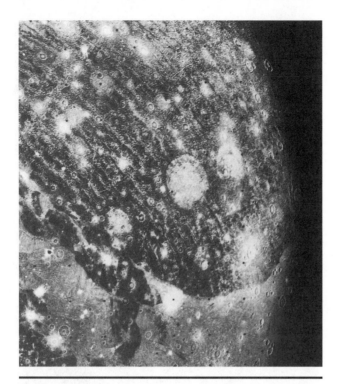

Figure 10-45. Contrast between an old, dark, cratered surface and younger, grooved terrain on Ganymede. The old surface preserved arcs that are probably part of a multiring basin system similar to those seen on Callisto (compare Figure 9-15c). Bright ovals, or palimpsests, are probably old craters filled in by ice. (NASA, Voyager 2)

Figure 10-46. Crater flattening on Europa. (**a**) The 26-km crater, Pwyll, sits on a young ice plain but has a filled-in and flattened appearance. (**b**) This larger impact structure produced a bullseye of concentric fractures, but no well-pronounced crater. The bright patch in the center may mark where water erupted and filled a central depression. Bright streak across the feature is a still younger fracture. (NASA, Galileo, courtesy Elizabeth Alvarez)

lithosphere. This process may have flattened or obliterated many old, large craters on Ganymede, Callisto, and other icy worlds while leaving smaller surface features more intact. Parmentier and Head (1981) calculated profiles of craters in relaxing viscous ice layers and matched observed craters. Large craters (with their greater initial elevation differences) flatten more quickly than small craters do, as measured in proportion to their initial depths. Craters 10 and 100 km across were found to relax in 30 Gy and 30 My, respectively, indicating that craters above a few tens of kilometers may flatten within the history of the satellites. On Europa, where the few fresh craters stand out from the ice-plain background, even the relatively fresh craters have a curiously flattened or subdued appearance, different from fresh craters on the moon, Mercury, or Mars. Figure 10-46 shows two examples. Pwyll, the 26-km diameter featured in Figure 10-46a, is fresh enough that its ejected ray material can be traced for hundreds of kilometers across Europa's plains, yet the crater itself has only a vestigial rim and a filled-in appearance. In Figure 10-46b, the 140-km impact feature looks more like a shatter mark in glass than a classic crater. These characteristics may have to do with infilling and the thinness of the icy lithosphere on Europa.

The whole question of the driving mechanism and mechanics of the fracturing and resurfacing of the icy satellites is a major area for future research. Figure 10-47 emphasizes that similar fracturing is seen on the Saturn satellites as well as those of Jupiter. Although most of the examples above involve near-linear swarms of fractures, Figure 10-47 shows two lens-shaped fracture systems associated with resurfacing on Ganymede and Saturn's bright icy moon, Enceladus. The two patterns are remarkably similar.

a

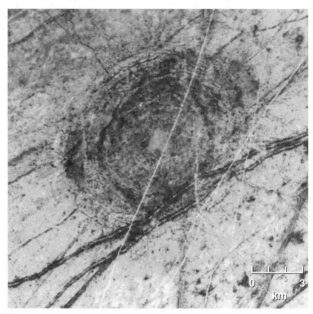

b

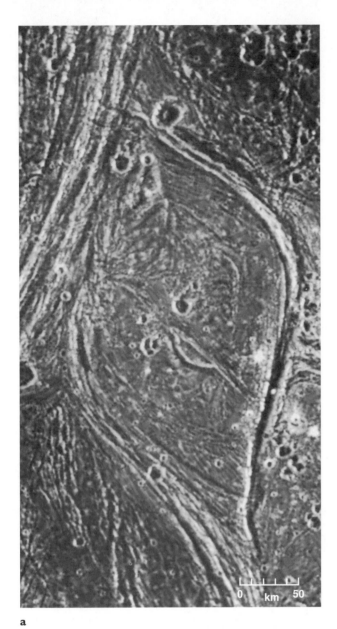

a

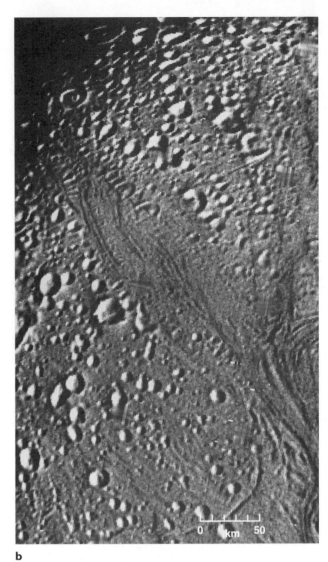

b

Figure 10-47. Lens-shaped fracturing and resurfaced areas on two icy satellites. (**a**) This region of Ganymede lies between cratered regions and swaths of linear fractures. (**b**) Similar area on Enceladus is a lens-shaped system where the ancient, cratered surface has been resurfaced. (NASA, Galileo and Voyager)

In a pre-Voyager attempt to explain the sodium glow around Io (see Chapter 2) and other Io peculiarities, Fanale, Johnson, and Matson (1974) raised an interesting point about the surfaces produced by heating and partially melting ice-rich or carbonaceous-like worlds. They pointed out that modest heating of a carbonaceous-chondrite-like body would produce water containing dissolved minerals; evaporation or loss of the water would leave behind evaporite deposits rich in salt (sodium chloride) and sulfur compounds such as those in Figure 10-35. In support of their model, Fanale, Johnson, and Matson noted that gypsum ($CaSO_4$ $2H_2O$), epsomite ($MgSO_4$ $\times$ H_2O, or magnesium sulfate hydrated with variable amounts of water), and bloedite ($MgSO_4$ Na_2SO_4 H_2O) have been identified in carbonaceous chondrites. (Recall that sulfur is also concentrated in apparent evaporite residues in Martian soils.)

Why are evaporites not more common among satellites of the outer solar system? Apparently, most water reaching the surfaces of these satellites froze too rapidly to leave evaporite deposits behind. Only on Io was there such repeated and intense heating that virtually all surface water was eventually driven away, leaving a sulfurous crust. Nonetheless, some regions of some bodies might have been reheated enough to produce local concentrations of evaporite minerals.

At the opposite extreme, bodies that never melted significantly never experienced differentiation of the H_2O

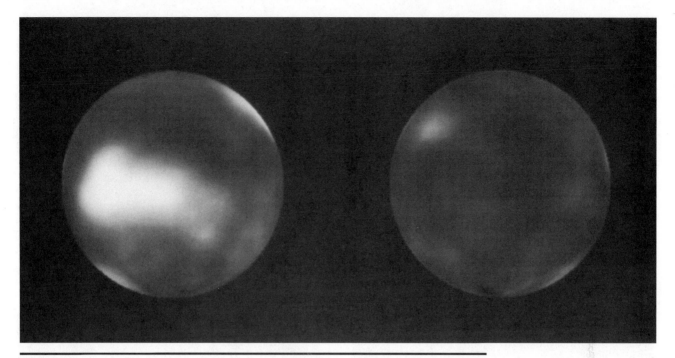

Figure 10-48. Surface features of Titan are dimly visible through the haze in these derived from Hubble Space Telescope images made in near infrared light that penetrates Titan's haze. The leading hemisphere has a large, bright region (reflective in near infrared, possibly cleaner ice; the trailing hemisphere is dark). (HST-derived maps, courtesy Peter Smith, University of Arizona)

and the dark carbonaceous dust and hence show the dark coloration and spectrum of the dust. Because modest differences in heating produce gross differences in geochemical differentiation of such materials, the surface chemistries of the outer solar system worlds may be very indicative of their past histories.

Titan

The known properties of the atmosphere guarantee an exotic, if still unknown, surface. As remarked in Chapter 2, the atmosphere is likely to produce rains and snows of methane-based and organic compounds. Hydrocarbon oceans were once thought possible. Based on these considerations, scientists have characterized Titan in colorful terms. Voyager scientist Donald Hunten commented that octane would be synthesized in the atmosphere and that he visualized Titan "raining frozen gasoline." Voyager scientist Von Eshelman characterized the surface as "a bizarre, murky swamp (with) hydrocarbon muck." Owen (1982) pictures good surface visibility (occasionally marred by methane rainstorms) but a low midday light level resembling that of a full-moonlit night on Earth. He suggests minimal winds. He also remarked that because of the virtual invisibility of celestial objects (obscured by the low methane clouds and mid-level smog) and the absence of a magnetic field, navigating on any methane lakes or seas would be difficult!

Peter Smith (University of Arizona) and colleagues in 1994 used the Hubble Space Telescope to photograph Titan in near-infrared wavelengths where the atmosphere is partly transparent; the images showed a highly reflective feature on the leading side, and a darker surface on the trailing side (Figure 10-48; Smith and others, 1996). This indicates that Titan's surface is not covered by a universal, uniform ocean of liquid, though it still could have local seas or lakes. Follow-up observations at the near-infrared "spectral window" wavelengths revealed spectral features of frozen H_2O on the surface. Thus, Griffith and others (2003) were able to conclude that "despite the hundreds of meters of organic liquids and solids (previously) hypothesized to exist on Titan's surface, its icy bedrock lies extensively exposed."

The Cassini spacecraft was successfully launched to Saturn and Titan in 1997, and a parachute probe from Cassini is designed to send back data about the surface of Titan in 2004.

The Moon

The planetary surface whose history is best known is probably the moon. In the beginning, there was intense cratering (Figure 6-6), repeatedly pulverizing the thin lithosphere that was trying to form on a magma ocean surface. After a few hundred million years, a heavily cratered surface emerged, underlain by a thickening lithosphere (Figure 9-4). Figure 10-49 shows the evolution of this surface seen from an imaginary point in space. In Figure 10-49a, we see the heavily cratered pre-Imbrium uplands, about 4200 My ago, in the region that was to become Mare

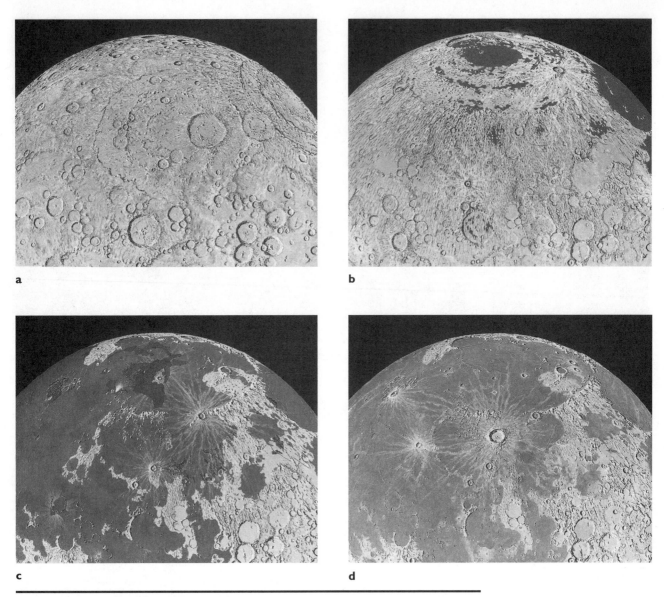

Figure 10-49. Imaginary views of the evolution of the Imbrium basin region of the moon, from prebasin to present configuration. See text for further explanation. (Paintings by Don Davis, U.S. Geological Survey, courtesy D. E. Wilhelms)

Imbrium. Volcanism during this period produced lava flows of KREEP basalt, found by astronauts in the lunar highlands. The surfaces of those flows were rapidly brecciated by the intense bombardment. Figure 10-49b shows the same scene about 3900 My ago, soon after the enormous impact that formed the huge Imbrium basin. The Imbrium ring system and ejecta are prominent. Already some lava flows from the partially molten zone below the lithosphere have erupted through the ring fractures created by the impact, producing an appearance like that of the Orientale basin today. In Figure 10-49c, about 3,400 My ago, the mare-forming eruptions are continuing. Mare Imbrium and Oceanus Procellarum nearly have their present configurations (see Figure 2-10). In Figure 10-49d, mare flooding has ceased, and recent craters, such as

Copernicus, have formed with their impressive ray systems. The scene shows the present appearance.

Figure 10-50 gives some loosely corresponding surface views. Figure 10-50a is an imaginary view of a fresh mare lava flow of fluid, pahoehoelike basalt, based on the terrestrial appearance of such flows. This is a view not encountered on the present-day moon because volcanism mostly stopped 3,000 My ago, and cratering has beaten the old flows into rubble. At the site of the Apollo photo of Figure 10-50b, an impact has punctured a thin regolith cover and exhumed boulders of intact lava. More typical of the moon today is Figure 10-50c, which shows a vista of a regolith-covered mare with upland mountains in the background and astronauts' tracks in the powdery soil.

a

b

c

Figure 10-50. Representations of different stages of lunar surface evolution. (**a**) Imaginary view of fresh pahoehoelike lava flow prior to pulverization by impact gardening. Such fresh flows might have been visible 3,500 My ago. (Painting by Hiroki Morinoue) (**b**) Boulders excavated by cratering near the edge of Mare Serenitatis. (NASA, Apollo 17) (**c**) Powdery soil at Apollo 15 landing site, on the edge of Mare Imbrium is typical of mature regolith. (NASA, Apollo 15)

The search for the most ancient surface materials of the moon has been frustrating. The uplands are the oldest areas, and many uplands show evidence of ancient lava flows almost covered with crater ejecta-debris. Apollo scientists targeted the Apollo 16 mission to one of these areas, hoping to collect ancient upland lava samples. Instead, the astronauts found highly brecciated material with ages typically 4,000 to 4,400 My. Many lunar scientists then interpreted the upland's intercrater plains as sheets of ejecta that swept across the uplands from the ancient basin-forming impacts, perhaps in the form of immense base surges (see Taylor, 1975, for a summary).

However, many small, dark halo impact craters in the intercrater plains penetrate a light-colored veneer and throw out darker ejecta (Schultz and Spudis, 1979). The ejecta has been shown spectrally to be basaltic (Hawke, Spudis, and Clark, 1985). Similarly, the Galileo spacecraft, flying past the moon's far side, discovered that darker upland plains generally have a more basaltic spectrum than brighter uplands. Probably many upland plains mark ancient lava flows, possibly the KREEP basalts. Lava plains that formed before a certain date, around 4,000 My ago, were forming during the intense bombardment and received a substantial dusting of bright ejecta, hiding their dark color and making them look like bright uplands.

Thus, while the moon looks sharply divided into two surface units—bright uplands and dark mare patches making the "man in the moon"—the physical evolution of these units may not be so distinct. The "true uplands" are the light-colored anorthositic rock that formed the early lithosphere about 4,400 My ago. Dark basaltic lavas that erupted on that surface prior to about 4,200 My ago were more or less chewed up by the intense early meteoritic bombardment (Figure 6-6) and converted back into rugged, light-colored uplands. Thick lava flows that formed about 4,000 My ago caught the tail end of the bombardment and received perhaps 3 to 20 times the amount of cratering as the dark lunar maria; they were thus masked by a veneer of overlapping rays and ejecta sheets. Only the lava flows formed after about 3,800 My retained their familiar dark coloration. Because most of the moon's surface formed either before 4,000 or after 3,800 My ago, most of the surface is either bright upland or dark mare, with few intermediate areas.

Does lunar volcanism occur today? Most astronomers for several generations have regarded the moon as a dead world. Indeed, this belief accounted for a certain lack of interest in the moon by astronomers until the middle 1900s. A few amateur astronomers had reported changes in certain lunar structures, but none of the reports were documented well enough to be taken seriously by most professional astronomers. However, on November 19, 1958, the Soviet astronomer Kozyrev (1959) obtained a photographic spectrum indicating an emission of gas in the lunar crater Alphonsus. Next, on two occasions in 1963, a number of observers at Lowell Observatory saw red glowing spots in and near the crater Aristarchus

(Greenacre, 1963). Such events came to be known as **lunar transient phenomena.** Other historical examples, less well verified, are also known.

A statistical study suggests that lunar transient phenomena are most frequently reported when the moon passes through its perigee point, when lunar tidal forces are largest (Middlehurst and Chapman, 1968). Apollo seismic observations showed that a particular type of moonquake also concentrates at this time. Observations by orbiting astronauts and their instruments turned up no evidence for active eruptions, but measurements of radioactive radon-222 gas, produced by uranium decay, showed peaks over Aristarchus, Grimaldi, and the edges of maria—possibly indicating gas leakage through fractures in these areas (Taylor, 1975). (Radon emissions were correlated with a major Japanese earthquake in 1978; Wakita and others, 1980). In short, significant lava eruptions seem unlikely at the present time, but gas emissions or perhaps even small ash ventings may occur occasionally.

The lunar surface is far from fully known. This was underscored in 1998 when the inexpensive Lunar Prospector spacecraft flew to the moon and discovered deposits of H_2O ice mixed into the soil in permanently shaded regions near the two poles. The detection was made by monitoring neutrons ejected from the surface by cosmic ray bombardment; their energies are modified by the presence of ice. For some years, geochemist James Arnold had argued that such ice would condense and accumulate during comet impacts. The ice mixing ratio in the soil is estimated at 0.3 to 1%.

So there is much still to see on the moon. Because Apollo landings were targeted on rather flat places for safety reasons, we have not yet seen surface vistas at the rims of large craters, along major fault scarps such as Straight Wall, at possible eruption sites, or among domes. A few places are young enough not to have been smoothed by regolith formation. Figure 10-51 shows an example of a strikingly rugged terrain that must have a spectacular appearance from the ground.

Mercury

As mentioned in earlier chapters, three flybys by Mariner 10 in 1974 and 1975 show that Mercury is superficially similar to the moon in having cratered uplands, multiring basins, and plains with the morphological features of lunar mare lavas. The Mariner findings and surface interpretation have been reviewed by Strom (1979).

A major difference between the surfaces of Mercury and the moon is the presence of more intercrater plains on Mercury, somewhat resembling those on the moon. These can be seen outside the Caloris basin by comparing Figures 9-15a and 9-15b. A possible interpretation is that because Mercury is the larger planet, its cooling took longer and its asthenospheric source for volcanic eruptions lasted longer, as seen in Figures 8-24 and 8-25. Also, as seen in

a   b

Figure 10-51. Interior of the 90-km crater Tycho, one of the youngest and roughest extensive regions on the moon. Lunar samples and crater counts suggest ages of 100 to 270 My. (**a**) Tycho and surroundings. (**b**) Close-up of rough floor northeast of the central peak along the wall shadow. Width 6.5 km; smallest features about 10 m. (NASA, Orbiter 5)

Table 6-1, Mercury's cratering rate probably averaged about twice the rate of the moon. Thus, while a Mercurian lava plain that formed 3,700 My ago would have escaped the early intense bombardment and would not have been converted into cratered uplands, it would have accumulated more cratering than a 3,700 My-old lunar mare. At present, it would be masked by a veneer of rays and secondary ejecta, thus creating the upland plain appearance seen on the photos.

Spectral studies suggest a roughly lunar-surface composition, with a feldspar-rich crust and volcanic lavas (Robinson and Lucey, 1997). However, some studies suggest less basaltic material than the moon, and less iron than might be expected from Mercury's high density (Jeanloz and others, 1995; see also review by Clark, 1997). The contraction evidenced by thrust faults on the surface (Figure 8-25), might have inhibited open fractures and discouraged basaltic magma access to the surface.

In spite of these differences, an astronaut plunked down at a random spot on Mercury might have a hard time telling visually (without looking at the sun, which is 2.5 times larger and 6.5 times brighter) that he or she was not on the moon.

Mars

Mars, being still more massive than Mercury, had still more prolonged volcanism, which resurfaced many of the early cratered surfaces and built spectacular volcanoes. Continuing volcanism on Mars produced basaltic flows and built up the largest known volcanic edifices in the solar system, such as the towering, youthful-looking Olympus Mons and neighboring volcanoes of the high Tharsis lava plains. These volcanoes are not known to be active, but neither are they known to be extinct! As we saw earlier, the rocks so far sampled from Mars mostly testify widespread basaltic volcanism. About half of Mars is covered by such young lava plains, especially the Tharsis plains. The other half is old, cratered uplands, and an important question is the nature of the rock units there. A significant question is whether internal differentiation has produced more siliceous rock types and whether surface deposition processes have produced conglomerates and other sedimentary rock types. One rock sampled at the Mars Pathfinder site, where the Ares Vallis channel carried floodwaters and debris out of the uplands, was marginally andesitic, and some were suspected conglomerates

a

b

c

d

e

f

Figure 10-52. Six landscapes on the surface of Venus, photographed by Russian Venera landers. Parts of the landers appear in the bottom of views **a**, **b**, and **f**. (**a**) First photo from the surface of Venus, showing loose boulders near Venera 9. (**b**)–(**d**) Boulders, gravel, and platey outcrops near Venera 13. (**e**)–(**f**) Sheetlike rock outcrop and loose (darker) soil near Venera 14. (Courtesy of C. P. Florensky and A. Basilevsky, Vernadsky Institute, Moscow)

(Matijevic and others, 1997). Does Mars approach the geological complexity of Earth, or is it more like the moon with a little wind to blow the dust around?

Important in shaping the special personality of Mars are ancient water flows and modern atmospheric processes that have created many surface features not seen on the worlds discussed above. We expand on issues about the ancient water flow in Chapter 13. In modern times, winds move materials around, creating transient bright and dark

markings, dune fields, stratified deposits, wind streaks, and deposits in crater floors.

This leads to the interesting problem of whether, and where, bedrock deposits are exposed on Mars. Mars Global Surveyor photos at resolutions of 10 m or so have shown that drifts of wind-deposited dust blanket much of Mars. Three out of three landing sites so far showed a mix of scattered boulders but not bedrock. Two out of three of these showed sizable dune deposits. Where will we find

Figure 10-53. Weathered pahoehoe basaltic plains resembling the landscapes of Venus are found on Mauna Kea volcano in Hawaii. High overcast provides diffuse lighting that heightens the resemblance to Venus. (Photo by author)

bedrock outcrops that show the geology of Martian rockforming units *in situ*? Are all the large-scale dark markings seen from Earth merely products of wind deposition, or do some mark regions of bedrock exposure? So far, the best chances to find such exposures may be in the difficult terrain of canyon walls and faulted valleys. Mars Global Surveyor has photographed many such strata, which probably include basalts as well as windblown and water deposited sediments averaging tens of meters thick.

Venus

The surface of Venus may be still more complex. We know that the atmosphere produces an extraordinary environment with temperatures around 750 K (891° F) day and night and with pressure 90 times that on Earth's surface. The surface air is "heavy," having a density about 50 times that of air on Earth, and 5% that of water. Though Venus is completely overcast by clouds, the bottom of the dense cloud layer is normally about 44 km above the surface, and the lower atmosphere is surprisingly bright and clear, something like a very cloudy day on Earth.

Wind velocities measured at or near the surface by three Soviet probes (Veneras 8, 9, and 10) and two American probes (Pioneer "Day" and "North") were very low, about 1 ± 1 m/s. The winds account for the low dust found in the surface atmosphere by the Russian probes (Keldysh, 1977). White (1981) calculated the dust transport and found ratios of wind saltation of dust grains on Venus:Earth:Mars to be 1:10:250. Nonetheless, Magellan radar mapping surprised observers with images of scattered dunes and windblown streaks of ejecta from impact craters (Greeley and others, 1991).

Photos from Venus's surface made by Venera landers (Figure 10-52) show different kinds of rocky landscapes,

ranging from gravel and loose angular boulders up to about 0.5 m across, to smoother surfaces, apparently with sheetlike outcrops of platey rock. Strikingly similar landscapes are seen in lightly eroded volcanic terrain on Earth, especially on weathered pahoehoe flows (Figure 10-53).

Soviet VEGA 1 and 2 missions, on their way to Halley's comet, also dropped landers that carried no cameras but measured compositions.

All seven Soviet landers that gave composition data (Veneras 8, 9, 10, 13, and 14, and VEGA 1 and 2) landed on plains near the equator, some on the fringes of uplands. Two types of composition measurement were used. Gamma-ray spectroscopy (Venera 8, 9, 10, VEGA 1, 2) measured the amount of radioactive potassium, uranium, and thorium in surface materials, grossly testing whether the rock is basaltic or granitic. X-ray fluorescence (Venera 13, 12, VEGA 2) measured the content of silicon, titanium, aluminum, iron, manganese, magnesium, calcium, potassium, sulfur, and chlorine in 1 cm^3 drilled from beneath the lander. At five sites, the composition resembled tholeitic basalt—the types of basalts found in Earth's oceanic crust, extruded from the upper mantle, beneath the thin seafloor crust. The Venera 13 site also gave a basaltic composition of a high-potassium type known as leucitic or subalkaline basalt. The Venera 8 site was the most provocative. Situated about 5,000 km east of Phoebe Regio, probably on a somewhat raised area that seems to be in the midst of volcanic flows in the Magellan imagery, it has a granitic composition, possibly a syenite type related to the Earth's continents (Saunders and others, 1991). If this assumption is correct, Venus may be only a little less differentiated from Earth in terms of measured rock chemistry, as we would expect on basis of size. In general, the dominance of basalts is consistent with the picture of Venus presented in Chapter 9, where lack of plate-active

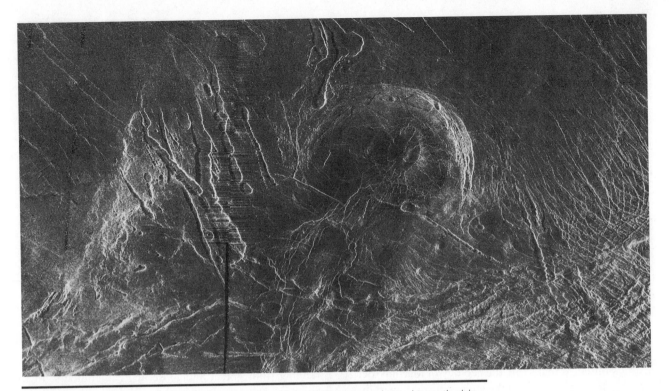

Figure 10-54. Volcanic terrain of Venus at the edge of Lakshmi Planum, an elevated smooth plain. Partly formed circular feature is probably a volcanic collapsed caldera, named Siddons. To its left are lava flow channels and possible collapsed lava tubes (compare Figure 10-19). At lower right is intensely fractured deformed terrain. (NASA, Magellan radar image)

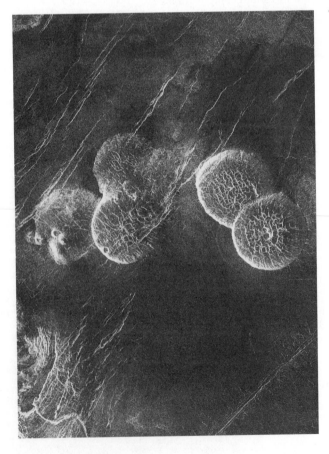

Figure 10-55. Unusual circular domes on the east side of Alpha Regio, Venus. These features average about 20–25 km across and about 750 m high. They may be pancakelike eruptions of viscous lava. Cinder cones can be seen at the left edge of the left dome. Tectonic fractures (bright lines) cross the area. (NASA, Magellan radar image)

tive surface on Earth. (Photo by author)

Figure 10-56. The most representative surface on Earth. (Photo by author)

tectonics has inhibited crustal recycling and granitic continent buildup (Saunders and others, 1991). After examining the evidence, Magellan researchers affirmed the general law that basaltic volcanism is the leitmotif of terrestrial planet geology (Figures 10-54 and 10-55).

How old are Venus's volcanic flows? Most researchers suspect that volcanism is active on Venus today, but no one has proved it. Among the evidence are rapid fluctuations in SO_2 gas content that might come from volcanic exhalations (Esposito, 1984) and (controversial) indications of concentrations of lightning around volcanic peaks (measured from radio signals). Crater counts, divided by the estimated cratering rate, suggest that various units of Venus span ages ranging from 800 My to less than 100 My, and thus support the conclusion that Venus is volcanically active today (Phillips and others, 1991).

Earth

Figure 10-56 reminds us that our most common "landscape" is unique in the solar system—and is not the one on which we spend most of our time. Our range of environments—from the dark depths of the oceans to arid deserts and Everest's airy peak—barely begins to match the variety found among other worlds, small and large. The astonishing individuality found among these places—from Hektor's dumbbell shape and Europa's ice-buried oceans and Titan's methane smog to Io's erupting volcanoes and Mars's deserted dunes and Earth's rain forests—emphasizes that each planetary body is its own special world, to a greater extent than we would have guessed a decade or two ago.

endogenic processes
exogenic processes
volcanism
tectonics
cumulates
aa
pahoehoe
fluidized eruption
nuée ardente
ash
cinders
tuff
welded tuff
ignimbrite
structural geology
cinder cone
spatter cone
scoria cone
hornito
shield volcano
composite volcano
crater
caldera
lava tube
pit crater
diatreme
maar
dark halo crater

volcano-tectonic sink
tectonism
fault
thrust fault
strike-slip fault
normal fault
graben
lineament
mare
wrinkle ridge
dome
rille
sinuous rille
linear rille
crater chain
slumping
creep
saltation
wind streaks
dune field
geochemical cycle
Urey reaction
red bed
evaporites
playas
duricrust
palimpsest
lunar transient phenomenon

PROBLEMS

1. (a) Why do terrestrial lava flows have different textures, such as aa and pahoehoe? (b) Why do lava flows found on the moon today lack these textures?

2. (a) What does the absence of large lunar volcanic mountains and the paucity of lunar cinder cones imply about the fluidity of lunar lavas? (b) Why are maars unlikely on the moon?

3. (a) What causes volcanic eruptions? (b) Do volcanic areas necessarily have totally molten zones (with the temperature exceeding the melting point of all rock minerals) below them within 100 km of the surface?

4. Why were lava eruptions on the moon (or on other planets) 3.5 Gy ago most likely to occur in or near recently formed, large impact basins?

5. (a) Why are Martian channels unlikely to form under present Martian conditions? (b) What past conditions might have produced them? (c) What do Martian polar strata imply about the Martian climate's history?

6. In the context of past and future evolution of life on Earth or on other planets, what is the philosophical importance of evidence about the existence or nonexistence of dramatic global climate fluctuations on Mars or other planets?

7. Describe the history and current location of H_2O, if any, on (a) Mars, (b) Venus, and (c) Io.

8. How are the surfaces and histories of Phobos, Deimos, J6 Himalia, J7 Elara, S9 Phoebe, belt asteroid 10 Hygiea, and Trojan 624 Hektor and certain other asteroids possibly related?

9. Name at least one planet besides Earth and at least one satellite that could have once had large oceans of liquid water. Cite evidence.

10. Suppose future astronauts colonize planetary bodies. Discuss how they might obtain building stone, H_2O, iron, sulfur, molecular hydrogen, and molecular oxygen on (a) the moon, (b) asteroid 1580 Betulia, (c) asteroid 16 Psyche, (d) Mars, (e) Ganymede, and (f) Io.

11. Smith and others (1979b) discuss a 20-km-thick crust of sulfur and silicates on Io, with perhaps a 1-km-deep layer of liquid sulfur involved in the eruptions. Johnson and others (1979) used the geometry of the erupting plumes to calculate an average resurfacing rate all over Io of 3×10^{-4} to 0.1 cm/y. The radius of Io is 1820 km. If a layer roughly 10 km deep is assumed to cycle through the eruptions, how long does it take for all this material to go through the cycle of eruption, burial, remelting, and reeruption? Has most of the material been through the cycle less than once, once, or more than once?

12. Volcanologist A. Rittmann (1962) discussed nine terrestrial eruptions with eruptive rates of about 0.03 to 3,000 km³/y (for individual volcanoes). Taking these as lower and upper limits on the eruption rates of Io volcanoes and assuming that there are eight eruptions on the average at all times, calculate the cycle time for the upper 10 km of Io's crust. Compare this with the result in problem 11.

13. (a) Mars is at prehelion. A polar ice field with dirty ice having an albedo of 40% is located so that the sun is 10° above the horizon. The emissivity of the ice-soil mixture is 0.5, and half the sunlight gets through the Martian atmosphere at the slant angle mentioned. Calculate the equilibrium temperature of the Martian surface. (b) A landslide occurs, exposing a cliff face with an 80° slope, which is fully exposed to the sun. Calculate the equilibrium temperature of the ice-soil mixture in the cliff face. (c) Comment on the future erosion of the cliff area.

Jupiter's swirling clouds demonstrate complex atmospheric processes. Various chemical compounds give the clouds different hues, and atmospheric turbulence affects the clouds' motions. Bright clouds are higher; dark regions tend to be lower formations. The large oval area, bottom, is the Great Red Spot, a storm system several times Earth's diameter. (NASA, Voyager 1 image made during flyby in 1979)

CHAPTER ELEVEN

Planetary Atmospheres

When studying planetary atmospheres, we want to learn their origin and evolution, the principles that determine their structure, and their meteorological conditions. Atmospheric physicists range from those concerned with present surface conditions to those concerned with long-term climatic history.

One vigorous area of research concerns interactions between atoms and magnetic fields at the fringe of the atmosphere, where the solar wind interacts with planetary gas. This study, sometimes called simply **particles and fields,** has expanded greatly with spacecraft measures near all the planets out to Neptune, but it is beyond the scope of this book.

Another growing area is research on climatic effects of natural or artificial atmospheric changes, such as volcanic ash in the stratosphere, industrial CO_2 increases, or chemical damage to the ozone layer. Planets offer natural laboratories for these studies; for example, the atmospheres of Venus and Mars, being dominantly CO_2, allow us to calibrate our understanding of the CO_2 greenhouse effect. Also, terrestrial ice ages and evidence for major climate changes on Mars provoke us to wonder about the drivers of climate change on Earth and all planets.

Origin of Planetary Atmospheres

Primitive Atmospheres

As was discussed in Chapter 5, the planets were immersed in a hydrogen-rich nebula when they formed. This tenuous nebula had a nearly insignificant pressure as planets began to grow, but the gas must have concentrated in the gravitational fields of the primeval planets once they reached the size range somewhere between Venus and Uranus. This concentration eventually formed thick **primitive atmospheres** around the giant planets, and thinner such atmospheres around the larger terrestrial planets. "Primitive" in this sense refers to starting with nearly solar composition except that the light gases, such as hydrogen (H atoms or H_2 molecules) would have tended to float to the top of the atmosphere and escape into the solar nebula (see the later section on atmospheric escape), whereas heavier gases—such as argon (Ar), neon (Ne), methane (CH_4), and ammonia (NH_3)—would have been more readily retained. Chemically, an H-rich atmosphere would be reducing, not oxidizing. Once the solar nebula cleared, leakage of any primitive atmosphere from terrestrial planets would have been rapid.

Inert Gases as Tracers of Early Conditions

Could our present atmosphere simply be a remnant of such a primitive atmosphere? No. Consider the amount of neon (Ne) in our present atmosphere. Neon is a heavy inert gas, and none is produced by radioactive decay. Because it is heavy, virtually none has escaped into space, and because it is inert, it has not chemically bonded into the crustal rocks or oceans. Thus, the Ne in our present atmosphere is a good tracer of the primitive atmosphere. By adding the proportionate amounts of other solar nebula gases to the existing Ne, we can calculate the corresponding mass of the primitive atmosphere, which turns out to be only 0.9% of the mass of the present atmosphere (Walker, 1977, p. 182). The conclusion is that today's atmosphere is much too extensive to be a remnant of the primitive atmosphere.

The above calculation suggests two possibilities: (1) The primitive atmospheres of terrestrial planets were small, only a fraction of the present atmospheres of Earth, Venus, and Mars; or (2) the primitive atmospheres may have been massive initially but were swept away from the planets as the nebula was cleared by violent T-Tauri-phase solar winds. The second idea has been invoked to help explain chemical properties of the planets and the presence of captured satellites.

The argon in the atmospheres of Venus and Mars complicated this analysis when it was measured by probes in the 1970s. Argon has several isotopes. The surprises were associated with argon 36 and argon 38, which are not radiogenic and thus are remnant tracers of the primitive atmosphere in the same way as neon, discussed above. Prior to the Venus measurement, some researchers had theorized that planetesimals in the vicinity of Venus were so much hotter than those of Earth that less argon was adsorbed on their surfaces. Also, a hot Venus might have retained less of its primitive atmosphere than Earth. Venus was thus predicted to have less argon 36 and argon 38 than Earth. The Venus measurements, combined with Viking Mars measurements, established just the reverse trend: concentrations (in units of 10^{-12} g of argon 36 and argon 38 per gram of planet mass) on Venus, Earth, and Mars of about 5,000, 46, and 0.2, respectively (Pollack and Black, 1979). A possible explanation is that the temperature in the nebula did not rise as fast as thought toward the sun, and so the nebular gas pressure and density increased sharply enough toward the sun to drive these amounts of argon into the planetesimals. As an alternative possibility, the closer the planetesimals were to the sun, the more solar wind argon atoms were trapped in the planetesimal materials. The matter remains controversial.

In any case, although 1950s researchers assumed that a hydrogen-rich primitive atmosphere was present during much of Earth's early history and might have been the atmosphere in which life evolved, the current view is that the hydrogen-rich primitive atmosphere was very transient and was soon replaced by a secondary atmosphere of volcanically emitted gases, before life evolved.

Secondary Atmospheres of Earth, Mars, and Venus

Secondary atmospheres are atmospheres that have been produced or significantly altered by gases exhausted from the planetary interior, as symbolized in Figure 11-1. Although the giant planets have gravitational fields strong enough to retain their primitive atmospheres, the atmospheres of the smaller planets are mostly secondary.

What kinds of gases were added by volcanoes? In a classic paper, Rubey (1951) showed that many of the gases now in Earth's atmosphere came from this source. Rubey started by making an inventory of existing volatiles at Earth's surface (such as water in the oceans, carbon dioxide in carbonate rocks, nitrogen in the air, and so on). With some assumptions about the primitive atmosphere, he calculated the relative amounts of gases that had to have been emitted by Earth in the last few billion years to explain today's atmospheric and surface volatile inventory. Table 11-1 page 320, shows the rough agreement between gases coming out of volcanoes and the gases required by Rubey's analysis. For example, the main gases needed are not the gases now in the atmosphere, but water to make the oceans and carbon dioxide to allow the formation of carbonate rocks. Strikingly, the main gases emitted by volcanism are indeed CO_2 and water vapor (H_2O). Thus, volcanic gases appear to account for much of Earth's present surface volatiles and secondary atmosphere.

Some controversy remains about the transition from the original primitive atmosphere. Was there ever a substantial, hydrogen-rich, primitive atmosphere with strongly reducing chemical properties? Or was it immediately swamped by a more oxidizing atmosphere of CO_2 and water vapor? An important extension of Rubey's work was

a study by Holland (1962), who showed that metallic iron in the early, undifferentiated Earth would have combined with and removed oxygen from the first volcanic gases as they ascended to the surface, making them rich in H_2, H_2O, CO, and H_2S. As a result of this study, researchers for some years assumed that the original primitive atmosphere and even the early secondary atmosphere were strongly reducing and rich in hydrogen compounds. Today, researchers are more inclined to believe that reducing conditions were very short-lived and were soon replaced by a secondary atmosphere neither strongly reducing nor strongly oxidizing, but perhaps rich in CO_2 and N_2.

In any case, it is a striking fact that today's oxygen-rich conditions were not present during the first third of Earth's history. They evolved slowly because of two processes. First, dissociation of H_2O molecules into H and O occurred in the upper atmosphere when H_2O molecules were struck and split by energetic ultraviolet sunlight. Second, photosynthesis began after green plants evolved. Analysis of ancient sediments show that the sediments deposited before about 2,500 My ago formed in oxygen-poor environments and that the O_2 began to increase rapidly after about 2,000 My ago. Fossils from this time show the emergence of photosynthetic plants that gave off O_2 (Walker, 1977).

Comparative Planetology of the Atmospheres of Venus, Earth, and Mars

These ideas of atmosphere evolution help explain relationships among the atmospheres of Venus, Earth, and Mars. At first glance, they appear entirely different: Venus has a massive CO_2 atmosphere; Earth has N_2-O_2; and Mars has a thin CO_2 atmosphere. But Table 11-2 page 320, shows that, contrary to our first impression, Venus, Earth, and Mars have similar total inventories of volatiles, once we count the volatiles bound up in rocks, ice, and permafrost. Note especially that Venus and Earth have nearly identical total CO_2 inventories! This is a beautiful tie-in with Table 11-1; that table shows that secondary atmospheres created by volcanism should be mostly H_2O and CO_2, with traces of other materials such as N_2. As noted in the last chapter, the difference is that on Earth, CO_2 dissolved into oceans and the Urey reaction drove it into the form of carbonate rocks; on Venus, this could not happen because of the lack of oceans. Although we think of nitrogen as being important on Earth (it dominates our atmosphere), Earth and Venus have similar small amounts of nitrogen; nitrogen looks important to us only because our massive CO_2 atmosphere has disappeared into carbonate rocks, leaving the nitrogen behind to dominate the air we breathe.

Argon on Venus supports this picture. Argon 40 is a radioactive decay product, mostly incorporated in rock-forming minerals; its concentration in the atmosphere thus measures not a primitive atmosphere but rather the amount of argon created inside Venus by decay of potassium and

a

Figure 11-1. Evolution of a secondary atmosphere occurs as volcanic gases, mostly carbon dioxide and water, are emitted from planetary interiors. This occurs during both large-scale eruptions and small-scale venting. (**a**) The plume from a volcanic eruption of Mauna Ulu volcano, Hawaii. (Photo by Boone Morrison) (**b**) Local emission from a volcanic fissure. The light-colored deposits are sulfur condensates. (Kilauea volcano, Hawaii; photo by author)

b

then outgassed during volcanism. It is present on Venus within a factor 4 of the relative abundance on Earth, giving evidence of volcanic outgassing on both planets. The lower abundance on Venus suggests that the volcanic outgassing there has been somewhat less efficient than that on Earth.

The amounts of H_2O and N_2 on the two planets may have depended strongly on initial bulk compositions, and the history of water on Venus is still a research issue. For example, Venus may have contained less H_2O than Earth at the outset because of its closer position to the sun and the consequent loss of volatiles from planetesimals at that

TABLE 11-1 Gases Added to the Atmosphere by Volcanic Outgassing of Earth (percent composition by weight)

Gas	Observed from eruptions that may tap the mantle (Hawaiian volcanoes)[a]	Observed from continental fumaroles and geysers	Volatiles outgassed according to calculation by Rubey (1951)
H_2O	57.8	99.4	92.8
CO_2[b]	23.5	0.33	5.1
Cl_2	0.1	0.12	1.7
N_2	5.7	0.05	0.24
S_2	12.6	0.03	0.13
Others	<1	<1	<1
Total	100	100	100

[a] Figures from Hawaiian volcanoes. This source, which lies above a thin oceanic crust, is the most primitive, deep-seated magma available. (Continental fumaroles, on the other hand, represent material that has been chemically reprocessed and contaminated with groundwater, etc.)

[b] Plus small amounts of carbon monoxide (CO).

TABLE 11-2 Inventories of Selected Outgassed Volatiles on Terrestrial Planets (10^{-9} kg/kg planetary mass)

Volatile	Venus	Earth	Mars
H_2O			
Atmosphere	60[a]	3[b]	0.02
Oceans and polar caps	—	250 000[a,b]	5 000?[a,c]
Crust	160 000?[c]	30 000[b]	10 000?[a,c]
Total	160 000?	280 000	15 000?
CO_2			
Atmosphere	100 000[a,c]	0.4[b]	50[a,d]
Polar caps	—	—	10[d]
Crust	—	100 000[a,b]	>900?[e]
Total	100 000	100 000	>1000?
N_2			
Atmosphere	2 000[a]	2 000[a]	300[a]
^{40}Ar			
Atmosphere	4[a]	11[a]	0.5[a]

Note: Entries with two references are averages.

[a] Pollack and Black (1979).

[b] Walker (1977).

[c] Khodakovsky and others (1979).

[d] Hess, Henry, and Tillman (1979).

[e] Fanale and Cannon (1979) estimate 6×10^{17} kg of CO_2 adsorbed in nontronite clays in Martian polar layered terrain and associated regolith, substantially exceeding the CO_2 in the atmosphere or polar caps.

distance. Others argue that water would have been delivered to Venus by comet impacts, regardless of the initial amounts. Also, as discussed later in this chapter, the higher temperatures of Venus probably contributed to H_2O dissociation and loss from the atmosphere.

Measurements of the D/H (deuterium/hydrogen) ratio by a Pioneer probe in the atmosphere of Venus shed light on this. Donahue and others (1982) studied D/H and titled their paper "Venus was Wet." D/H/ gives a measure of past hydrogen content, because deuterium is twice as heavy as H and hence escapes more slowly into space. Thus, given a present-day D/H measure and assumed primordial values based on solar and other data, we can calculate backward to past values and estimate the total past H that passed through Venus's atmosphere. This, in turn, implies the total H_2O, as most H would have united with O. The results suggest that Venus outgassed the equivalent of at least 0.3% of the water in Earth's oceans, and possibly a full ocean's worth.

The possibility that primordial Venus did have oceans of water is supported by an obscure fact of Venus geology. Venus has long, sinuous rilles, or channel formations, that appear to have been cut by lava flows. Calculations suggest that silicate lavas hot enough to cut channels would cool much too fast to carve channels over the long distances observed. For that reason, lower temperature carbonatite lavas, of molten carbonates and salts, have been suggested, with the idea that these lavas formed from massive evaporite deposits of carbonates and salts, left by the evaporated primeval oceans of Venus (Kargel, 1997; Grinspoon, 1997).

Study the figures for Mars in Table 11-2. With the smallest mass and internal heating, Mars may have produced the least gas per kilogram, but the proportions may still have matched those in Table 11-1. The volatiles of Mars are difficult to inventory for two reasons. First, the amounts of carbonate rocks and H_2O permafrost below the surface are uncertain. If water was more abundant in the past, much CO_2 could have been tied up in carbonate rocks, as occurred on Earth (Anders and Owen, 1977; Pollack, 1979). Second, Viking probe measurements gave evidence that Mars once had a more massive atmosphere, much of which leaked off into space because of the low gravity (McElroy, Yung, and Nier, 1976).

Structure and Condensates of Planetary Atmospheres

The same principles discussed in the beginning of Chapter 9 can be used to compute the structure of a planetary atmosphere. The hydrostatic equation $dP = -\rho\,g\,dz$ applies to both planetary atmospheres and interiors because both behave like fluid systems. (The minus sign appears here because z is chosen to increase upward as P decreases.)

Temperature Structure

In Chapter 9, when we tried to compute the internal structure of a planet, we saw that the equation of state became important. The same applies to atmospheres except that here the equation of state is much simpler, namely the **ideal gas law,** which applies to any gas. The ideal gas law relates the pressure P to the density ρ, the temperature T, and the composition, which in the case of gases can be indicated simply by the mean molecular weight μ. The ideal gas law reads

$$P = \frac{\rho}{\mu M_H} kT$$

where M_H is the mass of a hydrogen atom and k the Boltzmann constant (see Table 2-2).

Suppose that we now try to compute a crude model of the atmosphere by using the hydrostatic equation and measurements of the surface P, ρ, and T. Following the same scheme as in Chapter 9, we could divide the atmosphere into 100 layers, each 1 km thick, and try to compute the pressure and other conditions at each layer. The atmosphere is so thin that we can treat g as a constant, to a first approximation.

Using the hydrostatic equation, we insert the known surface values of ρ, g and $dz = 1$ km. We compute the change in pressure between the ground and the top of the first layer. We now have P_1, the pressure 1 km above the ground.

Now we are stuck, because we have no way of getting the density at that level to use in our next calculation for the second kilometer. If we had measures of ρ at all levels we could proceed. Or if we had measures of T at all levels, we could proceed by using the ideal gas law to compute ρ, as we would then know P, T, and the composition (assumed to be uniform).

In other words, to compute a reliable model of an atmosphere, given the conditions at any one level, we have two alternatives: We must either (1) have the values of P, ρ, or T or (2) make some simplifying assumption about the behavior of one of these.

One of the simplest and conceptually most useful assumptions is that the temperature is constant at all levels (that is, that we have an **isothermal atmosphere**). At first sight, this assumption seems a gross mistake, as we know that the air gets noticeably cooler as we ascend even a modest mountain. But the changes in atmospheric temperature are only some tens of Kelvin, out of about 400 K—a change of only a few percentage points rather than a substantial change. Model atmospheres assumed to be isothermal are among the simplest to calculate (see the accompanying mathematical notes).

One useful concept that comes from such considerations is the **scale height,** or vertical distance over which the density decreases by a specified factor, usually 0.368 (which is $1/e$, a value that comes out of the mathematical

A great deal can be learned about atmospheres by simply combining the hydrostatic equation and the ideal gas law, under the assumption that the atmosphere is isothermal. By substituting the ideal gas law's expression for ρ into the hydrostatic equation and by assuming T is constant, we have

$$dP = \frac{-\mu M_H P g}{kT} dz$$

This differential equation is readily solved. Dividing by P, we have on the left dP/P, which coincidentally is identical to $d \ln P$. Integrating this logarithmic form from the surface ($z = 0$) to any arbitrary height, we find that

$$P = P_0 \exp\left(\frac{-\mu M_H g}{kT} z\right)$$

where P_0 is the pressure at the ground level. We thus have solved for the pressure structure P as a function of z.

Two concepts are immediately evident. First, the density structure in an isothermal atmosphere is just the same as the pressure structure, because, by the ideal gas law with T constant, we have $P/P_0 = \rho/\rho_0$. The second concept is the concept of scale height, defined as shown:

$$\text{scale height} = H = \frac{kT}{\mu M_H g}$$

This expression was used to calculate the scale heights quoted in the text. Note that scale height depends on T and hence changes with the height in real (nonisothermal) atmospheres.

We thus have the simple expression

$$P = P_0 e^{-z/H}$$

We can see that P falls by $1/e$ for each increase in height by H. The expression gives the pressure and hence density at all heights in any reasonably isothermal part of an atmosphere.

---

theory). Sometimes the decimal scale height is used instead (the height over which the density decreases by a factor of 1/10). Due to combinations of temperature, gas composition, and gravity, scale heights do not vary much in the solar system. At Earth's surface the scale height is about 8 km, on Venus about 15 km, on Mars about 16 km, on Jupiter, Saturn, Uranus, and Neptune near the 1 bar pressure level, about 22, 44, 27, and 22 km, respectively. On Pluto the scale height is about 14 km and on Titan, roughly 21 km. Generally, lower gravity results in a much more extended atmosphere with a larger scale height and a lower mean density, but cold temperatures contract the atmosphere and reduce the scale height. Such figures emphasize that Earth probably has the most compressed atmosphere, and that we live within easy hiking distance of what is, for practical purposes, interplanetary space.

More sophisticated models of atmospheres quickly make us deal with the fact that real atmospheres are not isothermal. Upper atmospheres, especially, are hotter and more extended. In fact, if we could create an isothermal atmosphere and then let nature take its course, the atmosphere would soon develop its own temperature structure. Three major sources of heat energy causing this new distribution of temperature are (1) diurnally (day–night) varying inputs of radiation from the sun (visible radiation) and from the planet (infrared radiation), (2) condensation or evaporation of certain atmospheric constituents (heat input governed by the latent heat of condensation), and (3) heating of the top of the atmosphere by solar wind and radiation interactions. To see how these sources affect atmospheric structure, we must first discuss how energy is distributed through the atmosphere.

Rayleigh Scattering, Blue Skies, and Pink Skies

Around 1500, Leonardo da Vinci, in a book on painting, clearly described the fact that distant objects look more blue, and he surmised that blue light must be added to a beam of light traversing daylit air. In 1868, English physicist John Tyndall noticed the blue glow of a light beam as it passed through a liquid suspension of tiny colloidal particles whose diameters were less than the 0.5-μm wavelength of visual light. He correctly concluded that the blue color of the sky and the blue light added by the air must be caused by sunlight interacting with similar particles in the air, called **aerosols**. By 1871, another English physicist, John Strutt, also known as Lord Rayleigh, showed that light interacts with these small particles in a way called **scattering**. If light encounters particles smaller than its own wavelength, it does not reflect off them in the normal way, but rather is scattered in various directions in such a way that the intensity of the scattered light increases dramatically with shorter wavelengths.* That is, the scattered light looks much bluer than the original light. This type of scattering is called **Rayleigh scattering**. (Not widely known is that Leonardo da Vinci described it qualitatively.) For a few years, physicists thought the sky was blue only because of the presence of tiny water droplets, salt crystals, or other aerosols (Young, 1982). But in 1899, Lord Rayleigh discovered that even air molecules cause Rayleigh scattering. Even a sky without aerosol particles would be blue.

---

*The intensity of the scattered light is proportional to (wavelength)$^{-4}$.

Figure 11-2. Mars Pathfinder's 24th sunset on Mars. The sun is poised over the horizon. Dust absorption at low elevation causes darkening along the horizon and creates a glowing patch above the sun, which can also be seen at sunset during dust storms on Earth, due to light scattering through the dust. The color of the sunset glow is more neutral grey than the daytime pink sky of Mars because of the scattering properties of the dust grains. (NASA, courtesy Sara Smith, Pathfinder Project)

If we look at the clear daytime sky in a direction away from the sun, we see blue light scattered off gas molecules and tiny aerosol particles. If we look at the sun, we see the unscattered portion of the light, from which some of the blue has been removed. Thus, if we look at the setting sun, the sunlight has come along such a long path through the atmosphere that most of the blue light has been removed, and the setting sun looks vivid red in color. If we look at the sky on a foggy day, the air is full of moderate-size water droplets—particles larger than a wavelength. Hence, Rayleigh scattering does not occur and we see the neutral whitish color of sunlight scattered and reflected throughout the sky.

The nature of gas molecules and aerosol particles in other planetary atmospheres is very important to understanding how sunlight interacts with these atmospheres and how they look from above or from the surface.

On Mars, the air is thin so that there is less Rayleigh scattering by gas molecules. Until July 4, 1976, scientists invoked this fact to predict that the Martian sky would be very dark blue, like the Earth's sky at an altitude of 100,000 ft. On July 4, 1976, the first Viking probe landed and its cameras revealed a bright, pink or peach-colored sky.* What was wrong with the prediction? So much fine

(μm-scale) reddish dust from the surface is suspended in the Martian air that the scattered light from these larger-than-a-wavelength particles acquires their color, giving the sky its pinkish-tan cast. The Martian sky in the direction of sunrise and sunset, as shown in Figure 11-2, shows a darker band near the horizon because of the greater concentration of dust near the surface.

Radiative Transfer of Heat Energy

The theoretical treatment of the radiative transfer of heat is exceedingly complex and is the basis of much of astrophysics. We can only touch on it here. (It was mentioned briefly in Chapter 9 with regard to heat transfer in planetary interiors.)

The keys to the theory of radiative transfer are the concepts of opacity and optical depth. **Opacity** is a measure of how much radiation is absorbed* over a given distance through the atmosphere. In a dense fog on Earth, the opacity is very high, and we can see perhaps 10 m. On a clear day, we can see 100 km. The **optical depth** helps to express the effect of opacity. The optical depth is a dimensionless number that expresses the amount of radiation lost from a beam of light. The optical depth is defined so that at optical depth 1, most of the light has been lost out of the original beam. Optical depth 10 implies that virtually no light is getting through. Consider a beam of light approaching an atmosphere from space. At the top, the optical depth is zero. If the atmosphere is cloudy, like that of Venus, optical depth 1 may occur near the cloud tops. But if the atmosphere is clear, like much of Earth's, then

---

*Actually the pink sky was revealed on July 5. Researchers had expected a blue sky, dark because of the thin air. The actual sky color is very delicate and on the first released photos, engineers tweaked the controls to give the expected bluish tint to the sky. Hours later, scientists recognized that more red light was being recorded in the sky than blue light, and subsequent photos were corrected. Subsequent missions looked for dust-free conditions that might reveal the darker blue sky, but generally there is so much dust in the lower atmosphere that the sky in most places is nearly always pink. The sky toward the sun is more white, because of forward-scattering effects. After sunset and before sunrise, the sun illuminates the higher, cleaner atmosphere, and the dusk sky takes on a bluish tinge when the sun is just below the horizon.

---

*Or scattered. Scattering and absorption are two distinct mechanisms for removing energy from a beam of light. For simplicity, we will minimize the difference between these mechanisms, although the difference is crucial in advanced radiative transfer theory.

we may have an optical depth of only 0.5 or some other fraction at the ground.

The subtle point about opacity and optical depth is that they vary with the wavelength (color) of the radiation. If we happen to pick a wavelength that some atmospheric constituent, say carbon dioxide, strongly absorbs, the opacity will be very high at that precise wavelength even though it may be nearly zero at nearby wavelengths. In other words, if we looked through a filter of precisely that color, the atmosphere would look entirely opaque, but if we looked through a filter of a slightly different color, the atmosphere would appear transparent. To say it still another way, optical depth 1 would occur high in the atmosphere at the absorbing wavelength but low in the atmosphere (or not at all) at a nearby wavelength.

An atmosphere could be quite transparent in the visible wavelengths but opaque at infrared wavelengths. This is just the case on Earth, as shown by Figure 11-3. The atmosphere is moderately clear at visual wavelengths, but water vapor absorbs much of the infrared. Earth-based astronomers trying to detect diagnostic infrared spectral absorptions by certain minerals or ices on other planets must therefore observe from very high mountains or spacecraft above most of Earth's water vapor.

Another example of absorption of certain wavelengths, this time by methane, is dramatically seen in Figure 11-4, page 326.

To see why an atmosphere develops an irregular temperature profile, suppose we could somehow create a perfectly isothermal atmosphere. Let us examine the consequences with respect to radiative transfer theory. Consider two cases. In the first case, suppose the atmosphere was absolutely transparent at all wavelengths. Then the radiation from the sun would come through the atmosphere without any absorption. It would strike and heat the ground, causing the ground to radiate in the infrared, as in Figure 11-5a, page 326. The infrared radiation would also escape into space without affecting the atmosphere, as the atmosphere has no opacity at infrared wavelengths. Thus, there would be no radiative influence of the sunlight on the atmosphere. No radiative energy would be absorbed at any wavelength or at any level in the atmosphere, and so the atmosphere would remain cold and isothermal (neglecting conduction effects, where the atmosphere contacts the heated ground). This is like the condition on mountain ski slopes, where the air is dry and cold, but the sunlight feels strong and warming.

For our second case, suppose there is a small amount of opacity at various wavelengths. Some incoming sunlight and some outgoing infrared radiation would be absorbed. Thus, the sunlight would heat the atmosphere. Sunlight can cause heating only if there is some opacity causing radiative energy to be absorbed (Figure 11-5b).

To see one reason the temperature structure of an atmosphere can be quite irregular, let us consider a still more realistic example. Suppose the various absorbing constituents are arranged in layers. Take Earth as an example.

Ozone (O_3), which strongly absorbs ultraviolet sunlight, is created and concentrated by photochemical reactions only in a layer at high altitudes. Water vapor, which is a very strong absorber of infrared, is created primarily at low levels by the evaporation of surface water. If these two constituents are the main absorbers, some of the incoming sunlight will be absorbed at high altitudes, heating the upper atmosphere, and most of the outgoing planetary infrared radiation will be absorbed at low levels, causing the air temperature near the surface to be higher than that at intermediate altitudes. This in fact is the situation on Earth.

The Greenhouse Effect

The **greenhouse effect** is a strong heating of the lower atmosphere of a planet due to selective absorptions in the infrared. We can easily understand it by reviewing the above model. Suppose, as is the case on Earth, that the atmosphere is *relatively* transparent in the visible wavelengths but has some strong absorbing medium in the infrared, such as water vapor or carbon dioxide. The surface of the planet is heated to a few hundred Kelvin by whatever solar radiation gets through. Any surface heated to such temperatures will radiate in the infrared, and this outgoing infrared radiation will get absorbed by the atmosphere. What happens to that energy? The atmosphere cannot keep absorbing infrared energy indefinitely without getting warmer, so the atmosphere heats up and itself radiates in all directions. The lower atmosphere and the ground heat up, radiating still more infrared, until the amount of infrared energy escaping from the top of the atmosphere is equal to the amount of visible solar energy coming in. Only then is equilibrium achieved.

But the new equilibrium may leave the ground level much hotter than the few hundred Kelvin we initially expected from the solar input. This greenhouse effect is named after the same phenomenon that occurs in a greenhouse, where the glass panes are analogous to our infrared absorber. Glass lets in sunlight, but it blocks the outgoing infrared and traps the warm air, causing the greenhouse to get warmer until enough infrared escapes to reach equilibrium.

The greenhouse effect is familiar in nature on Earth. For example, during the cloudy night, when much water vapor is in the air, the clouds and humid lower atmosphere absorb the outgoing infrared radiation and the air is likely to stay warm; during a sparklingly clear, dry night, the temperature may drop rapidly because the infrared radiation from the ground goes right out through the atmosphere. For this reason, cool nights are common in dry deserts.

As deduced by Carl Sagan as early as 1962, the greenhouse effect probably accounts for the 750-K surface temperature of Venus. The 90 bar of CO_2 on Venus creates a greenhouse far exceeding that of the 1 bar of O_2 and N_2 on Earth. The greenhouse effect is probably crucial to the different histories of Venus, Earth, and perhaps even Mars.

a

b

Figure 11-3. Different opacities of Earth's atmosphere at visual wavelengths and infrared wavelengths are vividly shown in these two pictures. (**a**) Image at visual wavelengths (0.1–44 μm). Land masses of Africa and Europe, with scattered opaque clouds. (**b**) Image at infrared wavelengths (5.7–7.1 μm). The water vapor absorption bands show diffuse, moist cloud masses; the image does not penetrate to the ground. (European Meteorological Satellite images)

Figure 11-4. Photographs of Saturn made with films and filters sensitive to different colors of light. Left = blue (450 nm), the most common view. Right = wavelength absorbed by methane gas (897.5 nm); abundant methane in the atmosphere absorbs most of the light at this wavelength making the globe dark. The rings remain bright, being covered with ice that reflects sunlight at this wavelength. (Lunar and Planetary Laboratory, University of Arizona)

Study the **phase diagram** in Figure 11-6, which shows the pressures and temperatures producing stability for different phases of H_2O. The combination of temperature and pressure at which any volatile can exist simultaneously as gas, liquid, and solid (instead of freezing, melting, or subliming into one of these forms) is called the **triple point.** For water, it is P = 6.1 mbar and T = 273 K. Curiously, the pressure on most of Mars, except at the lowest spots, is less than 6.1 mbar, or so close that liquid isn't stable on Mars, except in the lowest canyons. In today's regime, liquid water erupted onto Mars would freeze or evaporate very quickly. This is why a slightly higher ancient pressure—allowing more stable liquid—would be very interesting in terms of explaining the dry river channels!

The temperature is low enough on Mars that, as the secondary atmosphere built up there, volcanically emitted H_2O probably condensed directly as ice, preventing an H_2O greenhouse effect, which appears when about 1 mbar of H_2O is available. Earth's initially warm temperature allowed H_2O to condense as liquid. On Venus, the strong CO_2 greenhouse, plus the increasing H_2O greenhouse and Venus's initially warmer condition, raised the temperature so much that the evolutionary path may have missed the liquid regime altogether, and gaseous water molecules eventually dissociated and/or reacted with other materials.

Figure 11-6 emphasizes that the creation of an Earth-like planet is a somewhat delicate matter. According to this view, if Earth had been only about 5% closer to the sun, it, too, would have suffered Venus's fate; water would not have condensed, CO_2 would have built up, an enormous greenhouse effect would have ensued, and life would not have evolved (Rasool and de Bergh, 1970). If Earth had been further from the sun, water might have remained frozen. Conversely, and intriguingly, a slight heat input to Mars might trigger a substantial greenhouse effect, causing dramatic warming.

Condensable Substances—Moist Atmospheres

Meteorologists consider an atmosphere **moist** when it contains substances that can condense into liquid or solid form. On Earth, water vapor is the familiar example, condensing to form cloud layers composed of water droplets or ice crystals. Such cloud layers may affect planetary

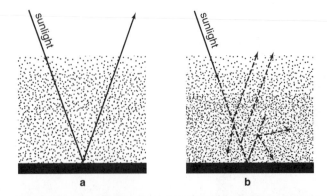

Figure 11-5. Transmission of sunlight through (**a**) a hypothetical transparent atmosphere and (**b**) a more realistic atmosphere with absorption and remission of radiation.

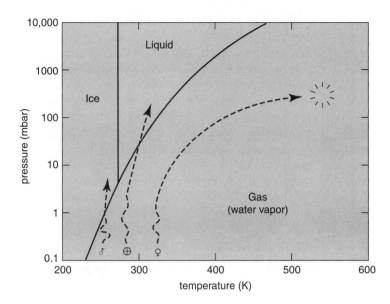

Figure 11-6. Phase diagram for water, showing phases existing at various temperature/pressure combinations. Dashed lines are schematic evolutionary tracks for H_2O on Mars, Earth, and Venus. Wiggles represent hypothetical climatic oscillations. The H_2O-caused greenhouse warming sets in at about 1 mbar of water vapor pressure.

atmospheric temperatures in two ways: (1) Clouds can interrupt the flow of radiation with an abrupt layer of high opacity and (2) enough heat can be released when some substances condense to affect the atmospheric structure. This is the case with water vapor on Earth; indeed, water vapor is so important that a model of the lower Earth atmosphere will be substantially wrong if it does not take water vapor into account.

For example, suppose an imaginary parcel of air has convected upward to a certain level (review the descrip-

tion of convection in Chapter 9). If there were no water vapor, the air parcel would tend toward a certain equilibrium temperature and pressure. If the parcel were moist to start with, the cooling might make water vapor condense, creating a puffy **cumulus cloud**, liberating heat and causing the parcel to tend toward a different equilibrium temperature than if it were dry. Thus, condensable substances change the temperature structure of atmospheres.

The temperature structure is often expressed in terms of the **temperature gradient,** which is the change in temperature per kilometer as we ascend into the atmosphere. Because this quantity is negative near Earth's surface, meteorologists often use for convenience the **lapse rate,** which is simply the negative of the temperature gradient. Thus on Earth, the typical near-surface temperature gradient is -6.5 K/km, whereas the lapse rate is 6.5 K/km. On Venus, the lapse rate near the surface is about 8 K/km (Seiff and others, 1979), and on Mars, about 3 K/km (Seiff and Kirk, 1977), according to measurements by landers.

The Sulfuric Acid Condensate Clouds of Venus

An extraterrestrial example of condensate clouds occurs on Venus, where the clouds were made from droplets of sulfuric acid (H_2SO_4, commonly used as battery acid) rather than water. The 1978 parachute probes of Venus from the Pioneer spacecraft confirmed earlier evidence that Venus's clouds are H_2SO_4 and showed that the clouds are concentrated in a layer from 48 to 58 km above the surface. The H_2SO_4 droplets in this region have diameters of about 1 to 10 μm. Above this cloud deck, a haze of 1- to 3-μm droplets extends upward to at least 68 km (Figure 11-7). Below the cloud deck, a haze of 1- to 2-μm droplets extends downward to a sharp cutoff at 31 km. Below that the atmosphere is dense but clear (Knollenberg and Hunten, 1979a, 1979b).

Figure 11-7. When Venus passes between Earth and the sun, sunlight backlights the atmosphere. A thin, high layer of haze scatters sunlight so that the atmosphere is visible all around the disk. This was the first evidence for an atmosphere on Venus, discovered in 1761. The sun is far out of the picture to the upper left. (New Mexico State University Observatory: courtesy B. A. Smith)

Theoretical work combined with other Pioneer observations of Venus have revealed how these clouds were probably formed. At altitudes of around 80 km, under the influence of solar ultraviolet, chemical reactions among minor atmospheric constituents produce small droplets of H_2SO_4, which then begin to fall slowly. Between altitudes of 48 and 58 km is a region with convective updrafts and downdrafts; thus, the particles are likely to be caught there, growing slowly during a lifetime of a few months. This is the main cloud deck (Toon, 1979; Knollenberg and Hunten, 1979a, 1979b). Droplets that get too big eventually sink or drop out of this layer into the warmer region below.

Whereas temperatures range between 288 and 364 K (59°F to 196°F) in the cloud deck, they reach 493 K (428°F) about 20 km below the clouds, and so the droplets rapidly evaporate as they fall out of the cloud. This phenomenon explains both the smaller particle size observed in the sulfuric acid "drizzle" falling out of the cloud and the termination of the drizzle below 31 km. Condensation and movement of the cloud particles involve a heat flux about one-fourth of the solar flux at that altitude. The clouds are believed to be stratuslike forms, not puffy cumulus clouds (Knollenberg and Hunten, 1979b).

Martian Clouds

The cycle of water condensation, freezing, and evaporation on Mars is important to understanding the Martian environment and its history in ancient times. We examine the past conditions in more detail in Chapter 13. On present-day Mars, however, areas warmed by the morning sun emit water vapor into the air. About half an hour after dawn, this vapor freezes to form clouds or ground fogs of water ice crystals in some regions. Ground fog clinging to the floors of channels and canyons is common. Higher clouds form, as on Earth, where air gets lifted over elevations, such as the slopes of volcanoes in the Tharsis region. Poleward of latitudes about 65°, winter conditions lead to freezing of even the CO_2 at about 148 K. This freezing creates a winter **polar hood** of CO_2 clouds and haze (CO_2 ice particles) hanging over the polar regions.

Such features were recorded a century ago by Earth-based telescopic observers who classified Martian clouds into three groups. **White clouds** are probably mostly H_2O ice crystals and may include crystal hazes of CO_2 ice around the poles and at altitudes of 50 km, where CO_2 ice can condense (Briggs and others, 1977). **Blue clouds** or blue hazes are clouds prominent on blue-sensitive photos. They, too, are probably composed of H_2O and CO_2 ice crystals, but they may involve smaller particles that scatter blue light. Kuiper (1964) simulated blue haze with tiny ice particles. Both kinds of clouds are common at dawn and dusk on Mars. Mars orbiters have photographed many kinds of Martian clouds at close range, including an un-usual cyclonic storm system reminiscent of those on Earth (Figure 11-8). The two Viking landers saw virtually no clouds from the surface, but Mars Pathfinder finally obtained beautiful views of high altitude cirrus-like clouds in the pre-dawn sky (Figure 11-9).

Yellow clouds, which are prominent in earthbased red and yellow photos, are dust storms composed of dust particles a few micrometers across. These often form in local areas and can spread to form **global dust storms,** especially during summer in the southern hemisphere. An example is seen in Figure 11-10b, page 330.

Dynamics of Planetary Atmospheres

Why do winds blow? Why shouldn't they blow themselves out and remain calm forever? In this section, we show that winds are a consequence of solar radiation input and that patterns of winds and clouds are associated with planetary rotation.

Vertical Mixing by Convection

Convection offers one way to drive atmospheric circulation, both on local and regional scales. Recall from the discussion of convection in Chapter 9 that convective flow will start in a medium if the temperature gradient becomes **superadiabatic**—that is, steeper than that experienced by an adiabatically rising parcel.

Thus we see one way that winds could start. Suppose we magically calmed all the winds on a planet. Sunlight strikes the planet. Perhaps a certain patch of dark ground gets quite hot and heats the parcel of air above it by conduction. That parcel expands and rises, causing new air to rush in horizontally to fill the potential vacuum. Thus, even if you magically calmed an entire atmosphere, local winds would establish themselves as convection began. If the temperature gradient happens to be superadiabatic, the process is self-regenerating and a substantial convective uplift may begin. Air passengers experience substantial choppiness as their plane flies through the ascending and descending currents.

Martian Dust Storms

This process of local heating not only causes winds but is probably also a key to the summer onset of global dust storms in the southern hemisphere of Mars. Mars passes through perihelion about a month before the beginning of summer in the south, so that maximum solar insolation occurs south of the equator in late spring. The high eccentricity of the planet causes it to be 17% closer to the sun at perihelion than at aphelion, so the insolation difference between northern and southern summers is substantial.

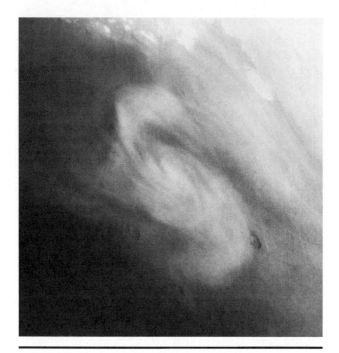

Figure 11-8. A cyclonic storm at 65 N latitude on Mars, associated with the boundary between cold air over the polar ice cap (outlying white frost patches, upper right) and warmer air from the mid-latitudes. The season is mid-summer, and the clouds are water ice crystals condensed from water vapor that evaporated from warm soil during the summer. Winds near the storm edge are estimated at 26 m/s (58 mph). The frost-filled crater Korolev, with a 90-km diameter, is in the upper right. (NASA, Viking 1)

Figure 11-9. The first views of clouds from the Martian surface were made by the Mars Pathfinder lander in 1997. The clouds in the pre-dawn sky are illuminated by the sun, about 40 min before sunrise. They had a pinkish color and moved in from the northeast at about 24 km/h, at a height of about 16 km (52,000 ft). (NASA, courtesy Sara Smith, Pathfinder Project)

The summer sun heats the Martian ground, the ground heats the adjacent air, and the warm air rises. This drives upward-swirling air masses into the atmosphere, resembling dust devils in the Southwestern American deserts. Martian dust devils have been photographed from orbiters (Figure 11-11, p. 330). Also, Metzger (1998) identified Martian dust devils on the horizon in Pathfinder lander images. Once airborne, the dust absorbs sunlight and makes the air still warmer, adding to convective turbulence and causing a positive feedback effect. Thus, the storms, once started, often spread in a day or so into massive dust clouds (Figures 11-10 and 11-11). Some die out after a few days; others eventually carry dust across much of the globe (Golitsyn, 1973; Gierasch and Goody, 1973).

Global Circulation of the Planetary Atmospheres

What controls global patterns of circulation on different planets? Suppose again that we could magically stop all winds. Stagnant air at the equator would then be heated and tend to rise, whereas cold air at the poles would tend to sink. This movement would set up a circulation pattern with air moving down at the poles toward the equator and then upward in the equatorial regions. This very simple model does not accurately predict planetary circulation patterns (although southward-moving weather patterns are familiar in the United States, especially during winter), but it does show that the continuous input of solar energy is a continuous driving force that stirs the atmosphere.

On real worlds, several factors complicate the flow of air:

1. The **Coriolis force** is an apparent force resulting from a planet's rotation. The equator of Earth is moving eastward at about 464 m/s (100 mph). Therefore, air masses moving away from the slow-moving poles appear to lag behind and shift westward while poleward-moving air pulls ahead to the east. The Coriolis deflections produce the spiral motions (with opposite directions in the north and south hemispheres) associated with cyclonic storms, in which air tries to rush radially into low-pressure regions. Several examples are prominent in Figure 11-12, page 331; Figure 11-8 showed that the same effect occurs on Mars.

2. Land masses and their relief affect airflow, both on a small scale (Figure 11-12) and on a continental scale.

3. Fluid mechanics systems, depending on their dimensions and other properties, tend to set up turbulence and cellular patterns. For example, if you try to push water

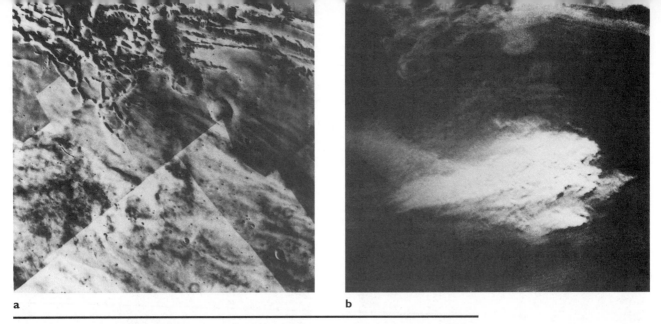

a   b

Figure 11-10. Development of a Martian dust storm. (**a**) Typical view of high plains south of Valles Marineris canyons (top) shows scattered water ice clouds in early winter. (**b**) Mid-spring view (same as in Figure 11-10a) shows a turbulent dust cloud 12 km high and 600 km wide (the width of Colorado) covering the same area. The canyons are filled with low dust clouds. The date is about 35 d before perihelion passage, when strong solar heating warms the ground at southern latitudes and promotes convection, raising dust. (NASA, Viking)

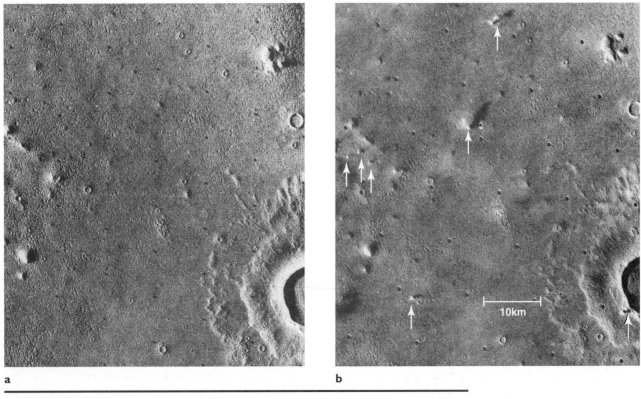

a   b

Figure 11-11. Dust devils on Mars. (**a**) Orbital view on a normal afternoon in the desert northwest of Olympus Mons volcano. Crater at lower right is surrounded by lobes of ejecta possibly formed by impact that melted permafrost, splashing out slurrylike mud ejects. The area is about the size of Rhode Island. (**b**) On another afternoon, in summer, transient bright features (arrows) cast shadows several kilometers long. These features are believed to be huge dust devils. On Earth, dust devils are created when sun heats the ground; air near the ground warms and rises in a swirling motion, carrying eddies of dust into the air. In some years, when dust devils inject enough dust into the air, vast dust storms develop on Mars. (NASA Viking Orbiter photos)

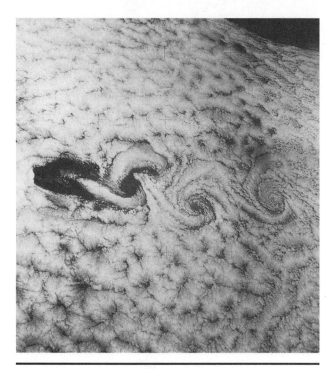

Figure 11-12. Series of cyclonic eddies produced as airflow is disturbed by Guadalupe Island, off Baja California. (NASA, Skylab 3)

water vapor are concentrated here and also because this is where the atmosphere can be heated by contact with the ground, often producing superadiabatic lapse rates. At higher altitudes, the air is thinner and more transparent to radiation, and little energy is absorbed in each kilometer of height. At such altitudes, therefore, there would be a much lower lapse rate, as was the case in our nonabsorbing isothermal atmosphere.

This intuitive reasoning is correct in that the planetary atmospheres generally do have low-altitude regions of high lapse rates, turbulent motion, and cloud-forming activity and higher regions of nearly uniform temperature, smooth airflow, and lack of turbulence. This structure is shown for Earth in Figure 11-13, page 332.

The region in which the temperature gradient approaches zero is known as the **tropopause;** it is quite abrupt in the terrestrial atmosphere. It lies at a height of about 18 km at the equator and as low as 7 km at higher latitudes. The region below the tropopause is the **troposphere** (from a Greek root referring to change or mixing). It is characterized by greater lapse rates, more turbulent motions, and chemical mixing. The region above is the **stratosphere,** characterized by a more nearly isothermal temperature profile and smoother motions. Winds near the tropopause are often laminar, or uniform, and of high speed. These winds are known on Earth as the **jet stream.** Air travelers will recall that once a jet airliner gets well above the turbulent cloud layer, the flight becomes much smoother. This is because the aircraft has entered the lower stratosphere.

Temperature Irregularities and the Mesosphere

In general, the region from the tropopause to the higher realm of increasing temperature is called the **mesosphere.** There is some confusion between this term and the term *stratosphere* when applied in general planetology, because the atmospheres of different planets may have different patterns of temperature irregularities. Chamberlain (1962) proposed that the term *stratosphere* be abolished altogether in planetological use, although here we have followed the terrestrial use: The stratosphere is the nearly isothermal part of the mesosphere bordering and immediately above the troposphere, and it may in principle constitute the entire mesosphere.

The best example of complex structure in a mesosphere is the terrestrial **ozone layer.** As the solar ultraviolet comes down through the atmosphere, it is absorbed by oxygen molecules (O_2), dissociating them into pairs of single oxygen atoms. The oxygen atoms may combine with other oxygen molecules to form ozone molecules (O_3). The ozone itself absorbs ultraviolet sunlight, especially in certain color regions known as the **Hartley-Huggins bands.** If this high-energy ultraviolet radiation were not stopped, it would be lethal to life on Earth as we know it.

around in a bathtub with your hand, much of the motion you impart to the water ends up in turbulent eddies instead of a smooth flow. It is extremely difficult to predict the details of the water motion. In the same way, no simple model of a global atmosphere can predict the details of atmospheric motions visible on a planet from day to day. (Recall the discussion of turbulence in Chapter 3.)

Atmospheric Levels and Upper Atmospheres

The principles discussed above can be applied to any part of any planetary atmosphere. However, certain additional phenomena enter the picture, especially in the upper atmospheres of the planets, because the upper atmospheres interact directly with the cosmic environment (for example, with the solar wind). The terminology used to discuss the general structure of planetary atmospheres at different levels derives from Earth's atmosphere and is based primarily on temperature effects. We begin with the lowest levels.

Troposphere, Tropopause, and Stratosphere

In a general planetary atmosphere (but particularly in the case of Earth), one might expect the lapse rate to be greatest near the ground because infrared absorbers such as

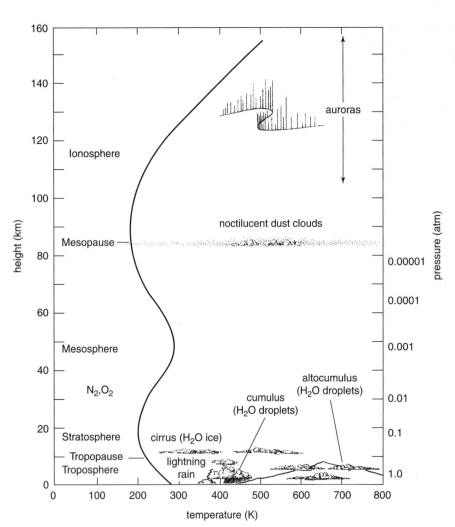

Figure 11-13. Structure and features of Earth's atmosphere.

In the region far enough down for oxygen molecules to be sufficiently dense, these energy absorptions heat the air significantly. The heating effect is maximized at about 50 km up, where there is a temperature maximum about 100 K greater than the tropopause temperature. The compositions and radiative properties of other planetary atmospheres may cause many variations in mesosphere structure (see the end of this chapter).

Thermospheres, Ionospheres, and Exospheres

In their outer reaches, the planetary atmospheres grade into the interplanetary gas with its high-energy particles. The **thermosphere** is defined as the outer region where the temperature increases with height. An important characteristic of the thermosphere is that in the rarefied gas, molecules can move substantial distances before colliding with other molecules. Thus the **mean free path**, or typical distance between collisions of molecules, becomes much longer than in the denser lower atmosphere; separation of gases by diffusion becomes important, whereas in the lower atmospheres, high density and turbulent mass motions ensure mixing. Molecules of the lighter gases, especially hydrogen and helium, have higher velocities and thus can rise to great heights on nearly ballistic trajectories. This behavior produces a degree of stratification; the lightest gases concentrate in the highest layers.

The planets are cold objects moving through the hot interplanetary gas. The solar wind sweeps out past the planets with an average velocity of 600 km/s near Earth. As solar wind atomic particles collide with the Earth's upper atmosphere, they heat the atmospheric boundaries. The thermosphere is also heated because the atmospheric gases absorb solar ultraviolet and x-ray photons, thus raising thermosphere temperatures.

Rocket and satellite observations have shown that the temperature of the terrestrial thermosphere increases from a cold 180 K at 80 km to high, variable temperatures, from 700 to 1,800 K, above 200 km, depending on solar activity (that is, the sunspot cycle), the time of day, and other factors. The primary cause of such high temperatures is absorption of solar ultraviolet radiation at altitudes between 80 and 200 km. High-energy particles from the solar wind may penetrate the planetary magnetic field and interact with the high atmosphere, generating additional heating.

The **ionosphere** is a portion of the thermosphere in which charged particles, or ions, are abundant. These ions are negatively charged electrons and positively charged atoms or molecules from which the electrons have been knocked off by photons of solar ultraviolet. These photons may impart so much energy that the electrons are knocked off with much higher velocities than the local equilibrium thermal velocities characterizing the neutral molecules. Indeed, measurements show that equilibrium does not exist between the radiation, the ions, and the neutral particles. Observed **electron temperatures** are often as much as twice the gas temperatures defined by the neutral particles.

Ions are so numerous that the ionosphere is an electrically conducting layer. Hence, it drastically affects the passage of radio waves. Prior to 1901 scientists confidently predicted that radio would be useful only over short ranges, as electromagnetic radiation travels in straight lines. Earth's ionosphere was discovered in 1901 when Marconi astonished the world by beaming radio signals across the Atlantic. Two physicists, Kennelly and Heaviside, independently but simultaneously explained Marconi's discovery by postulating a conducting, ionized layer high in Earth's atmosphere. Once called the Kennelly-Heaviside layer, the ionosphere absorbs and reflects radio waves, allowing "bounce beaming" over long distances.

Earth's ionosphere ranges from about 80 to 300 km in altitude, and it contains two main maxima in electron density: the lower E region (90 to 120 km) and the upper F region (150 to 300 km). During the daytime, the latter often has two maxima, F_1 and F_2. All the layers, especially the uppermost F_2 region, are highly variable and are affected by the time of day, the 11-y sunspot cycle, magnetic storms, and so on. Electron densities in the F_2 region reach as much as 10^6 electrons/cm^3.

Similar ionosphere structures were found by Pioneer and Viking probes on Venus and Mars, with ion densities reaching 10^5 ions/cm^3 at altitudes of 140 and 130 km, respectively (von Zahn and others, 1979).

The outer edge of the thermosphere is the **exosphere,** a level above which at least half the upward-moving molecules do not hit another molecule, instead following a long trajectory into space and possibly escaping. The exosphere temperature is very important to the evolution of the atmosphere. If the exosphere temperature is high, exosphere molecules will have high velocity and can escape into space. Certain molecules that are efficient radiators in the infrared may serve as thermostats that keep the thermosphere temperature down. "Thermostat molecules" are thus crucial to the atmosphere's history. The absence of such molecules forces the temperature up until outward radiation offsets the incoming solar radiation.

An example of the effect of thermostat molecules occurs in the exosphere of Mars. Without a good infrared radiator, its temperature would be about 2,000 K. Chamberlain (1962) pointed out that CO, produced from the abundant CO_2 on the planet, would be a good radiator. Chamberlain's original calculations suggested an exosphere temperature as low as 1,100 K, and subsequent theorists suggested still lower values of only a few hundred Kelvin. Viking entry probes measured an exosphere temperature of 210 K near an altitude of 175 km (Hanson, Sanatani, and Zuccaro, 1977). Similarly, Pioneer entry probes on Venus found an exosphere temperature of 285 K at an altitude of about 165 km (von Zahn and others, 1979). Because exosphere temperatures control escape rates into space, even short-term changes in exosphere temperatures, perhaps resulting from changes in the quantity of exosphere thermostat molecules, could create important episodes of enhanced atmosphere escape rates.

Escape of Planetary Atmospheres

The theory of gravitational escape of gases from planets was developed by Jeans (1916). If one of the upward-moving molecules in the exosphere does not hit another molecule, it will follow a ballistic path—an elliptical orbit arcing above the atmosphere and eventually returning. If the molecule should happen to be one of the fastest-moving molecules and have a velocity greater than the escape velocity for that particular altitude, it will follow a parabolic or hyperbolic orbit and escape from the planet (see Chapter 3). Every time this happens, the planet's atmosphere permanently loses one molecule.

Jeans, and later Spitzer, calculated the rate at which this happens. Each molecular species (H_2, N_2, O_2 CO_2, and so on) may be treated separately. The principle known as the **equipartition of energy** states that, on the average, each kind of molecular species has the same energy as any other kind. To have the same energy, the small ones must move fast, and the big ones must move slowly. Therefore, on the average, the light hydrogen molecules will move faster than the heavier helium molecules, which will move faster than the still heavier oxygen molecules, and so on.

The fastest-moving molecular species, hydrogen, will have the most molecules moving faster than the escape velocity and will be the first type of gas to be depleted in the atmosphere of any planet. Because of its strong

The expression for the lifetime of a planetary atmosphere is too complex to derive here. However, a rule of thumb emerges from the fact that the escape time depends rather critically on the escape velocity of the planet, for which an expression is derived in Chapter 3. This rule of thumb can be expressed as follows: An atmosphere will escape from a planet in a time less than the age of the solar system if

$$V \geqslant f v_{esc}$$

where

V = root-mean-square molecular velocity
   $= \sqrt{3kT/m}$
f = constant, about 0.25
v_{esc} = escape velocity from planet
m = molecular mass for given molecular species
k = Boltzmann constant

Sharonov (1958) writes a more general rule of thumb to take into account that the exosphere temperature will tend to decline roughly with the square root of the distance from the sun, to the first order (according to simple laws of radiative heating). Combining this with the equation and definitions above, Sharonov concludes that the ability to retain an atmosphere might be called the relative atmosphere retentivity, $\mathcal{R}$, which is defined as the ratio

$$\mathcal{R} = \frac{v_{esc}}{V} \propto v_{esc} a^{1/4}$$

where a is the distance of the planet or satellite from the sun. Planets are listed by relative retentivity $\mathcal{R}$ in Table 11-3.

Spitzer (1952) derived the time t to reduce atmospheric density of $1/e$ of its original value by thermal escape (neglecting nonthermal effects such as photodissociation and solar wind):

$$t = \frac{\sqrt{6\pi}}{3g} V \frac{e^{Y}}{Y}$$

where

g = surface gravity of planet $= \dfrac{GM}{r^2}$
Y = $\dfrac{GMm}{kT_x R}$
G = Newton's gravitational constant
M = planetary mass
R = planetary radius
T_x = exosphere temperature
m = molecular mass
   = molecular weight $\times$ mass of H atom

This expression applies to an idealized isothermal atmosphere. However, with a more realistic temperature structure, escape can be as much as 10^5–10^6 times faster. This discrepancy shows the importance of understanding exosphere temperature structure.

---

gravitational field, Jupiter has apparently retained much of its original hydrogen, whereas Earth—with its smaller gravitational attraction and lower escape velocity—has lost its hydrogen and helium but retained most heavier gases.

Four phenomena complicate this picture:

1. In any molecular species there is a distribution of speeds. For example, in the Martian exosphere, the *mean* hydrogen velocity would be about 2.3 km/s, about half of the 5-km/s escape velocity of Mars. Nonetheless, a certain fraction of H atoms would move faster than 5 km/s, facilitating a slow but steady escape.

2. Solar ultraviolet radiation may break up certain molecules, a process called **photodissociation.** For instance, H_2O (molecular weight 18) may be too heavy to escape, but if it breaks into H and OH in the exosphere, the H may rapidly escape from the planet. Even the OH may break apart. H_2O may thus be rapidly depleted as H leaks off and O oxidizes other materials.

3. If the high-speed atoms and molecules of the solar wind collide with exosphere molecules, they may kick the target molecules out of the atmosphere or at least dissociate them. Today's solar wind thus helps deplete atmospheres, and a primeval super solar wind might have been important in stripping primitive atmospheres off planets.

4. Heavy molecules tend to concentrate in the lower atmosphere; thus, their escape rate from the exosphere is limited by the rate at which they can diffuse upward.

These ideas apply to Venus and its missing H_2O. Any original H_2O was too heavy to escape in the age of Venus. But if H_2O broke into H and O atoms, the H could escape more easily. However, the low 285 K exosphere temperature found by Pioneer entry probes implies 20 Gy even for hydrogen to escape! Perhaps the exosphere was hotter in the past. At 1,000 K, the escape time could be much less than the age of Venus. Some evidence favors this. The ratio of deuterium, or the heavy isotope of hydrogen, to ordinary hydrogen, usually written D/H, is much higher on Venus than on Earth. The usual interpretation is that when H_2O broke into H and O, the lighter H atoms escaped much faster than the heavier D atoms, leaving a concentration of D behind. Grinspoon (1987) suggests an alternate model, in which Venus was initially dry, and the current D/H was built up by water provided by comet impacts. Certainly the question of the original water con-

TABLE 11-3 Planets Listed in Order of Atmospheric Retentivity

Planet	Relative retentivity R[a]	Surface pressure[b]	Constituents (% by no. molecules or volume)
Jupiter	250	High	86 H_2, 13 He, 0.07 CH_4
Saturn	170	High	97 H_2, 3 He, 0.05 CH_4
Neptune	147	High	>74 H_2, <25 He, >1 CH_4 at depth
Uranus	120	High	83 H_2, 15 He, 2 CH_4 at depth
Earth	30	1000	78 N_2, 21 O_2, 1 Ar
Venus	25	90 000	96 CO_2, 3 N_2, 0.1? H_2O
Mars	15	10	95 CO_2, 3 N_2, 1.6 Ar
Titan	12	1500	>90 N_2, 1–10 CH_4, ? Ar
Ganymede	11	$<10^{-8}$	
Triton	10?	0.016	N_2, CH_4
Io	10	$\sim 10^{-7}$?	? SO_2
Mercury	9	<0.3	
Europa	8	—	
Pluto	7	0.01?[c]	? CH_4, ? N_2
Callisto	6	—	
Moon	6	—	
Rhea	3	—	
Iapetus	3	—	
Dione	2	—	

[a]A measure based on escape velocity and exosphere temperature, discussed in the accompanying mathematical notes.

[b]Figures in millibars.

[c]At perihelion.

Sources: Orton (1978); Hanel and others (1979); Oyama and others (1979); Owen and others (1977); Cruikshank and Silvaggio (1979, 1980); Owen (1982); Broadfoot and others (1979); Kumar (1979); Belton, Hunten, and McElroy (1967); Ingersoll (1981); Barbato and Ayer (1981); Stone and Miner (1986, 1989); Ingersoll (1990); Niemann and others (1996).

tents and sources on Venus and Earth, and the relation to Mars, is a major issue in planetary science (see reviews by Pepin, 1991; Grinspoon, 1997).

Summary: Atmospheres of Selected Worlds

Table 11-3 summarizes planetary atmospheres with the planets listed in order of ability to retain atmospheric gases, based on the simplified considerations in the adjacent mathematical notes on atmosphere escape. A dramatic transition from air-rich to airless worlds occurs around the size range of Titan and Ganymede.

A note is in order about the units used to report atmospheric compositions. Table 11-3 lists fractional numbers of molecules of each species, or **mixing ratios.** Meteorologists commonly specify composition by mixing ratios rather than by, say, percentage by weight (the units usually used in discussing planetary interiors). Telescopic observers usually report compositions in terms of the equivalent amount of gas traversed by the light beam on its way to the telescope. This is the gas traversed either above the surface or above some opaque cloud layer. The unit often used is the **kilometer-atmosphere,** the amount of gas traversed in a 1-km path under standard atmospheric conditions, usually taken as 20° C and 1 atm (1 bar) of pressure. Frequently this same unit is defined as the **kilometer-amagat,** where *amagat* refers to a slightly more rigorously defined density equal to 0.0446 g mole/liter. A gram-mole is the number of grams equal to the molecular weight of the gas species.

The following sections summarize historical and modern studies for each planet.

Venus

The atmosphere of Venus was discovered by a Russian scientist, Lomonosov, during a solar transit of Venus in 1761. Lomonosov observed sunlight backlighting this atmosphere, as seen in Figure 11-7. The composition remained a matter of controversy for more than two centuries. When the earliest observers realized that Venus was covered by clouds, they assumed that they were seeing dense water vapor clouds like those of Earth. Science fiction writers depicted Venus as a rainy, steamy swamp populated by dinosaurs or some fictional equivalent. However, careful observations during the 1950s and early 1960s failed to give evidence of substantial amounts of water

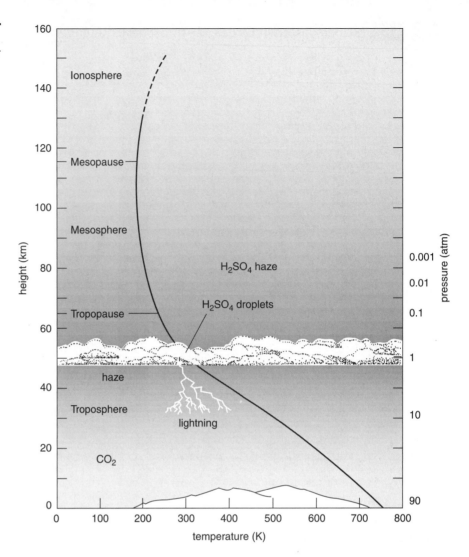

Figure 11-14. Structure and features of Venus's atmosphere, based on Pioneer and other measures.

vapor. A variety of (sometimes exotic) theoretical models of the atmosphere followed: models with planetwide oil fields; models in which all the water was tied up in low-level clouds, ice crystals, and so on; and models with tremendous winds and blowing dust that would generate the high surface temperatures by friction. After some controversy, most investigators correctly agreed that the clouds and general Venus meteorology are not dominated by effects of water, as the observed amount of water vapor above the clouds is very small.

As early as 1928, American astronomer Frank Ross announced the discovery of hazy cloud patterns visible on ultraviolet photos but virtually invisible to the eye. Chapter 2 detailed unsuccessful attempts to determine Venus's rotation period by tracking these patterns from Earth. These attempts led only to discovery of a west-to-east, 4-day cloud circulation but did not reveal the planet's rotation.

The flight of Mariner 2 past Venus in 1962, together with nearly simultaneous observations from Earth, led to the discovery that Venus is very hot, with surface temperatures initially estimated at 400 to 700 K (Sagan, 1962), which Sagan correctly attributed to a greenhouse effect.

Chapters 2 and 10 detailed the discoveries of the ensuing series of Soviet probes that were parachuted into the atmosphere, culminating with the first successful surface measurements by Venera 7 in 1970. This probe transmitted 23 minutes of data from the surface before failing and reported a temperature of 748 K and a pressure of 90 bar! A further breakthrough came in 1972–1973 as several scientists independently identified the Venus clouds as being made of 2-μm droplets of 78% to 90% sulfuric acid.

The Pioneer Venus mission parachuted four probes through Venus's atmosphere on December 19, 1978; the Soviet VEGA missions in 1985 put two balloons adrift in Venus's atmosphere, which gathered data for 46 hours each. These dramatic missions added numerous details to the above results. These included detection of minor constituents such as 180 ppm of SO_2, 67 ppm of Ar, and 20 ppm of CO, in addition to the major gases listed in Table 11-3 (Oyama and others, 1979). The atmospheric structure derived from Pioneer and other measurements is shown in Figure 11-14.

Daily monitoring of the Venusian clouds by space vehicles in the 1970s revealed strong westward cloud

a   b

Figure 11-15. Westward circulation of Venus's atmosphere is shown in these two views taken 5 h apart. In **b**, the features have shifted west (left), as seen from dusky features, left center. The planet itself takes 243 d to turn, but the 200-mph winds at the cloud level blow the clouds around the planet in about 4 d. These ultraviolet images enhance the cloud features; in visible light the disk is almost blank. (NASA, Pioneer Venus Orbiter)

circulation at about 100 m/s, as shown in Figure 11-15. Interestingly, such photos confirmed controversial features that had been dimly glimpsed and sketched by visual observers of previous decades: a semipermanent "cusp cap," or polar cap of bright haze, often bordered by a darker band, seen in Figure 11-15. Figure 11-16 shows the polar haze cap as it would appear from above. Such features are prominent only in ultraviolet light. Spacecraft studies showed that the bright cap and bright bands are high hazes composed of 0.25-μm particles, smaller than the 1- to 10-μm particles in the main cloud deck described earlier in this chapter.

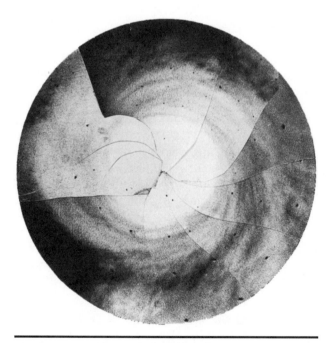

Figure 11-16. Photo-mosaic indicating the spiral structure of Venus's south polar cloud cap and its typical darker border, as seen from above the pole. (NASA, Pioneer images processed by Sanjay Limaye, Space Science and Engineering Center, University of Wisconsin)

A curious feature of Venus that has not been accounted for is the **ashen light,** which is a faint glow of the planet's dark hemisphere. It has been seen by a number of visual observers, particularly near inferior conjunction, when the dark side of the planet is directed toward Earth. Theoreticians have been unable to account for it, in spite of the fact that it has been reported even in telescopes with apertures as small as 30 cm. Usually described as tinted with a coppery color, it may be (if real) a high-altitude phenomenon similar to the terrestrial aurora or airglow (Levine, 1969).

Airglow in the form of certain spectral emission lines has been found on the dark side of Venus by Soviet and American spacecraft, but the glow is too faint to account for ashen light. Venera 11 and 12 probes in 1978 and the subsequent analysis of Pioneer data obtained at about the same time suggested lightning discharges on Venus. However, lightning detectors on the two 1985 VEGA balloons detected no lightning, and the importance of lightning on Venus is uncertain.

As described earlier, the past history of water and other volatiles in Venus's atmosphere is another major area of current research.

Earth

Figure 11-13 summarized our atmosphere's structure. A principal point emphasized in this chapter is the evolution of our atmosphere from a possible primitive atmosphere rich in hydrogen compounds, through an early secondary atmosphere poor in oxygen, to the modern oxygen-rich atmosphere. Water vapor and carbon dioxide were the most important gases added by volcanic degassing to form the secondary atmosphere. Most of the oxygen was added by biological processes. (See also next chapter.)

In both practical and academic terms, a major current issue is the question of climate change on Earth. Measurements of historical samples, as well as air measurements throughout this century, clearly show the increase in CO_2 since the industrial revolution. Models have predicted climate changes, including a probable global warming, as a

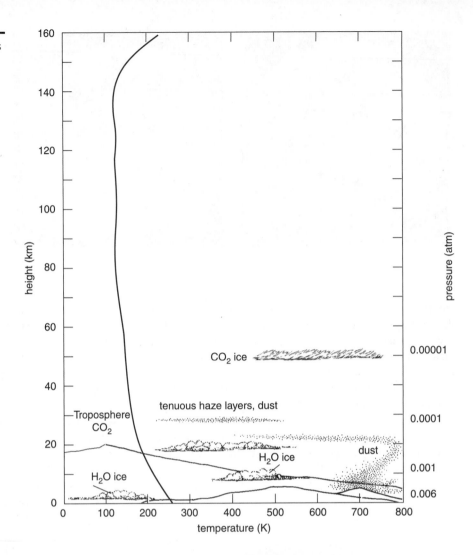

Figure 11-17. Structure and features of Mars's atmosphere, based on Viking and other measures.

result, probably becoming apparent in the decades around 2000. Increasingly, international scientists have concluded that this warming is beginning to be detected amidst the "noise" of natural year-to-year climate variation. The scientific challenge is to understand the mechanisms and the relation to natural climate variations. For example, short-term coolings, for a few years, have been measured when large volcanic eruptions spewed fine ash into the high atmosphere, blocking a bit of the incoming sunlight. On a larger scale, when a natural cooling cycle set in around A.D. 1100–1200, Viking sailors recorded increasing ice in their north Atlantic trade routes and abandoned their North American and Greenland settlements, and prehistoric native societies in the central and southwest United States experienced turmoil and restructuring. Yet the increasing number of record-breaking warm years in the last decades suggest that human-driven climate change is beginning. The social challenge is whether humans can agree internationally on what to do about our impacts on our planetary system.

Mars

Chapter 2 chronicled the drastic decline in estimates of Martian atmospheric density, from Herschel's estimate of

an Earth-like environment to modern measurements of 6 mbar surface pressure. Even at 6 mbar, Mars has a dynamic atmosphere, with cloud formation, dust storms, and seasonal transformation of ice at one polar cap to water vapor, which transfers through the atmosphere to freeze out at the winter polar ice caps. The atmospheric structure is diagrammed in Figure 11-17. The major problem of modern Martian atmospheric research is the question of how much water flowed on the ancient surface, and thus whether the ancient climate was different. Was this difference a profoundly denser or warmer early atmosphere, or did it reflect only ephemeral changes, such as short-term melting of subsurface permafrost in localized geothermal events? What does the climate change of Mars teach us about climate change episodes in Earth's history? We take up these questions in the discussion of Mars in the final chapter.

Jupiter and the Giant Planets

Earth-based spectra of Jupiter and Saturn long ago indicated absorptions by methane and ammonia (Kuiper, 1952), but scientists suspected that hydrogen and helium (which are harder to detect by spectra from Earth) were also retained by the strong gravities of these planets.

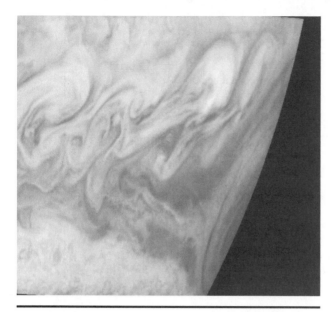

Figure 11-18. Turbulent clouds of Jupiter. The Great Red Spot can be seen on the horizon at the right. The bright, white clouds are highest; the red and darker clouds are lower formations. (NASA, courtesy Elizabeth Alvarez, Galileo Project)

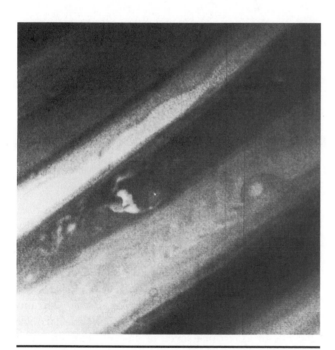

Figure 11-19. Small-scale cloud details in the atmosphere of Saturn. Bright oval (right) and spots in the dark belt are convective clouds. A wavelike longitudinal cloud stretches along the bright zone at the top. Smallest cloud features are about 175 km across. North is toward upper left. (NASA, Voyager 2)

Figure 11-20. Convective structure of belts and zones deduced from spacecraft measures and cloud motions. Zones are bright upwelling clouds; belts are descending areas exposing lower, warmer regions. Patterns acquire greater complexity due to shearing motions caused by Coriolis forces.

Molecular hydrogen was eventually detected by a series of Earth-based observations (reviewed by Owen, 1970). On Jupiter and Saturn, respectively, the Voyagers measured 81% and 88% by mass of H_2, very close to the solar hydrogen content of 79% by mass. The helium percentages for Jupiter, Saturn, and the sun were measured at 18%, 11%, and 20%, respectively (Ingersoll, 1981). The variations in helium abundance are somewhat puzzling, as helium is chemically inactive and might have been expected to be preserved in equal proportions; this is an area of current research. On both Jupiter and Saturn, the Voyagers also identified many trace constituents composing less than 1% of the mass. These included ammonia (NH_3), methane (CH_4), phosphine (PH_3), ethane (C_2H_6), and acetylene (C_2H_2) (Hannel and others, 1979; 1981).

The ever-changing, colorful cloud patterns, especially on Jupiter, have remarkably intricate structure, as shown by the chapter opening figure and Figures 11-18 and 11-19. Dark **belts**, bright **zones,** and a few prominent features such as Jupiter's Great Red Spot are relatively stable, but whole belts and zones change intensity from year to year, as can be monitored with a backyard telescope. As sketched in Figure 11-20, Voyager temperature maps revealed that the bright zones are high, upwelling clouds in the cold, upper atmosphere. The darker, more colorful belts are regions where we see deeper, warmer levels. The strong eastward and westward jet streams, moving at hundreds of meters per second relative to each other at different latitudes and altitudes, cause shearing of clouds on both planets, as described in Chapter 2. The Red Spot is one such turbulent eddy, circulating in a counterclockwise direction. Low temperatures in the Red Spot show that it lies higher than surrounding clouds. White oval disturbances occur on both Jupiter and Saturn; Voyager 2 observed a Saturnian example measuring 5,000 × 7,000 km and circulating with 100 m/s winds.

The rapid changes in the cloud markings are the first clue to the structure of giant planet atmospheres. The

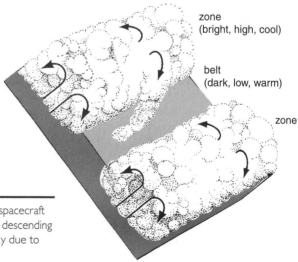

zone
(bright, high, cool)

belt
(dark, low, warm)

zone

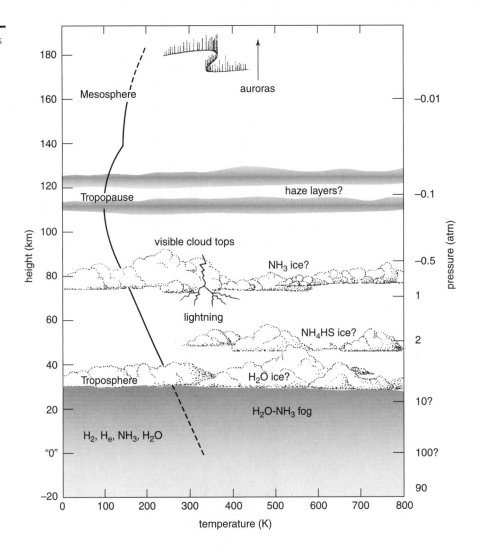

Figure 11-21. Structure and features of Jupiter's atmosphere, based on Voyager, Galileo, and other measures.

changes suggest convection driving ascent from below. Convection implies that the temperature gradient is adiabatic or slightly superadiabatic. These ideas led to early models of the temperature structure of Jupiter's atmosphere, based on Kuiper's (1952) calculation of an adiabatic temperature gradient of −3 K/km below the clouds. Voyager data (Figure 11-21) confirmed this reasoning, measuring a temperature gradient closer to −2 K/km. Figure 11-22 shows a similar atmospheric structure diagram for Saturn.

The source of cloud colors on Jupiter and Saturn is uncertain. Prinn and Owen (1976) showed that reactions among the major and minor constituents, such as sulfur and phosphorous, can produce colored compounds such as ammonium polysulfides ($[NH_4]_xS_y$), hydrogen polysulfides (H_xS_y), and phosphorous crystals (P_4). Their tones of yellow, orange, brown, and red may explain the colors. Other compounds, such as organic molecules, have also been proposed.

Voyager discovered remarkable visual phenomena in Jupiter's nighttime atmosphere: aurora and lightning. Both of these were recorded photographically in visual wave-

lengths as shown in Figure 11-23, page 342.* Ultraviolet auroral emissions near the poles were also detected. Voyager instruments recorded similar ultraviolet aurora within 12° of the poles of Saturn.

The atmospheres of Uranus and Neptune show some similarities and some differences when compared with those of Jupiter and Saturn. Most pronounced is the blue color. This stems from a combination of two effects. Above the clouds is a thick layer of gaseous haze in which Rayleigh scattering produces a sky-blue color. Perhaps more important, methane gas absorbs red and orange light, producing an overall blue tone with a faint greenish tinge.

---

*Interestingly, "frequent and powerful thunderstorms in the Jovian atmosphere," causing lightning at 10,000 times the terrestrial rate, had been predicted as early as 1975 by Bar-Nun, based on his hypothesis that observed Jovian acetylene was synthesized by lightning. Ironically, Lewis (1980) later concluded that the acetylene is not efficiently produced by the lightning, so that the prediction of lightning, though correct, has questionable status. The regard for prediction in science depends not only on the success, but also on the additional work the prediction stimulates, and the solidity of the original reasoning.

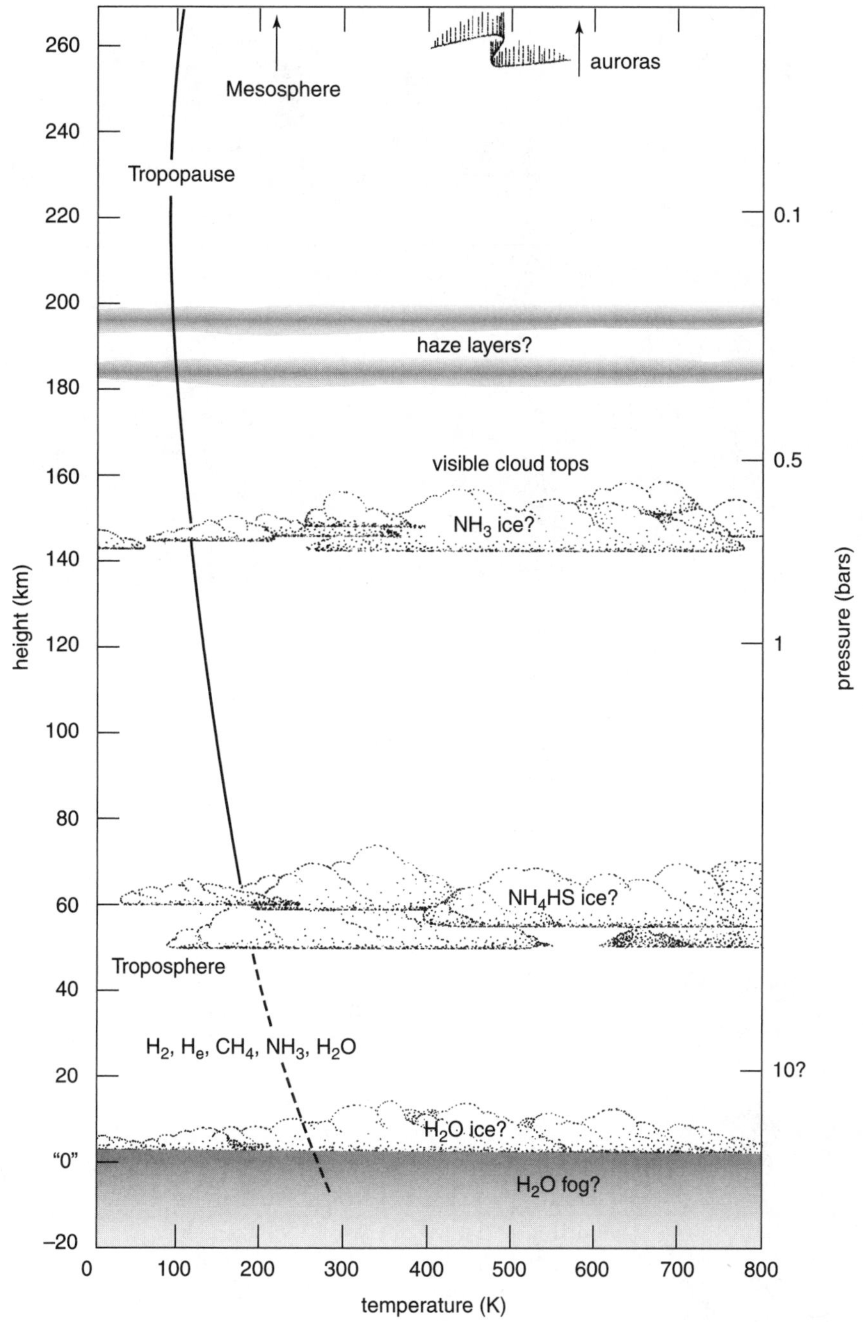

Figure 11-22. Structure and features of Saturn's atmosphere, based on Voyager and other measures.

Voyager showed Uranus to have a nearly featureless blue globe, with only the palest features visible through the haze. Recall that Uranus has little excess internal radiation, compared to its absorbed sunlight, so that there is little energy input from below to drive big storm systems. Neptune, in contrast, turned out to have some pronounced dark belts, a few bright white high-level clouds, several small dark ovals, and a Great Dark Spot 30,000 km across, 2½ times bigger than Earth (Figure 11-24). It is similar in behavior to Jupiter's Great Red Spot; both have counterclockwise (anticyclonic) rotation.

Wind speeds were measured on various cloud features. Unlike the atmospheres of Jupiter and Saturn, the atmospheres of Uranus and Neptune circulate slower at equatorial latitudes than the rotation of the planet itself; winds oppose the rotation direction. The poles and equators of Uranus and Neptune share similar temperatures, implying that considerable global circulation must occur to equalize the temperatures.

All four giants have atmospheres that are very roughly ¾ H_2 by mass and ¼ He, with traces of other gases. Declining temperatures cause more ammonia (NH_3) to freeze out in the form of cloud crystals as we move from Jupiter

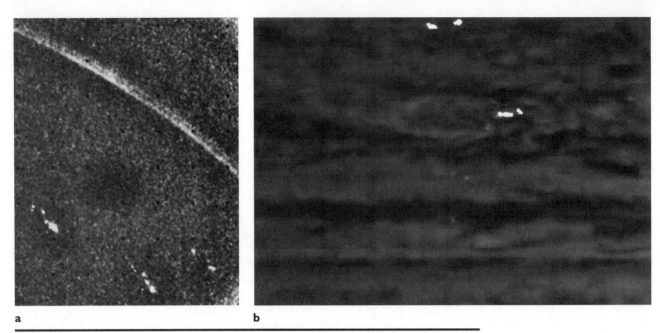

a   b

Figure 11-23. Lights in Jupiter's night sky, seen from above. (**a**) This three-minute exposure shows aurora over north pole (bright limb, across top) and bright lightning flashes among the clouds. The aurora is brighter than Earth's; the lightning is comparable to superbolts near the tops of terrestrial tropical thunderstorms. (NASA, 1979 Voyager 1 image) (**b**) One minute exposure of Jupiter's night side, illuminated by the moonlight of Io, shows lightning flashes among the cloud bands. The flashes are roughly 100 times brighter than typical terrestrial lightning. (NASA, 1997 Galileo image courtesy Elizabeth Alvarez)

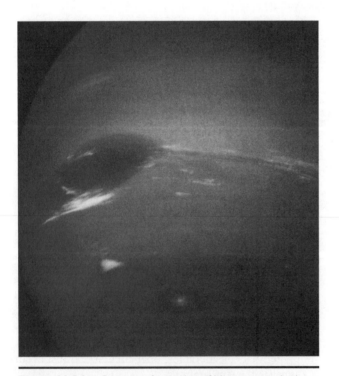

Figure 11-24. Close-up of a portion of Neptune surprised Voyager scientists by showing prominent cloud patterns in its atmosphere. Dark oval clouds, such as the prominent "Great Dark Spot," mark circulating storm systems, capped by white condensed clouds higher in the atmosphere. (NASA Voyager 2 photo)

to Neptune, and the methane (CH_4) absorption becomes more pronounced, helping to create the trend toward bluer color from Jupiter to Neptune.

Titan

Titan is one of the most fascinating of the atmosphere-rich worlds, partly because its N_2-rich atmosphere resembles Earth's, and partly because its haze cloaks the surface in mystery (Figure 11-25). Kuiper's 1944 discovery of absorption bands of CH_4 in the spectrum marked the first discovery of an atmosphere on a satellite, and Titan remains the only satellite with a dense atmosphere.

Geochemical theory prior to the 1980 flyby of Voyager 1 suggested that large amounts of N_2 might have been produced by volcanic activity and resultant dissociation of NH_3 gas (Atreya, Donahue, and Kuhn, 1978), but N_2 could not be confirmed from Earth because it lacks absorptions in accessible parts of the spectrum. Voyager data confirmed that the atmosphere is at least 90% N_2 by volume, and that CH_4 is only the number two constituent. Voyager data also revealed the striking fact that the surface atmospheric pressure on Titan is 1.50 ± 0.02 bars—closer to the terrestrial value than found on any other world (Stone and Miner, 1981; Owen, 1990)!

Polarization and photometric studies led to the 1973 discovery of the haze layer obscuring Titan's surface

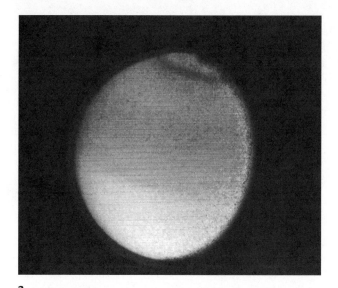

a

b

Figure 11-25. Saturn's giant Moon, Titan, about as big as Mercury. The smoggy, reddish-tan atmosphere of nitrogen is colored by organic compounds. (**a**) At gibbous lighting, a dark band can be seen at 60° N latitude, and the southern hemisphere is slightly brighter than the north. Scientists speculate the pattern may change with the seasons. (**b**) At crescent lighting, the illuminated atmosphere can be traced all the way around the moon. (NASA Voyager images)

(Veverka, 1973; Zellner, 1973). The haze is a photochemical smog produced when sunlight interacts with the CH_4 and other compounds in the atmosphere. Podolak, Noy, and Bar-Nun (1979) theorized that the haze consists of particles of polyethylene, a polymer of the gas ethylene (C_2H_4), which has been detected in Titan. Voyager 1 detected about 1% of methane (CH_4) by volume in the mid-atmosphere, and smaller traces of ethane (C_2H_6), acetylene (C_2H_2), ethylene (C_2H_4), hydrogen cyanide (HCN), and possibly methylacetylene (C_3H_4) and propane (C_3H_8) (Hanel and others, 1981). These compounds support the idea that the haze layer is a colored photochemical smog. Various researchers, such as Smith and others (1996) realized that the gases in Titan's haze are less opaque at near infrared wavelengths than at visible wavelengths, and they used the Hubble Space Telescope to pierce the clouds as described in the previous chapter (Figure 10-48).

Figure 11-27 summarizes the atmospheric structure. The measured surface temperature of Titan, a frigid 94 K ($-290°$F), is strikingly near the triple point temperature for methane, 90.7 K. Methane's triple point pressure is 117 mbar, reached some tens of km above the surface. This is why methane may exist as ice, liquid, or gas in Titan's lower atmosphere, just as water does in Earth's lower atmosphere, where the environment has a similar relation to the triple point for H_2O (273.01 K; 6.1 mbar). That is why researchers expect CH_4 rain, CH_4 snow, and perhaps CH_4 rivers or lakes on the surface (Figure 11-26). Nonetheless, spectroscopy reveals H_2O ice on the surface (Griffith and others, 2003; see also Chapter 10). The Cassini probe,

Huygens, should have an interesting journey as it parachutes onto Titan in 2004!

Small Worlds and Thin Atmospheres

None of the other worlds in the solar system have atmospheres that would be noticeable to an astronaut. Io is deeply embedded in Jupiter's magnetic field, and although volcanic vents provide some gas in the vicinity of Io, more atoms are probably knocked off the surface by high-energy particles in Jupiter's field. Many of the remarkable phenomena involving Io's thin atmosphere are described in Chapter 2. Pioneer 10 detected a thin ionosphere around Io in 1973, first confirming the presence of gas, which had been suspected from ground-based observations. Then, in 1974, astronomer Robert A. Brown used ground-based observations to discover an extended cloud of sodium (Na), sulfur (S), and other atoms that extends many Io diameters from Io and stretches in both directions around the orbit. These atoms emit a remarkable auroral glow in this cloud, as seen in Figure 11-28, page 345. The emission in the well-known yellow "D" line of the Na spectrum is especially strong. This glow is brightest when Io is moving toward or away from the sun at the two points of elongation (as seen from the inner solar system). The reason is that the glow is stimulated by solar photons that excite the Na atoms. The solar spectrum lacks photons of the D wavelength because of a deep absorption of the sodium D line by the solar atmosphere. Only when Io moves toward or away from the sun is the incoming solar light Doppler shifted enough that the sodium atoms are

a

b

Figure 11-26. Hypothetical scenes on Titan. (**a**) Above the main haze or cloud layer, Saturn may be visible in a hazy blue sky. (Painting by author) (**b**) Below the clouds, it is fairly dark. Temperature and pressure conditions suggest snowstorms and rainstorms of solid and liquid methane compounds, with possible rivers of liquid methane and ethane. (Painting by Ron Miller)

Figure 11-27. Structure and features of Titan's atmosphere, based on Voyager and other data. Note that the scale differs from the other atmosphere diagrams in this chapter in order to show the high haze layer of tiny particles.

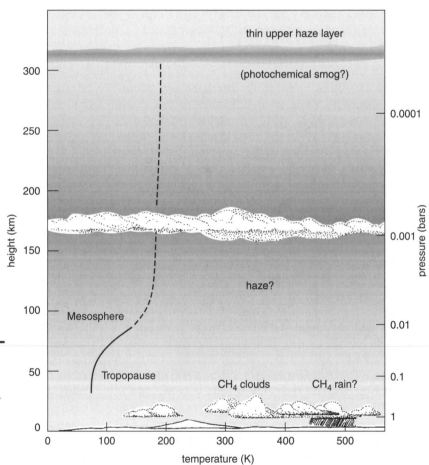

illuminated by solar light with the D line wavelength. The yellow sodium glow has been reported to have a brightness several times that of the faintest auroras visible to the eye on Earth. Thus, it should be visible to an observer on Io when Jupiter is half lit by the sun.

Another phenomenon involving volatiles on Io is the probable deposition of sulfur dioxide (SO_2) frost from gases produced by the volcanic eruptions. Whitish SO_2 frost deposits appear likely to form under relatively cold conditions in Io and may be the explanation of the bright-

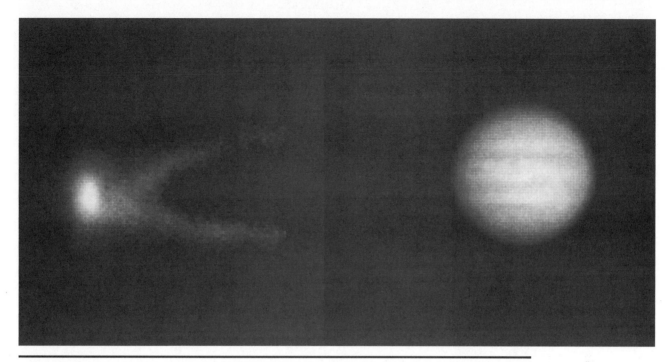

Figure 11-28. Part of the glowing ring of gas atoms blown off Io and extending around Jupiter (right) near Io's orbit. The ring is tipped, relative to Jupiter's equator, because the atoms spread out by moving with Jupiter's rotating magnetic field, which is tipped relative to Jupiter itself. The diamond ring effect is caused by looking tangentially along ribbon-like sheets that extend north and south along magnetic field lines. Gas temperatures in the toroidal ring range from 20,000 to 400,000 K. The image is a 10-minute exposure made in the light of glowing, ionized sulfur (SII) atoms; the image of Jupiter was a separate exposure inserted to show scale and orientation. (Jet Propulsion Laboratory, courtesy John Trauger)

ening *sometimes* observed for 10 to 15 minutes after Io comes out of eclipse (see Chapter 2). If eruptions are providing SO_2 at the time of the eclipse, the effect might be strong; when the volcanoes are less active, the effect might not occur. Veverka and others (1981) monitored three Io eclipses from Voyager spacecraft and saw only a 3% brightening of the south polar region for 3 minutes after the eclipse ended. The phenomenon remains intriguing.

Energetic solar photons, cosmic ray particles, and/or energetic particles from planetary magnetospheres hitting the ices of other small moons can produce extremely thin "atmospheres" of gas, or ionospheres of charged particles, around other small bodies. The Galileo spacecraft observed an ionosphere of charged particles around Europa, with a scale height of about 240 km, indicating a very thin atmosphere possibly made of H_2O or O_2 at temperatures of 340 to 600 K, respectively (Kliore and others, 1997). The temperatures, much higher than the surface temperature, are created by the high energy dislodging of the atoms from the surface ices.

Similarly, Galileo scientists detected atomic hydrogen atoms in a very thin atmosphere escaping from Ganymede. The discovery implied that the H atoms are remnants of H_2O knocked off the surface ice, probably by energetic solar UV photons, and that the heavier oxygen atoms are left behind in a thin atmosphere around Ganymede. If so, Colorado researcher Charles Barth, of the Galileo team, speculated that some of the oxygen atoms may be re-trapped in the surface ice. Ozone (O_3) has been directly detected, trapped in the Ganymede ice (Noll and others, 1996). Barth speculated that Ganymede may have very large oxygen resources, comparable to the total inventory of oxygen in Earth's atmosphere—an interesting possible resource for future explorers.

Voyager photos showed polar frost caps on Ganymede (Figure 8-29, upper right), probably thin deposits of H_2O ice. The deposit must be thin because underlying surface structure and albedo variations are seen. The frost could stand out from background ices, not only because of ice and soil compositional differences, but also because frost with different grain sizes can have different albedos. The cap edge lies at about 40° to 45° latitude, close to the calculated limiting latitude below which solar warming could sublime an H_2O frost cap (Smith and others, 1979). Once volatiles migrate to a polar region where they find cold surfaces to condense on, they may be fairly permanently trapped, like the polar H_2O ice cap of Mars. A similar migration and condensation of H_2O molecules released by comet impacts on the moon, may explain lunar polar ice deposits. Many of the outer satellites, if volatiles were once emitted on their surfaces, may have polar deposits and overlapping polar caps of CH_4, H_2O, and other frozen materials. On the satellites with higher obliquities, seasonal sublimation of such materials could provide very thin atmospheres whose pressures would equal the vapor pressures of these frosts under the ambient temperature conditions at the frost deposit sites.

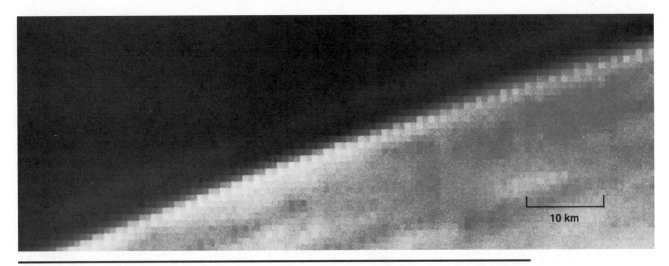

Figure 11-29. Triton's surprising haze layer. Although the atmosphere of Triton has a surface pressure only 0.02 mbar, Voyager photos revealed a haze layer about 3 km thick, floating about 3 km above the surface. The haze may be a condensate, or may be residual smoke from the eruptive vents on the surface. (NASA Voyager 2 photo)

Noll and others (1997) found O_3 in the ice of Saturn satellites Dione and Rhea, which orbit within the magnetosphere of Saturn. They suggest that energetic particles in the magnetic field break up H_2O molecules, as more directly observed on Ganymede, providing a thin gas envelope around the moons and also building up quantities of oxygen and ozone trapped in the ice.

Pluto is an even smaller icy world with atmospheric traces. Pluto is so far from the sun that the active ice is methane, which would be "burnt" away by the sun at the solar distance of Ganymede and Europa. When Cruikshank and Silvaggio (1980) found CH_4 ice on Pluto's surface, they calculated that it should sublime to maintain a thin atmosphere whose surface pressure would change markedly as Pluto moved from aphelion to perihelion. Current data suggest that the surface pressure could range from 0.001 to as high as 0.01 mbar during Pluto's 248-year "year" (compare with Mars's 6 mbar).

As remarked in the last chapter, Triton and Pluto may be linked in having similar origins as interplanetary bodies. Indeed, as we consider atmospheric processes on small icy bodies, we can see more clearly than ever that modern studies are blurring the classic distinctions between the behavior of planets, moons, and comets.

Consider three bodies, seemingly dissimilar: the "planet" Pluto (or rather the Pluto/Charon system), the moon Triton, and the "asteroid/comet" Chiron. Each has its own unique properties, but look at the links in the behavior of ices and gases.

In Pluto's case, a strange atmospheric transfer has transpired between Pluto and its moon Charon. During Pluto's "summer" at perihelion, methane ice sublimes into a gas on both bodies, but Charon is so small that the gas molecules can escape—essentially turning Charon into a comet. Nearby Pluto has stronger gravity, and so it not only retains its own methane but captures some of that leaking

off Charon. As a result, Charon's surface is nearly denuded of methane, leaving "rock-solid" H_2O ice, which is not volatile at those temperatures. Spectra reveal Charon's water-ice surface, in striking contrast to the surfaces of Pluto and Triton.

In the case of Triton, the high orbit inclination leaves first one pole in sunlight, then the other—reminiscent of Uranus. As a result, nitrogen and perhaps methane ice sublime from the sunny pole into the atmosphere, only to recondense at the cold opposite cap. Current data suggest a thin atmosphere of N_2 with additional CO and perhaps Ar, with a total pressure around 0.02 mbar. The southern cap of Triton has been exposed to the sun for several decades, but Voyager photographed a bright cap covering the polar region. Darker surface exists at lower latitudes, but a faint frost is seen at the terminator, where new condensation seems to be occurring (Smith and others, 1989). Thin clouds and haze layers were also seen, as shown in Figure 11-29. The seasonal production of hazes and transfer of material from one pole to the other, together with the active eruptions, probably explain some spectral changes that have been observed on Triton.

Chiron is another distant interplanetary body whose character is affected by ice sublimation. As remarked in Chapter 2, Chiron was cataloged as asteroid 2060 but was later seen acting like a comet. Meech and Belton (1990) outlined a unique effect that makes Chiron produce not only gas but a "dust atmosphere" in its coma, different from other comets. Because Chiron is so much bigger than other comets, its gravity is stronger. In addition, the ice that is subliming (CO_2?) probably has a much larger molecular weight than the H_2O that powers comets at lower solar distance; therefore, the molecular velocities are slower. The combination of high gravity and low gas velocity means that many dislodged dust particles move too slowly to escape and be blown into a comet tail by the solar wind. Thus, each comet outburst produces a long-lived "atmosphere" of fine dust around Chiron itself, which only gradually settles back as the production weakens.

In summary, the solar system contains a wider range of atmospheric phenomena than once expected. These range from the vast primitive, hydrogen-rich atmospheres of the giant planets, to the more familiar secondary atmospheres of terrestrial planets, and beyond, to the tenuous atmospheres of small icy bodies. The latter lead us to a blurry distinction between traditional atmospheric phenomena and purely cometary phenomena associated with ice sublimation.

CONCEPTS

particles and fields	cumulus cloud	ozone layer
primitive atmosphere	temperature gradient	Hartley-Huggins bands
secondary atmosphere	lapse rate	thermosphere
ideal gas law	polar hood	mean free path
isothermal atmosphere	white clouds (on Mars)	ionosphere
scale height	blue clouds (on Mars)	electron temperature
aerosols	yellow clouds (on Mars)	exosphere
scattering	global dust storms	equipartition of energy
Rayleigh scattering	superadiabatic	photodissociation
opacity	Coriolis force	mixing ratio
optical depth	tropopause	kilometer-atmosphere
greenhouse effect	troposphere	kilometer-amagat
phase diagram	stratosphere	ashen light
triple point for H_2O	jet stream	belt
moist (atmosphere)	mesosphere	zone

PROBLEMS

1. Describe how compositions of atmospheres evolve with time. Start with the primitive composition and then discuss the production of secondary atmospheres and thermal escape.

2. Why is a cloudy night likely to be warmer than a clear night if other conditions are the same? Why is a night in a clear, high desert colder than a night on the beach if the noontime temperatures are the same? Relate these events to the greenhouse effect.

3. Why does Titan, which resembles a terrestrial planet, have a nitrogen-methane atmosphere when terrestrial planets have nitrogen-carbon dioxide-water vapor atmospheres (or volatile inventories)?

4. Describe the surface climate on an imaginary planet in Earth's orbit with an atmosphere that has white clouds of optical depth 10 in the visual region of the spectrum but optical depth 0.1 in most of the infrared.

5. Why might one expect the atmosphere of Mars to have changed more in the past than Venus's or Earth's? Cite evidence that this happened.

6. Why do low-pressure storm systems develop cyclonic flow patterns, and why are the patterns in the two hemispheres mirror images of each other?

7. Describe the annual cycle of events at one of the poles of Mars.

8. Why is exosphere temperature critically important to an atmosphere's evolution?

ADVANCED PROBLEMS

9. How much does 1 m³ of the air around you weigh (as expressed in kg)?

10. If the nominal surface pressure on Mars is 7 mbar, calculate the pressure in a canyon 4 km below this level and at the top of a volcano 20 km above this level.

11. Which planet is better screened against meteorites, a Titan-size planet with 1-bar surface pressure or an Earth-size planet with 1-bar surface pressure?

12. (a) Calculate an atmospheric structure for an ancient Mars with an opaque cloud deck at 4 km altitude with albedo 40%. Assume dry CO_2 composition and ignore greenhouse effects. Calculation should show temperature trends above and below the clouds. (b) Describe how results would change with a greenhouse effect caused by CO_2 and large amounts of water vapor.

13. (a) Calculate thermal escape times for H, He, and H_2O on Mars with exosphere temperatures of 210 K and 1,000 K. Comment on the volatile history on Mars if transient events ever heated its exosphere to 1,000 K for 1 My. (b) Why is the calculation inadequate to fully explain the escape of Martian volatiles?

The sun and the sea are two components usually believed necessary to create an environment in which biochemical reactions can lead to the formation of life. Sunlight or some other source of energy is required to encourage the necessary chemical reactions, and liquid water to provide a medium in which complex organisms can evolve. (Photo by author)

CHAPTER TWELVE

Life: Its History and Occurrence

One of the most intriguing questions about our cosmic and human condition is whether planets elsewhere in the universe harbor what we are pleased to call "intelligent life." Extraterrestrial life must either exist or not exist. Either case has striking consequences. The nineteenth-century Scottish writer Thomas Carlyle sardonically said that other worlds offer "a sad spectacle. If they be inhabited, what a scope for misery and folly. If they be not inhabited, what a waste of space."

What can planetary science say about these two possibilities? As discussed in Chapter 4, the 1990s for the first time have brought direct observational evidence that planetary systems probably exist around a significant fraction of other stars. Even if only 1% of stars have planets, there could be a billion planetary systems in our galaxy alone. Chapters 4 and 6 showed that some interstellar dust and some meteorites contain **organic molecules** (C-H based compounds) formed in space. Chapters 8–10 revealed the possibility of a warm ocean under the ice of Europa, and organic materials in the nitrogen atmosphere of Titan—provocative environments for studying the evolution of organic compounds. The sterile soil of Mars's surface limits the chemical possibilities of life there, yet the 1996 announcement of seemingly Martian hydrocarbons and even microbe-like structures within carbonate globules inside rocks from Mars—though controversial—re-opened the debate about whether life might have evolved in ancient times on Mars.

Some workers call this field **exobiology,** the study of biological processes in the universe in general, primarily off the Earth. Critics have charged that exobiology is a subject without any known subject matter, but experiments have produced interesting results about biochemical evolution in extraterrestrial (or early terrestrial) conditions. In this chapter, we consider the issue in several steps. First, we discuss what we mean by *life* and what conditions life would require to exist on planets. Second, we review the long process that led to the evolution of life on Earth, as best we understand it. Third, we look more carefully at the availability of planets that could support a similar evolution of life elsewhere in the universe. Finally, we try to estimate the probability that intelligent life actually exists elsewhere and whether we might contact it, or it, us.

The Nature of Life

What do we mean by "life"? **Life** is not a status but a process—a series of chemical reactions using carbon-based molecules, by which matter is taken into a system and used to assist the system's growth and reproduction, with waste products being expelled. The system in which these processes occur is the cell. All known living things are composed of one or more cells.* A **cell** is in essence a container filled with an intricate fluid array of organic and inorganic molecules (protoplasm). This container is generally surrounded, protected, and defined by a cell wall typically 5–10 nm thick.

The elements most prominent in the organic molecules are carbon, hydrogen, oxygen, and nitrogen—all common in planetary matter. Phosphorus, also important to life (in small amounts), is also widely available. Carbon is especially critical because it can combine to make long chains of atoms—large molecules that encourage the complicated chemistry of genetics, reproduction, and so on. That is why the term **organic chemistry** (as well as *organic molecules*) refers not specifically to life-forms but more generally to all chemistry (and molecules) involving carbon, and more specifically the repeated C-H molecular bond.

We often make a mistake of thinking of ourselves as static beings instead of dynamic systems. We casually assume that we are constant entities, as if our identities solely depended on the form of our bodies. But our bodies today are not the same ones we had seven years ago. Hardly a cell is still alive that was part of that body. This dynamic conception of life is a far cry from the conception only a few generations back, when bodies were thought of as semipermanent machines whose parts gradually wore out. Even our seemingly inert skeletons are living and

---

*Viruses might be an exception. They are simpler than cells, yet can reproduce themselves using materials from host cells. Biologists disagree on whether viruses should be considered a form of life.

Figure 12-1. Broths rich in amino acids and complex organic molecules are likely to have formed in long-lived tidewater pools on ancient Earth. Evaporation of the water would have concentrated the remaining organic materials, allowing complex reactions. The resulting products could have "fertilized" oceans with protobiological materials or even the first living organisms. (Photo by author)

changing, always replacing their cells. We *must* keep changing, because the cells must keep processing new materials to stay alive. When the processing stops, that is the condition we call death, and the cellular material immediately begins to break down into its molecules and molecular fragments.

The nature of living beings is illustrated in an analogy made by the Russian biochemist A. I. Oparin (1962). Consider a bucket that has water pouring in at the top from a tap and flowing out at the same rate through a tap in the bottom. The water level in the bucket stays constant, and a casual observer would call it a "bucket of water." But it is not like an ordinary bucket standing full of water. The water at an instant is not the same water as at any other instant, yet the outward appearance is constant. We are like buckets with water and nutrients and air flowing through us, but with other, much more complex attributes, such as the ability to reproduce and to be affected by genetic changes that let us evolve from generation to generation.

This metaphor gives us some clue to the kinds of processes involved in the origin of life. We are looking for a process in which complex carbon-based molecules can enter cell systems that (1) draw material from the incoming molecules to create new molecules, (2) incorporate the new molecules into new structures, (3) eject unused material, and (4) reproduce themselves.

Scientists therefore usually choose to define life by these specific carbon-based processes. Often at this point people ask, "What about some unknown chemistry based on other elements, such as silicon, that can form big mol-

ecules? Or what about some unknown form of consciousness?"* The answer is that we have never observed or experimented with such life-forms, so we can say nothing substantive about them. If we admit they are plausible, we simply increase the probability of life or consciousness in the universe. But researchers usually restrict their discussion to forms on which chemical and behavioral data are available.

Whatever other concepts we invent—civilization, religion, technology, art, war, love, communication—it is the chemical processes of life that define us, just as they define the spiders, sea urchins, elephants, moths, amoebas, redwoods, and all the other incredibly varied living creatures around us. To judge whether life may exist on other planets—whether other planets are already "taken"—we must find out how those processes got started on Earth.

The Origin of Life on Earth

Most researchers have assumed that in addition to the life-forming elements—primarily carbon (C), hydrogen (H), oxygen (O), and nitrogen (N)—water was crucial to the development of life on Earth. Several facts suggest this:

---

*For example, in *The Black Cloud*, astronomer Fred Hoyle imagined an interstellar nebular cloud with matter and electromagnetic fields organized in such a way that it developed a consciousness or will of its own.

1. Most of our body weight is made up of water. The percentage is even higher for plants.

2. Body fluids, like tears, have a saltiness similar to that of the oceans. The oceans are our heritage. As embryos, we at first are immersed in fluid and develop bodies more like fish than like mammals—for example, we have gills.

3. Organisms deprived of liquid water quickly die. And dead organisms are shriveled and dried.

4. A watery medium is usually involved in reproduction. Male fish release sperm into the water to fertilize eggs nearby; amphibians began the conquest of land but had to return to water to lay eggs; land animals developed a system to fertilize egg cells inside the body, and/or lay eggs with a shell that protected the watery medium.

5. Taking a more theoretical view, water provides a medium in which organic molecules can be suspended and can interact—a likely place for life to begin.

So a good starting question is, what was the early history of C, H, O, N, and H_2O on Earth? The solar nebula was rich in these materials. As is clear from Chapter 5, it included hydrogen molecules, ammonia, methane, water, and perhaps some nitrogen molecules (respectively, H_2, NH_3, CH_4, H_2O, and N_2). Note which atoms were involved: C, H, O, and N—the elements of life! As discussed in Chapter 7, comets contain particles so rich in these materials that they are called CHON particles.

Probably within 100 My years after Earth formed, heat from internal radioactivity and impacts melted parts of Earth's interior, unleashing volcanic activity. Volcanic activity released gases, especially water vapor (H_2O). In the first 500 My, the intense early bombardment included large impacts capable of driving off masses of atmosphere or vaporizing oceans. (Recall from Chapter 9 that the moon has 1,000-km-scale impact basins, formed during this period.) Yet by about 4,000 My ago there were probably bodies of surface water under an atmosphere that was rich in hydrogen, ammonia, methane, water, nitrogen, and carbon dioxide, and large impacts capable of vaporizing oceans were becoming less frequent.

Chemical reactions occurring naturally in such an environment could have produced building blocks of life, as has been proved by laboratory experiments. In the 1950s, chemist Stanley L. Miller, working with Nobel laureate chemist Harold Urey, put a gaseous mixture of hydrogen, ammonia, methane, and water vapor (to represent the primitive atmosphere) over a pool of liquid water (primitive seas) and passed electric sparks through it (simulating energy sources such as lightning). After several days, the pool began to darken. The water now contained a solution of **amino acids,** the class of molecules that join to form **proteins,** the huge molecules in cells. This so-called **Miller experiment** proved that the complex organic chemicals necessary for life could be built in a natural environment (Miller, 1955). An example of a reaction in the Miller experiment is the production of the amino acid glycine:

$$NH_3 + 2CH_4 + 2H_2O \xrightarrow{\text{energy}}$$
$$\text{ammonia} \quad \text{methane} \quad \text{water}$$
$$C_2H_5O_2N + 5H_2$$
$$\text{glycine} \quad \text{hydrogen}$$

As discussed in Chapter 11, later geochemical studies suggested that Earth's early atmosphere may not have been so primitive—that is, rich in hydrogen based compounds—as Miller and Urey thought. It was probably more rich in volcanic gases such as CO_2, H_2O vapor, CO, and N_2. Solar ultraviolet light striking such an atmosphere tends to produce water and hydrogen cyanide (HCN). Further experiments showed that a Miller-like reaction produces amino acids even under these conditions (Goldsmith and Owen, 1979):

$$3HCN + 2H_2O \rightarrow C_2H_5O_2N + CN_2H_2$$
$$\text{hydrogen} \quad \text{water} \quad \text{glycine} \quad \text{cyanamide}$$
$$\text{cyanide}$$

These later experiments also showed that many kinds of energy sources—including ultraviolet light from the sun, volcanic activity, and even meteorite impacts—can produce the same reactions. Thus, the building blocks of life *would* have formed on early Earth and other planetary bodies. This conclusion was dramatically confirmed in the 1960s by the discovery of extraterrestrial amino acids (but not fossils or life forms) in several carbonaceous chondrite meteorites (Kvenvolden and others, 1970; Cronin, Pizzarello, and Moore, 1979; Engel and Nagy, 1982). Amino acids can form in either a "right-handed" or "left-handed" symmetric molecular structure, but on Earth, some biological or prebiological process led to strong preponderance of only one of the two structural options (called a **nonracemic mixture**). The carbonaceous chondrite amino acids were proven to be extraterrestrial by showing their molecular structures had equal proportions of the two structures (called a **racemic mixture**).

From such results, researchers have concluded life's organic building blocks form easily, and that Earth's oceans were well fertilized with molecular organic raw materials by 4,000 My ago or earlier.

What happened next? The classic view, developed in the 1950s and 1960s, is that such materials became concentrated in isolated tidewater ponds (Figure 12-1), as seawater swept into the pools and then evaporated, leaving heavy molecules behind to interact and form complex substances. Eventually, complex molecules like DNA evolved. DNA has the ability to split apart, with each fragment then picking up new atoms that recreate the original.

From Organic Molecules to Living Cells

The next steps are less certain. The important thing is that somehow, whether in tidal pools or other environments, organic materials become concentrated enough that cell-

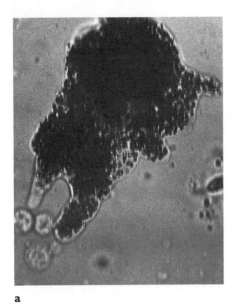

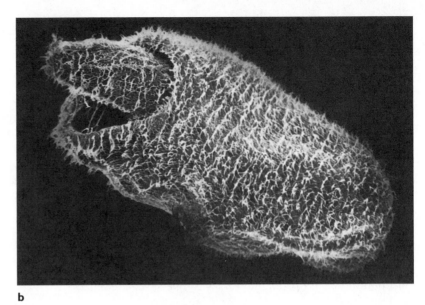

a

b

Figure 12-2. The importance of a fluid medium for primitive evolution is suggested by these photos of single-celled organisms engulfing food from surrounding fluid. (**a**) An amoeba flows to surround a nearby food particle. (Optical microscope photo, S. L. Wolfe) (**b**) The protozoan Woodruffia ingests a paramecium. (Electron microscope photo, T. K. Golder)

like structures formed, enclosing still higher concentrations of organics. Eventually these structures enclosed self-reproducing molecules, but initially they may not have been biological. For example, Florida biologist S. W. Fox has shown that simple heating of dry amino acids (as might happen on a dry planet such as present-day Mars) can create protein molecules. When water is added, these proteins assume the shape of round, cell-like objects called **proteinoids,** which take in material from the surrounding liquid, grow by attaching to each other, and divide. Though they are not considered living, they resemble bacteria so much that experts have trouble telling the difference in microscope views.

Possibly related to proteinoids are objects discovered in the 1930s by Dutch chemist H. G. Bungenberg de Jong. When proteins are mixed in solution with other complex molecules, both sets of substances spontaneously accumulate into cell-sized clusters called **coacervates.** The remaining fluid is almost entirely free of complex organic molecules (Oparin, 1962).

The next step toward recognizable life-forms is still more uncertain, but many biologists theorize that cell-like proteinoids and coacervates concentrated in moist environments, or in primeval pools of "organic broth," began reacting with surrounding fluids and with each other, accumulating more molecules and growing more complex, as suggested by Figure 12-2. Presumably, at some point, proteinoid-like globules containing DNA organized into self-reproducing *cells,* marking the origin of life. Details of this step remain murky!

The smallest living cells are generally of the order of 100 nm across (0.1 micron), but much larger cells are

common. The genetic properties that define the cell are in the molecular subgroups along with the giant DNA molecules. The simplest and most primitive cells, called **prokaryote** cells (from Greek roots for "before the nucleus"), were the first to appear around 3,500 to 3,700 My ago. These cells contain DNA molecules, but not in well-defined nuclei. They formed two broad types of life: bacteria and blue-green algae. Bacteria include life-forms called archaebacteria, which, according to DNA evidence, may have been among the first life-forms on Earth—the parents of us all. They may have been better-suited to the oxygen-poor early earth and can still be found in oxygen-poor or extreme environments. They include forms that manufacture CH_4 from CO_2; forms that live in high-salt environments; and "extremophiles" that live in hot-water geothermal vents. Some forms of blue-green algae, the other prokaryotes, produce oxygen and thus began to change Earth's atmosphere. Around 2,000 My ago, a second type of cell appeared, called **eukaryote** ("true nucleus"). In these cells, DNA is located in a well-defined central body called the nucleus. They are the cells of all advanced life-forms, including ourselves.

The underlying assumption of the classic "tidepool hypothesis" is that life prospers best today under what we think of as "moderate" temperatures and pressures, and so that must have been best for life to originate. However, research in the 1980s and 1990s has made it possible to study the degree to which genetic code-carrying DNA in various life-forms is different from one life-form to another, and this knowledge has changed our thinking somewhat. As the molecular linkages were traced down to the level of single-celled bacterial microbes, it turned out that all life-forms have linkages that seem to branch out from the archaebacteria, including extremophiles, which live in what we think of as difficult environments. A classic example is microbial life living in the high-temperature "smoker" vents on the sea-floor, where mineral-rich columns of water at hundreds of degrees, with a smoky black appearance, are released from geothermal vents. An interesting feature of such life is that it ultimately depends not primarily on sunlight, as do most familiar life-forms, whose food chains trace back to organic products produced by photosynthesis. Rather, they collect their biological materials from the minerals produced in the hot water of the vent. This realization revised the notion that sunlight must be life's main energy source (Karl, Wirsen, and Jannasch, 1980). Other extremophile microbes live in geothermal hot springs at locations such as Yellowstone Park. All branches of the tree of life, as mapped by genetic code research at the molecular level, seem to connect back to a tree trunk that involves extremophile archaebacteria. This discovery has led to suggestions that the first complex self-reproducing molecules, which began to build the first single-celled life-forms on Earth, did so in geothermal environments rather than in balmy tidepools.

If this is true, what does it suggest about the possibility for biochemistry at hypothetical geothermal vents on the floor of the buried oceans of Europa, or geothermal vents that melt ice deposits under the haze of Titan?

Whatever the processes, microscopic cellular life arose on Earth, probably between about 4,000 and 3,500 My ago, because fossil remains have been discovered in rocks after that period. In one location in western Greenland, 3,700-My-old rocks show evidence of biological carbon isotope chemistry (Nisbet, 1980). In a sequence from another site in Greenland, 3,700-My-old rocks do not show traces of ancient life, but rocks younger than about 3,100 My do. Also, 3,400-My-old rocks found in South Africa in 1977 contained fossil methane-producing bacteria (Gould, 1978).

Among the best evidence of early life are fossil remains of **stromatolites,** colonies of blue-green algae with a cabbagelike structure. Stromatolites, being soft material, do not leave fossils as clear as those of the hard-bodied creatures that evolved much later, such as shellfish. Still, 2,700- to 3,500-My-old stromatolites are known from Canada, Zimbabwe, and Australia (Walter, Buick, and Dunlop, 1980).

An interesting factor is that the intense early meteorite bombardment, prior to about 4,200 My ago, may have delayed biological evolution until things settled down. Anyway, life-forms such as bacteria and blue-green algae apparently were evolving rapidly by 3,500 My ago. Life arose "as soon as it could; perhaps it was as inevitable as quartz or feldspar" (Gould, 1978).

Earth was not a passive backdrop. Craters formed and filled. Ocean and land configurations changed. The atmosphere evolved chemically. Oxygen content in the initial atmosphere was small, probably only a few percentage points. Oxygen is chemically very reactive and would disappear rapidly in reactions that oxidize rock minerals unless it was replaced by some other process. That process is photosynthesis in plants. This process may have begun in the blue-green algae, but it continued as plants evolved. The oxygen content began to rise dramatically, probably around 2,500 My ago. Empirical evidence for this is that rocks formed and buried earlier than that are suboxidized, or formed under more oxygen-poor conditions than exist today (Goldsmith and Owen, 1979). Oxidized "red beds," where iron minerals have been oxidized to red colors, are rare before 2,000 My ago and common afterward (Walker, 1977).

As soon as some oxygen became available, solar ultraviolet light broke some O_2 molecules, and the free pair of O's joined with other O_2 molecules to make **ozone** (O_3). This led to production of the **ozone layer** high in Earth's atmosphere, which absorbs nearly all the incoming ultraviolet. Formation of this layer was an important step, because too much ultraviolet breaks up complex molecules and thus could prevent life—the apparent fate of surface biochemistry on Mars. Thus, paradoxically, the early presence of solar ultraviolet may have helped life start by providing energy to break up old molecules and form new ones, but its absence later allowed life to evolve safely on the surface.

Living things could hardly be unaffected by all these changes. Whereas the earliest life-forms developed and existed without oxygen, life now had to adjust to the dangerous reactive substance, oxygen. Such environmental changes favored evolution: When an accidental change in the DNA—a mutation—happened to produce traits well adapted to the new environment, the organism was likely to live long enough to have more offspring, promoting retention of the new trait and thus evolution of new species. This process is called **natural selection.**

It is strange to realize that prior to 2,500 My ago, Earth would have seemed to be an alien planet. The early atmosphere was rich in CO_2, like the atmospheres of Mars and Venus. The land was still barren. Lack of vegetation produced extreme erosion. Some areas must have looked like today's deserts or like Mars. Some areas were moist and washed by rains, but instead of luxurious forests there were only eroded gullies, bare dirt, and grand canyons. Brown vistas stretched to the sea.

Most life was still in the oceans, with soft bodies; it rarely produced fossils. Stromatolites increased in abundance 2,500 to 2,000 My ago, living at the boundary of water and rock along seacoasts. Today they are found only in rare, oxygen-poor salt marshes on some seacoasts. They produce oxygen by photosynthesis, and their abundance around 2,000 My ago abetted the upswing in oxygen (Walker, 1977; Goldsmith and Owen, 1979).

One sign of the adaptability of life, once it evolved, is the rapid proliferation of advanced species, as indicated in the geological time scale shown in Table 9-5. Half the available time was used to go from complex molecules to algae and bacteria, but it took only the last 12% of Earth history to go from the first hard-bodied sea creatures to humans (Figure 12-3). Biologists have attributed evolution in general and this rapid proliferation in particular to natural selection and competition of species, slow processes working entirely *within* the biosphere. However, as we see in a moment, it may also involve sudden changes induced from *outside* the biosphere. Details of natural selection are still being studied. Prototypes of new species may have evolved in obscure ecological niches and then emerged rapidly when sudden environmental changes (such as ice ages) caused extinction of earlier species better suited for earlier conditions.

Whatever the specific mechanism, we can say from the fossil record that Earth experienced evolution from nonliving organic chemicals to small organisms in a few billion years, and that these organisms evolved in less than a billion years to species with enough intelligence to begin fooling around with tools.

What Mars Tells Us

Because Mars is the most Earth-like planet, and one with evidence of past running water (see also next chapter), it has long been viewed as the planet that might first reveal additional secrets about the evolution or existence of life beyond Earth.

The Viking Landers' Experiments

Related discoveries have been made by Viking landers on Mars, which tested soil samples scooped from the surface (Figure 12-4). Some scientists expected the cameras on the two Viking landers to show primitive life-forms, such as lichens, on the surface. They were surprised when the landers revealed not only a barren surface, but also no organic molecules in the soil, to a level of a few parts per billion! The soil is very sterile. In retrospect, it seems clear that the surface soils of Mars have been sterilized by being exposed to solar ultraviolet rays. This radiation bathes the planet's surface because Mars lacks Earth's protective

Figure 12-3. We tend to think of trilobites as extremely ancient life because they date from near the beginning of the reliable record of easily recognized fossils. But in fact, they date from only the last 13% of Earth's history, and mark only the last part of the development of life. (Photo by Gayle Hartmann and Joyce Rehm)

a   b

Figure 12-4. When two Viking probes landed on Mars in 1976, their sampler arms scooped up soil for compositional and biological tests inside the spacecraft. Here, the arm pushes aside a frothy-textured rock (vesicular basalt?) to gather soil from under it. Protected soil in such locations would have less irradiation by solar ultraviolet light, expected to harm Martian organics. The soil was sterile. (NASA, Viking 2 image, October 9, 1976)

ozone layer. Recall from earlier chapters that solar wind particles and photons, especially UV photons, have enough energy to stimulate reactions among molecules on icy satellites. The breakup of CH_4 and related compounds may actually create molecular fragments with the C-H bond and stimulate production of organics. Yet at the same time, solar UV photons have enough energy to break apart organic molecules, whether on Mars or in our own skin. Mars's present surface has no sources for new C-H bonds, and the Martian dust, frequently moved by the wind, has probably been sufficiently irradiated to destroy organic molecules that might have formed in the past.

The Viking landers had four tests for life on Mars: tests for organic molecules in the soil, and three experiments designed to look for chemical signs of metabolic activity in it—for example, by adding nutrients and looking for chemical byproducts of life (Figure 12-4). The first experiment revealed sterility at both landing sites, but, paradoxically the other experiments gave some positive results! Most scientists now interpret those results to unfamiliar reactive chemical states in Martian minerals, produced by the UV radiation. Nutrient addition led to the production of organic molecules and the release of CO_2 gas, but this result is now attributed to the reactive states produced in the soil by ultraviolet radiation. Indeed, laboratory experiments in 1977 duplicated many Viking results using simulated Martian soil (from oxide minerals exposed to ultraviolet light and CO_2). The main significance of these results may be to show ways in which chemical reactions on a lifeless planet can synthesize organic molecules by utilizing atmospheric CO_2 under certain conditions (Horowitz, 1977).

The Viking instruments were tested in the barren dry valleys of Antarctica—a relatively Mars-like environment—and they confirmed that the soil there was sterile. All the more surprise, then, when Florida State University biologist E. I. Friedmann reported in 1978 that microorganisms such as bacteria, algae, and fungi thrive in cracks and pore spaces *inside* rocks in the dry valleys of Antarctica! Though the rock may be in a frozen area, sunlight heats the rock itself and creates a microscopic niche suitable for life and protected from the harsher external conditions. Life, seemingly inexhaustibly accommodating, fills the niche. In addition, terrestrial microbes have also been found living inside fractures in the massive Columbia River basalt lava flows, and at depths of kilometers under the ground. The mass of microbial life in the upper crust is probably greater than the mass of living organisms on Earth's surface. Such findings led to suggestions that the Viking landers had not ruled out the possibilities of life in other Martian environments, such as underground.

Fossil Microbes in Martian Rocks?

As discussed in Chapter 6, the handful of so-called SNC meteorites are rocks knocked off Mars. A scientific sensa-

tion occurred in 1996 when NASA researcher David McKay and eight colleagues (1996) announced that they had found evidence of possible fossil microbes in fractures inside one of these rocks. Because the discovery of alien life is one of the holy grails of science, this announcement was closely scrutinized by various teams and an unfortunately bitter controversy soon erupted. Facts that are not in dispute include these: (1) The Martian rock is very old, having formed about 4,500 My ago; (2) its fractures contain globules of carbonate (more than in other, younger Mars rocks) indicating that liquid water percolated through the rock in the past; (3) organic molecules called polycyclic aromatic hydrocarbons (PAHs) were found inside the rock on fracture surfaces, concentrated in and/or around the carbonate globules; (4) other organic molecules, including amino acids, are present; (5) peculiar microbe-like structures were photographed. McKay and his team used electron microscope images to reveal the latter small structures, 20 to about 200 hundred nm in size. They look vaguely organic, and some look like certain terrestrial microbes (Figure 12-5), including some found inside certain terrestrial basalts. These structures were not concentrated toward the rock's surface, as might be expected from terrestrial contamination. McKay's team emphasized, even in their initial paper, that "none of these observations in itself is conclusive for the existence of past life," but that the ensemble of traits suggested the structures might be fossil microbes that had existed in the ancient past when water ran on Mars. Other teams objected, concluding that the rock had been contaminated on Earth, or that the structures were common mineral forms which, seen edge on, looked like microbes. An interesting argument was raised that while the larger structures are in the size range of terrestrial microbes, the smallest structures, about 20 nm across, could not be microbes because the minimum 5 nm thickness for each cell wall leaves no more than 10 nm diameter for the cell interior, which would be too small to carry enough DNA molecules to transmit genetic information! A consequence of this suggestion, which has not been adequately pursued, is that these structures might represent not remnants of living organisms, but rather remnants of early organized structures of nonliving material on the road toward life.

In some forums, the attacks on the reality of "Martian microbes" bordered on criticism of scientific conduct; because the issue of alien life is a major issue, some groups felt that extraordinary proof should be required before any publication was made. In my own view, the early results of McKay and his colleagues were interesting enough that the scientific community should have been alerted and publication was proper—but like many issues in exobiology, that gets into the issue of what is the "minimum publishable unit" of work to make a scientific paper. At this writing, neither view is established, but the original paper, like the best scientific papers, has engendered a flood of new work and new understandings. Among recent studies,

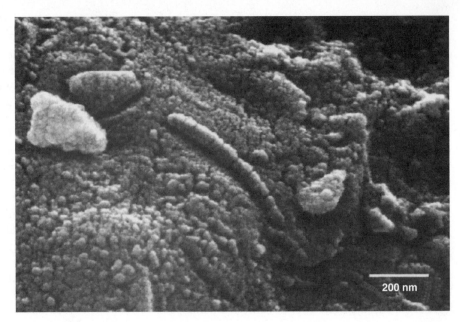

Figure 12-5. Example of the microbe-like structures found in carbonate globules inside cracks in the 4,500-My-old Martian meteorite, Allan Hills 84001. Controversy has swirled around whether these are fossil Martian organisms. (NASA, courtesy Everett Gibson)

Bada and others (1998) concluded that most amino acids they detected had racemic (handedness) properties matching terrestrial contaminants, though a small amount of amino acids "native" to the rock could not be ruled out. Similarly, Jull and others (1998) concluded that although most of the carbon had isotope ratios consistent with terrestrial contamination, some of the carbon in the carbonate globules is extraterrestrial. The problem is pinning down the compositions of the organics and the properties of the "microfossil" structures themselves, inside the carbonate globules. Meanwhile, research continues on in-rock terrestrial microbes, which have been found as far as 2.8 km below the surface, at temperatures as high as 75° C (167° F) (Fredrickson and Onstott, 1996).

Planets Outside the Solar System

Most scientists believe that if planetary surfaces exist with the necessary conditions—liquid water and C-H-O-N chemicals—life is likely to evolve on them. And advanced species will probably appear eventually. But are there any such planetary surfaces elsewhere in the solar system or the universe?

Planets have indeed been found orbiting other stars, but the orbital properties found so far are mostly unlike our own, habitable system (see Chapters 4 and 5). An informed guess would be that between 1% and 30% of all stars have planetary companions, but techniques so far are inadequate to detect Earth-sized planets in systems like ours.

However, even if planets exist near some other stars, there is no guarantee that they are habitable. Astronomers have proposed several conditions for a habitable planet:

1. The central star should not be more than about 1.5 M_{sun} so that it will last long enough for substantial life to evolve (at least 2 Gy), and so that it will not kill evolving life with too much ultraviolet radiation.

2. The central star should be at least 0.3 M_{sun} to be warm enough to create a reasonably large orbital zone in which a planet could retain liquid water.

3. The planet must orbit at the right distance from the star, so that liquid water will neither evaporate nor permanently freeze.

4. The planet's orbit must be almost circular to keep it at the proper distance from the star and prevent overly drastic seasonal changes.

5. The planet's gravity must be strong enough to hold a substantial atmosphere.

In work ahead of its time, Dole (1964) and Huang (1965) reviewed these criteria and sketched out conditions for habitable planets in stable orbits (Figure 12-6). Dole concluded that roughly 6% of the stars, mostly from 0.9 to 1.0 M_{sun}, have habitable planets. However, the figure remains highly uncertain.

Has Life Evolved Elsewhere?

If habitable planets exist and if life evolves readily under habitable conditions, is life abundant throughout the universe? There are additional factors to consider. For one thing, planetary environments change with time, so that today's habitable planet may not be habitable tomorrow. Can life survive such changes?

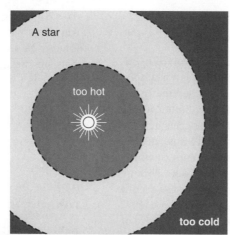

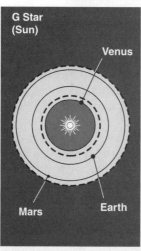

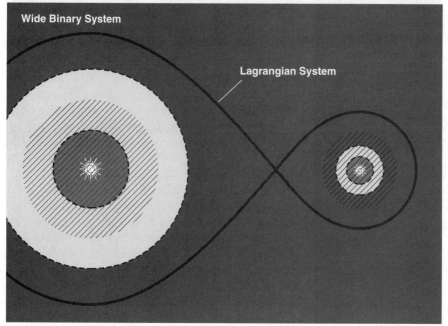

Figure 12-6. Doughnut-shaped light areas show the relative sizes of zones where water would remain liquid in planetary systems around stars of various spectral types. In a binary system (bottom), the situation is complicated because orbits outside the hatched regions experience large perturbations, causing evolution away from circular orbits, in turn causing temperature extremes.

Effects of Planetary and Astronomical Processes on Biological Evolution

Basic planetary or stellar processes may be involved in encouraging or hindering biological evolution. For example, Earth's fossil records indicate an episode called the "great dying" at the end of the Permian Period about 250 My ago. As many as 96% of all land and sea species became extinct in a few My or less (Raup, 1979). This is only the biggest of the so-called **mass extinction** events, in which large percentages of existing species and genera became extinct—usually marking the breaks in the fossil record by which geological periods, such as the Cretaceous and Tertiary, are defined. Raup and Sepkoski's (1986) study of 11,800 genera of marine animals during the last 270 My shows these statistics: 61% of existing genera disappeared during the Permian great dying 250 My ago; 49% of existing genera, including dinosaurs, became extinct at the end of the Cretaceous Period 65 My ago; 43% died out at the end of the Triassic Period 210 My ago; 30% disappeared at the end of the Jurassic Period about 140 My ago; and smaller extinctions grading down into the noise level of evolution. How can sudden mass extinctions be explained? Many possible effects have been suggested:

1. Interior convection could have caused plate tectonic crustal splitting, hence changing sea levels, ocean currents, wind patterns, and seasonal extremes (for example, Gartner and McGuirk, 1980). This process has been responsible for isolating land masses such as Australia and allowing different species to evolve there.

2. Volcanic eruptions may have spewed enough dust into the high atmosphere to reduce the sunlight reaching the surface. For example, widespread sunlight decreases of up to 25% occurred after the 1883 Krakatoa (Sumatra), 1912 Katmai (Alaska), and 1982 El Chichon (Mexico) eruptions. Larger, rarer eruptions might have cut sunlight for some years, causing the decline of some species and the ascendancy of new ones.

3. A more profound volcano tectonic episode might accompany the establishing of a mantle plume in Earth's interior. Certain stages of plume evolution might be marked by millions of years of intense volcanism, perhaps related to the hypothetical resurfacing events on Venus around 600 My ago.

4. Slight changes in the tip of the planetary axis to the plane of the ecliptic, caused by gravitational forces, are believed to have caused major climatic changes, such as the ice ages (for example, Imbrie and Imbrie, 1980).

5. Slight changes in the sun's radiation may have changed climates. Even in the course of our own civilization there has been some evidence for this. In 1645–1715, for example, records show that sunspot numbers were probably unusually low, the solar corona was observed to be pale, and solar flare activity was probably low (Eddy, 1976). Tree-ring data suggest possible climate variation during this period. Changes in *total* solar radiation are known to be tiny, but sunspot and solar flare activity has effects on the upper atmosphere that may propagate to lower levels.

6. Irradiation from a sufficiently nearby supernova could have affected the climate or directly damaged organisms.

Impacts as a Driver of Biological Evolution on Earth

A radical breakthrough on the whole question of mass extinctions came in the 1980s. The second-biggest extinction, defining the end of the Cretaceous and beginning of the Tertiary Period 65 My ago, was virtually proven to be the result of an asteroid or comet impact. This event is known as the K-T extinction event, a name derived from geologic abbreviations for the Cretaceous and Tertiary. But it was more profound than that name indicates: It ended the entire Mesozoic Era and initiated the Cenozoic Era, in which we live. Alvarez and others (1980) first presented evidence for the asteroid impact idea after measuring excesses in the iridium (Ir) content of 65-My-old limestones; these iridium contents turned out to be 20 to 160 times that of limestones of other ages. Because Ir is enriched in meteorites relative to Earth, they suggested, as an explanation, that a 6- to 14-km-diameter asteroid struck Earth and ejected massive amounts of meteoroid-enriched dust into the atmosphere. This hypothesis has had a tremendous impact on the paleontological community and caused a burst of new research. Within months, other investigators confirmed the Ir excess in 65-My-old sediments from many parts of the world. Additional evidence in the 65-My-old boundary layer includes shocked quartz grains produced only in impacts; spherules of once-melted glassy material produced in impacts; tsunami deposits centered around the Gulf of Mexico and the Caribbean; soot in an amount that would require burning of as much as twice the total forest mass on Earth today.

The effects of the impact are hard to contemplate. We know that some 75% of species that had evolved (notably dinosaurs) became extinct within a few million years, and new species (notably mammals) became dominant (Russell, 1982). However, detailed studies of the fossil record show that this did not happen all at once, and various species were affected differently. No land animals larger than 25 kg survived, but freshwater animals and plants were much less affected (Russell, 1982).

At first, paleontologists were loath to accept a catastrophic impact. However, the clincher came when researcher Alan Hildebrand and others identified a huge, 65 My-old impact crater, about 180 km across, buried under sediments straddling the Yucatan coast (Hildebrand and Boynton, 1991).

Researchers have studied impact processes to explain the observations. First, the dust thrown into the stratosphere absorbed sunlight and is calculated to have caused 3 to 6 mo of daytime darkness, consistent with a reported extinction of some 49% of genera of light-dependent floating marine organisms, but only 20% of (less light-dependent) land genera (Kerr, 1981; Ahrens and O'Keefe, 1982). Asteroid or comet impact may also damage the ozone (O_3) layer; Turco and others (1981) found that the Tunguska explosion of 1908 generated as much as 30 million T of nitrogen oxide (NO) in the stratosphere, reacting to deplete up to 45% of the O_3 in the northern hemisphere, consistent with measures of 1909–1911. If the impact occurred at least partly in the ocean, giant tsunamis would have devastated coastal areas. Still more gruesome, cratering expert Jay Melosh calculated that debris falling back into the atmosphere would create a worldwide radiant heat pulse from the sky, broiling exposed life-forms and possibly explaining the forest fires. Creatures in burrows, underwater, or in broad regions of rainstorms and snowstorms might survive this heat, consistent with complexities in the extinction patterns.

Interestingly, the next largest extinction event, 210 My ago, probably coincides with the creation of the 100-km Manicougan crater in Canada. Craters smaller than, say, 50 km may cause local effects that are lost in the noise of biological evolution. Craters larger than the 100–200 diameter range, big enough to lead to mass extinctions, can be expected to form randomly in 70 to 250 My intervals (see Figure 6-5; see also Gehrels, 1994; and Hartmann and Miller, 1991).

Strangely, the largest extinction event, the "great dying" at the end of the Permian Period, 250 My ago, shows no evidence of an impact, and its cause remains unknown. Perhaps it involved one of the effects listed above, or an impact whose evidence has been destroyed by plate tectonic activity.

If Earth (or any other planet) suffers catastrophic impacts that wipe out species every few hundred My, this is a radical change in classic Darwinism, which pictures evolution as spurred only by competition among species within environments that change slowly due to terrestrial forces. Disasters too large could hinder evolution of life itself; but impact-disasters in the right size range would wipe out many "well-adjusted" species, reducing competition and creating huge environmental niches to be filled by new species. Such events would prune the tree of life to allow new growth.

Evolutionary Processes on Other Worlds?

One of our problems is that we just don't know whether Earth—our statistic of one—has experienced rare special effects that have promoted life and intelligence. The mere presence of our unusually large moon is such an effect: Its tides helped produce tidal pools, where organic molecules could concentrate, and its tidal forces have helped hold Earth's axis in a fixed position, preventing axis excursions (such as probably happened on Mars) that could cause drastic climatological oscillations (Goldsmith and Owen, 1979).

Perhaps it is no coincidence, then, that the time scale of biological change, at least in the last quarter of Earth's history, has been comparable to the time scale of geological change. Life's response to changing environments is to change itself, on time scales as short as a million years. Old species die and new species prosper under the new conditions. We must not underestimate the possibility, however, that environmental changes on still shorter time scales prevented life from evolving on many or most planets.

Adaptability and Diversity of Life

The great variety of ancient and modern species on Earth and the variety of environments in which they have thrived suggest that, given time, life could also evolve to fit a wide range of conditions on other planets. Even humans have a remarkable adaptability. We thrive from equatorial wet jungles to dry deserts to arctic plains to Andean summits, where air pressure is barely half that at sea level. During the last ice age, we survived by migrating.

But the limits for survival of simpler life forms are even wider. Microflora are known in super-cooled Antarctic ponds that remain liquid at 228 K ($-49°$ F) because of dissolved calcium salts. Extremophile bacteria in Yellowstone Hot Springs prosper at a temperature of 363 K ($194°$ F), and exist in deep sea vents at 383 K ($230°$ F) (Fredrickson and Onstott, 1996). This is a 68% range of variation in temperature. In laboratory experiments, common bacteria have survived in liquid cultures at least 24 h in CO_2 atmospheres at 433 K ($320°$ F), extending the range to a 90% variation in temperature.

Habitable pressure regimes show an even greater range. Bacteria exist at altitudes where the atmospheric pressure is only 0.2 bar, advanced organisms live in ocean depths with pressures of hundreds of bars, and microbes several kilometers down in the crust experience even higher pressures. The pressure range exceeds a factor of a thousand.

Experiments with terrestrial life-forms in nonterrestrial environments have shown grass seeds germinating in atmospheres of C, H, O, and N compounds, insects with normal behavior at pressures as low as 100 to 160 mbar, insects surviving brief exposure to Martian surface pressures, and bacteria surviving in CO_2 at conditions between those of Earth and Venus (Siegel, 1970).

Conversely, alien organisms might survive terrestrial conditions and have devastating effects on Earth. There are historical examples of similar events. The plague caused by Asian bacteria introduced into Europe in the 1300s killed about a quarter of all Europeans, and as many as three-quarters of the inhabitants in some areas. Diseases introduced into Hawaii after the first European contact in

1778 killed about half of all Hawaiians within 50 years. Some 95% of the natives of Guam were wiped out by disease within a century of continued European contact (Underwood, 1975). For these reasons, early Apollo astronauts were quarantined until it was clear that they carried no lunar organisms. And future samples from Mars may be analyzed only in space labs, well above Earth's atmosphere.

Recently, public awareness has increased about the sudden appearance of deadly viruses that could, in principle, spread rapidly through human populations, disrupting civilization (Preston, 1994). Viruses are large molecules that, when injected into specific types of cells, have the ability to use the cell to reproduce the virus itself. Some of these appear to be bred in the humanlike primates of Africa, and then make the jump to humans. In 1979, astrophysicist Fred Hoyle and biochemist N. C. Wickramasinghe presented data that viral outbreaks have appeared historically in random spots and spread; they hypothesized that viral molecules, synthesized in space, might arrive in local regions in cometary or meteoritic impacts or atmospheric breakup events. This proposal generated little acceptance among biologists but still has intriguing aspects.

The whole subject is frightening! Change and evolution in life populations are caused not only by life's adaptations to new environments but also by the invasion and destruction of some populations by others. With these facts as well as cultural competition in mind, anthropologist D. K. Stern (1975) has remarked, "It is likely that the meeting of two alien civilizations will lead to the subordination of one by the other," a somber view at odds with popular concepts of benign aliens organized in a cheerfully democratic galactic federation.

Appearance of Alien Life

If alien life did evolve, could we expect it to evolve through any stages similar to our own? Opinion is divided on this point. MIT physicist Philip Morrison (1973) emphasized the **convergence** effect, whereby species with similar capabilities in similar habitats evolve to look alike. For example, three different species of large sea creatures are "designed" for fast swimming in the ocean and look alike: the extinct reptile Ichthyosaur; the shark, which is a fish; and the mammal that returned to the sea, the dolphin. Similarly, Morrison speculated, aliens living on planetary surfaces in gaseous atmospheres and using "intelligence" to manipulate their surroundings with tools might well have bilateral symmetry, appendages used as hands, a pair of eyes designed (like ours but unlike those of most animals) for stereo vision, and so on.

On the other hand, the paleontologist, George Gaylord Simpson (1964, 1973) argued that although life is likely to start, the long chain of environmental changes and evolutionary steps that produced humans is unlikely even to be approximated elsewhere, so that there is likely to be a "nonprevalence of humanoids." Simpson labeled

the whole attempt to estimate probabilities of alien life as nearly meaningless because of our lack of experiments or examples.

Physicist W. G. Pollard (1979) countered that we do have examples. He notes that about 180 My ago, Australia broke off Gondwanaland and can be thought of metaphorically as an Earth-like planet, "A," where evolution continued independently from a primarily reptilian stock. Similarly, South America broke off Africa 130 My ago and can be viewed as an independent planet "S." Independent evolution also continued on planet "E" (Earth, made up of Africa and adjoining land). In the same period of 130 to 180 My ago, independent evolution diverged, rather than converging on humans. Humans appeared only on E, primates developed on S, and marsupials on A. Parker (1977) argues that humans were unlikely to appear by continued evolution on A. Humanlike creatures, appearing on E only about 4 My ago, have existed for only 0.1% of Earth's history.

In summary, natural selection seems to produce species capable of occupying any habitable environment. Thus, we should not be surprised if life has evolved on another planet. But this life may look very strange to us, even if it displays recognizable intelligence. After all, if mushrooms and corals and woolly mammoths and Venus flytraps all evolved on one planet, how much greater may be the differences between life-forms on two different planets? Feathers and fur and sex and seeds and symphonies may be products of Earth only.

Effects of Technological Evolution on Biological Evolution

If life will evolve when the right conditions exist, and if those conditions probably do exist elsewhere, then what can we predict about life elsewhere? Should we predict intelligence and civilization? What do these terms mean? Should we assume that other civilizations will achieve space flight or interstellar communication? Should we assume they might visit us some day? This raises the question of technology and its role in the evolution of life. With our single inhabited world as an example, our answers to these questions are uncertain. Most writers on the subject assume that "intelligence" means, among other things, use of tools to modify the environment; hence, technology. Intelligence leads to technology, which leads to civilization.

But does technology also end civilization? As we have seen, while limited environmental change can be helpful, environmental change that is too much or too fast can be fatal! As Pulitzer Prize-winning naturalist René Dubos points out, we are umbilical to Earth, and if we alter our planet too much before acquiring an ability to leave it, we are finished. For this reason, consideration of our own case leads to the conclusion that the development of technology may actually mark the end of civilization on some worlds. There are numerous examples related to this point.

First, it is hardly wild speculation, as we see numbers of nominally moral, intelligent technologists in various regions of our own world spending entire careers devising weapons solely to deal death to our own species. In the past, war did not threaten our whole species, because conflicts involved only a small percentage of the world. But today, nuclear, biological, and other types of weapons could involve the whole world. For example, radioactive strontium 90 produced by nuclear explosions was blasted freely into the atmosphere before the nuclear test-ban treaty of 1963. It has a half-life of 28 y. A year after the first H-bomb tests by the United States in the Pacific, strontium 90 deposits in American soil increased soil radioactivity by 0.5%. A few weeks after a 1976 Chinese nuclear test, measurable airborne radioactive debris fell onto the United States. Clearly, a sufficiently massive nuclear exchange could devastate not only local target regions, but also broader civilization and future life, whose genetic pool would be exposed to high radiation levels for decades. The struggle faced by all intelligent cultures is whether viable political/social mechanisms to settle disputes can be developed before technology's offspring—weapons of mass destruction—come into play.

Second, we have also proved that disasters could happen by mistake. As our technology assumes a planetary scale, so do our accidents. Problems as diverse as nuclear power and aerosol spray cans illustrate the issue. Earth has limited fossil fuel reserves, yet planned nuclear power plants will produce radioactive plutonium wastes, among the most toxic of known materials. Although safe when sealed, some kilograms of plutonium dust accidentally spilled into the air could devastate life in whole states. Yet many kilograms are already being processed, and several governments promote greater dependence on nuclear power plants until other energy forms become available.

Third, technical accidents are not the only danger: We have seen our society spawn deliberate terrorists and madmen. With grams of stolen plutonium 238, suitcase-scale bombs, and/or biological weapons, such characters could threaten whole cities.

Fourth, ordinary consumer technology offers another example. In 1974, several scientists realized from theoretical calculations that cholorflourocarbons (CFCs such as the commercial product "freon"), used in aerosol cans and air conditioning systems, damage the ozone layer. This recognition came in part from the study by geochemists and planetary astronomers of chemical processes in the atmospheres of Venus and Mars. After a 1975 report by the National Academy of Sciences, confirming the ozone danger, moves were made in 1976 and 1977 by American regulatory agencies to phase out CFC production, including phase-out of CFC aerosol propellants by 1979. The predicted damage to the ozone layer from CFC chemicals was first detected in winter seasons over Antarctica, where a natural winter "hole" in the ozone layer broadened dramatically during the 1980s and early 1990s. Satellite data in the early 1990s showed that the winter breakdown of ozone had begun at high *northern* latitudes as well. The record winter ozone depletion over Antarctica occurred in 1993 (when volcanic acidic particles from the 1991 eruption of a Philippine volcano added to the effect); the 1995–1997 depletions seem to have stabilized at a slightly smaller effect. Most scientists attribute the widening of the ozone hole to CFCs, and its stabilization to the phase-out of CFCs. If the phaseout continues, experts in this field predict a decline in ozone damage by about 2010 (Monastersky, 1997). The ozone issue seems to be a case whereby scientists and international cooperation have dealt with an environmental problem in the nick of time. But we should be haunted by the words of Harvard planetary physicist Michael McElroy, who performed some of the earliest Freon calculations: "What the hell else has slipped by?" To put it another way, the CFC problem could be viewed as an experimental proof that a species could fatally flaw its own planet by explosive emergence of intelligence and technology that damages the planet before being recognized.

A fifth example of rapid human-caused change is the increase in carbon dioxide (CO_2) in our atmosphere due to fossil fuel burning, forest burning, and industry. As shown in Figure 5-12, a CO_2 increase of more than 20% has been clearly observed in data taken since the industrial revolution. Here, again, is an example of a planetary civilization altering its own world on a time scale of centuries—which could be too fast for a culture to alter events and save itself.

In summary, if humanity is any example, long-term survival of a planetary culture is not assured. Although human culture has been around less than a tenth of a percent of the history of our planet, we are already beginning to have brushes with global disaster, such as nuclear war, terrorism, ozone layer damage, and global climate change. In fact, we are already living in an ongoing mass extinction, caused by human activity (Myers, 1997). Thus, we can speculate that if evolution produces intelligent societies that remain tied to one planet, many of them may last only a fraction of a percent of the age of the universe—in which case there is little chance that a given culture will be around at the same time we are.

What Ends Civilization?

The question of whether alien cultures (or our own) could produce interstellar communication is closely tied to the question of how long civilizations last, on a cosmic time scale. The paragraphs above listed five examples of the proposition that development of technology is so explosive that intelligent species are likely to damage their own environment irreparably before they realize the consequences of their own activity.

But natural and social forces also disrupt cultures. Historians have argued for centuries about the causes of decline in major ancient societies, such as Rome or the Mayans. Some may fizzle from internal social exhaustion

Criterion	Fraction	
	Lower limit (?)	Upper limit (?)
1. Stars having planets	10^{-2}	0.3
2. Criterion 1: stars ever having habitable conditions on at least one planet	10^{-1}	0.7
3. Criterion 2: planets on which habitable conditions last long enough for life to evolve	10^{-1}	1
4. Criterion 3: planets on which life evolves	10^{-1}	1
5. Criterion 4: planets on which habitable conditions last long enough for intelligence to evolve	10^{-3}	0.9
6. Criterion 5: planets on which intelligence evolves	10^{-1}	1
7. Criterion 6: planets on which intelligent life endures	10^{-7}	10^{-1}
8. Fraction of duration of intelligent life during which it retains an interest in contact with Earth-like civilization	10^{-3}	1
Fraction of all stars with planets that bear intelligent life	10^{-19}	2×10^{-2}
Implication: Distance to nearest civilization	3×10^{8} light years	15 light years

or decay. Growing evidence suggests natural climate change can be a factor. The collapse of "high cultures" in the American Southwest and Midwest around 1100–1300 coincided with major climate changes; by 1400, beautifully painted pottery had been replaced by plainware and large permanent architecture in many areas gave way to brush huts. Similarly, 1997 drill-core discoveries of extreme dust deposition in the Gulf of Oman reveal that the collapse of the Akkadian Empire from Iraq to Turkey coincides with a 3-century extreme drought from 2200 to 1900 B.C. We tend to believe in continual progress and the idea that a modern democratic society, once achieved, is a self-sustaining accomplishment, but history shows that civilization is by no means a permanent condition.

As an important consideration in all this, if our own global civilization fell into new Dark Ages now, it would be very difficult for our descendants to climb back out of the pit. Re-emergence of a new technical infrastructure, with water delivery, manufactured products, and so on, would be hampered by the fact that *we* have already consumed most of the easily accessible coal, oil, ores, and other fuels and resources needed for a second industrial revolution. There is a defensible proposition that a planetary species gets only one shot at spreading its own first intelligent species among the planets and stars.

Thus, we have identified a **cultural hurdle** that we (and perhaps intelligent species on any planet) must surmount: the transition from scattered, competing nation-states with the capability to damage the planet to a some sort of stable planetary or interplanetary technological society of intelligence and imagination. "Planetary" may not mean global government, but it does mean global communication—the global village—and a way of monitoring for possible danger signs. Perhaps we and some other cultures will cross the hurdle to this sort of enduring culture. Perhaps some cultures have completed the further transition from being planetary cultures to being interplanetary, thus ensuring their survival against ecological disaster on any one planet. Such cultures might last and be detectable for billions of years instead of thousands.

Our conclusion so far is that a fraction of the stars have planets, a fraction of the planets ought to be inhabited, and a fraction of the inhabited planets ought to have either civilizations or relics of destroyed civilizations. The next question is how to assess the prospects that extraterrestrial carbon-based life as we know it exists today.

Alien Life in the Solar System?

According to our ground rules, a habitable planet has (or has had) liquid water and an environment in which complex carbon-based organic chemistry can proceed—and these conditions must have lasted long enough for a specified degree of complexity to have evolved. If the specified degree is to be technology-using intelligence, then we can probably rule out everybody in the solar system but Earth. But if we specify only microbial life, we cannot rule out so many bodies. As already noted, meteorites indicate that amino acids have been synthesized in carbonaceous asteroids (and elsewhere?) beyond Earth, and some researchers have already proposed a case for ancient microbial life on Mars. As for present-day Mars, most researchers agree that the surface dust layers are sterile. However, Levin and Stratt (1981) asserted that a Viking-class instrument failed to detect organic material in a certain Antarctic soil sample in which another Viking-class instrument did detect microorganisms! They argued that Viking may even have detected primitive life-forms when its experiments showed production of organic molecules from nutrients. Most analysts doubt this view, but a number have suggested that although the surface of Mars is probably sterile, liquid water and microbial life could possibly be hidden at the base of the Martian permafrost layer. Mars thus remains an active issue for exobiologists.

In addition, the Galileo spacecraft during the 1990s has strengthened the case that Jupiter's moon Europa may have a liquid water ocean beneath its crust of young ice. The surface conditions on Saturn's moon, Titan, are still uncertain, but organic molecules appear abundant, and local geothermal heat sources may be present.

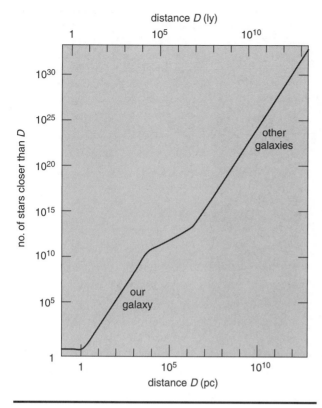

distance *D* (ly)

no. of stars closer than *D*

our galaxy

other galaxies

distance *D* (pc)

Figure 12-7. Distance in parsecs (bottom) and light-years (top) required to encounter the number of stars indicated at left. If one star in 10^9 has life-bearing planets, the nearest one might be within a few thousand parsecs.

Alien Life among the Stars?

The next step is to consider alien life on planets near other stars. There is no direct evidence, but various scientists have publicized a method for considering the possibilities (Sagan, 1973). The logic is to try to estimate a sequence of various fractions: the fraction of stars having planets, of those planets that are habitable, of those habitable planets where conditions remain favorable long enough for life to evolve, of those planets where life does evolve, of those where intelligence evolves, and of the planet's life during which intelligence lasts. The product of those fractions would be the fraction of stars harboring intelligent civilizations. The formalism is sometimes called the **Drake equation,** after American astronomer Frank Drake, who developed the scheme.

Table 12-1 shows optimistic (upper limit) and pessimistic (lower limit) estimates of the various fractions, and the consequent estimates of the upper and lower limits on the fraction of stars that might have civilizations today. The optimistic figure is a few percent; the pessimistic case is only one star in 10^{19}. Note the 17 order of magnitude uncertainty! Given these figures, how far away would the civilization be? Figure 12-7 shows the answer by plotting the radial distance required to include a given number of stars.

In the first case, the nearest civilization might be only 15 light years (LY) away—amazingly close. At the speed of light, that civilization might be reached in only one generation. In the pessimistic case, the closest civilization would probably not be in our own galaxy, but roughly 300 million LY away in a distant galaxy. Nonetheless, given the innumerable galaxies, it is hard to avoid the conclusion—even with the most pessimistic view—that millions or billions of technological civilizations exist outside the solar system. If our reasoning is right, then at this moment intelligent creatures may be pursuing their own ends in unknown places under unknown suns.

Where Are They?

If the universe is full of other civilizations, why haven't we seen them or heard from them?

One answer might be that we *are* being visited—witness the flying saucer reports. But although some UFO reports are intriguing, verifiable evidence for alien spaceships is abysmally poor. The largest scientific study of the UFO phenomenon (Condon and others, 1969), on which I was a field investigator, showed, in my opinion, no evidence that any UFOs are extraterrestrial. Some interesting phenomena, possibly atmospheric, were recorded, but many other seemingly sensational cases, especially photographic ones, were proved to be frauds or honest mistakes. The fact that the layperson hears most of these cases through the conduit of the popular press proved to be one of the main causes of misinformation. Many reports of seemingly strange events are given front page treatment on the first day; a few days later many are explained, but this is scarcely news and is relegated to a back page, if printed at all. The same reports, including photographs established as frauds by various investigators, may be picked up and run again by tabloids, who merchandise pseudoscience at supermarket checkout stands. Also, as is clearer every year, our culture is astoundingly good at developing and propagating subcultures devoted to belief in ludicrous misinformation. For these reasons, most people think that the UFO issue is much more mysterious than it really is.

Yet, for just the same reasons, it is extremely difficult to prove that NO alien spacecraft have ever been seen! No one has established that alien visits are *not* occurring today, but if they are, they are sufficiently rare that we have no evidence of them, in spite of enormous amounts of "surveillance" equipment, from Air Force satellites to home video cameras. The verifiable fact that no obvious historical visits have occurred at least shows that space is not chock-full of alien spacecraft trying to make contact with us.

A second answer to the question—"where are they?"—may be that the pessimistic figures are right, and that the nearest civilizations are in distant galaxies. Even their radio messages, if any, could be 10 My old or more by the time we receive them, and their spaceships would be unlikely to reach Earth if limited to speeds less than that of light, as current physics requires. But most investigators

Figure 12-8. Even if Earth has been visited on 10,000 occasions by alien expeditions, the last visit would probably have been too long ago to have left any record likely to be identified by historical or archaeological records, unless large, prominent artifacts were deliberately left. (Painting by Jim Nichols)

of the problem place the probability well above our lower limit. At a 1971 Soviet-American conference on this problem, the favored estimate came out to be one civilization per 10^5 stars. This would put a million civilizations in our galaxy and the nearest civilizations only a few hundred light years away. Why, then, are there no frequent contacts with Earth?

A third possible answer is that biological evolution need not produce creatures who have a desire to build civilizations or travel through space. Not even all human societies evolve toward technology. Are humans fated to be explorers, bridge builders, and businesspeople rather than artists, athletes, or daydreamers? Is the stereotyped aggressive Westerner more representative of the essence of humanity than the stereotyped contemplative Easterner? May not our aggressive technocracy be just one type of *cultural activity* rather than a universally achieved stage of *biological* evolution? Historically, many patterns we once assumed to be biologically imposed have turned out to be merely culturally conditioned.

If humanity is not predestined to develop a technological civilization, how much less certain is the course of social development on other worlds. It is absurdly anthropocentric to suppose that beings on other planets would resemble us psychologically or socially. Consider again the variety of highly evolved life-forms on our planet alone. Ants live in ordered societies that do not appear to regard individual survival as important. Dolphins communicate and have brains that seem almost comparable to ours, but they have no manipulative organs and hence no technology.

Perhaps we cannot expect aliens to be motivated by emotions that mean much to us. We certainly cannot expect, as happens in many early grade-B science fiction movies, that humanlike aliens will walk out of saucers and invite us to join their democratically constituted United Planets, a galactic organization structured by documents that are curiously reminiscent of the U.S. Constitution. So why assume that other civilizations might even care to try to visit or contact us?*

Evolutionary Clocks and the Explorative Interval

This brings us to a fourth and perhaps the most important possible answer for why we have no evidence of aliens: We may be farther from aliens in evolutionary time than in physical space. Biological evolution is so persistently experimental that even if another planet started evolving at exactly the same time as ours, and even if its biochemistry produced creatures like us, those creatures are not likely to be in a phase of evolution similar to ours. If the evolutionary "clocks" on the two planets got only 0.02% out of synchronization, the other group would be 1 My ahead of us or behind us, in evolutionary terms. Thus, even in the unlikely event that other planets produce civilizations recognizable to us, we might have to contact one of those civilizations in a very narrow time interval—just the right 0.01% of the planet's history—to find any society with recognizable common interests.

---

*Woody Allen (1980, p. 37) summed up much of this more succinctly: "If saucers come from outer space, why have their pilots not attempted to make contact with us, instead of hovering mysteriously over deserted areas? My own theory is that for creatures from another solar system, 'hovering' may be a socially acceptable mode of relating. . . . It should also be recalled that when we talk of 'life' on other planets we are frequently referring to amino acids, which are never very gregarious, even at parties."

Figure 12-9. Lines as long as a few miles in the Peruvian desert were claimed as alien landing sites requiring stupendous technologies beyond our means. Ground studies show that these lines were made by native people around 1,800 y ago and required technology no more stupendous than a rake to sweep aside dark desert pavement stones from the underlying bright dust. The lines are currently threatened by off-road vehicles, whose tracks show scale. (Photo by author)

Evolution may pass through only a brief **explorative interval,** during which societies on one planet would care to reach other planets; beyond that stage communication or space exploration might be no more attractive than a national program on our part to communicate meaningfully with chimpanzees, ants, or dolphins. To be sure, a few of our scholars try this, but they "contact" an infinitesimal fraction of these creatures. What fraction of the anthills or dolphin schools have we humans tried to "contact"? By the same token, our solar system might be ignored by advanced aliens. Aliens a million years ahead of us might be no more interested in us than we are in ants.

How long might an explorative interval last? We have used tools for about 2 My, and we assume that we will have progressed far beyond current technology within another million years, if we survive. Our explorative interval might be a few million years, then, or less than 0.1% of the history of the planet. This is the basis for the lower factor listed in criterion 8 in Table 12-1.

If civilizations are 500 LY apart, then interstellar voyages and message exchanges could take a millennium or more. There might be little incentive for the effort. Large spaceships in which many generations could live and die during interstellar voyages—self-contained "planets" that are a staple of science fiction stories—have also been hypothesized in published models of interstellar colonization by other species (Kuiper and Morris, 1977). In any case, several authors have concluded that, given the plausible number of ships in transit at any one time from all

civilizations, a visit to a specific random planet, such as Earth, would be rare.

This brings us to a fifth possible answer to the question of alien visits: They may have happened in the remote past. This is the "ancient astronaut" hypothesis, popularized in several pseudo-science books. Figure 12-8 illustrates a problem with this idea. For example, even if 10,000 visits to Earth occurred at random intervals in Earth's history, they would still average 460,000 y apart! Given the rate of terrestrial erosion by water and glaciation, the prospects for finding evidence of such ancient visits are poor! As for historical visits, there is no good evidence. The Soviet and American collaborators I. S. Shklovskii and Carl Sagan (1966) surveyed archaeological and mythological literature even before the hypothesis was a popular fad and found no compelling evidence for ancient astronauts. Neither Earth, the moon, Mars, nor Venus is littered with ancient alien artifacts, and not a single mysterious artifact has been publicly advanced as physical evidence of ancient astronauts. As illustrated by Figure 12-9 popular pseudo-science books have greatly exaggerated the mystery surrounding many archaeological structures and artifacts (Story, 1976).

Radio Communication

Possibly our period of isolation may be nearing an end. For most of a century, we have been broadcasting radio signals among ourselves. Already our unintentional but weak alert is more than 70 LY out from Earth. Aliens may one day pick up our signals and send radio messages (or an expedition?) in return. Radio astronomers are therefore still conducting modest programs to listen for such messages with large radio telescopes.

In one such search, for example, radio astronomers listened for broadcasts near 1420 MHz (the 21-cm-wavelength that marks an astronomically important radiation from neutral hydrogen atoms in space), targeting all solar-type stars not known to be members of multiple star systems, within about 82 LY (see Table 4-2); this included 185 stars. No artificial broadcasts were identified (Horowitz, 1978). Goldsmith and Owen (1979) summarize other searches involving more than 600 stars. All were negative.* Larger listening instruments have been proposed.

---

*The NASA search for extraterrestrial intelligence by radio listening was halted in late 1981 as the result of an amendment attached to the 1982 federal budget by Senator Proxmire, explicitly forbidding expenditures for this purpose, on the grounds that it is a "ridiculous waste." My own view, in contrast, is that such listening puts either positive or negative boundary conditions on one of the most fundamental questions about our cosmic surroundings: Are technological civilizations like ours making radio transmissions?

A message from 1,000 LY away would come from a civilization 1,000 y in the past, and no answers to our questions could come back for 2,000 y. Such communication would be unlike dialogue, but, as physicist P. Morrison has pointed out, more like our receipt of "messages" (books, letters, plays, and artworks) from ancient civilizations such as Greece.

Meanwhile, we have sent a few messages of our own. The Pioneer 10 spacecraft, which flew by Jupiter and left the solar system in 1973, carried a plaque designed to convey our appearance and location to possible alien discoverers of the derelict spacecraft. The first radio message was a test message sent from the large radio telescope at Arecibo, Puerto Rico, in 1974, beamed toward globular star cluster M 13, 27,000 LY away.*

---

*Astronomers in these projects were deluged with letters ranging from support to complaints about the nudity of the human figures on the Pioneer 10 plaque. The radio astronomers promptly received a telegram: "Message received. Help is on the way—M 13." Its authenticity might be questioned, because the round trip radio signal travel time to M 13 is 54,000 y.

SUMMARY

The discovery of firm evidence for alien civilizations, or even alien life-forms, could be a pivotal development in human history, as suggested in Figure 12-10. Yet questions of exobiology, and especially of intelligent life on other worlds, leave us with a mystery. Experimental evidence suggests that life is likely to start on other planets if liquid water, energy, and the right chemicals are present. Astronomical evidence suggests, but does not prove, that habitable planets ought to exist elsewhere in the universe. Biological evidence shows that life is adaptable and that species can evolve to fit different environments, from ocean depths to low-pressure atmospheres of different compositions.

While the limited evidence indicates that other life-forms should exist, there is no evidence that they do, or that they have tried to communicate with us. We can only speculate about the reasons. Perhaps they are too far away. Perhaps most civilizations destroy themselves before successfully exploring the universe. Perhaps evolution carries them beyond a stage where they would care to communicate with us. Perhaps they are unrecognizable.

Arthur C. Clarke has remarked that any technology much advanced beyond your own looks like magic to you. Perhaps we are too limited by our own concept of civilization. After all, one creature's civilization may be another's chaos, as shown by Mohandas Gandhi's remark when asked what he thought of Western civilization: He said it wouldn't be such a bad idea. It seems likely that our first contact with aliens (if they exist) might be as incomprehensible as the dramatized contact in Clarke's novel and the Kubrick-Clarke film, *2001: A Space Odyssey*.

Clearly, we have been reduced to speculation by a lack of facts. Indeed, the whole field of exobiology has been criticized as a science without any subject matter. Exobiology recalls Mark Twain's famous remark, "There is something fascinating about science. One gets such wholesale returns of conjecture from such a trifling investment of fact." In the same vein, exobiology spokesman Philip Morrison (as quoted by Simpson, 1973) has admitted of exobiology, "Here is a body of literature whose ratio of results/papers is lower than any other." The only way to reduce the conjecture and increase the proportion of fact is to pursue research in many related fields—physics, chemistry, geology, meteorology, biology—and listen to the skies with radio telescopes. There may be surprises waiting out there.

a

b

Figure 12-10. Discovery of positive evidence for extraterrestrial civilizations could be a dramatic and pivotal event in human history. This could occur by (**a**) humans going out and finding such evidence (discovery of alien obelisk on the moon, as shown in the MGM release *2001—A Space Odyssey,* 1968 Metro-Goldwyn-Mayer Inc.) or (**b**) such evidence arriving on Earth (arrival of the "mother ship" as shown in *Close Encounters of the Third Kind,* 1977 Columbia Pictures Industries, Inc.)

CONCEPTS

organic molecules
exobiology
life
cell
organic chemistry
amino acid
protein
Miller experiment
nonracemic mixture
racemic mixture
proteinoid
coacervate

prokaryote
eukaryote
stromatolite
ozone
ozone layer
natural selection
mass extinction
convergence
cultural hurdle
Drake equation
explorative interval

PROBLEMS

1. Compare Alpha Centauri, Barnard's star, and Sirius in terms of the probability of detecting radio broadcasts from intelligent creatures in these systems (see Table 4-2).

2. Describe several ways in which Earth's internal evolution has affected the evolution of life on Earth.

3. Describe several ways in which extraterrestrial events, such as solar or stellar evolution, might have affected evolution of life on Earth or other planets.

4. Describe ways in which technology could affect the survival of intelligent life on Earth or on other planets. Construct scenarios of (a) the possible destruction of life and (b) guaranteeing the survival of life. (*Hint:* In part [b] consider the impact of space travel.)

5. In your own opinion, what would be the long-range consequence of each of the following:
(a) Arrival of an alien spacecraft and visitors in a prominent place, such as the United Nations building
(b) Discovery of radio signals arriving from a planet about 10 LY distant and asking for two-way communication
(c) Proof (by some unspecified means) that life existed *nowhere* else in the observable universe

6. Given the assumption that our technology has the potential for creating planetwide changes in the environment, defend the proposition that *if* life on other worlds produces technologies like ours, then that life is likely either to have become extinct or to be widely dispersed among many planets by means of space travel.

7. In view of the devastation wrought on many terrestrial cultures by contact with more technologically advanced cultures, which would you say is the safest course: (1) aggressive broadcasting of radio signals to show where we are in hopes of attracting friendly contacts, (2) careful listening with large radio receivers to see whether there are any signs of intelligent life in space, and (3) neither broadcasting nor listening, but just waiting to see what happens?
(a) How would it affect the results of our listening program if other intelligent species had reached the first, second, or third conclusion?
(b) If we listen at many frequencies and pick up no artificial signals, does that prove that life has not evolved elsewhere in the universe?

8. If an Earth-sized planet (diameter 12,000 km) was circling a star 1 parsec away (3×10^{16} m):

(a) What would be its angular diameter when seen from Earth?

(b) Could this be resolved by existing telescopes?

(c) If the planet orbited 1 AU (1.5×10^{11} cm) from its stars, what would be its maximum angular separation from the star as seen from Earth?

(d) Could this angle be resolved?

9. Derive expressions for the inner and outer radii of a toroidal "zone of habitability" around any star of luminosity L (in units of solar luminosity) if an inhabited planet orbiting the star has an effective Bond albedo A and emissivity *e*, a circular orbit, and liquid water somewhere on its surface in order to support life.

A new frontier. Martian vista inside Gusev crater, photographed from Spirit lander in 2004, shows an austerely beautiful desert with distant hills and foreground sand drifts. (NASA, JPL)

CHAPTER THIRTEEN

Martian Epilogue: Applying Planetary Science on a New Frontier

Planetary science has exploded in subject matter, with probes exploring worlds as diverse as Titan and the asteroids, and more proposed from Mercury to Pluto. Yet nowhere is the excitement more palpable than in our exploration of Mars. The previous chapters were deliberately terse about Mars because we have saved a broader discussion of Martian phenomena for this final chapter. In this way, Mars can serve as a case study for application of the language and general principles developed in this book.

Mars is the most exciting frontier for planetary exploration in the solar system because it is the planet that speaks to us most directly about our own condition. Its ice caps, permafrost, clouds, 24-hour day, volcanoes, dunes, canyons, and other phenomena are familiar. Although Mars is a cold desert, summer afternoon temperatures can rise above freezing. Mars is the planet where human researchers may most easily learn to live off the land, and the planet that might give us the best natural lab to understand the early history of our own world and the conditions that led to life.

Mars offers provocative research problems that can be grouped under two broad mysteries: climate change and biology. To explore these mysteries, a flotilla of missions to Mars is planned for the next decade from the United States and other countries. Launches come every two years during favorable launch windows. To understand the goals of these missions, we will review general knowledge about Mars and its present conditions. These offer anchor points for considering the evolutionary history of the planet. Always, behind these discussions, lurks the question: What can Mars tell us about Earth and the story of life on our own planet?

Brief Review of Present Martian Features

Mars offers an incredibly rich variety of phenomena to help us understand the evolution of planets. As we have seen in earlier chapters, about half of Mars consists of ancient, heavily cratered uplands, cut by channels that appear to be dry beds of ancient rivers. Roughly the other half of Mars consists of plains, less cratered than the uplands and obviously younger. These plains are marked by many lava flows and about half a dozen enormous shield volcanoes, some reaching to heights of 24 km above the plains. The volcanoes have very few impact craters and appear to be very young, possibly still active. Water from the upland rivers appears to have emptied into the plains, although the plains themselves generally show few river channels, suggesting that the period of water flow was earlier than many of the volcanic flows. The giant canyon, Valles Marineris, is long enough to cross the United States and appears to be Mars's single large tectonic rift canyon comparable to the Red Sea or the Gulf of California in scale (see map, Figure 2-14).

As shown in the chapter opening image, all this topography is veiled by a pattern of bright and dark features, which appear to be primarily windblown dust deposits. The patterns are not well correlated with the topography. The shapes of the dark markings change slightly from year to year, as was shown in Figure 10-29. Streaky markings, connected with prevailing wind directions, are common and probably helped spawn the erroneous theory of thin, linear "canals" on Mars. Massive dune fields are also common (see Figure 10-32), and the view from the very first Martian lander, Viking 1 in 1976, showed prominent, graceful dunes (Figure 13-1).

Mars has complicated seasonal phenomena, similar to those of Earth. In the winter hemisphere, the polar ice cap grows and extends to latitudes of about 65° in the north and 57° in the south (comparable to the latitude of Hudson's Bay and Scotland on Earth). The winter cap grows as CO_2 frost is deposited on the surface (recall that solid CO_2 is familiar to us as "**dry ice**"). Figure 13-2 shows a winter morning frost deposit on the surface at the Viking 2 site, at 48° N latitude. In summer, the ice caps shrink back to a smaller size, seen in Figure 13-3.

Summer in the southern hemisphere currently coincides with Mars's passage through perihelion in its orbit, because Mars's orbit is more elliptical than Earth's; thus, the southern summer is noticeably warmer than the northern summer. Heating of the ground and air near the ground tends to initiate dust devils, as was shown in Figure 11-11. This is especially true in the southern hemisphere in summer. In some Martian years, these spread

a   b

Figure 13-1. The dune fields of Mars, photographed under back-lighting (**a**) and front lighting. (**b**) Rocks in the foreground range up to about a foot in length. Whether these are actively moving dunes or stabilized remnants of eroded sediment cover is uncertain. (NASA, Viking 1 lander)

from local features into vast regional and even global dust storms, as was indicated in Figure 11-10. The mechanism involves atmospheric instability and winds, as the airborne dust absorbs the sun and heats the upper atmosphere, while shading the surface and cooling the lower atmosphere.

The existence of the channels raises the intriguing question of water's history on Mars. At present, the water is concentrated in three "reservoirs": (1) the permanent polar ice caps, (2) underground ice layers some number of meters below the surface, especially at high latitudes, and (3) water of hydration in soil minerals. As we will see, many of the questions of Martian climate change and biology are intimately mixed with the questions of water's history on the planet.

The Martian Geological Eras

U.S. Geological Survey scientists, in creating geologic stratigraphic maps of Mars, created a subdivision of Martian time with three major eras, analogous to the major divisions of the geologic timescale on Earth (Tanaka, 1986). The eras were defined by major geological processes that dominated them, and their divisions were defined by crater densities. Later analysts attempted to use the crater density to fix the dates of each era (Hartmann and Neukum, 2001). Combining these studies, we have the following descriptions of the Martian chronology:

- **Noachian Era:** Earliest period of Martian history, believed to be characterized by a thicker early atmosphere, high rates of cratering, heat flow, volcanism, fluvial activity, and probable glacial activity. (Handy mnemonic: think of Noah and the flood). Climate may have been either warmer or

colder (see later part of this chapter). It lasted from planet formation 4.5 Gy ago until about 3.5 Gy. The latter date is fairly well constrained because it has to fall within the end of the early intense cratering; the uncertainty on the end date may be a few hundred My.

- **Hesperian Era:** Era of transition from Noachian conditions to modern, Amazonian conditions. In many of the major outflow channels, the last water flows and erosion episodes probably dated from this period. It probably lasted from about 3.5 Gy ago until roughly 3.3 to 2.9 Gy. The closing date is uncertain and values as young as 2 Gy are not completely excluded.

- **Amazonian Era:** The modern, dry, dusty, mostly frozen era. Lasted from roughly 3.3 or 2.9 Gy ago until the present. Although geologic activity has declined on Mars, evidence has been found of active geology (lava flows, water release) in this era.

Origin and Compositional Questions

Because Mars is farther than Earth from the sun, it probably accreted more water- or ice-bearing planetesimals than Earth did. The Kr-Xe chemistry of Mars suggests that Mars got a much higher proportion of CI carbonaceous chondrite matter than Earth did, which would have produced a large inventory of water and volatiles in the Martian parent material (Pepin, 1991). The influx of carbonaceous bodies into Mars might be also related to the capture of the two blackish (carbonaceous?) moons.

a

b

Figure 13-2. Summer and winter on Mars. (**a**) Viking 2 view of rocky Utopia plain as it appears during most of the 687-d Martian year. (**b**) For about 100 d during the Martian winter at the same site, a partial layer of frost formed, especially in the mornings. At least some of the frost persisted at higher temperatures than needed for pure CO_2 ice. It may have been H_2O frost. Such frost may reach the ground when H_2O ice and/or CO_2 ice condense on dust particles and then precipitate out of the atmosphere. (NASA)

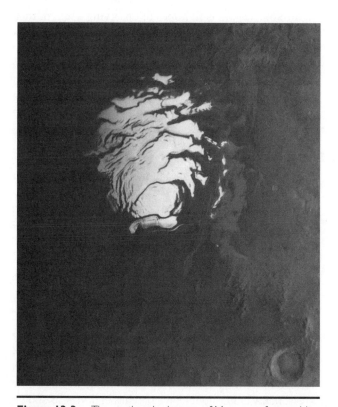

Figure 13-3. The south polar ice cap of Mars seen from orbit during the summer. This small cap, about 360 km across, probably consists of a mixture of H_2O and CO_2 ices. In the winter, condensation of CO_2 frost expands the cap to 10 times this size. Spiral breaks mark warmer, sun-facing ridges and valley walls in the layered sedimentary deposits around the pole. Night side of the planet is at left. (NASA and U.S. Geological Survey, courtesy Tammy Ruck and Larry Soderblom)

Thus, most theorists assume there is no problem *a priori* in Mars having a large initial amount of water, sufficient to explain the proposed river-forming activity and the channels described in Chapter 2.

Interior Structure and Related Considerations

The 1997 Pathfinder and Mars Global Surveyor missions provided an improved moment of inertia for Mars (0.366), (Golombek and others, 1997; Folkner and others, 1997) and found an iron core of radius 1520 to 1840 km with at least the outer part molten (Yoder and others, 2003). The existence of a core confirms the expected melting and differentiation, and is consistent with the extensive volcanism.

How active is the interior today? Is volcanism a present-day phenomenon on Mars? Probably. The surfaces of the large volcanoes, like Nix Olympica, are so sparsely cratered that their ages must lie in the last 10% of geologic time, and the 28 known Martian meteorite rocks include lava-like samples with ages of only 170 My and 330 My. It seems unlikely, therefore, that internal activity has simply shut off forever.

The Viking 2 lander carried the first successful seismometer experiment to Mars in 1976 (a duplicate seismometer on the Viking 1 lander failed to be deployed). During five months of operation, no large earthquakes were detected, though one possible event of Richter magnitude 3 was recorded (Anderson and others, 1977). Martian winds, shaking the lander that contained the seismometer, provided a high "seismic" noise level during much of the period. Analysis indicates that Mars is probably less seismically active than Earth and perhaps close to the moon in its level of activity. Future seismometry on Mars may be the means of clarifying the activity level of the youthful-looking giant volcanoes of Mars.

Calculated thermal models of Mars differ, reflecting uncertainty about the internal properties (Spohn and others, 2001). Figure 8-26 gave an example consistent with known facts. If there was ever magma ocean, it quickly solidified into an early crust of which we may still see remnants. This is consistent with the existence of one Martian meteorite which is a piece of the primordial crust—an igneous cumulate rock with a measured age of 4.5 Gy. The mantle reached the melting point of iron before the silicates melted; sluggish iron drainage probably formed a core before a well-developed asthenosphere of molten silicates could appear. Most calculations indicate that molten silicates appeared no less than 400 km down about 0.5 to 3 Gy ago. This magma may have been the source of the vast, sparsely cratered lava plains and huge volcanic cones (see Figure 2-15), no more than a few hundred My old.

The suggestion that the lithosphere has always been thick explains why well-developed plate tectonics did not appear, as evidenced by lack of folded mountains and the scarcity of major rift valleys. A period of expansion by 5 to 10 km during the heating (Solomon, 1976), may explain the many fractured areas and the appearance of the enormous rift canyon system, Valles Marineris (Valley of Mariner, named after the spacecraft that discovered it). A portion of this rift, and collapsed terrain at one end, can be seen in Figure 13-4.

The huge volcanoes of Mars lead to interesting ideas about Martian mantle plumes and mantle evolution in general. The plains of the western hemisphere of Mars (longitude 0° to 180°) are dominated by the Tharsis dome, a broad area of youthful lava flows and volcanic features, which rises about 8 km above the mean surface level (see map, Figure 2-14). The Tharsis dome is surmounted by the huge volcano, Olympus Mons (Figures 10-7 and 10-8), and other shield volcanoes rising over 20 km above the mean surface (Figure 13-5). This may mean that one or more mantle plumes have been active under that region, and in the absence of plate tectonics, the crust was fixed over the plume and the lavas kept accumulating in one area (Spohn and others, 2001). In contrast, on Earth, plate drift has carried the crust over plume sources, accounting for volcanic island chains, such as Hawaii. Even the total volcanic rock accumulations of the Hawaiian islands amount to less than a small percent of the volume of Olympus Mons, implying a very active or long-lived Martian source.

The importance of volcanism as a source of most crustal rock is shown both by Martian meteorites and by sampling of rocks on the surface of Mars. All 28 Martian meteoritic rocks known in 2004, from about four to eight sites on Mars, are igneous, and as noted above, many are ordinary basaltic lavas. As shown in Figure 13-6, the 1997 Mars Pathfinder rover, Sojourner, sampled rocks at the

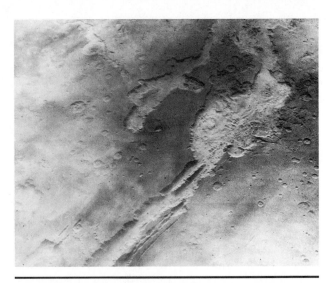

Figure 13-4. East end of the Valles Marineris, an apparent rift canyon system enlarged by collapse and erosion. North is at upper left. The linear canyons are about 150 km wide, opening into an oval region of collapsed terrain. Clouds obscure parts of the surface at left and lower right. Figure 10-17 shows a closeup of the north rim of the canyon and faults parallel to it. (NASA Viking 1 Orbiter)

mouth of Ares Vallis in hopes of observing a mix of rock types washed down from the old highlands, but the six rocks whose compositions were measured turned out to be basalts and a marginally more silica-rich rock type, andesite. The average composition matches Earth's mean crustal composition (see Table 9-3). Landers in 2004 are sampling compositions of rocks in Gusev crater and in Terra Meridiani, but the compositions have not been announced as of this writing. Rocks in Gusev crater appear visually similar to rocks at earlier landing sites (Figure 13-7), but in Terra Meridiani the lander found the first known bedrock outcrop, with centimeter-scale layering, which may suggest sedimentary rock, formed by wind deposition or water deposition (see discussion on p. 386).

Dust Storms and Dust Mantling

At a given spot on Mars today, an observer would be more conscious of dust mobility than of volcanoes as a shaper of the landscape. Isolated small dunes or drifts are common (chapter opener). Winds drop dune fields in many low areas; many crater floors have patches of dunes, as shown in Figure 10-32. The largest dune field in the solar system stretches around the north polar ice cap. When the Mars Global Surveyor orbiter arrived at Mars in 1997 with high resolution cameras, researchers noted that when one can see details smaller than 50–80 m, the visual impression of the surface changes, and dust deposits dominate many views. Armies of dunes march across the landscape and

a

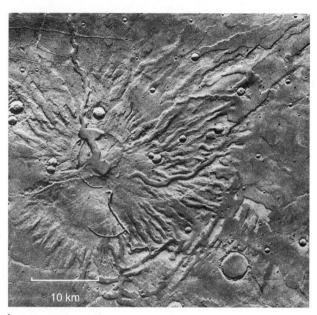

b

Figure 13-5. Mid-size Martian shield volcanoes. (**a**) Ceraunius Tholus extends the alignment of three larger volcanoes in the Tharsis region. The flanks are crossed by lava flow channels. The largest has dumped a delta-like lava flow onto the floor of an oval impact crater, formed by a low-angle impact. (**b**) Tyrrhena Patera is an older volcano on a plain in the Martian southern highlands. Its flanks are strongly dissected by channels. Geologists have debated whether these are all lava flow channels or if some could be caused by water runoff. The size is about a tenth that of Olympus Mons. (NASA, Viking)

massive dust drifts engulf small craters, reminiscent of snow drifts after a storm on Earth. As was shown in Figure 10-30, a given dust storm can leave long streaks of deposited material, either lighter or darker than the background. A later storm can leave a different pattern. This process, on a larger scale, explains the changes once cited by telescopic observers on Earth as evidence for vegetation.

The dust particles lifted in Martian storms are very small, about 1 μm across. Calculations suggest that high winds of at least 50 m/s (110 mph) are needed to raise dust in the air of Mars, contrasting with only about 8 m/s needed on Earth (Mutch and others, 1976, p. 242). Such winds could easily occur in localized "twisters" or dust devils. Viking Landers 1 and 2, at 22° and 48° N latitude, respectively, measured arrival of dust at their positions from storms that started in the southern hemisphere. The arrival at the Viking 1 site was accompanied by increased winds with gusts up to 26 m/s (58 mph). The Pathfinder lander observed winds mostly less than 10 m/s (22 mph). Reanalysis of Pathfinder surface photos suggested that dust devils may have been visible on the horizon in one or two images (Metzger, 1998).

At a local scale, the changes from year to year are small. Although the Viking landers observed essentially

Figure 13-6. The Sojourner rover backs up against a boulder to measure its composition. The rover's aft end carried a composition measuring device, which bombarded the rock with alpha particles from a radioactive source and measured particles and x-rays scattered back off the rock. This boulder had basaltic composition. The rover landed on Mars in 1997 with the Mars Pathfinder. (NASA, courtesy Sara Smith, Pathfinder project)

Figure 13-7. Fisheye wide-angle view from Spirit rover in 2004 shows arm reaching out to sample composition of a boulder. Rover wheels can be seen in lower corners. The view is on the floor of a large crater, named Gusev. (NASA, JPL)

throughout a Martian year, they observed only minuscule changes in the dune structure or topographic features of the landscape. Viking 1 observed a centimeter-scale slip in a dune face, changing the soil from bright to dark material (Figure 13-8). The larger boulder in the same images has a mantle of fine red dust on its top, deposited by filtering vertically down out of the atmosphere. Mars Pathfinder gave evidence of several soil types, including very fine bright drift material (Figure 13-9).

As the Sojourner rover moved, dust accumulated on its solar panel; the rate of obscuration of the solar cells was independent of vehicle motion, indicating the dust was settling out of the air. The measured accumulation was 3 micrograms/cm² per day averaged over 30 days, or a layer about 6 mm deep in a millennium, if the dust kept accumulating in one place, such as a deep crater floor.

Water on Present-Day Mars

If Mars were simply a volcanic, moon-like world with wind to blow the dust around, it would not be nearly so interesting as it is. However, one of space exploration's most provocative discoveries is that Mars's history involved abundant water.

As mentioned earlier, the water today is found in three forms: chemically bonded in minerals in the soil, frozen in ice deposits at the two poles, and frozen in per-

mafrost beneath the surface. The first two forms were discovered spectroscopically from Earth and confirmed by Viking. Viking lander soil samples released 0.1% to 1% by weight of water after being heated to 620–770 K. The subsurface ice was confirmed when the Mars Odyssey orbiter in 2001 detected 30% to 50% ice in the top two meters of soil (Boynton and others, 2002). It is indicated in other areas by topographic features resembling terrestrial collapses, formed when ice melts and water drains out of an area (Carr and Schaber, 1977), and also by "terrain softening" in which many landforms at latitudes above 30°–40° seem to have softer-than-normal contours, as if modified by ice flow or melting (Squyres and Carr, 1986; Squyres and others, 1992).

Condensation of Water and Carbon Dioxide on Mars

Mars has an important cycle of condensation involving the abundant CO_2 and sparse H_2O in its atmosphere. Many of the Martian phenomena described above can be understood with the phase diagram shown in Figure 13-10. The solid curves show the conditions for condensation of CO_2 and H_2O. The **triple point for H_2O,** the only condition under which ice, liquid water, and water vapor can coexist, is marked at $P = 6.1$ mbar and $T = 273$ K. Curve *a* marks the daily range of conditions for a representative high-altitude region on Mars, such as the Tharsis volcanoes. Normally, H_2O would be frozen and CO_2 gaseous at such a location. Since $P < 6.1$ mbar, the warmest days do

a   b

Figure 13-8. Changes in fine-scale dune structure at the foot of a 1-m boulder on Mars occurred sometime during the 112-d interval between these two pictures. A 30-cm tongue of dust slipped from the dune face, exposing darker dust. Dust atop the boulder settled out of the air. (NASA, Viking 1)

Figure 13-9. Drifts of bright dust against darker soil and rock can be seen in this view looking south across the Ares Vallis attach plains. The faint hill on the horizon is a knob 31 km away, seen through the hazy Martian air. (NASA, courtesy Sara Smith, Pathfinder project)

not produce stable liquid water, but rather cause ice to sublime directly into water vapor. Cold nights or winter days could cause condensation of CO_2 gas into CO_2 frost. Curve b shows conditions in a low-altitude region, such as the floor of Valles Marineris. Here, $P > 6.1$ mbar, so that a balmy summer Martian afternoon at 35°F might produce small amounts of liquid water in the soil in those regions. As the ambient air pressure is not provided by water vapor but by dry CO_2 gas, the humidity is low and this water will evaporate very rapidly into the dry air.

Figure 13-2 showed a consequence of Figure 13-10: the dramatic result of winter's onset at the modest latitude of 44° N on Mars. The first image is a view of the rocky plain as it appeared soon after Viking 2's landing in summer; the second image shows the scene during a 100-d winter period when frost formed at night on the surface. The complex mechanism of such frost deposition probably involves dust particles blown into the atmosphere by dust storms. These act as condensation nuclei for H_2O ice, but the particles still do not grow big enough to fall rapidly out of the atmosphere. However, as winter nights grow colder, the CO_2 ice regime on Figure 13-10 is reached, causing CO_2 ice to condense onto the dust/water particles, giving them a heavy enough coating to precipitate onto the ground (Pollack, 1979). The daytime sun, however, warms the CO_2 frost and sublimes it back into the atmosphere, leaving a frost of water-ice-coated dust grains, which is probably what is seen in Figure 13-2b.

In Figure 13-10, the conditions on Earth are shown in the upper right, where we see transitions from the solid to liquid H_2O regimes. Since Earth's atmosphere is not pure H_2O, the liquid water is not entirely stable, but evaporates into the air. If enough liquid water is available, an equilibrium water vapor pressure ("100% humidity") is reached.

Curve c shows a hypothetical condition for ancient Mars if the air pressure then was higher than it is today. In

Figure 13-10. Phase diagram for a planetary environment with carbon dioxide and water, accounting for various Martian phenomena. Curves *a* and *b* show conditions at high and low surface elevations on Mars, respectively. See the text for further discussion. (After a diagram by Mutch and others, 1976)

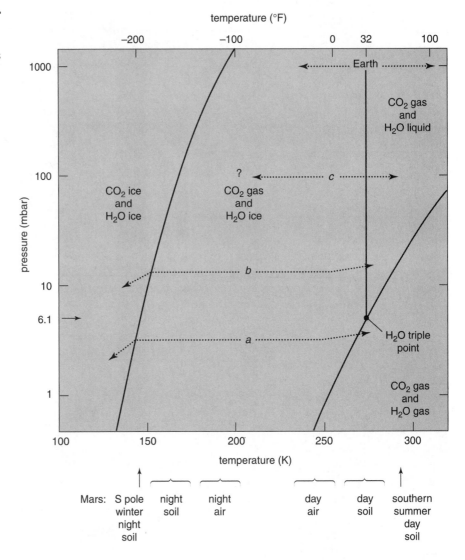

Processes at the Martian Polar Caps

The coldest Martian regions are the winter poles, currently tipped away from the sun by 25° and shrouded in darkness for half a Martian year, or 11 Earth months. During winter, the CO_2 that freezes out of the atmosphere at 148 K (−193°F) forms a transient broad polar cap, analogous to our water-snow winter cap that stretches across Canada and the northern United States. So much CO_2 freezes out of the atmosphere that the total air pressure varies by 26% seasonally, as a result of the annual shuttling of CO_2 back and forth from one winter polar cap to the other (Hess and others, 1979). CO_2 snow has been detected falling out of the atmosphere, and the depth of the transient CO_2 cap reaches about 2 m (Smith and others, 2001).

The cycle of polar events is complex, since there are two condensates, H_2O and CO_2. The cycle can be followed in Figure 13-10. Because the polar temperature is almost always below the freezing point for water, even in full sunlight, there is a small, permanent polar ice cap of frozen H_2O at each cap, shown in Figure 13-3. These permanent caps are slightly off center at each pole and reach about 10° from the north pole, but only about 6° from the south (where summer is warmer than in the north).

Adding to the complexity, the Martian obliquity varies from near 0° to ~45° on a timescale of 10–15 My. (Earth's obliquity, in contrast, is stabilized near 23.5° by lunar gravitational forces.) When the summer pole leans 40°–45° toward the sun, much H_2O ice burns off (Jakosky and Carr, 1985). The resulting water vapor condenses on dust grains suspended in the atmosphere, snowing out at winter latitudes ~40°–90°. Every 10–15 My, this creates a new mantle of ice and dust draped across the topography (Costard and others, 2001). Even though the ice sublimes each summer, a net accumulation of ice seems to occur at high obliquity. At lower obliquity a net loss of ice occurs, leaving a young blanket of dust tens of meters thick covering older topography at upper latitudes. Such mantles have been seen in Mars Global Surveyor images (Mustard and others, 2001). These phenomena lead

this case, a greenhouse effect might have warmed the air, and liquid water would have been stable in many regions on warmer days.

to questions about the large-scale history of water on the planet.

Martian Mystery No. 1: The Channels, the History of Water, and Climate Change

Martian science today is driven by two major mysteries that have profound consequences: the history of Martian water and the possibility of Martian life.

In view of the instability of liquid water on modern Mars, planetary scientists were astonished in 1972 when photographs from Mariner 9 revealed many well-preserved dry riverbeds. These came to be called **channels**—not to be confused with the largely imaginary "canals" popularized by Lowell around 1900 (see Chapter 2). Excellent examples of channels with tributary systems were shown in Figure 2-15.

Because Mars is a dry, dusty planet, rivers of water were unexpected. Perplexed researchers considered other origins for the channels, such as wind-eroded features, lava flow channels, tectonic grooves, or channels carved by liquid carbon dioxide. But detailed photos showed that Martian channels have all the properties of terrestrial riverbeds. They usually get wider and deeper downstream, like terrestrial rivers, but unlike lava flow channels. They often have tributaries, terraced banks, and interlacing, or "braided" sub-channels on their floors. (Braided channels often form on Earth during the last stages of a flood as the final flows of water wind among islands on a river floor.) They seem unlike what might be expected from liquid carbon dioxide, which requires 5 bars of pressure, and seems more likely to explode into the Martian near-vacuum than to flow in tidy rivers across the surface.

Dry riverbeds present a serious problem for understanding the planet. In the first place, the current temperature is mostly below the 0°C needed for water ice to melt (though daytime temperatures often rise above the −40°C at which salt-rich brines can be liquid). Second, in the upper elevations of the Martian surface, the current atmospheric pressure is less than the 6.1 mbar needed for water to remain liquid (Figure 13-10); water in those areas would spontaneously boil away into the gas phase. Even in the lower-elevation regions, water would evaporate rapidly. Combining these two ideas, we can say that the most likely description of a Martian river under modern conditions would be a catastrophic flood of briny water that pours out large volumes fast enough to surge across the surface in a few days or weeks, eroding rapidly into loosely consolidated soils, and whose surface freezes, protecting the liquid water underneath from evaporation. This protective ice-cover would allow flow for some period of time.

Supporting these ideas, most Martian meteoritic rocks found on Earth show signs of having been exposed to at least small amounts of water on Mars. Mineral deposits inside their cracks, left by water, prove that the Martian water was briny, with a composition similar to terrestrial sea water (Bridges and others, 2001). Such brines form readily in deserts when water runs across soils or percolates through them, dissolving salts.

Empirical rules have been derived by terrestrial hydrologists to describe relations of slope, sinuosity, rate of flow, etc. among terrestrial water-flow systems, and these have been applied to Mars (Baker, 1982). The general result is that the larger (kilometer-wide) Martian riverbed systems, called **outflow channels** (Figure 2-15), were indeed catastrophic releases of large amounts of water. Consistent with this theory, some of the larger outflow channels originate in discrete box-canyon-like areas of collapsed or "chaotic" terrain, as shown in Figure 13-11. Geologists interpret them as places where underground ice melted in ice-rich soils, causing the surface to collapse into the muddy substrate. This squeezed out the water, which flowed downhill out of the area, eroding a channel.

Closer examination of Mariner and Viking orbiter photos in the 1970s showed various kinds of Martian water flow features. As shown in Figure 13-12, some systems are not large, single outflow channels, but rather finer-scaled dendritic patterns of tributaries, hundreds of meters wide, often feeding into kilometer-scale channels, but spread over wide areas. These are called **valley networks**. They look like runoff systems, and led to controversial suggestions of widely dispersed rainfall on ancient Mars (Masursky and others, 1977)—a radical suggestion for a planet whose present atmosphere and surface can't sustain much liquid water! Other researchers proposed that such systems formed not by rain but by **headward sapping**, a process observed on Earth. If a layer of underground ice or an aquifer is exposed by erosion in a river bank, its melting, sublimation, or erosion may cause collapse and undercutting of the channel wall (Carr, 1996). A stubby side channel then erodes its way back from the main channels, creating a stubby side valley. As shown in Figure 13-13, some Martian channel systems fit this pattern very well.

In many equatorial areas, waters from southern upland channels have emptied onto the low northern plains, creating vast areas sculpted by water flow into flow-line ridges and teardrop streamlined islands (Figure 13-14). Similar streamlined flow features can be seen in dry riverbeds in terrestrial deserts (Figure 13-15).

Some Martian channels indicate distinct episodes of flow. For example, Figure 13-16 shows a high-resolution view of Nanedi Vallis. This winding channel has oxbows, or cut off bends, typical of a river that has established itself, meandering on a flat plain (Figure 13-16a). The high-resolution view shows three episodes of erosion: the main channel, a bench (upper center of Figure 13-16b) formed when the river shifted to the right side of the image, and

Figure 13-11. Martian channel emanating from a collapsed "box canyon" containing chaotic terrain. Note striated flow deposits on channel floor. Restricted area of source indicates that the origin of the flowing liquid was associated with the formation of the chaotic terrain. (NASA, Viking mosaic)

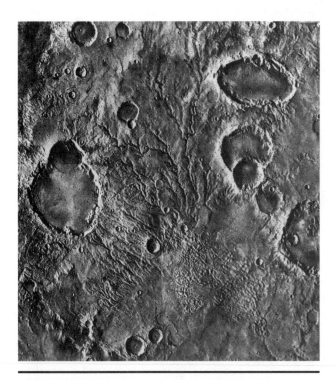

Figure 13-12. Martian channels and tributaries in the cratered uplands northeast of Argyre basin. Width of picture 310 km; main channels about 2 km wide. (NASA, Viking)

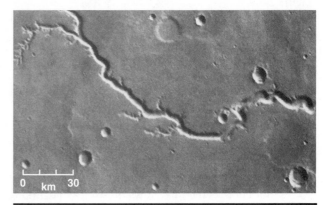

Figure 13-13. Nirgal Vallis, a Martian channel system with stubby tributaries. The tributaries may be cut outward from the channel by headward sapping (see text) (NASA, Viking Orbiter)

an intriguing, final narrow streambed running down the middle of the broader channel. The final streambed marks a late, final episode of flow. The trouble is that we don't know whether these three stages were formed a day apart (during a giant flood) or 100 My apart (during three separate episodes of water flow). The walls show resistant layers, which might require years of cutting if they are rock, but could have been cut in hours if the layering is merely dry mud or weak sediments. So the mystery re-

mains—did repeated episodes of flooding occur, eons apart, or does each river channel mark a single catastrophic flood?

The youngest lava plains, like the recent lavas of the Tharsis region, are mostly uncut by river channel systems. This fits the general idea that most large channel systems are old. An exception is Marte Vallis, which was shown in Figure 13-14b, a large channel system cut into unusually young lavas of Elysium Planitia, just west of Tharsis. Very few impact craters dot its floor. From crater counts, Burr and others (2001) and Berman and Hartmann (2001) found that the background lavas are very young, probably less than 100 My old in many cases, and that the river channel itself may be less than 20 My old—formed in the last percent of the planet's history! What was the source of water in the midst of this volcanic plain? The recent lava eruptions may have melted local underground ice reserves, causing massive water release.

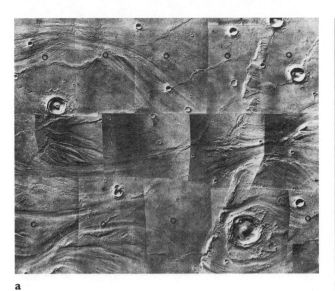

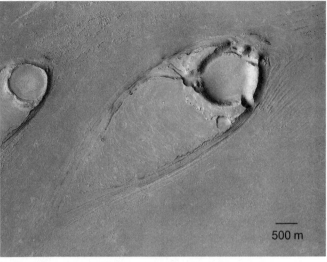

a

b

Figure 13-14. River erosion on Mars. (**a**) Chryse Planitia is modified by massive water flow. Erosion streamlines appear throughout this 200–250 km region. Flow was from west (left) where outflow channels from Figure 2–13 empty into the plain from the highlands. (NASA Viking mosaic) (**b**) Teardrop-shaped "islands" formed when water in Marte Vallis encountered craters, flowed around them, and left streamlined landforms. Note scarcity of impact craters post-dating the river floor. (Mars Global Surveyor; Malin Space Science Systems, NASA, JPL)

Figure 13-15. Terrestrial arroyo in Death Valley. Note terraces and braided deposits on the floor, as found in some Martian channels. In many Martian channels, however, such features at this scale are masked by dust drifts. (Photo by author)

Ancient Martian Seas and Lakes?

Certain workers, notably Parker (in the 1980s and in his 1994 Ph.D. dissertation), Baker (1982), and Clifford and Parker (2001), emphasized the large volumes of water that emptied from riverbeds in the southern uplands into the lower northern plains. Parker, in particular, claimed ancient shorelines could be seen around the edge of Martian lowlands, resembling ancient shorelines on Earth (for example at prehistoric Lake Bonneville in Utah). He said these marked coastlines of ancient Martian seas. Later altimetry data with 1-meter precision, from Mars Global

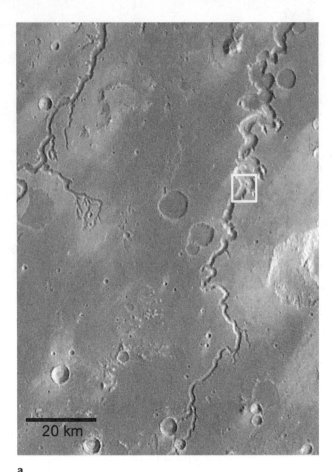

a

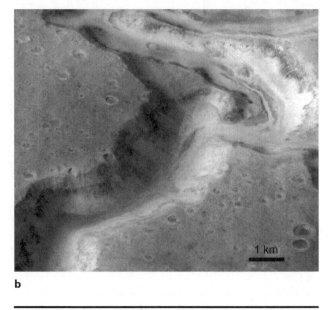

b

Figure 13-16. Possible evidence of a long-time duration of water flow on Mars? (**a**) Nanedi Vallis is a meandering channel with cutoff loops, or "oxbow" formations, which are often evidence of long, slow flow. Upper end of box is the region of **b.** (NASA, Viking orbiter) (**b**) Oblique close-up shows a portion of the winding channel with a much smaller channel down the center, marking a final flow. It is partly obscured by dust drifts. Outcrops halfway down the walls may mark resistant rock layers cut through by the flow. (NASA and Malin Space Science Systems, Mars Global Surveyor)

Surveyor, confirmed that one of Parker's shorelines does fall accurately along a constant-altitude contour—a prerequisite for an ocean surface. Clifford and Parker (2001) also argued that Mars Global Surveyor images at few-meter resolution confirmed some of the ancient shoreline deposits. Nonetheless, the whole subject of ancient Martian oceans, and the size of the largest bodies of water that ever ponded on Mars, remains controversial. On the other hand, deposition of apparent sedimentary layers in some Martian craters does make it seem likely that ancient Martian lakes, at least, existed.

Consideration of the channels, water erosion systems, and possible ancient seas re-emphasizes the great mystery of Mars's history of liquid water and its relation to the possibility of life.

Long-Term Climate Change and the History of Water

The discovery of ancient Martian aqueous features led to speculation that the early Noachian/Hesperian climate—probably 4.5 Gy to about 3.0 or 2.5 Gy ago—was different than today's climate. Various workers have analyzed how Mars's climate could have changed, and this is still an exciting topic of current research.

The first approach seemed obvious—if early Mars had rivers, the climate must have been warmer, in order to keep the ice melted. It was usually assumed that this early warm climate was associated with greenhouse warming due to a thicker initial CO_2 atmosphere. Studies of the Martian atmosphere imply that the total of volatiles degassed by Martian volcanism was more than is now visible, and allowed for high early pressures and abundant water (Plescia, 1993; see also Table 12-2). As early as 1981, Rossabacher and Judson (1981) concluded that if you add up the total H_2O in the present polar caps, in the present atmosphere, and lost into space, you account for only about 10% of what was probably emitted from Mars—and that the rest is stored as underground ice (or **ground ice** in geological parlance) in a **cryosphere**, the Martian global layer of **permafrost** (permanently frozen subsurface soil and rock layers).

Chemical studies allow estimates of the total initial atmosphere, and particular volatiles such as water. For example, McElroy, Yung, and Nier (1976) studied present nitrogen isotope ratios; noting that the lighter N isotopes escape faster than heavier ones, they concluded that as much as 30 mbar of N_2 was degassed into the atmosphere in the past. Other Viking studies of nitrogen and other isotopic systems infer early atmosphere pressures from 140 mbar to more than 1000 mbar, if the gas was all in the atmosphere at one time. For example, Anders and Owen (1977) theorized 140 mbar or more of CO_2 along with an amount of water vapor equivalent to a global ocean 9 m deep—that is, a water layer 9 m deep if spread over a spherical Mars. Donahue (1995) studied deuterium/hydrogen ratios in Martian meteorites; given that H escapes faster than D, the present ratio gives a measure of the total hydrogen, and by extension the total H and water vapor already lost into space. Donahue inferred a total crustal reservoir of water equivalent to a global ocean several hundred meters deep.

To summarize, Noachian Mars probably experienced an atmosphere with pressure in the range of $1/10$ Earth's to about the same as Earth's. Such pressures, of course, would have allowed liquid water, but the question remained whether the temperatures would have been high enough to avoid freezing.

Early models of the greenhouse warming from such atmospheres were not encouraging. The CO_2 greenhouse effect did not seem adequate to maintain Martian surface temperatures above the freezing point. However, Penn State researcher James Kasting (1991) pointed out that the postulated early Martian atmosphere would have been cloudy, and that the clouds would have held in the heat. Kasting's models suggested early Mars might have been barely warm enough to sustain briny liquid water rivers and even lakes or seas. In this scenario, the Martian atmosphere gradually leaked into space due to the low gravity, the greenhouse warming abated, and the water was sequestered as polar ice, underground ice, and water of hydration in Martian minerals.

Eventually, researchers realized that a different model of early Mars might be tenable. In this view, early Mars did have a thicker atmosphere, but the climate was colder, not warmer! A major factor in this idea is that stellar evolution models consistently show that the early sun, in its first Gy or so, was fainter than at present, slowly brightening to its present condition. This is typical of main sequence stars, and it means that the early planets should have received less solar heating. But how could a colder early Mars have rivers?

The idea in this scenario is that with a colder climate, the top of any underground ice layer would have been closer to the Martian surface. Another effect then comes into play. Early Mars (and Earth) had more radioactive material and hence higher heat flow from the interior than modern Mars. (Note that heat flow from the interior is a negligible factor in atmospheric climate, which is dominated by the sun, but it's a major factor in controlling subsurface temperatures and temperature gradient.) The higher early heat flow ensured that the bottom of any early underground ice layer would also have been closer to the surface.

In other words, on a cold early Mars, the whole underground ice layer—top to bottom—would have been closer to the surface. Thus, any slight local warming, due to volcanism or geothermal activity, would have melted subsurface ice. This, in turn, would cause collapse of the surface (explaining chaotic terrain) and release sudden,

short-lived, catastrophic volumes of water, which would rapidly erode riverbeds in loosely consolidated soils (explaining outflow channels). In this view, early Martian rivers were not long-term, stable, or steady flows on a balmy Mars, but were mostly short-lived briny floods that raced across the ground, cut channels, and ponded in craters and other low areas, eventually sinking into the ground and freezing and/or evaporating. Thus, a colder early Mars—paradoxically—could have favored creation of river valleys and short-lived lakes or even seas.

The nature of the earliest Martian climate—and the presence of large bodies of water—remains controversial. However, it seems clear that Noachian Mars, until about 3.5 Gy ago, had thicker atmosphere, warmer underground temperatures, more water flow, and higher rates of cratering, volcanism, and glaciation than today. Hesperian Mars, from about 3.5 Gy to about 2 to 3 Gy ago, marked a time of transition toward today's drier, dustier, less active Mars.

Hillside Gullies, Glaciers, and Recent Climate Changes

The subject of Martian water and its history has been complicated by amazing discoveries from the Mars Global Surveyor's high-resolution camera, suggesting the possibility of very recent cycles of water activity and/or climate change. In other words, not all Martian water activity is associated with ancient thick atmospheres of Noachian time. Malin and Edgett (2001) discovered that many Martian crater walls and hillsides, especially at latitudes around 40° to 70°, have what seem to be water-erosion gullies running down their faces. The gullies usually originate in blocky layers of strata at clifftops, range around 10–20 m wide and a few hundred meters in length, and have triangular fans of debris at their bases. Examples are shown in Figure 13-17a.

Virtually perfect duplicates of these gullies have been found in Iceland (Figure 13-17b), Greenland, and arctic Canada (Hartmann and others, 2003). The similarities have convinced most workers that the Martian gullies are formed by water drainage on the cliffs, although the source of the water remains uncertain. Early work centered on the idea that the water may originate in underground aquifers, exposed in the blocky layers on Martian cliffsides. More recent work centers on possible deposition of ice-rich layers during periods of high obliquity every 10–15 My (Costard and others, 2001; Mustard and others, 2001; see below). The Martian gullies seem most similar to gullies called **debris flows** on Earth, in which soils in a talus slope at the angle of repose become saturated with water (of whatever source); the water lubricates the soils and a slight disturbance can trigger a failure of cohesion, in which a muddy slurry breaks loose, flows down the slope, cuts its own gully as it goes, and deposits a fan of mud and rocks at the base of the slope.

The Martian gullies and their background hillsides are young and uncratered. Some of the debris fans appear

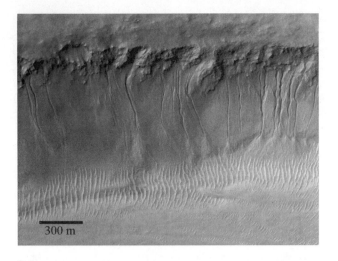

a   b

Figure 13-17. Hillside gullies on Mars and Earth. (**a**) Martian gullies on the walls of Nirgal Vallis, at 30° S latitude. A regional view of Nirgal Vallis is shown in Figure 13-13. (Mars Global Surveyor; NASA, JPL, Malin Space Science Systems, and Planetary Science Institute) (**b**) Near-duplicate gullies, created by water-induced debris flows, on basaltic hillsides in Iceland. (Photo by author)

to be deposited on top of sand dunes, which would normally be considered young, transient features. These aspects suggest to most analysts that the gullies are remarkably young geologic features, from the last few percent of Martian geologic time.

The situation got even more interesting with the discovery of Martian glacier-like features, with linkages to hillside gullies, as found independently on the Mars Global Surveyor images by researchers such as Daniel C. Berman, John Arfstrom, and myself (Hartmann et al., 2003). Figure 13-18 shows examples of an astonishing glacier-like tongue on a pole-facing crater wall at about 40° S latitude. Glacier-like chevron-patterned masses fill valleys on the south wall of the same crater. A number of observers believe these may be flowing masses of ice-rich soil (known on Earth as **rock glaciers**). In other words, arid Mars may have had glaciers at moderate latitudes in the recent past!

Obliquity Cycles and Recent Martian Climate Variations

The discovery of the young, water-flow gullies on Mars triggered its own avalanche (or debris flow?) of studies on water, ice, and climate effects at upper latitudes in 2001–2004. As mentioned above, recent work suggests climate changes due to the change in obliquity (axial tilt), and there are a number of clues that something odd has happened in recent Martian time:

- Ever since Mariner 9 mapping in 1972, researchers have puzzled over the structure of the Martian polar deposits, a mass of material deposited around the poles. These deposits are composed of layer upon layer of thin sediments. The fact that

the deposits are built up of distinct layers suggests cycles of deposition, not continuous deposition. **Unconformities** (discontinuous contacts between strata with different tilts) indicate long passages of time between episodes of peak deposition (Cutts and others, 1976).

- Earth itself has short term climate cycles. For example, the ice ages occur sporadically on 10^4–10^5 y timescales. North America as far south as Chicago and Pittsburgh was buried under an ice sheet 18,000 y ago. Rather abrupt changes have occurred. Earth's ice ages are probably controlled mainly by periodic changes in precession (21,000-y dominant periodicity) and obliquity (41,000-y dominant periodicity) (Hays and others, 1976). This establishes the principle that astronomical factors should be included when considering any planet's climate history.

- As mentioned earlier, Mars turns out to have short term periodic changes in obliquity that may control recent climate and ice deposition. These have been well analyzed. They include 0.12 My and 1.2 My cycles, combining into larger-scale cycles of about 10–15 My, in which the polar axial tilt changes from near 0° to more than 40° (Ward and others, 1975; Costard and others, 2001). More recent French calculations indicate excursions to 80°, and the most common values around 30° to 50°. Today's Mars may be anomalous!

The exciting idea developed around 2001–2004 is that every 10–15 My, when Mars nods 40°–45° toward the sun, massive ice deposits accumulate at high latitudes. The sun shines hotly on the summer pole, ice burns off the

Figure 13-18. Enigmatic, youthful Martian flow-like feature on a crater wall at 38° S latitude. Crater rim is near the top, crater floor is near the bottom. Some researchers interpret this feature as a glacier or rock-glacier.

Mars in extremely recent geologic time (well within the last 1% of the planet's history), and that the Martian climatic environment has gone through substantial short-term cycles of variation. At the same time, the peculiarities of obliquity have been used to suggest an interesting long-term climate change.

Obliquity Cycles and Pre-Tharsis Mars

A remarkable effect was found by Ward and others (1979) who showed that the present range of obliquity is influenced by the topographic bulge of the Tharsis volcanic dome, which covers much of one hemisphere. The bulge affects the moment of inertia. Subtracting the bulge, so as to model a pre-Tharsis, these researchers found an obliquity range larger than that at present, and more recent work has suggested a pre-Tharsis obliquity range higher than in recent time. Such high maxima of obliquity may have exaggerated the effects of summer polar cap burn-off and winter creation of ice deposits at modest latitudes. On the other hand, much of the CO_2 and H_2O of Mars may have been emitted by volcanoes as the huge volume of Tharsis lavas accumulated. This in turn may have been a factor in any Noachian greenhouse effect. Thus, the history of Tharsis may have been critical to the story of Martian climatic change.

Searching for Ancient Water and Ice—The Secret of Mars

The outflow channels, valley networks, hillside gullies, possible glaciers, and polar H_2O ice caps make it clear that Mars has had a rich fluvial history. What happened to the water? Some of it may have broken into H and OH in the high atmosphere and escaped into space, but there is much direct observational evidence that the Martian cryosphere, or permanently frozen layer, contains large reservoirs of ice.

- **Patterned ground** is terrain with interconnecting patterns of polygon- or lattice-shaped furrows and hummocks. On Earth, it is typically produced by expansion and contraction of ice-rich underground soil during freeze-thaw cycles. An example is seen in the right part of Figure 9-16. Even furrows seen on the ground at the Viking 2 landing (Figure 10-36) may be part of patterned ground. Patterns of 10–100 m scale on Earth stem from 100 m-scale depths of ice, and similar scales are seen on Mars. But 10-km scale patterns are also seen on Mars and may indicate much deeper ground ice.

- In brilliant and fundamental research in 1980, Russian geologist Ruslan Kuzmin and, independently, U.S. geologist Joseph Boyce analyzed **rampart craters**—the ones whose ejecta are a splotch of muddy slurry (see review by Squyres and oth-

pole, water vapor content increases 50–100× in the atmosphere, and ice condenses on dust particles in the cold, all-day darkness of the winter latitudes poleward of about 45° latitude. The resulting snowflakes precipitate onto the ground and leave a 10–30 m deep layer of dusty ice (Mustard and others, 2001; Costard and others, 2001). Half a year later, the midnight sun shines all day on these latitudes. Curiously enough, as shown by the French team of Francois Costard and his colleagues (2001), the places on Mars that receive the maximum total solar **insolation,** or total solar energy input, under these summer conditions are pole-facing slopes at moderately high latitudes—and this is just the location where some of the hillside gullies seem most concentrated (although that reported concentration on pole-facing slopes has been contested). So the idea is that the ice—subjected to summer sun, perhaps partly shielded from sublimation under CO_2 frost or a water ice mantle—melts, saturates the pole-facing hillsides, and causes debris flows and gullies.

The exciting ramifications of these discoveries are that liquid water has been an active agent on dry, frozen

ers, 1992). The interpretation was that rampart craters form when projectiles make craters deep enough to penetrate into a subsurface layer of ground ice. Kuzmin and Boyce showed that the higher the latitude, the shallower the crater needed to penetrate to ice and throughout mud. In other words, not only does massive Martian ground ice exist, but the higher the latitude, the closer it is to the surface. Mapping of such patterns by Squyres and his colleagues (1992) suggests depths of 400 m or so to the top of ground ice at low latitudes, and less than 100 m at high latitudes.

- Consistent with this, instruments on Mars Odyssey in 2001 discovered abundant H_2O ice—30% to 50%—in the top 2 m of soil around 65° latitude, both north and south (Boynton and others, 2002).

- Permanent caps of H_2O ice have been found at both the north pole and at the south pole (although the southern water-cap is usually masked by CO_2 frost).

- When water runs across desert soils, it dissolves soluble minerals such as NaCl and other salts, and carbonates and sulfates. When the water then ponds in desert lakes and evaporates, it leaves these materials behind. These minerals, as a class, are called **evaporites.** This is why water in "hard water" areas leaves bathtub rings of white carbonates, and is also why salts can build up in irrigated soil until farming becomes impossible. The salts and other minerals tend to bind loose soils into crumbly layers. The Viking landers discovered such bonded soils, which gained the name **duricrust** (Figure 10-36, lower right corner). The loose dust on Mars was found to contain some evaporite minerals, including 100× the sulfur content of the average of terrestrial crustal minerals (Clark and others, 1977; Rieder and others, 1997). Furthermore, the duricrust was found to be 20%–50% richer in sulfur than the loose dust (Mutch and others, 1977). These results were interpreted as supporting the idea that much of the Martian surface has been exposed to water. (The Viking 1 landing site, indeed, was on the floor of Chryse basin, into which several outflow channels empty.)

- Most Martian meteorite rocks contain salts and carbonates, and analysis has shown that the water necessary to leave such evaporites would be very similar in composition to terrestrial seawater. The Martian rocks need not have been immersed for long periods; transient wetting for only days might suffice. In the case of one Martian igneous rock, the 1300-My-old meteorite Lafayette, water-alteration minerals were abundant enough to be dated by two different labs, both of which found

that the water exposure occurred 670 My ago or somewhat less. These findings are interpreted as indicating that Martian rocks have been exposed to briny water in sporadic episodes postdating the formation of the rock itself (Shih, C.-Y. and others, 1998).

As Mars Global Surveyor and Mars Odyssey mapping of Mars began in 1997 and 2001, respectively, researchers hoped that their orbital spectrometers would map deposits of evaporite minerals of carbonates and salts, which would be the "smoking gun" evidence of ancient lakebeds or seafloors. By 2003, no sizeable (~km scale) deposits had been found. Contrary to some press reports, this does not necessarily prove that Martian lakes or seas never existed, but rather that none are now exposed. If they existed 4 Gy or 3 Gy ago, for example, they may have been covered (and preserved?) by wind-blown dust or sediments, and we might have to be very lucky to find one that has been recently exhumed in pristine condition.

For these reasons, the Spirit and Opportunity Martian rovers (as well as an unsuccessful European lander) were targeted in 2004 to land in low spots where water may have ponded in the past. (The European mission, however, delivered a successful orbiter that has taken awesome photos, and also will search for underground ice layers.) The idea was to look for concentrations of carbonates, salts, or other evaporite minerals that would prove that a lake had formed and evaporated. Spirit landed in a large crater, Gusev, into which an ancient river channel emptied, as seen in Figure 13-19a. The landscape showed scattered boulders, somewhat smaller than at the previous sites (Figure 13-19b). The onboard spectrometer detected some carbonates in the soils but the concentrations do not appear to be unusually high. Any ancient lakebed on Gusev crater may be covered with typical, windblown, wind-mixed soils.

The Opportunity rover landed in an unusual region—a nondescript depression where earlier orbiting spectrometers had indicated a deposit of the iron oxide mineral, grey hematite, sometimes formed by geothermal evaporation of water. (This mineral has also been used to make attractive, shiny, black jewelry.) This rover revealed a landscape totally different from the previous four landing sites! As seen in Figure 13-20a, the photos showed a grey, sandy soil containing hematite, with no scattered boulders. By chance, the rover landed in a small crater about 25 m across, and in the inner walls of this crater was the first discovered outcrop of bedrock on Mars—light-toned, finely layered sediments (Figure 13-20b). The rover drove at once to this outcrop, and close-up photography showed ripple-mark textures typical of sediment deposition under flowing water. Chemical analysis indicated whopping 20%–40% concentrations of sulfates in these rather weak rocks. The initial conclusion: These rocks were once soaked by, or formed under, Martian lake waters.

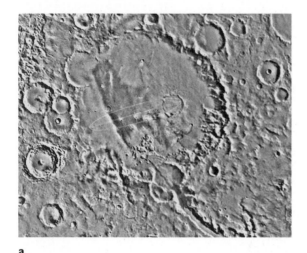

a

Figure 13-19. (**a**) A river channel (bottom) has cut into 150-km crater Gusev, indicating that water must have ponded there in the past. (Mars Global Surveyor; NASA, JPL, Malin Space Science Systems) (**b**) Spirit rover leaves tracks across the Gusev landscape. Lander platform is in distance. (NASA, JPL)

b

Martian Mystery No. 2: Did Life Ever Form on Mars?

Tied to the question of early water is the question of whether life got started on early Mars. This question is now a driver for much of international Mars exploration. It is a perfect scientific question, because either answer is profound. If we find that life did start on Mars (even if it went extinct), that will be the first time in history that we know life formed on another planet, and therefore that we are probably not alone in the universe. In spite of all the aliens we have seen in science fiction movies, we still don't really know whether we are alone! On the other hand, if we go to Mars and find that—in spite of early water—life never evolved on early Mars, then maybe something is wrong with our ideas about the origin of life. Maybe we are more alone than we thought.

If no life or biochemistry ever even started on Mars, we have some right to feel surprised. After all, early Mars—according to current understanding—was rather like early Earth. As we saw in Chapter 11, early Earth had

a

b

Figure 13-20 (**a**) Opportunity rover landed in a 25-m crater in a landscape with sandy gravel, free of boulders, but photos showed the first known layered bedrock outcrops, seen here exposed in the crater's inner rim. (NASA) (**b**) Closeup of rocks shows fine, centimeter-scale layering (left). The rocks contain 20%–40% sulfates; the hematite at the site is concentrated in the tiny spherules (lower corners), interpreted as concretions. Both observations imply the site was once saturated or ponded with Martian water.

Figure 13-21. Spirit rover reached the rim of impact crater Bonneville 66 Martian days after landing nearby. Exploration of such craters in some regions of Mars might reveal ancient underground strata layers, revealed in cross-section by the impact excavation. (NASA, JPL, Cornell)

a CO_2-rich atmosphere with minimal O_2, and both planets had water to some degree. In other words, early Earth was rather similar to early Mars, and if one but not the other produced life, we should want to know why!

Chapter 12 reviewed the controversial evidence that fossil microbes exist in the Martian meteoritic rock ALH 84001. Whether these turn out to be actual Martian fossils, most investigators believe we need to explore Mars itself, by robot or in person, to get the real story of whether life evolved on Mars and, if so, whether it died out. One objective will be to seek more Noachian sedimentary deposits that could reveal whether early life forms existed, or at least whether early conditions were conducive to life. Another objective will be to access underground strata, to see if they contain microbial life or organic material (Figure 13-21). Solar ultraviolet and particle radiation on the present surface will destroy organic molecules and sterilize present soils in a matter of days, but terrestrial bacteria have evolved to exist below the surface, and some seafloor species around so-called "smoker" geothermal vents apparently exist without dependence on sunlight. Thus, life seems excluded from the Martian surface, but life or records of it might exist underground. Still another objective will be to drill down into the Martian ground ice deposits, and better yet into underground aquifers of liquid water (if they exist) to see whether there is evidence of Martian microbes or organic materials associated with Martian water.

A challenge is to do all this without contaminating Mars with terrestrial microbes. Mars offers at least partial answers to profound questions about the distribution of life in the universe. We don't want to lose the opportunity by contaminating Mars so badly that we can't answer the question of whether life evolved there—or the more subtle question of whether life transferred from Earth to Mars long ago via meteoritic rock fragments. A first test of whether any organic material, or microbes, are native to Mars will be to study racemic ratios of organic molecules—that is, the ratio of left-handed to right-handed structures. A 50-50 ratio is the signature of non-biological synthesis. A markedly different ratio characterizes terrestrial biological materials, and a still different ratio on Mars could prove that organics there originated in native Martian organisms.

To summarize, the next-door-neighbor planet, Mars, offers our best and most immediate chance to learn about the frequency of life in the universe.

SUMMARY

Mars is a planet rich in "big picture" scientific issues. Many of the issues revolve around the role of water on Mars: How were river channels cut on a planet now dry and frozen? Possible solutions involve simply a thicker initial atmosphere, or transient volcanic outbursts of gas, or a greenhouse effect produced by more CO_2 and water vapor, or climatic oscillations involving changes in obliquity and planetary moment of inertia.

Regardless of the outcome of controversies about fossil microbes in Mars meteorite ALH 84001, buried in its soils may be treasures: the answers to questions humans have asked for many years about the nature and stability of planets, the origins of life on Earth, and the probabilities that organic chemistry has produced life throughout the universe. Martian landers are continuing the quest to learn if lakes or seas existed on Mars, and if they spawned simple life forms.

CONCEPTS

dry ice
triple point for H_2O
channels
outflow channels
valley networks
headward sapping
ground ice
cryosphere
permafrost

debris flows
rock glaciers
unconformities
insolation
patterned ground
rampart craters
evaporites
duricrust

Planetary Data Table

Object[b]	Semi-major axis (AU for planets & asteroids; 10³ km for satellites)	Revolution period (days unless marked otherwise)	Orbit inclination (degrees; with respect to ecliptic for planets & asteroids; w.r.t. planet equator for satellites)	Orbit eccentricity	Obliquity (degrees)	Rotation period (days; R = retrograde)	Diameter (km)
Sun	—	—	—	—	7.25	25.4	1,391,400
Mercury	0.387 AU	87.97	7.00	0.206	0	58.65	4878
Venus	0.723 AU	224.70	3.39	0.007	178	243.0 R	12,104
Earth	1.000 AU	365.26	0.00	0.017	23.4	1.00	12,756
*Moon	384.4	27.32	18 to 29	0.055	6.7	27.32	3476
Mars	1.524 AU	686.98	1.85	0.093	25.0	1.026	6787
*Phobos	9.38	0.319	1.0	0.018	~0	0.319	27 × 19
*Deimos	23.50	1.262	2.8	0.002	~0	1.262	15 × 11
Largest Belt Asteroids							
1 Ceres	2.768 AU	4.61y	10.60	0.077	?	0.378	914
4 Vesta	2.362 AU	3.63y	7.14	0.090	?	0.223	500
2 Pallas	2.773 AU	4.62y	34.80	0.233	?	0.328	522
10 Hygiea	3.138 AU	5.56y	3.84	0.118	?	~0.75	443
511 Davida	3.181 AU	5.67y	15.90	0.172	?	0.363	336
Jupiter	5.203 AU	11.86y	1.30	0.048	3.08	0.410	142,800
J16 Metis	128	0.295	0.0	0.0	~0?	0.30	40
J15 Adrastea	129	0.298	0.0	0.0	~0?	0.29	24 × 16
J5 Amalthea	181.5	0.498	0.45	0.003	~0?	0.498	270 × 155
J14 Thebe	222	0.674	0.8?	0.01	~0?	0.674	100
*J1 Io	422	1.769	0.03	0.000	~0	1.769	3630
*J2 Eruopa	671	3.551	0.46	0.000	~0	3.551	3130
*J3 Ganymede	1071	7.155	0.18	0.002	~0	7.155	5280
*J4 Callisto	1884	16.689	0.25	0.008	~0	16.689	4840
J13 Leda	11,110	240	26.7	0.146	?	?	~16
J6 Himalia	11,470	250.6	27.6	0.158	?	?	~180
J10 Lysithea	11,710	260	29.0	0.12	?	?	~40
J7 Elara	11,740	260.1	24.8	0.207	?	?	~80
J12 Ananke	20,700	617	147	0.169	?	?	~30
J11 Carme	22,350	692	163	0.207	?	?	~44
J8 Pasiphae	23,300	735	147	0.40	?	?	~35
J9 Sinope	23,700	758	156	0.275	?	?	~20
Trojan Asteroids							
624 Hektor	5.12	11.58y	18	0.02	~80	0.29	~300 × 150
911 Agamemnon	5.15	11.69y	22	0.07	?	0.33	~148

[a]Data come from various sources including reports on Jupiter and Saturn systems by Voyager team members (*Science* 1 June 1979; 23 Nov. 1979; 10 April 1981; 29 January 1982); tabulations in *Proc. Lun. Plan. Sci. Conf. 1979, 9*: inside covers; S. R. Taylor, *Planetary Science: A Lunar Perspective,* Houston: Lunar and Planetary Science Institute (1982); J. K. Beatty, et al. (editors), *The New Solar System,* Cambridge: Sky Publishing Corp. (1990); R. Binzel, T. Gehrels, M. Matthews (editors), *Asteroids II,* Tucson: University of Arizona Press (1989); and additional scientific reports.

Radius (km)	Mass (kg)	Mean density (kg/m³)	Visual geometric albedo	Escape velocity at surface (m/s)	Surface materials (known or probable)	Atmosphere (Main constituents)
695,700	1.99 (30)	1410	—	617,800	Ionized H and He gas	H + He
2439	3.30 (23)	5420	0.12	4300	Igneous rock (basaltic?)	None
6052	4.87 (24)	5250	0.59	10,400	Basaltic (+ granitic?) rock	CO_2
6378	5.98 (24)	5520	0.39	11,200	Water; basaltic + granitic rock	$N_2 + O_2$
1738	7.35 (22)	3340	0.11	2380	Anorthositic + basaltic rock + dust	None
3398	6.42 (23)	3940	0.15	5000	Basaltic rock + dust	CO_2
14 × 10	9.6 (15)	~1900	0.05	11	Carbonaceous rock + dust	None
8 × 6	2.0 (15)	~2100	0.05	6	Carbonaceous rock + dust	None
510	?	?	0.06	~650	Carbonaceous rock or dust	None
274	~2.5 (20)	~2900	0.24	~350	Basaltic rock or dust	None
269	?	?	0.08	~340	Silicate rock	None
222	?	?	0.05	~280	Carbonaceous rock or dust	None
171	?	?	0.04	~220	Carbonaceous rock or dust	None
71,400	1.90 (27)	1314	0.44	59,500	?	$H_2 + He$
20	?	?	low	~20	Rock	None
12 × 8	?	?	0.04	~20	Rock	None
135 × 78	?	?	0.05	~120	Rock + sulfurous coating?	None
50	?	?	~0.10	~50	Rock	None
1820	8.89 (22)	3530	0.63	2560	Sulfur compounds	Thin SO_2
1565	4.79 (22)	3030	0.64	2040	H_2O ice	None
2640	1.48 (23)	1930	0.43	2740	H_2O ice + carbonaceous dirt	None
2420	1.08 (23)	1790	0.17	2420	H_2O ice + carbonaceous dirt	None
~8	?	?	?	~4	Carbonaceous dirt + H_2O ice?	None
~90	?	?	0.03	~90	Carbonaceous dirt + H_2O ice?	None
~20	?	?	?	~10	Carbonaceous dirt + H_2O ice?	None
~40	?	?	0.03	~30	Carbonaceous dirt + H_2O ice?	None
~15	?	?	?	~8	Carbonaceous dirt + H_2O ice?	None
~22	?	?	?	~12	Carbonaceous dirt + H_2O ice?	None
~35	?	?	?	~14	Carbonaceous dirt + H_2O ice?	None
~20	?	?	?	~10	Carbonaceous dirt + H_2O ice?	None
~150 × 50	?	?	0.02	~120	Carbonaceous dirt + H_2O ice?	None
~74	?	?	0.04	~80	Carbonaceous dirt + H_2O ice?	None

(continued)

[b]A traditional letter/number designation scheme is indicated for some of the satellites. Generally, the larger ones discovered before the Voyager flights were given the first, lowest numbers, and others were numbered in order of distance from the planet.

*These are the most-discussed satellites and most important in terms of their unique features. Students should know some of the interesting characteristics of each of these moons.

Object[b]	Semi-major axis (AU for planets & asteroids; 10^3 km for satellites)	Revolution period (days unless marked otherwise)	Orbit inclination (degrees; with respect to ecliptic for planets & asteroids; w.r.t. planet equator for satellites)	Orbit eccentricity	Obliquity (degrees)	Rotation period (days; R = retrograde)	Diameter (km)
Saturn	9.539	29.46y	2.49	0.056	26.7	0.426	120,660
S18 Pan	133.6	0.576	0	0	?	?	~20
S15 Atlas	137.6	0.601	0.3	0.002	?	?	50 × 20
S16 Prometheus	139.4	0.613	0.0	0.003	?	?	140 × 80
S17 Pandora	141.7	0.629	0.05	0.004	?	?	110 × 70
*S11 Epimetheus	151.4	0.694	0.34	0.009	?	?	140 × 100
*S10 Janus	151.4	0.695	0.14	0.007	?	?	220 × 160
*S1 Mimas	185.54	0.942	1.5	0.020	~0	0.942	390
*S2 Enceladus	238.04	1.370	0.0	0.004	~0	1.370	500
*S3 Tethys	294.67	1.888	1.1	0.000	~0	1.888	1048
S13 Calypso[c]	294.67 (Trail)	1.888	~1?	0.0	?	?	34 × 22
S14 Telesto[c]	294.67 (Lead)	1.888	~1?	0.0	?	?	34 × 26
S4 Dione	377	2.737	0.0	0.002	~0	2.74	1120
S12 Helene	377	2.74	0.15	0.005	?	?	36 × 30
*S5 Rhea	527	4.518	0.4	0.001	~0	4.52	1530
*S6 Titan	1222	15.94	0.3	0.029	~0?	15.9?	5150
*S7 Hyperion	1484	21.28	~0.5	0.104	?	?	350 × 200
*S8 Iapetus	3564	79.33	14.72	0.028	~0	79.33	1440
*S9 Phoebe	12,930	550.4	150	0.163	?	0.4?	220
2060 Chiron	13.70	50.7y	6.92	0.379	?	0.25	200
Uranus	19.18	84.01y	0.77	0.047	97.9	0.75	51,118
U6 Cordelia	49.7	0.335	~0.14	~0	0?	?	~30
U7 Ophelia	53.8	0.376	~0.09	~0.01	0?	?	~30
U8 Bianca	59.2	0.435	~0.16	~0	0?	?	~40
U9 Cressida	61.8	0.464	~0.04	~0	0?	?	~70
U10 Desdemona	62.7	0.474	~0.16	~0	0?	?	~60
U11 Juliet	64.4	0.493	~0.06	~0	0?	?	~80
U12 Portia	66.1	0.513	~0.09	~0	0?	?	~110
U13 Rosalind	69.9	0.558	~0.28	~0	0?	?	~60
U14 Belinda	75.3	0.624	~0.03	~0	0?	?	~70
U15 Puck	86.0	0.762	~0.31	~0	0?	?	150
*U5 Miranda	129.8	1.414	3.40	0.00	0	1.414	470
*U1 Ariel	191.2	2.520	0.00	0.00	0	2.520	1160
*U2 Umbriel	266.0	4.144	0.00	0.00	0	4.144	1170
*U3 Titania	435.8	8.706	0.00	0.00	0	8.706	1580
*U4 Oberon	582.6	13.463	0.00	0.00	0	13.463	1520
Neptune	30.06	164.8y	1.77	0.009	29.6	0.80	49,528
N3 Naiad	48.0	0.296	~0.0	~0	0?	?	~50
N4 Thalassa	50.0	0.312	~4.5	~0	0?	?	~80
N5 Despina	52.2	0.333	~0.0	~0	0?	?	~180
N6 Galatea	62.0	0.429	~0.0	~0	0?	?	~150
N7 Larissa	73.6	0.554	~0.0	~0	0?	?	~190
*N8 Proteus	117.6	1.121	~0.0	~0	0?	?	~400
*N1 Triton	354.8	5.877	157	0.00	0	5.877	2700
*N2 Nereid	5513.4	360.16	29	0.75	?	?	~340
Pluto	39.53	247.7y	17.15	0.248	122.5	6.39	2300
*P1 Charon	19.6	6.39	98.8		0	6.39	1190

[a] Data come from various sources including reports on Jupiter and Saturn systems by Voyager team members (*Science* 1 June 1979; 23 Nov. 1979; 10 April 1981; 29 January 1982); tabulations in *Proc. Lun. Plan. Sci. Conf. 1979, 9*: inside covers; S. R. Taylor, *Planetary Science: A Lunar Perspective*, Houston: Lunar and Planetary Science Institute (1982); J. K. Beatty, et al. (editors), *The New Solar System*, Cambridge: Sky Publishing Corp. (1990); R. Binzel, T. Gehrels, M. Matthews (editors), *Asteroids II*, Tucson: University of Arizona Press (1989); and additional scientific reports.

[b] A traditional letter/number designation scheme is indicated for some of the satellites. Generally, the larger ones discovered before the Voyager flights were given the first, lowest numbers, and others were numbered in order of distance from the planet.

other objects in the solar system.[a]

Radius (km)	Mass (kg)	Mean density (kg/m^3)	Visual geometric albedo	Escape velocity at surface (m/s)	Surface materials (known or probable)	Atmosphere (Main constituents)
60,330	5.69 (26)	690	0.46	35,600	?	H_2 + He
~10	?	?	?	~10	H_2O ice?	None
25 × 10	?	?	0.4	15	Dirty H_2O ice?	None
70 × 40	?	?	0.6	50	H_2O ice?	None
55 × 35	?	?	0.6	35	H_2O ice?	None
70 × 50	?	?	0.5	45	Dirty H_2O ice?	None
110 × 80	?		0.5	75	Dirty H_2O ice?	None
196	4.5 (19)	1200	0.6	170	H_2O ice	None
250	8.4 (19)	1200	0.95	205	H_2O ice	None?
530	7.5 (20)	1210	0.7	435	H_2O ice	None
17 × 11	?	?	0.8	11	H_2O ice?	None
17 × 13	?	?	0.6	12	H_2O ice?	None
560	1.05 (21)	1430	0.5	500	Dirty H_2O ice	None
18 × 15	?	?	0.5	14	Dirty H_2O ice?	None
765	2.49 (21)	1330	0.6	660	H_2O ice	None
2575	1.35 (23)	1880	0.2	2640	Liquid CH_4 or ices?	N_2
205 × 110		?	0.3	160	Dirty ice?	None
730	1.88 (21)	1160	0.05 (L), 0.5 (T)[d]	590	H_2O-icy dirt(L); H_2O ice (T)[d]	None
110	?	?	0.06	90	Ice + carbonaceous dirt?	None
100	?	2000?	~0.04		Carbonaceous dirt + ice	Coma
25,559	8.68 (25)	1290	0.56	21,300	?	H_2 + He
~15	?	?	<0.1	~10	Carbonaceous dirt + ice (?)	None
~15	?	?	<0.1	~10	Carbonaceous dirt + ice (?)	None
~20	?	?	<0.1	~20	Carbonaceous dirt + ice (?)	None
~35	?	?	<0.1	~30	Carbonaceous dirt + ice (?)	None
~30	?	?	<0.1	~30	Carbonaceous dirt + ice (?)	None
~40	?	?	<0.1	~40	Carbonaceous dirt + ice (?)	None
~55	?	?	<0.1	~50	Carbonaceous dirt + ice (?)	None
~30	?	?	<0.1	~30	Carbonaceous dirt + ice (?)	None
~35	?	?	<0.1	~30	Carbonaceous dirt + ice (?)	None
75	?	?	0.07	~70	Carbonaceous dirt + ice (?)	None
235	6.89 (19)	1350	0.34	200	H_2O ice	None
580	1.26 (21)	1660	0.40	560	H_2O ice	None
585	1.33 (21)	1510	0.19	537	H_2O ice	None
790	3.48 (21)	1680	0.28	770	H_2O ice	None
760	3.03 (21)	1580	0.24	710	H_2O ice	None
24,764	1.02 (26)	1640	0.51	23,300	?	H_2 + He
~25	?	?	~0.06	~25	Carbonaceous dirt + ice (?)	None
~40	?	?	~0.06	~40	Carbonaceous dirt + ice (?)	None
~90	?	?	~0.06	~90	Carbonaceous dirt + ice (?)	None
~75	?	?	0.054	~75	Carbonaceous dirt + ice (?)	None
~95	?	?	0.056	~95	Carbonaceous dirt + ice (?)	None
~200	?	?	0.060	~200	Carbonaceous dirt + ice (?)	None
1350	2.14 (22)	2070	~0.75	1452	N_2 ice, CH_4 ice	N_2, CH_4
~170	?	?	0.14	~170	Carbonaceous dirt + ice (?)	None
1150	1.29 (22)	2030	~0.5	1100	CH_4 ice, (N_2 ice?)	Thin CH_4
595	1.77 (21)	~2000	~0.4	~590	H_2O ice	None

[c] Calypso and Telesto are in L_S and L_T Lagrangian point positions, respectively, relative to Tethys.

[d] L = Leading hemisphere. T = Trailing hemisphere.

*These are the most-discussed satellites and most important in terms of their unique features. Students should know some of the interesting characteristics of each of these moons.

References

Chapter One

Boorstein, D. J. 1983. *The Discoverers.* New York: Random House.

Crisswell, D. R. (ed.) 1977. *New Moons.* Houston: Lunar and Planetary Institute.

Dingle, H. 1973. "Philosophical Aspects of Cosmology." *Astronomy 1*:162.

Dyson, F. 1969. "Human Consequences of the Exploration of Space." *Bull. Atomic Scientists,* September, p. 8.

Gaffey, M. J., and T. McCord, 1977. "Mining the Asteroids." *Mercury 6:1* (No. 6, Nov.-Dec.).

Hartmann, W. K., R. Miller, and P. Lee. 1984. *Out of the Cradle.* New York: Workman Publishing Co.

Kuhn, T. S. 1962. *The Structure of Scientific Revolutions.* Chicago: University of Chicago Press.

Kulkami, S. R. 1997. "Brown Dwarfs: A Possible Missing Link Between Stars and Planets." *Science 276*:350–1354.

Lewis, J., M. S. Matthews, and M. L. Guerrieri (ed.) 1993. *Resources of Near-Earth Space.* Tucson: University of Arizona Press.

Lewis, J. S., and R. A. Lewis 1987. *Space Resources.* New York: Columbia Univ. Press.

Nicks, O. W. (ed.) 1970. *This Island Earth.* Washington, D.C.: *NASA SP-250.*

Novak, B. 1980. *Nature and Culture.* New York: Oxford University Press.

Stern, S. Alan. 2002. "Journey to the Farthest Planet." *Sci. Am. 286,* No. 5 March, pp. 56–63.

Udall, M. K. 1969. "The Environment—What You Can Do." *Congressman's Report 9,* May 20.

Wetherill, G. 1979. "Apollo Objects." *Sci. Am.,* March, p. 54.

Chapter Two

Alexander, A. F. O. 1962. *The Planet Saturn.* New York: Macmillan.

Anderson, L., and J. Fix. 1973 "Pluto: New Photometry and a Determination of the Axis of Rotation." *Icarus 20*:279.

Anderson, J. D., W. Sjögren, and G. Schubert 1996. "Galileo Gravity Results and the Internal Structure of Io." *Science 272*:709–712.

Beatty, J. K., and others, eds. 1981. *The New Solar System.* Cambridge, Mass.: Sky Publishing.

Beatty, J. K. and S. J. Goldman 1994. "The Great Crash of 1994: A First Report." *Sky and Telescope 88*:18–23.

Beatty, J. K., and others (ed.). 1981. *The New Solar System.* Cambridge, MA: Sky Publishing.

Betz, A. L., E. C. Sutton, R. McLaren, and C. McAlavy. 1977. "Laser Heterodyne Spectroscopy." In *Proceedings of the Symposium on Planetary Atmospheres,* ed. A. V. Jones. Ottawa: Royal Society of Canada.

Bigg, E. K. 1964. "Influence of the Satellite Io on Jupiter's Decametric Emission." *Nature 203*:1008.

Binder, A. B., and D. P. Cruikshank. 1964. "Evidence for an Atmosphere on Io." *Icarus 3*:299.

Binzel, R. P. 1990. "Pluto." *Sci. Am. 262*:50.

Brown, R. A. 1973. "Optical Line Emission from Io." In *Exploration of the Planetary System,* ed. A. Woszczyk and C. Iwaniszewska. Boston: Reidel.

Carusi, A., G. Valsecchi, and L. Kresak. 1981. "Orbital Patterns at Close Encounters." Paper presented at International Astronomical Union Colloquium 61 on Comets, Tucson.

Chapman, C. R. 1968. "The Discovery of Jupiter's Red Spot." *Sky and Telescope* 35:276.

———. 1977. *The Inner Planets.* New York: Scribner's.

C. R. 1968. "The Discovery of Jupiter's Red Spot." *Sky and Telescope* 35:276.

———. 1977. *The Inner Planets.* New York: Scribner's.

Christy, J. W., and R. Harrington. 1978. "The Satellite of Pluto." *Astron. J. 83:*1005.

Clark, R. N. 1980. "Ganymede, Europa, Callisto, and Saturn's Rings: Compositional Analysis from Reflectance Spectroscopy." *Icarus* 44:388.

Cruikshank, D. P., C. Pilcher, and D. Morrison. 1976. "Pluto: Evidence for Methane Frost." *Science* 194:835.

Cruikshank, D. P., and C. R. Chapman. 1967. "Mercury's Rotation and Visual Observations." *Sky and Telescope* 34:24.

Cuzzi, J. N., and J. B. Pollack. 1978. "Saturn's Rings" Partial Composition and Size Distribution as Constrained by Microwave Observations." *Icarus* 33:233.

Dyce, R. B., G. H. Pettengill, and I. I. Shapiro. 1967. "Radar Determinations of the Rotations of Venus and Mercury." *Astron. J.* 72:351.

Elliot, J. L., E. Dunham, and D. Mink. 1977. "The Rings of Uranus." *Nature* 267:328.

Fanale, F., T. Johnson, and D. Matson. 1974 "Io: A Surface Evaporite Deposit?" *Science* 186:922.

Florensky, C. P., A. Bailevsky, and A. Pronin. 1977. "The First Panoramas of the Venusian Surface: Geological-Morphological Analysis of Pictures." *COSPAR Space Research* 17:645.

Focas, J. H., and A. Dollfus. 1969. "Optical Characteristics and Thickness of Saturn's Rings Observed on the Ring Plane in 1966." *Astron. Astrophys.* 1:251.

Gladman, B. J. and others. (2001). "Discovery of 12 Satellites of Saturn Exhibiting Orbital Clustering." *Nature* 412:163–166.

Goldreich, P., and S. J. Peale. 1967. "Spin-Orbit Coupling in the Solar System II: The Resonant Rotation of Venus." *Astron. J.* 72:662.

Goldreich, P., and S. Soter. 1966. "Q in the Solar System." *Icarus* 5:375.

Goldreich, P., N. Murray, P. Longaretti, and D. Banfield. 1989. "Neptune's Story." *Science* 245:500.

Greenberg, R., D. Davis, W. Hartmann, and C. Chapman. 1977. "Size Distribution of Particles in Planetary Rings." *Icarus* 30:769.

Guerin, P. 1970. "The New Ring of Saturn." *Sky and Telescope* 40:88.

Hartmann, W. K. 1980. "Surface Evolution of Two-Component Stone/Ice Bodies." *Icarus* 44:441.

Hartmann, W. K. 2003. *A Traveler's Guide to Mars.* New York: Workman Publishing Co.

Hayes, S. H., and M. Belton. 1977. "The Rotational Periods of Uranus and Neptune." *Icarus* 32:383.

Hoyt, W. G. 1976. *Lowell and Mars.* Tucson: University of Arizona Press.

Ingersoll, Andrew P. 1990. "Atmospheres of the Giant Planets." In *The New Solar System.* Cambridge, Mass.: Sky Publishing Corp.

Jewitt, D. 1979. "Discovery of a New Jupiter Satellite." *Science* 206:951.

Keeler, J. E. 1895. "The Spectroscopic Proof of the Meteoric Constitution of Saturn's Rings." *Astrophys. J. 1:*416.

Keldysh, M. V. 1977. "Venus Exploration with the Venera 9 and Venera 10 Spacecraft." *Icarus 30:*605.

Kowal, C. T. 1978. "Letter to the editor." *Sky and Telescope* 55:195.

Kuiper, G. P. 1944. "Titan, A Satellite with an Atmosphere." *Astrophys. J. 100:*378.

———. 1956. "The Formation of the Planets, I, II, III." *J. Roy. Astron. Soc. Can. 50,* 57:105.

———. 1957. "Further Studies on the Origin of Pluto." *Astrophys. J.* 125:287.

Kuiper, G. P., D. P. Cruikshank, and U. Fink. 1970. "Letter to the editor." *Sky and Telescope* 39:80.

Larson, S. 1977. "Faint Inner Satellites of Saturn." Presented at the 8th annual meeting of the Division for Planetary Sciences, American Astronomical Society, Honolulu, Jan.

Lebofsky, L. A. 1975. "Stability of Frosts in the Solar System." *Icarus* 25:205.

Lebofsky, L. A., T. V. Johnson, and T. B. McCord. 1970. "Saturn's Rings: Spectral Reflectivity and Compositional Implications." *Icarus* 13:226.

Luu, Jane X., and D. Jewitt. 1988. "A Two-part Search for Slow-moving Objects." *Astron. J.* 95:1256.

Lyttleton, R. A. 1936. "On the Possible Results of an Encounter of Pluto with the Neptunian System." *Monthly Notices Roy. Astron. Soc.* 97:108.

———. 1968. *Mysteries of the Solar System.* New York: Oxford University Press.

Macy, W., and W. Sinton. 1977. "Detection of Methane and Ethane Emission on Neptune but not on Uranus." *Astrophys. J.* 218:L74.

McCord, T. B. 1969. "Comparison of the Reflectivity and Color of Bright and Dark Regions on the Surface of Mars." *Astrophys. J.* 156:79.

McKinnon, W. B., and H. Melosh. 1980. "Evolution of Planetary Lithospheres: Evidence from Multiringed Structures on Callisto and Ganymede." *Icarus* 44:454.

Minton, R. B. 1973. "The Red Polar Caps of Io." *Comm. Lunar Planet. Lab.* 10:35.

Murray, B. C., M. Belton, G. Danielson, M. Davies, D. Gault, B. Hapke, B. O'Leary, R. Strom, V. Suomi, and N. Trask. 1974. "Venus: Atmospheric Motion and Structure from Mariner 10 Pictures." *Science* 183:1307.

Owen, T. C., and G. P. Kuiper. 1964. "A determination of the Composition and Surface Pressure of the Martian Atmosphere." *Comm. Lunar Planet. Lab.* 2:113.

Peek, B. M. 1958. *The Planet Jupiter.* New York: Macmillan.

Pettengill, G. H., and R. B. Dyce. 1965. "A Radar Determination of the Rotation of the Planet Mercury." *Nature* 206:1240.

Phillips, R., R. Grimm, and M. Malin. 1991. "Hotspot Evolution and the Global Tectonics of Venus." *Science* 252:651.

Pilcher, C. P., C. R. Chapman, L. A. Lebofsky, and H. H. Kieffer. 1970. "Saturn's Rings: Identification of Water Frost." *Science* 167:1372.

Pollack, J. B. 1973. "Mariner 9 Television Observation of Phobos and Deimos, 1." *J. Geophys. Res.* 78:4313.

———. 1975. "The Rings of Saturn." *Space Sci. Rev. 18:*3.

———. 1978. "Near-Infrared Spectra of the Galilean Satellites: Observations and Compositional Implications." *Icarus* 36:271.

Pollack, J. B., D. Strecker, F. Witteborn, E. Erickson, and B. Baldwin. 1978. "Properties of the Clouds of Venus, as Inferred from Airborne Observations of Its Near-IR Reflectivity Spectrum." *Icarus,* 34:28.

Pollack, J. B. 1973. "Mariner 9 Television Observation of Phobos and Deimos, 1." *J. Geophys. Res.* 78:4313.

————. 1978. "Near-Infrared Spectra of the Galilean Satellites: Observations and Compositional Implications." *Icarus* 36:271.

Prinn, R. G., and T. Owen. 1976. "Chemistry and Spectroscopy of the Jovian Atmosphere." In *Jupiter*, ed. T. Gehrels and M. Matthews. Tucson: University of Arizona Press.

Reese, E. J., and B. A. Smith. 1968. "Evidence of Vorticity in the Great Red Spot of Jupiter." *Icarus* 9:474.

Saunders, R. S., and G. H. Pettengill. 1991. "Magellan: Mission Summary." *Science* 252:247. (See also succeeding articles.)

Smith, B. A. 1967. "Rotation of Venus: Containing Contradictions." *Science* 158:114.

————. 1990. "The Voyager Encounters." In *The New Solar System*, ed. J. K. Beatty and A. Chaikin. Cambridge, Mass.: Sky Publishing Corp.

Smith, B. A., and 64 others. 1989. "Voyager 2 at Neptune: Imaging Science Results." *Science* 246:1422.

Smith, B. A., and Voyager Imaging Team. 1979. "The Jupiter System through the Eyes of Voyager 1." *Science* 204:951.

Smith, B. A., and others. 1981. "Encounter with Saturn: Voyager 1 Imaging Science Results." *Science* 212:163.

Soderblom, L. A. 1980. "The Galilean Moons of Jupiter." *Sc. Am.*, 242 No. 1, Jan., p. 88.

Stern, S. A., D. Weintraub, and M. Festou 1993. "Evidence for a Low Surface Temperature on Pluto from Millimeter-Wave Thermal Emission Measurements." *Science* 262:1713–1716.

Stern, S. A. 1992. "The Pluto-Charon System." *Ann. Rev. Astron. Astrophys.* 30:185–233.

Stone, E. C., and E. D. Miner. 1989. "The Voyager 2 Encounter with the Neptunian System (1989)." *Science* 246:1417.

Trafton, L., T. Parkinson, and W. Macy. 1974. "The Spatial Extent of Sodium Emission around Io." *Astrophys. J.* 190:L85.

Trafton, L. 1977. "Uranus' Rotational Period." *Icarus* 32:402.

Walker, M. F., and R. Hardie. 1955. "A Photometric Determination of the Rotational Period of Pluto." *Publ. Astron. Soc. Pacific* 67:224.

Witteborn, F., J. Bregman, and J. Pollack. 1979. "Io: An Intense Brightening near 5μm." *Science* 203:643.

Young, A. T. 1973. "Are The Clouds of Venus Sulfuric Acid?" *Icarus* 18:564.

Young, L. D. G., and A. T. Young. 1973. "Comment on 'The Composition of the Venus Cloud Tops in Light of Recent Spectroscopic Data.'" *Astrophys. J.* 179:L39.

Chapter Three

Aggarwal, H. R., and V. Oberbeck. 1974. "Roche Limit of a Solid Body." *Astrophys. J.* 191:577.

Biermann, L. 1951. "Kometenschweife und solare Korpuskularstrahlung." *Z. Astrophys.* 29:274.

Bottke, William F. and 4 others. 2001. Dynamical Spreading of Asteroid Families by the Yarkovsky Effect." *Science* 294:1693–1696.

Bruman, J. R. 1969. "A Lunar Libration Point Experiment." *Icarus* 10:197.

Burns, J. A., P. Lamy, and S. Soter. 1979. "Radiation Forces on Small Particles in the Solar System." *Icarus* 40:1.

Burns, J. A. 1981. "Planetary Rings." In *The New Solar System*, ed. J. K. Beatty and others. Cambridge, Mass.: Sky Publishing.

Columbo, G., P. Goldreich, and A. Harris. 1976. "Spiral Structure as an Explanation for the Asymmetric Brightness of Saturn's A Ring." *Nature* 264:344.

Cuzzi, J. 1978. "The Rings of Saturn." In *The Saturn System*, ed. D. Hunten and D. Morrison. Washington, D.C.: NASA CP 2068.

Darwin, G. H. 1962. *The Tides and Kindred Phenomena in the Solar System*. San Francisco: W. H. Freeman. (Originally published 1898.)

Dermott, S., and C. Murray. 1981. "The Dynamics of Tadpole and Horseshoe Orbits. I and II." *Icarus* 48:1 and 12.

Farinella, P., D. Vokrouhlicky, W. Hartmann, D. Davis, and S. Weidenschilling 1998. "Meteorite Delivery via Yarkovsky Orbital Drift." *Icarus,* in press.

Franklin, F. A., and G. Colombo. 1978. "On the Azimuthal Brightness Variations of Saturn's Rings." *Icarus* 33:279.

Froeschle, Cl. and R. Greenberg 1989. "Mean Motion Resonances." In *Asteroids II*, ed. R. Binzel, T. Gehrels, and M. Matthews. Tucson: University of Arizona Press.

Gehrels, T. 1977. "Some Interrelations of Asteroids, Trojans, and Satellites." In *Comets, Asteroids, Meteorites*, ed. A. Delsemme. Toledo: University of Toledo Press.

Goldreich, P., and S. Tremaine. 1978. "The Velocity Dispersion in Saturn's Rings." *Icarus* 34:227.

————. 1979. "Towards a Theory for the Uranian Rings." *Nature* 277:97.

Greenberg, R., D. Davis, W. Hartmann, and C. Chapman. 1977. "Size Distribution of Particles in Planetary Rings." *Icarus* 30:796.

Greenberg, R., and D. R. Davis. 1978. "Stability at Potential Maxima: The L^4 and L^5 Points of the Restricted Three Body Problem." *Amer. J. Phys.*, no. 7.

Greenberg, R. 1978. "Orbital Resonance in a Dissipative Medium." *Icarus* 33:62.

Greenberg, R., D. Davis, W. Hartmann, and C. Chapman. 1977. "Size Distribution of Particles in Planetary Rings." *Icarus* 30:796.

Harrington, R., and P. Seidelmann. 1981. "The Dynamics of the Saturnian Satellites 1980S1 and 1980S3." *Icarus* 47:97.

Hartmann, W. K. and 7 others. 1999. "Reviewing the Yarkovsky Effect: New Light in the Delivery of Stone and Iron Meteorites from the Asteroid Belt." *Meteoritics and Plan. Sci.* 34:161–167.

Johnson, R. D., and C. Holbrow. 1977. *Space Settlements—A Design Study*. Washington, D.C.: NASA SP-413.

Kaula, W. M., and A. Harris. 1975. "Dynamics of Lunar Origin and Orbital Evolution." *Rev. Geophys. Spa. Phys.* 3:363.

Kordylewski, K. 1961. "Summary of Letter to the Editor." *Sky and Telescope* 22:63.

Lago, B., and Z. Cazanave. 1979. "Possible Dynamical Evolution of the Rotation of Venus Since Formation." *Moon and Planets* 21:127.

Melosh, H. J. 1981. "Atmospheric Breakup of Terrestrial Impactors." In *Multi-Ring Basins*, ed. P. Schultz and R. Merrill. New York: Pergamon Press.

Mikkola, S. and 3 others. 1993. *Celest. Mech. Dynamical Astron.* 58:3–64.

Newton, I. 1962. *Principia*. A. Motte, trans. Berkeley: University of California Press. (Originally published 1687, translated 1729).

O'Neill, G. K. 1974. "The Colonization of Space." *Physics Today*, September, p. 32.

————. 1975. "Space Colonies and Energy Supply to the Earth." *Science* 10:943.

————. 1977. *The High Frontier*. New York: Morrow.

Öpik, E. J. 1951. "Collision Probabilities with the Planets and the Distribution of Interplanetary Matter." *Proc. Ray. Irish Acad.* 54:165.

Peale, S. J. 1976. "Orbital Resonance in the Solar System." *Ann. Rev. Astron. Astrophys.* 14:215.

Peterson, C. 1976. "A Source Mechanism for Meteorites: Controlled by the Yarkovsky Effect." *Icarus* 29:91–111.

Poynting, J. 1903. "Radiation in the Solar System: Its Effect on Temperature and Its Pressure on Small Bodies." *Monthly Notices Roy. Astron. Soc. 64*, app. 1.

Rietsma, H. J., R. Beebe, and B. Smith. 1976. "Azimuthal Brightness Variations in Saturn's Rings." *Astron. J.* 81:209.

Robertson, H. P. 1937. "Dynamical Effects of Radiation in the Solar System." *Monthly Notices Roy. Astron. Soc.* 997:423.

Roche, E. 1850. "La Figure d'une Masse Fluide soumise à l'attraction d'un pointe eloigne." *Mem. Acad. Montpellier 1 (Sciences):* 1847.

Roosen, R. G. 1968. "A Photographic Investigation of the Gegenschien and the Earth-Moon Libration Point L_5." *Icarus* 9:429.

Rubincam, D. P. 1995. "Asteroid Orbit Evolution Due to Thermal Drag." *J. Geophys. Research,* 100:1585–1594.

Scholl, H. and 4 others. 1989. "Secular Resonances." In *Asteroids II,* ed. R. Binzel, T. Gehrels, and M. Matthews. Tucson: Univ. Arizona Press.

Tittemore, W. C. 1990. "Chaotic Motion of Europa and Ganymede and the Ganymede-Callisto Dichotomy." *Science* 250: 263–267.

Weisel, W. E. 1981. "The Origin and Evolution of the Great Resonance in the Jovian Satellite System." *Astron. J.* 86:611.

Wiegert, Paul A., K. Innanen, and S. Mikkola 1997. "An Asteroidal Companion to Earth." *Nature* 387:685–686.

Williams, J. G. 1969. Ph.D. dissertation. University of California, Los Angeles.

Wyatt, S. P., and F. L. Whipple. 1950. "The Poynting-Robertson Effect on Meteor Orbits." *Astrophys. J.* 111:134.

Yoder, C.F. 1979. "How Tidal Heating in Io Drives the Galilean Orbital Resonance Locks." *Nature* 279:767.

Chapter Four

Aanestad, P. A., and S. Kenyon. 1978. "Temperature Fluctuations in Interstellar Dust Grains." In *Thermodynamics and Kinetics of Dust Formation in the Space Medium.* Houston: Lunar and Planetary Institute.

Abt, H. A. 1978. "The Binary Frequency Along the Main Sequence." In *Protostars and Planets,* ed. T. Gehrels and M. Matthews, Tucson: University of Arizona Press.

Allen, C. W. 1973. *Astrophysical Quantities.* London: Athlone Press.

Arny, T., and P. Weissman. 1973. "Interaction of Protostars in a Collapsing Cluster." *Astron. J.* 78:309.

Artymowicz, P. 1997. "Beta Pictoris: An Early Solar System?" *Ann. Rev. Earth Planet. Sci.* 25:175–219.

Bachiller, R. 1996. "Bipolar Molecular Outflows from Young Stars and Protostars." *Ann. Rev. Astron. Astrophys.* 34:111–154.

Batten, A. H. 1973. *Binary and Multiple Systems of Stars.* Oxford: Pergamon Press.

Black, D.C. and M. Matthews, ed. 1985. *Protostars and Planets II.* Tucson: University of Arizona Press.

Bodenheimer, P., and D. C. Black. 1978. "Numerical Calculations of Protostellar Hydrodynamic Collapse." In *Protostars and Planets,* ed. T. Gehrels and M. Matthews. Tucson: University of Arizona Press.

Bodenheimer, P. 1976. "Contraction Models for the Evolution of Jupiter." *Icarus* 29:165.

Böhm, K. H. 1978. "Herbig-Haro Objects and Their Interpretation." In *Protostars and Planets,* ed. T. Gehrels and M. Matthews. Tucson: University of Arizona Press.

Bok, B. J., C. Cordwell, and R. Cromwell. 1971. "Globules." In *Dark Nebulae, Globules, and Protostars,* ed. B. Lynds. Tucson: University of Arizona Press.

Bonsacl, W. K., and J. L. Greenstein. 1960. "The Abundance of Lithium in T Tauri Stars and Related Objects." *Astrophys. J.* 131:83.

Cameron, A. G. W. 1975. "The Origin and Evolution of the Solar System." *Sci. Am.* September, p. 32.

Campbell, B. 1989. "Search for Planetary-Mass Companions to Nearby Stars." *Highlights of Astronomy* 8:109.

Davidson, K., and M. Harwit. 1967. "Infrared and Radio Appearance of Cocoon Stars." *Astrophys J.* 133:443.

Donn, B. 1978. "Condensation Processes and the Formation of Cosmic Grains." In *Protostars and Planets,* ed. T. Gehrels and M. Matthews. Tucson: University of Arizona Press.

Fertel, J. F. "Silicon Monoxide Bands in Some Low-Temperature Stars." *Astrophys. J.* 159:L7.

Field, G. B. 1975. "The Composition of Interstellar Dust." In *The Dusty Universe,* ed. G. Field and A. Cameron. New York: Neale Watson Academic Publications.

———. 1978. "Conditions in Collapsing Clouds." In *Protostars and Planets,* ed. T. Gehrels and M. Matthews. Tucson: University of Arizona Press.

Fleck, R. C. 1978. "On the Origin of Close Binary and Planetary Systems." *Astrophys. J.* 225:198.

Gatewood, G. 1976. "On the Astrometric Detection of Neighboring Planetary Systems." *Icarus* 27:1.

Gatewood, G., and H. Eichhorn. 1973. "An Unsuccessful Search for a Planetary Companion of Barnard's Star." *Astron. J.* 78:769.

Gibson, E. G. 1973. *The Quiet Sun.* Washington, D.C.: NASA SP-303.

Greenberg, R., W. Hartmann, C. Chapman, and J. Wacker. 1978. "The Accretion of Planets from Planetesimals." In *Protostars and Planets,* ed. T. Gehrels and M. Matthews. Tucson: University of Arizona Press.

Greene, Thomas P. 2001. "Protostars." *Amer. Scientist,* 89:316–325.

Handbury, M. J., and I. P. Williams, 1976. "The Peculiar Binary System Epsilon Aurigae." *Astrophys. Spa. Sci.* 45:439.

Hansen, R., B. Jones, and D. Lin., 1983. "The Astrometric Position of T Tauri and the Nature of Its Companion." *Astrophys. Journ. Letters* 270:L27.

Harrington, R. 1977. "Planetary Orbits in Binary Stars." *Astron. J.* 82:753.

Harrington, R. S., and A. L. Behall. 1973. "The Mass Ratio of L726-8." *Astron. J.* 78:1096.

Hartmann, L., and S. J. Kenyon 1996. "The FU Orionis Phenomenon." *Ann. Rev. Astron. Astrophys.* 34:207–240.

Hartmann, W. K. 1970. "Growth of Planetesimals in Nebulae Surrounding Young Stars." In *Evolution stellaire avant la sequence principale.* Memoires de la Société Royale des Sciences de Liège. Collection in -8, 5th sér,19:215.

———. 1978. "Planet Formation: Mechanism of Early Growth." *Icarus* 33:50.

Henry, T. 1996. Brown Dwarfs Revealed—At Last! *Sky and Telescope* 91:24–28.

Henyey, L., R. LeLevier, and R. Levee. 1955. "The Early Phases of Stellar Evolution." *Publ. Astron. Soc. Pacific* 67:154.

Herbig, G. H. 1968. "The Structure and Spectrum of R Monocerotis." *Astrophys. J.* 152:439.

———. 1970. "Introductory Remarks." In *Evolution stellaire avant la sequence principale.* Memoires de la Société Royale des Sciences de Liège. Collection in -8 , 5th sér,19:13.

———. 1977. "Eruptive Phenomenon in Early Stellar Evolution." *Astrophys. J.* 217:693.

Herbst, W., and G. E. Assousa, 1978. "The Role of Supernovae in Star Formation and Spiral Structure." In *Protostars and Planets,* ed. T. Gehrels and M. Matthews. Tucson: University of Arizona Press.

Hills, J. G. 1977. "Exchange Collisions between Binary and Single Stars." *Astron J.* 82:606.

Hoxie, D. 1969. "The Structure and Evolution of Stars of Very Low Mass." Ph.D. dissertation, University of Arizona.

Hoyle, F., and N. Wickramasinghe. 1962. "Graphite Particles as Interstellar Grains." *Monthly Notices Roy. Astron. Soc.* 124:417.

Hyland, A., E. Becklin, G. Neugebauer, and G. Wallerstein. 1969. "Observations of the Infrared Object VY Canis Majoris." *Astrophys. J.* 158:618.

Imhoff, C. L. 1978. "T Tauri Star Evolution and Evidence for Planetary Formation." In *Protostars and Planets,* ed. T. Gehrels and M. Matthews. Tucson: University of Arizona Press.

Joy, A. H. 1945. "T Tauri Variable Stars." *Astrophys. J.* 102:168.

Kamijo, F. 1963. "A Theoretical Study on the Long Period Variable Stars. III. Formation of Solid and Liquid Particles in the Circumstellar Envelope." *Publ. Astron. Soc. Japan* 15:440.

Knacke, R. 1978. "Mineralogical Similarities between Interstellar Dust and Primitive Solar System Material." In *Protostars and Planets,* ed. T. Gehrels and M. Matthews. Tucson: University of Arizona Press.

Knacke, R., and J. E. Gaustad. 1969. "Possible Identification of Interstellar Silicate Absorption in the Infrared Spectrum of 119 Tauri." *Astrophys. J.* 155:L189.

Kuhi, L. V. 1966. "T Tauri Stars: A Short Review." *J. Roy. Astron. Soc. Can.* 60:1.

———. 1978. "Spectral Characteristics of T Tauri Stars." In *Protostars and Planets,* ed. T. Gehrels and M. Matthews. Tucson: University of Arizona Press.

Kuiper, G. P. 1955. "On the Origin of Binary Stars." *Publ. Astron. Soc. Pacific* 67:387.

Kulkami, S. R. 1997. "Brown Dwarfs: A Possible Missing Link between Stars and Planets." *Science* 276:1350–1354.

Kumar, S. S. 1964. "On the Nature of Planetary Companions of Stars." *Z. Astrophys.* 58:248.

Larimer, J. W., and E. Anders. 1967. "Chemical Fractionations in Meteorites. II. Abundance Patterns and Their Interpretation." *Geochim. Cosmochim. Acta* 31:1239.

Larimer, J. W. 1967. "Chemical Fractionations in Meteorites. I. Condensation of the Elements." *Geochim. Cosmochim. Acta* 31:1215.

Larson, R. B. 1968. "Numerical Calculations of the Dynamics of a Collapsing Protostar." *Monthly Notices Roy. Astron. Soc.* 145:271.

———. 1972. "The Evolution of Spherical Protostars with Masses of 0.25 to 10 $M^{\wedge\wedge}$." *Monthly Notices Roy. Astron. Soc.* 157:121.

———. 1978. "The Stellar State: Formation of Solar-Type Stars." In *Protostars and Planets,* ed. T. Gehrels and M. Matthews. Tucson: University of Arizona Press.

Low, F., and B. Smith. 1966. "Infrared Observations of a Preplanetary System." *Nature* 212:675.

Lucy, L. B. 1977. "A Numerical Approach to the Testing of the Fission Hypothesis." *Astron. J.* 82:1013.

Martin, E. L., R. Rebolo, and M. R. Zapatero-Osorio. 1997. "The Discovery of Brown Dwarfs." *Amer. Scientist,* 85:522–529.

Mendoza, E. E. V. 1966. "Infrared Photometry of T Tauri Stars and Related Objects." *Astrophys. J.* 143:1010.

Marzari, F. and S. J. Weidenschilling. 2002. "Eccentric Extra-Solar Planets: The Jumping Jupiter Model." *Icarus* 156:570–580.

Messenger, Scott, and 4 others. 2003. "Samples of Stars Beyond the Solar System: Silicate Grains in Interplanetary Dust." *Science* 300:105–108.

Ney, E. P. 1972. "Infrared Excesses in Supergiant Stars: Evidence for Silicates." *Publ. Astr. Soc. Pacific* 84:613.

Oort, J. H., and H. C. van de Hulst. 1946. "Gas and Smoke in Interstellar Space." *Bull. Astron. Inst. Neth.* 10:187.

Poveda, A. 1965. "The H-R Diagram of Young Clusters and the Formation of Planetary Systems." *Bol. Observ. Tonantz, Tacubaya* 4:15.

Probst, R. G. 1977. "Parallax, Orbit, and Mass of Ross 614." *Astron j.* 82:656.

Queloz, D. and M. Mayor 1997. *Science* 20:375.

Rayashi, C. 1961. "Stellar Evolution in Early Phases of Gravitational Contraction." *Publ. Astron. Soc. Japan* 13:450.

Reeves, H., and others. 1972. "On the Origin of Light Elements." Cal Tech Preprint OAP-296.

Roth, J. 1997. "Does 51 Pegasi's Planet Really Exist?" *Sky and Telescope* 93:24–25.

Rydgren, A. E. 1978. "Interpreting Infrared Observations of T Tauri Stars." In *Protostars and Planets,* ed. T. Gehrels and M. Matthews. Tucson: University of Arizona Press.

Serkowski, K. 1976. "Feasibility of a Search for Planets around Solar-Type Stars with a Polarimetric Velocity Meter." *Icarus* 27:13.

Simon, T., and H. M. Dyck. 1975. "Silicate Absorption at 18 µm in Two Peculiar Infrared Sources." *Nature* 253:101.

Srinivasan, B., and E. Anders. 1978. "Noble Gases in the Murchison Meteorite: Possible Relics of S-Process Nucleosynthesis." *Science* 201:51.

Strom, S. E. 1972. "Optical and Infrared Observations of Young Stellar Objects—an Informal Review." *Publ. Astron. Soc. Pacific* 84:745.

Ulrich, R. K. 1978. "The Status of T Tauri Models." In *Protostars and Planets,* ed. T. Gehrels and M. Matthews. Tucson: University of Arizona Press.

van de Kamp, P. 1963. "Astrometric Study of Barnard's Star with Planets Taken with the Sprul 24-inch Refractor." *Astron. J.* 68:515.

Ward, W. R. 1981. "Solar Nebula Dispersal and the Stability of the Planetary System, 1." *Icarus* 47:234.

Welin, G. 1978. "The FU Orionis Phenomenon." In *Protostars and Planets,* ed. T. Gehrels and M. Matthews. Tucson: University of Arizona Press.

Wilner, S. P., R. Puettner, and R. Russell. 1978. "Unidentified Infrared Spectral Features." In *Thermodynamics and Kinetics of Dust Formation in the Space Medium.* Houston: Lunar and Planetary Institute.

Wright, A. 1970. "Results of a Computer Program for Gravito-Gas Dynamic Collapse." In *Evolution stellaire avant la sequence principale.* Memoires de la Société Royale des Sciences de Liège. Collection in -8, 5th sér,19:75.

Chapter Five

Aggarwal, H. R., and V. Oberbeck. 1974. "Roche Limit of a Solid Body." *Astrophys. J.* 191:577.

Aitekeeva, Z. Z. 1968. "Anomalous Satellites of Planets." *Solar System Res.* 2:19.

Albee, A. L., A. J. Gancarz, and A. A. Chodos. 1973. "Metamorphism of Apollo 16 and 17 and Luna 20 Metaclastic Rocks at about 3.95 AE: Samples 61156, 64423, 14-2, 65015, 67483, 15-2, 76055, 22006, and 22007." *Proc. Lun. Sci. Conf.* 3:569–595.

Alfvén, H., and G. Arrhenius. 1976. *Evolution of the Solar System.* Washington, D.C.: NASA SP-345.

Alfvén, H. 1954. *The Origin of the Solar System.* London: Oxford Press.

Arnold, J. R. 1965. "The Origin of Meteorites as Small Bodies II." *Astrophys. J.* 141:1536.

Binder, A. B. 1974. "On the Origin of the Moon by Rotational Fission." *Moon* 11:53.

———. 1975. "On the Petrology and Structure of a Gravitationally Differentiated Moon of Fission Origin." *Moon* 13:431.

Blander, M., and J. Katz. 1967. "Condensation of Primordial Dust." *Geochim. Cosmochim. Acta* 31:1025.

Boynton, W. V. 1985. "Meteoritic Evidence Concerning Conditions in the Solar Nebula." In *Protostars and Planets II,* ed. D. Black and M. Matthews. Tucson: University of Arizona Press.

Bronshten, V. A. 1968. "Origin of Irregular Satellites of Jupiter." *Solar System Res. 1:*23.

Burns, J. A. 1973. "Where Are the Satellites of the Inner Planets?" *Nature 242:*23.

———. 1978. "The Dynamical Evolution and Origin of the Martian Moons." *Vistas in Astronomy 22:*193.

———. 1980. "A Model of the Jovian Ring." Paper presented at International Astronomical Union Colloquium 57, May 1980, Kona, Hawaii.

———. 1990. "Planetary Rings." In *The New Solar System,* ed. J. Beatty and A. Chaikin. Cambridge, Mass.: Sky Publishing Corp.

Cameron, A. G. W., and W. Ward. 1976. "The Origin of the Moon [abstract]." In *Lunar Science VII.* Houston: Lunar Science Institute.

Cameron, A. G. W., and M. Pine. 1973. "Numerical Models of the Primitive Solar Nebula." *Icarus 18:*377.

Cameron, A. G. W. 1962. "The Formation of the Sun and the Planets." *Icarus 1:*13.

———. 1975. "The Origin and Evolution of the Solar System." *Sci. Am.,*September, p. 32.

———. A. G. W. 1985. "Formation and Evolution of the Primitive Solar Nebula." In *Protostars and Planets II,* ed. D. Black and M. Matthews. Tucson: University of Arizona Press.

Cameron, A. G. W., and J. Truran. 1977. "The Supernova Trigger for Formation of the Solar System." *Icarus 30:*447.

Canup, R., and L. Esposito. 1996. Accretion of the Moon from an Impact-Generated Disk. *Icarus 119:*427–446.

Carlson, R. W., and G. Lugmair. 1988. "The Age of Ferroan Anorthosite 60025L Oldest Crust on a Young Moon." *Earth Planet. Sci. Lett. 90:*119.

Carusi, A., and G. Valsecchi. 1980. "Comets and Temporary Satellites of Jupiter." Paper presented at International Astronomical Union Colloquium 57, May 1980, Kona, Hawaii.

Chandrasekhar, S. 1946. "On a New Theory of Weizsäcker on the Origin of the Solar System." *Rev. Mod. Phys. 18:*94.

Christy, J., and R. Harrington. 1978. "The Satellite of Pluto." *Astron. J. 83:*1005.

Chyba, C., and P. D. Nicholson. 1987. Abstract, Division of Planetary Sciences of Am. Astronom. Soc., p. 821.

Clayton, R., L. Grossman, and T. Mayeda. 1973. "A Component of Primitive Nuclear Composition in Carbonaceous Meteorites." *Science 182:*485.

Clayton, R., T. Mayeda, and A. Rubin. 1984. "Oxygen Isotopic Compositions of Enstatite Chondrites and Aubrites." *J. Geophys. Res. Suppl. 89:*C245.

Consolmagno, G. J., and J. Lewis. 1977. "Preliminary Thermal History Models of Icy Satellites." In *Planetary Satellites,* ed. J. Burns. Tucson: University of Arizona Press.

Consolmagno, G. J. 181. "Io: Thermal Models and Chemical Evolution." *Icarus 47:*36.

Consolmagno, G., and J. Jokopii. 1978. "26Al and the Partial Ionization of the Solar Nebula." *Moon and Planets 19:*253.

Cruikshank, D. P., and D. Morrison. 1976. "The Alilean Satellites of Jupiter." *Sci. Am.,* May, p. 108.

Cunningham, C. J. 1988. Introduction to Asteroids: The Next Frontier. Richmond, VA: Willman-Bell, Inc.

Day, K. L., and B. Donn. 1978. "Condensation of Nonequilibrium Phases of Refractory Silicates from the Vapor." *Science 22:*307.

Dermott, S. F. 1968. "On the Origin of Commensurabilities in the Solar System I." *Monthly Notices Roy. Astron. Soc. 141:*349.

Durda, D. D. 1996. The Formation of Asteroidal Satellites in Catastrophic Collisions. Icarus 120:212–219.

Elliot, J. L., and others. 1978. "The Radii of Uranian Rings″, α, β, γ, δ, ϵ, η, 4, 5, and 6 from Their Occulations by SAO 158687." *Astron. J. 83:*980.

Gancarz, A. J., and G. Wasserberg. 1977. "Initial Pb of the Amìitsoq Gneiss, West Greenland, and Implications for the Age of the Earth." *Geochim. Cosmochim. Acta 41:*1283.

Gehrels, T., and others. 1980. "Imaging Photopolarimeter on Pioneer Saturn." *Science 207:*436.

Goettel, K. A., and S. S. Barshay. 1978. "The Chemical Equilibrium Model for Condensation in the Solar Nebula: Assumptions, Implications, and Limitations." In *The Origin of the Solar System,* ed. S. F. Dermott. New York: Wiley.

Goldreich, P., and W. Ward. 1973. "The Formation of Planetesimals." *Astrophys. J. 183:*1051.

Goldreich, P., and S. Tremaine. 1978a. "The Formation of the Cassini Division in Saturn's Rings." *Icarus 34:*240.

———. 1978b. "The Velocity of Dispersion in Saturn's Rings." *Icarus 34:*227.

Goldreich, P., and S. Tremaine. 1979. "Towards a Theory for the Uranian Rings." *Nature 277:*97–99.

Goldreich, P. 1965. "An Explanation of the Frequent Occurrence of Commensurable Mean Motions in the Solar System." *Monthly Notices Roy. Astron. Soc. 130:*159.

Greenberg, R., J. Wacker, W. Hartmann, and C. Chapman. 1978b. "Planetesimals to Planets: Numerical Simulation of Collisional Evolution." *Icarus 35:*1.

Greenberg, R. J. 1977. "Orbit-Orbit Resonances in the Solar System: Varieties and Similarities." *Vistas in Astronomy 21:*209.

Greenberg, R., W. Hartmann, C. Chapman, J. Wacker. 1978. "The Accretion of Planets from Planetesimals." In *Protostars and Planets,* ed. T. Gehrels and M. Matthews. Tucson: University of Arizona Press.

Grossman, L. 1973. "Refractory Trace Elements in Ca-Al-rich Inclusions in the Allende Chondrite." *Geochim. Cosmochim. Acta 37:*1119.

———. 1975. "The Most Primitive Objects in the Solar System: Carbonaceous Chondrites." *Sci. Am.,* February, p. 30.

———. 1977. "Chemical Frationation in the Solar Nebula." In *The Soviet-American Conferences on Cosmochemistry of the Moon and Planets,* ed. J. Pomeroy and N. Hubbard. Washington, D.C.: NASA SP-370, pt. 2.

Gurevich, L. E., and A. Lebedinskii. 1950. "Formation of the Planets." *Izv. AN SSSR, seriya fizich.,* 14:765.

Haggarty, S. E. 1978. "The Allende Meteorite: Solid Solution Characteristics." *Proc. Lun. Planet. Sci. Cont. 9:*1331.

Halliday, A. N., M. Rahkamper, D-C. Lee, and W. Yi 1997. *Earth and Planet. Sci. Lett. 142:*75.

Harris, A. W. 1977. "An Analytical Theory of Planetary Rotation Rates." *Icarus 31:*168.

———. 1978. "Satellite Formation II." *Icarus 34:*128.

Hartmann, W. K. 1979. "Diverse Puzzling Asteroids and a Possible Unified Explanation," in *Asteroids,* ed. T. Gehrels. Tucson: Univ. Arizona Press.

———. 1978. "Planet Formation: Mechanism of Early Growth." *Icarus 33:*50–61.

———. 1969. "Terrestrial, Lunar, and Interplanetary Rock Fragmentation." *Icarus 10:*201.

———. 1970. "Growth of Planetesimals in Nebulae Surrounding Young Stars." In *Evolution stellaire avant la sequence principale.* Mémoires de la Société Royale des Sciences de Liège. collection in -8 , 5th Ser.,*19:*215.

———. 1975. "Lunar Cataclysm: A Misconception?" *Icarus 24:*181.

————. 1976. "Planet Formation: Composition Mixing and Lunar Compositional Anomalies." *Icarus* 27:553.

————. 1977. "Large Planetesimals in the Early Solar System." In *Comets, Asteroids, Meteorites,* ed. A. Delsemme. Toledo, Ohio: University of Toledo Press.

————. 1978. "Planet Formation: Mechanism of Early Growth." *Icarus* 33:50.

————. 1987. "A Satellite-Asteroid Mystery and a Possible Early Flux of Scattered C-class Asteroids." *Icarus* 71:57.

————. 1990. "Additional Evidence about an Early Intense Flux of C Asteroids and the Origin of Phobos." *Icarus* 87:236.

Hartmann, W. K., and D. R. Davis. 1975. "Satellite-sized Planetesimals and Lunar Origin." *Icarus* 24:504.

Hartmann, W. K., and S. Vail. 1986. "Giant Impactors: Plausible Sizes and Populations." In *Origin of the Moon*, ed. W. Hartmann, R. Phillips, and G. Taylor), Lunar and Planetary Institute: Houston.

Hartmann, W. K., R. Phillips, and G. Taylor. 1986. *Origin of the Moon.* Houston: Lunar and Planetary Institute.

Hartmann, W. K., R. Strom, S. Weidenschilling, K. Blasius, A. Woronow, M. Dence, R. Grieve, J. Diaz, C. R. Chapman, E. M. Shoemaker, and K. Jones. 1981. "Cratering Chronology of the Planets." In *Basaltic Volcanism of the Planets,* ed. R. Merrill and others. New York: Pergamon Press.

Herbig, H. G. 1970. "VY Canis Majoris II. Interpretation of the Energy Distribution." *Contrib. Lick Observ. 317.*

Hoyle, F. 1963. "Formation of the Planets." In *Origin of the Solar System,* ed. R. Jastrow and A. Cameron. New York: Academic Press.

Hunten, D. M. 1979. "Capture of Phobos and Deimos by Proto-atmospheric Drag." *Icarus* 37:113.

Ida, S., R. Canup, and G. R. Stewart. 1997. "Lunar Accretion from an Impact-generated Disk." *Nature* 389:53–357.

Isaacman, R., and C. Sagan. 1977. "Computer Simulations of Planetary Accretion Dynamics: Sensitivity to Initial Conditions." *Icarus* 31:510.

Jewitt, D. C., and G. E. Danielson. 1981. "The Jovian Ring." *J. Geophys. Res.* 86:8691.

Jewitt, D., G. E. Danielson, and S. Synnott. 1979. "Discovery of a New Jupiter Satellite." *Science* 206:951.

Jewitt, D., and P. Goldreich. 1980. "The Ring of Jupiter." Paper presented at the International Astronomical Union Colloquium 57, May 1980, Kona, Hawaii.

Kaula, W. M., and P. Bigeleisen. 1975. "Early Scattering by Jupiter and Its Collision Effects in the Terrestrial Zone." *Icarus* 25:18.

Kaula, W. M. 1968. *An Introduction to Planetary Physics.* New York: Wiley.

Kerridge, J. F., and J. Vedder. 1972. "Accretionary Processes in the Early Solar System: An Experimental Approach." *Science* 177:161.

Kortenkamp, Stephen J., G. W. Wetherill, and Satoshi Inaba. 2001. "Runaway Growth of Planetary Embryos Facilitated by Massive Bodies in a Protoplanetary Disk," *Science* 293:1127–1129.

Kuhl, L. V. 1966. "T Tauri Stars: A Short Review." *J. Roy. Astron. Soc. Can.* 60:1.

————. 1978. "Rotation of Pre-main-sequence Stars in NGC2264 and the Solar Angular Momentum Problem." *Moon and Planets* 19:199.

Kuiper, G. P. 1951. "On the Origin of the Solar System." In *Astrophysics,* ed. J. A. Hynek. New York: McGraw-Hill.

————. 1956a. "The Formation of the Planets, I, II, III." *J. Roy. Astron. Soc. Can.* 50:57, 105, 158.

————. 1956b. "On the Origin of the Satellites and the Trojans." *Vistas in Astronomy* 2:1631.

Kuiper, G. P., D. P. Cruikshank, and U. Fink. 1970. Letter in *Sky and Telescope* 39:80.

Lange, D. E., and J. Larimer. 1973. "Chondrules: An Origin by Impacts between Dust Grains." *Science* 182:920.

Larimer, J. W., and E. Anders. 1967. "Chemical Fractionations in Meteorites II. Abundance Patterns and Their Interpretations." *Geochim. Cosmochim. Acta* 31:1239.

Larimer, J. W. 1967. "Chemical Fractionations in Meteorites I. Condensation of the Elements." *Geochim. Cosmochim. Acta* 31:1215.

Larson, R. B. 1972. "The Evolution of Spherical Protostars with Masses 0.25 to 10 M^^." *Monthly Notices Roy. Astron. Soc.* 157:121.

Lebofsky, L. A., T. V. Johnson, and T. B. McCord. 1970. "Saturn's Rings: Spectral Reflectivity and Compositional Implications." *Icarus* 13:226.

Lee, J. T., B. Li, and O. K. Mauel (1997) "On the Signature of Local Element Synthesis." *Comments on Astrophysics* 18:335–345.

Lee, D-C., A. Halliday, G. Snyder, and L. Taylor (1997) "Age and Origin of the Moon." *Science* 278:1098–1103.

Lewis, J. S. 1972a. "Low Temperature Condensation from the Solar Nebula." *Icarus* 16:241.

————. 1972b. "Metal/silicate fractionation in the solar system." *Earth Planet. Sci. Lett. 15:*286.

————. 1974. "The Temperature Gradient in the Solar System." *Science* 186:440.

Lissauer, Jack J. and J. N. Cuzzi 1985. "Rings and Moons: Clues to Understanding the Solar Nebula." In *Protostars and Planets III,* ed. D. Black and M. Matthews. Tucson: University of Arizona Press.

Lord, H. C. 1965. "Molecular Equilibria and Condensation in a Solar Nebula and Cool Stellar Atmospheres." *Icarus* 4:279.

Luu, J. 1991. CCD Photometry and Spoectroscopy of Outer Jovian Satellites. *Bull. Amer. Astron. Soc.* 23:1171 (abstract).

Mason, B. 1962. *Meteorites.* New York: Wiley.

McCord, T. B. 1966. "The Dynamical Evolution of the Neptunian System." *Astron. J.* 71:585.

————. 1968. "The Loss of Retrograde Satellites in the Solar System." *J. Geophys. Res.* 73:1497.

McKinnon, W. B. (1984) "On the Origin of Triton and Pluto." *Nature* 311:355–358.

McKinnon, W. B. (1989) "Origin of the Pluto-Charon Binary. "*Astrophys. Journ.* 344:L41–44.

McSween, H. Y., Jr. 1989. "Chondritic Meteorites and the Formation of Planets." *American Scientist* 77:146.

Millis, R., and L. Wasserman. 1978. "The Occultation of BD -15 3969 by the Rings of Uranus." *Astron. J.* 83:933.

Mizuno, H. 1980. "Formation of Giant Planets." *Progress of Theor. Physics,* 64:544.

Neugebauer, G., and others. 1981. "Spectra of the Jovian Ring and Amalthea." *Astron. J.* 86:607.

Nicholson, P. D., and others. 1978. "The Rings of Uranus: Results of the 10 April 1978 Occultation." *Astron. J.* 83:1240.

Öpik, E. J. 1963. "Survival of Comet Nuclei and the Asteroids." *Adv. Astron. Astrophys.* 2:219.

————. 1966. "The Stray Bodies in the Solar System II. The Cometary Origin of Meteorites." *Adv. Astron. Astrophys.* 4:301.

Papanastassiou, D. A., and G. Wasserburg. 1976. "Rb-Sr Age of Troctolite 76535." *Proc. Lun. Sci. Conf.* 7:2035.

Peale, S. J. 1978. "An Observational Test for the Origin of the Titan-Hyperion Orbital Resonance." *Icarus* 36:240.

Pepin, R. O., and D. Phinney. 1975. "The Formation Interval of the Earth." *Lun. Sci. Conf. Abstracts* 6:682.

Pepin, R. O. 1991. "On the Origin and Early Evolution of Terrestrial Planet Atmospheres and Meteoritic Volatiles." *Icarus* 92:2.

Piironen, J. O. 1978. "The Profile of Saturn's Ring A as Derived from Stellar Occultations." *Moon and Planets* 19:61.

Pilcher, C., C. Chapman, L. Lebofsky, and H. Kieffer. 1970. "Saturn's Rings: Identification of Water Frost." *Science* 167:1372.

Podosek, F. A. 1970. "Dating of Meteorites by the High-Temperature Release of Iodine-Correlated Xe129." *Geochim. Cosmochim. Acta* 34:341.

Pollack, J. B., J. Burns, M. Tauber. 1979. "Gas Drag in Primordial Circumplanetary Envelopes: A Mechanism for Satellite Capture." *Icarus* 37:587.

Pollack, J. B. 1975. "The Rings of Saturn." *Space Sci. Rev.* 18:3.

———. 1978. "The Rings of Saturn." *Am. Scientist* 66:30.

Proveda, A. 1965. "H-R Diagram of Young Clusters and the Formation of Planetary Systems." *Bol. Observ. Tonantz. Tacubaya* 4:15.

Reynolds, J. G. 1960. "Determination of the Age of the Elements." *Phys. Rev. Lett.* 4:8.

Ringwood, A. E. 1970. "Origin of the Moon: The Precipitation Hypothesis." *Earth Planet. Sci. Lett.* 8:131.

Safronov, V. S. 1966. "Sizes of the Largest Bodies Falling onto the Planets during Their Formation." *Sov. Astron. -AJ* 9:987.

———. 1972. *Evolution of the Protoplanetary Cloud and Formation of the Earth and the Planets.* Jerusalem: Israel Program for Scientific Translations. (Originally published 1969 in Russian.)

Schonfeld, E. 1976. "Chronology of the Lunar Crust." *Proc. Lun. Sci. Conf.* 7:2093.

Schwartz, K. and G. Schubart. 1969. "The Early Despinning of the Sun." *Astrophys. Space Sci.* 5:444.

Slattery, W. L. 1978. "Protoplanetary Core Formation by Rain-out of Iron Drops." *Moon and Planets* 19:443.

Smith, P. K., and P. Buseck. 1981. "Graphitic Carbon in the Allende Meteorite: A Microstructural Study." *Science* 212:322.

Smoluchowski, R. 1978. "Width of a Planetary Ring System and C-ring of Saturn." *Nature* 274:669.

———. 1979. "Planetary Rings." In *Comments on Astrophys.* 8:69.

Spaute, D. and 3 others. (1992) "Accretional Evolution of a Planetesimal Swarm: 1. A New Simulation." *Icarus* 92:147–164.

Synnott, S. P. 1980. "Discovery of a 15th Satellite of Jupiter." Paper presented at the International Astronomical Union Colloquium 57, May 1980. Kona, Hawaii.

Taylor, S. R. 1975. *Lunar Science: A Post-Apollo View.* New York: Pergamon Press.

Taylor, S. R. 1992. *Solar System Evolution.* Cambridge, MA: Cambridge University Press.

Tholen, D., and B. Zellner. 1984. "Multicolor Photometry of Outer Jovian Satellites." *Icarus* 58:246–253.

Thompson, W. R., B. Murray, B. Khare, and C. Sagan. 1987. "Coloration and Darkening of Methane Clathrate and Other Ices by Charged Particle Irradiation: Application to the Outer Solar System." *J. Geophys. Res.* 92:14933–14947.

Tscharnuter, W. M. 1978. "Collapse of the Pre-Solar Nebula." *Moon and Planets* 19:229.

Urey, H. C. 1952. *The Planets: Their Origin and Development.* New Haven: Yale University Press.

Van Allen, J., and others. "Saturn's Magnetosphere, Rings, and Inner Satellites." *Science* 207:415.

Veverka, J., P. Thomas, and T. Duxbury. 1978. "The Puzzling Moons of Mars." *Sky and Telescope* 56:186.

Wanke, H., and G. Dreibus. 1986. "Geochemical Evidence for the Formation of the Moon by Impact-induced Fission of the Proto-Earth." In *Origin of the Moon,* ed. W. Hartmann, R. Phillips, and G. Taylor. Houston: Lunar and Planetary Institute.

Ward, W. R., and M. Reid. 1973. "Solar Tidal Friction and Satellite Loss." *Monthly Notices Roy. Astron. Soc.* 164:21.

Wasserburg, G. 1985. "Short-Lived Nuclei in the Early Solar System." In *Protostars and Planets II,* ed. D. Black and M. Matthews. Tucson: University of Arizona Press.

Wasson, J. T. 1974. *Meteorites.* New York: Springer-Verlag.

Watson K., B. C. Murray, and H. Brown. 1963. "The Stability of Volatiles in the Solar System." *Icarus* 1:317.

Weidenschilling, S. J. 1980. "Dust to Planetesimals: Settling and Coagulation in the Solar Nebula." *Icarus* 44:172.

Wetherill, G. W. 1975. "Late Heavy Bombardment of the Moon and Terrestrial Planets." *Proc. Lun. Sci. Conf.* 6:1539.

———. 1976. "The Role of Large Bodies in the Formation of the Earth and Moon." *Proc. Lun. Sci. Conf.* 7:3245.

———. 1977. "Evolution of the Earth's Planetesimal Swarm Subsequent to the Formation of the Earth and Moon." *Proc. Lun. Sci. Conf.* 8:1.

———. 1981. "The Formation of the Earth from Planetesimals." *Sci. Am.,* June, p. 163.

———. 1990. "Formation of the Earth." *Ann. Rev. Earth Planet. Sci.* 18:205.

Whipple, F. L. 1964. "History of the Solar System." *Proc. Natl. Acad. Sci.* 52:517.

Wise, D. U. 1966. "Origin of the Moon by Fission." In *The Earth-Moon System,* ed. B. Marsden and A. Cameron. New York: Plenum.

Wood, J. A. 1963. "On the Origin of Chondrules and Chondrites." *Icarus* 2:152.

———. 1985. "Meteoritic Constraints on Processes in the Solar Nebula." In *Protostars and Planets II,* ed. D. Black and M. Matthews. Tucson: University of Arizona Press.

Chapter Six

Anders, E. 1963. "On the Origin of Carbonaceous Chondrites." *Ann. N.Y. Acad. Sci.* 108:514.

———. 1978. "Most Stony Meteorites Come from the Asteroid Belt." In *Asteroids: An Exploration Assessment,* ed. D. Morrison and W. Wells. Washington, D.C.: NASA CP-2053.

Arnold, J. R. 1965. "The Origin of Meteorites as Small Bodies II." *Astrophys. J.* 141:1536.

Baldwin, R. B. 1949. *The Face of the Moon.* Chicago: University of Chicago Press.

———. 1963. *The Measure of the Moon.* Chicago: University of Chicago Press.

Ballabh, G., A. Bhatnagar, and N. Bhandari. 1978. "The Orbit of the Dhajala Meteorite." *Icarus* 33:361.

Binder, A., and others. 1977. "The Geology of the Viking I Lander Site." *J. Geophys. Res.* 82:4439.

Binzel, R. P. and S. Xu. 1993. "Chips off of Asteroid 4 Vesta: Evidence for the Parent Body of Basalt Achondrite Meteorites." *Science* 260:186–191.

Blander, M., and M. Abdel-Gawad. 1969. "The Origin of Meteorites and the Constrained Equilibrium Condensation Theory." *Geochim. Cosmochim. Acta* 33:710.

Bogard, D. D. 1976. "Fragment from a Young Asteroid?" *Astronomy* 4:14.

Bogard, D. D., and L. Husain. 1977. "Ar-40/Ar-39 Dating of Plainview Brecciated Chondrite; Evidence of Regolith Formation since 3.7 Aeons B.P." Paper presented at the 8th annual meeting of the Division of Planetary Sciences, American Astronomical Society, 19–22 January, Honolulu.

Bogard, D., and P. Johnson. 1983. "Martian Gases in an Antarctic Meteorite?" *Science* 221:651.

Bohn, R. B. and 3 others. 1994. "Infrared Spectroscopy of Triton and Pluto Ice Analogs: The Case for Saturated Hydrocarbons." *Icarus* 111:51–173.

Bunch, T. E. and R. S. Rajan. 1988. "Meteorite Regolithic Breccias." In *Meteorites and the Early Solar System,* ed. J. Kerridge and M. Matthews. Tucson: University of Arizona Press.

Chapman, C. R., and D. R. Davis. 1975. "Asteroid Collisional Evolution: Evidence for a Much Larger Early Population." *Science* 190:553.

Chapman, C. R., J. Williams, and W. Hartmann. 1978. "The Asteroids." *Ann. Rev. Astron. Astrophys.* 16:33.

Chapman, C. R., T. McCord, and C. Pieters. 1973. "Minor Planets and Related Objects X. Spectrophotometric Study of the Composition of 1685 Toro." *Astron. J.* 78:502.

Chapman, D., and H. Larson. 1963. "On the Lunar Origin of Tektites." *J. Geophys. Res.* 68:4305.

Chyba, C., P. Thomas, and K. Zahnle. 1993. "The 1908 Tunguska Explosion: Atmospheric Disruption of a Stony Asteroid." *Nature* 361:40–44.

Clepecha, Z. 1981. "Possible Meteorite Fall, Alps Region." *Scientific Event Alert Network Bulletin 6,* no. 10, p. 12.

Crabb, J., and L. Schultz. 1981. "Cosmic Ray Exposure Ages of Ordinary Chondrites and Their Significance for Parent Body Stratiography." *Geochim. Cosmochim. Acta.* 45:215.

Cruikshank, D., D. Tholen. W. Hartmann, J. Bell, and R. Brown. 1991. "Three Basaltic Earth-Approaching Asteroids and the Source of the Basaltic Meteorites." *Icarus* 89:1.

Davis, D. R., C. Chapman, R. Greenberg, S. Weidenschilling, and A. Harris. 1979. "Collisional Evolution of Asteroids: Populations, Rotations, and Velocities." In *Asteroids,* ed. T. Gehrels. Tucson: University of Arizona Press.

Dodd, R. T. 1965. "Preferred Orientation of Chondrules in Chondrites." *Icarus* 4:308.

———. 1971. "The Petrology of Chondrules in the Sharps Meteorite." *Contrib. Mineral. Petrol.* 31:201.

———. 1981. *Meteorites.* Cambridge: Cambridge University Press.

Dohnanyi, J. S. 1970. "On the Origin and Distribution of Meteoroids." *J. Geophys. Res.* 75:3468.

———. 1972. "Interplanetary Objects in Review: Statistics of Their Masses and Dynamics." *Icarus* 17:1.

DuFresne, E., and E. Anders. 1963. "Chemical Evolution of the Carbonaceous Chondrites." In *The Moon, Meteorites and Comets,* ed. B. M. Middlehurst and G. P. Kuiper. Chicago: University of Chicago Press.

Eberhardt, P., and D. Hess. 1960. "Helium in Stone Meteorites." *Astrophys. J.* 131:38.

Farinella, P., F. Marzari, D. Vokrouhlicky, W. K. Hartmann, D. R. Davis, and S. J. Weidenschilling. 1997. "Meteorite Delivery through Yarkovsky Orbital Drift." *Bull. Amer. Astron. Soc.* 1045.

Faul, H. 1966. "Tektites Are Terrestrial." *Science* 152:1341.

Feierberg, M., H. Larson, and C. Chapman. 1982. "Spectroscopic Evidence for Undifferentiated S-type Asteroids." *Astrophys. J.* 257:361–372.

Feierberg, M., and M. Drake. 1980. "The Meteorite-Asteroid Connection: The Infrared Spectra of Eucrites, Shergottites, and Vesta." *Science* 209:805.

Fredriksson, K. 1963. "Chondrules and the Meteorite Parent Bodies." *Trans. N.Y. Acad. Sci.* 25:756.

Gallant, Roy A. 1994. "Journey to Tunguska." *Sky and Telescope,* June, 38–43.

Ganapathy, R., R. Keays, J. Laul, and E. Anders. 1970. "Trace Elements in Apollo 11 Lunar Rocks: Implications of Meteorite Influx and Origin of Moon." *Proc. Apollo 11 Lunar Sic. Conf.* 2:1117.

Ganapathy, R., D. Brownlee, and P. Hodge. 1978. "Silicate Spherules from Deepsea Sediments: Confirmation of Extraterrestrial Origin." *Science* 201:1119.

Gladman, B. J. and 4 others. 1996. "The Exchange of Impact Ejecta between Terrestrial Planets." *Science* 271:1387–1392.

Goldstein, J. I., and J. M. Short. 1967. "The Iron Meteorites, Their Thermal History and Parent Bodies." *Geochim. Cosmochim. Acta* 31:1733.

Goplan, K., and G. Wetherill. 1971. "Rubidium-Strontium Studies on Black Hypersthene Chondrites." *J. Geophys. Res.* 76:8484.

Greenberg, R. and C. Chapman. 1984. "Asteroids and Meteorites: Origin of Stony-Iron Meteorites at Mantle-Core Boundaries." *Icarus* 57:267–279.

Hartmann, W. K., R. Strom, S. Weidenschilling, K. Blasius, A. Woronow, M. Dence, R. Grieve, J. Diaz, C. R. Chapman, E. M. Shoemaker, and K. Jones. 1981. "Cratering Chronology of the Planets." In *Basaltic Volcanism of the Planets,* ed. R. Merrill and others. New York: Pergamon Press.

Hartmann, W. K., and C. A. Wood. 1971. "Moon: Origin and Evolution of Multi-ring Basins." *The Moon* 3:3.

Hartmann, W. K., and G. P. Kuiper. 1962. "Concentric Systems Surrounding Lunar Basins." *Comm. Lunar Planet. Lab.* 1:51.

Hartmann, W. K., and D. J. Tholen. 1990. "Comet Nuclei and Trojan Asteroids, A New Link and a Possible Mechanism for Comet Splitting." *Icarus* 86:448.

Hartmann, W. K. 1970. "Lunar Cratering Chronology." *Icarus* 13:299.

———. 1975. "Lunar Cataclysm: A Misconception?" *Icarus* 24:181.

———. 1979. "A Special Class of Planetesimal Collisions and a Possible Confirmation." *Proc. Lun. Planet. Sci. Conf.* 10:1897.

———. 1980. "Dropping Stones in Magma Oceans: Effects of Early Lunar Cratering." In *Proceedings of Conference on the Lunar Highlands Crust,* ed. J. Papike and R. Merrill. New York: Pergamon Press.

Hartmann, W. K., and C. Impey. 1994. *Astronomy: The Cosmic Journey.* Belmont, CA: Wadsworth.

Hartmann, W. K., and L. Wilkening. 1981. "Chondrule-sized Spherules from an Explosion Crater." *Meteoritics* 15:299 (abstract).

Heineman, R., and L. Brady. 1929. "The Winona Meteorite." *Amer. J. Sci. 18* (5th ser.):477.

Herndon, J. M., and L. Wilkening. 1978. "Conclusions Derived from the Evidence of Accretion in Meteorites." In *Protostars and Planets,* ed. T. Gehrels and M. Matthews. Tucson: University of Arizona Press.

Heymann, D. 1978. "Solar Gases in Meteorites: The Origin of Chondrites and Cl Carbonaceous Chondrites." *Meteoritics* 13:291.

Housen, K. R., L. Wilkening, C. R. Chapman, and R. Greenberg. 1979. "Regolith Development and Evolution on Asteroids and the Moon." In *Asteroids,* ed. T. Gehrels. Tucson: University of Arizona Press.

Hunten, D. M. 1979. "Capture of Phobos and Deimos by Proto-atmospheric Drag." *Icarus* 37:113.

Ip, W. 1978. "Model Consideration of the Bombardment Event of the Asteroid Belt by the Planetesimals Scattered from the Jupiter Zone." *Icarus* 34:117.

Jacchia, L. G. 1974. "A Meteorite That Missed the Earth." *Sky and Telescope* 48:4.

Kaula, W., and P. Bigeleisen. 1975. "Early Scattering by Jupiter and Its Collision Effects in the Terrestrial Zone." *Icarus* 25:18.

Keay, C. S. 1980. "Anomalous Sounds from the Entry of Meteor Fireballs." *Science* 210:11.

Kerr, R. A. 2001. "Ancient Sky Rocks and an Unblemished Eros." *Science* 294:39.

Keiffer, S. W. 1975. "Droplet Chondrules." *Science* 189:333.

Kelly, W. R., and J. Larimer. 1977. "Chemical Fractionation in Meteorites VIII." *Geochim. Cosmochim. Acta 41*:93.

Kerridge, F., and M. Matthews, ed. 1988. *Meteorites and the Early Solar System.* Tucson: University of Arizona Press.

Khare, B. N. and 7 others. 1993. "Production and Optical Constants of Ice Tholin from Charged Particle Irradiation of (1:6) C2H6/H2O at 77 K." *Icarus 103*:290–300.

Krinov, E. L. 1966. *Giant Meteorites.* London: Pergamon Press.

———. 1960. *Principles of Meteorites.* Elmsford, New York: Pergamon Press.

Kvenvolden, K., and others. 1970. "Evidence for extraterrestrial amino acids and hydrocarbons in the Murchison meteorite." *Nature 228*:923.

Lauretta, Dante S., K. Lodders, and B. Fegley. 1997. "Experimental Simulations of Sulfide Formation in the Solar Nebula." *Science 277*:358–360.

Levin, B. J., A. Simonenko, and E. Anders. 1976. "Farmington Meteorite: A Fragment of an Apollo Asteroid?" *Icarus 28*:307.

Lewis, R. S., and others. 1979. "Stellar Condensates in Meteorites: Isotopic Evidence from Noble Gases." *Astrophys. J. 234*:L165.

Lorin, J., and P. Pellas. 1979. "Preirradiation History of Djermaia (H) Chondritic Breccia." *Icarus 40*:502.

Margot, J. L. and 7 others. 2002. "Binary Asteroids in the Near-Earth Population." *Science 296*:1445–1448.

Mason, B. 1962. *Meteorites.* New York: Wiley.

Masursky, H., G. Colton, and F. El-Bax, ed. 1978. *Apollo over the Moon.* Washington, D.C.: NASA SP-362.

McCord, T. B., J. Adams, and T. Johnson. 1970. "Asteroid Vesta: Spectral Reflectivity and Compositional Implications." *Science 168*:1445–1447.

McSween, H. Y. Jr. 1987. *Meteorites and Their Parent Planets.* Cambridge University Press.

Michel, Patrick and 3 others. 2001. "Collisions and Gravitational Reaccumulation: Forming Asteroid Families and Satellites," *Science 294*:1696–1700.

Migliorini, F. and 5 others. 1997. "Vesta Fragments from nu-6 and 3:1 Resonances: Implications for V-type Near-Earth Asteroids and Howeardite, Eucrite and Diogenite Meteorites." *Meteoritics and Plan. Sci. 32*:903–916.

Morrison, D., C. Chapman, and P. Slovic. 1994. "The Impact Hazard." In *Hazards due to Comets and Asteroids,* ed. T. Gehrels, Tucson: University of Arizona Press .

Nininger, H. H. 1952. *Out of the Sky.* Denver: University of Denver Press.

Nyquist, L. E. and 6 others. 1979. "Early Differentiation, Late Magmatism, and Recent Bombardment on the Shergottite Parent Planet." *Meteoritics 14*:502.

O'Keefe, J., ed. 1963. *Tektites.* Chicago: University of Chicago Press.

———. 1978. "The Tektite Problem." *Sci. Am.,* February, p. 116.

Öpik, E. J. 1963. "Survival of Comet Nuclei and the Asteroids." *Adv. Astron. Astrophys. 2*:219.

———. 1966. "The Stray Bodies in the Solar System II. The Cometary Origin of Meteorites." *Adv. Astron. Astrophys. 4*:301.

Pellegrino, C. R., and J. Stoff. 1979. "Organic Clues in Carbonaceous Meteorites." *Sky and Telescope 57*:330.

Pepin, R. O., and M. Carr. 1991. In *Mars,* ed. H. Kieffer and B. Jakosky. Tucson: University of Arizona Press.

Prior, G. T. 1920. "The Classification of Meteorites." *Mineral. Mag. 19*:51.

Richardson, S. M. 1978. "Vein Formation in the C1 Carbonaceous Chondrites." *Meteoritics 13*:141.

Ringwood, A. E. 1966. "Genesis of Chondritic Meteorites." *Rev. Geophys. Spa. Phys. 4*:113.

Sagan, C. 1975. "Kalliope and the Ka;aaba: The Origin of Meteorites." *Natural History 84*:8.

Schmitz, B. amd 3 others. 1997. "Accretion Rates of Meteorites and Cosmic Dust in the Early Ordovician." *Science 278*:88–90.

Scott, E. R. D. 1977. "Parent Bodies of Iron Meteorites." In *Comets, Asteroids, and Meteorites,* ed. A. Delsemme, p. 439. Toledo: University of Toledo Press.

———. 1978. "Tabulation of Meteorites." *Proc. Lun. Planet. Sci. Conf. 9*:pt. 2, endpaper.

Sears, Derek W. and 4 others. 1997. "The Metamorphic History of Eucrites and Eucrite-related Meteorites and the Case for Late Metamorphism." *Meteoritics and Plan. Sci. 32*:917–927.

Smith, P. P. K., and P. Buseck. 1981. "Graphitic Carbon in the Allende Meteorite: A Microstructural Study." *Science 212*:322.

Stoffler, D. 1997. "Minerals in the Deep Earth: A Message from the Asteroid Belt." *Science 278*:1576–1577.

Strazzulla, G. and R. Johnson. 1991. "Irradiation Effects on Comets and Cometary Debris." In *Comets in the Post-Halley Era,* ed. R. Newburn, M Neugebauer, and J. Rahe Dordrecht: Kluwer, pp. 243–275.

Suess, H. E. 1949. "Chemistry of the Formation of Planets." *Z. Electrochem. 53*:237.

Svetsov, V. V. 1996. "Total Ablation of the Debris from the 1908 Tunguska Explosion." *Nature 383*:697–699.

Taylor, S. R., and M. Kaye. 1969. "Genetic Significance of the Chemical Composition of Tektites: A Review." *Geochim. Cosmochim. Acta 7*:34.

Taylor, S. R. 1975. *Lunar Science: A Post-Apollo View.* New York: Pergamon Press.

Taylor, S. R., and D. Heymann. 1969. "Shock Reheating and the Gas Retention Ages of Chondrites." *Earth Planet, Sci. Lett. 7*:151.

Tera, F., D. Papanastassiou, and G. Wasserburg. 1974. "Isotopic Evidence for a Terminal Lunar Cataclysm." *Earth Planet, Sci. Lett. 22*:1.

Thomas, Peter C. and 5 others. 1997. "Impact Excavation on Asteroid 4 Vesta: Hubble Space Telescope Results." *Science 277*: 1492–1495.

Thompson, W. R., B. Murray, B. Khare, and C. Sagan. 1987. "Coloration and Darkening of Methane Clathrate and Other Ices by Charged Particle Irradiation: Application to the Outer Solar System." *J. Geophys. Res. 92*:14933–14947.

Uhlig, H. H. 1955. "Contribution of Metallurgy to the Study of Meteorites I. Structure of Metallic Meteorites, Their Composition and the Effect of Pressure." *Geochim. Cosmochim. Acta 6*:282.

Urey, H. C., and H. Craig. 1953. "The Composition of the Stone Meteorites and the Origin of the Meteorites." *Geochim. Cosmochim. Acta 4*:36.

Van Schmus, W. R., and J. A. Wood. 1967. "A Chemical-Petrologic Classification for the Chondritic Meteorites." *Geochim. Cosmochim. Acta 31*:747.

Veski, Siim and 4 others. 2001. "Ecological Catastrophe in Connection with the Impact of the Kaali Meteorite about 800–400 B.C. on the Island of Saaremaa, Estonia." *Meteoritics Plan. Sci. 36:* 1367–1375.

Vortman, L. J. 1977. "Craters from Surface Explosions and Energy Dependence—A Retrospective." In *Impact and Explosion Cratering,* ed. D. Roddy, R. Pepin, and R. Merrill. New York: Pergamon Press.

Wahl, W. 1952. "The Brecciated Stony Meteorites and Meteorites Containing Foreign Fragments." *Geochim. Cosmochim. Acta 2*:91.

Wasson. J. T. 1974. *Meteorites.* New York: Springer-Verlag.

———. 1985. *Meteorites: Their Record of Early Solar-system History.* New York: Freeman.

Wasson, J. T. and G. W. Wetherill. 1979. "Dynamical, Chemical, and Isotopic Evidence Regarding the Formation Locations of

Asteroids and Meteorites." In *Asteroids,* ed. T. Gehrels. Tucson: University of Arizona Press.

Weidenschilling, S. J., F. Marzari, and L. L. Hood. 1998. "The Origin of Chondrules at Jovian Resonances." *Science* 279:681–684.

Wetherill, G. W. 1975. "Late Heavy Bombardment of the Moon and Terrestrial Planets." *Proc. Lun. Sci. Conf.* 6:1539.

———. 1976. "Where Do the Meteorites Come From? A Reevaluation of the Earth-crossing Apollo Objects as Sources of Chondritic Meteorites." *Geochim. Cosmochim. Acta* 40:1297.

———. 1977a. "Evolution of the Earth's Planetesimal Swarm Subsequent to the Formation of the Earth and Moon." *Proc. Lun. Sci. Conf.* 8:1.

———. 1977b. "Pre-mare Cratering and Early Solar System History." In *The Soviet-American Conference on Cosmochemistry of the Moon and Planets,* ed. J. Pomeroy and N. Hubbard. Washington: NASA SP-378.

Whipple, F. L. 1966. "Chondrules: Suggestion Concerning the Origin." *Science* 153:54.

Whitaker, E. 1978. "Galileo's Lunar Observations and the Dating of the Composition of Siderus Nuncius." *J. Hist. Astron.* 9:155.

Wilhelms, D. E., C. Hodges, and R. Pike. 1977. "Nested-Crater Model of Lunar Ringed Basins." In *Impact and Explosion Cratering,* ed. D. Roddy and others. New York: Pergamon Press.

Wilk, H. B. 1956. "The Chemical Composition of Some Stony Meteorites." *Geochim. Cosmochim. Acta* 9:279.

Wilkening, L. 1977. "Meteorites in Meteorites: Evidence for Mixing among the Asteroids." In *Comets, Asteroids, Meteorites,* ed., A. H. Delsemme. Toledo, Ohio: University of Toledo Press.

———. 1978. "Carbonaceous Chondritic Material in the Solar System." *Naturwissenschaften* 65:73.

Williams, J. G. 1973. "Meteorites from the Asteroid Belt?" *EOS, Trans. Am. Geophys. Union* 54:233 (abstract).

Wood, J. A. 1963. "Physics and Chemistry of Meteorites." In *The Moon, Meteorites, and Comets,* ed. B. M. Middlehurst and G. P. Kuiper. Chicago: University of Chicago Press.

Wood, J. A., and H. McSween. 1977. "Chondrules as Condensation Products." In *Comets, Meteorites, Asteroids,* ed. A. H. Delsemme. Toledo, Ohio: University of Toledo Press.

Zolensky, M., and H. McSween. 1988. "Aqueous Alteration." In *Meteorites and the Early Solar System,* ed. J. Kerridge and M. Matthews. Tucson: University of Arizona Press.

Chapter Seven

Anders, E. 1964. "Origin, Age, and Composition of Meteorites." *Space Sci. Rev.* 4:582.

———. 1965. "Fragmentation History of Asteroids." *Icarus* 4:399.

Arnold, J. R. 1965. "The Origin of Meteorites as Small Bodies." *Astrophys. J.* 141:1536.

———. 1970. "Asteroid Families and Jet Streams." *Astron. J.* 14:1235.

Belton, M. and others. 1992. "Galileo Encounter with 951 Gaspra: First Pictures of an Asteroid." *Science* 257:1647.

Bierman, L. 1967. "Theoretical Considerations of Small Particles in Interplanetary Space." In *The Zodiacal Light and the Interplanetary Medium* (ed. J. Weinberg), Washington, DC: NASA SP-150, p. 279.

Binzel, Richard. In press. Trojan, Hilda, and Cybele Asteroids: New Light Curve Observations. *Icarus* 95:222–238.

Binzel, R. P. and 6 others. 1993. "Discovery of a Main-Belt Asteroid Resembling Ordinary Chondrite Meteorites." *Science* 262:1541–1543.

Binzel, R. P., and T. van Flandern. 1979. "Minor Planets: The Discovery of Minor Satellites." *Science* 203:903.

Bottke, W. F. and H. J. Melosh 1996 "Binary Asteroids and the Formation of Doublet Craters." *Icarus* 124:372–391.

Bowell, E., C. Chapman, J. Gradie, D. Morrison, and B. Zellner. 1978. "Taxonomy of Asteroids." *Icarus* 35:313.

Brandt, John C. 1990. "Comets." In *The New Solar System.* Cambridge, Sky Publishing Corp.

Brouwer, D. 1951. "Secular Variations of the Orbital Elements of Minor Planets." *Astron. J.* 56:9.

Brown, Peter G. and 21 others. 2001. "The Fall, Recovery, Orbit, and Composition of the Tagish Lake Meteorite: A New Type of Carbonaceous Chondrite." *Science* 290:320–325.

Brown, R. H., D. Cruikshank, Y. Pendleton, and G. Veeder. 1997. "Surface Composition of Kuiper Belt Object 1993SC." *Science* 276:937–939.

Brownlee, D., R. Rajan, and D. Tomandl. 1977. "A Chemical and Textural Comparison between Carbonaceous Chondrites and Interplanetary Dust." In *Comets, Asteroids, Meteorites,* ed. A. Delsemme. Toledo, Ohio: University of Toledo Press.

Chapman, C. R. 1975. "The Nature of Asteroids." *Sci. Am.,* January, p. 24.

———. 1979. "The Asteroids: Nature, Interrelations, Origin, and Evolution." In *Asteroids,* ed. T. Gehrels. Tucson: University of Arizona Press.

———. 1996 "S type Asteroids, Ordinary Chondrites, and Space Weathering: The Evidence from Galileo's Fly-bys of Gaspra and Ida." *Meteoritics and Plan. Sci.* 31:699–725.

Chapman, C. R., and D. Davis. 1975. "Asteroid Collisional Evolution: Evidence for a Much Larger Early Population." *Science* 190:553.

Chapman, C. R., T. McCord, and C. Pieters. 1973. "Minor Planets and Related Objects X. Spectrophotometric Study of the Composition of 1685 Toro." *Astron. J.* 78:502.

Chapman, C. R., J. Williams, and W. Hartmann. 1978. "The Asteroids." *Ann. Rev. Astron. Astrophys.* 16:33.

Clark, R. N. 1980. "The Spectral Reflectance of Water-Mineral Mixtures in Low Temperature." *J. Geophys. Res.* 86:3074.

Clayton, R. N., T. Mayeda, and A. Rubin. 1984. "Oxygen Isotopic Compositions of Enstatite Chondrites and Aubrites." *Journ. Geophys. Res. Suppl.* 89:C245-9 (Proc. Lun. Plan. Sci. Conf. 15).

Criswell, D. R., ed. 1978. *New Moons.* Houston: Lunar and Planetary Institute.

Cruikshank, D., W. Hartmann, and D. Tholen. 1985. "Colour, Albedo, and Nucleus Size of Halley's Comet." *Nature* 315: 122–124.

Cruikshank, D., D. Tholen, W. Hartmann, J. Bell, and R. Brown. 1991. "Three Basaltic Earth-Approaching Asteroids and the Source of the Basaltic Meteorites." *Icarus* 89:1.

Davies, J. G., and A. Lovell. 1955. "The Giacobinid Meteor Stream." *Monthly Notices Roy. Astron. Soc.* 15:23.

Davies, J. G., and W. Turski. 1962. "The Formation of the Giacobinid Meteor Stream." *Monthly Notices of the Roy. Astron. Soc.* 123:459.

Davis, D. R., C. Chapman, R. Greenberg, S. Weidenschilling, and A. Harris. 1979. "Collisional Evolution of Asteroids: Populations, Rotations, and Velocities." In *Asteroids,* ed. T. Gehrels. Tucson: University of Arizona Press.

Davis, D. R. 1999. "The Collisional History of Asteroid 253 Mathilde." *Icarus* 140:49–52.

Davis, D.R., and C. Chapman. 1977. "The Collisional Evolution of Asteroid Composition Classes." *Lunar Plan. Sci. Conf.* 8 (Houston: Lunar and Planetary Institute), pp. 224–226 (abstract).

Davis, D. R. and P. Farinella. 1997. "Collisional Evolution of Edgeworth-Kuiper Belt Objects." *Icarus* 125:50–60.

Degewij, J., J. Gradie, T. Lebertre, W. Wisniewski, and B. Zellner. 1977. Abstract from 150th meeting, American Astronomy Society.

Dunlap, J. L., and T. Gehrels. 1969. "Minor Planets III: Light-curves of a Trojan Asteroid." *Astron. J.* 74:797.

Durda, D. D. 1996. "The Formation of Asteroid Satellites in Catastrophic Collisions," *Icarus* 120:212–219.

Everhart, E. 1977. "The Evolution of Comet Orbits as Perturbed by Uranus and Neptune." In *Comets, Asteroids, Meteorites,* ed. A. Delsemme. Toledo, Ohio: University of Toledo Press.

Everhart, E. 1979. "Chaotic Orbits in the Solar System." In *Asteroids,* ed. T. Gehrels. Tucson: University of Arizona Press.

Fanale, Fraser and J. Salvail 1990. "Evolution of the Water Regime of Phobos." *Icarus* 88:380–395.

French, Linda, M. Kramer, and H. Vargass. 1986. "Rotation Properties of Trojan Asteroids." *Bull. Am. Astron. Soc.* 18:796–797.

Gaffey, M. J. 1978. "Asteroid Surface Materials: Mineralogical Characterizations from Reflectance Spectra." *Space Sci. Rev. 21:* 555.

Gaffey, M. J., and T. McCord. 1977. "Mining the Asteroids." 6, no. 6, p. 1.

Gault, D. E., E. M. Shoemaker, and H. J. Moore. 1963. "Spray Ejected from the Lunar Surface by Meteoroid Impact." NASA Technical Note Dd 767.

Gradie, J., and E. Tedesco. 1982. "Compositional Structure of the Asteroid Belt." *Science* 215:1405.

Gradie, J., and J. Veverka. 1980. "The Composition of the Trojan Asteroids." *Nature* 283:840.

Harris, A. W. 1977. "An Analytical Theory of Planetary Rotation Rates." *Icarus* 31:168.

Harris, A. W., and J. A. Burns. 1979. "Asteroid Rotation I. Tabulation and Analysis of Data." *Icarus* 40:115.

Hartmann, W. K. 1980. "Surface Evolution of Two-Component Stone/Ice Bodies in the Jupiter Region." *Icarus* 44:441–453.

Hartmann, W. K. 1990. "Additional Evidence about an Early Intense Flux of C Asteroids and the Origin of Phobos." *Icarus* 87:236–240.

Hartmann, W. K. and D. Tholen 1990. "Comet Nuclei and Trojan Asteroids: A New Link and a Possible Mechanism for Comet Splittings." *Icarus* 86:448–454.

Hartmann, W. K. and 4 others. 1988. "Trojan and Hilda Asteroid Lightcurves." *Icarus* 73:487–498.

Hiroi, Takahiro, M. Zolensky, and Carlé Pieters. 2001. "The Tagish Lake Meteorite: A Possible Sample from a D-Type Asteroid." *Science* 293:2234–2236.

Jacchia, L. G., F. Verniani, and R. E. Briggs. 1967. "An Analysis of 413 Precisely Reduced Photographic Meteors." *Smithsonian Contrib. Astrophys.* 10, no. 1.

Jewitt, D. C. and 3 others. 1997. "Measurements of $^{12}C/^{13}C$, $^{14}N/^{15}N$, and $^{32}S/^{34}S$ Ratios in Comet Hale-Bopp (C/1995 O1)." *Science* 278:90–93.

Liou, J.-C. and R. Malhotra 1997. "Depletion of the Outer Asteroid Belt." *Science* 275:375–377.

Luu, Jane X., D. Jewitt and C. Trujillo. 2000. "Water Ice in 2060 Chiron and its Implications for Centaurs and Kuiper Belt Objects." *Astrophys. J. Letters.*

Magri, C. and 4 others. 2001. "Radar Constraints on Asteroid Regolith Compositions Using 433 Eros as Ground Truth." *Meteoritics and Planet. Sci.* 36:1697.

Margot, J. L. and M. E. Brown. 2003. A Low-Density M-Type Asteroid in the Main Belt. *Science 300:*1939–1942.

McKay, C. P. and W. J. Borucki 1997. "Organic Synthesis in Experimental Impact Shocks." *Science* 276:390–392.

McLean, D. R. 1967. "The Leonid Meteor Shower of November 17, 1966." *Comm. Lunar Planet, Lab.* 6:43.

Meech, K. and M. Belton. 1990. "The Atmosphere of 2060 Chiron." *Astron. Journ.* 100:1323.

Merline, W. J. and 5 others. 2002. "Asteroids Do Have Satellites." In *Asteroids III,* ed. W. Bottke and 3 others. Tucson: University of Arizona Press.

Millman, P. M. 1979. "Interplanetary Dust." *Naturwissenschaften* 66:134.

Oort, J. H. 1950. "The Structure of the Cloud of Comets Surrounding the Solar System and a Hypothesis Concerning its Origin." *Bull. Astron. Inst. Netherlands* 11:91.

———. 1963. "Empirical Data on the Origin of Comets." In *The Moon, Meteorites, and Comets,* ed. B. Middlehurst and G. Kuiper. Chicago: University of Chicago Press.

Öpik, E. J. 1963. "The Stray Bodies in the Solar System I. Survival of Cometary Nuclei and the Asteroids." *Adv. Astron. Astrophys.* 2:219.

———. 1966. "The Stray Bodies in the Solar System II. The Cometary Origin of Meteorites." *Adv. Astron. Astrophys.* 4:301.

Piotrowski, S. 1953. "The Collisions of Asteroids." *Acta. Astron.* A5:115.

Pizzarello, Sandra and 6 others. 2001. "The Organic Content of the Tagish Lake Meteorite." *Science* 293:2236–2239.

Pohn, H. A. 1966. "Observations of a Double Nucleus in Comet Ikeya-Seki." *Sky and Telescope* 31:376.

Sagdeev, R. V., and others, 1966. "Encounters with Comet Halley—The First Results." *Nature* 321:259.

Schmitz, B., B. Peucker-Ehrenbrink, M. Lindstrom, and M. Tassinari. 1997. "Accretion Rates of Meteorites and Cosmic Dust in the Early Ordovician." *Science* 278:88–90.

Sekanina, Z. 1977. "Differential Nongravitational Forces in the Motions of the Split Comets." In *Comets, Meteorites, and Asteroids,* ed. A. Delsemme. Toledo, Ohio: University of Toledo Press.

Smith, Peter H. and 25 others. 1997. "Results from the Mars Pathfinder Camera." *Science* 278:1758–1765.

Sullivan, R. J. and 3 others. 2002. "Asteroid Geology from Galileo and NEAR Shoemaker Data." In *Asteroids III,* ed. W. Bottke and 3 others. Tucson: University of Arizona Press.

Taylor, S. R. 1988. "Planetary Compositions." In *Meteorites and the Early Solar System,* ed. J. Kerridge and M. Matthews. Tucson: University of Arizona Press.

Tholen, David, and M. Barucci. 1989. "Asteroid Taxonomy." In *Asteroids II,* ed. R. Binzel and others. Tucson: University of Arizona Press.

Verniani, F. 1969. "Structure and Fragmentation of Meteoroids." *Space Sci. Rev.* 10:230.

Wasson, J. T. 1974. *Meteorites.* New York: Springer-Verlag.

Weidenschilling, S. J. 1997. "The Origin of Comets in the Solar Nebula: A Unified Model." *Icarus* 127:290–306.

Whipple, F. 1967. "On Maintaining the Meteoritic Complex." In *Zodiacal Light and the Interplanetary Medium,* ed. J. Weinberg. Washington, D.C.: NASA SP-150.

Yeomans, D. K. and 12 others. 1997. "Estimating the Mass of Asteroid 253 Mathilde from Tracking Data During the NEAR Flyby." *Science* 278:2106–2109.

Chapter Eight

Anderson, D. L., and others. 1977. "Seismology on Mars." *J. Geophys. Res.* 82:4524.

Anderson, J. W. Sjögren, and G. Schubert 1996. "Galileo Gravity Results and the Internal Structure of Io." *Science* 272:709–712.

Bacon, F. 1620. *Novum Organum.* London.

Basu, A. R., and others. 1981. "Eastern Indian 3800-My-old Crust and Early Mantle Differentiation." *Science* 212:1502.

Ballabh, G., A. Bhatnagar, and N. Bhandari. 1978. "The Orbit of the Dhaiala Meteorite." *Icarus* 33:361.

Ben-Avraham, Z. 1981. "The Movement of Continents." *Am. Scientist* 69:291.

Biermann, L. 1951. "Komentenschweife und solare Korpuskularstrahlung." *Z. Astrophys.* 29:274.

Brandt, J. C., and R. D. Chapman. 1981. *Introduction to Comets.* Cambridge: Cambridge University Press.

Brower, K. 1978. *The Starship and the Canoe.* New York: Holt, Rinehart & Winston.

Cameron, A. G. W. 1973. "Accumulation Processes in the Primitive Solar Nebula." *Icarus* 18:407.

Carr, M. H. and 21 others. 1998. "Evidence from Galileo for a Subsurface Ocean on Europa." To be published in *Nature.*

Ceplecha, A., R. Rajche, and L. Sehnal. 1959. "New Czechoslovak Meteorite, Luhy." *Bull. Astron. Inst. Czech.* 10:147.

Clark, Pamela E. 1997. "Mercury: Planet of Fire and Ice (Part 2)." *The Mercury Messenger,* Issue 9, December (Houston: Lunar and Planetary Institute).

Cole, G. H. A. 1978. *The Structure of Planets.* New York: Crane, Russak.

Consolmagno, G. J. 1981. Io: Thermal Models and Chemical Evolution. *Icarus* 47:36.

Consolmagno, G., and J. Lewis. 1977. "Preliminary Thermal History Models of Icy Satellites." In *Planetary Satellites,* ed. J. Burns. Tucson: University of Arizona Press.

———. 1978. "The Evolution of Icy Satellite Interiors and Surfaces." *Icarus* 34:280.

Cox, K. G. 1978. "Kimberlite Pipes." *Sci. Am.,* April, p. 120.

Criswell, D. R., ed. 1977. *New Moons.* Houston: Lunar and Planetary Institute.

Delsemme, A. H. 1977. "The Pristine Nature of Comets." In *Comets, Asteroids, Meteorites,* ed. A. Delsemme. Toledo, Ohio: University of Toledo Press.

DeMarcus, W. C. 1958. "The Constitution of Jupiter and Saturn." *Astron. J.* 63:2.

Dermott, S. F., R. Malhotra, and C. D. Murray. 1988. "Dynamics of the Uranian and Saturian Satellite Systems: A Chaotic Route to Melting Miranda?" *Icarus* 76:295–334.

Dewey, J. F., and J. Bird. 1970. "Mountain Belts and the New Global Tectonics." *J. Geophys. Res.* 75:2625.

Fairbridge, R. W. 1972. *The Encyclopedia of Geochemistry and Environmental Sciences.* New York: Van Nostrand Reinhold.

Fay, T., and W. Wisniewski. 1978. "The Light Curve of the Nucleus of Comet d'Arrest." *Icarus* 34:1.

Fayet, G. 1910. "Recherches concernant les excentricités des comètes." *Paris Mém.* 26:A.1 a A.134.

Fernández, J. A. 1978. "Mass Removed by the Outer Planets in the Early Solar System." *Icarus* 34:173.

Folkner, W. M. and 4 others. 1997. "Interior Structure and Seasonal Mass Redistribution of Mars from Radio Tracking of Mars Pathfinder." *Science* 278:1749–1752.

Frey, H. 1977. "Origin of the Earth's Ocean Basins." *Icarus* 32:235.

Fricker, P., R. Reynolds, and A. Summers. 1974. "On the Thermal Evolution of the Terrestrial Planet." *Moon* 9:211.

Goodwin, A. M. 1976. "Giant Impacting and the Development of Continental Crust." In *The Early History of the Earth,* ed. B. F. Windley. New York: Wiley.

Guttenberg, B. 1926. "Untersuchungen zu Frage, bis zu welcher tiefe die erde Kristallin ist." *Z. Geophysik* 2:24.

Halliday, A. N., M. Rahkamper, D-C. Lee, and W. Yi. 1997. *Earth and Planet. Sci. Lett.* 142:75.

Hartmann, W., D. Cruikshank, and J. Degewij. 182. "Remote Comets and Related Bodies: VJHK Colorimetry and Surface Materials." *Icarus,* in press.

Hartmann, W. K., and D. J. Tholen. 1990. "Comet Nuclei and Trojan Asteroids, A New Link and a Possible Mechanism for Comet Splitting." *Icarus* 86:448.

Herbert, F., M. Drake, and C. Sonett. 1978. "Geophysical and Geochemical Evolution of the Lunar Magma Ocean." *Proc. Lun. Sci. Conf.* 9:249.

Hsui, A., and M. Toksöz. 1977. "Thermal Evolution of Planetary Size Bodies." *Proc. Lun. Sci. Conf.* 8:447.

Hubbard, W. 1970. "Structure of Jupiter: Chemical Composition, Contraction and Rotation." *Astrophys. J.* 162:687.

———. 1977. "The Jovian Surface Condition and Cooling Rate." *Icarus* 30:305.

———. 1981. "Interiors of the Giant Planets." *Science* 214:145.

———. 1990. "Interiors of the Giant Planets." In *The New Solar System,* ed. J. Beatty and A. Chaikin. Cambridge, Mass.: Sky Publishing Corp.

Humes, D. H., J. Alvarez, R. O'Neal, and W. Kinar. 1974. "The Interplanetary and Near-Jupiter Meteoroid Environments." *J. Geophys. Res.* 79:3677.

Jacobsen, S. B., and G. Wasserburg. 1979. "The Mean Age of the Mantle and Crustal Reservoirs." *J. Geophys. Res.* 84:7411.

Jeanloz, R., and 3 others. 1995. "Evidence for a Basalt-Free Surface on Mercury and Implications for Internal Heat." *Science* 268:1455–1457.

Jeffreys, H. 1924. "On the Internal Constitution of the Earth." *Monthly Notices Roy. Astron. Soc.* 84:534.

———. 1962. *The Earth.* New York: Cambridge University Press.

Johnson, T. V. and 5 others. 1988. "Io: Evidence for Silicate Volcanism in 1986." *Science* 242:1280–1283.

Johnson, T. V. 2000. "Jupiter and its Moons." *Sci. Am.* 282:40–49 (Feb.).

Kargel, J. S. 1998. "The Salt of Europa." *Science* 280:1211–1212.

Kaula, W. M. 1968. *An Introduction to Planetary Physics.* New York: Wiley.

———. 1990. "Venus: A Contrast in Evolution to Earth." *Science* 247:1191.

Kaula, W. M. 1995. "Venus Reconsidered." *Science* 270:1460–1462.

Khurana, K. K. and 4 others. 1997. "Absence of an Internal Magnetic Field at Callisto." *Nature* 387:262–264.

Klepeis, J. E. and 3 others. 1991. "Hydrogen-helium Mixtures at Megabar Pressures: Implications for Jupiter and Saturn." *Science* 254:986–989.

Kresak, L. 1977. "Asteroid vs. Comet Discrimination from Orbital Data." In *Comets, Asteroids, Meteorites,* ed. A. Delsemme. Toledo, Ohio: University of Toledo Press.

Krinov, E. L. 1963. "The Tunguska and Sikhote-Alin Meteorites." In *The Moon, Meteorites, and Comets,* eds. B. M. Middlehurst and G. P. Kuiper. Chicago: University of Chicago Press.

Kuiper, G. P. 1952. "Planetary Atmospheres and Their Origin." In *The Atmospheres of the Earth and Planets,* ed. G. Kuiper. Chicago: University of Chicago Press.

Latham, G. V. 1971. "Lunar Seismology." *EOS, Trans. Am. Geophys. Union* 52:162.

Lebofsky, L. A. 1975. "Stability of Frosts in the Inner Solar System." *Icarus* 25:205.

Levin, B. J., A. Simonenko, and E. Anders. 1976. "Farmington Meteorite: A Fragment of an Apollo Asteroid?" *Icarus* 28:307.

Lewis, J. S. 1971a. "Satellites of the Outer Planets: Their Physical and Chemical Nature." *Icarus* 15:174.

———. 1971b. "Satellites of the Outer Planets: Thermal Models." *Science* 172:1127.

Marcialis, R., and R. Greenberg. 1987. "Warming of Miranda during Chaotic Rotation." *Nature* 38:227.

Marsh, B. D. 1979. "Island-Arc Volcanism." *Am Scientist* 67:161.

Mason, B. 1966. *Principles of Geochemistry.* NY: John Wiley & Sons, Inc.

McCord, T. B., and 11 others. 1988. "Salts on Europa's Surface Detected by Galileo's Near Infrared Mapping Spectrometer." *Science* 280:1242–1245.

McCrosky, R. E. 1967. "Orbits of Photographic Meteors." *Smithsonian Astrophys. Observ. Spec. Rept.* 252.

Menard, H. W. 1969. "The Deep-Ocean Floor." *Sci Am.*, September, p. 127.

Mohorovi, I. A. 1909. *Jb. Met. Oserv. Agram.* 9:1. (Zagreb, Croatia).

Muller, P. M., and W. L. Sjogren. 1968. "Mascons: Lunar Mass Concentrations." *Science 161*:680.

Nelson, R. M. 1997 "Mercury: The Forgotten Planet." *Scientific Amer.* 277:No. 5, 56–67 (November).

Nierenberg, W. A. 1978. "The Deep Sea Drilling Project After Ten Years." *Am. Scientist 66*:20

Niven, L., and J. Pournelle. 1977. *Lucifer's Hammer.* New York: Fawcett Crest.

Oberg, J. 1977. "Tunguska: Collision with a Comet." *Astronomy* 5:18.

Öpik, E. J. 1962. "Jupiter: Chemical Composition, Structure, and Origin of a Giant Planet." *Icarus 1*:200.

Oppenheimer, M. 1978. "What Are Comets Made Of?" *Natural History,* March, p. 42.

Pappalardo, R. T. and 9 others. 1998. "Geological Evidence for Solid-State Convection in Europan Ice Shell." *Nature 391*:365.

Peale, S. J., P. Cassen, and R. Reynolds. 1979. "Melting of Io by Tidal Dissipation." *Science 203*:892.

———. 1980. "Tidal Dissipation, Orbital Evolution, and the Nature of Saturn's Inner Moons." *Icarus 43*:65.

Pieters, C. 1982. *Science 315*:59.

Pieters, C. 1986. *Rev. Geophys.* 24:557.

Pinet, P. C., S. Chevrel, and P. Martin. 1993. "Copernicus; A Regional Probe of the Lunar Interior." *Science 260*:797–801.

Phillips, R. J. and Vicki L. Hansen. 1998. "Geological Evolution of Venus: Rises, Plains, Plumes, and Plateaus." *Science 279*: 1492–1497.

Podolak, M. 1978. "Models of Saturn's Interior: Evidence for Phase Separation." *Icarus 33*:342.

Poisson, S. D. 1829. *Mem. Acad. Sci. Paris 8*:623.

Pollack, James B. 1985. "Formation of the Giant Planets and Their Satellite-Ring Systems: An Overview." In *Protostars and Planets II,* ed. D. Black and M. Matthews. Tucson: University of Arizona Press.

Prockter, L. M. and R. Pappalardo. 2000. "Folds on Europa: Implications for Crustal Cycling and Accommodation of Extension." *Science 289*:941–944.

Rayleigh, Lord (J. W. Strutt). 1887. "On Waves Propagated along the Plane Surface on an Elastic Solid." *Proc. London Math Soc. (1), 17*:4.

Reasenberg, R. D. 1977. "The Moment of Inertia and Isostasy of Mars." *J. Geophys. Res. 82*:369.

Richter, N. B. 1978. "Periodic Comet Schwassmann-Wachmann I: One of the Most Interesting and Puzzling Objects of Our Solar System." International Astronomical Union Commission 15, *Circular Letter* no. 1.

Ridpath, I. 1977. "The Tunguska Mystery—Solved?" *Astronomy* 5:22 (Dec.).

Russell, H. N. 1920. "On the Origin of Periodic Comets." *Astron. J. 33*:49.

Schiaparelli, G. B. 1866. *Bull. Meteorol. Observ. Coll. Romano 5*:10.

Schubert, G., D. Stevenson, and K. Ellsworth. 1981. "Internal Structures of the Galilean Satellites." *Icarus 47*:46.

Shoemaker, E. M. 1996. In *Proceedings of Conference on Europa,* San Juan Capistrano, CA: San Juan Institute.

Slattery, W., W. DeCampli, and A. Cameron. 1980. "Protoplanetary Core Formation by Rainout of Minerals." *Moon and Planets* 23:381.

Smith, B. A., and 29 others. 1979. "The Galilean Satellites and Jupiter: Voyager 2 Imaging Science Results." *Science 206*:927–950.

Smith, B. A. and others. 1982. "A New Look at the Saturn System: The Voyager 2 Images." *Science 215*:504.

Smith, B. A., and 39 others. 1986. "Voyager 2 in the Uranian System: Imaging Science Results." *Science 233*:43.

Soberman, R., S. Neste, and K. Lichtenfeld. 1974. "Optical Measurement of Interplanetary Particulates from Pioneer 10." *J. Geophys. Res. 79*:3685.

Solomon, S. 1976. "Some Aspects of Core Formation in Mercury." *Icarus 28*:509.

———. 1979. "Formation, History, and Energetics of Cores in Terrestrial Planets." *Phys. Earth Planet. Interiors 19*:168.

Solomon, S., and J. Head. 1979. "Vertical Movement in Mare Basins." *J. Geophys. Res. 84*:1667.

Stevenson, D. J. 1981. "Models of the Earth's Core." *Science 214*:611.

Stevenson, D. J. 1996. "When Galileo Met Ganymede." *Nature 384*:511–512.

Stoffler, D. 1997. "Minerals in the Deep Earth: A Message from the Asteroid Belt." *Science 278*:1576–1577.

Stone, E. C., and E. Miner. 1989. "The Voyager 2 Encounter with the Neptunian System." *Science 246*:1417.

Strom, R., N. Trask, and J. Guest. 1975. "Tectonism and Volcanism on Mercury." *J. Geophys. Res. 80*:2478.

Taylor, D. J. 1965. "Spectrophotometry of Jupiter's 3400-10,000 Å Spectrum and a Bolometric Albedo for Jupiter." *Icarus 4*:362.

Taylor, S. R. 1975. *Lunar Science: A Post-Apollo View.* NY: Pergamon Press.

Taylor, S. R. 1982. *Planetary Science: A Lunar Perspective.* Houston: Lunar and Planetary Institute.

Taylor, S. R. 1992. *Solar System Evolution.* Cambridge, MA: Cambridge University Press.

Tazieff, H. 1964. *When the Earth Trembles.* London: Hart-Davis.

Tittemore, W. C. 1990. "Chaotic Motion of Europa and Ganymede and the Ganymede-Callisto Dichotomy." *Science 250*: 263–267.

Toksöz, M., and A. Hsui. 1978. "Thermal History and Evolution of Mars." *Icarus 34*:537.

Turtle, E. P. and E. Pierazzo. 2001. "Thickness of a Europan Ice Shell from Impact Crater Simulations." *Science 294*:1326–1328.

Urey, H. C. 1952. *The Planets: Their Origin and Development.* New Haven: Yale University Press.

Van Allen, J. A. 1961. "The Geomagnetically Trapped Corpuscular Radiation." In *Science in Space,* eds. L. V. Berkner and H. Olishaw. New York: McGraw-Hill.

Van Flandern, T. 1977. "A Former Major Planet of the Solar System." In *Comets, Asteroids, Meteorites,* ed. A. Delsemme. Toledo, Ohio: University of Toledo Press.

Van Woerkom, A. J. J. 1948. "On the Origin of Comets." *Bull. Astron. Inst. Neth. 10*:445.

Veeder, G. J. and 4 others. 1994. "Io's Heat Flow from Infrared Radiometry: 1983–1993." *Journ. Geophys. Res. 99*:17095–17162.

Vsekhsvyatsky, S. K. 1977. "Comets and the Cosmogony of the Solar System." In *Comets, Asteroids, Meteorites,* ed. A. Delsemme. Toledo, Ohio: University of Toledo Press.

Warner, J. L., and D. A. Morrison. 1978. "Planetary Tectonics I: The Role of Water." *Lun. Planet. Sci. abstracts 9*:1217.

Wasson, J. T. 1974. *Meteorites.* New York: Springer-Verlag.

Watson, K., B. C. Murray, and H. Brown. 1963. "The Stability of Volatiles in the Solar System." *Icarus 1*:317.

Weissman, Paul R. 1985. "The Origin of Comets: Implications for Planetary Formation." In *Protostars and Planets II*, ed. D. Black and M. Matthews. Tucson: University of Arizona Press.

Whipple, F. 1950. "A Comet Model I. The Acceleration of Comet Encke." *Astrophys. J.* 111:375.

———. 1963. "On the Structure of the Cometary Nucleus." In *The Moon, Meteorites, and Comets*, ed. B. M. Middlehurst and G. P. Kuiper. Chicago: University of Chicago Press.

———. 1975. "Do Comets Play a Role in Galactic Chemistry and γ-ray Bursts?" *Astron. J. 80*:525.

———. 1980. "The Spin of Comets." *Sci. Am.*, March, p. 124. (Reprinted in *Comets*, ed. J. Brandt. San Francisco: W. H. Freeman.)

Wilde, S. A. and 3 others. 2001. "Evidence from Detrital Zircons for the Existence of Continental Crust and Oceans on Earth 4.4 Gyr ago." *Nature 409*:175–178.

Wildt, R. 1938. "On the State of Matter in the Interior of the Planets." *Astrophys. J. 85*:508.

Windley, B. F. 1976. "New Tectonic Models for the Evolution of Archaean Continents and Oceans." In *The Early History of the Earth*, ed. B. F. Windley. New York: Wiley.

Wu, S. C. 1978. "Mars Synthetic Topographic Mapping." *Icarus* 33:417.

Yoder, C. F. 1979. "How Tidal Heating in Io Drives the Galilean Orbital Resonance Locks." *Nature 279*:767.

Yoder, C. F. and 4 others. 2003. "Fluid Core Size of Mars from Detection of the Solar Tide." *Science 300*:299–304.

Chapter Nine

Arnold, J. 1977. "Lunar versus Asteroidal Resources." In *New Moons*, ed. D. Criswell. Houston: Lunar and Planetary Institute.

Baldwin, R. B. 1949. *The Face of the Moon*. Chicago: University of Chicago Press.

———. 1963. *The Measure of the Moon*. Chicago: University of Chicago Press.

Basilevsky, A. T. 1989. "The Planet Next Door." *Sky and Telescope*, April, p. 36.

Beatty, J. K. 1982. "Venus: The Mystery Continues." *Sky and Telescope 63*:134.

Bell, J. F., D. Cruikshank, and M. Gaffey. 1985. *Icarus 61*:192.

Bowen, N. L. 1928. "The Evolution of the Igneous Rocks." Princeton: Princeton University Press. Reprint. New York: Dover, 1956.

Carey, W. 1962. "The Scale of Geotectonic Phenomena." *J. Geol. Soc. India 3*:97.

Carr, M. H., and others. 1977. "Martian Impact Craters and Emplacement of Ejecta by Surface Flow." *J. Geophys. Res. 82*:4055.

Clark, B. C., and 9 others. 1977. "The Viking X-Ray Fluorescence Experiment: Analytical Methods and Early Results." *J. Geophys. Res. 82*:4577–4594.

Clark, Beth Ellen and 3 others. 2002. "Asteroid Space Weathering and Regolith Evolution." In *Asteroids III*, ed. W. Bottke and 3 others. Tucson: University of Arizona Press.

Clark, R. N. 1980. "Ganymede, Europa, Callisto, and Saturn's Rings: Compositional Analysis from Reflectance Spectroscopy." *Icarus 44*:388.

Cruikshank, D. P. 1979. "The Surfaces and Interiors of Saturn's Satellites." *Rev. Geophys. Spa Phys. 17*:165.

Cruikshank, D. P., and D. Morrison. 1990. "Icy Bodies of the Outer Solar System." In *The New Solar System*, ed. J. Beatty and A. Chaikin. Cambridge, Mass: Sky Publishing Corp.

Cruikshank, D. P., and P. M. Silvaggio. 1979. "Triton: A Satellite with an Atmosphere." *Astrophys. J. 233*:1016.

———. 1980. "The Surface and Atmosphere of Pluto." *Icarus 41*:96.

Delsemme, A. H. 1977. "The Pristine Nature of Comets." In *Comets, Asteroids, Meteorites*, ed. A. Delsemme. Toledo, Ohio: University of Toledo Press.

Dodd, Robert T. 1981. *Meteorites*. Cambridge University Press.

Erikson, E., D. Goorvitch, J. Simpson, and D. Strecker. 1978. "Far Infrared Spectrophotometry of Jupiter and Saturn." *Icarus 35*:61.

Flint, R. F., and B. J. Skinner. 1977. *Physical Geology*, 2nd ed. New York: Wiley.

Ganapathy, R., and others. 1970. "Trace Elements in Apollo 11 Lunar Rocks: Implications for Meteorite Influx and Origin of the Moon." *Proc. Apollo 11 Lunar Sci. Conf. 1*:1117.

Gault, D. E., E. D. Heitowitt, and H. J. Moore. 1964. "Some Observations of Hypervelocity Impacts with Porous Media." In *The Lunar Surface Layer*, ed. J. W. Salisbury and P. E. Glaser. New York: Academic Press.

Geikie, A. 1905. *The Founders of Geology*, 2nd ed. New York: MacMillan. Reprint. New York: Dover.

Gibson, E. G. 1973. *The Quiet Sun*. Washington, D.C.: NASA SP-303.

Grinspoon, D. H. 1987. "Was Venus Wet? Deuterium Reconsidered." *Science 238*:1702.

Hapke, B. 1965. "Effects of a Simulated Solar Wind on the Photometric Properties of Rocks and Powders." *Ann. N.Y. Acad. Sci. 123*:711.

Hartmann, W. K. 1966. "Martian Cratering." *Icarus 5*:565.

———. 1973. "Ancient Lunar Mega-Regolith and Subsurface Structure." *Icarus 18*:634.

———. 1980a. "Dropping Stone in Magma Oceans: Effects of Early Lunar Cratering." In *Proc. Conf. Lunar Highlands Crust*, ed. J. Papike and R. Merrill. New York: Pergamon Press.

———. 1980b. "Surface Evolution of Two-Component Stone/Ice Bodies in the Jupiter Region." *Icarus 44*:441.

Hartmann, W. K., and G. P. Kuiper. 1962. "Concentric Structures Surrounding Lunar Basins." *Comm. Lunar Planet. Lab. 1*:51.

Hartmann, W. K., and C. A. Wood. 1971. "Moon: Origin of Evolution of Multi-Ring Basins." *The Moon 3*:3.

Hanel, R., and others. 1979. "Infrared Observations of the Jovian System from Voyager 1." *Science 204*:972.

Herbert, F., and others. 1977. "Some Constraints on the Thermal History of the Lunar Magma Ocean." *Proc. Lun. Sci. Conf. 8*:573.

Herbert, F., and C. Sonett. 1979. "Electromagnetic Heating of Minor Planets in the Early Solar system." In *Asteroids*, ed. T. Gehrels. Tucson: University of Arizona Press.

Hooke, R. 1705. "Lectures and Discourses on Earthquakes." *Posthumous Works*. London.

Hurley, P. M. 1968. "The Confirmation of Continental Drift." *Sci. Am.*, April, p. 52.

Lofgren, G. E., and others. 1981. "Petrology and Chemistry of Terrestrial, Lunar, and Meteoritic Basalts." In *Basaltic Volcanism on the Terrestrial Planets*, ed. Basaltic Volcanism Study Project. New York: Pergamon Press.

Lupo, M. J., and J. Lewis. 1979. "Mass-Radius Relationships in Icy Satellites." *Icarus 40*:157.

Mason, B. 1958. *Principles of Geochemistry*. New York: John Wiley & Sons, Inc.

———. 1962. *Meteorites*. New York: John Wiley & Sons, Inc.

Mason, B. 1966. *Principles of Geochemistry*. NY: John Wiley & Sons, Inc.

Masursky, H., and others. 1980. "Pioneer Venus Radar Results: Altimetry and Surface Properties." *J. Geophys. res. 85*:8232.

Masursky, H., E. Eliason, R. Jordan, G. Pettingill, P. Ford, and G. McGill. 1979. "Pioneer Venus Radar Observations-The First Year." *Bull. Amer. Astron. Soc. 11*:549 (abstract).

McCracken, C., and M. Dubin. 1964. "Dust Bombardment on the Lunar Surface." In *The Lunar Surface Layer*, ed. J. Salisbury and P. Glaser. New York: Academic Press.

Melosh, H. J. 1989. *Impact Cratering: A Geologic Process.* NY: Oxford Univ. Press.

Morrison, D. A., and J. Warner. 1978. "Planetary Tectonics II. A Gravitational Effect." *Lun. Planet. Sci. Abstracts* 9:769.

Murthy, V. R., and H. T. Hall. 1972. "The Origin and Chemical Composition of the Earth's Core." *Phys. Earth Planet. Interiors* 6:123.

Mutch, T. A., R. Arvidson, A. Binder, E. Guinness, and E. Morris. 1977. "The Geology of the Viking Lander 2 Site." *J. Geophys. Res.* 82:4452.

Oldham, R. D. 1990. *Phil. Trans. Roy. Soc. London* A194:135.

———. 1906. "The Constitution of the Earth." *Quart. J. Geol. So.* 62:456.

Owen, B. B., and J. E. Martin. 1966. "The Selection, Orientation, and Mounting of Diamonds for Use as Bridgman Anvils." *J. Sci. Instr.* 43:197.

Parker, E. N. 1970. "The Origin of Magnetic Fields." *Astrophys. J.* 160:383.

Phillips, R., R. Grimm, and M. Malin. 1991. "Hot-Spot Evolution and the Global Tectonics of Venus." *Science* 252:651.

Phillips, R., W. Kaula, G. McGill, and M. Malin. 1981. "Tectonics and Evolution of Venus." *Science* 212:879.

Pieters, C. 1982. "Copernicus Crater Central Peak: Lunar Mountain of Unique Composition." *Science* 215:59.

Podolak, M., and A. Cameron. 1974. "Possible Formation of Meteoritic Chondrules and Inclusions in the Precollapse Jovian Protoplanetary Atmosphere." *Icarus* 22:123.

Podolak, M., and R. Reynolds. 1981. "On the Structure and Composition of Uranus and Neptune." *Icarus* 46:40.

Ramsey, W. H. 1967. "On the Constitutions of Uranus and Neptune." *Planet. Space Sci.* 15:1609.

Ringwood, A. E. 1956. "The Olivine-Spinel Transition in the Earth's Mantle." *Nature* 178:1303.

Ringwood, A. E., and A. Major. 1966. "High-Pressure Transformations in Pyroxenes." *Earth Planet. Sci. Lett.* 1:351.

Ritsema, A. R. 1954. "A Statistical Study of the Seismicity of the Earth." *Meterol. Geophys. Serv., Indonesia, Verhandl.* 46.

Rittmann, A. 1962. *Volcanoes and Their Activity.* New York: John Wiley & Sons.

Saunders, R. S., and M. Malin. 1977. "Geologic Interpretation of New Observations of the Surface of Venus." *Geophys. Res. Lett.* 4:547.

Scarf, F. L., and others. 1981. "Jupiter Tail Phenomena Upstream from Saturn." *Nature* 292:585.

Sharp, R. P. 1971. "The Surface of Mars 2. Uncratered Terrains." *J. Geophys. Res.* 76:331.

Shoemaker, E. M. 1965. "Preliminary Analysis of the Fine Structure of the Lunar Surface." In *Ranger VII, part II, Experimenters' Analyses and Interpretations.* Pasadena, CA: Jet Propulsion Laboratory, JPL TR 32–700.

Shoemaker, E. M., and R. Hackman. 1962. "Stratigraphic Basis for a Lunar Timescale." In *The Moon*, ed. Z. Kopal and Z. Mikhailov. London: Academic Press.

Short, N. M. 1975. *Planetary Geology.* Englewood Cliffs, NJ: Prentice-Hall.

Short, N. M., and M. Forman. 1972. "Thickness of Crater Impact Ejecta on the Lunar Surface." *Mod. Geol.* 3:69.

Sinton, W. M. 1979. "Io: Confirmation of 5-μm Outbursts and a Possible Explanation." *Bull. Amer. Astron. Soc.* 11:585 (abstract).

Smith, B. A., and others (Voyager Imaging Team). 1979. "The Jupiter System through the Eyes of Voyager I." *Science* 204:951.

Smith, B. A., and others (Voyager Imaging Team). 1982. "A New Look at the Saturn System: The Voyager 2 Images." *Science* 215:504.

Smith, B. A. 1979. "Crustal Structure and Volcanic Mechanisms." *Bull. Amer. Astron. Soc.* 11:585 (abstract)

Soderblom, L., and L. Lebofsky. 1972. "Technique for Rapid Determination of Relative Ages of Lunar Areas from Orbital Photography." *J. Geophys. Res.* 77:279.

Solomon, S. C., and J. Longhi. 1977. "Magma Oceanography 1. Thermal Evolution." *Proc. Lun. Sci. Conf.* 8:583.

Sonett, C., B. Smith, D. Colburn, G. Schubert, and K. Schwartz. 1971. "Preliminary Assets of the Lunar Lithospheric Thermal Gradient, Heat Flux, Deep Temperature and Compositional Gradation." Paper presented at the Lunar Science Conference, Houston.

Sonett, C. P., D. Colburn, K. Schwartz, and K. Keil. 1970. "The Melting of Asteroidal-sized Bodies by Unipolar Dynamo Induction from a Primordial T Tauri Sun." *Astrophys. Spa. Sci.* 7:446.

Soter, S. 1974. Paper presented at Satellite Colloquium, Cornell University, August.

Stone, E. C., and E. D. Miner. 1989. "The Voyager 2 Encounter with the Neptunian System." *Science* 246:1417.

Strazzulla, G. 1986. "Organic Material from Phoebe to Iapetus." *Icarus* 66:397.

Taylor, S. R. 1975. *Lunar Science: A Post-Apollo View.* NY: Pergamon Press.

Taylor, S. R. 1982. *Planetary Science: A Lunar Perspective.* Lunar and Planetary Science Institute.

———. 1990. "Magnetospheres, Cosmic Rays, and the Interplanetary Medium." In *The New Solar System*, ed. J. Beatty and A. Chaikin. Cambridge, MA: Sky Publishing.

Toulmin, T., and others. 1977. "Geochemical and Mineralogical Interpretation of the Viking Inorganic Chemical Results." *J. Geophys. Res.* 82:4625.

Vilas, F., S. Larson, K. Stockstill, and M. Gaffey. 1996. "Unraveling the Zebra: Clues to the Iapetus Dark Material Composition." *Icarus* 124:262–267.

Vine, F. J., and D. Matthews. 1963. "Magnetic Anomalies over Ocean Ridges." *Nature* 199:947.

Vortman, L. J. 1977. "Craters from Surface Explosions and Energy Dependence—A Retrospective View." In *Impact and Explosion Cratering*, ed. D. Roddy, R. Pepin, and R. Merrill. New York: Pergamon Press.

Warren, P. H., and J. Wasson. 1979. "The Compositional-Petrographic Search for Pristine Nonmare Rocks, 3rd Voray." *Proc. Lunar Planet. Sci. Conf.* 10:583.

Warren, P. H., and Wasson, J. T. 1979. "Effects of Pressure on the Crystallization of a 'Chondritic' Magma Ocean and Implications for the Bulk Composition of the Moon." *Proc. Lun. Planet. Sci. Conf.* 10:2051.

Wasson, J. T. 1974. *Meteorites.* New York: Springer-Verlag.

Whitaker, E. A. 1978. "Galileo's Lunar Observations and the Dating of the Composition of the Sidereus Nuncius." *J. Hist. Astron.* 9:155.

Wildt, R. 1972. "Hydrogen Planets and High-Pressure Physics." *Phys. Earth Planet. Interiors* 6:1.

Wilhelms, D. E., C. Hodges, and R. Pike. 1977. "Nested-Crater Model of Lunar Ringed Basins." In *Impact and Explosion Cratering*, ed. D. Roddy, R. Pepin, and R. Merrill. New York: Pergamon Press.

Williams, R. J. 1976. "Hydrogen Resources for the Moon." In *Lunar Utilization*, ed. D. Criswell. Houston: Lunar Science Institute.

Wilson, P. D., and C. Sagan. 1995. "Spectrophotometry and Organic Matter on Iapetus, 1. Composition Models." *J. Geophys. Res. (Planets)* 100:7531–7537.

Wood, J. A. 1977. "A Survey of Lunar Rock Types and Comparison of the Crusts of Earth and Moon." In *The Soviet-American Conference on Geochemistry of the Moon and Planets,* ed. J. Pomeroy and N. Hubbard. Washington, D.C.: NASA SP-370.

Wyllie, P. J. 1971. *The Dynamic Earth: Textbook in Geosciences.* New York: Wiley.

Wyllie, P. J. 1971. *The Dynamic Earth.* New York: Wiley.

Chapter Ten

Adams, J. B. 1968. "Lunar and Martian Surfaces: Petrologic Significance of Absorption Bands in the Near Infrared." *Science* 159:1453–1455.

Arnold, J. 1977. "Lunar versus Asteroidal Resources." In *New Moons,* ed. D. Criswell. Houston: Lunar and Planetary Institute.

Baldwin, R. B. 1949. *The Face of the Moon.* Chicago: University of Chicago Press.

Binder, A. B. 1966. "Mariner IV: Analysis of Preliminary Photographs." *Science* 152:1053.

Binder, A. B. 1977. "The Geology of the Viking 1 Lander Site." *J. Geophys. Res.* 82:4439.

Binder, A. B., and D. Cruikshank. 1966. "Lithological and Mineralogical Investigation of the Surface of Mars." *Icarus* 5:521.

Burns, B. 1982. "Cratering Analysis of the Surface of Venus as Mapped by 12.6-cm Radar." Thesis (NAIC 167). Arecibo, P.R.: Natl. Astron. and Ionosph. Center.

Burns, J. A., ed. 1977. *Planetary Satellites.* Tucson: University of Arizona Press.

Chapman, C. R., and K. Jones. 1977. "Cratering and Obliteration History of Mars." *Ann. Rev. Earth Planet. Sci.* 5:515.

Clark, B. C., and others. 1977. "The Viking X-ray Fluorescence Experiment: Analytical Methods and Early Results." *J. Geophys. Res.* 82:4577.

Connes, P., J. Connes, W. Benedict, and L. Kaplan. 1967. "Traces of HCl and HF in the Atmosphere of Venus." *Astrophys. J.* 147:1230.

Conrath, B., and 15 others. 1989. "Infrared Observations of the Neptune System." *Science* 246:1454–1459.

Cruikshank, D. P. 1979. "The Surfaces and Interiors of Saturn's Satellites." *Rev. Geophys. Space Phys.* 17:165.

Cruikshank, D. P., and 6 others. 1983. "The Dark Side of Iapetus." *Icarus* 53:90.

Cruikshank, D. P., and 6 others. 1991. "Solid C/N Bearing Material on Outer Solar System Body." *Icarus* 94:345.

Cruikshank, D., and P. Silvaggio. 1980. "The Surface and Atmosphere of Pluto." *Icarus* 41:2367.

Cruikshank, D. P., R. Brown, L. Giver, and A. Tokunaga. 1989. "Triton: Do We See to the Surface?" *Science* 245:283.

Cutts, J. A., and others. 1976. "North Polar Region of Mars: Imaging Results from Viking 2." *Science* 194:1329–1337.

Cutts, J. A., K. Blasius, and W. J. Roberts. 1979. "Evolution of Martian Polar Landscapes: Interplay of Long-Term Variations of Perennial Ice Cover and Dust Storm Intensity." *J. Geophys. Res.* 84:2975.

Eaton, J., and J. Murata. 1960. "How Volcanoes Grow." *Science* 132:925.

Epstein, S., and H. P. Taylor, Jr. 1970. "$^{18}O/^{16}O$, $^{30}Si/^{28}Si$, D/H, and $^{13}C/^{12}C$ Studies of Lunar Rocks and Minerals." *Science* 167:533.

Esposito, L. W. 1984. "Sulfur Dioxide: Episodic Injection Shows Evidence for Active Venus Volcanism." *Science* 223:1072.

Fanale, F., T. Johnson, and D. Matson. 1974. "Io: A Surface Evaporite Deposit?" *Science* 186:922.

Fanale, F., and J. Salvail. 1990. "Evolution of the Water Regime of Phobos." *Icarus* 88:383.

Fink, U., and others. 1980. "Detection of a CH_4 Atmosphere on Pluto." *Icarus* 44:62.

Fink, U., and 4 others. 1997. "The Steep Red Spectrum of 1992AD: An Asteroid Covered with Organic Material?" *Icarus* 97:145–149.

Goldreich, P., and 4 others. 1989. "Neptune's Story." *Science* 245:500.

Greeley, R., and others. 1974. "Wind Tunnel Simulations of Light and Dark Streaks on Mars." *Science* 183:847.

———. 1991. "Wind Streaks on Venus: Magellan Results." Abstracts, 23rd Meeting, Div. of Plan. Sci. of Am. Astr. Soc.:122.

Greeley, R., R. Papson, and J. Veverka. 1978. "Crater Streaks in the Chryse Planetia Region of Mars: Early Viking Results." *Icarus* 34:556.

Greenacre, J. A. 1963. "A Recent Observation of Lunar Color Phenomena." *Sky and Telescope* 26:316. [See also: *Sky and Telescope* 27:3.]

Griffith, Caitlin A. and 4 others. 2003. "Evidence for the Exposure of Water Ice on Titan's Surface." *Science* 300:628–630.

Hanel, R., and 12 others. 1986. "Infrared Observations of the Uranian System." *Science* 233:70–74.

Hanel, R., and others. 1979a. "Infrared Observations of the Jovian System from Voyager 2." *Science* 206:952–956.

———. 1979b. "Infrared Observations of the Jovian System from Voyager 1." *Science* 204:972–976.

———. 1982. "Infrared Observations of the Saturnian System from Voyager 2." *Science* 215:544.

Hargraves, R., and others. 1977. "The Viking Magnetic Properties Experiment: Primary Mission Results." *J. Geophys. Res.* 82:4547.

Hartmann, W. K. 1980. "Surface Evolution of Two-Component Stone/Ice Bodies in the Jupiter Region." *Icarus* 44:441.

Hartmann, W. K. 1982. *Astronomy: The Cosmic Journey,* 2nd ed. Belmont, CA.: Wadsworth.

———. 1990. "Additional Evidence about an Early Intense Flux of C Asteroids and the Origin of Phobos." *Icarus* 87:236.

Hartmann, W. K. and 10 others. 1981. "Chronology of Planetary Volcanism by Comparative Studies of Planetary Cratering." In *Basaltic Volcanism on the Terrestrial Planets,* ed. Basaltic Volcanism Study Project. New York: Pergamon Press.

Hawke, B. R., P. Spudis, and P. Clark. 1985. "The Origin of Selected Lunar Geochemical Anomalies." *Earth, Moon and Planets,* pp. 257–273.

Hunten, D. 1973. "The Escape of Light Gases from Planetary Atmospheres." *J. Atmosph. Sci.* 30:1481.

Hviid, S. F., and 12 others. 1997. "Magnetic Properties Experiment on the Mars Pathfinder Lander, Preliminary Results." *Science* 278:1768–1770.

Johnson, T. V. 2000. "Jupiter and its Moons." *Sci. Amer.* 282:40–49 (Feb.)

Johnson, T. V., and 5 others. 1988. "Io: Evidence for Silicate Volcanism in 1986." *Science* 242:1280.

Keldysh, M. V. 1977. "Venus Exploration with the Venera 9 and Venera 10 Spacecraft." *Icarus* 30:605.

KenKnight, C., D. Rosenberg, and G. Wehner. 1967. "Parameters of the Optical Properties of the Lunar Surface Powder in Relation to Solar Wind Bombardment." *J. Geophys. Res.* 72:3105.

Kieffer, H. H. 1977. "Thermal and Albedo Mapping of Mars during the Viking Primary Mission." *J. Geophys. Res.* 82:4249.

King, T., and 4 others. 1992. "Evidence for Ammonium-bearing Minerals on Ceres." *Science* 255:1551.

Kozyrev, N. 1959. "Observation of a Volcanic Process on the Moon." *Sky and Telescope* 18:184.

Lewis, J. S. 1968. "An Estimate of the Surface Condition of Venus." *Icarus* 8:434.

McCord, T. B., and J. Adams. 1969. "Spectral Reflectivity of Mars." *Science 163*:1058–1060.

Middlehurst, B., and W. Chapman. 1968. "Tidal Cycles and Lunar Event Mechanisms." *Astron. J. 73*:192.

Milton, D. J. 1967. "Slopes on the Moon." *Science 156*:1135.

Morrison, D. 1977. "Radiometry of Satellites and of the Rings of Saturn." In *Planetary Satellites*, ed. J. A. Burns. Tucson: University of Arizona Press.

Morrison, D. A., and J. L. Warner. 1978. "Planetary Tectonics II. A Gravitational Effect." *Lun. Planet. Sci. Abstracts 9*.

Mueller, B., D. Tholen, W. Hartmann, D. Cruikshank. 1992. "Extraordinary Colors of Asteroid 5145." *Icarus 97*:150–154.

Mutch, T. A. 1970. *Geology of the Moon*. Princeton: Princeton University Press.

Nakamura, N., and M. Tatsumoto. 1977. "The History of the Apollo 17 Station 7 Boulder." *Proc. Lun. Sci. Conf. 8*:2301.

Neukum, G., and others. 1975. "A Study of Lunar Impact Crater Size Distributions." *Moon 12*:201.

Orton, G. 1978. "Planetary Atmospheres." *Proc. Lunar Planet. Sci. Conf. 9*:endpaper.

Owen, T. 1993. "Surface Ices and the Atmospheric Composition of Pluto." In *The New Solar System* (ed. J. Beatty), Cambridge, MA: Sky Publishing Corporation.

Parmentier, E., and J. Head. 1981. "Viscous Relaxation of Impact Craters on Icy Planetary Surfaces: Determination of Viscosity Variation with Depth." *Icarus 47*:100.

Phillips, R. J., and 6 others. 1991. "Impact Craters on Venus: Initial Analysis from Magellan." *Science 252*:288.

Pollack, J. 1977. "Phobos and Deimos." In *Planetary Satellites*, ed. J. Burns. Tucson: University of Arizona Press.

Redman, A. 1962. *Volcanoes and Their Activity*. New York: Wiley.

Rittmann, A. 1962. *Volcanoes and Their Activity*. New York: Wiley.

Sagan, C. 1979. "Sulfur Flows on Io." *Nature 280*:750.

Sagan, C., and J. B. Pollack. 1969. "Windblown Dust on Mars." *Nature 223*:791.

Saunders, R. S., and 5 others. 1991. "An Overview of Venus Geology." *Science 252*:249.

Salisbury, J., and V. Smalley. 1964. "The Lunar Surface Layer." In *The Lunar Surface Layer*, ed. J. Salisbury and P. Glaser. New York: Academic Press, Inc.

Schultz, P. H., and P. Spudis. 1979. "Evidence for Ancient Mare Volcanism." *Proc. Lun. Planet. Sci. Conf. 10*:2899.

Seiff, A., and others. 1979. "Structure of the Atmosphere of Venus up to 110 Kilometers: Preliminary Results from the Four Pioneer Venus Entry Probes." *Science 203*:787–790.

Shoemaker, E. M., and others. 1970. "Lunar Regolith at Tranquility Base." *Science 167*:452.

Sill, G. T. 1973. "Sulfuric Acid in the Venus Clouds." *Comm. Lunar Planet. Lab. 9*:191.

Sinton, W. M. 1979. "Io: Confirmation of 5 μm Outbursts and a Possible Explanation." *Bull. Amer. Astron. Soc. 11*:598 (abstract).

———. 1981. "The Thermal Emission Spectrum of Io and a Determination of the Heat Flux from Its Hot Spots." *J. Geophys. Res. 86*:3122.

Smith, B. A., and others (Voyager Imaging Team). 1979. "The Galilean Satellites of Jupiter: Voyager 2 Imaging Science Results." *Science 206*:927.

Smith, B. A., and others (Voyager Imaging Team). 1982. "A New Look at the Saturn System: The Voyager 2 Images." *Science 215*:504.

Smith, B. A., and 39 others. 1986. "Voyager 2 in the Uranian System: Imaging Science Results." *Science 233*:43.

Smith, B. A., and 64 others. 1989. "Voyager 2 at Neptune: Imaging Science Results." *Science 246*:1422.

Squyres, S. W. 1979. "The Evolution of Dust Deposits in the Martian North Polar Region." *Icarus 40*:244.

Strom, R. G. 1979. "Mercury: A Post-Mariner 10 Assessment." *Space Sci. Rev. 24*:3.

Strom, R. B. 1987. "The Solar System Cratering Record: Voyager 2 Results at Uranus and Implications for the Origin of Impacting Objects." *Icarus 70*:517.

Taylor, F. W. 1979. "Infrared Remote Sounding of the Middle Atmosphere of Venus from the Pioneer Orbiter." *Science 203*:779.

Taylor, S. R. 1975. *Lunar Science: A Post-Apollo View*. New York: Pergamon Press.

Van Houten, F. B. 1973. "Origin of Red Beds: A Review— 1961–1972." *Ann. Rev. Earth Planet. Sci. 1*:39.

Veverka, J., P. Thomas, and T. Duxbury. 1978. "The Puzzling Moons of Mars." *Sky and Telescope 56*:186.

Veverka, J., P. Thomas, and R. Greeley. 1977. "A Study of the Variable Features on Mars during the Viking Primary Mission." *J. Geophys. Res. 82*:4167.

Wakita, H., Y. Nakamura, K. Notsu, M. Noguchi, and T. Asada. 1980. "Random Anomaly: A Possible Precursor of the 1978 Izu-Oshima-Kinkai Earthquake." *Science 207*:882.

White, B. R. 1981. "Venusian Saltation." *Icarus 46*:226.

Wilson, L., and J. Head. 1981. "Ascent and Emplacement of Basaltic Magma on the Earth and Moon." *J. Geophys. Res. 86*:2971.

Witteborn, F., J. Gregman, and J. Pollack. 1979. "Io: An Intense Brightening Near 5 μm." *Science 203*:643.

Wood, J. A. 1977. "A Survey of Lunar Rock Types and Comparison of the Crusts of Earth and Moon." In *The Soviet-American Conference on Cosmochemistry of the Moon and Planets*, ed. J. Pomeroy and N. Hubbard. Washington, D.C.: NASA SP-370, p. 35.

Woronow, A. 1978. "A General Cratering History Model and Its Implications for the Lunar Highlands." *Icarus 34*:76.

Young, A. T. 1973. "Are the Clouds of Venus Sulfuric Acid?" *Icarus 18*:564.

Chapter Eleven

Anders, E., and T. Owen. 1977. "Mars and Earth: Origin and Abundance of Volatiles." *Science 198*:453.

Bagnold, R. A. 1941. *The Physics of Blown Sand and Desert Dunes*. London: Chapman and Hall.

Barbato, J. P., and E. Ayer. 1981. *Atmospheres*. New York: Pergamon Press.

Belton, M., D. Hunten, and M. McElroy. 1967. "A Search for an Atmosphere on Mercury." *Astrophys. J. 150*:1111.

Binzel, R. P. 1990. "Pluto." *Scientific Am. 262*:50.

Breed, C. S. 1978. "Terrestrial Analogs of the Hellespontus Dunes, Mars." *Icarus 30*:326.

Briggs, G., and others. 1977. "Martian Dynamical Phenomena During June-November 1976: Viking Orbiter Imaging Results." *J. Geophys. Res. 82*:4121.

Broadfoot, A., and others. 1980. "Extreme Ultraviolet Observations from Voyager 1 Encounter with Jupiter." *Science 204*:979.

Broadfoot, A. L., and 16 others. 1979. "Extreme UV Observations from Voyager 1 Encounter with Jupiter." *Science 294*:979–982.

Carr, M., H. Masursky, R. Strom, and R. Terrile. 1979. "Volcanic Features of Io." *Nature 280*:729.

Chamberlain, J. W. 1962. "Upper Atmospheres of the Planets." *Astrophys. J. 136*. 382.

Cruikshank, D. P., and P. Silvaggio. 1979. "Triton: A Satellite with an Atmosphere." *Astrophys. J. 233*:1016.

———. 1980. "The Surface and Atmosphere of Pluto." *Icarus 41*:96.

Dermott, S., R. Malhotra, and C. Murray. 1988. "Dynamics of the Uranian and Saturnian Satellite Systems: A Chaotic Route for Melting Miranda?" *Icarus 76:*295.

Farmer, C. B., and P. Doms. 1979. "Global Seasonal Variation of Water Vapor on Mars and the Implications for Permafrost." *J. Geophys. Res. 84:*2881.

Fielder, G. 1961. *Structure of the Moon's Surface.* Elmsford, N.Y.: Pergamon Press.

Gierasch, P., and R. Goody. 1973. "A Model of a Martian Great Dust Storm." *J. Atmos. Sci. 30:*169.

Golitsyn, G. 1973. "On the Martian Dust Storms." *Icarus 18:*113.

Gradie, J., P. Thomas, and J. Veverka. 1980. "The Surface Composition of Amalthea." *Icarus 44:*373.

Greenberg, R., and R. Marcialis. 1987. Div. Planet. Sci., Am. Astron. Soc. Abstracts, p. 820.

Griffith, Caitlin, and 4 others. 2003. "Evidence for the Exposure of Water Ice on Titan's Surface." *Science 300:*628–630.

Hanel, R., and others. 1979. "Infrared Observations of the Jovian System from Voyager 1." *Science 204:*972.

———. 1981. "Infrared Observations of the Saturnian System from Voyager 1." *Science 212:*192.

Hanson, W. B., S. Santani, and D. Zuccaro. 1977. "The Martian Ionosphere as Observed by the Viking Retarding Potential Analyzers." *J. Geophys. Res. 82:*4351.

Hartmann, W. K. 1963. "Radial Structures Surrounding Lunar Basins I." *Comm. Lunar Planet. Lab. 2:*1.

———. 1964. "Radial Structures Surrounding Lunar Basins II." *Comm. Lunar Planet. Lab. 2:*175.

Hess, S. L., and others. 1977. "Meteorological Results from the Surface of Mars: Viking 1 and 2." *J. Geophys. Res. 82:*4559–4574.

Holland, H. d. 1962. "Model for the Evolution of the Earth's Atmosphere." In *Petrologic Studies.* New York: Geological Society of America.

Ingersoll, A. 1981, 1990. "Jupiter and Saturn." In *The New Solar System,* ed. J. Beatty, B. O'Leary, and A. Chaikin (3rd ed., 1990). Cambridge, Mass.: Sky Publishing Corp.

Johnson, T. V., A. Cook, C. Sagan, and L. Soderblom. 1979. "Volcanic Resurfacing Rates and Implications for Volatiles on Io." *Nature 280:*746.

Khodakovsky, I., and others. 1979. "Venus: Preliminary Prediction of the Mineral Composition of Surface Rocks." *Icarus 39:*352.

Kieffer, H., and others. 1976. "Martian North Polar Summer Temperatures: Dirty Water Ice." *Science 194:*1341.

Knollenberg, R., and D. Hunten. 1979a. "Clouds of Venus: Particle Size Distribution Measurements." *Science 203:*792.

———. 1979b. "Clouds of Venus: A Preliminary Assessment of Microstructure." *Science 205:*70.

Kuiper, G. P. 1944. "Titan: A Satellite with an Atmosphere." *Astrophys. J. 100:*378.

———. 1952. "Planetary Atmospheres and Their Origin." In *Atmospheres of the Earth and Planets,* ed. G. P. Kuiper. Chicago: University of Chicago Press.

Kuiper, G. P. 1964. "Infrared Spectra of Stars and Planets IV. The Spectrum of Mars, 1- 2.5 micron, and the Structure of its Atmosphere." *Comm. Lunar Plan. Lab. 2:*79.

Kumar, S. 1979. "The Stability of an SO2 Atmosphere on Io." *Nature 280:*758.

MacDonald, G. A. 1972. *Volcanoes.* Englewood Cliffs, NJ: Prentice Hall.

Marcialis, R., and R. Greenberg. 1987. "Warming of Miranda during Chaotic / Rotation." *Nature 32:*227.

Masursky, H., and others. 1979. "Pioneer Venus Radar Observations—The First Year." *Bull. Amer. Astron. Soc. 11:*549.

———. 1980. "Pioneer Venus Radar Results: Geology from Images and Altimetry." *J. Geophys. Res. 85:*8232.

McCauley, J., B. Smith, and L. Soderblom. 1979. "Erosional Scarps on Io." *Nature 280:*736.

McElroy, M. B., Y. Yung, and A. Nier. 1976. "Isotopic Composition of Nitrogen: Implications for the Past History of Mars' Atmosphere." *Science 194:*70.

Meech, K. and M. Belton. 1990. "The Atmosphere of 2060 Chiron." *Astron. Journ. 100:*1323.

Morrison, D., D. Cruikshank, and J. Burns. 1977. "Introducing the Satellites." In *Planetary Satellites,* ed. J. A. Burns. Tucson: University of Arizona Press.

Mutch, T. A., and others. 1976. *The Geology of Mars.* Princeton: Princeton University Press.

Nieman, H. B., and 12 others. 1996. The Galileo Probe Mass Spectrometer: Composition of Jupiter's Atmosphere. Science 272: 846–849.

Orton, G. 1978. "Data Tables." *Proc. Lun. Planet. Sci. Conf. 9:* endpapers.

Owen, T. 1970. "The Atmosphere of Jupiter." *Science 167:*1675.

———. 1982. "Titan." *Sc. Amer.,* February, p. 98.

Owen, T., and others. 1977. "The Composition of the Atmosphere at the Surface of Mars." *J. Geophys Res. 82:*4635.

Oyama, V., and others. 1979. "Venus Lower Atmosphere Composition: Analysis by Gas Chromatography." *Science 203:*802. (Revision supplied by NASA)

Pang, K., J. Rhoads, A. Lane, and J. Ajello. 1980. "Spectral Evidence for a Carbonaceous Chondrite Surface Composition on Deimos." *Nature 283:*277.

Peale, S. J., P. Cassen, and R. Reynolds. 1979. "Melting of Io by Tidal Dissipation." *Science 203:*892.

Pettengill, G. H., and others. 1979. "Venus: Preliminary Topographic and Surface Imaging Results from the Pioneer Orbiter." *Science 205:*90–93.

Podolak, M., N. Noy, and A. Bar-Nun. 1979. "Photochemical Aerosols in Titan's Atmospheres." *Icarus 40:*193.

Pollack, J. B. 1979. "Climatic Change on the Terrestrial Planets." *Icarus 37:*479.

Pollack, J. B., and D. Black. 1979. "Implications of the Gas Compositional Measurements of Pioneer Venus for the Original Planetary Atmospheres." *Science 205:*56.

Prinn, R. G., and T. Owen. 1976. "Chemistry and Spectroscopy of the Jovian Atmosphere." In *Jupiter,* ed. T. Gehrels. Tucson: University of Arizona Press.

Rasool, S. I., and C. de Bergh. 1970. "The Runaway Greenhouse and the Accumulation of CO_2 in the Venus Atmosphere." *Nature 226:*1037.

Ross, F. 1928. "Photographs of Venus." *Astrophys. J. 68:*57.

Rubey, W. W. 1951. "Geologic History of Sea Water." *Bull. Geol. Soc. Am. 62:*1111. Reprinted in *The Origin and Evolution of Atmospheres and Oceans,* ed. P. Brancazio and A. Cameron. New York: Wiley, 1964.

Sagan, C. 1962. "Structure of the Lower Atmosphere of Venus." *Icarus 1:*151.

Saunders, R. S., and M. Malin. 1977. "Geologic Interpretation of New Observations of the Surface of Venus." *Geophys. Res. Lett. 4:*547.

Seiff, A., and D. Kirk. 1977. "Structure of the Atmosphere of Mars in Summer at Mid-Latitudes." *J. Geophys. Res. 82:*4364.

Seiff, A., and others. 1979. "Thermal Contrast in the Atmosphere of Venus: Initial Appraisal from Pioneer Venus Probe Data." *Science 205:*46.

Smith, B. A., and others. 1979. "The Galilean Satellites and Jupiter: Voyager 2 Imaging Science Results." *Science 206:*927.

Smith, B. A., and others (Voyager Imaging Team). 1979a. "The Galilean Satellites and Jupiter Voyager 2 Imaging Science Results." *Science* 206:927.

———. 1979b. "The Role of SO_2 in Volcanism on Io." *Nature* 280:738–743.

———. 1981. "Encounter with Saturn: Voyager 1 Imaging Science Results." *Science* 212:163.

Smith, B. A., and 39 others. 1986. "Voyager 2 in the Uranian System; Imaging Science Results." *Science, 233*:43.

Soter, S. 1971. "The Dust Belts of Mars." Ph.D. dissertation, Cornell University.

Stone, E. C., and E. D. Miner. 1981. "Voyager 1 Encounter with the Saturnian System." *Science* 212:159.

———. 1986. "The Voyager 2 Encounter with the Uranian System." *Science* 233:39.

———. 1989. "The Voyager 2 Encounter with the Neptunian System." *Science* 246:1417.

Strom, R. G. 1964. "Tectonic Maps of the Moon I." *Comm. Lunar Planet. Lab.* 2:205.

Toon, O. B. 1979. "A Physical-Chemical Model of the Venus Clouds." *Bull. Amer. Astron. Soc.* 22:544 (abstract).

Urey, H. C. 1952. *The Planets: Their Origin and Development.* New Haven: Yale University Press.

Veverka, J. 1973. "Titan: Polarimetric Evidence for an Optically Thick Atmosphere?" *Icarus* 18:657.

Veverka, J., and others. 1981. "Voyager Search for Post-eclipse Brightening on Io." *Icarus* 47:60.

von Zahn, V., and others. 1979. "Venus Thermosphere: In Situ Composition Measurements, the Temperature Profile, and the Homopause Altitude." *Science* 203:768.

Walker, J. C. G. 1977. *Evolution of the Atmosphere.* New York: Macmillan.

Wise, D. U., M. Golombek, and G. McGill. 1979. "Tharsis Province of Mars: Geologic Sequence Geometry, and a Deformation Mechanism." *Icarus* 38:456.

Chapter Twelve

Allen, W. 1980. *Side Effects.* New York: Ballantine Books.

Alvarez, L. W., W. Alvarez, F. Asaro, and H. Michel. 1980. "Extraterrestrial Cause for the Cretaceous-Tertiary Extinction." *Science* 208:1095.

Atreya, S., T. Donahue, and W. Kuhn. 1978. "Evolution of a Nitrogen Atmosphere on Titan." *Science* 201:611.

Bar-Nun, A. 1975. "Thunderstorms on Jupiter." *Icarus* 24:86.

Brown, H. 1948. "Rare Gases and the Formation of the Earth's Atmosphere." In *The Atmospheres of the Earth and Planets,* ed. G. P. Kuiper. Chicago: University of Chicago Press.

Cess, R., V. Ramanathan, and T. Owen. 1980. "The Martian Paleoclimate and Enhanced Atmospheric CO_2." *Icarus* 41:159.

Chapman, C. R., and K. Jones. 1977. "Cratering and Obliteration History of Mars." *Ann. Rev. Earth Planet. Sci.* 5:515.

Combes, M., and others. 1981. "Upper Limit of the Gaseous CH_4 Abundance on Triton." *Icarus, 47*:139.

Condon, E. U., and others. 1969. *Scientific Study of Unidentified Flying Objects.* New York: Dutton.

Connes, P., J. Connes, W. Benedict, and L. Kaplan. 1967. "Traces of CHl and HF in the Atmosphere of Venus." *Astrophys. J.* 147:1230.

Cronin, F. R., S. Pizzarello, and C. B. Moore. 1979. "Amino Acids in an Antarctic Carbonaceous Chondrite." *Science* 206:335.

Cutts, J., and others. 1976. "North Polar Region of Mars; Imaging results from Viking 2." *Science* 194:1329.

Dole, S. H. 1964. *Habitable Planets for Man.* Waltham, Mass.: Blaisdell.

Engel, M., and B. Nagy. 1982. "Distribution and Enantiometric Composition of Amino Acids in the Murchison Meteorite." *Nature* 296:837.

Finale, F., and W. Cannon. 1979. "Mars: CO_2 Absorption and Capillary Condensation on Clays—Significance for Volatile Storage and Atmospheric History." *J. Geophys. Res.* 84:8404.

Fink, U., and others. 1980. "Detection of a CH4 Atmosphere on Pluto." *Icarus* 44:62.

Fink, U., and 4 others. 1992. "The Steep Red Spectrum of 1992 AD: An Asteroid Covered with Organic Material?" *Icarus* 97: 145–149.

Frederickson, J. K., and T. C. Onstott. 1996. Microbes Deep Inside the Earth. *Scientific Amer.* 275 (October):68–73.

Gartner, S., and J. McGuirk. 1980. "Terminal Cretaceous Extinction Scenario for a Catastrophe." *Science* 206:1272.

Goldsmith, D., and T. Owen. 1979. *The Search for Life in the Universe.* Menlo Park: Benjamin/Cummings.

Gould, S. J. 1978. "An Early Start." *Natural History* 87 (February):10.

Grinspoon, D. H. 1987. "Was Venus Wet?: Deuterium Reconsidered." *Science* 238:1702.

Hartmann, W. 1973. "Martian Cratering 4. Mariner 9 Initial Analysis of Cratering Chronology." *J. Geophys. Res.* 78:4096.

———. 1978. "Martian Cratering 5: Toward an Empirical Martian Chronology, and Its Implications." *Geophys. Res. Lett.* 5:450.

Hartmann, W. K., and C. A. Wood. 1971. "Moon: Origin and Evolution of Multi-Ring Basins." *The Moon 3*:3.

Hartmann, W. K., and Ron Miller, 1991. *The History of Earth.* New York: Workman Publishing Co.

Hays, J., J. Imbrie, and N. Shackleton. 1976. "Variations in the Earth's Orbit: Pacemaker of the Ice Ages." *Science* 194:1121.

Hildebrand, A. R., and W. Boynton. 1991. "Cretaceous Ground Zero." *Natural History,* June, p. 47.

Horowitz, N. H. 1977. "The Search for Life on Mars." *Sci. Am.,* November, p. 52.

Horowitz, P. 1978. "A Search for Ultra-Narrowband Signals of Extraterrestrial Origin." *Science* 201:733.

Hunten, D. M. 1977. "Titan's Atmosphere and Surface." In *Planetary Satellites,* ed. J. Burns. Tucson: University of Arizona Press.

Imbrie, J., and J. Z. Imbrie. 1980. "Modeling the Climatic Response to Orbital Variations." *Science* 207:943.

Jeans, J. H. 1916. *Dynamical Theory of Gases.* Cambridge: Cambridge University Press.

Karl, D., C. Wirsen, and H. Jannasch. 1980. "Deep-sea Primary Production at the Galápagos Hydrothermal Vents." *Science* 207:1345.

Kerr, R. A. 1981. "Impact Looks Real, the Catastrophe Smaller." *Science* 214:896.

Khodakovsky, I. L., and others. 1979. "Venus: Preliminary Prediction of the Mineral Composition of Surface Rocks." *Icarus* 39:352.

Ksanfomality, L. V. 1980. "Discovery of Frequent Lightning Discharges on Clouds of Venus." *Nature* 284:244.

Kuiper, T., and M. Morris. 1977. "Searching for Extraterrestrial Civilizations." *Science* 196:616.

Kvenvolden, K., and others. 1970. "Evidence for Extraterrestrial Amino Acids and Hydrocarbons in the Murchison Meteorite." *Nature* 228:923.

Levin, G., and P. Straat. 1981. "A Search for a Nonbiological Explanation of the Viking Labeled Release Life Detection Experiment." *Icarus* 45:494.

Levine, J. S. 1969. "The Ashen Light: An Auroral Phenomenon on Venus." *Planet Space Sci.* 17:1081.

Lewis, J. S. 1980. "Lightning Synthesis of Organic Compounds on Jupiter." *Icarus* 43:85.

Lewis, J. S., and R. G. Prinn. 1970. "Jupiter's Clouds: Structure and Composition." *Science* 169:472.

Malin, M. C. 1976. "Age of Martian Channels." *J. Geophys. Res.* 81:4825.

Masursky, H., and others. 1977. "Classification and Time of Formation of Martian Channels Based on Viking Data." *J. Geophys. Res.* 82:4016.

McElroy, M. B., Y. Yung, and A. Nier. 1976. "Isotopic Composition of Nitrogen: Implications for the Past History of Mars' Atmosphere." *Science* 194:70.

Meech, K. and M. Belton. 1990. "The Atmosphere of 2060 Chiron." *Astron. Journ.* 100:1323.

Mueller, B., D. Tholen, W. Hartmann, D. Cruikshank. 1992. "Extraordinary Colors of Asteroidal Object 5145." *Icarus* 97:150–154.

Murray, B., W. Ward, and S. Yeung. 1973. "Periodic Insolation Variations on Mars." *Science* 80:638.

Mutch, T., R. Arvidson, J. Head, K. Jones, and R. S. Saunders. 1976. *The Geology of Mars.* Princeton: Princeton University Press.

Myers, N. 1997. Mass Extinction and Evolution. *Science* 278:597–598.

Nisbet, E. G. 1980. "Achaean Stromatolites and the Search for the Earliest Life." *Nature* 284:395.

Oparin, A. I. 1962. *Life: Its Nature, Origin, and Development.* New York: Academic Press.

Parker, P. 1977. "An Ecological Comparison of Marsupial and Placental Patterns of Reproduction." In *The Biology of Marsupials,* ed. B. Stonehouse and D. Gilmore. London: Macmillan.

Pepin, R. O. 1991. "On the Origin and Early Evolution of Terrestrial Planet Atmospheres and Meteoritic Volatiles." *Icarus* 92:2.

Pollard, W. G. 1979. "The Prevalence of Earthlike Planets." *Am. Scientist* 67:653.

Rages, K., and J. Pollack. 1980. "Titan Aerosols: Optical Properties and Vertical Distribution." *Icarus* 41:119.

Raup, D. M. 1979. "Size of the Permo-Triassic Bottleneck and Its Evolutionary Implications." *Science* 206:217.

Raup, D. M., and J. Sepkoski, Jr., 1986. "Periodic Extinction of Families and Genera." *Science* 231:833.

Russell, D. A. 1982. "The Mass Extinctions of the Late Mesozoic." *Sci. Am.,* January, p. 58.

Sagan, C., ed. 1973. *Communication with Extraterrestrial Intelligence.* Cambridge, Mass.: MIT Press.

Sharonov, V. V. 1958. *The Nature of Planets.* Trans. Israel Program for Scientific Translations. Washington, D.C.: Department of Commerce, Office of Technical Services.

Shklovskii, I. S., and C. Sagan. 1966. *Intelligent Life in the Universe.* San Francisco: Holden-Day.

Siegel, S. M. 1970. "Experimental Biology of Extreme Environments and Its Significance for Space Bioscience 2." *Spaceflight* 12:256.

Simpson, G. G. 1964. "The Nonprevalence of Humanoids." *Science* 143:769.

———. 1973. "Added Comments on 'The Nonprevalence of Humanoids.'" In *Communication with Extraterrestrial Intelligence,* ed. C. Sagan. Cambridge, Mass.: MIT Press.

Smith, B. A., and 64 others. 1989. "Voyager 2 at Neptune: Imaging Science Results." *Science* 248:1422.

Spitzer, L. 1952. "The Terrestrial Atmosphere above 300 Kilometers." In *The Atmosphere of the Earth and Planets,* ed. G. P. Kuiper. Chicago: University of Chicago Press.

Stewart, A., and others. 1979. "Ultraviolet Spectroscopy of Venus: Initial Results from the Pioneer Venus Orbiter." *Science* 203:777.

Story, R. 1976. *The Space-Gods Revealed.* New York: Harper & Row.

Suess, H. E. 1949. "Die Haufigkeit der Edelgase auf der Erde und im Kosmos." *J. Geol.* 57:600.

Trafton, L. M. 1972. "On the Possible Detection of H2 in Titan's Atmosphere." *Astrophys. J.* 175:285.

Underwood, J. 1975. *Biocultural Interactions and Human Variations.* Dubuque, Iowa: W. C. Brown.

Walker, J. C. G. 1977. *Evolution of the Atmosphere.* New York: Macmillan.

Walter, M. R., R. Buick, and J. Dunlop. 1980. "Stromatolites 3400–3500 Myr Old from the North Pole Area, Western Australia." *Nature* 284:443.

Ward, W. R. 1974. "Climatic Variations on Mars 1. Astronomical Theory of Insolation." *J. Geophys. Res.* 79:3375.

Ward, W., J. Burns, and O. Toon. 1979. "Past Obliquity Oscillations of Mars: Role of the Tharsis Uplift." *J. Geophys. Res.* 84:243.

Young, A. T. 1982. "Rayleigh Scattering." *Phys. Today,* January, p. 42.

Zellner, B. 1973. "The Polarization of Titan." *Icarus* 18:661.

Chapter Thirteen

Ahrens, T. J., and J. D. O'Keefe. 1981. "Impact of an Asteroid or Comet on the Ocean and Extinction of Terrestrial Life." *Lunar Planet Sci. Abstracts* 13:3 (see also press abstracts, p. 1).

Anders, E., and T. Owen. 1977. "Mars and Earth: Origin and Abundance of Volatiles." *Science* 198:453.

Anderson, J. W. and 4 others 1997. "Gravitational Evidence for an Undifferentiated Callisto." *Nature* 387:264–266.

Baker, V. R. 1982. *The Channels of Mars.* Austin: University of Texas Press.

Berman, Daniel C. and W. K. Hartmann. 2002. "Recent Fluvial, Volcanic, and Tectonic Activity on the Cerberus Plains of Mars." *Icarus* 159:1–17.

Boynton, W. V. and 24 others. 2002. "Distribution of Hydrogen in the Near-Surface of Mars: Evidence for Subsurface Ice Deposits." *Science* 296:81–85.

Bowen, C. D. 1963. *Francis Bacon.* Boston: Little, Brown.

Bridges, J. C. and 5 others. 2001. "Alteration Assemblages in Martian Meteorites: Implications for Near-Surface Processes." In *Chronology and Evolution of Mars,* ed. R. Kallenbach, J. Geiss, and W. Hartmann (Bern: International Space Science Institute); *Space Sci. Rev.* 96:365–392.

Burr, Devon and 3 others. 2002. "Repeated Aqueous Flooding from the Cerberus Fossae: Evidence for Very Recently Extant, Deep Groundwater on Mars." *Icarus,* 159:53–73.

Carr, M. H. 1981. *The Surface of Mars.* New Haven: Yale University Press.

Carr, Michael H. 1996. *Water on Mars.* New York: Oxford University Press.

Carr, M., and G. Schaber. 1977. "Martian Permafrost Features." *J. Geophys. Res.* 82:4039.

Clark, B. C., and others. 1977. "The Viking X-ray Fluorescence Experiment: Analytical Methods and Early Results." *J. Geophys. Res.* 82:4577.

Clifford, Stephen M. 1993. "A Model for the Hydrologic and Climatic Behavior of Water on Mars." *Journal of Geophys. Research.* 98:10,973–11,016.

Clifford, Stephen M. and Timothy J. Parker. 2001. "The Evolution of the Martian Hydrosphere: Implications for the Fate of a Primordial Ocean and the Current State of the Northern Plains." *Icarus,* 154:40–79.

Costard, F. and 3 others. 2002. "Formations of Recent Martian Debris Flows by Melting of Near-Surface Ground Ice at High Obliquity." *Science* 295:110–113.

Costard, F., F. Forget, N. Mangold, and J.-P. Peulvast 2001. "Formation of Recent Martian Debris Flows by Melting on Near-Surface Ground Ice at High Obliquity." *Science* 295:110–113.

Donahue, T., and others. 1982. "Venus Was Wet: A Measurement of the Ratio of Deuterium to Hydrogen." *Science* 216:630.

Golombek, M. and 13 others. 1997. "Overview of the Mars Pathfinder Mission and Assessment of Landing Site Predictions." *Science* 278:1743–1748.

Gould, S. J. 1974. "The Great Dying." *Natural History* 83 (October):22.

Hartmann, W. K. 1978. "Martian Cratering V. Toward an Empirical Martian Chronology, and Its Implications." *Geophys. Res. Letters* 5:450.

Hartmann, W. K., and G. Neukum. 2001. "Cratering Chronology and Evolution of Mars." In *Chronology and Evolution of Mars,* Eds. R. Kallenbach, J. Geiss, and W. K. Hartmann. (Bern: International Space Science Institute); also *Space Sci. Rev.* 96:165–194.

Hoffert, M. I., and others. 1981. "Liquid Water on Mars: An Energy Balance Climate Model for CO2/H2O Atmospheres." *Icarus* 47:112.

Hoyle, F. 1972. *From Stonehenge to Modern Cosmology.* San Francisco: W. H. Freeman.

Hoyle, F., and N. C. Wickramasinghe. 1978. *Lifecloud.* New York: Harper & Row.

Jakosky, Bruce M. and Michael H. Carr. 1985. "Possible Precipitation of Ice at Low Latitudes of Mars During Periods of High Obliquity." *Nature* 315:559–561.

Kasting, James F. 1991. "CO_2 Condensation and the Climate of Early Mars." *Icarus* 94:1–13.

Komar, P. D. 1979. "Comparisons of the Hydraulics of Water Flows in Martian Outflow Channels with Flows of Similar Scale on Earth." *Icarus* 37:156.

Kyte, F. T., N. Z. Zhou, and J. Wasson. 1980. "Siderophile-Enriched Sediments from the Cretaceous-Tertiary Boundary." *Nature* 228:651.

Masursky, H., J. Boyce, A. Dial, G. Schaber, and M. Strobell. 1977. "Classification and Time of Formation of Martian Channels Based on Viking Data." *J. Geophys. Res.* 82:4016.

McElroy, M. B., Y. Yung, and A. Nier. 1976. "Isotopic Composition of Nitrogen: Implications for the Past History of Mars' Atmosphere." *Science* 194:70.

Metzger, S. M. 1998. Dust Devil Vortices at the Ares Allis MPF Landing Site. Lunar Planet. Sci. Conf 29 (Houston), abstract.

Miller, S. L. 1955. "Production of Some Organic Compounds under Possible Primitive Earth Conditions." *J. Amer. Chem. Soc.* 77:2351.

Morrison, P. 1973. "Discussion." In *Communication with Extraterrestrial Intelligence,* ed. C. Sagan. Cambridge, Mass.: MIT Press, p. 120.

Mustard, J. F., C. Cooper, and M. Rifkin. 2001. "Evidence for Recent Climate Change on Mars from the Identification of Youthful Near-Surface Ground Ice." *Nature* 412:411–413.

Mutch, T. A., and others.. 1977. "The Geology of the Viking Lander 2 Site." *J. Geophys. Res.* 82:4452.

Parker, T. J. and 4 others. 1993. "Coastal Geomorphology of the Martian Northern Plains." *J. Geophys. Research* 98:11,061–11,078.

Pieri, D. 1976. "Distribution of Small Channels on the Martian Surface." *Icarus* 37:351.

Pollack, J. B. 1979. "Climatic Change on the Terrestrial Planets." *Icarus* 37:479.

Plescia, J. 1990. "Recent Flood Lavas in the Elysium Region of Mars." *Icarus* 88:465–490.

Tanaka, K. L. 1986. "The Stratigraphy of Mars." *Proc. Lunar Planet. Sci. Conf. 17, Part 1. J. Geophys. Res. Supp.* 91:E139–158.

Rossabacher, L. A., and S. Judson. 1981. "Ground Ice on Mars: Inventory, Distribution, and Resulting Landforms." *Icarus* 45:39.

Sagan, C. 1970. "Life." In *Encyclopaedia Brittanica.* Chicago: Encyclopaedia Britannica, Inc.

Smith, David E., Maria Zuber, and G. A. Neumann. 2001. "Seasonal Variations of Snow Depth on Mars." *Science* 294:2141–2146.

Solomon, S. 1976. "Some Aspects of Core Formation in Mercury." *Icarus* 28:509.

Squyres, S. W. and M. H. Carr. 1986. "Geomorphic Evidence for Distribution of Ground Ice on Mars," *Science* 231:249–252.

Squyres, S. W. and 4 others. 1992. "Ice in the Martian Regolith." In *Mars,* ed. H. Kieffer and 3 others. Tucson: University of Arizona Press.

Toon, O. B., and others. 1980. "The Astronomical Theory of Climate Change on Mars." *Icarus* 44:552.

Turco, R., and others. 1981. "Tunguska Meteor Fall of 1908: Effects on Stratospheric Ozone." *Science* 212:19.

Wallace, D., and C. Sagan. 1979. "Evaporation of Ice in Planetary Atmospheres: Ice-Covered Rivers on Mars." *Icarus* 39:385.

Wright, I. P., M. Grady, and C. Pillinger. 1989. "Organic Materials in a Martian Meteorite." *Nature* 340:220.

Yoder, C. F. and 4 others. 2003. "Fluid Core Size of Mars from Detection of the Solar Tide." *Science* 300:299–304.

Index

Aa lava, 271, 272
Aberration (of light) 67
Absolute magnitude 172
Accretion
  of carbonaceous material 111
  chemical equilibrium model 111
  collisional (full-scale planets), 114–116
  collisional (planetesimals), 112–113
  heterogeneous accretion model 111
Achondrites, 141–142, 145
Adiabatic change/temperature gradient
  206
Aggregation
  of solar nebula dust, 107, 109, 111–112
  of T Tauri stars, 87–88
Albedo
  defined, 41, 43
  of asteroids, 158–159, 163, 164
  Bond albedo 43
  Charon 37
  of comets 43
  in diametral calculations 172
  Earth, 43, 295
  Galilean moons 295
  geometric albedo 43
  ion sputtering, effect of 265
  irradiation, effect of (Iapetus), 264, 265,
    266
  Jupiter 295
  Mercury, 43, 295
  Moon, 43, 295
  Neptune, 43, 295
  Pluto, 37, 295
  Saturn 295
  Uranus 295
  Venus, 43, 295
Aldrin, Edward "Buzz" 1
Alien life
  appearance of (convergence effect) 360
  contact with, 363–367
  the Drake equation 363
  estimated possibilities for, 362–363
  See also Life
Allende meteorite, 108, 110
Amalthea, cratering on 297–298
Amazonian Era (Mars), 372
Ammonia ice 238
Amphiboles 238
Angra dos Reis (ADOR) anomaly 145
Aphelion, defined 39
Apoapse/apapsis, defined, 39, 48
Apogee, defined 39
Apollo 11, 1
Apparition, defined 39
Argon (in Venus/Mars) 317
Ariel, surface appearance of 36
Armstrong, Neil A. 1

Ashen light (Venus) 337
Asteroids
  1 Ceres, 9, 12, 157, 167, 173, 178
  2 Pallas, 157, 173
  3 Juno, 157
  4 Vesta, 157, 167, 173
  8 Flora (family), 168
  44 Nysa, 164
  243 Ida, 169, 170, 173
  253 Mathilda, 115, 173
  433 Eros, 176
  434 Hungaria, 104
  624 Hektor, 171, 176
  951 Gaspra, 170
  1580 Betulia, 171
  1685 Toro, 162
  2060 Chiron, 154, 160–161, 164,
    346–347
  3971 Dionysius, 169
  4179 Toutatis, 169, 171–172
  4769 Castalia, 169
  5145 Pholus, 164
  albedos of, 158–159, 163, 164
  asteroid or comet?, 160–161
  Aten-type 158
  as captured moons, 167–168
  carbonaceous chondrites (C-type) 164
  centaurs, 159, 161–162, 164
  Chiron, cometary activity of 154
  collisions of. See
    Collisions/fragmentation
  coloration of, 164, 166, 167, 239
  compound shapes of, 169–171
  confusion of types/nomenclature for,
    157, 160
  defined 10
  densities of 173
  discovery of, 156–157
  Earth-crossing/Earth approaching, 158,
    160, 169, 172
  economic potential of, 3, 186–187
  elongated shapes of 176
  exploring/visiting 186
  first discovery of (Ceres) 12
  Hirayama families of, 168–169
  ice in, 161, 164, 166
  Kirkwood gaps 54, 167
  Kuiper belt. See Kuiper belt objects
  Lagrangian points for, 33, 54–57, 160
  letter/number code for 10
  Mathilda, meteorite impact on 115
  moonlets around, 169, 171
  number of, 177–178, 179
  and Oort cloud, 161, 162, 166–167
  Pluto considered as, 9, 299
  properties of (chart), 158–159
  reflectance spectrum of (Apollo) 162

  resonances 54
  rotations 176–177
  as scientific preserves 4
  shapes 176–179
  size distribution of, 177–178, 179
  spectral types of 163
  spectral types (taxonomy of), 162–163,
    173
  table of data 158–159
  Trojan, 55–56, 164
  Types C, D and P 164
  Yarkovsky effect on, 69
  See also Interplanetary bodies
Astroblemes 249
Astrogeology, 6, 41
Astrometry
  defined 52
  astrometic binary 90
Astronomical unit (AU)
  defined 12
  and Bode's Rule 12
Atmospheres, planetary. See Planetary
  atmospheres

Basalts
  basalt flows 277
  gabbros, 208, 241
  KREEP basalts, 243, 271, 306, 308
Basins, multiring, 255–256, 257, 285, 286
Beta Pictoris systems, 88–89, 95
Binary/multiple stars
  defined 90
  astrometric binary 90
  habitable planets, criteria for, 356–357
  and planetary formation 91
  possible origins of, 93–94
  single stars, rarity of 90
  spectroscopic binary 90
  star systems, survey of (chart) 91
  systems containing n members 92
  visual binary 90
  vs. double star 90
Blue skies (Rayleigh scattering), 322–323
Bode's Rule
  illustrated 12
  Neptune as failure of 35
Bok globules, 80–81
Boltzmann constant 41
Bow shock (solar wind), 201, 209
Bowen, Catherine Drinker
  comment on Copernican revolution 2
Brahe, Tycho 47
Braided runoff channels (Mars) 379
Breccia
  brecciated meteorites, 143, 151
  in sedimentary rocks 242
Broom stars 173

Brown dwarfs
  defined 5
  in multiple star systems, 92, 93

Caldera, volcanic
  defined 275
  on Io, 299–300
  in Mexico (MacDougal Crater) 277
Callisto
  albedo of 295
  crater counts and isochrons 263
  cratering/resurfacing on, 301–304
  multiring impact crater on 257
  photograph of 222, 257
  physical characteristics, 28, 222–223
  surface temperature of 295
  Voyager flyby 28
Cantaloupe terrain (Triton) 258
Carbon
  in cometary nuclei, 161, 173–174
  graphite 229
  ice, carbonaceous grains in, 110–111
  primordial biological carbon chemistry
    353
  production of 266
Carbon dioxide
  atmospheric increase of 361
  carbon dioxide ice 238
  condensation of (Mars), 371–373,
    376–378
  in Martian polar hood 328
  synthesis of organic molecules from 355
  and the Urey reaction 291
  and the Venusian atmosphere 292
Carbonaceous chondrites
  heat balance of 209
  origins/properties of, 110, 120, 138–139,
    141, 147, 174
  racemic amino acids in 351
  temperature vs. distance from sun 208
Carbonaceous condensates (from solar
    nebula) 110
Carbonate rocks
  and atmospheric CO2 291
Cassini, Giovanni
  Cassini's division (Saturn's rings) 30
  describes Great Red Spot 48
Celestial mechanics
  historical development, 47–48
  introduction 47
Centaurs, 159, 161–162
Ceres, discovery of 12
Channels, runoff
  braided 379
  headward sapping 379
  Martian, 21–22, 23, 291, 379–381,
    385–386
  outflow channels, 379
  valley networks, 379
Chaotic terrain (Mars) 21, 379
Charon
  origin of, 122, 124
  physical/orbital characteristics of 37
Chemical differentiation (during heating)
  lithophile/siderophile elements 135
Chiron
  atmosphere of, 346–347
  cometary activity of 154

CHON particles
  in comets, 174–175
  and early life, 351, 359
Chondrites
  carbonaceous, 110, 120, 138–139, 141,
    147, 174
  ordinary, gas-rich, 140–141
  as S-class asteroids 163
Chondrules
  description/formation of, 136–138
  See also Glasses
Cinders/cinder cones, 272, 273, 274, 290
Circular velocity, 49–50
Clouds, atmospheric
  on Io (sodium) 29
  on Jupiter, 24, 25, 26, 340
  Martian, 328, 329
  in moist atmospheres, 326–327
  on Neptune 37
  on Saturn, 340, 341
  sulfuric acid (Venus), 327–328
  on Venus, 13–15
Clouds, nebular
  dust/gas clouds (Monoceros) 76
  See also Interstellar material
Cocoon nebulae
  astrophysical observations of 95
  bipolar mass ejection from, 84, 86
  evidence for, telescopic, 82–83
  evidence for, theoretical 82
  evolutionary processes in, 94–95
  Herbig-Haro objects 84
  the nebular hypothesis, 81–82
  solar nebula as 102
  trapping protostar visible light, 87, 94
Collisions/fragmentation
  and asteroid compound shapes, 169–170
  asteroids, surface cratering of, 169–170
  collision velocities 114
  collisional accretion, 112–113, 114–116
  collisional mixing (brecciated
    meteorites), 143, 151
  collisions: history vs. mass, 178–179
  Hirayama families, 168–169
  influence on planetary size, distribution,
    115, 116
  laboratory collisions 151
  meteoroid collisions, dates of, 145–146
  of meteoroids/planetesimals, 152,
    169–170
  Neumann bands as evidence of 143
  summary discussion of 152
Coloration
  of asteroids, 164, 166, 167, 239
  of Io, 300–301
  of Kuiper belt objects, 164, 166, 167, 239
  of Mars, 18–19, 292–293
  and Solar wind 266
Comets
  carbonaceous chondrites in 147
  CHON particles in, 174–175
  comet or asteroid?, 160–161
  cometary meteors, 182–183
  defined, 10, 12, 173
  discovery of 156
  fragmentation of, 179–180
  Halley's 155, 171, 175
  Ikeya-Seki 155, 179

introduction to, 155–156, 173–174
jets from, 175–176
Kohoutek 175
Kuiper belt objects as nuclei 162
meteoroid ejection velocities 183
nucleus, composition of, 174–175
number and size distribution of 179
and Oort cloud, 165, 166–167
orbits of (typical), 156, 173
as pristine bodies, 185–186
Shoemaker-Levy 9, 26–27, 180
table of data 176
tails: appearance and composition, 174,
    175
West 180
Wild 2, 171
Commensurability
  defined, 53–54
  and resonance, 53–54
Comparative planetology 6
Condensation
  carbonaceous condensates 110
  of high-temperature refractories, 108,
    110
  ice, 110–111
  of nickel-iron and silicates, 108–109
  of olivine 109
  of planetary dust, 106–108, 109
Conjunction, defined 42
Continental drift/plate tectonics. See Plate
    tectonics/continental drift
Continental shield 212
Convection
  adiabatic change/temperature gradient
    206
  convection cells 206
Copernicus, Nicolaus
  biographical sketch of 47
  Copernican revolution 2
Cores, planetary
  defined 191
  evolution of 211
  of Galilean satellites 223
  of giant planets, 230–231
Coriolis force
  in atmospheric circulation 329
Corona, solar
  and solar wind 67
Cosmic ray exposure (CRE)
  meteorite age measurement, 143, 145
Craters, impact. See Impact craters
    (various)
Craters, volcanic. See Volcanism/volcanoes
Creationism 6
Crust, planetary
  density and crustal differentiation, 191,
    196–197
  description/evolution of, 211–212
  feldspars in, 109, 208
  formation of (Earth) 212
  See also Plate tectonics/continental drift;
    Seismology/earthquakes

Deimos
  crater counts and isochrons 263
  cratering on, 170, 294, 297
  density of 173
  discovery of 23

photographs of 170, 266
space weather is on, 266
Density
  of asteroids 173
  and crustal differentiation, 191, 196–197
  of ice 191
  of interplanetary bodies, 173, 195–196
  mean density, defined 191
  population density 3
  solar nebula, density/pressure of,
    104–105
D/H ratio (Venus) 321
Diatremes, 273, 278
Dione, physical properties of 33
DNA
  archaebacteria, DNA branching from,
    352–353
  in Martian rocks 355
  and natural selection 353
Domes
  on the Moon, 281, 283
  on Venus 312
Doppler effect
  defined 52
  and red shift 56
Drake equation (alien life) 363
Dune fields (Mars), 288–291, 372,
  374–376, 377
Duricrust (Mars), 292, 386
Dust
  dust clouds, nonlucent 332
  dust disks (Beta Pictoris systems), 88–89
  dust storms (Mars), 328–329, 330,
    374–376
  in evolution of T Tauri stars, 87–88
  orbital dynamics of, 103–104
  planetary condensation of, 106–110
Dyson, Freeman
  comment on the human condition 2

Earth
  atmosphere of, 15, 321–322, 332,
    337–338
  continental drift/plate tectonics,
    214–217
  core, formation of 211
  crater counts and isochrons 262
  cross-section of 212
  crust, formation of 212
  deep drilling projects 200
  faulting on. See Faults/faulting
  feldspar in, 109, 208
  formation date of 100
  Galileo photograph of 18
  geochemical cycle on 291
  global warming of, 337–338
  heat balance of 209
  interior dynamics, introduction to,
    212–214
  major historical eruptions, 203, 204
  mantle, formation of, 211–212
  mass asymmetry of 199
  moment of inertia of 198
  mountain belt, cross-section of 215
  ocean ridge, cross-section of 215
  Pangea, breakup of, 214, 216
  relief map of 16
  stratigraphic studies, 258–260

surface heat flow 199
tectonic collision zone, cross-section of
  217
thermal history of 217
volcanic outgassing on, 318–321, 337,
  338
  See also Heating, internal; Planetary
    atmospheres; Seismology/earthquakes
Earth-crossing asteroids, 158, 160, 169,
  172, 176
Earthquakes. See Seismology/earthquakes
Eclipse
  defined (umbral/penumbral) 39
  dynamics of 42
Electromagnetism
  electromagnetic induction heating 207
Electron temperatures (atmospheric) 333
Ellipse, geometry of 48
Ellipticity, dynamical
  of planets/planetary bodies 198
Elongation, defined 42
Enceladus
  cratering/resurfacing on, 301–304
  as "missing link" 298
  surface appearance of 34
  water/ice volcanism on, 33, 34
  photographs of 34, 304
Endogenic processes 271
Environment, damage to 1
Epimethius 170
Equations of state
  for planetary interiors 192
Equipartition energy (atmospheric),
  333–334
Escape velocity, 50–51
Escape/retentivity (atmospheric),
  333–335
Eucrites 141
Europa
  albedo of 295
  crater counts and isochrons 263
  cratering/resurfacing on, 301–304
  graben on 281
  physical characteristics, 28–29, 224–227
  surface temperature of 295
  photographs of 26, 28, 225–226
Evaporites
  in Death Valley 292
  formation of 242
  on Mars 292, 386
  scarcity on outer satellites 304
Evolution, stellar. See H-R diagram; Star
  formation
Evolution, theory of
  as pragmatic, predictive discipline 7
  vs. creationism 6
  See also Life
Exobiology
  defined 6
  See also Alien life
Exosphere 333
Extinction events, 358–359
Extraterrestrial civilizations. See Alien life
Extremophile bacteria
  alien organisms as, 359–360
  archaebacteria 352
  in sea-floor smoker vents 353
  in Yellowstone hot springs 359

Faults/faulting
  graben (normal faults), 215, 279, 280, 281
  lineaments, 279, 285
  strike/slip faults, 279, 280
  thrust faulting, mechanism of 221
  thrust faulting (on Mercury), 220–221
  thrust faults, 215, 279, 280
  See also Plate tectonics/continental drift;
    Seismology/earthquakes
Feldspars, 109, 208, 237–238
Fireball
  defined 130
  "Tagish Lake," 164
Formation age
  defined 99
  of planetary bodies 100
  See also Solar system, formation of
Formation interval 101
Fracturing and resurfacing
  on Europa and Ganymede, 302–303
  on Ganymede and Enceladus 304
  of Saturn satellites 298
Freon (CFC) ecological disaster 361
FU Orionis stars 88
Fusion crust (meteorites) 132

Gabbros, 208, 241
Galilean satellites
  albedo of 295
  condensation sequence of 119
  structures of 223
  surface temperatures of 295
  See also Callisto; Europa; Ganymede; Io
Galilei, Galileo, 47–48
Ganymede
  albedo of 295
  atmosphere of, 345–346
  crater counts and isochrons 263
  cratering/resurfacing on, 301–304
  impact craters on, 129, 263, 302–303
  photographs of, 129, 224, 302–304
  physical characteristics, 28, 223–224
  surface temperature of 295
Gas-rich chondrites, 140–141
Gegenschein 184
Genesis rocks 246
Geochemical cycle (Earth) 291
Geothite 239
Giant planets
  structure of, 230–231
  See also specific planets
Giga (prefix) 18
Glasses
  production of 264
  structure of 237
  See also Chondrules
Glassy spherules (chondrules), 136–138
Global warming, 337–338
Glomar Challenger 200
Gondwanaland 214
Graben (normal faults)
  defined, 279, 281
  on Europa 281
  on Mars 281
  schematic cross-sections of, 215, 280
Granites
  formation of 241
  rhyolite 241

Graphite
  forms of 239
  See also Carbon
Gravitation
  Gravitational constant (G) 41
  and rocket thrust 54
  vs. weight 54
Gravitational collapse
  introduction to, 78–79
  and evolution of universe, 79–80
  graphical analysis of 80
  and Jupiter-size bodies 80
  of planetesimals, 113–114
  protoclusters 79
  protostars, 79, 83
  virial theorem (Jeans theorem) 79
Great Red Spot, 24, 25, 26
Greenhouse effect, 324, 326

Habitable planets, criteria for, 356–357
Harmonic law (Kepler) 48
Hartley-Huggins bands, 331–332
Hawaiian volcanoes
  Kilauea, 6, 273, 275, 277, 280
  Mauna Loa 276
Heating, internal
  convection, planetary 206
  Earth, thermal history of 217
  electromagnetic induction heating 207
  in Europa 29
  heat balance, planetary, 208–209
  heat flow/temperature gradient, surface,
    199–200
  heat transport, theory of 206
  initial melting, heat sources for,
    206–208
  in Io 29
  Mars, thermal history of 222
  measured heat flow (solar system) 199
  Mercury, thermal history of, 219–220
  Moon, thermal history of, 219–220
  summary discussion of 232
  temperature vs. distance from sun 208
  thermal gradients 206
  Venus, thermal history of 217
  See also Tidal heating;
    Volcanism/volcanoes
Hematite 239
Henyey track (H-R diagram) 87
Herbig-Haro objects 84
Hertzsprung, Ejnar. *See* H-R diagram
Hesperian Era (Mars), 372
Hirayama families, 168–169
Horseshoe orbits 57
Hoyle, Fred 2
H-R diagram
  calculated evolutionary tracks on 87
  Henyey track 87
  Hyashi track, 86–87
  introduction to, 84, 86
  main sequence stars, 86, 87
  pre-main sequence stars, 86, 87
  See also Star formation
Hyashi track (H-R diagram), 86–87
Hydrogen
  hydrogen atom, mass of 41
  liquid, 230–231
  metallic, 195, 230–231
Hyperion, 32–33

Iapetus
  albedo of 295
  dark side of, 32, 265, 266
  photographs of 32, 365
  surface temperature of 295
Ice/water
  in aethenosphere 202
  in asteroids, 161, 164, 166
  on Callisto, 28, 222–223, 301–304
  in carbonaceous-soil admixtures, 27–28
  in cometary nuclei, 111, 161, 174–175,
    238
  condensation from solar nebula, 110–111
  condensation of (Mars), 373, 376–378
  on Enceladus, 33, 34, 111, 227–228
  on Europa, 28–29, 111, 224–227,
    301–304
  frozen, sublimation of 27
  on Ganymede, 28, 223–224, 301–304
  icy satellites, overview of, 221–222
  importance to life, 348, 352
  Jupiter, trace amounts on, 23–24
  Martian polar icecaps, 22, 370, 372,
    378–379, 386
  Martian runoff channels, riverbeds,
    21–22, 23, 291, 379–381, 385–386
  Martian water, ancient, 381–383,
    385–386
  Martian water, present, 376–378
  mean density of (ice) 191
  on Mimas 33
  on the Moon 308
  salt-and-pepper model 28
  on Saturn's moons, 227–229
  in Saturn's rings, 30, 111
  on Tethys, Dione, Rhea, 33, 111, 229
  on Triton, 258, 298
Igneous rocks
  basalts (gabbros), 208, 241
  classification scheme of (chart) 240
  description, occurrence of 239
  fractionization/reaction series of 242
  grain size of (phaneritic/aphanitic), 239,
    241
  granites, 241–242
Impact craters, physics of
  base surge, 253, 255
  basic crater features, 251–255
  basic impact event, description of,
    249–251
  bedrock, fracturing of, 252, 255
  central peaks, 252, 254, 255
  crater counts and isochrons, 261,
    262–263
  crater diameter, factors affecting, 251, 252
  crater floor, description of 255
  crater retention ages 261
  crater saturation equilibrium, 260–261
  DL dating system, 261, 264
  ejecta blanket, 251–253, 255
  ejecta blankets, lobate, 256, 258
  explosive test crater, 253, 254
  glasses, production of 264
  irradiation, darkening by, 264, 265, 266
  micrometeorite effects 264
  multiring basins, 255–256, 257, 285, 286
  peak rings, 255, 256
  primary impact craters, features of,
    251–253

rampart craters (Mars), 385–386
rays from, 253, 255
secondary impact craters 255
secondary mounds in, 252, 255
simple vs. complex craters 255
sputtering (by low-energy ions) 265
stratigraphic studies (Earth-moon),
  258–260
as tools for surface dating 258
Impact craters (Earth)
  astroblemes, defined 249
  Brent Crater, Ontario 129
  consequences of, 2–3
  crater pairs 169
  dates of 250
  discovery of 249
  discussion of 267
  El Paso fireball (1996) 132
  impact energy (megaton equivalent) 134
  impact rate, Earth-Moon system, 133, 134
  impact rates, present-day, 2–3, 133, 134
  impact rates, primeval, 133, 135
  Meteor Crater, Arizona 253
  as origin of tectonic plates 217
  selected craters (table) 250
  table of data 250
  Tunguska event (Siberia), 149–150
  Utah near miss (1972) 130
  See also Impact craters (other bodies);
    Impact craters, physics of
Impact craters (other bodies)
  crater counts and isochrons, 261–264
  cratering from, 127–128
  Europa 303
  Ganymede 129, 302–303
  impact rates, other planets, 133–134, 135
  Jupiter (Shoemaker-Levy 9), 26–27
  Mars, 129, 169
  Mathilda 115
  Mercury, 13, 128
  Mimas 115
  Miranda 128
  on moon. *See* Moon, surface features of
  Phobos, 115, 128
  Saturn moons 128
  Tethys 115
  Venus 15
  See also Impact craters (Earth); Impact
    craters, physics of
Inclination, orbital
  defined, 39, 41, 49
  of Jovian satellites, 25–26
  of Mercury and Venus 13
  and retrograde orbital motion 26
  See also Obliquity; Orbits, planetary
Inclination, rotational. *See* Obliquity
Infrared spectroscopy
  of Beta Pictoris dust disk 89
  infrared reflectance spectroscopy 162
  of interstellar molecules and dust, 77–78
  pyroxene absorption bands 162
  Wien's law, 82, 85
  See also Spectroscopy; Interplanetary
    bodies
Infrared stars
  in evolutionary sequence 83
  proplyds, 84, 85
  R Mons 82
  spectrum of (typical), 82, 84

Interplanetary bodies
    asteroids, spectral types (taxonomic
        classes) of, 162–163, 173
    density of 173
    history and distribution of 165
    infrared absorption spectroscopy in,
        162–163
    irregular shapes of, 176–177, 178
    pristine bodies, 185–186
    rotation and shape of, 176–177, 178
    Types C, D and P asteroids 163
    zonal grouping of, 163–164
    See also Asteroids; Comets; Solar system,
        formation of; Interplanetary bodies
Interstellar material
    composition of, 75, 77
    infrared spectroscopy of, 77–78
    interstellar dust, formation of, 77–78
    nebular dust/gas clouds (Monoceros) 76
Invariable plane, 39, 41
Io
    atmosphere of, 343–345
    coloration of, 300–301
    photographs of 8, 26, 28, 228, 229–301
    seen against Jupiter's clouds 119
    sodium clouds on, 29, 304
    sulfur eruptions on 227
    tidal heating on 227
    volcanism on, 223, 227, 228, 299–301
Ionization (solar nebula) 106
Ionosphere 333
Irradiation
    and criteria for life 356
    darkening by (albedo), 264, 265, 266
    and life on earth 358
    and life on Mars 353
Isochrons, 261, 262–263

Jeans theorem (virial theorem)
    of gravitational collapse, 79, 81
Jovian satellites. See Jupiter, satellites of
Jupiter
    albedo of 295
    atmosphere of, 23–24, 338–340, 342
    clouds/winds on, 24, 25, 26, 316
    as cooled brown dwarf 5
    disturbances on 24
    formation by gravitational collapse 80
    Great Red Spot, 24, 25, 26
    heat balance of 209
    mass of 23
    metallic hydrogen on, 231–232
    moment of inertia of 198
    orbital resonance with 54
    outer moons of 121
    probes to 24
    radio emissions from 29
    ring system of, 61, 62
    rotational periods, clouds (System I, II)
        24
    rotational periods, planetary 24
    satellite capture by, 26, 119–120
    Shoemaker-Levy 9 impact on, 26–27
    structure of, 230–231
    surface heat flow/temperature, 199, 295
    See also Jupiter, satellites of
Jupiter, satellites of
    Amalthea 25
    Galilean satellites, 24, 223

small moons (inner group), 25–26
small moons (outer group) 26
See also Callisto; Ganymede; Europa; Io
Jupiter mass (MJupiter)
    as planetary upper size limit 5

Kaaba (Muslim shrine) 130
Kepler, Johannes
    dates of 47
    Kepler's Laws, 48–49, 50
Kilauea (volcano), 6, 273, 275, 277, 280
Kilometer-atmosphere/kilometer-amagat
    334
Kimberlite pipes 200
Kirkwood gaps (in asteroid belt), 54, 69,
    167
KREEP basalts
    defined 243
    lunar, 306, 308
    origin of 271
Kuiper belt objects
    coloration of, 164, 166, 167, 239
    as inactive comet nuclei 162
    Kuiper belt as stable asteroid location,
        128, 129
    Oort Cloud contrast with, 161, 162
    orbital inclination of 161
    as plutinos 161
    Pluto and Triton as, 9, 299
    properties of 159
    satellites of, 169

Lagrangian points, 33, 54–57, 160, 357
Laurasia 214
Lava tubes 278
Law of areas (Kepler) 48
Lead (on Venus) 292
Life
    introduction: the nature of life,
        349–350
    alien life, appearance of 360
    alien life, contact with, 363–367
    alien life, estimated possibilities for,
        362–363
    alien life, the Drake equation 363
    amino acids, 351, 352
    archaebacteria, DNA branching from,
        352–353
    archaebacteria, extremophile 352
    cellular life, beginnings of 353
    CO2, synthesis of organic molecules
        from 355
    coacervates 352
    convergence effect 360
    diversity/adaptability of, 359–360
    early life, chemistry of, 350–351
    evolution, theory of, 6, 7
    extinction events, 358–359
    fluid medium, importance of 352
    freon (CFC) ecological disaster 361
    habitable planets, criteria for, 356–357
    on Mars, 387, 389
    on Mars (Viking Lander), 354–355
    in Martian rocks 355
    meteorite impacts, effects of, 358–359
    Miller experiment 351
    outside the solar system 356
    oxygen-rich vs. CO2 environments 353
    ozone layer, significance of 353

plagues and viruses, 359–360
planetary/astronomical processes, effect
    of 358
prokaryote/eukaryote cells 352
proliferation and natural selection,
    353–354
proteins/proteinoids, 351, 352
racemic/nonracemic mixtures 351
in sea-floor smoker vents 353
stromatolites (blue-green algae) 353
technology, effect of, 360–361
the tidepool hypothesis 352
trilobites 354
Light, velocity of 41
Limonite 239
Lineaments, 279, 285
Lithophile elements 135
Lithosphere
    evolution/composition of, 244–246
    formation of 208
    genesis rocks in 246
    and plate tectonics/earthquakes 204
    primeval lithosphere, evolution of,
        248–249
    on Venus 218
Lowell, Percival, 19–20
Luminosity (of Sun) 41

Maar 278
Maghemite 239
Magma
    fractionization/reaction series of 242
    magma ocean (lunar) 207
Magnetic braking
    of Sun's rotation 106
Magnetic fields
    dipole field, 209, 210
    on Europa 28
    Jupiter's (effect on Io) 29
    lines of force 209
    magnetic coupling 106
    Mars 221
    moon (Earth's) 220
    reversal of, 209–210
    in solar nebula, 105–106
    of Sun 106
    surface 211
    typical planetary field 210
    See also Magnetism, planetary
Magnetism, planetary
    introduction to 209
    electromagnetic induction heating
        207
    magnetic correction 209
    magnetosphere, magnetopause 209
    origins of (dynamo theory), 210–211
    paleomagnetism 210
    solar wind, magnetic interactions
        with 209
    See also Magnetic fields
Magnetite 239
Magnetosphere, magnetopause 209
Main sequence stars, 86, 87
Malthus, Thomas 3
Mantle (Earth)
    formation of, 211–212
    mantle plumes, 215, 217
    materials of (Earth/moon) 243
Mare, lunar, 252, 281

Mars
  albedo of 295
  Apollinaris Patera 276
  argon on 317
  bedrock on, 310–311
  canals on, 19–20
  chaotic terrain on, 21, 379
  climate variations on, 382–385
  clouds on, 328, 329
  drawings of (early) 21
  dry riverbeds on 23
  dune fields on, 288–291, 372, 374–376, 377
  dust storms on, 328–329, 330, 374–376, 379
  evaporite minerals on (duricrust), 292, 386
  expected impact of Phobos on 118
  geological eras of, 372
  graben on 281
  ice/water condensation on, 373, 376–378
  impact craters on, 129, 258, 263
  instability of liquid on 326
  isochrons, 261, 262–263
  lakebeds on, 386, 388
  landings on (Pathfinder/Sojourner) 4
  life on, 354–356, 387, 389
  magnetic field of 221
  map of, 22, 370
  Marsquakes, 204–205
  meteorites from, 148–149, 374
  moment of inertia of, 198, 372
  obliquity cycles, 384–385
  Olympus Mons (shield volcano), 273–274
  Opportunity (rover), 386, 388
  outgassing on 385
  Pathfinder (lander), 4, 373
  patterned ground on 385
  pink sky (Rayleigh scattering) 323
  polar hood on 328
  polar icecaps on, 22, 370, 372, 378–379, 386
  rampart craters on 385–386
  red coloration of, 18–19, 292–293
  runoff channels on, 21–22, 23, 291, 379–381, 385–386
  saltation on, 287, 288
  satellites of, 23, 24, 119–120, 121
  soil composition of, 243–244
  Sojourner (rover), 4, 374, 375
  Spirit (rover), 376, 386, 387
  stratified terrain on 290
  surface characteristics of, 18–22, 121, 270, 309–311
  surface chemistry of, 292–293
  surface temperature of 295
  thermal history of 222
  view of (Global Surveyor) 22
  Viking landers on, 22–23, 371–373
  volcanism on, 221, 273–274, 320, 373–374, 375
  water on (ancient), 381–383, 385–386
  water on (present), 376–378
  windblown deposits on, 287–291
Mascons
  in planetary interiors, 198–199
Mass asymmetry
  of planets/planetary bodies 199

Mathilda
  density of 173
  meteorite impact on 115
Mauna Loa (volcano) 276
Mauna Ulu (volcano) 319
Mean density, defined 191
Mercury
  albedo of 295
  crater counts and isochrons 262
  impact craters on, 256, 257, 308–309
  Mariner 10 map of 14
  orbit, angular separation of 13
  orbital period of 13
  photomosaic of 13
  physical description of, 12–13
  rotational period of 13
  similarity to Moon 309
  surface temperature of 295
  thermal history of, 219–220
  thrust faulting on, 220–221
Mesosiderite stony iron meteorites 142
Mesosphere 331
Metamorphic rocks 243
Meteorite ages
  age vs. meteorite type 144
  Angra dos Reis (ADOR) anomaly 145
  BABI point, 144–145
  cosmic ray exposure (CRE) age, 143, 145
  as evidence of solar system collisions, 145–146
  gas retention age 143
  rubidium-strontium system of measurement, 144–145
  and Yarkovsky effect 145
Meteorite classification
  introduction to, 134–136
  achondrites, 141–142
  carbonaceous chondrites, 138–139, 141
  carbonaceous inclusions in (Allende meteorite), 108, 110
  chondrites 140
  chondrules, 136–138
  classification chart 136
  gas-rich chondrites, 140–141
  iron (Widmanstätten pattern), 142–143
  stony irons (pallasite, mesosiderites) 142
Meteorite falls
  brightness 132
  defined 130
  sounds of 132
  temperature of (fusion crust), 132, 133
  trains 132
  velocity of, 132, 133
  See also Impact craters (Earth); Impact craters (other bodies)
Meteorite impacts
  effects on life, 358–359
  See also Impact craters (Earth)
Meteorites
  ages of. See Meteorite ages
  Antarctic meteorite fields, 131–132
  brecciated (collisional mixing) 143
  classification of. See Meteorite classification
  find, defined 130
  formation age of 100
  Kaaba (Muslim shrine) 130
  lunar, 148–149
  Martian, 148–149, 374

  meteorite, defined 130
  meteorite studies, history of, 130–131
  meteoritic complex, defined 127
  Neumann bands in 143
  olivine in 109
  origins/orbits of, 147–148
  parent body, defined 130
  tektites 146
  Thomas Jefferson comment regarding 131
  type distribution 148
  Winona stone 131
  See also Impact craters (Earth); Impact craters (other bodies)
Meteoroids
  collisions of. See Collisions/fragmentation
  meteoroid, defined 130
  meteoroid flux 132
  Yarkovsky effect on, 68, 70
  See also Meteorite falls; Meteorites; Planetesimals
Meteors
  arrival times for 181
  cometary meteors, 182–183
  defined 130
  Leonid swarm, 181–182
  meteor showers, 181–182
  meteoroid swarm 182
  radiant (source direction) 181
Methane ice 238
Micas 238
Micrometeorites
  effects of 264
Miller experiment 351
Mimas 33
Minerals
  amphiboles 238
  classification by composition 237
  clays 239
  defined 237
  feldspars, 109, 208, 237–238
  graphite 239
  iron oxides 239
  micas 238
  pyroxenes 238
  quartz 238
  troilite, 109, 239
MIR space station 3
Miranda
  fracturing of, 228, 230
  impact craters on 128
  internal heating of, 35, 228
  surface appearance of 36
Mohole Project 200
Mohorovicic discontinuity 202
Moments of inertia
  planetary bodies 198
Montmorillonite 239
Moon, properties of
  anorthosites 219, 241
  Apollo landing on, 1, 3
  formation age of 100
  KREEP basalt on, 306, 308
  magnetic field of 220
  map of 19
  mass asymmetry of 199
  meteorites from, 148–149
  moment of inertia of 198

moonquakes, 204–205
norites 246
olivine on 200
origin of, 18, 121–122, 123
physical characteristics of, 17–18, 236
seismic velocity profile of 219
siderophile elements, deficiency of 122
stratigraphic studies, 258–260
structure of 219
surface heat flow 199
surface layers, cross-section of 248
thermal history of, 219–220
tidal effects on, 58–59
troctolites 246
volcanism on, 18, 308
See also Moon, Surface features of
Moon, surface features of
cinder cone on 274
crater chains 284
crater counts and isochrons 262
domes, 281, 283
Hadley Rille 284
ice on 308
impact craters, 20, 126, 128, 252–257
lineament grid system 285
lunar transient phenomena 308
magma ocean (primordial) 207
mare, 281, 282, 286
radial/concentric ridges 285
regolith on, 17, 246–248
rilles, 282–284, 286
surface evolution of, 305–308
Tycho crater 309
wrinkle ridges, 281, 282
See also Moon, properties of
Multiple stars. *See* Binary/multiple stars
Multiring basins, 255–256, 257, 285, 286

Nebulae
disk-shaped 82
nebula infall (FU Orionis stars) 88
nebular hypothesis, 81–82
See also cocoon nebulae; Solar nebula
Neptune
albedo of 295
atmosphere of, 37, 341–342
Bode's Rule, failure of 35
clouds/storms on 37
expected impact of Triton on 37
heat balance of 209
moment of inertia of 198
orbital perturbations of (Planet "X"),
37, 39
ring systems of, 37, 63
satellites of 37
structure of 231
surface heat flow/temperature, 199, 295
Neumann bands 143
Newton, Isaac
dates of, 47, 48
Newton's Laws, 48, 49
Noahician Era (Mars), 372
Nodes, orbital, 129–130
Nontronite 239
Nuclear waste, 2–3
Nuée ardente (glowing cloud) 272

Oberon, size/brightness of 36
Oblateness (dynamic/geometric) 198

Obliquity
of solar system planets, 39, 41
Occultation, defined 42
Oceanic ridges. *See* Ridges, oceanic
Olivine
composition of, 200, 238–239
condensation of 109
crystal structure of 237
on the moon 200
olivine-spinel transition 195
Olympus Mons (Martian volcano),
273–274
Oort cloud
contrast with Kuiper belt objects, 161,
162
and evolution of planetary bodies 165
perturbation of bodies in, 166–167
Opportunity (Mars rover), 386, 388
Opposition, defined 42
Optical depth/opacity (atmospheric),
323–324, 325
Orbital period
of mercury 13
rotational/orbital resonance 13
Orbits, planetary
angular separation of 13
circular velocity, 49–50
escape velocity, 50–51
general arrangement of 9
horseshoe orbits 57
parabolic velocity 51
perturbations and resonances, 53–54
retrograde vs. prograde 14
three-body problem, 52–53
tidal effects on, 58–59
See also Inclination, orbital
Orbits, satellite
effect of tidal forces on 119
outer Jupiter satellites, 118–119, 120
retrograde vs. prograde, 118–119
Outer Solar System
simplified overview 27
See also Jupiter, satellites of; Neptune;
Pluto; Saturn
Outgassing
on Mars, 382–383
volcanic, atmospheric effects of,
318–321, 337, 338
Ozone layer 331
significance to life 353

Pahoehoe lava
characteristics of, 271–272, 311
on the Moon 307
Paleomagnetism 210
Palimpsest cratering 302
Pallas, density of 173
Pallasite stony iron meteorites 142
Pangèa, 214, 216
Parabolic velocity 51
Parent body, defined 130
Pathfinder (Mars lander) 4, 373
Periapse/periapsis, 39, 48
Perigee, defined 39
Perihelion, defined 39
petrology, defined 237
Phobos
as captured satellite, 120, 121
crater counts and isochrons 263

density of 120, 173
discovery of 23
expected impact on Mars 118
fractures and internal structures, 98, 298
ice in, 298
impact craters on, 128, 294, 296–297
penumbra of 43
photographs of, 18, 24, 98, 115, 121, 172
retrograde motion of 118
Phoebe, 32, 120
Photodissociation (atmospheric) 334
Physics, defined 5
Pink sky (Mars) 323
Plagues and viruses, 359–360
Planck's constant 41
Planet "X," 37, 38
Planetary atmospheres
argon in (Venus/Mars) 317
atmospheric escape/retentivity, 333–335
carbon dioxide (Venus) 292
carbonate rocks and atmospheric CO_2 291
Chiron, 346–347
climate changes (Martian), 382–385
convective mixing 328
D/H ratio (Venus) 321
Earth, 15, 321–322, 332, 337–338
Earth, volcanic outgassing on, 318–321,
337–338
electron temperatures in 333
equipartition energy, 333–334
exosphere 333
Ganymede, 345–346
global circulation: coriolis force in 329
greenhouse effect, 324, 326
Hartley-Huggins bands, 331–332
ideal gas law, 321, 322
inert gases in primordial atmospheres,
317–318
Io, 343–345
ionosphere 333
irregular temperature profile in 324
isothermal atmosphere 321
Jupiter, 23–24, 338–340, 342
kilometer-atmosphere/kilometer-amagat
334
Mars 338
Mesosphere 331
moist atmospheres/clouds, 326–327
Neptune, 37, 341–342
optical depth/opacity, 323–324, 325
ozone layer 331
phase diagram for (generic), 326, 327
phase diagram for (Mars), 377–378
photodissociation 334
Pluto 346
radiative heat transfer in, 323–324
Rayleigh scattering and blue skies,
322–323
secondary atmospheres (volcanic
outgassing), 318–321
selective wavelength absorption (Saturn)
326
stratosphere 331
thermosphere, physics of, 332–333
Titan, 33–34, 342–343, 344
triple point in, 326, 376–378
Triton, 37, 346
troposphere/tropopause 331
Uranus, 35, 340–341

Planetary interiors
  computer models of (hydrostatic
    equations) 192
  density and crustal differentiation, 191,
    196–197
  density-mass diagram, 195–196
  dynamical ellipticity of 198
  equations of state, 191–192
  high-pressure lab tests 195
  internal pressure: mathematical notes
    on 194
  mascons in, 198–199
  mass asymmetry of 199
  mean density 191
  melting in, 193–194
  moments of inertia, theoretical/actual 198
  olivine-spinel transition 195
  plastic flow (rheidity), 192–193
  pressure ionization 195
  solid state phase changes, 194–195
  terminology, state changes in 195
Planetary science
  comparative planetology 6
  defined 5
  as integrative discipline 5
  U.S. implementation of, 5–6
Planetesimals
  aggregation of, 107, 109
  collisional accretion of, 112–113
  collisions of *See* Collisions/fragmentation
  compositions of 152
  defined, 10, 102
  gravitational collapse of, 113–114
  lifetimes of, 128, 129
  orbital capture of, 127, 167–168
  orbital nodes of, 129–130
  sweep-up of (in solar system) 117
  See also Asteroids; Kuiper belt objects;
    Meteorites
Planets/planetary bodies
  accretion models for 111
  around other stars, 89–93, 356–358,
    362–363
  collisional accretion of, 114–116
  defined, 5, 9–10
  density-mass diagram, 195–196
  dynamical ellipticity of 198
  Earth-sized alien planets 93
  extrasolar, 89–93, 356–358, 362–363
  gross composition of 111
  habitable planets, criteria for, 356–357
  mascons in 199
  mass asymmetry of 199
  moments of inertia, theoretical/actual 198
  nebular mass required for formation,
    102–103
  orbits, general arrangement of 9
  planetary processes, effect on life 358
  planetary systems, possible origins of,
    93–94
  protoplanet, defined 102
  relative sizes of (terrestrial) 11
  rotational periods of, 176–177, 178
  surface heat flow/temperature gradient,
    199–200, 293–294, 295, 297
  surface material 296
  surface temperatures, 293–294, 295, 297
  symbols for 12
  telescopic appearance of, 39, 40

upper size limit for 5
  See also Heating, internal; Planetary
    interiors; Solar nebula; Solar system
Plate tectonics/continental drift
  on Earth, 214–217
  on Ganymede 223
  Mid-Atlantic ridge (and Pangea),
    214–216
  rift faulting (Iceland) 215
  and rift valleys, 214, 216
  and sea-floor spreading 214
  See also Faults/faulting;
    Seismology/earthquakes
Pleiades 74
Plurality of worlds 19
Plutinos (Kuiper belt) 161
Pluto
  albedo of, 37, 295
  atmosphere of 346
  as Kuiper belt asteroid, 9, 299
  physical/orbital characteristics of 39
Population density 3
Poynting-Robertson effect
  mathematical expression for 68
  and orbital ring decay, 117–118
  physics of, 67–68
  time scales for 68
  vs. particle size 70
Prefixes, SI system of 43
Pristine bodies, 185–186
Probes, spacecraft
  to Jupiter 24
  to Mars, 22–23, 386
  to Venus, 14, 15, 131
Prograde motion
  defined 14
  satellites, effect on 118
Proplyds, 84, 85
Protoclusters 79
Protoplanet, defined 102
Protostars
  defined 79
  in evolutionary sequence (chart) 83
  on H-R diagram, 86–87
  visible light from, 87, 94
Pyroxenes 238

Quartz 238

Racemic/nonracemic mixtures 351
Radial velocity 51
Radiation pressure
  discussion of, 65, 67
  mathematical notes on 66
  vs. particle size, 67, 70
Radiative heat transfer, 323–324
Radio emissions (Jupiter) 29
Rampart craters (Mars) 385–386
Rayleigh scattering
  and appearance of Uranus 35
  and blue skies, 322–323
  and pink Martian sky 323
Red shift 56
Regolith
  lunar, 17, 246–248
  megaregolith, 247–248
  slumping/creep of, 281, 285–287
Resonance, rotational/orbital
  and commensurability, 53–54

Kirkwood gaps, 54, 167
  of Mercury 13
  in ring systems, 60–64
  of Saturn's satellites 227
Retrograde motion
  defined 14
  of Jovian satellites 26
  satellites, effect on, 118–119
  of Triton and Phobos 118
  of Uranus 35
Revolution vs. rotation, 41, 43
Reynolds number 69
Rhea, 33, 260
Rheidity (plastic flow)
  description of, 192–193
  on Europa 227
Ridges, lunar, 281, 282, 285
Ridges, oceanic
  cross-section of 215
  Mid-Atlantic ridge (and Pangea),
    214–216
  and sea-floor spreading 214
Rifts
  Mid-Atlantic ridge (and Pangea),
    214–216
  rift faulting (Iceland) 215
  on satellites 228–230
  and rift valleys, 214, 216
  rift zone (Hawaii) 296
  and sea-floor spreading 214
Rilles
  lunar, 282–284
  terrestrial (Hawaii) 283
Ring systems
  Cassini's division in 30
  as clues to planetary origin, 117–118
  Jupiter, 61, 62, 117
  Neptune 37
  resonances in, 60–64
  Roche's limit in, 60–61, 62, 63
  Saturn, 30–32, 46, 117
  Uranus, 35, 117–118
Roche's limit
  and breakup of spherical bodies, 59, 60
  and ring systems, 60–61, 62, 63, 118
Rocket thrust
  vs. gravitation 54
Rocks, planetary
  composition of (by weight) 245
  elemental abundances in 244
  survey of 243
  See also specific rock types
Rotational period
  of Mercury 13
  of planets and irregular shapes, 176–177,
    178
  rotational/orbital resonance 13
  of Venus 14
Rubidium-strontium system
  meteorite/rock age measurement,
    144–145
Russell, Henry N. *See* H-R diagram

Salt-and-pepper model 28
Saltation (Mars), 287, 288
Satellites
  cratering, 127–129, 261–264
  defined 10
  icy satellites, overview of, 221–222

Jovian, orbital inclination of, 25–26
of Jupiter, 24–26, 118–119, 120
Lagrangian points for, 33, 54–57, 160, 357
letter/number code for 10
of Kuiper belt objects, 169
of Mars, 23, 24, 119–120, 121
Moon, origin of, 121–122, 123
of Neptune, 37, 120
origins of (capture), 119–121
origins of (catastrophic), 122–124
prograde satellites, 118–119
retrograde orbits, effects of 118
of Saturn, 32–34, 120
temporary capture by Jupiter 26
tidal effects on orbits 119
of Uranus, 35, 36
See also other major satellites by name
Saturation equilibrium (impact craters), 260–261
Saturn
albedo of 295
atmosphere of, 340, 341
atmospheric wavelength absorption 326
heat balance of 209
moment of inertia of 198
principal features of (visual) 30
ring system of, 30–32, 46, 61–65, 117–118
satellites of, 32–34, 56, 63, 170, 227–229
shepherding satellites of, 62, 64
structure of 231
surface heat flow/temperature, 199, 295
See also Titan
Science as synthesis 4
S-Class asteroids 163
Sedimentary rocks
breccia 242
evaporites, 242, 292
geochemical differentiation in 242
shale, sandstone 242
and weathering 242
Seismology/earthquakes
aethenosphere, 202, 203–204
depth distribution of, 204–205
earthquakes, origins of, 202–203
epicenter 202
faults. *See* Faults/faulting
lithosphere, 204, 208
major historical earthquakes, 203, 204
Mohorovicic discontinuity 202
Moon, seismic velocity profile of 219
passive/active seismology 203
plates, movement of, 203–204
P-waves and S-waves, 201–202
San Francisco earthquake (1906) 201
seismic wave velocities 202
seismometer, 200–201
table of earthquakes 203
See also Plate tectonics/continental drift
Seven of Nine
transformation of 709
See also Voyager
Shadows (umbra, penumbra), 39, 43
Shale, sandstone 242
Shepherding satellites (Saturn), 62, 64
Shield volcanoes
Apollinaris Patera (Mars) 276
defined 273

Mauna Loa (Hawaii), 276, 277
maximum height of 275
Olympus Mons (Mars), 273–274
Shoemaker-Levy 9
fragmentation of 180
impact on Jupiter, 26–27, 118
Siderophile elements 135
Single stars, rarity of 90
Smog, photochemical (Titan) 34
Smoker vents, sea-floor 353
SNC meteorites 148
Socioenvironmental issues, 3–4
Sodium clouds (Io), 29, 304
Sojourner (Mars rover) 4, 374, 375
Solar constant, 39, 41
Solar nebula
collisional accretion in (planetesimals), 112–113
composition of, 102–103
condensation processes within, 106–111
cooling of 105
density/pressure of, 104–105
dust, aggregation of, 111–112
gravitational collapse (planetesimals), 113–114
ionization of 106
magnetic effects in, 105–106
mass required for planet formation, 102–103
shape of, 103–105
sweep-up of planetesimals 117
See also Solar system, formation of
Solar system, formation of
characteristics requiring explanation 102
collisional accretion (full-size planets), 114–116
date/duration of 99
planets, gross composition of 111
ring systems as clues to, 117–118
solidification/formation ages of, 99–100
supernova prior to, 100–101
See also Solar nebula
Solar wind
bow shock effect, 209, 210
and carbon/colored organics production 266
effect vs. particle size 70
extent of 67
introduction to 67
planetary magnetic interactions with 209
and solar corona 67
Solidification age 99
Space science, defined 6
Space weathering 266
Spallation 265
Spectroscopy
defined 52
spectral types (of asteroids) 163
spectroscopic binary 90
spectrum-luminosity diagram. *See* H-R diagram
See also Infrared spectroscopy
Spherical frontier, living on, 2–3
Spirit (Mars rover), 376, 386, 387
Sputtering (by low-energy ions) 265
Star formation
Beta Pictoris systems, 88–89, 95
Bok globules, 80–81

cocoon nebulae, evolutionary processes in, 94–95
evidence of 75
FU Orionis stars 88
sequence of (schematic view) 83
T Tauri stars, 87–88
See also Cocoon nebulae; Gravitational collapse; H-R diagram; Infrared stars; Interstellar material
Star systems
habitable planets, criteria for, 356–357
nearest (table) 91
star, defined 5
See also Binary/multiple stars
Stefan-Boltzmann constant 41
Stellar evolution. *See* H-R diagram; Star formation
Stoke's law 114
Stony iron meteorites (pallasite, mesosiderites) 142
Stony meteorites. *See* Achondrites; Carbonaceous chondrites; Chondrites
Stratosphere 331
Sulfur eruptions (Io) 227
Sulfuric acid clouds (Venus), 327–328, 336
Sun
luminosity of 41
magnetic field of 106
mass of 41
moment of inertia of 198
rotation, magnetic braking of 106
solar constant, 39, 41
solar wind, 67, 70
surface heat flow 199
See also Solar nebula
Supernova
and solar system formation, 100–101
Surface heat flow
Earth 199
Jupiter, 199, 295
Moon 199
Neptune, 199, 295
radiative heat transfer, 323–324
Saturn, 199, 295
Sun 199
temperature gradients, 199–200, 297
Uranus, 199, 295
See also Planetary atmospheres
Symbols, planetary 12

T Tauri stars
evolution of, 87–88
rotation/magnetic braking of 106
Tectonics. *See* Faults/faulting; Plate tectonics/continental drift; Seismology/earthquakes
Tektites 146
Temperature
of meteorite falls (fusion crust), 132, 133
of planetary surfaces, 199–200, 293–294, 295, 297
vs. wavelength (Wien's law), 82, 85
See also Surface heat flow
Terminology, state changes in 195
Terms, planetary astronomy 42
Terrestrial planets
defined 10
relative sizes of 11
Tethys, 33, 115, 229

Thermosphere, physics of, 332–333
Three-body problem, 52–53
Tidal effects
  body tides 57
  on Comet Shoemaker-Levy 118
  importance of 70
  mathematical notes on 58
  on Moon's rotation, orbit, 58–59
  ocean tides, 57–58
  Roche's limit, 59, 60
  on satellite orbits 119
  tidal bulges, 57, 59
  See also Tidal heating
Tidal heating
  Enceladus 33
  Europa 224
  Ganymede 223
  Io, 29, 59–60, 227
  Miranda 35
  Saturnian satellites, 227–228
  Triton 37
  See also Heating, internal
Titan
  albedo of 295
  atmosphere of, 33–34, 342–343, 344
  photochemical smog on 34
  photographs of 305, 343–344
  surface features of 305, 342–344
  surface temperature of 295
Titania
  icy surface of 229
  relative size/brightness of 36
Transient phenomena, lunar 308
Transit, defined 42
Trenches, oceanic 214
Triple point (atmospheric), 326,
  376–378
Triton
  atmosphere of, 37, 346
  as captured satellite, 120–121
  dark volcanic plumes on, 298–299
  expected impact on Neptune 37
  icy "cantaloupe terrain" on 258
  internal heating of 37
  as Kuiper asteroid 299
  photographs of 38, 346, 358, 398, 399
  retrograde orbit of, 37, 118
Troilite, 109, 239
Trojan asteroids, 55–56, 160, 167
Troposphere/tropopause 331
Tuff
  tuff rings, 278, 279
  volcanic tuff 272
Tunguska event (Siberia), 149–150
Turbulence, 69–70
Tycho. *See* Brahe

Ultraviolet imaging (Venus) 15
Umbra/penumbra, 39, 43
Umbriel, size/brightness of 36
Uplands (Mercury) 13
Uranus
  albedo of 295
  appearance of 35
  atmosphere of, 35, 340–342
  discovery of 34
  heat balance of 209
  moment of inertia 198
  ring system of, 35, 63
  rotational axis, inclination, 35, 36
  rotational period 35
  satellites of, 35, 36, 228–230
  structure of 231
  surface heat flow/temperature, 199, 295
Urey reaction 291

Velocity
  circular, 49–50
  escape, 50–51
  of light 41
  of meteorite falls, 132, 133
  parabolic 51
  Radial 51
  velocity equation 56
Venera probes (of Venus) 14
Venus
  albedo of 295
  argon on 317
  ashen light around 337
  atmosphere, structure/features,
    335–337
  clouds/winds on, 13–15
  crater counts and isochrons 262
  D/H ratio on 321
  as evening/morning star 13
  impact crater on, 15, 129, 256
  lead on 292
  physical description, 14–15
  relief map of 16
  retrograde rotation of 14
  sulfuric acid clouds on, 327–328, 336
  surface characteristics of, 311–313
  surface chemistry of 292
  surface temperature of 295
  thermal history of 217
  ultraviolet image of 15
  Venera probes, 14, 311
  volcanism on, 15, 217–219, 318, 320
Viking landers (Mars), 22–23, 371–373
Virial theorem (Jeans theorem)
  of gravitational collapse, 79, 81
Volcanism/volcanoes
  aa lava, 271–272

ash/cinders from, 272, 273, 274
caldera, 275, 277
causes of, 278–279
cinder cones, 273, 274, 290
cumulates 271
dark halo craters 278
diatremes (craters), 273, 278
on Enceladus, 227–228
fire fountain eruption 280
fluidized eruptions 272
on Io, 29, 223, 227, 228, 299–301
lava tubes 278
maars (craters) 278
major historical eruptions, 203, 204
on Mars, 221, 273–274, 320, 373–374,
  375
maximum height (shield volcanoes) 275
on the Moon, 18, 308
Nuée ardente (glowing cloud) 272
rocks produced by, 271–272
shield volcanoes, 273–276
on Triton, 298–299
tuff 272
tuff rings, 278, 279
on Venus, 15, 318, 319
volcanic landforms, schematic of 273
volcanic outgassing, atmospheric effects
  of, 318–321, 337, 338
volcano-tectonic sink 278
See also Heating, internal
Vorticity, example of 70
Voyager
  Callisto flyby 28
  Io photograph by 8
  See also Seven of Nine

Water. *See* Ice/water
Watt (power)
  joule-second equivalent of 41
Weight vs. gravity 54
Widmanstätten pattern (iron meteorites),
  142–143
Winds, planetary
  on Jupiter, 24, 25, 26, 316
  Mars, windblown deposits on, 287–291
  on Venus, 13–15
  See also Solar wind
World, defined 12
Wrinkle ridges (lunar), 281, 282

Yarkovsky effect
  defined 68
  on meter-scale bodies, 69, 70
  and stony vs. iron meteorites 145

Zodiacal light, 183–185